A Course in
Real Analysis

SECOND EDITION

A Course in
Real Analysis

SECOND EDITION

John N. McDonald

School of Mathematical and Statistical Sciences
Arizona State University

Neil A. Weiss

School of Mathematical and Statistical Sciences
Arizona State University

Biographies by Carol A. Weiss

AMSTERDAM ▪ BOSTON ▪ HEIDELBERG ▪ LONDON
NEW YORK ▪ OXFORD ▪ PARIS ▪ SAN DIEGO
SAN FRANCISCO ▪ SINGAPORE ▪ SYDNEY ▪ TOKYO

Academic Press is an imprint of Elsevier

Academic Press is an imprint of Elsevier
225 Wyman Street, Waltham, MA 02451, USA
The Boulevard, Langford Lane, Kidlington, Oxford, OX5 1GB, UK

Library of Congress Cataloging-in-Publication Data

McDonald, John N.
  A course in real analysis / John McDonald, Neil A. Weiss ; biographies by Carol A. Weiss. – 2nd ed.
     p. cm.
  Includes bibliographical references and index.
  ISBN 978-0-12-387774-1 (hardback)
1. Mathematical analysis. I. Weiss, N. A. (Neil A.) II. Title.

  QA300.M38 2012
  515–dc23   2011037823

British Library Cataloguing-in-Publication Data
A catalogue record for this book is available from the British Library.

For information on all Academic Press publications visit our website at: www.elsevierdirect.com

To Pat and Carol

Photo Credits

Contents

About the Authors:
John N. McDonald

After receiving his Ph.D. in mathematics from Rutgers University, John N. McDonald joined the faculty in the Department of Mathematics (now the School of Mathematical and Statistical Sciences) at Arizona State University, where he attained the rank of full professor. McDonald has taught a wide range of mathematics courses, including calculus, linear algebra, differential equations, real analysis, complex analysis, and functional analysis.

Known by colleagues and students alike as an excellent instructor, McDonald was honored by his department with the *Charles Wexler Teaching Award*. He also serves as a mentor in the prestigious Joaquin Bustoz Math-Science Honors Program, an intense academic program that provides motivated students an opportunity to commence university mathematics and science studies prior to graduating high school.

McDonald has numerous research publications, which span the areas of complex analysis, functional analysis, harmonic analysis, and probability theory. He is also a former Managing Editor of the *Rocky Mountain Journal of Mathematics*.

McDonald and his wife, Pat, have four children and six grandchildren. In addition to spending time with his family, he enjoys music, film, and staying physically fit through jogging and other exercising.

About the Authors:
Neil A. Weiss

Neil A. Weiss received his Ph.D. from UCLA and sub-sequently accepted an assistant-professor position at Arizona State University (ASU), where he was ulti-mately promoted to the rank of full professor. Weiss has taught mathematics, probability, statistics, and operations research from the freshman level to the advanced graduate level.

In recognition of his excellence in teaching, he received the *Dean's Quality Teaching Award* from the ASU College of Liberal Arts and Sciences. He has also been runner-up twice for the *Charles Wexler Teaching Award* in the ASU School of Mathematical and Statistical Sciences. Weiss's comprehensive knowledge and experience ensures that his texts are mathematically accurate, as well as pedagogically sound.

Weiss has published research papers in both theoretical and applied mathematics, including probability, engineering, operations research, numerical analysis, and psychology. He has also published several teaching-related papers.

In addition to his numerous research publications, Weiss has authored or coauthored books in real analysis, probability, statistics, and finite mathematics. His texts—well known for their precision, readability, and pedagogical excellence—are used worldwide.

In his spare time, Weiss enjoys walking and studying and practicing meditation. He is married and has two sons and three grandchildren.

Preface

This book is about real analysis, but it is not an ordinary real analysis book. Written with the student in mind, it incorporates pedagogical techniques not often found in books at this level.

In brief, *A Course in Real Analysis* is a modern graduate-level or advanced-undergraduate-level textbook about real analysis that engages its readers with motivation of key concepts, hundreds of examples, over 1300 exercises, and applications to probability and statistics, Fourier analysis, wavelets, measurable dynamical systems, Hausdorff measure, and fractals.

What Makes This Book Unique

A Course in Real Analysis contains many features that are unique for a real analysis text. Here are some of those features.

Motivation of key concepts. All key concepts are motivated. The importance of and rationale behind ideas such as measurable functions, measurable sets, and Lebesgue integration are made transparent.

Detailed theoretical discussion. Detailed proofs of most results (i.e., lemmas, theorems, corollaries, and propositions) are provided. To fully engage the reader, proofs or parts of proofs are sometimes assigned as exercises.

Illustrative examples. Following most definitions and results, one or more examples, most of which consist of several parts, are presented that illustrate the concept or result in order to solidify it in the reader's mind and provide a concrete frame of reference.

Abundant and varied exercises. The book contains over 1300 exercises, not including parts, that vary widely with regard to application and level.

Applications. Diverse applications appear throughout the text, some as examples and others as entire sections or chapters. For instance, applications to probability theory are ubiquitous. Other applications include those to statistics, Fourier analysis, wavelets, measurable dynamical systems, Hausdorff measure, and fractals.

Careful referencing. As an aid to effective use of the book, references (including page numbers) to definitions, examples, exercises, and results are consistently provided. Additionally, post-referenced exercises are marked with a star (★); all such exercises are strongly recommended for solution by the reader.

Biographies. Each chapter begins with a biography of a famous mathematician. In addition to being of general interest, these biographies help the reader obtain a perspective on how real analysis and its applications have developed.

New to the Second Edition

In this second edition of *A Course in Real Analysis*, besides fine tuning the material from the first edition, several significant revisions have been made, many of which are based on feedback from instructors, students, and reviewers. Some of the most important revisions are as follows.

Chapter Splits. To make the chapters more uniform, long chapters in the first edition have been split into two chapters in the second edition.

Riemann Integrability. A detailed proof of the fact that a bounded function on a closed bounded interval is Riemann integrable if and only if its set of points of discontinuity has measure zero has been supplied.

Extensions to Measures. The treatment of extensions to measures has been simplified and improved by stating and proving a result that shows that the collection of all Cartesian products of sets formed from a finite number of semi-algebras is also a semialgebra.

Classical Change-of-Variable Formula. A subsection has been included that applies the general change-of-variable formula for measurable transformations to get a result that contains as a special case the classical change-of-variable formula in Euclidean n-space.

Self-Adjoint Operators. A section on self-adjoint operators on Hilbert space has been added that includes the statement and proof of the spectral theorem for compact self-adjoint operators.

The Schwartz Class. A subsection that discusses the Schwartz class—the collection of all complex-valued rapidly decreasing functions—and the behavior of the Fourier transform on that class has been provided.

Hausdorff Measure and Fractals. Another applications chapter has been added that includes a discussion of outer measure and metric outer measure, Hausdorff measure, Hausdorff dimension and topological dimension, and fractals.

Organization

A Course in Real Analysis offers considerable flexibility in the choice of material to cover. The following list is a brief explanation of the organization of the text.

- Chapters 1 and 2 present prerequisite material that may be review for many students but provide a common ground for all readers. At the option of the instructor, these two chapters can be covered either briefly or in detail; they can also be assigned to the students for independent reading.

- Chapters 3–6 present the elements of measure and integration by first discussing the Lebesgue theory on the line (Chapters 3 and 4) and then the abstract theory (Chapters 5 and 6). This material is prerequisite to all subsequent chapters.

- Chapter 7 presents an introduction to probability theory that includes the mathematical model for probability, random variables, expectation, and laws of large numbers. Although optional, this chapter is recommended as it provides a myriad of examples and applications for other topics.

- In Chapters 8 and 9, differentiation of functions and differentiation of measures are discussed, respectively. Topics examined include differentiability, bounded variation, absolute continuity of functions, signed and complex measures, the Radon-Nikodym theorem, decomposition of measures, and measurable transformations.

- Chapters 10 and 11 provide the fundamentals of topological and metric spaces. These chapters can be covered relatively quickly when the students have a background in topology from other courses. In addition to topics traditionally found in an introduction to topology, a discussion of weak topologies and function spaces is included.

- Completeness, compactness, and approximation comprise the topics for Chapter 12. Examined therein are the Baire category theorem, contractions of complete metric spaces, compactness in function and product spaces, and the Stone-Weierstrass theorem.

- Hilbert spaces and the classical Banach spaces are presented in Chapter 13. Among other things, bases and duality in Hilbert space, completeness and duality of $\mathcal{L}^p$-spaces, and duality in spaces of continuous functions are discussed.

- The basic theory of normed and locally convex spaces is introduced in Chapter 14. Topics include the Hahn-Banach theorem, linear operators on Banach spaces, fundamental properties of locally convex spaces, and the Krein-Milman theorem.

- Chapter 15 provides applications of previous chapters to harmonic analysis. The elements of Fourier series and transforms and the $\mathcal{L}^2$-theory of the Fourier transform are examined. In addition, an introduction to wavelets and the wavelet transform is presented.

- Chapter 16 introduces measurable dynamical systems and includes a discussion of ergodic theorems, isomorphisms of measurable dynamical systems, and entropy.

- Chapter 17, which is new to this edition, presents outer measure and measurability, Hausdorff measure, Hausdorff dimension and topological dimension, and an introduction to fractals.

Figure P.1 on the next page summarizes the preceding discussion and depicts the interdependence among chapters. In the flowchart, the prerequisites for a given chapter consist of all chapters that have a path leading to that chapter.

Acknowledgments

It is our pleasure to thank the following reviewers of the first and second editions of the book. Their comments and suggestions resulted in significant improvements to the text.

Bruce A. Barnes University of Oregon	Todd Kemp University of California, San Diego
Dennis D. Berkey Boston University	Yon-Seo Kim University of Chicago
Courtney Coleman Harvey Mudd College	Michael Klass University of California, Berkeley
Peter Duren University of Michigan	Enno Lenzmann University of Copenhagen
Wilfrid Gangbo Georgia Institute of Technology	Mara D. Neusel Texas Tech University
Maria Girardi University of South Carolina	Duong H. Phong Columbia University
Sigurdur Helgason Massachusetts Institute of Technology	Bert Schreiber Wayne State University

Our special thanks go to Bruce Barnes, who undertook a detailed reading and critiquing of the entire manuscript. We also thank the graduate students who furnished valuable feedback, in particular, Mohammed Alhodaly, Hamed Alsulami, Jimmy Mopecha, Lynn Tobin, and, especially, Jim Andrews, Trent Buskirk, Menassie Ephrem, Ken Peterson, John Williams, and Xiangrong Yin.

We thank Arizona State University for its support and those chairs of the ASU Mathematics Department who provided encouragement for the project: Rosemary Renaut, Christian Ringhofer, Nevin Savage, and William T. Trotter.

We also thank all of those at Elsevier/Academic Press for helping make this book a reality, in particular, Jeff Freeland, Katy Morrissey, Patricia Osborn, Marilyn Rash, and Lauren Schultz Yuhasz. These people ensured that the process of publishing the second edition of the book went smoothly and efficiently.

Our appreciation goes as well to our proofreaders, Cindy Scott and Carol Weiss. Finally, we would like to express our heartfelt thanks to Carol Weiss. Apart from writing the text, she was involved in every aspect of development and production. Moreover, Carol researched and wrote the biographies.

J.N.M
N.A.W.

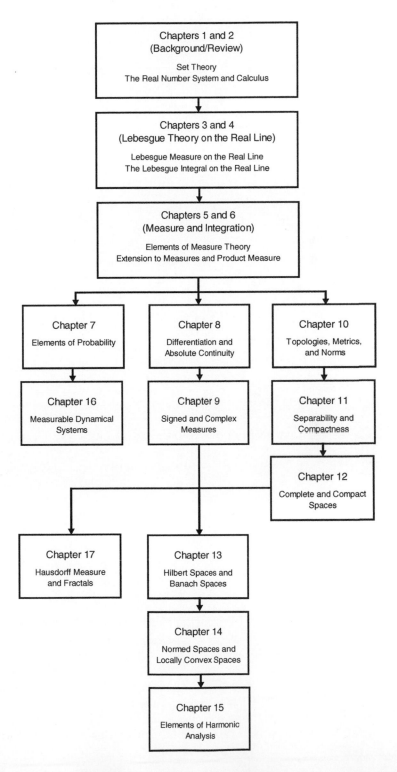

FIGURE P.1 Interdependence among chapters

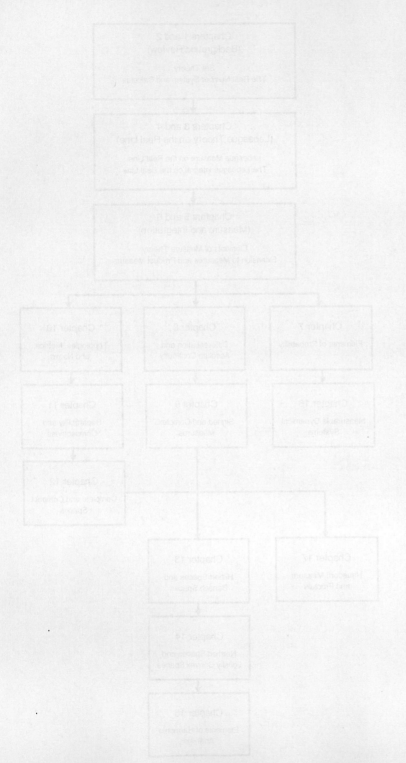

FIGURE P.1 Interdependence among chapters.

PART ONE

□

Set Theory, Real Numbers, and Calculus

Georg Cantor
(1845–1918)

Georg Cantor was born on March 3, 1845, in St. Petersburg, Russia, the son of Georg Waldemar Cantor, a successful merchant and Marie Böhm, whose family included renowned violinists.

Cantor's father refused to allow Cantor to become a violinist so, being interested in mathematics, he entered the University of Zurich in 1862 and transferred to the University of Berlin in 1863 after the death of his father. He received his doctorate in mathematics from the University of Berlin in 1867, having studied under Weierstrass, Kummer, and Kronecker. In 1869, he accepted a teaching position at the University of Halle and became a full professor in 1879.

Cantor wanted to obtain a professorship at the University of Berlin, where both pay and prestige were higher, but Kronecker, believing that much of Cantor's work (particularly his "transfinite numbers") was unsound, stood firmly in Cantor's path.

Others, however, acknowledged Cantor's genius. Mittag-Leffler published Cantor's writings in *Acta Mathematica*. Cantor was an honorary member of the London Mathematical Society and received honorary doctorates from both Christiania and St. Andrews. Hilbert said Cantor's work was "...the finest product of mathematical genius and one of the supreme achievements of purely intellectual human activity."

Known as the founder of set theory, Cantor also made fundamental contributions to classical analysis. Many concepts in modern mathematics bear his name, among which are Cantor series and Cantor sets; he also developed the first usable definition of the continuum. In 1872, he published a paper on trigonometric series in which he used fundamental series to introduce real numbers. Today these fundamental series are called Cauchy sequences.

The controversy surrounding his work took a heavy toll on Cantor; beginning in 1884, bouts of deep depression drove him often to a sanitarium. Cantor died in a psychiatric clinic at the University of Halle on January 6, 1918.

1

Set Theory

In this chapter, we will introduce the fundamentals of set theory. Although some readers may be familiar with much of the material, we present this chapter as a way to provide a common ground for all readers of the text.

We will first discuss basic definitions and properties of sets. Next we will explore relationships between functions and sets, discuss Cartesian products, and introduce countability. Finally, we will examine algebras, σ-algebras, and monotone classes — special collections of sets that play a prominent role in analysis and measure theory.

1.1 BASIC DEFINITIONS AND PROPERTIES

A **set** is a collection of elements. If A is a set and x is an element (member, point) of A, then we write $x \in A$; $x \notin A$ means that x is not an element of A and, in general, we use "/" to signify negation. The symbol $\emptyset$ denotes the **empty set,** a set containing no elements.

Let A and B be sets. If every element of A is an element of B, then A is said to be a **subset** of B, denoted $A \subset B$ or $B \supset A$. Two sets A and B are **equal** if they contain the same elements — in other words, if $A \subset B$ and $B \subset A$. If $A \subset B$ but $B \not\subset A$, then we say that A is a **proper subset** of B.

EXAMPLE 1.1   *Illustrates Sets and Subsets*

In this text, the following sets play a fundamental role:

$$\mathbb{C} = \text{collection of complex numbers}$$
$$\mathcal{R} = \text{collection of real numbers}$$

$$Q = \text{collection of rational numbers}$$
$$\mathcal{Z} = \text{collection of integers}$$

and

$$\mathcal{N} = \text{collection of positive integers}$$

Note that $\mathcal{N} \subset \mathcal{Z} \subset Q \subset \mathcal{R} \subset \mathbb{C}$ or, equivalently, $\mathbb{C} \supset \mathcal{R} \supset Q \supset \mathcal{Z} \supset \mathcal{N}$. But, $\mathbb{C} \not\subset \mathcal{R} \not\subset Q \not\subset \mathcal{Z} \not\subset \mathcal{N}$ or, equivalently, $\mathcal{N} \not\supset \mathcal{Z} \not\supset Q \not\supset \mathcal{R} \not\supset \mathbb{C}$. □

We will use the notation $\{a\}$ to denote the set consisting of the element a; $\{a, b\}$ to denote the set consisting of the elements a and b; $\{a, b, c\}$ to denote the set consisting of the elements a, b, and c; and so on.

Let Ω be a set. Subsets of Ω are frequently defined in terms of properties that its elements must satisfy. If $P(x)$ is some proposition concerning x, then $\{x \in \Omega : P(x)\}$ is the collection of elements $x \in \Omega$ such that $P(x)$ is true. For example, $\{x \in \mathcal{N} : x^2 < 5\} = \{1, 2\}$. When no confusion is possible, we will sometimes abbreviate $\{x \in \Omega : P(x)\}$ to $\{x : P(x)\}$.

Of particular importance in real analysis are **intervals** of real numbers. The notation and terminology associated with these subsets of $\mathcal{R}$ are presented in the following definition.

DEFINITION 1.1   Intervals of Real Numbers

Let a and b be real numbers such that $a < b$. The **bounded intervals** with *endpoints* a and b are as follows:

$$(a, b) = \{x \in \mathcal{R} : a < x < b\}$$
$$[a, b) = \{x \in \mathcal{R} : a \leq x < b\}$$
$$(a, b] = \{x \in \mathcal{R} : a < x \leq b\}$$
$$[a, b] = \{x \in \mathcal{R} : a \leq x \leq b\}.$$

The **unbounded intervals** are as follows:

$$(a, \infty) = \{x \in \mathcal{R} : x > a\}$$
$$[a, \infty) = \{x \in \mathcal{R} : x \geq a\}$$
$$(-\infty, b) = \{x \in \mathcal{R} : x < b\}$$
$$(-\infty, b] = \{x \in \mathcal{R} : x \leq b\}$$
$$(-\infty, \infty) = \{x \in \mathcal{R}\} = \mathcal{R}.$$

Complement, Intersection, and Union

We will now discuss three fundamental operations on sets — complement, intersection, and union. In what follows, we will assume that all sets under consideration are subsets of some fixed set Ω, often referred to as the *universal set*. The set of all subsets of Ω is called the **power set** of Ω and is denoted by $\mathcal{P}(\Omega)$. Thus, $A \subset \Omega$ if and only if $A \in \mathcal{P}(\Omega)$.

Let A and B be subsets of Ω. The **complement** of A, denoted A^c, is the set of elements of Ω that do not belong to A. Thus,

$$A^c = \{\, x : x \notin A \,\}.$$

The **intersection** of A and B, denoted $A \cap B$, is the set of elements of Ω that belong to both A and B. Thus,

$$A \cap B = \{\, x : x \in A \text{ and } x \in B \,\}.$$

The **union** of A and B, denoted $A \cup B$, is the set of elements of Ω that belong to either A or B (or both); in other words, those elements that belong to at least one of A and B. Thus,

$$A \cup B = \{\, x : x \in A \text{ or } x \in B \,\}.$$

Two important relationships among the three set operations of union, intersection, and complement are given in the following proposition, known as **De Morgan's laws.**

□ □ □ **PROPOSITION 1.1 De Morgan's Laws**

Let A and B be subsets of Ω. Then,
a) $(A \cup B)^c = A^c \cap B^c$.
b) $(A \cap B)^c = A^c \cup B^c$.

PROOF We prove part (a) and leave the proof of part (b) as an exercise for the reader. Suppose that $x \in (A \cup B)^c$. Then $x \notin A \cup B$ so that $x \notin A$ and $x \notin B$. But then $x \in A^c$ and $x \in B^c$, which implies that $x \in A^c \cap B^c$. Thus, $(A \cup B)^c \subset A^c \cap B^c$.

Conversely, suppose that $x \in A^c \cap B^c$. Then $x \in A^c$ and $x \in B^c$ so that $x \notin A$ and $x \notin B$. But then $x \notin A \cup B$, which implies that $x \in (A \cup B)^c$. Thus, $A^c \cap B^c \subset (A \cup B)^c$.

We have now shown that $(A \cup B)^c \subset A^c \cap B^c$ and $A^c \cap B^c \subset (A \cup B)^c$. This means that $(A \cup B)^c = A^c \cap B^c$. ■

The following proposition shows that intersection and union obey the **distributive laws.** The proof is left to the reader as an exercise.

□ □ □ **PROPOSITION 1.2 Distributive Laws**

Let A, B, and C be subsets of Ω. Then,
a) $A \cap (B \cup C) = (A \cap B) \cup (A \cap C)$.
b) $A \cup (B \cap C) = (A \cup B) \cap (A \cup C)$.

Relative Complement and Symmetric Difference

Several set operations can be derived from the three basic operations of complement, intersection, and union. Two of the most important such operations are **relative complement** and **symmetric difference.** The definitions of these two set operations follow.

DEFINITION 1.2   Relative Complement

Let A and B be subsets of Ω. Then the **complement of A relative to B**, denoted $\boldsymbol{B \setminus A}$, is the set of all elements belonging to B that do not belong to A. Thus,

$$B \setminus A = \{\, x : x \in B \text{ and } x \notin A \,\}.$$

In particular, we have that $A^c = \Omega \setminus A$; that is, the (absolute) complement of A is the complement of A relative to Ω.

Note: Clearly, we have $B \setminus A = B \cap A^c$.

DEFINITION 1.3   Symmetric Difference

Let A and B be subsets of Ω. Then the **symmetric difference of A and B**, denoted $\boldsymbol{A \triangle B}$, is the set of all elements belonging to either A or B but not both A and B; in other words, those elements that belong to exactly one of A and B. Thus,

$$A \triangle B = \{\, x : x \in A \text{ or } x \in B, \text{ and } x \notin A \cap B \,\}.$$

Note: It is easy to see that $A \triangle B = (A \setminus B) \cup (B \setminus A)$. We leave the verification to the reader as an exercise.

More on Set Operations

Exercises 1.1 and 1.2 discuss several properties of union and intersection. Among those properties are the following two:

$$A \cap (B \cap C) = (A \cap B) \cap C \qquad \text{and} \qquad A \cup (B \cup C) = (A \cup B) \cup C.$$

The two sets in the first equality consist of all elements that belong to A, B, and C, which we write as $A \cap B \cap C$. Thus,

$$A \cap B \cap C = \{\, x : x \in A \text{ and } x \in B \text{ and } x \in C \,\}.$$

The two sets in the second equality consist of all elements that belong to at least one of A, B, and C, which we write as $A \cup B \cup C$. Thus,

$$A \cup B \cup C = \{\, x : x \in A \text{ or } x \in B \text{ or } x \in C \,\}.$$

We can generalize the notions of intersection and union to arbitrary collections of sets.

DEFINITION 1.4   Intersection and Union

Let $\mathcal{C}$ be a collection of subsets of Ω, that is, $\mathcal{C} \subset \mathcal{P}(\Omega)$.

a) The **intersection** of $\mathcal{C}$, denoted $\bigcap_{A \in \mathcal{C}} A$, is the set of elements of Ω that belong to each set in the collection $\mathcal{C}$. Thus,

$$\bigcap_{A \in \mathcal{C}} A = \{\, x : x \in A \text{ for all } A \in \mathcal{C} \,\}.$$

b) The **union** of $\mathcal{C}$, denoted $\bigcup_{A \in \mathcal{C}} A$, is the set of elements of Ω that belong to at least one of the sets in the collection $\mathcal{C}$. Thus,

$$\bigcup_{A \in \mathcal{C}} A = \{\, x : x \in A \text{ for some } A \in \mathcal{C} \,\}.$$

EXAMPLE 1.2   Illustrates Definition 1.4

Let $\Omega = \mathcal{R}$ and $\mathcal{C} = \{\, [0, 1/n] : n \in \mathcal{N} \,\}$. Then

$$\bigcap_{A \in \mathcal{C}} A = \{0\} \qquad \text{and} \qquad \bigcup_{A \in \mathcal{C}} A = [0, 1],$$

as the reader can easily verify.   □

De Morgan's laws and the distributive laws hold for any collection of subsets. These are stated in the following two propositions whose proofs are left to the reader as exercises.

□ □ □   PROPOSITION 1.3 De Morgan's Laws

Let $\mathcal{C}$ be a collection of subsets of Ω. Then,

a) $\left(\bigcap_{A \in \mathcal{C}} A \right)^c = \bigcup_{A \in \mathcal{C}} A^c.$

b) $\left(\bigcup_{A \in \mathcal{C}} A \right)^c = \bigcap_{A \in \mathcal{C}} A^c.$

□ □ □   PROPOSITION 1.4 Distributive Laws

Let $\mathcal{C}$ be a collection of subsets of Ω and B a subset of Ω. Then,

a) $B \cap \left(\bigcup_{A \in \mathcal{C}} A \right) = \bigcup_{A \in \mathcal{C}} (B \cap A).$

b) $B \cup \left(\bigcap_{A \in \mathcal{C}} A \right) = \bigcap_{A \in \mathcal{C}} (B \cup A).$

Indexed Collections of Sets

Suppose that I is a set and that to each $\iota \in I$ there corresponds a unique subset A_ι of Ω. Then we have an **indexed collection** of subsets of Ω, indexed by I. We denote such a collection by $\{A_\iota\}_{\iota \in I}$.

In case $I = \{1, 2, \ldots, N\}$, the indexed collection is denoted by $\{A_n\}_{n=1}^{N}$ and is called a **finite sequence** of sets. Similarly, if $I = \mathcal{N}$, the indexed collection is denoted by $\{A_n\}_{n=1}^{\infty}$ and is called an **infinite sequence** of sets. In both of these cases, we say that the indexed collection is a **sequence** of sets, and we often write $\{A_n\}_n$ to represent either a finite or infinite sequence of sets.

For an indexed collection of sets $\{A_\iota\}_{\iota \in I}$, we denote the intersection and union of the collection by $\bigcap_{\iota \in I} A_\iota$ and $\bigcup_{\iota \in I} A_\iota$, respectively. Thus,

$$\bigcap_{\iota \in I} A_\iota = \{\, x : x \in A_\iota \text{ for all } \iota \in I \,\}$$

and

$$\bigcup_{\iota \in I} A_\iota = \{\, x : x \in A_\iota \text{ for some } \iota \in I \,\}.$$

In case $I = \{1, 2, \ldots, N\}$, we use the notations $\bigcap_{n=1}^{N} A_n$ and $\bigcup_{n=1}^{N} A_n$, respectively, for the intersection and union of the indexed collection. Similarly, if $I = \mathcal{N}$, we use the notations $\bigcap_{n=1}^{\infty} A_n$ and $\bigcup_{n=1}^{\infty} A_n$, respectively, for the intersection and union of the indexed collection. For example, if we let $A_n = (0, 1/n]$ for each $n \in \mathcal{N}$, then

$$\bigcap_{n=1}^{\infty} A_n = \emptyset \quad \text{and} \quad \bigcup_{n=1}^{\infty} A_n = (0, 1].$$

Disjoint Collections of Sets

An essential concept in analysis is that of **disjoint sets.** Two sets are disjoint if they have no elements in common. More generally, we have the following definition.

DEFINITION 1.5 **Disjoint and Pairwise Disjoint**

Two subsets A and B of Ω are said to be **disjoint** if $A \cap B = \emptyset$.

A collection $\mathcal{C}$ of subsets of Ω is said to be **pairwise disjoint** if each two distinct members of $\mathcal{C}$ are disjoint. If $\mathcal{C}$ is a pairwise disjoint collection, we often say the members of $\mathcal{C}$ are pairwise disjoint sets.

An indexed collection $\{A_\iota\}_{\iota \in I}$ of subsets of Ω is said to be **pairwise disjoint** if $A_\iota \cap A_\jmath = \emptyset$ whenever $\iota \neq \jmath$. In case $I = \{1, 2, \ldots, N\}$ or $I = \mathcal{N}$ and the indexed collection is pairwise disjoint, we say that we have a **pairwise disjoint sequence** of subsets of Ω.

EXAMPLE 1.3   *Illustrates Definition 1.5*

Let $\Omega = \mathcal{R}$.

a) The sets $\mathcal{Z}$ and $(0,1)$ are disjoint, since $\mathcal{Z} \cap (0,1) = \emptyset$.

b) The sets $\mathcal{Z}$ and $[0,1]$ are not disjoint, since $\mathcal{Z} \cap [0,1] = \{0,1\} \neq \emptyset$.

c) The indexed collection $\{[n-1,n)\}_{n=1}^{\infty}$ of subsets of $\mathcal{R}$ is pairwise disjoint because
$$[m-1,m) \cap [n-1,n) = \emptyset, \qquad m \neq n.$$

d) The indexed collection $\{[n-1,n]\}_{n=1}^{\infty}$ of subsets of $\mathcal{R}$ is not pairwise disjoint, because, for instance, $[0,1] \cap [1,2] = \{1\} \neq \emptyset$. Note, however, that the intersection of all the members of the collection is in fact empty, that is, $\bigcap_{n=1}^{\infty} [n-1,n] = \emptyset$. This shows that for a collection of sets to be pairwise disjoint it is not sufficient for the intersection of that collection to be empty. Is it necessary? □

Exercises for Section 1.1

1.1 Let A, B, and C be subsets of Ω. Prove each of the following.

a) $A \cup B = B \cup A$

b) $A \cup \emptyset = A$

c) $A \cup (B \cup C) = (A \cup B) \cup C$

d) $A \subset A \cup B$

e) $A = A \cup B$ if and only if $B \subset A$

1.2 Let A, B, and C be subsets of Ω. Prove each of the following.

a) $A \cap B = B \cap A$

b) $A \cap \emptyset = \emptyset$

c) $A \cap (B \cap C) = (A \cap B) \cap C$

d) $A \supset A \cap B$

e) $A = A \cap B$ if and only if $B \supset A$

1.3 Let A and B be subsets of Ω. Verify each of the following statements.

a) $A = (A \cap B) \cup (A \cap B^c)$

b) $A \cap B = \emptyset \Rightarrow A \subset B^c$

c) $A \subset B \Rightarrow B^c \subset A^c$

1.4 Let A and B be subsets of Ω. Prove that

a) $A \setminus B = A \cap B^c$.

b) $A \triangle B = (A \setminus B) \cup (B \setminus A)$.

1.5 Let A, B, and C be subsets of Ω. Establish each of the following facts.

a) $A \triangle (B \triangle C) = (A \triangle B) \triangle C$

b) $A \triangle \Omega = A^c$

c) $A \triangle \emptyset = A$

d) $A \triangle A = \emptyset$

1.6 Let A, B, and C be subsets of Ω.

a) Prove that $A \cap (B \triangle C) = (A \cap B) \triangle (A \cap C)$.

b) What is the relationship between $A \cup (B \triangle C)$ and $(A \cup B) \triangle (A \cup C)$?

c) Precisely when does $A \cup (B \triangle C) = (A \cup B) \triangle (A \cup C)$?

1.7 Let A and B be subsets of Ω. Show that $A = A \triangle B$ if and only if $B = \emptyset$.

1.8 Let $\{A_n\}_{n=1}^{\infty}$ be an infinite sequence of subsets of Ω.

a) Prove that
$$\bigcup_{n=1}^{\infty} \left(\bigcap_{k=n}^{\infty} A_k \right) \subset \bigcap_{n=1}^{\infty} \left(\bigcup_{k=n}^{\infty} A_k \right).$$

The set on the left is called the **limit inferior** of $\{A_n\}_{n=1}^{\infty}$ and is denoted by $\liminf_{n\to\infty} A_n$; the set on the right is called the **limit superior** of $\{A_n\}_{n=1}^{\infty}$ and is denoted by $\limsup_{n\to\infty} A_n$.

b) Describe in words the limit inferior and limit superior of $\{A_n\}_{n=1}^{\infty}$, and use that description to interpret the relation in part (a).

c) Let $\Omega = \mathcal{R}$ and define

$$A_n = \begin{cases} [0, 1 + 1/n], & \text{if } n \text{ is an even positive integer;} \\ [-1 - 1/n, 0], & \text{if } n \text{ is an odd positive integer.} \end{cases}$$

Determine $\liminf_{n\to\infty} A_n$ and $\limsup_{n\to\infty} A_n$.

1.9 Prove the general form of De Morgan's laws, Proposition 1.3 on page 7.

1.10 Prove the general form of the distributive laws for sets, Proposition 1.4 on page 7.

1.11 Let $\mathcal{C}$ be a collection of subsets of Ω and B a subset of Ω. Prove each of the following facts.

a) If $B \subset \bigcup_{A \in \mathcal{C}} A$, then $B = \bigcup_{A \in \mathcal{C}} (A \cap B)$.

b) If $\bigcup_{A \in \mathcal{C}} A = \Omega$, then $E = \bigcup_{A \in \mathcal{C}} (A \cap E)$ for each subset E of Ω.

c) If $\mathcal{C}$ is pairwise disjoint, then so is the collection $\{A \cap E : A \in \mathcal{C}\}$ for each subset E of Ω.

d) We say that $\mathcal{C}$ is a **partition** of Ω if it is pairwise disjoint and its union is Ω. Conclude from parts (b) and (c) that if $\mathcal{C}$ is a partition of Ω, then each subset E of Ω can be expressed as a disjoint union of the collection of sets $\{A \cap E : A \in \mathcal{C}\}$.

1.12 There is a slight distinction between the notions of pairwise disjoint for nonindexed collections of sets and indexed collections of sets, namely, an indexed collection of sets $\{A_\iota\}_{\iota \in I}$ can fail to be pairwise disjoint even though the collection, $\mathcal{C} = \{A_\iota : \iota \in I\}$, is pairwise disjoint. Provide an example that illustrates this fact.

1.13 Give an example of a collection $\mathcal{C}$ of sets that is not pairwise disjoint, has at least four members, and is such that any three distinct members of $\mathcal{C}$ have an empty intersection.

1.2 FUNCTIONS AND SETS

Suppose that Ω and Λ are sets. A **function** (*mapping, transformation*) from Ω to Λ is a rule that assigns to each element $x \in \Omega$ a unique element $f(x) \in \Lambda$.[†] We call $f(x)$ the **value of f at x** or the **image of x under f.**

To indicate that f is a function from Ω to Λ, we often write $f: \Omega \to \Lambda$. The set Ω is called the **domain** of f. The set $\{f(x) : x \in \Omega\}$ is called the **range** of f. We note that, in general, the range of f will be a proper subset of Λ. Two further concepts important in the study of functions are given in Definition 1.6.

DEFINITION 1.6 **One-to-One and Onto**

Let f be a function from Ω to Λ.

a) f is said to be **one-to-one** (or *injective*) if distinct elements of Ω have distinct images; that is, if $f(x_1) = f(x_2)$ implies that $x_1 = x_2$.

[†] We will take an intuitive approach to functions; that is, we will not use the definition based on ordered pairs.

b) f is said to be **onto** (or *surjective*) if each element of Λ is the image of some element of Ω; that is, for each $y \in \Lambda$, there is an $x \in \Omega$ such that $y = f(x)$. Thus, f is onto if and only if the range of f equals Λ.

If a function is both one-to-one and onto, then we can invert the function by using the rule that assigns to each element in the range the unique element in the domain of which it is the image. More precisely, we have the following definition.

DEFINITION 1.7 **Inverse of a Function**

Suppose that $f: \Omega \to \Lambda$ is one-to-one and onto. For $y \in \Lambda$, let $f^{-1}(y)$ be the unique $x \in \Omega$ such that $y = f(x)$. The function $f^{-1}: \Lambda \to \Omega$ so defined is called the **inverse** of the function f.

EXAMPLE 1.4 *Illustrates Definition 1.7*

Define $f: [0, 1] \to [2, 5]$ by $f(x) = 3x^2 + 2$. Then f is one-to-one and onto. As the reader can verify, the inverse of the function f, $f^{-1}: [2, 5] \to [0, 1]$, is given by $f^{-1}(y) = \sqrt{(y - 2)/3}$. □

Let f be a function from Ω to Λ and g be a function from Λ to Γ. Then we can define a function from Ω to Γ by first applying f and then applying g to that result. Here is a formal definition.

DEFINITION 1.8 **Composition of Functions**

Let $f: \Omega \to \Lambda$ and $g: \Lambda \to \Gamma$. Then the **composition** of g with f, denoted $\boldsymbol{g \circ f}$, is the function $g \circ f: \Omega \to \Gamma$ defined by

$$(g \circ f)(x) = g(f(x)).$$

EXAMPLE 1.5 *Illustrates Definition 1.8*

Define $f: \mathcal{R} \to [0, \infty)$ by $f(x) = x^2$ and $g: [0, \infty) \to \mathcal{R}$ by $g(y) = \sqrt{y}$. Then the composition of g with f, $g \circ f: \mathcal{R} \to \mathcal{R}$, is given by

$$(g \circ f)(x) = g(f(x)) = g(x^2) = \sqrt{x^2} = |x|.$$

In this case, we can also consider the composition going the other way, that is, the composition of f with g, $f \circ g: [0, \infty) \to [0, \infty)$, which is given by

$$(f \circ g)(y) = f(g(y)) = f(\sqrt{y}) = (\sqrt{y})^2 = y.$$ □

Sometimes we have a function defined on a set that we want to restrict to a subset of that set. To be specific, suppose $f: \Omega \to \Lambda$ and that $A \subset \Omega$. From f we can obtain a function from A to Λ, called the **restriction** of f to A, denoted $f_{|A}$, and defined by $f_{|A}(x) = f(x)$ for $x \in A$.

Sequences and Subsequences

Sequences constitute an important class of functions. An **infinite sequence** is a function whose domain is the set of positive integers, $\mathcal{N}$. If s is an infinite sequence, then $s(n)$ is called the **nth term** of the sequence and is usually denoted s_n. For ease in notation, we use $\{s_n\}_{n=1}^{\infty}$ to denote both the infinite sequence whose nth term is s_n and the range of the sequence, that is, $\{\, s_n : n \in \mathcal{N} \,\}$; context will determine which meaning is intended.

We will also have occasion to consider finite sequences. A **finite sequence** of length N is a function whose domain is the first N positive integers, $\{1, 2, \ldots, N\}$. As for infinite sequences, if s is a finite sequence, then $s(n)$ is called the **nth term** of the sequence and is usually denoted s_n. For ease in notation, we use $\{s_n\}_{n=1}^{N}$ to denote both the finite sequence of length N whose nth term is s_n and the range of the sequence, that is, $\{\, s_n : n = 1,\, 2,\, \ldots,\, N \,\}$; context will determine which meaning is intended.

We use the term **sequence** to refer to both infinite and finite sequences. The notation $\{s_n\}_n$ represents a sequence that may be finite or infinite and whose nth term is s_n. If the range of a sequence $\{s_n\}_n$ is a subset of a set Ω, then we say that $\{s_n\}_n$ is a **sequence of Ω** or a **sequence of elements of Ω**.

EXAMPLE 1.6   *Illustrates Sequences*

a) The sequence $\{3^{-n}\}_{n=1}^{\infty}$ is an infinite sequence of $\mathcal{R}$.

b) A sequence $\{A_n\}_n$ of subsets of a set Ω, as defined on page 8, is a sequence of $\mathcal{P}(\Omega)$, the set of all subsets of Ω.   ◻

Let $\{s_n\}_{n=1}^{\infty}$ be an infinite sequence and $\{n_k\}_{k=1}^{\infty}$ an infinite sequence of positive integers such that $n_1 < n_2 < \cdots$. Then the sequence $\{s_{n_k}\}_{k=1}^{\infty}$ is said to be a (infinite) **subsequence** of $\{s_n\}_{n=1}^{\infty}$. We note that a subsequence of a sequence is the composition of two functions.

To illustrate subsequences, let $\{s_n\}_{n=1}^{\infty}$ be the sequence in part (a) of Example 1.6 and let $n_k = 2k$. Then $s_{n_k} = 3^{-n_k} = 3^{-2k} = 9^{-k}$. In other words, the subsequence $\{s_{n_k}\}_{k=1}^{\infty}$ is the sequence $\{9^{-n}\}_{n=1}^{\infty}$.

Subsequences of finite sequences are defined similarly to those for infinite sequences. We leave the details to the reader.

Images and Inverse Images

Let $f : \Omega \to \Lambda$. If $A \subset \Omega$, then we define

$$f(A) = \{\, f(x) : x \in A \,\},$$

called the **image of A under f**.

If $B \subset \Lambda$, then we define

$$f^{-1}(B) = \{\, x \in \Omega : f(x) \in B \,\},$$

called the **inverse image of B under f**.

The next two propositions relate set operations and functions. We state the results in terms of indexed collections because that is generally what we deal with.[†] The proofs of the propositions are left to the reader as exercises.

□ □ □ **PROPOSITION 1.5**

Let $f : \Omega \to \Lambda$, $A \subset \Omega$, and $\{A_\iota\}_{\iota \in I}$ an indexed collection of subsets of Ω. Then

a) $f\left(\bigcup_{\iota \in I} A_\iota\right) = \bigcup_{\iota \in I} f(A_\iota)$

and

b) $f\left(\bigcap_{\iota \in I} A_\iota\right) \subset \bigcap_{\iota \in I} f(A_\iota)$.

If f is one-to-one, then

c) $f\left(\bigcap_{\iota \in I} A_\iota\right) = \bigcap_{\iota \in I} f(A_\iota)$

and

d) $f(A^c) \subset \left(f(A)\right)^c$.

And, if f is onto, then

e) $f(A^c) \supset \left(f(A)\right)^c$.

□ □ □ **PROPOSITION 1.6**

Let $f : \Omega \to \Lambda$, $B \subset \Lambda$, and $\{B_\iota\}_{\iota \in I}$ an indexed collection of subsets of Λ. Then

a) $f^{-1}\left(\bigcup_{\iota \in I} B_\iota\right) = \bigcup_{\iota \in I} f^{-1}(B_\iota)$,

b) $f^{-1}\left(\bigcap_{\iota \in I} B_\iota\right) = \bigcap_{\iota \in I} f^{-1}(B_\iota)$,

and

c) $f^{-1}(B^c) = \left(f^{-1}(B)\right)^c$.

[†] Actually, we are not losing generality, as any collection of subsets is an indexed collection that is indexed by the collection itself.

The Axiom of Choice and Zorn's Lemma

Many of the results that we will discuss in this text require more than the axioms of elementary set theory. Rather, they depend in addition on an axiom called the **axiom of choice,** which is independent of (i.e., cannot be derived from) the axioms of elementary set theory.

Roughly speaking, the axiom of choice asserts that given a collection of nonempty sets, it is possible to select an element from each set in the collection. More precisely, we have the following statement.

Axiom of Choice

Suppose that $\mathcal{C}$ is a collection of nonempty sets. Then there exists a function $f \colon \mathcal{C} \to \bigcup_{A \in \mathcal{C}} A$ such that $f(A) \in A$ for each $A \in \mathcal{C}$.

Although most mathematicians use the axiom of choice without hesitation, some employ it only when they cannot obtain a proof without it, and others consider it unacceptable. In this text, we will apply the axiom of choice freely, both tacitly and explicitly.

There are several important equivalences to the axiom of choice. We will discuss only one, namely, *Zorn's lemma*. In preparation for stating Zorn's lemma, we make the following definition.

DEFINITION 1.9 **Partial Ordering; Partially Ordered Set**

Let Ω be a set. A relation $\prec$ on Ω is said to be a **partial ordering** if for all $x, y, z \in \Omega$,
a) $x \prec x$ *[reflexive]*.
b) $x \prec y$ and $y \prec x$ implies $x = y$ *[antisymmetric]*.
c) $x \prec y$ and $y \prec z$ implies $x \prec z$ *[transitive]*.
The pair $(\Omega, \prec)$ is called a **partially ordered set.**

EXAMPLE 1.7 *Illustrates Definition 1.9*

a) We have that $\leq$ is a partial ordering on $\mathcal{R}$ and, hence, $(\mathcal{R}, \leq)$ is a partially ordered set.
b) Let Ω be a set. Then $\subset$ is a partial ordering on $\mathcal{P}(\Omega)$ and, hence, $(\mathcal{P}(\Omega), \subset)$ is a partially ordered set.   □

Let $(\Omega, \prec)$ be a partially ordered set. A subset C of Ω is called a **chain** if for each $x, y \in C$, either $x \prec y$ or $y \prec x$. An element $u \in \Omega$ is called an **upper bound** for a subset A of Ω if $x \prec u$ for all $x \in A$. An element $m \in \Omega$ is called a **maximal element** of Ω if $x \in \Omega$ and $m \prec x$ implies that $x = m$.

With the preceding definitions in mind, we can now state Zorn's lemma which, as we mentioned earlier, is equivalent to the axiom of choice.

Zorn's Lemma

Let $(\Omega, \prec)$ be a nonempty partially ordered set with the property that each chain has an upper bound. Then Ω has a maximal element.

Applications of both the axiom of choice and Zorn's lemma appear throughout the text.

Cartesian Products

Next we will introduce Cartesian products. First we define the Cartesian product of a finite number of sets.

DEFINITION 1.10 **Cartesian Product of a Finite Number of Sets**

Let A and B be two sets. Then the **Cartesian product** of A and B (in that order), denoted $\boldsymbol{A \times B}$, is the set of all ordered pairs (a, b), where $a \in A$ and $b \in B$. Thus,

$$A \times B = \{ (a, b) : a \in A, \ b \in B \}.$$

More generally, if A_1, A_2, ..., A_n are sets, then the Cartesian product of those n sets, denoted $\boldsymbol{A_1 \times A_2 \times \cdots \times A_n}$ or $\boldsymbol{\times_{k=1}^{n} A_k}$, is the set of all ordered n-tuples $(a_1, a_2, \ldots, a_n)$, where $a_k \in A_k$ for $k = 1, 2, \ldots, n$. Thus,

$$\underset{k=1}{\overset{n}{\times}} A_k = \{ (a_1, a_2, \ldots, a_n) : a_k \in A_k, \ 1 \le k \le n \}.$$

An important special case of Cartesian product occurs when all of the sets in the product are identical. If $A_k = A$ for $1 \le k \le n$, where A is some set, then we write A^n for the Cartesian product. In other words,

$$A^n = \underbrace{A \times A \times \cdots \times A}_{n \text{ times}}$$

EXAMPLE 1.8 *Illustrates Definition 1.10*

a) If at least one of A and B are empty, then so is $A \times B$.
b) Let Γ and Λ be two sets, $A \subset \Gamma$ and $B \subset \Lambda$. Then the subset $A \times B$ of $\Gamma \times \Lambda$ is called a **rectangle**. Note that, in general, not every subset of $\Gamma \times \Lambda$ is a rectangle.
c) The set $\mathcal{R}^n$ is called **Euclidean n-space**.
d) The set $\mathbb{C}^n$ is called **unitary n-space**. □

We can generalize the Cartesian product to any collection of sets. This is done as follows.

DEFINITION 1.11 **Cartesian Product of a Collection of Sets**

Let $\{A_\iota\}_{\iota \in I}$ be an indexed collection of sets. The **Cartesian product** of the collection, denoted $\times_{\iota \in I} A_\iota$, is the set of all functions x on I such that $x(\iota)$ is an element of A_ι for each $\iota \in I$. Thus,

$$\underset{\iota \in I}{\times} A_\iota = \left\{ x \colon I \to \bigcup_{\iota \in I} A_\iota \; : \; x(\iota) \in A_\iota, \; \iota \in I \right\}.$$

We call $x(\iota)$ the ι**th coordinate** of x and usually denote it by x_ι.

If $A_\iota = \emptyset$ for some $\iota \in I$, then $\times_{\iota \in I} A_\iota = \emptyset$. Conversely, in view of the axiom of choice, if $A_\iota \neq \emptyset$ for all $\iota \in I$, then $\times_{\iota \in I} A_\iota \neq \emptyset$.

An important special case of Cartesian product occurs when all of the sets in the product are identical. Suppose that $A_\iota = A$ for all $\iota \in I$, where A is some set. Then we write A^I for the Cartesian product. Thus, A^I is the set of all functions from I to A.

EXAMPLE 1.9 *Illustrates Definition 1.11*

a) If $I = \{1, 2, \ldots, n\}$, then we use the notation $\times_{k=1}^{n} A_k$ for the Cartesian product.
b) If $I = \{1, 2, \ldots, n\}$, then we write A^n in place of $A^{\{1,2,\ldots,n\}}$. Thus, A^n denotes the set of all sequences of length n of elements of A.
c) If $I = \mathcal{N}$, we use the notation $\times_{n=1}^{\infty} A_n$ for the Cartesian product.
d) If $I = \mathcal{N}$, then we sometimes write A^∞ in place of $A^\mathcal{N}$. Thus, A^∞ denotes the set of all infinite sequences of elements of A.
e) $\mathcal{R}^{[0,1]}$ is the set of all real-valued functions on $[0, 1]$.
f) $\mathbb{C}^\mathcal{R}$ is the set of all complex-valued functions on $\mathcal{R}$. □

We seemingly have two different definitions of the Cartesian product of a finite number of sets, one given by Definition 1.10 and the other by Definition 1.11. However, the appropriate identification shows that the difference is only apparent, as we now show.

Indeed, let $A_1, A_2, \ldots, A_n$ be n sets. By Definition 1.10, $\times_{k=1}^{n} A_k$ is the set of all ordered n-tuples $(a_1, a_2, \ldots, a_n)$, where $a_k \in A_k$ for $1 \leq k \leq n$. On the other hand, by Definition 1.11, $\times_{k=1}^{n} A_k$ is the set of all functions x on $\{1, 2, \ldots, n\}$ such that $x_k \in A_k$ for $1 \leq k \leq n$. If we identify each such function x with the ordered n-tuple $(x_1, x_2, \ldots, x_n)$, then we obtain a 1-1 correspondence between the Cartesian product $\times_{k=1}^{n} A_k$ as defined in Definition 1.11 and the Cartesian product $\times_{k=1}^{n} A_k$ as defined in Definition 1.10.

We will follow conventional notation and use the ordered n-tuple interpretation of the Cartesian product of a finite number of sets. Thus, for example, we construe $\mathcal{R}^n$ as the set of all ordered n-tuples of real numbers, realizing, however, that it can also be interpreted as the set of all sequences of length n of real numbers.

Exercises for Section 1.2

1.14 Suppose that $f:\Omega \to \Lambda$ is one-to-one and onto. Prove that $f^{-1}(f(x)) = x$ for all $x \in \Omega$ and $f(f^{-1}(y)) = y$ for all $y \in \Lambda$.

1.15 Let $f:\Omega \to \Lambda$.
a) Prove that f is one-to-one if and only if there is a function $g:\Lambda \to \Omega$ such that $(g \circ f)(x) = x$ for all $x \in \Omega$.
b) Prove that f is onto if and only if there is a function $g:\Lambda \to \Omega$ with $(f \circ g)(y) = y$ for all $y \in \Lambda$. *Hint:* The axiom of choice is needed for the "only if" part.
c) Suppose there is a function $g:\Lambda \to \Omega$ such that $(g \circ f)(x) = x$ for all $x \in \Omega$ and $(f \circ g)(y) = y$ for all $y \in \Lambda$. Prove that $g = f^{-1}$.

1.16 Let $\{s_n\}_{n=1}^{\infty}$ be an infinite sequence of elements of Ω and $\{s_{n_k}\}_{k=1}^{\infty}$ a subsequence of $\{s_n\}_{n=1}^{\infty}$. Interpret the subsequence as a composition of two functions.

1.17 Suppose that $f:\Omega \to \Lambda$ is one-to-one and onto. Show that for $B \subset \Lambda$, the two definitions of $f^{-1}(B)$ are consistent; that is, the image of B under f^{-1} equals the inverse image of B under f.

1.18 Prove Proposition 1.5 on page 13.

1.19 Refer to Proposition 1.5 on page 13.
a) Show that the assumption of one-to-one cannot be dropped for parts (c) and (d).
b) Show that the assumption of onto cannot be dropped for part (e).

1.20 Prove Proposition 1.6 on page 13.

1.21 Let $f:\Omega \to \Lambda$, $A \subset \Omega$, and $B \subset \Lambda$.
a) Show that $f(f^{-1}(B)) \subset B$ and that equality holds if f is onto.
b) Show that $f^{-1}(f(A)) \supset A$ and that equality holds if f is one-to-one.

1.22 Show that the axiom of choice is equivalent to the following statement: If $\{A_\iota\}_{\iota \in I}$ is any indexed collection of nonempty sets, then $\bigtimes_{\iota \in I} A_\iota \neq \emptyset$.

1.23 Let Ω be a nonempty set. Construct a one-to-one function from $\mathcal{P}(\Omega)$ onto $\{0, 1\}^{\Omega}$.

1.24 Let Ω be a nonempty set. Prove that there is no function from Ω onto $\mathcal{P}(\Omega)$. *Hint:* Suppose to the contrary that such a function, say, f, exists. Let $A = \{x \in \Omega : x \notin f(x)\}$.

1.3 EQUIVALENCE OF SETS; COUNTABILITY

We see from Proposition 1.5 on page 13, that if $f:\Omega \to \Lambda$ is one-to-one and onto, then, from a set theoretic point of view, Ω and Λ are equivalent because, under those circumstances, the set operations are preserved by f. Thus we can think of f as simply renaming the elements of Ω according to the rule $x \to f(x)$. If $f:\Omega \to \Lambda$ is one-to-one and onto, then it is called a **1-1 correspondence** (or *bijective function*).

Keeping the previous paragraph in mind, we now define set equivalence. Suppose that A and B are any two sets. Let us write $A \sim B$ if there is a 1-1 correspondence from A to B. We leave it as an exercise for the reader to show that for any three sets A, B, and C,

- $A \sim A$ *[reflexive]*.

- $A \sim B$ implies $B \sim A$ *[symmetric]*.

- $A \sim B$ and $B \sim C$ implies $A \sim C$ *[transitive]*.

In view of these facts, we make the following definition.

DEFINITION 1.12   Equivalence of Sets

Two sets are said to be **equivalent** if there is a 1-1 correspondence from one to the other.

Finite, Infinite, Countable, and Uncountable Sets

Using the concept of equivalence of two sets, we can now present definitions regarding the "size" of a set in the sense of how many elements it contains.

DEFINITION 1.13   Finite, Infinite, Countable, and Uncountable

Let A be a set. We say that
a) A is **finite** if it is either empty or equivalent to the first N positive integers for some $N \in \mathcal{N}$. In the former case, A is said to consist of 0 elements and, in the latter case, N elements.
b) A is **infinite** if it is not finite.
c) A is **countably infinite** if it is equivalent to $\mathcal{N}$.
d) A is **countable** if it is either finite or countably infinite.
e) A is **uncountable** if it is not countable.

EXAMPLE 1.10   Illustrates Definition 1.13

a) The set of all integers $\mathcal{Z}$ is countably infinite. Indeed, the function $f : \mathcal{N} \to \mathcal{Z}$ defined by

$$f(n) = \begin{cases} n/2, & n \text{ even}; \\ -(n-1)/2, & n \text{ odd}, \end{cases}$$

is a 1-1 correspondence from $\mathcal{N}$ to $\mathcal{Z}$.
b) Any (nondegenerate) interval of $\mathcal{R}$ is uncountable. One proof of this fact is presented in Exercise 1.26 and another in Exercise 3.34. In particular, $\mathcal{R}$ is uncountable.
c) Define $f : \mathcal{N}^2 \to \mathcal{N}$ by $f(m, n) = 2^{m-1}(2n - 1)$. Then it can be shown (see Exercise 1.27) that f is a 1-1 correspondence. Consequently, $\mathcal{N}^2$ is countably infinite and, hence, countable. ◻

We can express countability in terms of sequences. If A is countably infinite, then, by definition, there is a 1-1 correspondence $f : \mathcal{N} \to A$. If we let $s_n = f(n)$, then the infinite sequence $\{s_n\}_{n=1}^{\infty}$ is called an **enumeration** of A. Similarly, if A is a finite nonempty set, then, by definition, there is an $N \in \mathcal{N}$ and a 1-1 correspondence $f : \{1, 2, \ldots, N\} \to A$. If we let $s_n = f(n)$, then the finite sequence $\{s_n\}_{n=1}^{N}$ is called an enumeration of A.

In particular, we see that if a set is countably infinite, then it is the range of an infinite sequence (but not conversely); and that if a set is finite and nonempty, then it is the range of a finite sequence (and conversely). The following proposition is also quite useful.

□ □ □ PROPOSITION 1.7

A nonempty set is countable if and only if it is the range of an infinite sequence.

PROOF Suppose A is countable. Then, by definition, it is either finite or countably infinite. If it is countably infinite, then, by definition, it is equivalent to $\mathcal{N}$, which means there is a one-to-one and onto function f from $\mathcal{N}$ to A. Letting $s_n = f(n)$, we have that A is the range of the infinite sequence $\{s_n\}_{n=1}^{\infty}$.

If A is finite (and nonempty), then, by definition, there is an $N \in \mathcal{N}$ such that A is equivalent to the first N positive integers. Let g be a one-to-one and onto function from $\{1, 2, \ldots, N\}$ to A. Select $x \in A$ and define the infinite sequence s by $s_n = g(n)$ for $n = 1, 2, \ldots, N$, and $s_n = x$ for $n > N$. Then A is the range of the infinite sequence $\{s_n\}_{n=1}^{\infty}$.

Conversely, suppose A is the range of an infinite sequence $\{s_n\}_{n=1}^{\infty}$. We claim that A is countable. If A is finite, there is nothing to prove. So, assume that A is infinite. We will construct a 1-1 correspondence from $\mathcal{N}$ to A thereby proving that A is countably infinite and, hence, countable.

Let $n_1 = 1$. Since A is infinite, $A \setminus \{s_{n_1}\} \neq \emptyset$. Therefore, because the range of $\{s_n\}_{n=1}^{\infty}$ is A, the set $\{n \in \mathcal{N} : s_n \neq s_1\}$ is not empty. Denote by n_2 the smallest integer in that set.[†] Note that $n_1 < n_2$.

Proceeding inductively, note again that since A is infinite, we have that $A \setminus \{s_{n_1}, s_{n_2}, \ldots, s_{n_k}\} \neq \emptyset$. Therefore, as the range of $\{s_n\}_{n=1}^{\infty}$ is A, the set $\{n \in \mathcal{N} : s_n \neq s_{n_j}, 1 \leq j \leq k\}$ is not empty. Denote by n_{k+1} the smallest integer in that set and note that $n_k < n_{k+1}$.

We claim that the function $f : \mathcal{N} \to A$ defined by $f(k) = s_{n_k}$ is a 1-1 correspondence. By construction, f is one-to-one. So it remains to show that f is onto. Let $x \in A$. Since the range of $\{s_n\}_{n=1}^{\infty}$ is A, the set $\{n \in \mathcal{N} : s_n = x\}$ is not empty. Let m be the smallest integer in that set. If $m = 1$, then $x = s_1 = s_{n_1} = f(1)$. Otherwise, let k be the smallest integer such that $m \leq n_k$. Because $s_n \neq x = s_m$ for $n < m$, we have that $s_m \neq s_{n_j}$ for $1 \leq j \leq k - 1$, which implies that $m \geq n_k$. Therefore, $m = n_k$ and, consequently, $x = s_{n_k} = f(k)$. ∎

Proposition 1.7 often provides an efficient method for proving that a set is countable. The next two propositions illustrate that fact.

□ □ □ PROPOSITION 1.8

A subset of a countable set is countable.

PROOF Let A be a countable set and $B \subset A$. We claim that B is countable. If $B = \emptyset$, there is nothing to prove; so, assume $B \neq \emptyset$. This implies that A is nonempty and, hence, by Proposition 1.7, A is the range of an infinite sequence $\{s_n\}_{n=1}^{\infty}$.

Choose $x \in B$ and let $t_n = s_n$ if $s_n \in B$ and $t_n = x$ if $s_n \notin B$. Then B is the range of the infinite sequence $\{t_n\}_{n=1}^{\infty}$. Applying Proposition 1.7 again, we conclude that B is countable. ∎

[†] Here and elsewhere in this proof we are using the **well-ordering principle:** *Each nonempty subset of positive integers has a smallest element.*

□ □ □ **PROPOSITION 1.9**

The image of a countable set is countable.

PROOF Let A be a countable set and f a function defined on A. By Proposition 1.7, A is the range of an infinite sequence $\{s_n\}_{n=1}^{\infty}$. For each $n \in \mathcal{N}$, define $t_n = f(s_n)$.

Now, let $y \in f(A)$. Then there is an $x \in A$ such that $f(x) = y$. Since A is the range of the infinite sequence $\{s_n\}_{n=1}^{\infty}$, there is an $n \in \mathcal{N}$ such that $s_n = x$. Therefore, $y = f(x) = f(s_n) = t_n$. This shows that $f(A)$ is the range of the infinite sequence $\{t_n\}_{n=1}^{\infty}$. Hence, by Proposition 1.7, $f(A)$ is countable. ■

□ □ □ **PROPOSITION 1.10**

A countable union of countable sets is countable.

PROOF Suppose that $\mathcal{C}$ is a countable collection of countable subsets of a set Ω, and let $A = \bigcup_{C \in \mathcal{C}} C$. We must prove that A is countable.

If $\mathcal{C}$ is empty, then its union is empty and hence countable. So, assume that $\mathcal{C}$ is a nonempty collection. Without loss of generality, we can also assume that each member of $\mathcal{C}$ is nonempty.

Since $\mathcal{C}$ is nonempty and countable, Proposition 1.7 implies that it is the range of an infinite sequence $\{A_n\}_{n=1}^{\infty}$ and, since each member of $\mathcal{C}$ is countable, Proposition 1.7 implies that each A_n is the range of an infinite sequence $\{x_m^{(n)}\}_{m=1}^{\infty}$.

Now, define $f: \mathcal{N}^2 \to A$ by $f(m, n) = x_m^{(n)}$ and note that f is onto. By Example 1.10(c) on page 18, $\mathcal{N}^2$ is countable. Therefore, by Proposition 1.9, A is countable, being the image of $\mathcal{N}^2$ under f. ■

In Example 1.10(c) we pointed out that $\mathcal{N}^2$, the Cartesian product of $\mathcal{N}$ with itself, is countable. More generally, we have the following fact.

□ □ □ **PROPOSITION 1.11**

The Cartesian product of two countable sets is countable.

PROOF Let A and B be two countable sets. If either A or B is empty, then so is $A \times B$. So assume that both A and B are nonempty. By Proposition 1.7, A and B are the range of infinite sequences, say, $\{a_n\}_{n=1}^{\infty}$ and $\{b_n\}_{n=1}^{\infty}$. Define $f: \mathcal{N}^2 \to A \times B$ by $f(m, n) = (a_m, b_n)$. Then f is onto and, consequently, because $\mathcal{N}^2$ is countable, Proposition 1.9 implies that $A \times B$ is countable. ■

We can easily extend Proposition 1.11 to any finite number of sets. This extension will be explored in the exercises.

□ □ □ **PROPOSITION 1.12**

The set Q of rational numbers is countable.

PROOF Example 1.10(a) on page 18 shows that $\mathcal{Z}$ is countable. Hence, by Proposition 1.11, so is $\mathcal{Z} \times \mathcal{N}$. Define $f: \mathcal{Z} \times \mathcal{N} \to Q$ by $f(z, n) = z/n$. Since f is onto, Proposition 1.9 implies that Q is countable. ■

Exercises for Section 1.3

Note: A ★ denotes an exercise that will be subsequently referenced.

1.25　If A and B are sets, write $A \sim B$ if there is a 1-1 correspondence from A to B. Prove that $\sim$ is reflexive, symmetric, and transitive. In other words, if A, B, and C are sets, show that the following properties hold.
a) $A \sim A$ ［*reflexive*］
b) $A \sim B$ implies $B \sim A$ ［*symmetric*］
c) $A \sim B$ and $B \sim C$ implies $A \sim C$ ［*transitive*］

★1.26　In this exercise, we will prove that any (nondegenerate) interval of $\mathcal{R}$ is uncountable.
a) Show that the interval $[0, 1)$ is uncountable. *Hint:* Suppose to the contrary that $[0, 1)$ is countable and let $\{x_n\}_{n=1}^{\infty}$ be an enumeration of its elements. For each $n \in \mathcal{N}$, let $0.d_{n1}d_{n2}\ldots$ denote the unique decimal expansion of x_n not containing only finitely many digits differing from 9. Then consider the number $0.a_1 a_2 \ldots$, where $a_n = 1$ if $d_{nn} = 0$ and $a_n = 0$ otherwise.
b) Use part (a) to conclude that $(0, 1)$ is uncountable.
c) Use part (b) to show that any bounded interval of the form (a, b) is uncountable. *Hint:* Construct a one-to-one and onto function from $(0, 1)$ to (a, b).
d) Use part (c) to conclude that any interval is uncountable. In particular, $\mathcal{R}$ is uncountable.

1.27　Refer to Example 1.10(c) on page 18. Prove that the function $f : \mathcal{N}^2 \to \mathcal{N}$ defined by $f(m, n) = 2^{m-1}(2n - 1)$ is a 1-1 correspondence.

★1.28　Prove that any infinite set contains a countably infinite subset.

1.29　Let A be a set. Prove that the following statements are equivalent.
a) A is infinite.
b) There is a one-to-one function $f : A \to A$ that is not onto.
c) There is an onto function $g : A \to A$ that is not one-to-one.

1.30　Suppose $f : A \to B$ is one-to-one and that B is countable. Prove that A is countable.

1.31　Prove that the Cartesian product of a finite number of countable sets is countable.

1.32　In Proposition 1.11, we proved that the Cartesian product of two countable sets is countable and, in Exercise 1.31, we showed that the Cartesian product of a finite number of countable sets is countable. Is it true, in general, that the Cartesian product of a countable number of countable sets is countable?

★1.33　Let Ω be a set. A relation $\equiv$ on Ω is said to be an **equivalence relation** if for all $x, y, z \in \Omega$,

- $x \equiv x$ ［*reflexive*］
- $x \equiv y$ implies $y \equiv x$ ［*symmetric*］
- $x \equiv y$ and $y \equiv z$ implies $x \equiv z$ ［*transitive*］

a) Give three examples of equivalence relations.
b) Give three examples of relations that are not equivalence relations.

1.34　Refer to Exercise 1.33. Let Ω be a nonempty set and $\equiv$ an equivalence relation on Ω. For each $x \in \Omega$, define $E_x = \{ y \in \Omega : y \equiv x \}$. And let $\mathcal{C} = \{ E_x : x \in \Omega \}$. Each member of $\mathcal{C}$ is called an **equivalence class** of Ω under $\equiv$.
a) Show that for each $x, y \in \Omega$, either $E_x \cap E_y = \emptyset$ or $E_x = E_y$.
b) Prove that $\Omega = \bigcup_{A \in \mathcal{C}} A$.
c) Conclude that $\equiv$ partitions Ω into disjoint equivalence classes; that is, Ω is a disjoint union of the equivalence classes under $\equiv$.

1.35　Let a and b be real numbers such that $a < b$. Prove that the intervals (a, b) and $[a, b]$ are equivalent.

1.36 Prove the **Schröder-Bernstein theorem:** Suppose that A and B are sets and that there are one-to-one functions $f\colon A \to B$ and $g\colon B \to A$. Then $A \sim B$. Proceed as follows. Define

$$\tau(E) = g(f(E)^c)^c, \qquad E \subset A.$$

a) Show that if $E \subset F \subset A$, then $\tau(E) \subset \tau(F)$.
b) Let $\mathcal{C} = \{ E \subset A : E \subset \tau(E) \}$ and set $G = \bigcup_{E \in \mathcal{C}} E$. Prove that $\tau(G) = G$ and, hence, that $G^c = g(f(G)^c)$. In particular, G^c is a subset of the range of g.
c) Define $h\colon A \to B$ by

$$h(x) = \begin{cases} f(x), & \text{if } x \in G; \\ g^{-1}(x), & \text{if } x \notin G. \end{cases}$$

Prove that h is a 1-1 correspondence.

1.4 ALGEBRAS, σ-ALGEBRAS, AND MONOTONE CLASSES

In set theory, as in other branches of mathematics, it is important to distinguish collections that are closed under the relevant operations.[†] For example, in linear algebra, the relevant operations are vector addition and scalar multiplication. Subsets of vector spaces closed under those operations are called subspaces and receive intensive study because of their significance.

Algebras

The three basic operations in set theory are union, intersection, and complementation. A nonempty collection of sets closed under these operations is called an **algebra of sets.** Thus, we make the following definition.

DEFINITION 1.14 **Algebra of Sets**

Let Ω be a set. A nonempty collection $\mathcal{A}_0$ of subsets of Ω is called an **algebra** if the following two conditions are satisfied:
a) $A \in \mathcal{A}_0$ implies $A^c \in \mathcal{A}_0$.
b) $A, B \in \mathcal{A}_0$ implies $A \cup B \in \mathcal{A}_0$.

Conspicuous by its absence in Definition 1.14 is closure under intersection. However, it is easy to show that this property follows from the two stated in the definition—an algebra is necessarily closed under intersection; that is, if $\mathcal{A}_0$ is an algebra and $A, B \in \mathcal{A}_0$, then $A \cap B \in \mathcal{A}_0$. We leave the proof of that fact to the reader as an exercise, as we do the proofs of the following two facts.

- An algebra is closed under finite unions and intersections; that is, if $\mathcal{A}_0$ is an algebra and $A_k \in \mathcal{A}_0$ for $k = 1, 2, \ldots, n$, then $\bigcup_{k=1}^n A_k \in \mathcal{A}_0$ and $\bigcap_{k=1}^n A_k \in \mathcal{A}_0$.

- A nonempty collection of subsets of Ω is an algebra if it is closed under complementation and intersection.

[†] Roughly speaking, a collection (set) $\mathcal{C}$ is closed under an operation if whenever the operation is applied to elements of $\mathcal{C}$, the resulting element also belongs to $\mathcal{C}$.

EXAMPLE 1.11   *Illustrates Definition 1.14*

Let Ω be a nonempty set. It is easy to see that each of the following is an algebra of subsets of Ω:
a) the power set $\mathcal{P}(\Omega)$, that is, the set of all subsets of Ω;
b) the trivial algebra $\{\emptyset, \Omega\}$; and
c) $\{\emptyset, A, A^c, \Omega\}$, where A is a nonempty proper subset of Ω. $\Box$

Next we will prove that the union of a sequence of members of an algebra can always be expressed as a disjoint union of members of the algebra. More precisely, we have the following useful proposition.

$\Box\ \Box\ \Box$ **PROPOSITION 1.13**

Let $\mathcal{A}_0$ be an algebra of subsets of Ω and $\{A_n\}_n$ a sequence of sets in $\mathcal{A}_0$ (i.e., $A_n \in \mathcal{A}_0$ for each n). Then there is a pairwise disjoint sequence $\{B_n\}_n$ of sets in $\mathcal{A}_0$ such that $\bigcup_n A_n = \bigcup_n B_n$.

PROOF   The proof uses a process that we will refer to informally as *disjointizing*. Let $B_1 = A_1$ and, for $n \geq 2$, let $B_n = A_n \setminus (\bigcup_{k=1}^{n-1} A_k)$.

First we prove that $B_n \in \mathcal{A}_0$ for each n. Let $C_n = \bigcup_{k=1}^{n-1} A_k$. As $\mathcal{A}_0$ is an algebra, we have, in turn, that $C_n \in \mathcal{A}_0$ (because $\mathcal{A}_0$ is closed under finite unions), that $C_n^c \in \mathcal{A}_0$ (because $\mathcal{A}_0$ is closed under complementation), and that $B_n = A_n \setminus C_n = A_n \cap C_n^c \in \mathcal{A}_0$ (because $\mathcal{A}_0$ is closed under intersection).

Next we show that $\bigcup_n A_n = \bigcup_n B_n$. Since $B_n \subset A_n$ for each n, it is clear that $\bigcup_n A_n \supset \bigcup_n B_n$. To show the reverse inclusion, let $x \in \bigcup_n A_n$. Then $x \in A_n$ for some n. Let m be the smallest such n. If $m = 1$, then $x \in A_1 = B_1$. If $m \geq 2$, we have $x \in A_m$ and $x \notin A_k$ for $k < m$. This result means that $x \in A_m$ but $x \notin \bigcup_{k=1}^{m-1} A_k$; in other words,

$$x \in A_m \setminus \left(\bigcup_{k=1}^{m-1} A_k \right) = B_m \subset \bigcup_n B_n.$$

Thus, $\bigcup_n A_n \subset \bigcup_n B_n$.   ∎

It is useful to know that given a collection of subsets, there is a smallest algebra containing the collection. We state this fact formally in the following proposition whose proof is left to the reader as an exercise.

$\Box\ \Box\ \Box$ **PROPOSITION 1.14**

Let $\mathcal{C}$ be a nonempty collection of subsets of Ω. Then there is a smallest algebra of subsets of Ω containing $\mathcal{C}$.

The smallest algebra containing a collection $\mathcal{C}$ of subsets of Ω is called the **algebra generated** by $\mathcal{C}$ and is denoted $\sigma_0(\mathcal{C})$. Thus, $\sigma_0(\mathcal{C})$ is an algebra of subsets of Ω; $\mathcal{C} \subset \sigma_0(\mathcal{C})$; and if $\mathcal{A}_0$ is an algebra of subsets of Ω such that $\mathcal{C} \subset \mathcal{A}_0$, then $\mathcal{A}_0 \supset \sigma_0(\mathcal{C})$.

As a simple example, let A be a nonempty proper subset of a set Ω. Then $\sigma_0(\{A\}) = \{\emptyset, A, A^c, \Omega\}$.

σ-Algebras

As we have seen, an algebra of sets is closed under finite unions (and intersections). For the purposes of modern mathematics, a stronger condition is usually required, namely, closure under countably-infinite unions (and intersections). Hence, we make the following definition.

DEFINITION 1.15 **σ-Algebra of Sets**

Let Ω be a set. A nonempty collection $\mathcal{A}$ of subsets of Ω is called a **σ-algebra** if the following two conditions are satisfied:

a) $A \in \mathcal{A}$ implies $A^c \in \mathcal{A}$

b) $\{A_n\}_n \subset \mathcal{A}$ implies $\bigcup_n A_n \in \mathcal{A}$

Using the same type of argument used for algebras, we can show that a σ-algebra is necessarily closed under countable intersections; that is, if $\mathcal{A}$ is a σ-algebra and $\{A_n\}_n \subset \mathcal{A}$, then $\bigcap_n A_n \in \mathcal{A}$. We leave the proof of this fact to the reader as an exercise. We also leave it to the reader to prove that a nonempty collection of subsets of Ω is a σ-algebra if it is closed under complementation and countable intersections.

EXAMPLE 1.12 *Illustrates Definition 1.15*

a) Clearly, any σ-algebra is an algebra. However, the converse is not true. See the exercises for several examples.

b) The three algebras given in Example 1.11 are also σ-algebras. □

Additional examples of σ-algebras are presented in the exercises. We will also encounter several σ-algebras in future chapters; for instance, in Chapter 3, we will discuss the σ-algebra of Borel sets and the σ-algebra of Lebesgue measurable sets.

It is useful to know that given a collection of subsets, there is a smallest σ-algebra containing the collection. We state this fact formally in the following proposition whose proof is left to the reader as an exercise.

□ □ □ **PROPOSITION 1.15**

Let $\mathcal{C}$ be a nonempty collection of subsets of Ω. Then there is a smallest σ-algebra of subsets of Ω containing $\mathcal{C}$.

The smallest σ-algebra containing a collection $\mathcal{C}$ of subsets of Ω is called the **σ-algebra generated** by $\mathcal{C}$ and is denoted $\sigma(\mathcal{C})$. Thus, $\sigma(\mathcal{C})$ is a σ-algebra of subsets of Ω; $\mathcal{C} \subset \sigma(\mathcal{C})$; and if $\mathcal{A}$ is a σ-algebra of subsets of Ω such that $\mathcal{C} \subset \mathcal{A}$, then $\mathcal{A} \supset \sigma(\mathcal{C})$.

As a simple example, let A be a nonempty proper subset of a set Ω. Then $\sigma(\{A\}) = \{\emptyset, A, A^c, \Omega\}$.

Monotone Classes and the Monotone Class Theorem

Besides algebras and σ-algebras, we also need to consider monotone classes. Here is the definition of a monotone class.

DEFINITION 1.16   Monotone Class

Let Ω be a set. A nonempty collection $\mathcal{D}$ of subsets of Ω is called a **monotone class** if it satisfies the following two conditions:

a) $\{D_n\}_{n=1}^{\infty} \subset \mathcal{D}$ and $D_1 \subset D_2 \subset \cdots$ implies $\bigcup_{n=1}^{\infty} D_n \in \mathcal{D}$.

b) $\{D_n\}_{n=1}^{\infty} \subset \mathcal{D}$ and $D_1 \supset D_2 \supset \cdots$ implies $\bigcap_{n=1}^{\infty} D_n \in \mathcal{D}$.

Let $\{A_n\}_{n=1}^{\infty}$ be a sequence of subsets of Ω. If $A_1 \subset A_2 \subset \cdots$, then the sequence is said to be **monotone nondecreasing** or, more simply, **nondecreasing**. If $A_1 \supset A_2 \supset \cdots$, then the sequence is said to be **monotone nonincreasing** or, more simply, **nonincreasing**. A sequence of subsets is called **monotone** if it is either monotone nondecreasing or monotone nonincreasing. Using this terminology, we see that a monotone class is a collection of subsets that is closed under unions of nondecreasing sequences and intersections of nonincreasing sequences.

EXAMPLE 1.13   *Illustrates Definition 1.16*

a) Any σ-algebra is a monotone class.

b) Let $A \subset \Omega$ and $\mathcal{D} = \{A\}$. Then, trivially, $\mathcal{D}$ is a monotone class. Note, however, that it is not a σ-algebra.   ◻

For us, the most important result regarding monotone classes is the following theorem, known as the **monotone class theorem.**

◻ ◻ ◻   **THEOREM 1.1**   Monotone Class Theorem

Let Ω be a set and $\mathcal{A}_0$ an algebra of subsets of Ω. Let $\mathcal{D}$ be a collection of subsets of Ω such that $\mathcal{D} \supset \mathcal{A}_0$ and $\mathcal{D}$ is a monotone class. Then $\mathcal{D} \supset \sigma(\mathcal{A}_0)$, the σ-algebra generated by $\mathcal{A}_0$.

PROOF   Let $\mathcal{F}$ be the smallest monotone class that contains $\mathcal{A}_0$. (Exercise 1.53 guarantees the existence of $\mathcal{F}$.) We claim that $\mathcal{F} = \sigma(\mathcal{A}_0)$. Since every σ-algebra is a monotone class, we have $\sigma(\mathcal{A}_0) \supset \mathcal{F}$. If we can show $\mathcal{F}$ is a σ-algebra, that will imply $\sigma(\mathcal{A}_0) \subset \mathcal{F}$, and the desired equality will follow. First we show that $\mathcal{F}$ is an algebra.

Suppose that $A \in \mathcal{A}_0$, and let $\mathcal{E} = \{ F \in \mathcal{F} : A \cup F \in \mathcal{F} \}$. We will show that $\mathcal{E}$ is a monotone class containing $\mathcal{A}_0$. Since $\mathcal{A}_0$ is an algebra and $\mathcal{F} \supset \mathcal{A}_0$, it follows that $\mathcal{E} \supset \mathcal{A}_0$. Now suppose that $\{E_n\}_n \subset \mathcal{E}$ and $E_1 \subset E_2 \subset \cdots$. Because $\{E_n\}_n \subset \mathcal{F}$ and $\mathcal{F}$ is a monotone class, $\bigcup_n E_n \in \mathcal{F}$. And, because $\mathcal{F}$ is a monotone class, $\{A \cup E_n\}_n \subset \mathcal{F}$, and $A \cup E_1 \subset A \cup E_2 \subset \cdots$, we have that $A \cup \left(\bigcup_n E_n \right) = \bigcup_n (A \cup E_n) \in \mathcal{F}$. Therefore, $\bigcup_n E_n \in \mathcal{E}$. Similarly, $\mathcal{E}$ is closed under intersections of nonincreasing sequences. So $\mathcal{E}$ is a monotone class containing $\mathcal{A}_0$ and, consequently, $\mathcal{E} \supset \mathcal{F}$. But, by definition, $\mathcal{E} \subset \mathcal{F}$. Thus, $\mathcal{E} = \mathcal{F}$. In other words, $A \cup F \in \mathcal{F}$ for all $A \in \mathcal{A}_0$ and $F \in \mathcal{F}$.

Now suppose that $G \in \mathcal{F}$, and let $\mathcal{G} = \{ F \in \mathcal{F} : F \cup G \in \mathcal{F} \}$. We will show that $\mathcal{G}$ is a monotone class containing $\mathcal{A}_0$. From the previous paragraph, we know that $\mathcal{G} \supset \mathcal{A}_0$ and, using the same argument as in that paragraph, we can show that $\mathcal{G}$ is a monotone class. This implies that $\mathcal{G} = \mathcal{F}$. In other words, $F \cup G \in \mathcal{F}$ for all $F, G \in \mathcal{F}$. Hence, $\mathcal{F}$ is closed under union.

Next we show that $\mathcal{F}$ is closed under complementation. To that end, let $\mathcal{H} = \{ F \in \mathcal{F} : F^c \in \mathcal{F} \}$. Because $\mathcal{A}_0$ is an algebra and $\mathcal{F} \supset \mathcal{A}_0$, it follows that $\mathcal{H} \supset \mathcal{A}_0$. Also, because $\mathcal{F}$ is a monotone class, it is easy to see that $\mathcal{H}$ is a monotone class. Therefore, $\mathcal{H} = \mathcal{F}$; that is, $\mathcal{F}$ is closed under complementation.

We have now shown that $\mathcal{F}$ is an algebra of sets. To show that it is a σ-algebra, we need only prove that it is closed under countably-infinite unions. So let $\{F_n\}_{n=1}^{\infty} \subset \mathcal{F}$. For $n \in \mathcal{N}$, let $E_n = \bigcup_{k=1}^{n} F_k$. Then $E_1 \subset E_2 \subset \cdots$ and $\bigcup_{n=1}^{\infty} E_n = \bigcup_{n=1}^{\infty} F_n$. Since $\mathcal{F}$ is an algebra, $\{E_n\}_{n=1}^{\infty} \subset \mathcal{F}$ and, therefore, since $\mathcal{F}$ is a monotone class, $\bigcup_{n=1}^{\infty} F_n = \bigcup_{n=1}^{\infty} E_n \in \mathcal{F}$. Hence, $\mathcal{F}$ is a σ-algebra.

We now know that $\mathcal{F} = \sigma(\mathcal{A}_0)$. That is, the smallest monotone class that contains $\mathcal{A}_0$ is $\sigma(\mathcal{A}_0)$. Because $\mathcal{D}$ is a monotone class that contains $\mathcal{A}_0$, it must be that $\mathcal{D} \supset \sigma(\mathcal{A}_0)$. ∎

In proving the monotone class theorem, we showed that *the smallest monotone class that contains an algebra of subsets is the σ-algebra generated by the algebra.* That result is important in its own right.

Exercises for Section 1.4

1.37 Suppose that $\mathcal{A}_0$ is an algebra.
a) Show that $\mathcal{A}_0$ is closed under intersection; that is, $A, B \in \mathcal{A}_0$ implies $A \cap B \in \mathcal{A}_0$.
b) Show that $\emptyset \in \mathcal{A}_0$.

1.38 Prove that an algebra is closed under finite unions and intersections.

1.39 Show that if a collection of subsets of Ω is closed under complementation and intersection, then it is an algebra.

1.40 Let Ω be an infinite set and $\mathcal{D} = \{ A \subset \Omega : A \text{ is finite or } A^c \text{ is finite} \}$. Prove that $\mathcal{D}$ is an algebra.

1.41 This exercise generalizes Example 1.11(c). Suppose that $\{A_k\}_{k=1}^{n}$ is a pairwise disjoint finite sequence of nonempty subsets of Ω whose union is Ω. Let $\mathcal{D}$ be the collection of all finite (including empty) unions of members of $\{A_k\}_{k=1}^{n}$. Prove that $\mathcal{D}$ is an algebra.

1.42 Let $\mathcal{C}$ denote the collection of all intervals of $\mathcal{R}$, including degenerate intervals of the form $[a, a]$ and (a, a). And let $\mathcal{D}$ be the collection of finite disjoint unions of members of $\mathcal{C}$. Prove that $\mathcal{D}$ is an algebra. *Hint:* First show that $\mathcal{D}$ is closed under intersection and then under complementation.

1.43 Let Ω be a set. A nonempty collection $\mathcal{S}$ of subsets of Ω is called a **semialgebra** if the following conditions hold:

- $A, B \in \mathcal{S}$ implies $A \cap B \in \mathcal{S}$.
- $A \in \mathcal{S}$ implies that either $A^c = \emptyset$ or there is a pairwise disjoint finite sequence $\{A_k\}_{k=1}^{n}$ of members of $\mathcal{S}$ such that $A^c = \bigcup_{k=1}^{n} A_k$.

In words, $\mathcal{S}$ is a semialgebra if it is closed under intersection and the complement of each member of $\mathcal{S}$ is a finite (possibly empty) disjoint union of members of $\mathcal{S}$.

a) Show that any algebra is a semialgebra.

b) Give two examples of semialgebras that are not algebras.

c) Let $\{A_k\}_{k=1}^n$ be a pairwise disjoint finite sequence of nonempty subsets of Ω whose union is Ω. Set $\mathcal{S} = \{\emptyset\} \cup \{A_k : 1 \le k \le n\}$. Prove that $\mathcal{S}$ is a semialgebra.

d) Let $\mathcal{C}$ denote the collection of all intervals of $\mathcal{R}$, including degenerate intervals of the form $[a, a]$ and (a, a). Show that $\mathcal{C}$ is a semialgebra.

e) Let $\mathcal{S}$ be a semialgebra and $\mathcal{D}$ the collection consisting of the empty set and all finite disjoint unions of members of $\mathcal{S}$. Prove that $\mathcal{D}$ is an algebra. *Hint:* First show that $\mathcal{D}$ is closed under intersection and then under complementation.

1.44 Prove Proposition 1.14 on page 23: Let $\mathcal{C}$ be a nonempty collection of subsets of Ω. Then there is a smallest algebra of subsets of Ω containing $\mathcal{C}$. *Hint:* Consider the collection of all algebras of subsets of Ω that contain $\mathcal{C}$.

1.45 Refer to Exercise 1.43. In part (e) of that exercise, we proved that the collection $\mathcal{D}$ consisting of the empty set and all finite disjoint unions of members of a semialgebra, $\mathcal{S}$, constitutes an algebra. Show that $\mathcal{D}$ is the algebra generated by $\mathcal{S}$.

1.46 Prove each of the following facts.

a) A σ-algebra is closed under countable intersections; that is, if $\mathcal{A}$ is a σ-algebra and $\{A_n\}_n \subset \mathcal{A}$, then $\bigcap_n A_n \in \mathcal{A}$.

b) A nonempty collection of subsets of Ω is a σ-algebra if it is closed under complementation and countable intersections.

1.47 In this exercise, we will provide two examples of algebras that are not σ-algebras.

a) Prove that the collection $\mathcal{D}$ defined in Exercise 1.40, although an algebra, is not a σ-algebra.

b) Prove that the collection $\mathcal{D}$ defined in Exercise 1.42, although an algebra, is not a σ-algebra.

1.48 Show that the collection $\mathcal{D}$ defined in Exercise 1.41 is a σ-algebra.

1.49 Suppose that $\{A_n\}_{n=1}^\infty$ is a pairwise disjoint sequence of nonempty subsets of Ω whose union is Ω.

a) Prove that the collection $\mathcal{D}$ of countable (including empty) unions of members of $\{A_n\}_{n=1}^\infty$ is a σ-algebra.

b) Prove that the collection $\mathcal{E}$ of finite (including empty) unions of members of $\{A_n\}_{n=1}^\infty$ is not an algebra and, hence, not a σ-algebra.

1.50 Prove Proposition 1.15 on page 24: Let $\mathcal{C}$ be a nonempty collection of subsets of Ω. Then there is a smallest σ-algebra of subsets of Ω containing $\mathcal{C}$. *Hint:* Consider the collection of all σ-algebras of subsets of Ω that contain $\mathcal{C}$.

1.51 Refer to Exercise 1.49, where $\{A_n\}_{n=1}^\infty$ is a pairwise disjoint sequence of nonempty subsets of Ω whose union is Ω. In part (a) of that exercise, we proved that the collection $\mathcal{D}$ of countable (including empty) unions of members of $\{A_n\}_{n=1}^\infty$ is a σ-algebra. Show that $\mathcal{D}$ is the σ-algebra generated by that sequence.

1.52 Let Ω be a set. Prove that an algebra of subsets of Ω is a σ-algebra if and only if it is a monotone class.

1.53 Let $\mathcal{C}$ be a nonempty collection of subsets of Ω. Prove that there is a smallest monotone class of subsets of Ω containing $\mathcal{C}$.

Georg Friedrich Bernhard Riemann (1826–1866)

Bernhard Riemann was born on September 17, 1826, in Breselenz, Germany. He was the second of six children of Friedrich Bernhard Riemann, who was a poor Lutheran pastor, and Charlotte Ebell, a housewife.

Although extremely shy, Riemann showed exceptional mathematical aptitude from an early age, such as fantastic calculation abilities. In 1840, Riemann entered directly into the third class at the Lyceum (middle school) in Hannover. He lived with his grandmother until her death in 1842, when he moved to the Johanneum Gymnasium in Lüneburg.

Riemann enrolled in Göttingen University in 1846 to study theology in accord with his father's wishes. However, he soon also attended mathematical lectures and persuaded his father to allow him to study mathematics. Despite the presence of Gauss, Göttingen had only a simple mathematics curriculum. Consequently, in 1847, Riemann entered Berlin University, where both Jacobi and Dirichlet had great influence on him.

W. E. Weber's return to Göttingen University sparked an improvement in the mathematical climate there and, in 1849, Riemann also returned to Göttingen. There, in 1851, he earned his Ph.D. with his thesis on complex function theory and Riemann surfaces.

Riemann continued his studies, submitting papers on Fourier series and geometry to qualify to become an unpaid lecturer. The mathematical tools that Riemann developed in his geometry paper were used by Albert Einstein in his theory of relativity. Riemann's first lectures were on partial differential equations in relation to physics, which were reprinted for 80 years after his death. At last, in 1857, Riemann was appointed Assistant Professor (with pay!) at Göttingen and Professor 2 years later.

In 1862, Riemann married Elise Koch and also became quite ill with tuberculosis. He spent most of the next four years trying to regain his health in the more hospitable climate of Italy. But, in Selasca, Italy, on July 20, 1866, Riemann succumbed to tuberculosis at the age of 39.

2

The Real Number System and Calculus

As further preparation for our study of real analysis, we will present in this chapter several topics often encountered in previous mathematics courses. But, again, although some readers may be familiar with much of the material, we present this chapter as a way to provide a common ground for all readers of the text.

We will first discuss the real number system and the extended real number system. Next we will investigate sequences of real numbers, exploring, in particular, cluster points and limits of such sequences. Then we will introduce open and closed sets of real numbers and examine some of their basic properties. In the final sections of this chapter, we will present continuous functions and the Riemann integral with an eye toward remedying some of the deficiencies experienced in using these classical concepts in modern analysis.

2.1 THE REAL NUMBER SYSTEM

Although it is mathematically satisfying to construct the real numbers $\mathcal{R}$ from "scratch," such a construction would be an aside to the main thrust of this text. Thus, we will not endeavor to present a construction of the real numbers.[†] Instead, we will briefly review the main properties of the real number system, specifically, three groups of axioms that together characterize that system.

---

† Readers interested in a construction of the real numbers are referred to Cohen and Ehrlich's *The Structure of the Real Number System* (New York: D. Van Nostrand Reinhold, 1963).

Axioms for the Real Number System

The first group of axioms for the real number system consists of the **field axioms.** These axioms provide the basic properties of the real numbers relative to the two binary operations of addition ($+$) and multiplication ($\cdot$). We follow convention in using juxtaposition to indicate multiplication when convenient.

Field Axioms

Let $x, y, z \in \mathcal{R}$. Then we have:

(F1) $x + y = y + x$ and $xy = yx$.   (**commutative**)

(F2) $(x + y) + z = x + (y + z)$ and $(xy)z = x(yz)$.   (**associative**)

(F3) $x(y + z) = xy + xz$.   (**distributive**)

(F4)   There exist $0, 1 \in \mathcal{R}$ with $0 \neq 1$, such that for each $x \in \mathcal{R}$, $x + 0 = x$ and $x \cdot 1 = x$.   (**identities**)

(F5)   For each $x \in \mathcal{R}$, there is a $-x \in \mathcal{R}$ such that $x + (-x) = 0$ and, if $x \neq 0$, there is an $x^{-1} \in \mathcal{R}$ such that $xx^{-1} = 1$.   (**inverses**)

Because of (F2), $x + y + z$ is defined unambiguously, as is any finite sum; likewise, xyz is defined unambiguously, as is any finite product. If $x_1, x_2, \ldots, x_n$ are real numbers, then we use following notation:

$$\sum_{k=1}^{n} x_k = x_1 + x_2 + \cdots + x_n \qquad \text{and} \qquad \prod_{k=1}^{n} x_k = x_1 x_2 \cdots x_n.$$

Also, regarding (F5), we will usually write $y - x$ for $y + (-x)$ and often write y/x or $\frac{y}{x}$ for yx^{-1}.

The second group of axioms consists of the **order axioms.** These axioms provide the basic properties of the real numbers relative to the less-than ($<$) ordering.

Order Axioms [†]

Let $x, y, z \in \mathcal{R}$. Then we have:

(O1) $x < y$ and $y < z$ implies $x < z$.   (**transitive**)

(O2) $x < y$ implies $x + z < y + z$.

(O3) $x < y$ and $0 < z$ implies $xz < yz$.

(O4)   Exactly one of $x = y$, $x < y$, and $y < x$ holds.   (**trichotomous**)

Note: We will also employ the following notation: $x \leq y$ means that $x < y$ or $x = y$; $x > y$ means that $y < x$; and $x \geq y$ means that $y \leq x$.

The third group of axioms actually consists of one axiom, called the **completeness axiom** or the **least upper-bound axiom.** In preparation for stating that axiom, we first introduce some terminology.

[†] The order axioms can also be stated in terms of the positive real numbers. See Exercise 2.1.

Let A be a nonempty subset of $\mathcal{R}$. A real number u is called an **upper bound** for A if $x \leq u$ for all $x \in A$. Note that not every subset of $\mathcal{R}$ has an upper bound, for example, neither $\mathcal{R}$ nor $\mathcal{N}$ has an upper bound. If a subset of $\mathcal{R}$ has an upper bound, then we say that it is **bounded above.**

A real number u is called a **least upper bound** or **supremum** for A if it is an upper bound for A and is smaller than or equal to any other upper bound for A. It is easy to see that a set can have at most one least upper bound. Also, by definition, a necessary condition for a subset of $\mathcal{R}$ to have a least upper bound is that it be bounded above. That this condition is sufficient is the content of the completeness axiom.

Completeness Axiom

A nonempty subset of real numbers that is bounded above has a least upper bound.

Let A be a nonempty subset of $\mathcal{R}$ that is bounded above. Then each of the following notations is used to denote the least upper bound of A:

$$\sup A, \qquad \sup_{x \in A} x, \qquad \text{or} \qquad \sup\{\, x : x \in A \,\}.$$

Similarly, we can define lower bound and greatest lower bound: Let A be a nonempty subset of $\mathcal{R}$. A real number ℓ is called a **lower bound** for A if $x \geq \ell$ for all $x \in A$. Note that not every subset of $\mathcal{R}$ has a lower bound, for example, neither $\mathcal{R}$ nor $\mathcal{Z}$ has a lower bound. If a subset of $\mathcal{R}$ has a lower bound, then we say that it is **bounded below.**

A real number ℓ is called a **greatest lower bound** or **infimum** for A if it is a lower bound for A and is greater than or equal to any other lower bound for A. It is easy to see that a set can have at most one greatest lower bound. Also, by definition, a necessary condition for a subset of $\mathcal{R}$ to have a greatest lower bound is that it be bounded below. That this condition is sufficient is a consequence of the completeness axiom. In other words, we have the following proposition whose proof is left to the reader as an exercise. (See Exercise 2.4.)

□ □ □ **PROPOSITION 2.1**

A nonempty subset of real numbers that is bounded below has a greatest lower bound.

Let A be a nonempty subset of $\mathcal{R}$ that is bounded below. Then each of the following notations is used to denote the greatest lower bound of A.

$$\inf A, \qquad \inf_{x \in A} x, \qquad \text{or} \qquad \inf\{\, x : x \in A \,\}.$$

EXAMPLE 2.1 *Illustrates Least Upper Bound and Greatest Lower Bound*

a) $\sup[0,1) = 1$ and $\inf[0,1) = 0$.
b) $\mathcal{N}$ has no least upper bound, but $\inf \mathcal{N} = 1$.
c) Let $A = \{\, x : x^2 < 2 \,\}$. Then $\sup_{x \in A} x = \sqrt{2}$ and $\inf_{x \in A} x = -\sqrt{2}$. □

An important consequence of the completeness axiom is that given any real number, we can find a positive integer exceeding that number. In other words, we have the following proposition, known as the **Archimedean principle.**

□ □ □ **PROPOSITION 2.2 Archimedean Principle**

For each $x \in \mathcal{R}$, there is an $n \in \mathcal{N}$ such that $n > x$.

PROOF Let $A = \{ m \in \mathcal{N} : m \leq x \}$. If $A = \emptyset$, then $1 > x$ and we are done. So, we can assume that A is nonempty. By definition, A is bounded above by x and, hence, by the completeness axiom, A has a least upper bound, say, u. Then $u - 1$ is not an upper bound for A and, hence, there is a $k \in A$ such that $k > u - 1$. Let $n = k + 1$ and note that $n \in \mathcal{N}$. Because $n > u$ and u is an upper bound for A, we have that $n \notin A$. And from this last result and the fact that $n \in \mathcal{N}$, we conclude that $n > x$. ∎

The next two propositions show that between any two real numbers there is both an irrational number and a rational number. We will find these two facts essential.

□ □ □ **PROPOSITION 2.3 Density of the Irrational Numbers**

Between any two real numbers there is an irrational number.

PROOF Let $a, b \in \mathcal{R}$ with $a < b$. In Chapter 1, we noted that the interval (a, b) is uncountable. (See Exercise 1.26 for a proof.) Since the set of rational numbers Q is countable (Proposition 1.12 on page 20), it follows from Proposition 1.8 on page 19 that any subset of Q is countable. If (a, b) contained no irrational number then it would be an uncountable subset of Q. ∎

□ □ □ **PROPOSITION 2.4 Density of the Rational Numbers**

Between any two real numbers there is a rational number.

PROOF Let $a, b \in \mathcal{R}$ with $a < b$. First assume that $a > 0$. By the Archimedean principle, there is an $n \in \mathcal{N}$ with $n > (b - a)^{-1}$. Note that $nb > nb - a > 1 + na - a \geq 1$; so, $nb > 1$.

Now, let $A = \{ k \in \mathcal{N} : k \geq nb \}$. By the Archimedean principle, $A \neq \emptyset$ and, therefore, by the well-ordering principle, A has a smallest member, say, j. As $nb > 1$, $j \geq 2$. This, in turn, implies that $j - 1 \in \mathcal{N}$ and, consequently, because j is the smallest member of A, we must have $j - 1 < nb$. Letting $m = j - 1$, we have that

$$a = b - (b - a) < \frac{m+1}{n} - \frac{1}{n} = \frac{m}{n} < b.$$

Letting $r = m/n$, we have that $r \in Q$ and $a < r < b$.

Next we remove the restriction that $a > 0$. Applying the Archimedean principle, we choose an $n \in \mathcal{N}$ such that $n > -a$. Then $n + a > 0$ and, so, by what we have already proved, there is an $r \in Q$ such that $n + a < r < n + b$. Then $r - n \in Q$ and $a < r - n < b$. ∎

The Extended Real Number System

It is convenient to enlarge the set of real numbers to the **extended real numbers,** which we denote by $\mathcal{R}^*$. This set is obtained by adding two distinct symbols, ∞ and $-\infty$, to the real numbers; thus, $\mathcal{R}^* = \mathcal{R} \cup \{-\infty, \infty\}$.

We extend the usual ordering of $\mathcal{R}$ to $\mathcal{R}^*$ by using the following definitions: $-\infty < \infty$ and $-\infty < x < \infty$ for all $x \in \mathcal{R}$. We also extend the binary operations of addition and multiplication to $\mathcal{R}^*$. In doing so, we make the convention that, for $x \in \mathcal{R}^*$, $x - \infty = x + (-\infty)$ and $x - (-\infty) = x + \infty$ in the sense that if one side of the equation is defined, then the other is defined likewise.

Now, for $x \in \mathcal{R}$, we define

$$x + \infty = \infty + x = \infty \quad \text{and} \quad x - \infty = -\infty + x = -\infty;$$

and

$$
\begin{aligned}
x \cdot \infty = \infty \cdot x = \infty \quad &\text{and} \quad x \cdot (-\infty) = (-\infty) \cdot x = -\infty, \quad &&\text{if } x > 0; \\
x \cdot \infty = \infty \cdot x = -\infty \quad &\text{and} \quad x \cdot (-\infty) = (-\infty) \cdot x = \infty, \quad &&\text{if } x < 0; \\
x \cdot \infty = \infty \cdot x = 0 \quad &\text{and} \quad x \cdot (-\infty) = (-\infty) \cdot x = 0, \quad &&\text{if } x = 0.
\end{aligned}
$$

Also, we define

$$
\begin{aligned}
\infty + \infty = \infty \quad &\text{and} \quad -\infty - \infty = -\infty; \\
\infty \cdot \infty = \infty \quad &\text{and} \quad (-\infty) \cdot (-\infty) = \infty;
\end{aligned}
$$

and

$$\infty \cdot (-\infty) = (-\infty) \cdot \infty = -\infty.$$

The expressions $\infty - \infty$ and $-\infty + \infty$ are left undefined because they cannot be defined in a way that is consistent with the rules of ordinary addition and multiplication. See Exercise 2.10.

In Definition 1.1 on page 4, we defined intervals of $\mathcal{R}$. We can extend that definition to intervals of $\mathcal{R}^*$ and, in fact, this extension simplifies the number of cases that need to be considered.

DEFINITION 2.1 **Intervals of $\mathcal{R}^*$**

Let a and b be extended real numbers such that $a < b$. Then the **intervals of $\mathcal{R}^*$** with *endpoints* a and b are as follows:

$$
\begin{aligned}
(a, b) &= \{\, x \in \mathcal{R}^* : a < x < b \,\} \\
[a, b) &= \{\, x \in \mathcal{R}^* : a \leq x < b \,\} \\
(a, b] &= \{\, x \in \mathcal{R}^* : a < x \leq b \,\} \\
[a, b] &= \{\, x \in \mathcal{R}^* : a \leq x \leq b \,\}
\end{aligned}
$$

If a and b are both in $\mathcal{R}$, then these are the **bounded intervals** of $\mathcal{R}$ as given in Definition 1.1. On the other hand, if either $a = -\infty$ or $b = \infty$, then the preceding four sets are **unbounded intervals.**

Note that in $\mathcal{R}^*$, every set is bounded above by ∞. Thus, every nonempty subset of $\mathcal{R}^*$ has a least upper bound — if it is bounded above in $\mathcal{R}$, then its least upper bound is also in $\mathcal{R}$; if it is not bounded above in $\mathcal{R}$, then its least upper bound is ∞. Since every member of $\mathcal{R}^*$ is vacuously an upper bound for $\emptyset$, we see that the empty set also has a least upper bound in $\mathcal{R}^*$, namely, $-\infty$. Similar remarks hold for greatest lower bounds. Thus, we have the following proposition.

□ □ □ **PROPOSITION 2.5**

Every subset A of $\mathcal{R}^*$ has both a least upper bound and greatest lower bound. We have the following:

a) *If $A = \emptyset$, then $\sup A = -\infty$ and $\inf A = \infty$.*

b) *If A is bounded above in $\mathcal{R}$, then $\sup A \in \mathcal{R}$; otherwise, $\sup A = \infty$.*

c) *If A is bounded below in $\mathcal{R}$, then $\inf A \in \mathcal{R}$; otherwise, $\inf A = -\infty$.*

EXAMPLE 2.2 *Illustrates Proposition 2.5*

a) $\inf \mathcal{N} = 1$ and $\sup \mathcal{N} = \infty$.

b) $\inf \mathcal{Z} = -\infty$ and $\sup \mathcal{Z} = \infty$.

c) If $A = \{1, 2, 3, \infty\}$, then $\inf A = 1$ and $\sup A = \infty$.

d) Suppose that I is an interval in $\mathcal{R}^*$ with endpoints a and b. Then (see Exercise 2.11) we have $\inf I = a$ and $\sup I = b$. □

Exercises for Section 2.1

Note: A ★ denotes an exercise that will be subsequently referenced.

2.1 The order axioms for the real number system can also be stated in terms of the positive real numbers as follows. Let $\mathcal{R}^+$ denote the subset of positive real numbers.

(O1′) $x, y \in \mathcal{R}^+$ implies $x + y \in \mathcal{R}^+$.

(O2′) $x, y \in \mathcal{R}^+$ implies $xy \in \mathcal{R}^+$.

(O3′) $x \in \mathcal{R}^+$ implies $-x \notin \mathcal{R}^+$.

(O4′) For each $x \in \mathcal{R}$, we have $x = 0$, $x \in \mathcal{R}^+$, or $-x \in \mathcal{R}^+$.

Prove that these four axioms are equivalent to the order axioms given on page 30. *Note:* Assuming the order axioms given on page 30, we define $\mathcal{R}^+ = \{\, x : x > 0 \,\}$; whereas, assuming the order axioms in this exercise, we define $x < y$ to mean $y - x \in \mathcal{R}^+$.

★2.2 The **absolute value** of a real number x, denoted $|x|$, is defined by

$$|x| = \begin{cases} x, & \text{if } x \geq 0; \\ -x, & \text{if } x < 0. \end{cases}$$

Let $x, y \in \mathcal{R}$. Prove each of the following facts.

a) $|-x| = |x|$

b) $|xy| = |x||y|$

c) $|x + y| \leq |x| + |y|$ *[triangle inequality]*

d) $\big||x| - |y|\big| \leq |x - y|$

★2.3 For $x, y \in \mathcal{R}$, we define the **maximum** of x and y to be the larger of those two numbers. We denote the maximum by $\max\{x, y\}$ or $x \vee y$. Thus,

$$x \vee y = \max\{x, y\} = \begin{cases} x, & \text{if } x \geq y; \\ y, & \text{if } x < y. \end{cases}$$

Similarly, we define the **minimum** of x and y to be the smaller of those two numbers. We denote the minimum by $\min\{x, y\}$ or $x \wedge y$. Thus,

$$x \wedge y = \min\{x, y\} = \begin{cases} y, & \text{if } x \geq y; \\ x, & \text{if } x < y. \end{cases}$$

Let $x, y \in \mathcal{R}$. Referring to Exercise 2.2, prove each of the following facts.

a) $|x| = x \vee -x$

b) $x \vee y = \frac{1}{2}(x + y + |x - y|)$

c) $x \wedge y = \frac{1}{2}(x + y - |x - y|)$

2.4 Suppose that A is bounded below. Prove that A has a greatest lower bound and that, in fact, $\inf A = -\sup\{-x : x \in A\}$.

2.5 Suppose that $A \subset B$. Prove the following.

a) If B is bounded above, then so is A and $\sup A \leq \sup B$.

b) If B is bounded below, then so is A and $\inf A \geq \inf B$.

2.6 Suppose that F is a finite nonempty subset of $\mathcal{R}$.

a) Prove that F is bounded above.

b) Prove that $\sup F \in F$. (We call this element of F the **maximum** of F, and denote it by $\max F$, $\max_{x \in F} x$, or $\max\{x : x \in F\}$.

c) Referring to Exercise 2.3, show that if $F = \{x, y\}$, then $\max F = x \vee y$.

d) Prove that F is bounded below.

e) Prove that $\inf F \in F$. (We call this element of F the **minimum** of F, and denote it by $\min F$, $\min_{x \in F} x$, or $\min\{x : x \in F\}$.

f) Referring to Exercise 2.3, show that if $F = \{x, y\}$, then $\min F = x \wedge y$.

2.7 Prove that any (nondegenerate) interval of real numbers contains infinitely many irrational numbers, in fact, uncountably many.

2.8 Prove that any (nondegenerate) interval of real numbers contains infinitely many rational numbers.

2.9 Let $x \in \mathcal{R}$ and set $A = \{z \in \mathcal{Z} : z \leq x\}$.

a) Prove that A is nonempty.

b) Explain why A has a least upper bound.

c) Prove that $\sup A \in A$ and, hence, that $\sup A$ is an integer.

d) The integer $\sup\{z \in \mathcal{Z} : z \leq x\}$ is called the **greatest integer** in x and is denoted by $[x]$. Prove that $[x] \leq x < [x] + 1$ or, equivalently, that $x - 1 < [x] \leq x$.

e) The function $f : \mathcal{R} \rightarrow \mathcal{Z}$ defined by $f(x) = [x]$ is called the **greatest integer function**. Prove that for each $z \in \mathcal{Z}$, $f(z + x) = z + f(x)$.

2.10 Show that $\infty - \infty$ cannot be defined in a way that is consistent with the rules of ordinary addition and multiplication.

2.11 Prove that if I is an interval in $\mathcal{R}^*$ with endpoints a and b, then $\inf I = a$ and $\sup I = b$.

2.2 SEQUENCES OF REAL NUMBERS

Recall from Chapter 1 (see page 12) that an infinite sequence is a function whose domain is the set of positive integers $\mathcal{N}$. In this section, we will study infinite sequences of real numbers. A **sequence of real numbers** is a sequence whose range is a subset of $\mathcal{R}$. To begin, we recall the following definition from calculus.

DEFINITION 2.2 **Convergent Sequence; Limit**

A sequence $\{x_n\}_{n=1}^{\infty}$ of real numbers is said to **converge** to the real number x if for each $\epsilon > 0$, there is an $N \in \mathcal{N}$ such that $|x - x_n| < \epsilon$ whenever $n \geq N$. In other words, the sequence converges to x if for each $\epsilon > 0$, all but finitely many terms of the sequence lie within ϵ of x. The number x is called the **limit** of the sequence $\{x_n\}_{n=1}^{\infty}$ and we write

$$\lim_{n\to\infty} x_n = x \qquad \text{or} \qquad x_n \to x, \text{ as } n \to \infty.$$

If a sequence converges, we say that it has a limit.

The sequence $\{(n-1)/n\}_{n=1}^{\infty}$ converges; its limit is 1: $\lim_{n\to\infty}(n-1)/n = 1$. On the other hand, it is easy to find sequences of real numbers that do not converge.

Consider, for instance, the sequence $\{(-1)^n\}_{n=1}^{\infty}$. This sequence does not converge because its terms oscillate between -1 and 1 and, hence, do not approach any single number. The sequence $\{n^2\}_{n=1}^{\infty}$ also does not converge but for an intrinsically different reason—its terms are becoming indefinitely large and, hence, do not approach a real number. If we would allow limits in $\mathcal{R}^*$, this latter sequence would converge to ∞.

It is convenient to permit convergence to extended real numbers and, in fact, to allow the sequences themselves to contain extended real numbers (i.e., to have range $\mathcal{R}^*$). Here we will discuss convergence to extended real numbers but will restrict ourselves to sequences of real numbers, leaving the generalization to sequences of extended real numbers to the reader.

DEFINITION 2.3   **Convergent Sequence (Extended Sense)**

A sequence $\{x_n\}_{n=1}^{\infty}$ of real numbers is said to **converge** in $\mathcal{R}^*$ if one of the following three conditions hold:

a) The sequence converges to a real number in the sense of Definition 2.2. In this situation, we say that the sequence **converges in $\mathcal{R}$** or that **the limit exists and is finite**.

b) For each $M \in \mathcal{R}$, there is an $N \in \mathcal{N}$ such that $x_n > M$ whenever $n \geq N$. In this situation, we say that the sequence **converges to ∞** and write $\lim_{n\to\infty} x_n = \infty$.

c) For each $M \in \mathcal{R}$, there is an $N \in \mathcal{N}$ such that $x_n < M$ whenever $n \geq N$. In this situation, we say that the sequence **converges to $-\infty$** and write $\lim_{n\to\infty} x_n = -\infty$.

Sequences, such as $\{n^2\}_{n=1}^{\infty}$ or $\{(n+1)/n\}_{n=1}^{\infty}$, whose terms never decrease with increasing n or never increase with increasing n, play an important role in analysis. More generally, let $\{x_n\}_{n=1}^{\infty}$ be a sequence of real numbers. If $x_1 \leq x_2 \leq \cdots$, then the sequence is said to be **nondecreasing**. If $x_1 \geq x_2 \geq \cdots$, then the sequence is said to be **nonincreasing**. A sequence of real numbers is called **monotone** if it is either nondecreasing or nonincreasing.

The next proposition, whose proof is left to the reader as an exercise, shows that any monotone sequence of real numbers has a limit (in $\mathcal{R}^*$). In stating this and other propositions, we use the terminology that a sequence is bounded above if its range is bounded above, and that the least upper bound of a sequence is the least upper bound of the range of the sequence. Similarly, we say that a sequence is bounded below if its range is bounded below, and that the greatest lower bound of a sequence is the greatest lower bound of the range of the sequence.

□ □ □ **PROPOSITION 2.6**

Any monotone sequence of real numbers converges in $\mathcal{R}^*$. In fact, we have the following:

a) If $\{x_n\}_{n=1}^{\infty}$ is nondecreasing, then

$$\lim_{n\to\infty} x_n = \sup\{x_n : n \in \mathcal{N}\}.$$

In particular, the limit exists and is finite if $\{x_n\}_{n=1}^{\infty}$ is bounded above and is ∞ otherwise.

b) If $\{x_n\}_{n=1}^{\infty}$ is nonincreasing, then

$$\lim_{n\to\infty} x_n = \inf\{x_n : n \in \mathcal{N}\}.$$

In particular, the limit exists and is finite if $\{x_n\}_{n=1}^{\infty}$ is bounded below and is $-\infty$ otherwise.

Cluster Points

By permitting sequences of real numbers to converge to extended real numbers, we have dealt with one type of nonconvergence of sequences, namely, when the terms of the sequence are becoming either indefinitely large or indefinitely small. The other type of nonconvergence occurs when the terms of the sequence do not approach any single number, either real or extended real. To analyze sequences of this type, we introduce the concept of a **cluster point.**

For a sequence of real numbers to converge to a real number x requires that for each $\epsilon > 0$, all but finitely many terms of the sequence lie within ϵ of x. Thus, we see that the terms of a sequence that converges in $\mathcal{R}$ are "clustering" around the limit of the sequence and no other number.

If we consider again the sequence $\{(-1)^n\}_{n=1}^{\infty}$, we see that it does not converge because some of the terms of the sequence are clustering around -1 and some are clustering around 1. That is, for each $\epsilon > 0$, infinitely many terms of the sequence lie within ϵ of -1 and infinitely many lie within ϵ of 1. This leads us to the following definition.

DEFINITION 2.4   **Cluster Point**

Let $\{x_n\}_{n=1}^{\infty}$ be a sequence of real numbers.

a) A real number x is said to be a **cluster point** of $\{x_n\}_{n=1}^{\infty}$ if for each $\epsilon > 0$ and $N \in \mathcal{N}$, there is an $n \geq N$ such that $|x - x_n| < \epsilon$.

b) ∞ is a cluster point of $\{x_n\}_{n=1}^{\infty}$ if for each $M \in \mathcal{R}$ and $N \in \mathcal{N}$, there is an $n \geq N$ such that $x_n > M$.

c) $-\infty$ is a cluster point of $\{x_n\}_{n=1}^{\infty}$ if for each $M \in \mathcal{R}$ and $N \in \mathcal{N}$, there is an $n \geq N$ such that $x_n < M$.

Remark: Because we are restricting ourselves to sequences of real numbers, the condition in part (b) of Definition 2.4 for ∞ to be a cluster point is equivalent to the following condition: For each $M \in \mathcal{R}$, there is an $n \in \mathcal{N}$ such that $x_n > M$; and, similarly, the condition in part (c) of the definition can be restated. However, it is better to use the definitions as stated in Definition 2.4 because they generalize properly to sequences of extended real numbers.

EXAMPLE 2.3 *Illustrates Definition 2.4*

a) As the reader can easily verify, the sequence $\{(-1)^n\}_{n=1}^{\infty}$ has two cluster points, namely, -1 and 1.

b) Consider the sequence $2, 1, 0, 2, 2, \frac{1}{2}, 2, 3, \frac{2}{3}, 2, 4, \frac{3}{4}, \ldots$, that is,

$$x_n = \begin{cases} (n-3)/n, & \text{if } n \equiv 0 \pmod 3; \\ 2, & \text{if } n \equiv 1 \pmod 3; \\ (n+1)/3, & \text{if } n \equiv 2 \pmod 3. \end{cases}$$

This sequence has three cluster points, namely, 1, 2, and ∞.

c) Let $\{r_n\}_{n=1}^{\infty}$ be an enumeration of the rational numbers. From the density of the rational numbers (Proposition 2.4 on page 32), it follows that every extended real number is a cluster point of the sequence $\{r_n\}_{n=1}^{\infty}$. We leave the details to the reader. ◻

The cluster points of a sequence of real numbers can be characterized as follows. (See Exercise 2.17.)

- A real number x is a cluster point if and only if for each $\epsilon > 0$, infinitely many terms of the sequence are within ϵ of x.

- ∞ is a cluster point of a sequence if and only if for each $M \in \mathcal{R}$, infinitely many terms of the sequence exceed M if and only if the sequence is unbounded above.

- $-\infty$ is a cluster point of a sequence if and only if for each $M \in \mathcal{R}$, infinitely many terms of the sequence are smaller than M if and only if the sequence is unbounded below.

All three sequences in Example 2.3 have more than one cluster point and none of those sequences converge. More generally, we have the following proposition.

◻ ◻ ◻ **PROPOSITION 2.7**

A convergent sequence has exactly one cluster point, namely, its limit. Thus, a sequence having more than one cluster point cannot converge.

PROOF Suppose that $\{x_n\}_{n=1}^{\infty}$ is a convergent sequence of real numbers, say, $x_n \to x$, as $n \to \infty$. We will prove that x is a cluster point of the sequence and that it is the only cluster point of the sequence. In doing so, we will assume that $x \in \mathcal{R}$ and leave the other two cases ($x = \infty$ and $x = -\infty$) to the reader.

To verify that x is a cluster point, let $\epsilon > 0$ and $N \in \mathcal{N}$. We must find an $n \geq N$ such that $|x - x_n| < \epsilon$. But, in fact, since $x_n \to x$, there is a $K \in \mathcal{N}$ such that $|x - x_n| < \epsilon$ for all $n \geq K$. Let $n = \max\{N, K\}$. Then $n \geq N$ and, because $n \geq K$, we have $|x - x_n| < \epsilon$. Thus, x is a cluster point of $\{x_n\}_{n=1}^{\infty}$.

Now we show that no real number different from x can be a cluster point of $\{x_n\}_{n=1}^{\infty}$. Let $y \in \mathcal{R}$ and $y \neq x$. Let $\epsilon = |y - x|/2$. Choose $N \in \mathcal{N}$ such that $n \geq N$ implies $|x - x_n| < \epsilon$. Then for $n \geq N$, we have

$$|y - x_n| \geq |y - x| - |x - x_n| > 2\epsilon - \epsilon = \epsilon$$

and, consequently, y is not a cluster point of $\{x_n\}_{n=1}^{\infty}$.

Next we show that ∞ is not a cluster point of $\{x_n\}_{n=1}^{\infty}$. Choose $N \in \mathcal{N}$ such that $|x - x_n| < 1$ for $n \geq N$. Then, letting $M = x + 1$, we have that $x_n < M$ for $n \geq N$. Thus, ∞ is not a cluster point of $\{x_n\}_{n=1}^{\infty}$.

Finally we show that $-\infty$ is not a cluster point of $\{x_n\}_{n=1}^{\infty}$. Choose $N \in \mathcal{N}$ such that $|x - x_n| < 1$ for $n \geq N$. Then, letting $M = x - 1$, we have $x_n > M$ for $n \geq N$. Thus, $-\infty$ is not a cluster point of $\{x_n\}_{n=1}^{\infty}$. ∎

Limit Superior and Limit Inferior

Two of the most important concepts associated with infinite sequences of real numbers are the limit superior and the limit inferior. Although a sequence of real numbers does not necessarily have a limit (even in $\mathcal{R}^*$), it always has both a limit superior and limit inferior. As we will see, these two extended real numbers are cluster points of the sequence, in fact, the largest and smallest cluster points, respectively.

First we introduce some convenient notation. For a sequence $\{x_n\}_{n=1}^{\infty}$ of real numbers, we write

$$\inf_n x_n = \inf\{\, x_n : n \in \mathcal{N} \,\}, \qquad \sup_{k \geq n} x_k = \sup\{\, x_k : k \geq n \,\},$$

$$\sup_n x_n = \sup\{\, x_n : n \in \mathcal{N} \,\}, \qquad \inf_{k \geq n} x_k = \inf\{\, x_k : k > n \,\}.$$

DEFINITION 2.5   Limit Superior and Limit Inferior

Let $\{x_n\}_{n=1}^{\infty}$ be a sequence of real numbers.

a) The **limit superior** of the sequence is the extended real number given by

$$\limsup_{n \to \infty} x_n = \inf_n \sup_{k \geq n} x_k.$$

b) The **limit inferior** of the sequence is the extended real number given by

$$\liminf_{n \to \infty} x_n = \sup_n \inf_{k \geq n} x_k.$$

Note: Notations for the limit superior and limit inferior other than the ones presented in Definition 2.5 are commonly used. They are:

$$\limsup_{n \to \infty} x_n = \limsup x_n = \overline{\lim_{n \to \infty}} \, x_n = \overline{\lim} \, x_n$$

and

$$\liminf_{n \to \infty} x_n = \liminf x_n = \underline{\lim_{n \to \infty}} \, x_n = \underline{\lim} \, x_n.$$

EXAMPLE 2.4 *Illustrates Definition 2.5*

Refer to Example 2.3 on page 38.

a) Let $x_n = (-1)^n$. Then, for each $n \in \mathcal{N}$, $\sup_{k \geq n} x_k = 1$ and $\inf_{k \geq n} x_k = -1$. Therefore,

$$\inf_n \sup_{k \geq n} x_k = 1 \qquad \text{and} \qquad \sup_n \inf_{k \geq n} x_k = -1.$$

In other words, $\limsup x_n = 1$ and $\liminf x_n = -1$.

b) Consider the sequence $2, 1, 0, 2, 2, \frac{1}{2}, 2, 3, \frac{2}{3}, 2, 4, \frac{3}{4}, \ldots$, that is,

$$x_n = \begin{cases} (n-3)/n, & \text{if } n \equiv 0 \pmod 3; \\ 2, & \text{if } n \equiv 1 \pmod 3; \\ (n+1)/3, & \text{if } n \equiv 2 \pmod 3. \end{cases}$$

Then, for each $n \in \mathcal{N}$, we have $\sup_{k \geq n} x_k = \infty$ and

$$\inf_{k \geq n} x_k = \begin{cases} (n-3)/n, & \text{if } n \equiv 0 \pmod 3; \\ (n-1)/(n+2), & \text{if } n \equiv 1 \pmod 3; \\ (n-2)/(n+1), & \text{if } n \equiv 2 \pmod 3. \end{cases}$$

Therefore,

$$\inf_n \sup_{k \geq n} x_k = \infty \qquad \text{and} \qquad \sup_n \inf_{k \geq n} x_k = 1.$$

In other words, $\limsup x_n = \infty$ and $\liminf x_n = 1$.

c) Let $\{r_n\}_{n=1}^{\infty}$ be an enumeration of the rational numbers. For each $n \in \mathcal{N}$, we have $\sup_{k \geq n} x_k = \infty$ and $\inf_{k \geq n} x_k = -\infty$. Hence,

$$\inf_n \sup_{k \geq n} x_k = \infty \qquad \text{and} \qquad \sup_n \inf_{k \geq n} x_k = -\infty.$$

In other words, $\limsup x_n = \infty$ and $\liminf x_n = -\infty$. □

It is helpful to note that the sequences $\{y_n\}_{n=1}^{\infty}$ and $\{z_n\}_{n=1}^{\infty}$ defined by $y_n = \sup_{k \geq n} x_k$ and $z_n = \inf_{k \geq n} x_k$ are, respectively, nonincreasing and nondecreasing. Consequently, by Proposition 2.6 on page 37, both sequences are convergent, converging to, respectively,

$$\inf_n y_n = \inf_n \sup_{k \geq n} x_k = \limsup_{n \to \infty} x_n$$

and

$$\sup_n z_n = \sup_n \inf_{k \geq n} x_k = \liminf_{n \to \infty} x_n.$$

In other words,

$$\limsup_{n \to \infty} x_n = \lim_{n \to \infty} \sup_{k \geq n} x_k \qquad \text{and} \qquad \liminf_{n \to \infty} x_n = \lim_{n \to \infty} \inf_{k \geq n} x_k.$$

The next two propositions characterize the limit superior and limit inferior of a sequence of real numbers, providing both mathematical and intuitive interpretations. We will prove the first part of the first proposition and leave the proofs of the remaining parts of both propositions to the reader as exercises.

□ □ □ **PROPOSITION 2.8**

Let $\{x_n\}_{n=1}^{\infty}$ be a sequence of real numbers. We have:

a) $\limsup x_n = x \in \mathcal{R}$ if and only if for each $\epsilon > 0$,
   (i) there is an $N \in \mathcal{N}$ such that $x_n \leq x + \epsilon$ for $n \geq N$, and
   (ii) for each $n \in \mathcal{N}$, there is an $m \geq n$ such that $x_m > x - \epsilon$;
   in other words, if and only if for each $\epsilon > 0$, infinitely many terms of the sequence are within ϵ of x and only finitely many are greater than $x + \epsilon$.

b) $\limsup x_n = \infty$ if and only if for each $M \in \mathcal{R}$ and $N \in \mathcal{N}$, there is an $n \geq N$ such that $x_n > M$; in other words, if and only if the sequence is unbounded above.

c) $\limsup x_n = -\infty$ if and only if $\lim_{n \to \infty} x_n = -\infty$.

PROOF We prove part (a) and leave the proofs of the remaining two parts to the reader as Exercise 2.30. Let $y_n = \sup_{k \geq n} x_k$ and recall that $\{y_n\}_{n=1}^{\infty}$ is nonincreasing and converges to $\limsup x_n$.

Suppose that $x \in \mathcal{R}$ and $\limsup_{n \to \infty} x_n = x$. Then $y_n \geq x$ for all $n \in \mathcal{N}$ and $y_n \to x$ as $n \to \infty$. Let $\epsilon > 0$ be given. Choose $N \in \mathcal{N}$ such that $n \geq N$ implies $y_n - x < \epsilon$. Then, for $n \geq N$, $x_n \leq y_n < x + \epsilon$. This establishes (i). To establish (ii), we note that if $n \in \mathcal{N}$, then $\sup_{k \geq n} x_k = y_n \geq x$ and, hence, $\sup_{k \geq n} x_k > x - \epsilon$. This means $x - \epsilon$ is not an upper bound for $\{x_k : k \geq n\}$; in other words, $x_m > x - \epsilon$ for some $m \geq n$.

Conversely, suppose that for each $\epsilon > 0$, (i) and (ii) hold. We must prove that $\limsup x_n = x$ or, equivalently, that $\lim_{n \to \infty} y_n = x$. Let $\epsilon > 0$ be given. By (i), we can choose $N \in \mathcal{N}$ such that $x_n \leq x + \epsilon$ for $n \geq N$. This implies that, for $n \geq N$, $y_n = \sup_{k \geq n} x_k \leq x + \epsilon$. By (ii), we know that for each $n \in \mathcal{N}$, there is an $m \geq n$ such that $x_m > x - \epsilon$, which implies that, for each $n \in \mathcal{N}$, $y_n = \sup_{k \geq n} x_k > x - \epsilon$. Thus, we have proved that, for each $\epsilon > 0$, there is an $N \in \mathcal{N}$ such that $|y_n - x| \leq \epsilon$ whenever $n \geq N$; thus, $\lim_{n \to \infty} y_n = x$. ∎

□ □ □ **PROPOSITION 2.9**

Let $\{x_n\}_{n=1}^{\infty}$ be a sequence of real numbers. We have:

a) $\liminf x_n = x \in \mathcal{R}$ if and only if for each $\epsilon > 0$,
   (i) there is an $N \in \mathcal{N}$ such that $x_n \geq x - \epsilon$ for $n \geq N$, and
   (ii) for each $n \in \mathcal{N}$, there is an $m \geq n$ such that $x_m < x + \epsilon$;
   in other words, if and only if for each $\epsilon > 0$, infinitely many terms of the sequence are within ϵ of x and only finitely many are less than $x - \epsilon$.

b) $\liminf x_n = -\infty$ if and only if for each $M \in \mathcal{R}$ and $N \in \mathcal{N}$, there is an $n \geq N$ such that $x_n < M$; in other words, if and only if the sequence is unbounded below.

c) $\liminf x_n = \infty$ if and only if $\lim_{n \to \infty} x_n = \infty$.

We mentioned earlier that the limit superior and limit inferior are, respectively, the largest and smallest cluster points of a sequence. This is illustrated by Examples 2.3 and 2.4 (pages 38 and 40) and is proved in our next proposition.

□ □ □ **PROPOSITION 2.10**

Let $\{x_n\}_{n=1}^{\infty}$ be a sequence of real numbers. Then,

a) $\limsup x_n$ is the largest cluster point of $\{x_n\}_{n=1}^\infty$.

b) $\liminf x_n$ is the smallest cluster point of $\{x_n\}_{n=1}^\infty$.

PROOF We prove part (a) and leave the proof of part (b) to the reader as an exercise. Let $x = \limsup x_n$. It follows immediately from the definition of cluster point (Definition 2.4 on page 37) and Proposition 2.8 that x is a cluster point of $\{x_n\}_{n=1}^\infty$.

It remains to prove that x is the largest cluster point of $\{x_n\}_{n=1}^\infty$. If $x = \infty$, there is nothing to prove. If $x = -\infty$, then Proposition 2.8(c) shows that $\lim_{n \to \infty} x_n = -\infty$. Therefore, by Proposition 2.7 on page 38, $-\infty$ is the only cluster point of $\{x_n\}_{n=1}^\infty$ and, hence, the largest.

So, we can assume that $x \in \mathcal{R}$. By Proposition 2.8(a), only finitely many terms of the sequence exceed $x + 1$; consequently, ∞ is not a cluster point. It therefore remains to prove that if $y \in \mathcal{R}$ and $y > x$, then y is not a cluster point of $\{x_n\}_{n=1}^\infty$. Let $\epsilon = (y - x)/2$. Applying Proposition 2.8(a) again, we know that only finitely many terms of $\{x_n\}_{n=1}^\infty$ exceed $x + \epsilon$ or, equivalently, $y - \epsilon$. This shows that y is not a cluster point. ∎

The following proposition is often useful. The sufficiency part of the proposition enables us to prove that a sequence converges without explicitly finding its limit, and the necessity part often makes it easy to show that a sequence does not converge.

☐ ☐ ☐ **PROPOSITION 2.11**

A necessary and sufficient condition for a sequence of real numbers to converge is that its limit superior and limit inferior are equal. In such cases, the sequence converges to the common value of the limit superior and limit inferior.

PROOF Suppose that $\{x_n\}_{n=1}^\infty$ converges. Then, by Proposition 2.7, the sequence has a unique cluster point, namely, its limit. Since the limit superior and limit inferior are both cluster points (Proposition 2.10), they must both equal the limit of the sequence and, hence, each other.

Conversely, suppose that $\limsup x_n = \liminf x_n$. Then, by Proposition 2.10, $\{x_n\}_{n=1}^\infty$ has exactly one cluster point, namely, the common value of the limit superior and limit inferior. Call that common value x. We claim $\lim_{n \to \infty} x_n = x$. If $x = -\infty$, the result is true by Proposition 2.8(c) on page 41, whereas if $x = \infty$, the result is true by Proposition 2.9(c) on page 41.

Hence, it remains to show that if $x \in \mathcal{R}$, then $\lim_{n \to \infty} x_n = x$. Let $\epsilon > 0$. Then, by Proposition 2.8(a), there is an $N_1 \in \mathcal{N}$ such that $x_n < x + \epsilon$ for $n \geq N_1$ and, by Proposition 2.9(a), there is an $N_2 \in \mathcal{N}$ such that $x_n > x - \epsilon$ for $n \geq N_2$. Set $N = \max\{N_1, N_2\}$. Then, for $n \geq N$, we have that $x - \epsilon < x_n < x + \epsilon$, that is, $|x - x_n| < \epsilon$. ∎

Proposition 2.7 states that a convergent sequence has exactly one cluster point, namely, its limit. It follows immediately from Propositions 2.10 and 2.11 that the converse is true. In other words, we have the following.

□ □ □ PROPOSITION 2.12

A sequence of real numbers converges if and only if it has exactly one cluster point. In such cases, the limit of the sequence is the unique cluster point.

Cauchy Sequences

Proposition 2.11 provides a criterion for determining whether a sequence of real numbers converges. A special case of that criterion is that a sequence converges in $\mathcal{R}$ if and only if its limit superior and limit inferior are equal and finite.

Another criterion for determining whether the limit of a sequence exists and is finite is the **Cauchy criterion**. Roughly speaking, a sequence is a Cauchy sequence if the terms of the sequence become closer and closer together as the sequence progresses. More precisely, we have the following definition.

DEFINITION 2.6   Cauchy Sequence

A sequence $\{x_n\}_{n=1}^{\infty}$ of real numbers is called a **Cauchy sequence** if for each $\epsilon > 0$, there is an $N \in \mathcal{N}$ such that $|x_n - x_m| < \epsilon$ whenever $n, m \geq N$.

With Definition 2.6 in mind, we now state and prove the Cauchy criterion for convergence of sequences of real numbers.

□ □ □ THEOREM 2.1 Cauchy Criterion

A sequence of real numbers converges in $\mathcal{R}$ if and only if it is a Cauchy sequence.

PROOF   Let $\{x_n\}_{n=1}^{\infty}$ be a sequence of real numbers. Suppose that the limit of the sequence exists and is finite, say, x. Let $\epsilon > 0$ be given. Then we can choose $N \in \mathcal{N}$ such that $n \geq N$ implies $|x - x_n| < \epsilon/2$. Therefore, if $n, m \geq N$, we have

$$|x_n - x_m| \leq |x_n - x| + |x - x_m| < \frac{\epsilon}{2} + \frac{\epsilon}{2} = \epsilon.$$

Thus, $\{x_n\}_{n=1}^{\infty}$ is a Cauchy sequence.

Conversely, suppose that $\{x_n\}_{n=1}^{\infty}$ is a Cauchy sequence. Let $\epsilon > 0$ be given. Then we can choose $N \in \mathcal{N}$ such that $|x_n - x_m| < \epsilon$ whenever $n, m \geq N$. In particular, we have that

$$x_N - \epsilon < x_n < x_N + \epsilon, \qquad n \geq N. \tag{2.1}$$

From (2.1) and Exercise 2.29(b) on page 45, we see that both $\limsup x_n$ and $\liminf x_n$ lie in the interval $[x_N - \epsilon, x_N + \epsilon]$. This shows that both $\limsup x_n$ and $\liminf x_n$ are finite and that

$$0 \leq \limsup x_n - \liminf x_n \leq 2\epsilon.$$

As $\epsilon > 0$ was chosen arbitrarily, we conclude that $\liminf x_n = \limsup x_n$ and that their common value is a real number. So, by Proposition 2.11, $\lim_{n \to \infty} x_n$ exists and is finite. ■

Exercises for Section 2.2

2.12 Prove that the limit of a sequence of real numbers, if it exists, must be unique.

2.13 Let $\{x_n\}_{n=1}^{\infty}$ and $\{y_n\}_{n=1}^{\infty}$ be two sequences of real numbers whose limits exist and are finite. Also, let $c \in \mathcal{R}$. Prove that each of the following holds.

a) $\lim_{n \to \infty} (x_n + y_n) = \lim_{n \to \infty} x_n + \lim_{n \to \infty} y_n$

b) $\lim_{n \to \infty} cx_n = c \cdot \lim_{n \to \infty} x_n$

c) $\lim_{n \to \infty} (x_n y_n) = \lim_{n \to \infty} x_n \cdot \lim_{n \to \infty} y_n$

2.14 Refer to Exercise 2.13. Decide under which conditions each of (a)–(c) holds if convergence is allowed in the extended sense, that is, in $\mathcal{R}^*$.

2.15 Let $\{x_n\}_{n=1}^{\infty}$ and $\{y_n\}_{n=1}^{\infty}$ be two convergent sequences of real numbers such that $x_n \leq y_n$ for n sufficiently large, that is, there is an $N \in \mathcal{N}$ such that $x_n \leq y_n$ for $n \geq N$. Prove that $\lim_{n \to \infty} x_n \leq \lim_{n \to \infty} y_n$.

2.16 Prove Proposition 2.6 on page 37.

2.17 Refer to Definition 2.4 on page 37. Let $\{x_n\}_{n=1}^{\infty}$ be a sequence of real numbers. Prove each of the following facts.

a) A real number x is a cluster point of $\{x_n\}_{n=1}^{\infty}$ if and only if for each $\epsilon > 0$, infinitely many terms of the sequence are within ϵ of x.

b) ∞ is a cluster point of $\{x_n\}_{n=1}^{\infty}$ if and only if for each $M \in \mathcal{R}$, infinitely many terms of the sequence exceed M if and only if the sequence is unbounded above.

c) $-\infty$ is a cluster point of $\{x_n\}_{n=1}^{\infty}$ if and only if for each $M \in \mathcal{R}$, infinitely many terms of the sequence are smaller than M if and only if the sequence is unbounded below.

2.18 Find the cluster points of each of the following sequences.

a) $\{1/n\}_{n=1}^{\infty}$

b) $\{1 + (-1)^n\}_{n=1}^{\infty}$

c) $\{\sin(n\pi/2)\}_{n=1}^{\infty}$

2.19 Consider the sequence $\{x_n\}_{n=1}^{\infty}$ defined by

$$x_n = \begin{cases} 1, & \text{if } n \equiv 0 \pmod 3; \\ 2, & \text{if } n \equiv 1 \pmod 3; \\ n + \frac{1}{n}, & \text{if } n \equiv 2 \pmod 3. \end{cases}$$

Determine the cluster points of the sequence.

2.20 Consider the sequence $\{x_n\}_{n=1}^{\infty}$ defined by

$$x_n = \begin{cases} n, & \text{if } n \text{ is odd}; \\ (n-1)/n, & \text{if } n \text{ is even}. \end{cases}$$

Determine the cluster points of the sequence.

2.21 Let $\{r_n\}_{n=1}^{\infty}$ be an enumeration of the rational numbers. Prove that every extended real number is a cluster point of this sequence.

2.22 Let r be a rational number, say, $r = p/q$ where p and q are integers with no common divisors. Define $x_n = nr - [nr]$, where $[x]$ denotes the greatest integer in x. Determine the cluster points of $\{x_n\}_{n=1}^{\infty}$.

2.23 Let c be an irrational number and define $x_n = nc - [nc]$, where $[x]$ denotes the greatest integer in x. Determine the cluster points of $\{x_n\}_{n=1}^{\infty}$ by proceeding as follows.

a) Show that the terms of the sequence are distinct, that is, $x_n = x_m$ implies $n = m$.

b) Prove that for each $\epsilon > 0$ and $N \in \mathcal{N}$, there is an $n \geq N$ such that $0 < x_n < \epsilon$. *Hint:* Use the Archimedean principle to choose an $m \in \mathcal{N}$ such that $1/m < \epsilon$. For $1 \leq k \leq m$, let $I_k = [\frac{k-1}{m}, \frac{k}{m})$ and note that the I_ks are disjoint, their union is $[0, 1)$, and each has length $1/m$. Now consider $\{ x_j : j = 1, N + 1, 2N + 1, \ldots, mN + 1 \}$ and observe that, by part (a), this set consists of $m + 1$ distinct numbers in $[0, 1)$.

c) Let $x \in [0, 1)$. Prove that for each $\epsilon > 0$ and $N \in \mathcal{N}$, there is an $n \geq N$ such that $|x - x_n| < \epsilon$. *Hint:* Choose an $m \in \mathcal{N}$ such that $1/m < \epsilon$. Apply part (b) to choose an $n \geq N$ such that $0 < x_n < 1/m$. Let k be the unique integer between 1 and m such that $(k - 1)/m \leq x < k/m$. Now let ℓ be the largest positive integer such that $\ell x_n < k/m$.

d) Obtain the cluster points of $\{x_n\}_{n=1}^{\infty}$.

2.24 Complete the proof of Proposition 2.7 on page 38 by showing that a sequence converging to ∞ or $-\infty$ has that value as its unique cluster point.

2.25 Prove that $\inf_{k \geq n} x_k \leq \sup_{k \geq m} x_k$ for all $n, m \in \mathcal{N}$, where $\{x_n\}_{n=1}^{\infty}$ is any sequence of real numbers.

2.26 Let $\{x_n\}_{n=1}^{\infty}$ be a sequence of real numbers and c a real number. Show that

a) $\limsup(c + x_n) = c + \limsup x_n$.

b) $\liminf(c + x_n) = c + \liminf x_n$.

c) $\limsup(cx_n) = \begin{cases} c \limsup x_n, & \text{if } c \geq 0; \\ c \liminf x_n, & \text{if } c < 0. \end{cases}$

d) $\liminf(cx_n) = \begin{cases} c \liminf x_n, & \text{if } c \geq 0; \\ c \limsup x_n, & \text{if } c < 0. \end{cases}$

Note that as special cases of parts (c) and (d), we have

$$\limsup(-x_n) = -\liminf x_n \qquad \text{and} \qquad \liminf(-x_n) = -\limsup x_n.$$

2.27 Let $\{x_n\}_{n=1}^{\infty}$ and $\{y_n\}_{n=1}^{\infty}$ be sequences of real numbers. Verify that each of the following holds, provided the right-hand side makes sense.

a) $\limsup(x_n + y_n) \leq \limsup x_n + \limsup y_n$.

b) $\limsup(x_n + y_n) \geq \limsup x_n + \liminf y_n$.

c) $\liminf(x_n + y_n) \geq \liminf x_n + \liminf y_n$.

d) $\liminf(x_n + y_n) \leq \limsup x_n + \liminf y_n$.

2.28 Let $\{x_n\}_{n=1}^{\infty}$ and $\{y_n\}_{n=1}^{\infty}$ be sequences of real numbers and assume that $\lim_{n \to \infty} y_n$ exists and is finite. Prove that

a) $\limsup(x_n + y_n) = \limsup x_n + \lim y_n$.

b) $\liminf(x_n + y_n) = \liminf x_n + \lim y_n$.

2.29 Let $\{x_n\}_{n=1}^{\infty}$ and $\{y_n\}_{n=1}^{\infty}$ be sequences of real numbers. Suppose $x_n \leq y_n$ for n sufficiently large; that is, there is an $N \in \mathcal{N}$ such that $x_n \leq y_n$ for $n \geq N$.

a) Prove that $\limsup x_n \leq \limsup y_n$ and $\liminf x_n \leq \liminf y_n$.

b) Suppose a and b are extended real numbers such that for n sufficiently large, $a \leq x_n \leq b$. Show that $a \leq \liminf x_n \leq \limsup x_n \leq b$.

2.30 Refer to Proposition 2.8 on page 41. Prove parts (b) and (c).

2.31 Prove Proposition 2.9 on page 41.

★2.32 Prove that $\lim_{n \to \infty} x_n = x$ if and only if every subsequence of $\{x_n\}_{n=1}^{\infty}$ has a subsequence that converges to x.

2.33 Prove that an extended real number is a cluster point of a sequence if and only if the sequence has a subsequence converging to that number. Conclude that the limit superior of a sequence is the limit of a subsequence of the sequence and likewise for the limit inferior.

2.34 Provide an example of a sequence of real numbers that converges in $\mathcal{R}^*$ but is not a Cauchy sequence.

★**2.35** Let $\{x_n\}_{n=1}^{\infty}$ be a sequence of real numbers. Define

$$a_n = \frac{x_1 + \cdots + x_n}{n} = \frac{1}{n}\sum_{k=1}^{n} x_k,$$

so that a_n is the arithmetic mean of the first n terms of $\{x_n\}_{n=1}^{\infty}$.

a) Prove that

$$\liminf_{n\to\infty} x_n \le \liminf_{n\to\infty} a_n \le \limsup_{n\to\infty} a_n \le \limsup_{n\to\infty} x_n.$$

b) Prove that if $\{x_n\}_{n=1}^{\infty}$ converges, so does $\{a_n\}_{n=1}^{\infty}$, and $\lim_{n\to\infty} a_n = \lim_{n\to\infty} x_n$.

c) Show that the converse of part (b) fails.

2.36 In this exercise, we will discuss **infinite series**. Let $\{x_n\}_{n=1}^{\infty}$ be a sequence of real numbers. The sequence $\{s_n\}_{n=1}^{\infty}$ defined by

$$s_n = \sum_{k=1}^{n} x_k, \qquad n \in \mathcal{N},$$

is called the *sequence of partial sums* of $\{x_n\}_{n=1}^{\infty}$. If the sequence $\{s_n\}_{n=1}^{\infty}$ converges to a real number, say, s, then we say that $\{x_n\}_{n=1}^{\infty}$ is *summable* to s or that the *infinite series* $\sum_{n=1}^{\infty} x_n$ *converges* to s, and we write

$$s = \sum_{n=1}^{\infty} x_n.$$

We also say that s is the *sum of the infinite series*. If the sequence $\{s_n\}_{n=1}^{\infty}$ does not converge to a real number, then we say that $\{x_n\}_{n=1}^{\infty}$ is not summable or that the infinite series $\sum_{n=1}^{\infty} x_n$ *diverges*. For brevity, we often write $\sum x_n$ in place of $\sum_{n=1}^{\infty} x_n$.

a) Prove that if $x_n \ge 0$ for each $n \in \mathcal{N}$, then either $\lim_{n\to\infty} s_n = \infty$ or $\sum x_n$ converges.

b) Show that if $\sum x_n$ converges, then $\lim_{n\to\infty} x_n = 0$.

c) Show that if $\sum x_n$ converges, then $\lim_{n\to\infty} \sum_{k=n}^{\infty} x_k = 0$.

d) Prove that if $\sum |x_n|$ converges, then so does $\sum x_n$. *Hint:* Use the Cauchy criterion.

★**2.37** In this exercise, we will consider **generalized sums**. Let I be a nonempty set and let $\{x_\iota\}_{\iota \in I}$ be an indexed collection of nonnegative real numbers, that is, $x_\iota \ge 0$ for each $\iota \in I$. Define

$$\sum_{\iota \in I} x_\iota = \sup\left\{ \sum_{\iota \in F} x_\iota : F \text{ finite, } F \subset I \right\}, \tag{2.2}$$

where each sum in the set on the right is the ordinary sum of a finite collection of real numbers.

a) Suppose that $I = \{1, \ldots, n\}$. Show that $\sum_{\iota \in I} x_\iota = \sum_{k=1}^{n} x_k$, where the term on the left is interpreted as in (2.2).

b) Suppose that $I = \mathcal{N}$. Show that $\sum_{\iota \in I} x_\iota = \sum_{n=1}^{\infty} x_n$, where the term on the left is interpreted as in (2.2) and the term on the right as the sum of the infinite series if it converges and ∞ otherwise.

c) Show that if $\sum_{\iota \in I} x_\iota < \infty$, then $\{\iota \in I : x_\iota > 0\}$ is countable. *Note:* This result is often applied in the following form: If $f : \Omega \to [0, \infty)$ is such that $\sum_{x \in \Omega} f(x) < \infty$, then $\{x : f(x) > 0\}$ is countable.

2.3 OPEN AND CLOSED SETS

In this section, we will discuss open and closed sets of real numbers. These sets not only play a significant role in classical analysis but, as we will see throughout this book, figure prominently in many areas of modern analysis.

We begin with the definition of an open set. Roughly speaking, a set O of real numbers is *open* if for each $x \in O$, we can remain in O by staying sufficiently close to x. More precisely, we have the following definition.

DEFINITION 2.7 **Open Set**

A subset $O \subset \mathcal{R}$ is said to be an **open set** if for each $x \in O$, there is an $r > 0$ such that $(x - r, x + r) \subset O$. In other words, O is open if for each $x \in O$, there is an $r > 0$ such that all numbers within r of x are also members of O.

EXAMPLE 2.5 *Illustrates Definition 2.7*

a) Any interval of the form (a, b), where $-\infty \le a < b \le \infty$, is an open set. Therefore, such intervals are called **open intervals.**

b) The interval $(0, 1]$ is not open, because $(1 - r, 1 + r) \not\subset (0, 1]$ for all $r > 0$. Similarly, neither $[0, 1)$ nor $[0, 1]$ are open sets.

c) Let K be a nonempty countable subset of $\mathcal{R}$. Then K is not open. Indeed, a nonempty open set must contain an open interval and such an interval is uncountable, as we know from Exercise 1.26 on page 21. In particular, then, $\mathcal{N}$, $\mathcal{Z}$, and Q are not open, and no nonempty finite set is open.

d) The set Q^c of irrational numbers is not open. If it were, then it would have to contain an open interval. Such an interval would contain no rational numbers, which is impossible by Proposition 2.4 on page 32. □

Our next theorem displays three fundamental properties of the collection of open sets. As we will see in Section 10.1, these three properties are precisely the ones needed to generalize the concept of open sets to other frameworks in the form of *topological spaces*.

□ □ □ **THEOREM ? ?**

a) $\mathcal{R}$ and $\emptyset$ are open sets.

b) If A and B are open sets, then so is $A \cap B$.

c) If $\{O_\iota\}_{\iota \in I}$ is a collection of open sets, then $\bigcup_{\iota \in I} O_\iota$ is open.

PROOF The proof of (a) is trivial. For (b), suppose A and B are open sets. We must show $A \cap B$ is open. Let $x \in A \cap B$. Then $x \in A$ and $x \in B$. As A and B are open, there exist $r_1, r_2 > 0$ such that $(x - r_1, x + r_1) \subset A$ and $(x - r_2, x + r_2) \subset B$. Let $r = \min\{r_1, r_2\}$. Then we have $(x - r, x + r) \subset A$ and $(x - r, x + r) \subset B$ so that $(x - r, x + r) \subset A \cap B$. Hence, $A \cap B$ is open.

For (c), suppose $\{O_\iota\}_{\iota \in I}$ is a collection of open sets and let $O = \bigcup_{\iota \in I} O_\iota$. We must show O is open. Let $x \in O$. Then $x \in O_\iota$ for some $\iota \in I$, say, ι_0. As $x \in O_{\iota_0}$ and O_{ι_0} is open, there is an $r > 0$ such that $(x - r, x + r) \subset O_{\iota_0}$. Consequently, because $O_{\iota_0} \subset O$, we have that $(x - r, x + r) \subset O$. Thus, O is open.   ∎

Theorem 2.2(b) shows that the intersection of two open sets is open. It follows easily by induction that the intersection of a finite number of open sets is open; that is, if O_k is an open set for $k = 1, 2, \ldots, n$, then $\bigcap_{k=1}^{n} O_k$ is also an open set. However, the extension to arbitrary (even countable) collections is not valid. Indeed, for each $n \in \mathcal{N}$, let $O_n = (-1/n, 1/n)$. Then each O_n is open but $\bigcap_{n=1}^{\infty} O_n = \{0\}$ is not open.

We have seen that if $a, b \in \mathcal{R}^*$ with $a < b$, then the interval (a, b) is an open set, called an open interval. It follows from Theorem 2.2(c) that unions of collections of open intervals are open sets. As the next proposition shows, all open sets are of this form.

□ □ □   **PROPOSITION 2.13**

Each open set O is a countable union of disjoint open intervals. The representation is unique in the sense that if $\mathcal{C}$ and $\mathcal{D}$ are two pairwise disjoint collections of open intervals whose union is O, then $\mathcal{C} = \mathcal{D}$.

PROOF   Let O be an open set. We first show that O can be expressed as a union of open intervals. The idea is this: For each $x \in O$, go as far as possible in either direction from x without leaving O; this will yield an open interval containing x and contained in O. The union of these open intervals will equal O.

More formally, let $x \in O$. Define

$$A_x = \{\, y : y < x \text{ and } (y, x) \subset O \,\} \quad \text{and} \quad B_x = \{\, z : z > x \text{ and } (x, z) \subset O \,\}.$$

The sets A_x and B_x are nonempty because O is open. Let $a_x = \inf A_x$ and $b_x = \sup B_x$. Note that $a_x < x < b_x$. Indeed, if $y \in A_x$, then $a_x \leq y < x$ and, so, $a_x < x$; similarly, we see that $x < b_x$. Also note that $a_x, b_x \notin O$. To verify this fact, suppose to the contrary that $a_x \in O$. Then $(a_x - r, a_x + r) \subset O$ for some $r > 0$, and we can always choose $r < x - a_x$. Because $a_x + r > a_x$, there is a $y \in A_x$ such that $y < a_x + r$ and, because $y \in A_x$, $(y, x) \subset O$. It follows that $(a_x - r, x) = (a_x - r, a_x + r) \cup (y, x) \subset O$ and, hence, that $a_x - r \in A_x$. But this is impossible because a_x is a lower bound for A_x. Thus, $a_x \notin O$ and, similarly, $b_x \notin O$.

Set $I_x = (a_x, b_x)$ and note that $x \in I_x$. We claim that $I_x \subset O$. Let $u \in I_x$; then $a_x < u < b_x$. Thus, we can choose $y \in A_x$ and $z \in B_x$ such that $y < u < z$. If $u \leq x$, then $u \in (y, x] \subset O$ and, if $u > x$, then $u \in (x, z) \subset O$. Hence, $I_x \subset O$. We can now conclude that $\bigcup_{x \in O} I_x \subset O$. However, as $x \in I_x$, $\bigcup_{x \in O} I_x \supset O$. Thus, $O = \bigcup_{x \in O} I_x$.

Next we show that either $I_x \cap I_y = \emptyset$ or $I_x = I_y$. So, suppose that $I_x \cap I_y \neq \emptyset$. Then $a_x < b_y$ and $a_y < b_x$. Since $a_x \notin O$, we have $a_x \notin (a_y, b_y)$ and, so, $a_x \leq a_y$. Similarly, $a_y \leq a_x$. Thus, $a_x = a_y$. Likewise, $b_x = b_y$. Hence, $I_x = I_y$.

Now, let $\mathcal{C} = \{\, I_x : x \in O \,\}$. Then, as we have seen, $\mathcal{C}$ is pairwise disjoint and $\bigcup_{A \in \mathcal{C}} A = O$. We claim that $\mathcal{C}$ is countable. Let $A \in \mathcal{C}$. Because A is an open interval, we can, by the density of the rational numbers, select a rational number $r_A \in A$. Define $f: \mathcal{C} \to Q$ by $f(A) = r_A$. This function is one-to-one because $\mathcal{C}$ is pairwise disjoint. Hence, $\mathcal{C}$ is equivalent to a subset of Q and, consequently, is countable.

We leave the proof of the uniqueness of the representation as an exercise for the reader. ∎

Closed Sets

Open sets constitute an important class of sets. Another important class of sets comprises the *closed sets*. To begin our discussion of closed sets, we make the following definition.

DEFINITION 2.8 **Limit Point, Closure**

Let $E \subset \mathcal{R}$. A real number x is called a **limit point**[†] of E if for each $\epsilon > 0$, there is a $y \in E$ such that $|y - x| < \epsilon$. The set of all limit points of E, denoted $\overline{E}$, is called the **closure** of E.

It is easy to see that each of the following two conditions is equivalent to x being a limit point of E (i.e., $x \in \overline{E}$).

- Each open interval containing x contains a member of E; that is, if I is an open interval such that $x \in I$, then $I \cap E \neq \emptyset$.

- There is a sequence $\{x_n\}_{n=1}^{\infty}$ of elements of E such that $\lim_{n \to \infty} x_n = x$, thus, the terminology x is a limit point of E.

EXAMPLE 2.6 *Illustrates Definition 2.8*

We leave the verification of each part that follows to the reader.
a) $\overline{\mathcal{R}} = \mathcal{R}$ and $\overline{\emptyset} = \emptyset$.
b) Let $a, b \in \mathcal{R}$ with $a < b$. Then $\overline{(a,b)} = \overline{[a,b)} = \overline{(a,b]} = \overline{[a,b]} = [a,b]$.
c) $\overline{\mathcal{N}} = \mathcal{N}$ and $\overline{\mathcal{Z}} = \mathcal{Z}$.
d) $\overline{Q} = \mathcal{R}$ and $\overline{Q^c} = \mathcal{R}$.
e) If A is a finite subset of $\mathcal{R}$, then $\overline{A} = A$. □

Note that every point of a set E is a limit point of E, that is, $E \subset \overline{E}$. However, the converse is not true — there may be limit points of E that do not belong to E. For instance, 1 is a limit point of $[0, 1)$ but does not belong to that set. If a set contains all its limit points, it is called *closed*.

[†] Some texts use the term **point of closure** instead of *limit point* and reserve the term *limit point* for a related concept.

DEFINITION 2.9 Closed Set

A subset $F \subset \mathcal{R}$ is said to be a **closed set** if $\overline{F} = F$, that is, if F contains all its limit points.

EXAMPLE 2.7 *Illustrates Definition 2.9*

Referring to Example 2.6, we conclude the following:

a) $\mathcal{R}$ and $\emptyset$ are closed sets. But we also know from Theorem 2.2(a) that $\mathcal{R}$ and $\emptyset$ are open sets. We leave it to the reader as an exercise to show that these are the only two subsets of $\mathcal{R}$ that are both open and closed. (See Exercise 2.46.)

b) The intervals of $\mathcal{R}$ that are closed are of the form $[a, b]$, $[a, \infty)$, and $(-\infty, b]$, where $a, b \in \mathcal{R}$. Such intervals are called **closed intervals.** *Note:* Intervals of the form $(a, b]$ and $[a, b)$, where $a, b \in \mathcal{R}$, are called **half-open intervals.** Degenerate intervals of the form $[a, a]$ are closed sets; degenerate intervals of the form (a, a), $(a, a]$, and $[a, a)$ are empty and, hence, both open and closed.

c) $\mathcal{N}$ and $\mathcal{Z}$ are closed.

d) Neither Q nor Q^c is closed.

e) Any finite subset of $\mathcal{R}$ is closed.

f) A set may be neither open nor closed; examples are Q, Q^c, and any half-open interval. ◻

The fundamental relationship between open and closed sets is elucidated by the following proposition. This proposition also provides a way to *define* closed sets in more general settings, which we do in Chapter 10.

◻ ◻ ◻ **PROPOSITION 2.14**

A set is open if and only if its complement is closed or, equivalently, a set is closed if and only if its complement is open.

PROOF Suppose that E^c is open. We will show that E is closed by proving that it contains all its limit points. So, assume that $x \notin E$, that is, $x \in E^c$. Since E^c is open, there is an $r > 0$ such that $(x - r, x + r) \subset E^c$. But then $(x - r, x + r)$ is an open interval about x containing no points of E; hence, $x \notin \overline{E}$. We have therefore shown that $\overline{E} \subset E$, as required.

Conversely, suppose that E is closed. If $x \in E^c$, then $x \notin E$ and, consequently, since E is closed, we have $x \notin \overline{E}$. Hence, there is an $\epsilon > 0$ such that $(x - \epsilon, x + \epsilon) \subset E^c$. We have thus shown that E^c is open. ∎

Open and Closed Sets of a Subset of $\mathcal{R}$

Frequently, our "universal set" will be a proper subset of $\mathcal{R}$. Therefore, we need to discuss open and closed sets of a subset $D \subset \mathcal{R}$. Observe that the following definition of an open set of D reduces to Definition 2.7 on page 47 when $D = \mathcal{R}$.

DEFINITION 2.10   **Open Set of D**

Let $D \subset \mathcal{R}$. A subset $G \subset D$ is said to be **open in D** if for each $x \in G$, there is an $r > 0$ such that $(x - r, x + r) \cap D \subset G$. Thus, G is an open subset of D if for each $x \in G$, there is an $r > 0$ such that all numbers within r of x that are members of D are also members of G.

EXAMPLE 2.8   *Illustrates Definition 2.10*

a) Let $D = [0, 2]$. Then the interval $[0, 1)$ is open in D. Note, however, that it is not open in $\mathcal{R}$.

b) Let $D = [0, 2]$. Then the interval $[0, 1]$ is not open in D because, for each $r > 0$, we have $(1 - r, 1 + r) \cap [0, 2] \not\subset [0, 1]$.

c) Let $D = \mathcal{N}$. Then every subset $A \subset \mathcal{N}$ is open in $\mathcal{N}$. Indeed, if $n \in A$, then $(n - \frac{1}{2}, n + \frac{1}{2}) \cap \mathcal{N} = \{n\} \subset A$.   □

The following theorem provides the relationship between open sets of a subset of $\mathcal{R}$ and open sets of $\mathcal{R}$.

□ □ □   THEOREM 2.3

Let $D \subset \mathcal{R}$. A set $G \subset D$ is open in D if and only if there is an open set O of $\mathcal{R}$ such that $G = D \cap O$. In other words, the open sets in D are precisely the open sets of $\mathcal{R}$ intersected with D.

PROOF   Suppose $G \subset D$ is open in D. Then, for each $x \in G$, there is an open interval I_x (open in $\mathcal{R}$) containing x such that $I_x \cap D \subset G$. Let $O = \bigcup_{x \in G} I_x$. Then, by Theorem 2.2(c), O is open in $\mathcal{R}$. We will show that $G = D \cap O$. If $x \in G$, then $x \in D$ and $x \in I_x \subset O$; thus, $G \subset D \cap O$. On the other hand, since $I_x \cap D \subset G$ for all $x \in G$, we have

$$D \cap O = D \cap \left(\bigcup_{x \in G} I_x \right) = \bigcup_{x \in G} (I_x \cap D) \subset G.$$

Hence, $G = D \cap O$, as required.

Conversely, suppose $G = D \cap O$ for some open set O of $\mathcal{R}$. If $x \in G$, then $x \in O$ and, hence, there is an $r > 0$ such that $(x - r, x + r) \subset O$. This, in turn, implies that $(x - r, x + r) \cap D \subset O \cap D = G$. Hence G is open in D.   ■

Limit points, closure, and closed sets in D are defined in a way analogous to that in $\mathcal{R}$. We leave the details to the reader in Exercise 2.52.

Open Sets in Euclidean n-Space

We also need the concept of open sets in Euclidean n-space. Let $x, y \in \mathcal{R}^n$, say, $x = (x_1, \ldots, x_n)$ and $y = (y_1, \ldots, y_n)$. As in calculus, the (Euclidean) **distance** between x and y, denoted $\boldsymbol{d(x, y)}$, is defined by

$$d(x, y) = \sqrt{(x_1 - y_1)^2 + \cdots + (x_n - y_n)^2}.$$

Also, for $r > 0$, the **open ball** of radius r centered at x, denoted $\boldsymbol{B_r(x)}$, is defined by

$$B_r(x) = \{\, y \in \mathcal{R}^n : d(x, y) < r \,\}.$$

In $\mathcal{R}$, $\mathcal{R}^2$, and $\mathcal{R}^3$, the open balls of radius r centered at x are, respectively, the open interval $(x - r, x + r)$, the open disk of radius r centered at x, and the open sphere of radius r centered at x.

With the above notation in mind, we can now define open sets in Euclidean n-space. Specifically, a subset $O \subset \mathcal{R}^n$ is said to be an **open set** if for each $x \in O$, there is an $r > 0$ such that $B_r(x) \subset O$. Note the following:

- Limit points, closure, and closed sets in $\mathcal{R}^n$ are defined in a way analogous to that in $\mathcal{R}$.

- If $D \subset \mathcal{R}^n$, then the open sets of D are defined as in Definition 2.10 on page 51, using $B_r(x)$ instead of $(x - r, x + r)$.

- The open sets of $\mathbb{C}$ are simply the open sets in $\mathcal{R}^2$ upon identifying $x + iy$ with (x, y).

An essential result in Euclidean n-space is called the **Heine-Borel theorem**. As this result is a special case of a more general theorem that we prove in Chapter 11, we state it here without proof.

Heine-Borel Theorem

A subset E of $\mathcal{R}^n$ is closed and bounded if and only if any collection $\mathcal{O}$ of open sets that covers E (i.e., $E \subset \bigcup_{O \in \mathcal{O}} O$) contains a finite subcollection that also covers E. Such a subset E is called **compact.**

Exercises for Section 2.3

2.38 Prove that the intersection of a finite number of opens sets is open; that is, if O_k is an open set for $k = 1, 2, \ldots, n$, then $\bigcap_{k=1}^{n} O_k$ is also an open set.

2.39 Prove the uniqueness portion of Proposition 2.13 on page 48.

2.40 Prove **Lindelöf's theorem:** Let $\mathcal{O}$ be a collection of open sets. Then there is a countable subcollection $\{O_n\}_n$ of $\mathcal{O}$ such that

$$\bigcup_{O \in \mathcal{O}} O = \bigcup_n O_n.$$

2.41 Let $E \subset \mathcal{R}$ and $x \in \mathcal{R}$. Prove that each of the following is equivalent.
   a) $x \in \overline{E}$ (i.e., x is a limit point of E).
   b) Each open interval containing x contains a member of E; that is, if I is an open interval such that $x \in I$, then $I \cap E \neq \emptyset$.
   c) Each open set containing x contains a member of E; that is, if O is an open set such that $x \in O$, then $O \cap E \neq \emptyset$.
   d) There is a sequence $\{x_n\}_{n=1}^{\infty}$ of elements of E such that $\lim_{n \to \infty} x_n = x$.

2.42 Refer to Example 2.6 on page 49. Verify each of the statements made in that example.

2.43 Let $E \subset \mathcal{R}$. A real number x is called an **accumulation point** of E if for each $\epsilon > 0$, there is a $y \in E$ such that $0 < |y - x| < \epsilon$. Prove that the following are equivalent.
   a) x is an accumulation point of E.

b) Each open interval containing x contains a member of E different from x; that is, if I is an open interval such that $x \in I$, then $I \cap (E \setminus \{x\}) \neq \emptyset$.

c) $x \in \overline{E \setminus \{x\}}$.

d) There exists a sequence $\{x_n\}_{n=1}^{\infty}$ of distinct elements of E such that $\lim_{n \to \infty} x_n = x$.

2.44 Refer to Exercise 2.43. Let E' denote the set of all accumulation points of a set $E \subset \mathcal{R}$.

a) Prove that E' is closed.

b) Prove that $\overline{E} = E \cup E'$.

★2.45 Refer to Exercise 2.43. Prove the **Bolzano-Weierstrass theorem:** Every bounded infinite subset of real numbers has an accumulation point. *Hint:* Use the fact that every infinite set contains a countably infinite subset (Exercise 1.28 on page 21).

2.46 Prove that $\mathcal{R}$ and $\emptyset$ are the only two subsets of $\mathcal{R}$ that are both open and closed. *Hint:* Let A be a nonempty proper open subset of $\mathcal{R}$. Choose $x \in A$ and let a_x and b_x be as in the proof of Proposition 2.13 on page 48.

2.47 Let A and B be subsets of $\mathcal{R}$. Establish each of the following facts.

a) $\overline{A \cup B} = \overline{A} \cup \overline{B}$.

b) $\overline{A \cap B} \subset \overline{A} \cap \overline{B}$. Provide an example to show that the reverse inclusion does not hold in general.

c) $\overline{A}$ is closed.

d) If A and B are closed, then so is $A \cup B$.

e) If $\{F_\iota\}_{\iota \in I}$ is a collection of closed sets, then $\bigcap_{\iota \in I} F_\iota$ is closed.

2.48 *True or False:* If $\{F_\iota\}_{\iota \in I}$ is a collection of closed sets, then $\bigcup_{\iota \in I} F_\iota$ is closed.

2.49 Let A and B be subsets of $\mathcal{R}$ and $\{A_\iota\}_{\iota \in I}$ be a collection of subsets of $\mathcal{R}$. Establish each of the following facts.

a) If $A \subset B$, then $\overline{A} \subset \overline{B}$.

b) If $A \subset B \subset \overline{A}$, then $\overline{A} = \overline{B}$.

c) We have

$$\bigcup_{\iota \in I} \overline{A_\iota} \subset \overline{\bigcup_{\iota \in I} A_\iota}.$$

d) Referring to part (c), can we, in general, replace "$\subset$" by "$=$"? If not, state a condition on I that assures the replacement is valid.

2.50 Let $D \subset \mathcal{R}$.

a) Suppose that D is an open subset of $\mathcal{R}$. Prove that a subset of D is open in D if and only if it is open in $\mathcal{R}$.

b) Show that the result of part (a) fails to hold without the assumption that D is an open subset of $\mathcal{R}$.

c) Prove that a subset of D is open in D if it is open in $\mathcal{R}$.

2.51 Let $D \subset \mathcal{R}$. Prove that the collection of open sets of D satisfies the three properties listed in Theorem 2.2 (page 47). That is,

a) D and $\emptyset$ are open in D.

b) If A and B are open in D, then so is $A \cap B$.

c) If $\{G_\iota\}_{\iota \in I}$ is a collection of sets open in D, then $\bigcup_{\iota \in I} G_\iota$ is open in D.

2.52 In this exercise, we will explore limit points, closure, and closed sets of a subset of $\mathcal{R}$. Let $D \subset \mathcal{R}$ and $E \subset D$.

a) Define a *limit point of E in D;* call such a limit point a D-limit point of E.

b) Define the *closure of E in D;* call it the D-closure of E.

c) Define E *is closed in D.*

d) Prove that E is closed in D if and only if $D \setminus E$ is open in D.

e) Prove that E is closed in D if and only if there is a closed set F of $\mathcal{R}$ such that $E = D \cap F$.

2.4 REAL-VALUED FUNCTIONS

A **real-valued function** is a function whose range is a subset of $\mathcal{R}$. If $f\colon \Omega \to \mathcal{R}$, then we say that f is a **real-valued function on Ω.** In this section, we will discuss real-valued functions and several concepts associated with them. Much of the section is concerned with real-valued functions whose domains are a subset of $\mathcal{R}$.

We begin by defining algebraic operations on real-valued functions. This is done *pointwise* as follows. Suppose that f and g are real-valued functions on Ω and that $\alpha \in \mathcal{R}$. Then we define the functions $f + g$, αf, and $f \cdot g$ on Ω by

$$(f + g)(x) = f(x) + g(x),$$
$$(\alpha f)(x) = \alpha f(x),$$
$$(f \cdot g)(x) = f(x)g(x),$$

for each $x \in \Omega$.

Continuous Functions

The most important functions in calculus are the continuous functions. They play a prominent role in modern analysis as well. Roughly speaking, a function f is continuous at x_0 if $f(x)$ can be made arbitrarily close to $f(x_0)$ by taking x sufficiently close to x_0. More precisely, we have the following definition for a real-valued function defined on a subset of $\mathcal{R}$.

DEFINITION 2.11 Continuous Function

Let $D \subset \mathcal{R}$, $f\colon D \to \mathcal{R}$, and $x_0 \in D$. We say that f is **continuous at x_0** if for each $\epsilon > 0$, there is a $\delta > 0$ such that $|f(x) - f(x_0)| < \epsilon$ whenever $x \in D$ and $|x - x_0| < \delta$. We say that f is **continuous on D** if it is continuous at every point of D. We denote by $C(D)$ the collection of all continuous functions on D.[†] For simplicity, and when no confusion will arise, we often write C for $C(\mathcal{R})$.

Note: If f is not continuous at x_0, then we say that f is **discontinuous** at x_0 or that x_0 is a **point of discontinuity** of f.

EXAMPLE 2.9 *Illustrates Definition 2.11*

a) Let $D = (0, \infty)$ and define $f(x) = 1/x$. Then f is continuous on D.
b) Let $D = \mathcal{R}$. Define $f(0) = 0$ and $f(x) = \sin(1/x)$ for $x \neq 0$. Then f is continuous except at 0.
c) Let $D = \mathcal{R}$ and define $f(x) = [x]$. Then f is continuous except at points of $\mathcal{Z}$.
d) Every function is continuous on $\mathcal{N}$. Indeed, let $f\colon \mathcal{N} \to \mathcal{R}$ and $x_0 \in \mathcal{N}$. Then $|f(x) - f(x_0)| = 0$ whenever $x \in \mathcal{N}$ and $|x - x_0| < 1$. ◻

[†] This notation is temporary and will be modified and generalized in Chapter 11.

An important property of the continuous functions on a subset D of $\mathcal{R}$ is that they form an **algebra of functions.** In other words, we have the following theorem whose proof is left to the reader as an exercise.

□ □ □ THEOREM 2.4

Let $D \subset \mathcal{R}$. Then the collection $C(D)$ of continuous functions on D is an algebra of functions. That is, if $f, g \in C(D)$ and $\alpha \in \mathcal{R}$, then

a) $f + g \in C(D)$.

b) $\alpha f \in C(D)$.

c) $f \cdot g \in C(D)$.

Our next theorem provides a relationship between continuous functions on D and the open sets in D.

□ □ □ THEOREM 2.5

Let $D \subset \mathcal{R}$ and $f: D \to \mathcal{R}$. Then f is continuous on D if and only if $f^{-1}(O)$ is open in D for each open set O in $\mathcal{R}$.

PROOF   Suppose f is continuous on D. Let O be an open set in $\mathcal{R}$ and $x_0 \in f^{-1}(O)$. Then $f(x_0) \in O$ and, consequently, because O is open, there is an $r > 0$ such that $(f(x_0) - r, f(x_0) + r) \subset O$. As f is continuous at x_0, there is a $\delta > 0$ such that $|f(x) - f(x_0)| < r$ whenever $|x - x_0| < \delta$ and $x \in D$. Therefore, we see that, if $x \in (x_0 - \delta, x_0 + \delta) \cap D$, then $f(x) \in (f(x_0) - r, f(x_0) + r) \subset O$ and, hence, $x \in f^{-1}(O)$. So, we have found a $\delta > 0$ such that $(x_0 - \delta, x_0 + \delta) \cap D \subset f^{-1}(O)$. It now follows that $f^{-1}(O)$ is open in D.

Conversely, suppose $f^{-1}(O)$ is open in D for each open set O in $\mathcal{R}$. Let $x_0 \in D$. We will prove that f is continuous at x_0. Let $\epsilon > 0$. The set $(f(x_0) - \epsilon, f(x_0) + \epsilon)$ is open in $\mathcal{R}$ and, so, $G = f^{-1}\big((f(x_0) - \epsilon, f(x_0) + \epsilon)\big)$ is open in D. As $x_0 \in G$ and G is open in D, there is an $r > 0$ such that $(x_0 - r, x_0 + r) \cap D \subset G$. Thus, if $x \in D$ and $|x - x_0| < r$, then $x \in G$ and, hence, $f(x) \in (f(x_0) - \epsilon, f(x_0) + \epsilon)$, that is, $|f(x) - f(x_0)| < \epsilon$.   ■

□ □ □ COROLLARY 2.1

A function $f: \mathcal{R} \to \mathcal{R}$ is continuous if and only if $f^{-1}(O)$ is open in $\mathcal{R}$ whenever O is open in $\mathcal{R}$.

We can restate Theorem 2.5 as follows: *A real-valued function on D is continuous if and only if the inverse image of each open set in $\mathcal{R}$ is open in D.* This relationship between continuous functions and open sets is significant because it provides a way for us to define continuity of functions in very general settings, as we will see in Chapter 10.

Monotone Functions

Functions defined on an interval that never decrease as x increases or never increase as x increases play a significant role in analysis.

DEFINITION 2.12   **Monotone Function**

Let f be a real-valued function defined on an interval I of real numbers. Then f is said to be

a) **nondecreasing** on I if $f(x) \le f(y)$ whenever $x, y \in I$ and $x < y$.
b) **nonincreasing** on I if $f(x) \ge f(y)$ whenever $x, y \in I$ and $x < y$.
c) **monotone** on I if it is either nondecreasing or nonincreasing on I.

Note: Some authors use the term *increasing* in place of *nondecreasing* and use the phrase *strictly increasing* to indicate that $f(x) < f(y)$ whenever $x < y$. We will use both "increasing" and "strictly increasing" to describe functions satisfying this latter condition, but will avoid both terms for functions that do not increase in the strict sense. Thus, for us, each of the terms "nondecreasing," "strictly increasing," and "increasing" applies equally well to the function $f(x) = x^3$; but we would only use the term "nondecreasing" to describe the function $f(x) = 1$. Similar remarks hold for the three terms *nonincreasing, strictly decreasing,* and *decreasing.*

EXAMPLE 2.10   *Illustrates Definition 2.12*

a) The function $f(x) = e^x$ is nondecreasing on any interval. It is also monotone and (strictly) increasing.
b) The function $f(x) = 1/x$ is nonincreasing on any interval not containing 0. It is also monotone and (strictly) decreasing on any such interval.
c) The function $f(x) = \sin x$ is nondecreasing on $[-\pi/2, \pi/2]$ and nonincreasing on $[\pi/2, 3\pi/2]$. However, it is not monotone on $[0, \pi]$.   ❑

Pointwise Limits

We know from Theorem 2.4 on page 55 that $C(D)$, the collection of real-valued continuous functions on a set $D \subset \mathcal{R}$, is an algebra of functions — it is closed under sums, multiples, and products. Being an algebra is a useful and important property for a collection of functions to have.

Another desirable property for a collection of functions is that it be *closed under pointwise limits.* This concept is relevant to any collection of real-valued functions that have a common domain, not just to those whose domain is some subset of $\mathcal{R}$. To begin, we define **pointwise convergence** of a sequence of real-valued functions.

DEFINITION 2.13   **Pointwise Convergence**

Let $\{f_n\}_{n=1}^{\infty}$ be a sequence of real-valued functions on a set Ω, that is, $f_n: \Omega \to \mathcal{R}$ for each $n \in \mathcal{N}$. Then we say that $\{f_n\}_{n=1}^{\infty}$ **converges pointwise** on Ω if for each $x \in \Omega$, the sequence $\{f_n(x)\}_{n=1}^{\infty}$ of real numbers converges in $\mathcal{R}$.

If $\{f_n\}_{n=1}^{\infty}$ converges pointwise on Ω, then we can define a function $f: \Omega \to \mathcal{R}$ by $f(x) = \lim_{n \to \infty} f_n(x)$. We say that the function f is the **pointwise limit** of the sequence of functions $\{f_n\}_{n=1}^{\infty}$ or that the sequence of functions $\{f_n\}_{n=1}^{\infty}$ **converges pointwise** to the function f. We write $f_n \to f$ **pointwise** to indicate pointwise convergence of $\{f_n\}_{n=1}^{\infty}$ to f.

EXAMPLE 2.11 *Illustrates Definition 2.13*

a) For each $n \in \mathcal{N}$, define $f_n: \mathcal{R} \to \mathcal{R}$ by $f_n(x) = (1 + x/n)^n$. Then $f_n \to f$ pointwise on $\mathcal{R}$, where $f(x) = e^x$.

b) Let $D \subset \mathcal{R}$. For each $n \in \mathcal{N}$, define $f_n: D \to \mathcal{R}$ by $f_n(x) = x^n$. If $D = [0, 1]$, then $f_n \to f$ pointwise, where

$$f(x) = \begin{cases} 0, & \text{if } 0 \le x < 1; \\ 1, & x = 1. \end{cases}$$

However, the sequence of functions $\{f_n\}_{n=1}^{\infty}$ fails to converge pointwise if $D = [-1, 1]$ since the sequence $\{(-1)^n\}_{n=1}^{\infty}$ does not converge; it also fails to converge pointwise if $D = [0, 2]$ since, for instance, the sequence $\{2^n\}_{n=1}^{\infty}$ does not converge in $\mathcal{R}$.

c) For each $n \in \mathcal{N}$, define $f_n: \mathcal{R} \to \mathcal{R}$ by

$$f_n(x) = \begin{cases} n^2 x, & \text{if } |x| < \frac{1}{n}; \\ 1/x, & \text{otherwise.} \end{cases}$$

Then $f_n \to f$ pointwise on $\mathcal{R}$, where $f(0) = 0$ and $f(x) = 1/x$ for $x \ne 0$.

d) Let $D \subset \mathcal{R}$. For each $n \in \mathcal{N}$, define $f_n: D \to \mathcal{R}$ by $f_n(x) = x/n$. Then $f_n \to 0$ pointwise on D, where 0 denotes the function identically equal to 0, that is, $f(x) = 0$ for all $x \in D$. □

DEFINITION 2.14 **Closure Under Pointwise Limits**

Let $\mathcal{F}$ be a collection of real-valued functions on Ω. We say that $\mathcal{F}$ is **closed under pointwise limits** if whenever $\{f_n\}_{n=1}^{\infty} \subset \mathcal{F}$ and $f_n \to f$ pointwise on Ω, then $f \in \mathcal{F}$.

Obviously, the collection of all real-valued functions on a set Ω is closed under pointwise limits. In particular, the collection of all real-valued functions on a subset D of $\mathcal{R}$ is closed under pointwise limits. However, in general, $C(D)$ is not closed under pointwise limits, as we see from parts (b) and (c) of Example 2.11.

The fact that $C(D)$ is not generally closed under pointwise limits will lead us naturally into a discussion of Borel measurable functions in Chapter 3. For now, we introduce a type of convergence stronger than pointwise convergence that does yield closure for $C(D)$. This type of convergence is called **uniform convergence**, a concept that is relevant to any sequence of real-valued functions that have a common domain.

DEFINITION 2.15   Uniform Convergence

Let $\{f_n\}_{n=1}^{\infty}$ be a sequence of real-valued functions on a set Ω. Then we say that $\{f_n\}_{n=1}^{\infty}$ **converges uniformly** to the real-valued function f on Ω, if for each $\epsilon > 0$, there is an $N \in \mathcal{N}$ such that $n \geq N$ implies $|f_n(x) - f(x)| < \epsilon$ for all $x \in \Omega$. We write $\boldsymbol{f_n \to f}$ **uniformly** to indicate uniform convergence of $\{f_n\}_{n=1}^{\infty}$ to f.

The "uniform" in *uniform convergence* refers to the fact that by taking n sufficiently large, $f_n(x)$ can be made arbitrarily close to $f(x)$ *for all $x \in \Omega$*, that is, *uniformly* over Ω.

Clearly, uniform convergence implies pointwise convergence. The converse is not true, however, as Example 2.11(b) shows. Uniform convergence also depends on Ω. For example, let $f_n(x) = x/n$ and $f(x) = 0$. If $\Omega = [0, 1]$, then $f_n \to f$ uniformly on Ω. But, if $\Omega = \mathcal{R}$, then $f_n \nrightarrow f$ uniformly on Ω, although it does so pointwise.

The next proposition verifies our contention that $C(D)$ is closed under uniform limits.

□ □ □   PROPOSITION 2.15

Let $D \subset \mathcal{R}$. Suppose that $\{f_n\}_{n=1}^{\infty} \subset C(D)$ and that $f_n \to f$ uniformly. Then $f \in C(D)$.

PROOF   Let $x_0 \in D$ and $\epsilon > 0$. Because $f_n \to f$ uniformly, we can choose $N \in \mathcal{N}$ such that $|f_N(x) - f(x)| < \epsilon/3$ for all $x \in D$. And, because f_N is continuous on D and, hence, at x_0, we can choose $\delta > 0$ such that $|f_N(x) - f_N(x_0)| < \epsilon/3$ whenever $x \in D$ and $|x - x_0| < \delta$. It follows that whenever $x \in D$ and $|x - x_0| < \delta$, we have

$$|f(x) - f(x_0)| \leq |f(x) - f_N(x)| + |f_N(x) - f_N(x_0)| + |f_N(x_0) - f(x_0)|$$
$$< \epsilon/3 + \epsilon/3 + \epsilon/3 = \epsilon,$$

Thus, f is continuous on D.   ∎

Monotone Sequences of Functions

As we will see beginning in Chapter 4, it is important to consider *monotone* sequences of functions. As for pointwise and uniform convergence, this concept is relevant to any sequence of real-valued functions that have a common domain.

DEFINITION 2.16   Monotone Sequence of Functions

Let $\{f_n\}_{n=1}^{\infty}$ be a sequence of real-valued functions on a set Ω. Then we say that $\{f_n\}_{n=1}^{\infty}$ is

a) **nondecreasing** if for each $x \in \Omega$, $\{f_n(x)\}_{n=1}^{\infty}$ is a nondecreasing sequence of real numbers.

b) **nonincreasing** if for each $x \in \Omega$, $\{f_n(x)\}_{n=1}^{\infty}$ is a nonincreasing sequence of real numbers.

c) **monotone** if it is either nondecreasing or nonincreasing.

EXAMPLE 2.12　*Illustrates Definition 2.16*

a) Let $D \subset \mathcal{R}$ and define $f_n: D \to \mathcal{R}$ by $f_n(x) = x^n$. Then $\{f_n\}_{n=1}^{\infty}$ is nonincreasing if $D = [0,1]$, nondecreasing if $D = [1,2]$, and not monotone if $D = [0,2]$ or $D = (-1,0]$.

b) Let $D \subset \mathcal{R}$ and define $f_n: D \to \mathcal{R}$ by $f_n(x) = x/n$. Then $\{f_n\}_{n=1}^{\infty}$ is nondecreasing if $D \subset (-\infty, 0]$, nonincreasing if $D \subset [0, \infty)$, but not monotone if D contains both positive and negative numbers.

c) Define $f_n: [0,1] \to \mathcal{R}$ by

$$f_n(x) = \begin{cases} n^2 x, & \text{if } 0 \le x < \frac{1}{2n}; \\ n - n^2 x, & \text{if } \frac{1}{2n} \le x < \frac{1}{n}; \\ 0, & \text{if } \frac{1}{n} \le x \le 1. \end{cases}$$

The sequence $\{f_n\}_{n=1}^{\infty}$ is not monotone.

d) Let $\{A_n\}_{n=1}^{\infty}$ be a sequence of subsets of a set Ω. Define $f_n: \Omega \to \mathcal{R}$ by $f_n(x) = 1$ if $x \in A_n$ and $f_n(x) = 0$ if $x \notin A_n$.

(i) $\{f_n\}_{n=1}^{\infty}$ is a nondecreasing sequence of functions if and only if $\{A_n\}_{n=1}^{\infty}$ is a (monotone) nondecreasing sequence of sets.

(ii) $\{f_n\}_{n=1}^{\infty}$ is a nonincreasing sequence of functions if and only if $\{A_n\}_{n=1}^{\infty}$ is a (monotone) nonincreasing sequence of sets.　□

Exercises for Section 2.4

2.53 Let f and g be real-valued functions on Ω. Write the pointwise definitions of $f \vee g$ (the maximum of f and g) and $f \wedge g$ (the minimum of f and g). Refer to Exercise 2.3 on page 34.

2.54 Prove Theorem 2.4 on page 55.

2.55 Show that if $f \in C(D)$, then so is $|f|$.

2.56 Prove that $C(D)$ is closed under maximums and minimums. That is, if $f, g \in C(D)$, then

a) $f \vee g \in C(D)$. *Hint:* Use Exercise 2.55, Theorem 2.4 on page 55, and Exercise 2.3(b) on page 34.

b) $f \wedge g \in C(D)$. *Hint:* Use Exercise 2.55, Theorem 2.4 on page 55, and Exercise 2.3(c) on page 34.

2.57 Verify each part of Example 2.10 on page 56.

2.58 Define $f: [0,3] \to \mathcal{R}$ by $f(x) = 2$ if $1 \le x < 2$ and $f(x) = 1$ otherwise. On which (nondegenerate) subintervals of $[0,3]$ is f

a) nondecreasing?

b) strictly increasing?

c) nonincreasing?

d) strictly decreasing?

e) monotone?

★2.59 Suppose that $f: (a, b) \to \mathcal{R}$ is nondecreasing. For $x \in (a, b)$, let

$$L_x = \{\, f(t) : a < t < x \,\} \quad \text{and} \quad R_x = \{\, f(t) : x < t < b \,\},$$

and define $f(x-) = \sup L_x$ and $f(x+) = \inf R_x$.
a) Show that $f(x-) \le f(x) \le f(x+)$ for all $x \in (a, b)$.
b) Prove that f is continuous at x if and only if $f(x-) = f(x+)$.
c) Prove that f has countably many discontinuities; that is, the set of points at which f is discontinuous is countable.
d) Deduce that a nonincreasing function on (a, b) has countably many discontinuities.

2.60 For each $n \in \mathcal{N}$, define $f_n: [0, 1] \to \mathcal{R}$ by $f_n(x) = x^n$. Also, define

$$f(x) = \begin{cases} 0, & \text{if } 0 \le x < 1; \\ 1, & x = 1. \end{cases}$$

Prove that $f_n \to f$ pointwise, but not uniformly, on $[0, 1]$.

2.61 Let $D \subset \mathcal{R}$. For each $n \in \mathcal{N}$, define $f: D \to \mathcal{R}$ by $f_n(x) = x/n$. Also, define $f: D \to \mathcal{R}$ by $f(x) = 0$ for all $x \in D$.
a) Show that if $D = [0, 1]$, then $f_n \to f$ uniformly.
b) Show that if $D = \mathcal{R}$, then $f_n \not\to f$ uniformly.
c) In part (b), is it possible for $\{f_n\}_{n=1}^\infty$ to converge uniformly to some function? Explain your answer.

2.62 Verify each part of Example 2.12 on page 59.

★2.63 In this exercise, we ask you to prove **Dini's theorem:** Suppose that $\{f_n\}_{n=1}^\infty$ is a monotone sequence of continuous functions defined on a closed bounded interval $[a, b]$. Further suppose that $\{f_n\}_{n=1}^\infty$ converges pointwise to the continuous function f. Prove that $f_n \to f$ uniformly on $[a, b]$ by applying the following steps.
a) Explain why we can assume without loss of generality that $\{f_n\}_{n=1}^\infty$ is nonincreasing and that $f = 0$.
b) Let $\epsilon > 0$. For each $n \in \mathcal{N}$, set $O_n^\epsilon = \{\, x \in [a, b] : f_n(x) < \epsilon \,\}$. Show that $\{O_n^\epsilon\}_{n=1}^\infty$ is a monotone nondecreasing sequence of open sets in $[a, b]$ whose union is $[a, b]$.
c) Use part (b) to prove that there is an $N \in \mathcal{N}$ such that $O_N^\epsilon = [a, b]$. *Hint:* Use the Heine-Borel theorem.
d) Use part (c) to conclude that $f_n \to 0$ uniformly.

2.64 Refer to Exercise 2.63. Show that the conclusion of Dini's theorem does not hold if we weaken the hypotheses of that theorem in any one of the following ways:
a) The interval on which the functions are defined is permitted to be any closed interval.
b) The interval on which the functions are defined is permitted to be any bounded interval.
c) The limiting function f is not restricted to being continuous.
d) The monotonicity requirement is dropped.

2.65 Refer to Example 2.11(a) on page 57. Prove that the convergence is uniform on any bounded subinterval of $[0, \infty)$. *Hint:* Apply the binomial theorem and Dini's theorem (Exercise 2.63).

2.66 Let $\{f_n\}_{n=1}^\infty$ be a sequence of real-valued functions on Ω.
a) We say that $\{f_n\}_{n=1}^\infty$ is **pointwise Cauchy** on Ω, if for each $x \in \Omega$, $\{f_n(x)\}_{n=1}^\infty$ is a Cauchy sequence. Prove that if $\{f_n\}_{n=1}^\infty$ is pointwise Cauchy on Ω, then it converges pointwise on Ω.
b) We say that $\{f_n\}_{n=1}^\infty$ is **uniformly Cauchy** on Ω, if for each $\epsilon > 0$, there is an $N \in \mathcal{N}$ such that $|f_n(x) - f_m(x)| < \epsilon$ whenever $m, n \ge N$ and $x \in \Omega$. Prove that if $\{f_n\}_{n=1}^\infty$ is uniformly Cauchy on Ω, then it converges uniformly on Ω.

2.5 THE CANTOR SET AND CANTOR FUNCTION

We next introduce a set and function, called the *Cantor set* and *Cantor function,* named in honor of Georg Cantor, that will serve as useful examples and counterexamples throughout the text.

Before discussing the Cantor set, we present the following proposition which states in part that for each integer $p \geq 2$, every number between 0 and 1 has a base-p expansion. The proof is left to the reader as an exercise. (See Exercise 2.67.)

□ □ □ **PROPOSITION 2.16 Base-p Expansion**

Let p be an integer greater than 1. Then for each $x \in [0, 1]$, there is a sequence $\{a_n\}_{n=1}^{\infty}$ of integers such that $0 \leq a_n \leq p - 1$ for all n and

$$x = \sum_{n=1}^{\infty} \frac{a_n}{p^n} = \frac{a_1}{p} + \frac{a_2}{p^2} + \frac{a_3}{p^3} + \cdots. \tag{2.3}$$

The sequence $\{a_n\}_{n=1}^{\infty}$ is unique unless $x \neq 1$ and is of the form q/p^m for some $q, m \in \mathcal{N}$, in which case there are exactly two such sequences, one that has only finitely many nonzero terms and one that has only finitely many terms different from $p - 1$.

Note: We use the notation

$$x = 0.a_1 a_2 a_3 \ldots \quad (p) \tag{2.4}$$

as a shorthand for Eq. (2.3).

EXAMPLE 2.13 **Illustrates Proposition 2.16**

a) For each $p \geq 2$, we have

$$0 = 0.000 \ldots \quad (p)$$

and

$$1 = 0.(p - 1)(p - 1)(p - 1) \ldots \quad (p).$$

b) The number $1/2$ has, respectively, the binary ($p = 2$), ternary ($p = 3$), and decimal ($p = 10$) expansions given by

$$0.1000 \ldots \quad (2),$$
$$0.1111 \ldots \quad (3),$$
$$0.5000 \ldots \quad (10).$$

As predicted by Proposition 2.16, $1/2$ also has a second binary expansion and decimal expansion. They are, respectively,

$$0.0111 \ldots \quad (2),$$
$$0.4999 \ldots \quad (10).$$

But the ternary expansion of $1/2$ is unique. □

The Cantor Set

We now construct the Cantor set, a subset of $[0, 1]$ obtained as follows.

Step 1: Delete the middle third open interval of $[0, 1]$, namely, $(1/3, 2/3)$. See Fig. 2.1.

FIGURE 2.1 Set remaining after Step 1.

Step 2: After the first step, there remain two closed intervals, namely, $[0, 1/3]$ and $[2/3, 1]$. Delete the middle third open interval from each of those two intervals, namely, $(1/9, 2/9)$ and $(7/9, 8/9)$. See Fig. 2.2.

FIGURE 2.2 Set remaining after Step 2.

$$\vdots$$

Step n: After the $(n-1)^{\text{st}}$ step, there remain 2^{n-1} closed intervals. Delete from each of these the middle third open interval.

$$\vdots$$

Continue this process inductively. For each $n \in \mathcal{N}$, let G_n denote the set removed at the nth step, P_n the set remaining after the nth step, $G = \bigcup_{n=1}^{\infty} G_n$, and $P = \bigcap_{n=1}^{\infty} P_n$. We have the following noteworthy facts:

- G_n is the union of 2^{n-1} disjoint open intervals, each of length $1/3^n$. In particular, G_n is an open set.

- P_n is the union of 2^n disjoint closed intervals, each of length $1/3^n$. In particular, P_n is a closed set.

- $P = [0, 1] \setminus G$ and, so, $P \cap G = \emptyset$ and $P \cup G = [0, 1]$.

- G is the disjoint union of all removed open intervals, the sum of whose lengths is $\sum_{n=1}^{\infty} 2^{n-1}/3^n = 1$. In particular, G is an open set.

- P is a closed set, being the intersection of closed sets. It contains no interval because, for each $n \in \mathcal{N}$, $P \subset P_n$, and P_n contains no interval whose length exceeds $1/3^n$. The set P is called the **Cantor set** or, sometimes, the **Cantor ternary set**.

As we have just seen, G, the complement in $[0,1]$ of the Cantor set, is a disjoint union of open intervals, the sum of whose lengths is 1. But the length of $[0,1]$ is also 1. Thus, from the point of view of length, the Cantor set appears to be "small." On the other hand, as we will see shortly, P is uncountable, so that from a cardinality point of view, the Cantor set is "large." These, among other properties of the Cantor set, make it useful for illustrating many subtle concepts.

We mentioned that the Cantor set is sometimes called the *Cantor ternary set*. The reason for this will now be revealed. We begin with the following lemma.

□ □ □ **LEMMA 2.1**

An interval (a,b) is one of the 2^{n-1} open intervals removed from $[0,1]$ at the nth step in the construction of the Cantor set if and only if a and b are of the form

$$a = 0.(2c_1)(2c_2)\ldots(2c_{n-1})1000\ldots \quad (3),$$
$$b = 0.(2c_1)(2c_2)\ldots(2c_{n-1})2000\ldots \quad (3),$$

(2.5)

where $c_k \in \{0,1\}$ for $1 \leq k \leq n-1$.

PROOF We proceed by induction. At the first step ($n = 1$) of the construction of the Cantor set, exactly one interval is removed, namely, $(1/3, 2/3)$. And we have that

$$\tfrac{1}{3} = 0.1000\ldots \quad (3),$$
$$\tfrac{2}{3} = 0.2000\ldots \quad (3),$$

which is of the form (2.5) with $n = 1$.

Proceeding inductively, we note that an open interval (a, b) is removed at the nth step if and only if $a = r + 1/3^n$ and $b = r + 2/3^n$, where r equals 0 or is the right endpoint of one of the open intervals removed on or before the $(n-1)^{\text{st}}$ step. If $r = 0$, then

$$a = \tfrac{1}{3^n} = 0.\underbrace{0\ldots0}_{n-1 \text{ times}}1000\ldots \quad (3),$$

$$b = \tfrac{2}{3^n} = 0.\underbrace{0\ldots0}_{n-1 \text{ times}}2000\ldots \quad (3),$$

which is of the form (2.5) with $c_k = 0$ for $1 \leq k \leq n-1$. Otherwise, we have by the induction assumption that there is a positive integer $m \leq n-1$ such that $r = 0.(2c_1)(2c_2)\ldots(2c_{m-1})2000\ldots \quad (3)$, where $c_k \in \{0,1\}, 1 \leq k \leq m-1$. Then we have

$$a = r + \tfrac{1}{3^n} = 0.(2c_1)(2c_2)\ldots(2c_{m-1})2\underbrace{0\ldots0}_{n-m-1 \text{ times}}1000\ldots \quad (3),$$

$$b = r + \tfrac{2}{3^n} = 0.(2c_1)(2c_2)\ldots(2c_{m-1})2\underbrace{0\ldots0}_{n-m-1 \text{ times}}2000\ldots \quad (3),$$

which is of the form (2.5) with $c_m = 1$ and $c_k = 0$ for $m+1 \leq k \leq n-1$. ∎

From Lemma 2.1, we can obtain the following proposition whose proof is left to the reader as an exercise. See Exercise 2.69.

☐ ☐ ☐ PROPOSITION 2.17

The Cantor set consists of all numbers in $[0,1]$ that have a ternary expansion without the digit 1.

The Cantor Function

Using Proposition 2.17, we can now define the Cantor function, which is a real-valued function on $[0,1]$. We begin by specifying its values on the Cantor set, in other words, by defining a function $f: P \to \mathcal{R}$.

Let $x \in P$. By Proposition 2.17, x has a (unique) ternary expansion without the digit 1, say,

$$x = 0.(2c_1)(2c_2)\dots \quad (3),$$

where $c_n \in \{0,1\}$ for each $n \in \mathcal{N}$. We define

$$f(x) = 0.c_1c_2\dots \quad (2).$$

☐ ☐ ☐ PROPOSITION 2.18

Let $f: P \to \mathcal{R}$ be as defined in the preceding text. Then the range of f is $[0,1]$.

PROOF From the definition of f, its range is a subset of $[0,1]$. To show that it is onto, let $y \in [0,1]$. By Proposition 2.16 on page 61, we can write $y = 0.d_1d_2\dots \quad (2)$, where $d_n \in \{0,1\}$ for each $n \in \mathcal{N}$. Let $x = 0.(2d_1)(2d_2)\dots \quad (3)$. From Proposition 2.17, we know that $x \in P$ and, by definition, $f(x) = y$. ∎

☐ ☐ ☐ COROLLARY 2.2

The Cantor set is uncountable.

PROOF From Proposition 2.18, we have $f(P) = [0,1]$. By Proposition 1.9 on page 20, the image of a countable set is countable. Thus, since $[0,1]$ is uncountable and is the image of P under f, P must be uncountable. ∎

Next we extend f to a function ψ on $[0,1]$. If $x \in P$, we define $\psi(x) = f(x)$. If $x \in [0,1] \setminus P$, then x is in exactly one of the open intervals (a,b) removed from $[0,1]$ in the construction of the Cantor set. By Lemma 2.1 on page 63, there is an $n \in \mathcal{N}$ such that

$$a = 0.(2c_1)(2c_2)\dots(2c_{n-1})1000\dots \quad (3),$$
$$b = 0.(2c_1)(2c_2)\dots(2c_{n-1})2000\dots \quad (3),$$

where $c_k \in \{0,1\}$ for $1 \le k \le n-1$. Now note that

$$f(a) = 0.c_1c_2\dots c_{n-1}0111\dots \quad (2),$$
$$f(b) = 0.c_1c_2\dots c_{n-1}1000\dots \quad (2),$$

and, hence, $f(a) = f(b)$. We define $\psi(x)$ to be the common value of $f(a)$ and $f(b)$.

The function ψ is called the **Cantor function** or **Lebesgue singular function**. Its graph is sketched in Fig. 2.3.

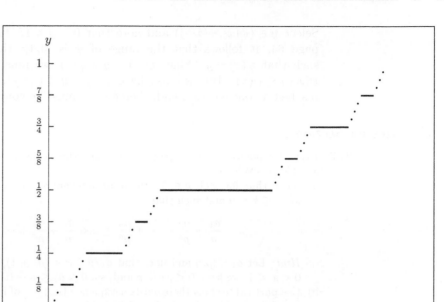

FIGURE 2.3 Sketch of the Cantor function.

In the next two propositions, we state two important properties of the Cantor function. One is that it is nondecreasing and the other is that it is continuous. The proof of the first proposition is left to the reader as an exercise.   (See Exercise 2.72.)

□ □ □   PROPOSITION 2.19

The Cantor function is nondecreasing.

□ □ □   PROPOSITION 2.20

The Cantor function is continuous.

PROOF   We will prove that ψ is continuous at each $x \in (0,1)$ and leave to the reader the proof of continuity at the endpoints of the interval. So, assume that $x \in (0,1)$. By Proposition 2.19, ψ is nondecreasing. Let

$$L_x = \{\, \psi(t) : 0 < t < x \,\} \qquad \text{and} \qquad R_x = \{\, \psi(t) : x < t < 1 \,\},$$

and let $\psi(x-) = \sup L_x$ and $\psi(x+) = \inf R_x$.

To show that ψ is continuous at x, it suffices, by Exercise 2.59 on page 60, to prove that $\psi(x-) = \psi(x+)$. Suppose that this is not the case. Then, again, by Exercise 2.59, either $\psi(x-) < \psi(x)$ or $\psi(x) < \psi(x+)$. We will consider the case where the latter holds true, realizing that a similar argument would ensue if the latter does not hold true but the former does.

As ψ is nondecreasing, we have

$$0 = \psi(0) \leq \psi(x) < \psi(x+) \leq \psi(1) = 1.$$

Select $y \in (\psi(x), \psi(x+))$ and note that $0 < y < 1$. From Proposition 2.18 on page 64, it follows that the range of ψ is $[0, 1]$; thus, there is a $z \in (0, 1)$ such that $\psi(z) = y$. Since $\psi(x) < y = \psi(z)$, we must have $x < z$ and, hence, $\psi(x+) \leq \psi(z)$. But we also have $\psi(z) = y < \psi(x+)$. Consequently, we have reached a contradiction and, therefore, ψ must be continuous at x. ∎

Exercises for Section 2.5

2.67 Prove Proposition 2.16 on page 61 by using the following steps. We can assume that $x \in (0, 1)$. (Why?)

a) Show that for each $n \in \mathcal{N}$, there are integers $a_1, a_2, \ldots, a_n$ with $0 \leq a_k \leq p - 1$ for $1 \leq k \leq n$ and such that

$$\frac{a_1}{p} + \frac{a_2}{p^2} + \cdots + \frac{a_n}{p^n} \leq x < \frac{a_1}{p} + \frac{a_2}{p^2} + \cdots + \frac{a_n}{p^n} + \frac{1}{p^n}.$$

Hint: Let $a_1 = [px]$ and note that $a_1/p \leq x < (a_1 + 1)/p$. Also note that, because $0 < x < 1$, we have $0 < px < p$ and, so, $0 \leq a_1 \leq p - 1$. Now use induction.

b) Use part (a) to show there exists a sequence $\{a_n\}_{n=1}^{\infty}$ of integers with $0 \leq a_n \leq p - 1$ for each $n \in \mathcal{N}$ and such that (2.3) holds.

c) Show that if x is of the form q/p^m, where $q, m \in \mathcal{N}$, then it has two base-p expansions, one that has only finitely many nonzero terms and one that has only finitely many terms different from $p - 1$. *Hint:* Use the Euclidean algorithm to write $q = b_1 p^{m-1} + b_2 p^{m-2} + \cdots + b_{m-1} p + b_m$, where $b_k \in \mathcal{Z}$ and $0 \leq b_k \leq p - 1$ for $k = 1, 2, \ldots, m$.

d) Prove that x can have at most two different base-p expansions and that, if it has two, it is of the form q/p^m, where $q, m \in \mathcal{N}$. *Hint:* Assume x has two different base-p expansions, say, $0.a_1 a_2 \ldots$ (p) and $0.b_1 b_2 \ldots$ (p). Let n be the first positive integer for which $a_k \neq b_k$ and assume without loss of generality that $a_n < b_n$. Show that this implies that $a_n = b_n - 1$ and that, for $k \geq n + 1$, $a_k = p - 1$ and $b_k = 0$.

2.68 Refer to Example 2.13 on page 61. For each part that follows, explain why we know from Proposition 2.16 that $1/2$ has

a) two binary expansions,

b) two decimal expansions,

c) a unique ternary expansion.

2.69 Prove Proposition 2.17: The Cantor set consists of all numbers in $[0, 1]$ that have a ternary expansion without the digit 1. Proceed as follows.

a) Show that $x \in G_n$ if and only if each of its ternary expansions is of the form

$$0.(2c_1)(2c_2) \ldots (2c_{n-1}) 1 a_{n+1} a_{n+2} \ldots \quad (3), \tag{2.6}$$

where $c_k \in \{0, 1\}$ for $1 \leq k \leq n - 1$ and not all the a_ks are 0 and not all are 2.

b) Use part (a) to show that G consists of all numbers in $(0, 1)$ that require a 1 in (each of) their ternary expansions.

c) Use part (b) to conclude that Proposition 2.17 holds.

2.70 Refer to the notation introduced on page 62.

a) Let I be any one of the 2^n closed intervals whose union is P_n. Prove that $I \cap P$ is uncountable.

b) Prove that for each $x \in P$ and $\delta > 0$, the set $(x - \delta, x + \delta) \cap P$ is uncountable.

2.71 Prove that the Cantor function ψ satisfies $\psi(x) = 2\psi\left(\frac{x}{3}\right)$ for all $x \in [0, 1]$.

2.72 Prove Proposition 2.19: The Cantor function ψ is a nondecreasing function on $[0, 1]$. *Hint:* Let $x, y \in [0, 1]$ with $x < y$. You must show $\psi(x) \leq \psi(y)$. To accomplish that, consider cases depending on whether x is a member of the Cantor set and whether y is a member of the Cantor set. First consider the case where both x and y are members of the Cantor set.

2.73 Complete the proof of Proposition 2.20 on page 65 by showing that the Cantor function is continuous at the endpoints of $[0, 1]$.

2.74 Let ψ denote the Cantor function and define

$$D = \left\{ \frac{\psi(x+h) - \psi(x)}{h} : x \in [0,1], \ h \neq 0 \right\}.$$

Show that $\inf D = 0$ and $\sup D = \infty$.

2.75 Generalize the technique used in the proof of Proposition 2.20 on page 65 to establish the following fact: If $f : [a, b] \to [c, d]$ is monotone and onto, then f is continuous on $[a, b]$.

2.6 THE RIEMANN INTEGRAL

In this section we will discuss the Riemann integral. We will define it in a way that motivates the definition of the Lebesgue integral, which is presented in Chapter 4.

In defining the Riemann integral, we need the concept of *characteristic function*. The characteristic function of a set indicates which elements are in the set and which are not. More precisely, we have the following definition.

DEFINITION 2.17 **Characteristic Function**

Let Ω be a set and $A \subset \Omega$. Then the **characteristic function** of A, denoted χ_A, is the real-valued function on Ω defined by

$$\chi_A(x) = \begin{cases} 1, & \text{if } x \in A; \\ 0, & \text{if } x \notin A. \end{cases}$$

The Riemann Integral

First we define the integral of a step function on a closed and bounded interval. A **step function** on an interval $[a, b]$ is a function of the form

$$h = \sum_{k=1}^{n} a_k \chi_{I_k}, \tag{2.7}$$

where $n \in \mathcal{N}$, $\{a_k\}_{k=1}^{n}$ is a sequence of real numbers, and $\{I_k\}_{k=1}^{n}$ is a finite sequence of pairwise disjoint intervals whose union is $[a, b]$. We permit degenerate intervals in this representation, that is, intervals of the form $[c, c] = \{c\}$ or $[c, c) = (c, c] = (c, c) = \emptyset$.

Let us denote by $\ell(I)$ the length of an interval I, where the length of a degenerate interval is defined to be 0. Then the *integral* of the step function in (2.7) is defined by

$$\int_a^b h(x) \, dx = \sum_{k=1}^{n} a_k \ell(I_k).$$

Note the following:

- The preceding equation defines the *integral* of a step function, not the *Riemann integral* of a step function. But we will see shortly that the two integrals agree.

- This definition of the integral of a step function is well-posed: If we can also write $h = \sum_{j=1}^{m} b_j \chi_{J_j}$, then $\sum_{j=1}^{m} b_j \ell(J_j) = \sum_{k=1}^{n} a_k \ell(I_k)$. See Exercise 2.77.

For example, let $[a, b] = [0, 1]$ and define

$$
h(x) = \begin{cases}
2, & 0 \leq x < \frac{1}{3}; \\
3, & \frac{1}{3} \leq x \leq \frac{1}{2}; \\
-4, & \frac{1}{2} < x \leq 1.
\end{cases}
$$

Then $h = 2\chi_{[0,1/3)} + 3\chi_{[1/3,1/2]} - 4\chi_{(1/2,1]}$ and we have

$$
\int_0^1 h(x)\, dx = 2\left(\frac{1}{3} - 0\right) + 3\left(\frac{1}{2} - \frac{1}{3}\right) - 4\left(1 - \frac{1}{2}\right) = -\frac{5}{6}.
$$

Next we define the upper and lower Riemann integrals of a bounded real-valued function. Let f be a bounded real-valued function on $[a, b]$, that is, $f : [a, b] \to \mathcal{R}$ and there is an $M \in \mathcal{R}$ such that $|f(x)| \leq M$ for all $x \in [a, b]$. Then the **upper Riemann integral** of f over $[a, b]$ is defined by

$$
\overline{\int_a^b} f(x)\, dx = \inf \left\{ \int_a^b h(x)\, dx : h \text{ a step function and } h \geq f \right\}.
$$

Similarly, the **lower Riemann integral** of f over $[a, b]$ is defined by

$$
\underline{\int_a^b} f(x)\, dx = \sup \left\{ \int_a^b h(x)\, dx : h \text{ a step function and } h \leq f \right\}.
$$

It is not too difficult to show that (see Exercise 2.79)

$$
\underline{\int_a^b} f(x)\, dx \leq \overline{\int_a^b} f(x)\, dx.
$$

If equality holds, then f is said to be Riemann integrable over $[a, b]$.

DEFINITION 2.18   Riemann Integrable; Riemann Integral

Let f be a bounded real-valued function on $[a, b]$. If

$$\underline{\int_a^b} f(x)\, dx = \overline{\int_a^b} f(x)\, dx,$$

then we say that f is **Riemann integrable** over $[a, b]$. In this case, the common value of the upper and lower Riemann integrals is called the **Riemann integral** of f over $[a, b]$ and is denoted by

$$\int_a^b f(x)\, dx.$$

We write $\boldsymbol{R([a, b])}$ for the collection of all Riemann integrable functions over the interval $[a, b]$.

EXAMPLE 2.14   *Illustrates Definition 2.18*

a) Let f be a step function on $[a, b]$. Then, see Exercise 2.80,

$$\overline{\int_a^b} f(x)\, dx = \int_a^b f(x)\, dx$$

and

$$\underline{\int_a^b} f(x)\, dx = \int_a^b f(x)\, dx,$$

where each integral on the right is interpreted as the integral of a step function. Thus, every step function on $[a, b]$ is Riemann integrable and, moreover, its Riemann integral equals its integral as a step function.

b) As we will discover later in this section, a continuous function on $[a, b]$ is Riemann integrable thereon.

c) Define $f(0) = 0$ and $f(x) = 1/x^2$ for $x \neq 0$. Then f is not Riemann integrable over a (closed and bounded) interval containing 0, as it is not bounded on such an interval. On the other hand, by part (b), f is Riemann integrable over a (closed and bounded) interval that does not contain 0, because it is continuous on such an interval.

d) Here is an example of a bounded function that is not Riemann integrable. For $x \in [0, 1]$, define $f(x) = \chi_Q(x)$. Now, a step function h that dominates f must satisfy $h(x) \geq 1$ except at a finite number of points. Because 1 (the function identically equal to 1) is a step function that dominates f, it follows that $\overline{\int_0^1} f(x)\, dx = 1$. Similarly, a step function h that is dominated by f must satisfy $h(x) \leq 0$ except at a finite number of points. Because 0 (the function identically equal to 0) is a step function that is dominated by f, it follows that $\underline{\int_0^1} f(x)\, dx = 0$. Hence, f is not Riemann integrable over $[0, 1]$.   □

Basic Properties of the Riemann Integral

The following theorem provides some fundamental properties of the Riemann integral. Its proof can be found in many advanced calculus and introductory real analysis texts.

◻ ◻ ◻ **THEOREM 2.6**

Suppose that $f, g \in R([a,b])$ and that $\alpha \in \mathcal{R}$.

a) If $a \leq c \leq b$, then $f \in R([a,c]) \cap R([c,b])$ and

$$\int_a^b f(x)\,dx = \int_a^c f(x)\,dx + \int_c^b f(x)\,dx.$$

b) If $f \leq g$, then

$$\int_a^b f(x)\,dx \leq \int_a^b g(x)\,dx.$$

c) $f + g \in R([a,b])$ and

$$\int_a^b (f+g)(x)\,dx = \int_a^b f(x)\,dx + \int_a^b g(x)\,dx.$$

d) $\alpha f \in R([a,b])$ and

$$\int_a^b (\alpha f)(x)\,dx = \alpha \int_a^b f(x)\,dx.$$

e) $|f| \in R([a,b])$ and

$$\left| \int_a^b f(x)\,dx \right| \leq \int_a^b |f(x)|\,dx.$$

Characterization of Riemann Integrable Functions

We stated earlier that a continuous function is Riemann integrable. Although there are noncontinuous functions that are Riemann integrable (e.g., step functions), we will soon see that a function is Riemann integrable if and only if it is "essentially continuous."

To make "essentially continuous" precise, we introduce the concept of sets of measure zero. Intuitively, these are sets without much content, although they may be large in the sense of cardinality.

DEFINITION 2.19 **Set of Measure Zero**

A subset E of $\mathcal{R}$ is said to have **measure zero** if for each $\epsilon > 0$, there exists a sequence $\{I_n\}_n$ of open intervals such that $E \subset \bigcup_n I_n$ and $\sum_n \ell(I_n) < \epsilon$.

The next proposition shows that a countable union of sets of measure zero also has measure zero. Its proof is left to the reader as an exercise.

□ □ □ **PROPOSITION 2.21**

Let $\{E_n\}_n$ be a sequence of subsets of $\mathcal{R}$, each having measure zero. Then $\bigcup_n E_n$ has measure zero.

EXAMPLE 2.15 *Illustrates Sets of Measure Zero*

a) A singleton set has measure zero; that is, if $x \in \mathcal{R}$, then $\{x\}$ has measure zero. Indeed, for $\epsilon > 0$, $I_\epsilon = (x - \epsilon/4, x + \epsilon/4)$ is an open interval, $\{x\} \subset I_\epsilon$, and $\ell(I_\epsilon) = \epsilon/2 < \epsilon$.

b) It follows immediately from part (a) and Proposition 2.21 that a countable subset of $\mathcal{R}$ has measure zero. In particular, a finite subset of $\mathcal{R}$ has measure zero and $\mathcal{N}$, $\mathcal{Z}$, and $\mathcal{Q}$ have measure zero.

c) The Cantor set has measure zero. (See Exercise 2.84). Since the Cantor set is uncountable (Corollary 2.2 on page 64), we see that although being countable is a sufficient condition for a subset of $\mathcal{R}$ to have measure zero, it is not necessary.

d) A (nondegenerate) interval does not have measure zero, as we ask you to show in Exercise 2.86. □

We are now in a position to state a continuity-type characterization of Riemann integrable functions. We begin with several lemmas whose proofs are left to the reader as exercises.

□ □ □ **LEMMA 2.2**

Let r and s be step functions on the interval $[a, b]$. If $r(x) = s(x)$ for all but a finite number of $x \in [a, b]$, then $\int_a^b s(x)\, dx = \int_a^b r(x)\, dx$.

□ □ □ **LEMMA 2.3**

Suppose that $\{I_j\}_{j=1}^N$ is a finite sequence of open intervals with $\sum_{j=1}^N \ell(I_j) < \epsilon$. Then there is a finite sequence of pairwise disjoint open intervals $\{J_k\}_{k=1}^L$ with $\bigcup_{k=1}^L J_k = \bigcup_{j=1}^N I_j$ and $\sum_{k=1}^L \ell(J_k) < \epsilon$.

□ □ □ **LEMMA 2.4**

Suppose that f and g are bounded functions on $[a, b]$ and that $f \geq g$. Then

$$\overline{\int_a^b} f(x)\, dx \geq \overline{\int_a^b} g(x)\, dx \qquad \text{and} \qquad \underline{\int_a^b} f(x)\, dx \geq \underline{\int_a^b} g(x)\, dx.$$

□ □ □ **LEMMA 2.5**

*Let f be a bounded function on $[a, b]$. Define the **upper envelope** and **lower envelope** of f by*

$$f^\star(x) = \inf_{\delta > 0} \sup_{|y-x| < \delta} f(y) \qquad \text{and} \qquad f_\star(x) = \sup_{\delta > 0} \inf_{|y-x| < \delta} f(y),$$

respectively. Then

a) $f_\star(x) \leq f(x) \leq f^\star(x)$ for all $x \in [a, b]$.

b) f is continuous at x if and only if $f^\star(x) = f_\star(x)$.

c) $\{x : f^\star(x) < t\}$ and $\{x : f_\star(x) > t\}$ are open in $[a, b]$ for each $t \in \mathcal{R}$.

d) $\{x : f^\star(x) \geq t\}$, $\{x : f_\star(x) \leq t\}$, and $\{x : f^\star(x) - f_\star(x) \geq t\}$ are closed for each $t \in \mathcal{R}$.

□ □ □ **LEMMA 2.6**

Let f be a bounded function on $[c, d]$. If $f^\star(x) - f_\star(x) < \epsilon$ for all $x \in [c, d]$, then there is a partition $c = x_0 < x_1 < \cdots < x_n = d$ of $[c, d]$ such that

$$\sup_{x \in [x_j, x_{j+1})} f(x) - \inf_{x \in [x_j, x_{j+1})} f(x) < \epsilon, \qquad 0 \leq j \leq n - 1.$$

Note: If $c = d$ (i.e., $[c, d]$ is degenerate), then Lemma 2.6 still holds, namely, vacuously with the "empty partition" given by $c = x_0 = d$.

□ □ □ **THEOREM 2.7**

Let f be bounded on $[a, b]$. Then f is Riemann integrable on $[a, b]$ if and only if its set of points of discontinuity has measure zero.

PROOF Suppose f is Riemann integrable on $[a, b]$. Let s be a step function with $s \geq f$. Then $s^\star$ is a step function and $s^\star \geq f^\star$. Referring to Lemma 2.2, we see that

$$\int_a^b s(x)\,dx = \int_a^b s^\star(x)\,dx \geq \overline{\int_a^b} f^\star(x)\,dx.$$

Thus, $\overline{\int_a^b} f(x)\,dx \geq \overline{\int_a^b} f^\star(x)\,dx$. Similarly, $\int_{\underline{a}}^b f_\star(x)\,dx \geq \int_{\underline{a}}^b f(x)\,dx$. Therefore, in view of Lemmas 2.4 and 2.5(a), and the assumed Riemann integrability of f, we have

$$\overline{\int_a^b} f_\star(x)\,dx \leq \overline{\int_a^b} f^\star(x)\,dx \leq \overline{\int_a^b} f(x)\,dx = \int_a^b f(x)\,dx \leq \int_{\underline{a}}^b f_\star(x)\,dx$$

and

$$\overline{\int_a^b} f^\star(x)\,dx \leq \int_{\underline{a}}^b f(x)\,dx = \int_a^b f(x)\,dx \leq \int_{\underline{a}}^b f_\star(x)\,dx \leq \int_{\underline{a}}^b f^\star(x)\,dx.$$

It now follows easily that $f^\star$ and $f_\star$ are Riemann integrable on $[a, b]$, that

$$\int_a^b f^\star(x)\,dx = \int_a^b f(x)\,dx = \int_a^b f_\star(x)\,dx,$$

and that

$$\int_a^b \left(f^\star(x) - f_\star(x) \right) dx = 0.$$

Let D denote the set of points of discontinuity of f. We want to show that D has measure zero. For convenience, set $g = f^* - f_*$. By Lemma 2.5, we have that $D = \{\, x : g(x) > 0 \,\}$. For each $n \in \mathcal{N}$, let $D_n = \{\, x : g(x) \geq 1/n \,\}$. Then $D = \bigcup_{n=1}^{\infty} D_n$. Hence, in view of Proposition 2.21 on page 71, we need only show that each D_n has measure zero.

To that end, let $\epsilon > 0$ be given. Because $\int_a^b g(x)\, dx = 0$, we can choose a step function $s \geq g$ such that $\int_a^b s(x)\, dx < \epsilon/2n$. We know that s has the form $s = \sum_i a_i \chi_{I_i}$, where $\{I_i\}_i$ is a finite sequence of pairwise disjoint intervals whose union is $[a, b]$. If $I_i \cap D_n \neq \emptyset$, then $a_i \geq 1/n$. Consequently,

$$\frac{1}{n} \sum_{I_i \cap D_n \neq \emptyset} \ell(I_i) \leq \int_a^b s(x)\, dx < \frac{\epsilon}{2n}.$$

Hence, $\sum_{I_i \cap D_n \neq \emptyset} \ell(I_i) < \epsilon/2$.

Now, the union of all I_is that have a nonempty intersection with D_n covers D_n. Choose finitely many open intervals J_k, $1 \leq k \leq N$, of total length less than $\epsilon/2$ that cover the endpoints of the I_is. Denoting the open interval with the same endpoints as I_i by I_i°, it follows that

$$D_n \subset \left(\bigcup_{k=1}^{N} J_k \right) \cup \left(\bigcup_{I_i \cap D_n \neq \emptyset} I_i^\circ \right).$$

However,

$$\sum_{k=1}^{N} \ell(J_k) + \sum_{I_i \cap D_n \neq \emptyset} \ell(I_i^\circ) = \sum_{k=1}^{N} \ell(J_k) + \sum_{I_i \cap D_n \neq \emptyset} \ell(I_i) < \frac{\epsilon}{2} + \frac{\epsilon}{2} = \epsilon.$$

We have now shown that, for each $\epsilon > 0$, there is a sequence of open intervals whose union covers D_n and the sum of whose lengths is less than ϵ. Consequently, each D_n has measure zero and, hence, so does D.

Conversely, suppose that the set D of points of discontinuity of f has measure zero. From Lemma 2.5(b), we know that $D = \{\, x \in [a, b] : f^*(x) - f_*(x) > 0 \,\}$. Set $M = \sup\{\, |f(x)| : x \in [a, b] \,\}$.

Let $\epsilon > 0$ be given. The set

$$D_\epsilon = \left\{\, x \in [a, b] : f^*(x) - f_*(x) \geq \frac{\epsilon}{2(b - a)} \,\right\}$$

is a subset of D and, hence, from Exercise 2.85, D_ϵ has measure zero. So, there is a sequence of open intervals $\{I_n\}_n$ such that $D_\epsilon \subset \bigcup_n I_n$ and $\sum_n \ell(I_n) < \epsilon/4M$. By Lemma 2.5(d), D_ϵ is closed, and clearly it is bounded. Hence, by the Heine-Borel theorem, there is an $N \in \mathcal{N}$ such that $D_\epsilon \subset \bigcup_{j=1}^{N} I_j$. Applying Lemma 2.3, we can choose a finite number of pairwise disjoint open intervals $\{J_k\}_{k=1}^{L}$ such that $D_\epsilon \subset \bigcup_{k=1}^{L} J_k$ and $\sum_{k=1}^{L} \ell(J_k) < \epsilon/4M$.

Now, the set $[a,b] \setminus \bigcup_{k=1}^{L} J_k$ is a finite pairwise disjoint union of closed intervals, say $\bigcup_{n=1}^{N} K_n$. Let $K_n = [c_n, d_n]$, and note that $K_n \subset D_\epsilon^c$ for $n = 1, 2, \ldots, N$. Applying Lemma 2.6, we can, for each $n = 1, 2, \ldots, N$, choose a partition $c_n = x_0^{(n)} < x_1^{(n)} < \cdots < x_{k_n}^{(n)} = d_n$ so that

$$\sup_{x \in \left[x_j^{(n)}, x_{j+1}^{(n)}\right)} f(x) - \inf_{x \in \left[x_j^{(n)}, x_{j+1}^{(n)}\right)} f(x) < \frac{\epsilon}{2(b-a)}, \qquad 0 \leq j \leq k_n - 1.$$

For $n = 1, \ldots, N$ and $0 \leq j \leq k_n - 1$, set

$$m_j^{(n)} = \inf_{x \in \left[x_j^{(n)}, x_{j+1}^{(n)}\right)} f(x) \qquad \text{and} \qquad M_j^{(n)} = \sup_{x \in \left[x_j^{(n)}, x_{j+1}^{(n)}\right)} f(x).$$

Now, define step functions s and r as follows:

$$s(x) = \begin{cases} M_j^{(n)}, & \text{if } x \in \left[x_j^{(n)}, x_{j+1}^{(n)}\right), \quad 0 \leq j \leq k_n - 1, \ 1 \leq n \leq N; \\ M, & \text{if } x \in \{d_n\} \cup \left(\bigcup_{k=1}^{L} J_k \cap [a,b]\right), \quad 1 \leq n \leq N. \end{cases}$$

and

$$r(x) = \begin{cases} m_j^{(n)}, & \text{if } x \in \left[x_j^{(n)}, x_{j+1}^{(n)}\right), \quad 0 \leq j \leq k_n - 1, \ 1 \leq n \leq N; \\ -M, & \text{if } x \in \{d_n\} \cup \left(\bigcup_{k=1}^{L} J_k \cap [a,b]\right), \quad 1 \leq n \leq N. \end{cases}$$

Then $s \geq f \geq r$ on $[a,b]$. Consequently,

$$0 \leq \overline{\int_a^b} f(x)\, dx - \underline{\int_a^b} f(x)\, dx \leq \int_a^b s(x)\, dx - \int_a^b r(x)\, dx$$

$$= \sum_{n=1}^{N} \sum_{j=1}^{k_n-1} M_j^{(n)} \left(x_{j+1}^{(n)} - x_j^{(n)}\right) + M \sum_{k=1}^{L} \ell(J_k \cap [a,b])$$

$$- \sum_{n=1}^{N} \sum_{j=1}^{k_n-1} m_j^{(n)} \left(x_{j+1}^{(n)} - x_j^{(n)}\right) + M \sum_{k=1}^{L} \ell(J_k \cap [a,b])$$

$$= \sum_{n=1}^{N} \sum_{j=1}^{k_n-1} \left(M_j^{(n)} - m_j^{(n)}\right) \left(x_{j+1}^{(n)} - x_j^{(n)}\right) + 2M \sum_{k=1}^{L} \ell(J_k \cap [a,b])$$

$$< \frac{\epsilon}{2(b-a)} \sum_{n=1}^{N} \ell(K_n) + 2M \sum_{k=1}^{L} \ell(J_k)$$

$$< \frac{\epsilon}{2(b-a)} \cdot (b-a) + 2M \cdot \frac{\epsilon}{4M} = \epsilon.$$

As ϵ is an arbitrary positive number, it follows that $\overline{\int_a^b} f(x)\, dx - \underline{\int_a^b} f(x)\, dx = 0$ and, hence, that f is Riemann integrable over $[a,b]$. ∎

□ □ □ **COROLLARY 2.3**

If f is continuous on $[a, b]$, then it is Riemann integrable thereon and

$$\int_a^b f(x)\, dx = \sup_h \int_a^b h(x)\, dx, \tag{2.8}$$

where the supremum is taken over all step functions h that are dominated by f.

Equation (2.8) will serve as motivation when, in Chapter 4, we define the Lebesgue integral of a measurable function.

EXAMPLE 2.16 *Illustrates Theorem 2.7*

a) From Theorem 2.7 and Example 2.15(b), we see that a bounded function on $[a, b]$ having countably many discontinuities is Riemann integrable.

b) Every monotone function on $[a, b]$ is Riemann integrable. This follows immediately from Exercise 2.59(c) on page 60 and part (a) here. □

Convergence Properties of the Riemann Integral

An important question in analysis is: If a sequence of functions converges pointwise, can the integral and limit be interchanged? For Riemann integration, this question can be stated as follows. Suppose that $\{f_n\}_{n=1}^\infty$ is a sequence of Riemann integrable functions on $[a, b]$ that converges pointwise to a Riemann integrable function f. Is it true that

$$\lim_{n\to\infty} \int_a^b f_n(x)\, dx = \int_a^b f(x)\, dx? \tag{2.9}$$

The answer, in general, is no. For example, let $f = 0$ (the function identically 0) and $f_n = n\chi_{(0,1/n)}$ for each $n \in \mathcal{N}$. Then $f_n \to f$ pointwise on $[0, 1]$. But $\int_0^1 f(x)\, dx = 0$ and $\int_0^1 f_n(x)\, dx = 1$ for each $n \in \mathcal{N}$. Therefore,

$$\lim_{n\to\infty} \int_0^1 f_n(x)\, dx = 1 \neq 0 = \int_0^1 f(x)\, dx.$$

Even if each f_n and f are continuous, the limit and the integral cannot, in general, be interchanged. As in the discussion of pointwise convergence of continuous functions, the concept of uniform convergence plays an important role here.

□ □ □ **THEOREM 2.8**

Suppose $\{f_n\}_{n=1}^\infty$ is a sequence of Riemann integrable functions on $[a, b]$ that converges uniformly to a function f. Then f is Riemann integrable over $[a, b]$ and

$$\lim_{n\to\infty} \int_a^b f_n(x)\, dx = \int_a^b f(x)\, dx.$$

PROOF   That f is Riemann integrable is left to the reader as an exercise. (See Exercise 2.96.)

Let $\epsilon > 0$ be given. Choose $N \in \mathcal{N}$ such that $|f(x) - f_n(x)| < \epsilon/(b-a)$ for all $x \in [a, b]$ whenever $n \geq N$. Using Theorem 2.6 on page 70, we have, for $n \geq N$,

$$\left| \int_a^b f(x)\, dx - \int_a^b f_n(x)\, dx \right| = \left| \int_a^b (f - f_n)(x)\, dx \right|$$

$$\leq \int_a^b |f(x) - f_n(x)|\, dx < \int_a^b \frac{\epsilon}{b-a}\, dx = \epsilon,$$

as required. ■

As Theorem 2.8 shows, uniform convergence is a sufficient condition for the interchange of limit and integral. It is not, however, a necessary condition. To see this, let $f_n(x) = x^n$ and

$$f(x) = \begin{cases} 1, & x = 1; \\ 0, & x \neq 1. \end{cases}$$

Then $f_n \to f$ pointwise, but not uniformly, on $[0, 1]$. Moreover,

$$\lim_{n \to \infty} \int_0^1 f_n(x)\, dx = \lim_{n \to \infty} \frac{1}{n+1} = 0 = \int_0^1 f(x)\, dx.$$

We have seen three important consequences of uniform convergence. It is a sufficient condition for

- the limit of a sequence of continuous functions to be continuous (Proposition 2.15 on page 58).

- the limit of a sequence of Riemann integrable functions to be Riemann integrable (Theorem 2.8).

- the interchange of limit and integral (Theorem 2.8).

Although uniform convergence has these and other desirable consequences, it is a very strong condition to place on a sequence of functions, especially when the common domain is the entire real line. This fact and a need to "integrate" non–Riemann integrable functions will lead us naturally to Lebesgue measurable functions and the Lebesgue integral in Chapter 4.

Exercises for Section 2.6

2.76 Let Ω be a set and $A, B \subset \Omega$. Prove the following facts.

a) $\chi_{A \cap B} = \chi_A \cdot \chi_B$.

b) If $A \cap B = \emptyset$, then $\chi_{A \cup B} = \chi_A + \chi_B$.

c) More generally than in part (b), if $\{C_n\}_n$ is a pairwise disjoint sequence of subsets of Ω, then $\chi_{\bigcup_n C_n} = \sum_n \chi_{C_n}$.

d) Obtain a general formula for $\chi_{A \cup B}$.

2.77 Show that the definition of the Riemann integral of a step function is well-posed by verifying the following fact. Suppose that $h = \sum_{k=1}^{n} a_k \chi_{I_k} = \sum_{j=1}^{m} b_j \chi_{J_j}$, where $n, m \in \mathcal{N}$, $\{a_k\}_{k=1}^{n}$ and $\{b_j\}_{j=1}^{m}$ are sequences of real numbers, and $\{I_k\}_{k=1}^{n}$ and $\{J_j\}_{j=1}^{m}$ are each a finite sequence of pairwise disjoint intervals whose union is $[a, b]$. Prove that $\sum_{k=1}^{n} a_k \ell(I_k) = \sum_{j=1}^{m} b_j \ell(J_j)$. *Hint:* First show that $I_k \cap J_j \neq \emptyset$ implies that $a_k = b_j$. Then show that $\sum_{k=1}^{n} a_k \ell(I_k) = \sum_{k=1}^{n} a_k \sum_{j=1}^{m} \ell(I_k \cap J_j)$.

2.78 Suppose that g and h are step functions on $[a, b]$ and that $g \leq h$ on $[a, b]$, that is, $g(x) \leq h(x)$ for all $x \in [a, b]$. Prove that $\int_a^b g(x)\, dx \leq \int_a^b h(x)\, dx$, where each integral is interpreted as the integral of a step function. *Hint:* See the hint given in Exercise 2.77.

2.79 Let f be a bounded function on $[a, b]$. Prove that

$$\underline{\int_a^b} f(x)\, dx \leq \overline{\int_a^b} f(x)\, dx.$$

Hint: Use Exercise 2.78.

2.80 Let f be a step function on $[a, b]$. Prove that

$$\overline{\int_a^b} f(x)\, dx = \int_a^b f(x)\, dx$$

and

$$\underline{\int_a^b} f(x)\, dx = \int_a^b f(x)\, dx,$$

where each integral on the right is interpreted as the integral of a step function. Thus, a step function on $[a, b]$ is Riemann integrable thereon and its Riemann integral equals its integral as a step function. *Hint:* Use Exercise 2.78.

2.81 Prove that a real-valued function f on $[a, b]$ is a step function if and only if there is a partition $a = x_0 < x_1 < \cdots < x_n = b$ of $[a, b]$ and real numbers $c_1, c_2, \ldots, c_n$ such that for each $1 \leq k \leq n$, we have $f(x) = c_k$ for $x_{k-1} < x < x_k$.

2.82 Show that the definitions of the upper and lower Riemann integrals, as presented on page 68, are equivalent to those usually encountered in advanced calculus and introductory real analysis courses. Specifically, let f be a bounded real-valued function on $[a, b]$. For a partition $a = x_0 < x_1 < \cdots < x_n = b$ of $[a, b]$, define, for $1 \leq k \leq n$, $I_k = [x_{k-1}, x_k]$, $M(f, I_k) = \sup\{f(x) : x \in I_k\}$, and $m(f, I_k) = \inf\{f(x) : x \in I_k\}$. Also define

$$U_a^b f = \inf\left\{ \sum_{k=1}^{n} M(f, I_k)(x_k - x_{k-1}) : a = x_0 < x_1 < \cdots < x_n = b \right\}$$

and

$$L_a^b f = \sup\left\{ \sum_{k=1}^{n} m(f, I_k)(x_k - x_{k-1}) : a = x_0 < x_1 < \cdots < x_n = b \right\}.$$

Prove that $U_a^b f = \overline{\int_a^b} f(x)\, dx$ and $L_a^b(f) = \underline{\int_a^b} f(x)\, dx$. *Hint:* Use Exercise 2.81.

2.83 Prove Proposition 2.21 on page 71.

2.84 Prove that the Cantor set P has measure zero. *Hint:* Recall that $P \subset P_n$ for each $n \in \mathcal{N}$, where P_n is the set remaining after the nth step in the construction of the Cantor set.

2.85 Show that a subset of a set of measure zero also has measure zero.

★**2.86** Prove that a nondegenerate interval does not have measure zero by proceeding as follows.

a) Let $a, b \in \mathcal{R}$ with $a < b$. Show that if $\{I_k\}_{k=1}^n$ is a finite sequence of open intervals whose union contains $[a, b]$, then $\sum_{k=1}^n \ell(I_k) > b - a$.

b) Deduce from part (a) that if $a < b$, then $[a, b]$ does not have measure zero. *Hint:* Use the Heine-Borel theorem.

c) Conclude from part (b) that a nondegenerate interval does not have measure zero.

2.87 Prove Lemma 2.2 on page 71.

2.88 Prove Lemma 2.3 on page 71.

2.89 Prove Lemma 2.4 on page 71.

2.90 Prove Lemma 2.5 on page 71.

2.91 Prove Lemma 2.6 on page 72.

2.92 Define $f: [0, 1] \to \mathcal{R}$ by

$$f(x) = \begin{cases} 0, & \text{if } x = 0 \text{ or } x \in [0, 1] \setminus Q; \\ 1/q, & \text{if } x \in (0, 1] \cap Q \text{ and } x = p/q \text{ in lowest terms.} \end{cases}$$

a) Show that the set of points of discontinuity of f is $(0, 1] \cap Q$.

b) Deduce from part (a) that f is Riemann integrable on $[0, 1]$.

c) Show that $\int_0^1 f(x)\, dx = 0$.

2.93 Find a function on $[0, 1]$ that has uncountably many points of discontinuity but is Riemann integrable. *Hint:* Do something with the Cantor set.

2.94 Construct a sequence of continuous functions on $[0, 1]$ that converges pointwise to a continuous function but for which the limit and integral cannot be interchanged.

2.95 Refer to Exercise 2.94. Is it possible to find such a sequence of functions if the sequence is required to be monotone?

2.96 Prove that if $\{f_n\}_{n=1}^\infty$ is a sequence of Riemann integrable functions on $[a, b]$ that converges uniformly to a function f, then f is Riemann integrable. Proceed as follows.

a) Show that f is bounded.

b) Show that the set of points of discontinuity of f has measure zero. *Hint:* Let E_n denote the set of points of discontinuity of f_n and set $E = \bigcup_{n=1}^\infty E_n$. Show that f is continuous at each point of $[a, b] \setminus E$.

2.97 Let $\{f_n\}_{n=1}^\infty$ be a sequence of Riemann integrable functions on $[0, 1]$ that converges pointwise to the function f. Construct an example showing that f need not be Riemann integrable on $[0, 1]$, even if $\{f_n\}_{n=1}^\infty$ is monotone and f is bounded.

PART TWO

□

Measure, Integration, and Differentiation

Émile Félix-Édouard-Justin Borel
(1871–1956)

Émile Borel was born at Saint-Affrique, France, on January 7, 1871, the third child of Honoré Borel, a Protestant minister, and Emilie Teissi-Solier. Émile had two older sisters who were 14 and 16 years old when he was born.

From an early age, Borel exhibited a strong proclivity for mathematics. At the age of 11, he went to live with his eldest sister, Madame Lebeau. This move allowed Borel to attend the Lycée at Montauben, where he showed extraordinary talents. Several years later, Borel went to Paris, where he took courses at the Lycée Louis-le-Grand in preparation for taking the examinations to enter the École Polytechnique and the École Normale. He ranked first in both of these examinations and could choose either of the two universities. In 1890, Borel entered the École Polytechnique in Paris, where he graduated in 1893. He received his doctorate from the École Normale Supérieure in 1894.

Borel's most important research was done in the 1890s when he worked on probability, infinitesimal calculus, divergent series, and measure theory. In 1896, he provided the proof of Picard's theorem, a proof that mathematicians had been seeking for nearly 20 years. Although John von Neumann is credited as the founder of game theory, Borel completed a series of papers on the subject between 1921 and 1927, thus being the first to define games of strategy.

After WW I, Borel developed an interest in politics, serving as Minister of the Navy from 1925–1940. He was arrested and briefly imprisoned by the Vichy regime in 1940, after which he worked in the Résistance. His honors included the Résistance medal in 1945, the Croix de Guerre in 1918, the Grand Cross of the Légion d'Honneur in 1950, and the first gold medal of the Centre National de la Recherche Scientifique in 1955.

Borel was appointed to the faculty of the École Normale Supérieure in 1896, held the Chair in Function Theory at the Sorbonne from 1909 until 1940, and was director and founding member of the Henri Poincaré Institute from 1928 until his death on February 3, 1956, in Paris.

□ 3 □

Lebesgue Measure on the Real Line

In Chapter 2, we discussed open sets, continuous functions, and the Riemann integral. Those classical concepts have served mathematics and its applications well. However, for the purposes of modern mathematics, a more general and sophisticated framework is required. In this chapter and the next, we will take the first steps toward obtaining that framework.

Here we will first expand the collection of continuous functions to the collection of *Borel measurable functions,* the smallest algebra of functions that contains the continuous functions and is closed under pointwise limits. In doing so, we will be led to consider the collection of *Borel sets,* the smallest σ-algebra of subsets of $\mathcal{R}$ that contains the open sets.

Next we will generalize the concept of length so that it applies to all Borel sets. In fact, that generalization will be to an even larger collection of sets than the Borel sets, namely, the *Lebesgue measurable sets.*

3.1 BOREL MEASURABLE FUNCTIONS AND BOREL SETS

In Chapter 2, we showed that the collection of continuous, real-valued functions forms an algebra but is not closed under pointwise limits. As this latter property is essential in modern mathematical analysis, we will enlarge the collection of continuous functions to a collection of functions that is closed under pointwise limits.

Specifically, we will consider the smallest algebra of (real-valued) functions that contains the continuous functions and is closed under pointwise limits. Such an algebra of functions exists — it is the intersection of all algebras of functions that contain the continuous functions and are closed under pointwise limits.

Note: The aforementioned intersection is not vacuous because the collection of all real-valued functions is an algebra that contains the continuous functions and is closed under pointwise limits.

As we will see presently, the condition of being an algebra is superfluous. That is, the smallest collection of functions that contains the continuous functions and is closed under pointwise limits is necessarily an algebra of functions. Thus, we make the following definition:

DEFINITION 3.1   Borel Measurable Functions

We denote by $\widehat{C}$ the smallest collection of real-valued functions on $\mathcal{R}$ that contains the collection of continuous functions and is closed under pointwise limits. The members of $\widehat{C}$ are called **Borel measurable functions.**

□ □ □   **THEOREM 3.1**

The collection $\widehat{C}$ of Borel measurable functions forms an algebra. That is, if f and g are Borel measurable and $\alpha \in \mathcal{R}$, then

a) $f + g$ is Borel measurable.
b) αf is Borel measurable.
c) $f \cdot g$ is Borel measurable.

PROOF   We prove only part (a); parts (b) and (c) are left as exercises. First of all, let $g \in C$ (the collection of continuous functions on $\mathcal{R}$) and set

$$\mathcal{D} = \{\, f \in \widehat{C} : f + g \in \widehat{C} \,\}.$$

If $f \in C$, then $f \in \widehat{C}$ and $f + g \in C \subset \widehat{C}$. Thus, $\mathcal{D} \supset C$. Now suppose that $\{f_n\}_{n=1}^{\infty} \subset \mathcal{D}$ and that $f_n \to f$ pointwise. Then $f_n \in \widehat{C}$ and $f_n + g \in \widehat{C}$ for all $n \in \mathcal{N}$ and $f_n + g \to f + g$ pointwise. Since $\widehat{C}$ is closed under pointwise limits, we conclude that $f \in \widehat{C}$ and $f + g \in \widehat{C}$. Hence, $f \in \mathcal{D}$. Therefore, we see that $\mathcal{D}$ is closed under pointwise limits.

The previous paragraph shows that $\mathcal{D}$ contains the continuous functions and is closed under pointwise limits. Since, by definition, $\widehat{C}$ is the smallest such collection of functions, it follows that $\mathcal{D} \supset \widehat{C}$. But, by the definition of $\mathcal{D}$, $\mathcal{D} \subset \widehat{C}$. Thus, $\mathcal{D} = \widehat{C}$; in other words, $f + g \in \widehat{C}$ whenever $f \in \widehat{C}$ and $g \in C$.

Next, let $f \in \widehat{C}$ and set

$$\mathcal{E} = \{\, g \in \widehat{C} : f + g \in \widehat{C} \,\}.$$

It follows from what we just proved that $\mathcal{E}$ contains the continuous functions. Moreover, the same argument that we used to show that $\mathcal{D}$ is closed under pointwise limits shows that $\mathcal{E}$ is closed under pointwise limits. Hence, $\mathcal{E} = \widehat{C}$; that is, $f + g \in \widehat{C}$ whenever $f \in \widehat{C}$ and $g \in \widehat{C}$.   ■

Borel Sets

In Chapter 2, we discovered that there is a natural association between the continuous functions and the collection of open sets: A function is continuous if and only if the inverse image of each open set is open. Now we ask which collection of sets corresponds naturally to the Borel measurable functions. As we will see, it is the collection of sets whose characteristic functions are Borel measurable functions.

DEFINITION 3.2   **Borel Sets**

A set $B \subset \mathcal{R}$ is called a **Borel set** if its characteristic function is Borel measurable. The collection of all Borel sets is denoted $\mathcal{B}$. So, $\mathcal{B} = \{ B \subset \mathcal{R} : \chi_B \in \widehat{C} \}$.

To begin, we will prove that the Borel sets form a σ-algebra of subsets of $\mathcal{R}$. In order to accomplish this, we will need several lemmas. The proof of the first lemma is left as an exercise for the reader (see Exercise 3.2).

□ □ □   **LEMMA 3.1**

Let h denote the absolute value function, that is $h(x) = |x|$, $x \in \mathcal{R}$. Then there is a sequence $\{p_n\}_{n=1}^{\infty}$ of polynomials such that $p_n \to h$ pointwise.

Next we introduce the notation used for the maximum and minimum of two functions and prove that the Borel measurable functions are closed under those two operations.

DEFINITION 3.3   **Maximum and Minimum of Two Functions**

Let f and g be real-valued functions on $\mathcal{R}$. Then we define $f \vee g = \max\{f, g\}$ and $f \wedge g = \min\{f, g\}$. That is,

$$(f \vee g)(x) = \max\{f(x), g(x)\}$$

and

$$(f \wedge g)(x) = \min\{f(x), g(x)\}.$$

□ □ □   **LEMMA 3.2**

If f and g are Borel measurable functions, then so are $f \vee g$ and $f \wedge g$. More generally, if f_1, f_2, ..., f_n are Borel measurable functions, then so are $f_1 \vee \cdots \vee f_n$ and $f_1 \wedge \cdots \wedge f_n$.

PROOF   We first note that the following identities hold:

$$f \vee g = \frac{1}{2}\big(f + g + |f - g|\big) \tag{3.1}$$

and

$$f \wedge g = \frac{1}{2}\big(f + g - |f - g|\big). \tag{3.2}$$

Next we show that $|F| \in \widehat{C}$ whenever $F \in \widehat{C}$. From Lemma 3.1, we can choose a sequence of polynomials $\{p_n\}_{n=1}^{\infty}$ such that $p_n(x) \to |x|$ for all $x \in \mathcal{R}$. Because $\widehat{C}$ is an algebra of functions (Theorem 3.1), it follows that $p_n \circ F \in \widehat{C}$ for all $n \in \mathcal{N}$. However, $p_n \circ F \to |F|$ pointwise and, consequently, as $\widehat{C}$ is closed under pointwise limits, $|F| \in \widehat{C}$.

Now suppose that $f, g \in \widehat{C}$. Then $|f - g| \in \widehat{C}$ (why?). Using (3.1) and (3.2) and the fact that $\widehat{C}$ is an algebra, we deduce that $f \vee g \in \widehat{C}$ and $f \wedge g \in \widehat{C}$. The remaining conclusions of the lemma follow by mathematical induction. ∎

□ □ □ **LEMMA 3.3**

If $\{f_n\}_{n=1}^{\infty}$ is a sequence of Borel measurable functions with $\{f_n(x)\}_{n=1}^{\infty}$ bounded for each $x \in \mathcal{R}$, then $\sup_n f_n$ and $\inf_n f_n$ are Borel measurable.

PROOF By Lemma 3.2, if $f_1, f_2, \ldots, f_n$ are Borel measurable, then so are $f_1 \vee \cdots \vee f_n$ and $f_1 \wedge \cdots \wedge f_n$. But,

$$\sup_n f_n = \lim_{n \to \infty} f_1 \vee \cdots \vee f_n$$

and

$$\inf_n f_n = \lim_{n \to \infty} f_1 \wedge \cdots \wedge f_n.$$

The lemma now follows because $\widehat{C}$ is closed under pointwise limits. ∎

Now that we have established Lemmas 3.1–3.3, we can prove that the collection $\mathcal{B}$ of Borel sets is a σ-algebra of subsets of $\mathcal{R}$.

□ □ □ **THEOREM 3.2**

The collection of Borel sets $\mathcal{B} = \{ B \subset \mathcal{R} : \chi_B \in \widehat{C} \}$ is a σ-algebra of subsets of $\mathcal{R}$.

PROOF We first show that the collection of Borel sets is closed under complementation. Assume $B \in \mathcal{B}$. Then, by definition, $\chi_B \in \widehat{C}$. Since $1 \in \widehat{C}$ and $\widehat{C}$ is an algebra, we conclude that $1 - \chi_B \in \widehat{C}$. But $1 - \chi_B = \chi_{B^c}$ and, consequently, $B^c \in \mathcal{B}$.

Next we show that the collection of Borel sets is closed under countable unions. Suppose that $B_n \in \mathcal{B}$, for $n \in \mathcal{N}$. Then $\chi_{B_n} \in \widehat{C}$ for $n \in \mathcal{N}$ and, therefore, by Lemma 3.3, $\sup_n \chi_{B_n} \in \widehat{C}$. But $\sup_n \chi_{B_n} = \chi_{\bigcup_{n=1}^{\infty} B_n}$. Hence, $\bigcup_{n=1}^{\infty} B_n \in \mathcal{B}$. ∎

Further Properties of Borel Sets and Borel Measurable Functions

It is left as an exercise for the reader to show that if O is an open set, then χ_O is a Borel measurable function. In other words, *every open set is a Borel set*. We will prove shortly that, in fact, $\mathcal{B}$ is the smallest σ-algebra that contains all the open sets.

But first, we will justify our contention that it is natural to associate the Borel sets with the Borel measurable functions, as we do the open sets with the continuous functions. Specifically, we will show that a function is Borel measurable if and only if the inverse image of each open set is a Borel set. In order to accomplish this, we will introduce another collection of functions which, as we will see, turns out to be identical to the collection of Borel measurable functions.

□ □ □ **LEMMA 3.4**

Let $\mathcal{F} = \{\, f : f^{-1}(O) \in \mathcal{B} \text{ for all open sets } O \,\}$. Then $\mathcal{F}$ contains the continuous functions.

PROOF   Suppose that f is a continuous function on $\mathcal{R}$. Then, by Theorem 2.5 on page 55, $f^{-1}(O)$ is open whenever O is open. But every open set is a Borel set (Exercise 3.3). Therefore, $f^{-1}(O) \in \mathcal{B}$ whenever O is open. This shows that $\mathcal{F}$ contains the continuous functions.   ∎

□ □ □ **LEMMA 3.5**

$f \in \mathcal{F}$ if either of the following conditions hold:
a) For each $a \in \mathcal{R}$, $f^{-1}\big((-\infty, a)\big) \in \mathcal{B}$.
b) For each $a \in \mathcal{R}$, $f^{-1}\big((a, \infty)\big) \in \mathcal{B}$.

PROOF   We will prove part (a). The proof of part (b) is similar and is left as an exercise. So, suppose that f satisfies the condition in part (a). We claim that $f \in \mathcal{F}$; that is, $f^{-1}(O) \in \mathcal{B}$ for all open sets O. Set

$$\mathcal{A} = \{\, A \subset \mathcal{R} : f^{-1}(A) \in \mathcal{B} \,\}.$$

Because $f^{-1}(A^c) = [f^{-1}(A)]^c$, $f^{-1}(\bigcup_n A_n) = \bigcup_n f^{-1}(A_n)$, and $\mathcal{B}$ is a σ-algebra, it follows that $\mathcal{A}$ is a σ-algebra.

Now, by assumption, $\mathcal{A}$ contains all sets of the form $(-\infty, \alpha)$, where $\alpha \in \mathcal{R}$. If $a \in \mathcal{R}$, then we can write $(-\infty, a] = \bigcap_{n=1}^{\infty}(-\infty, a + 1/n)$; therefore, $(-\infty, a] \in \mathcal{A}$ because $\mathcal{A}$ is a σ-algebra. This in turn implies that $(a, b) \in \mathcal{A}$ for each $a, b \in \mathcal{R}$, since $(a, b) = (-\infty, b) \cap (-\infty, a]^c$. It now follows easily that $\mathcal{A}$ contains all open intervals. But, by Proposition 2.13 on page 48, every open set is a countable union of open intervals. Consequently, $\mathcal{A}$ contains all open sets. This means that $f^{-1}(O) \in \mathcal{B}$ for all open sets O; that is, $f \in \mathcal{F}$.   ∎

□ □ □ **LEMMA 3.6**

$\mathcal{F}$ is closed under pointwise limits.

PROOF   Suppose that $\{g_n\}_{n=1}^{\infty} \subset \mathcal{F}$ and let $g = \sup_n g_n$. Then, for each $a \in \mathcal{R}$, we have $g^{-1}\big((a, \infty)\big) = \bigcup_{n=1}^{\infty} g_n^{-1}\big((a, \infty)\big) \in \mathcal{B}$. Therefore, by the preceding lemma, $\sup_n g_n \in \mathcal{F}$. Similarly, $\inf_n g_n \in \mathcal{F}$.

Now suppose that $\{f_n\}_{n=1}^{\infty} \subset \mathcal{F}$ and that $f_n \to f$ pointwise. Then, for each $x \in \mathcal{R}$, $\lim_{n \to \infty} f_n(x) = f(x)$ and, so, $\limsup_{n \to \infty} f_n(x) = f(x)$. Consequently, $f = \inf_n \{\sup_{k \geq n} f_k\}$. Let $g_n = \sup_{k \geq n} f_k$. Then the previous paragraph shows in turn that $g_n \in \mathcal{F}$ for all $n \in \mathcal{N}$ and $\inf_n g_n \in \mathcal{F}$. Hence, $f \in \mathcal{F}$.   ∎

□ □ □ **COROLLARY 3.1**

$\mathcal{F}$ contains the Borel measurable functions.

PROOF   By Lemmas 3.4 and 3.6, $\mathcal{F}$ contains the continuous functions and is closed under pointwise limits. Since the collection of Borel measurable functions $\widehat{C}$ is the smallest collection of functions that contains the continuous functions and is closed under pointwise limits, it must be that $\mathcal{F} \supset \widehat{C}$.   ∎

□ □ □   **LEMMA 3.7**

Let $f \in \mathcal{F}$. Then there is a sequence $\{f_n\}_{n=1}^{\infty}$ of Borel measurable functions such that $f_n \to f$ pointwise.

PROOF   First of all, note that if $a, b \in \mathcal{R}$, then $f^{-1}([a,b)) \in \mathcal{B}$ (why?). For $n \in \mathcal{N}$, let

$$E_{nk} = \left\{ x : \frac{k}{n} \le f(x) < \frac{k+1}{n} \right\} = f^{-1}\left(\left[\frac{k}{n}, \frac{k+1}{n} \right) \right)$$

for $k = 0, \pm 1, \pm 2, \ldots$. Then $E_{nk} \in \mathcal{B}$ and so $\chi_{E_{nk}} \in \widehat{C}$. Since $\widehat{C}$ is an algebra of functions and is closed under pointwise limits, the function

$$f_n = \sum_{k=-\infty}^{\infty} \frac{k}{n} \chi_{E_{nk}} = \lim_{k \to \infty} \sum_{j=-k}^{k} \frac{j}{n} \chi_{E_{nj}}$$

is in $\widehat{C}$. It is easy to see that $|f(x) - f_n(x)| < 1/n$ for all $x \in \mathcal{R}$. Hence $f_n \to f$ pointwise (in fact, the convergence is uniform).   ∎

Using the preceding results, it is now evident that $\mathcal{F}$ and the collection of Borel measurable functions are identical. Specifically, we have the following theorem.

□ □ □   **THEOREM 3.3**

A function f is Borel measurable if and only if the inverse image of each open set under f is a Borel set; that is, if and only if $f^{-1}(O) \in \mathcal{B}$ for all open sets O.

PROOF   By Corollary 3.1, $\mathcal{F} \supset \widehat{C}$. Conversely, suppose that $f \in \mathcal{F}$. Then, by Lemma 3.7, f is the pointwise limit of functions in $\widehat{C}$. Since $\widehat{C}$ is closed under pointwise limits, this implies $f \in \widehat{C}$. Thus, $\widehat{C} \supset \mathcal{F}$.   ∎

We mentioned previously that the collection of Borel sets $\mathcal{B}$ is the smallest σ-algebra of sets that contains the open sets. Here is a proof of that result.

□ □ □   **THEOREM 3.4**

The collection of Borel sets $\mathcal{B}$ is the smallest σ-algebra of subsets of $\mathcal{R}$ that contains all the open sets.

PROOF   We already know that $\mathcal{B}$ is a σ-algebra that contains all the open sets. Let $\mathcal{A}$ be any σ-algebra that contains all the open sets. We claim that $\mathcal{B} \subset \mathcal{A}$. Let $\mathcal{G} = \{ f : f^{-1}(O) \in \mathcal{A} \text{ for all open sets } O \}$. The arguments used for Lemmas 3.4 and 3.6 depend only on the fact that $\mathcal{B}$ is a σ-algebra containing the open sets. It follows that $\mathcal{G}$ contains the continuous functions and is closed under pointwise limits; thus, $\mathcal{G} \supset \widehat{C}$. This last fact implies that $\chi_B \in \mathcal{G}$ for all $B \in \mathcal{B}$. But then, for each $B \in \mathcal{B}$, $B = \chi_B^{-1}\left((1/2, 3/2) \right) \in \mathcal{A}$. Thus $\mathcal{B} \subset \mathcal{A}$.   ∎

Remarks: In many texts, the collection of Borel sets $\mathcal{B}$ is *defined* to be the smallest σ-algebra of sets that contains the open sets; and a function is *defined* to be Borel measurable if the inverse image of each open set is a Borel set. As we see from Theorems 3.3 and 3.4, the definitions presented here (Definitions 3.1 and 3.2) are equivalent to those. It seems more natural, though, to introduce the Borel measurable functions in a way that is motivated by a defect in the collection of continuous functions; namely, the defect of not being closed under pointwise limits. Once this introduction is accomplished, however, it may indeed be easier to think of Borel measurable functions via Theorem 3.3 and Borel sets via Theorem 3.4. Moreover, those characterizations will be used as a means to define Borel sets and Borel measurable functions in more general contexts.

Here now are several examples that illustrate Borel measurable functions and Borel sets. We have left some of the justifications as exercises for the reader.

EXAMPLE 3.1 *Illustrates Borel Measurable Functions and Borel Sets*

a) Because $\mathcal{B}$ is a σ-algebra containing the open sets, it follows that all open sets, closed sets, and intervals are Borel sets.

b) If B is a countable set, then $B \in \mathcal{B}$; in particular, $Q \in \mathcal{B}$. From this, it also follows that the set of irrational numbers $\mathcal{R} \setminus Q$ is in $\mathcal{B}$.

c) By definition, any continuous function is Borel measurable.

d) χ_Q is Borel measurable because $Q \in \mathcal{B}$. Indeed, if $B \in \mathcal{B}$, then χ_B is Borel measurable by the definition of $\mathcal{B}$.

e) If $B_1, \ldots, B_n \in \mathcal{B}$ and $\alpha_1, \ldots, \alpha_n \in \mathcal{R}$, then $f = \sum_{k=1}^{n} \alpha_k \chi_{B_k}$ is Borel measurable. This fact follows immediately from part (d) and the fact that $\widehat{C}$ is an algebra of functions. In particular, all step functions are Borel measurable.

f) Every monotone function is Borel measurable, as the reader can easily verify by applying Lemma 3.5.

g) A real-valued function f on $\mathcal{R}$ that is 0 except on a countable set is Borel measurable. Indeed, suppose K is countable and $f(x) = 0$ for $x \notin K$. Let $\{x_n\}_n$ be an enumeration of K. Then $f = \sum_n f(x_n)\chi_{\{x_n\}}$. If K is finite, then f is Borel measurable by parts (a) and (e). If K is infinite, then f is the pointwise limit of the Borel measurable functions $\sum_{k=1}^{n} f(x_k)\chi_{\{x_k\}}$, $n \in \mathcal{N}$, and, hence, is itself Borel measurable. □

Borel Measurable Functions and Borel Sets on Subsets of $\mathcal{R}$

We conclude this section with a brief discussion of Borel measurable functions and Borel sets when the underlying space is some subset $D \subset \mathcal{R}$. The pertinent definitions and theorems are obvious modifications of those discussed earlier.

DEFINITION 3.4 Borel Measurable Functions on D

We denote by $\widehat{C}(D)$ the smallest collection of real-valued functions on D that contains the continuous functions on D and is closed under pointwise limits. The members of $\widehat{C}(D)$ are called **Borel measurable functions on** D.

DEFINITION 3.5   Borel Sets of D

A set $B \subset D$ is called a **Borel set of D** if its characteristic function is a Borel measurable function on D. The collection of all Borel sets of D is denoted $\mathcal{B}(D)$. Thus, $\mathcal{B}(D) = \{\, B \subset D : \chi_B \in \widehat{C}(D) \,\}$.

Using arguments similar to those used earlier, we can obtain the following theorems:

□ □ □ **THEOREM 3.5**

A function f is Borel measurable on D if and only if the inverse image of each open set under f is a Borel set in D, that is, $f^{-1}(O) \in \mathcal{B}(D)$ for all open sets O.

□ □ □ **THEOREM 3.6**

The collection of Borel sets of D, $\mathcal{B}(D)$, is the smallest σ-algebra of subsets of D that contains all open sets in D.

An interesting and useful characterization of $\mathcal{B}(D)$ is given by the following theorem. Note the analogy with open sets in D (see Theorem 2.3 on page 51).

□ □ □ **THEOREM 3.7**

$B \in \mathcal{B}(D)$ if and only if there is an $A \in \mathcal{B}$ such that $B = D \cap A$. That is, the Borel sets of D are precisely the Borel sets (of $\mathcal{R}$) intersected with D.

PROOF   Let $\mathcal{A} = \{\, D \cap A : A \in \mathcal{B} \,\}$. We claim that $\mathcal{A} = \mathcal{B}(D)$. It is easy to see that $\mathcal{A}$ is a σ-algebra of subsets of D and, since $\mathcal{B}$ contains all open sets of $\mathcal{R}$, $\mathcal{A}$ contains all open sets of D. Thus, by Theorem 3.6, $\mathcal{A} \supset \mathcal{B}(D)$.

Now, let $\mathcal{C}$ be any σ-algebra of subsets of D that contains the open sets of D and set

$$\mathcal{D} = \{\, A \in \mathcal{B} : D \cap A \in \mathcal{C} \,\}.$$

Then $\mathcal{D}$ contains the open sets of $\mathcal{R}$ because $\mathcal{C}$ contains the open sets of D. Also, $\mathcal{D}$ is a σ-algebra because $\mathcal{B}$ and $\mathcal{C}$ are. Consequently, $\mathcal{D} \supset \mathcal{B}$. But, by definition, $\mathcal{D} \subset \mathcal{B}$. Hence, $\mathcal{D} = \mathcal{B}$. It now follows that $\mathcal{A} \subset \mathcal{C}$ and, since $\mathcal{C}$ was an arbitrary σ-algebra of subsets of D that contains the open sets, we conclude that $\mathcal{A} \subset \mathcal{B}(D)$. This last result and the previous paragraph show that $\mathcal{A} = \mathcal{B}(D)$. ∎

Exercises for Section 3.1

Note: A ★ denotes an exercise that will be subsequently referenced.

3.1   Prove parts (b) and (c) of Theorem 3.1 on page 82.

★3.2   Let $h(x) = |x|$. Prove Lemma 3.1 on page 83 by proceeding as follows:

a) Show that there exists a sequence of polynomials that converges uniformly to h on $[-1, 1]$. *Hint:* Consider the Taylor series expansion for $(1 - t)^{1/2}$ on $[0, 1]$.

b) Use part (a) to conclude that for each compact subset K of $\mathcal{R}$, there exists a sequence of polynomials that converges uniformly to h on K. *Hint:* If $b > 0$, we can write $|x| = b \cdot |x/b|$.

c) Use part (b) to conclude that there exists a sequence of polynomials that converges to h uniformly on each compact subset of $\mathcal{R}$.

d) Deduce Lemma 3.1 from part (c).

3.3 Prove that every open set is a Borel set by showing that for each open set O, χ_O is a Borel measurable function. *Hint:* Begin by showing that χ_I is Borel measurable for each open interval I.

3.4 Verify part (b) of Lemma 3.5 on page 85.

3.5 Show that f is Borel measurable if and only if $f^{-1}(B) \in \mathcal{B}$ for all Borel sets B.

3.6 Let D be a dense subset of $\mathcal{R}$. Show that f is Borel measurable if either of the following conditions holds:

a) For each $d \in D$, $f^{-1}\big((-\infty, d)\big) \in \mathcal{B}$.

b) For each $d \in D$, $f^{-1}\big((d, \infty)\big) \in \mathcal{B}$.

3.7 Show that all closed sets and all intervals are Borel sets.

3.8 Prove that every monotone function is Borel measurable.

3.9 Prove Theorems 3.5–3.7.

3.10 Verify (3.1) and (3.2) on page 83.

3.11 Show that any countable subset of $\mathcal{R}$ is a Borel set.

3.12 For subsets A and B of $\mathcal{R}$, define

$$A + B = \{\, a + b : a \in A \text{ and } b \in B \,\}.$$

Suppose that B is a Borel set. Prove that $A + B$ is a Borel set if A is

a) countable.   b) open.

3.13 Most functions encountered in a calculus course can be obtained from the identity function $i(x) = x$ by using the standard operations of algebra (sums, products, quotients, and the extraction of roots) together with the operation of passing to the limit in a sequence of functions. For example,

$$e^{x^{1/2}} = \lim_{n \to \infty} \sum_{k=0}^{n} \frac{(x^{1/2})^k}{k!}.$$

Explain why any function obtained using the forementioned operations is a Borel measurable function.

3.2 LEBESGUE OUTER MEASURE

In the previous section, we enlarged our basic collection of functions from the continuous functions to the Borel measurable functions. Although both of those collections of functions are algebras, the latter collection has the advantage of being closed under pointwise limits.

Our next goal is to extend the Riemann integral to an integral that applies to all Borel measurable functions. The extension is not trivial since there are Borel measurable functions that are not Riemann integrable. Indeed, as we learned in Theorem 2.7 on page 72, a bounded function is Riemann integrable if and only if it is continuous except on a set of measure zero. There are certainly Borel measurable functions that do not satisfy this last condition (e.g., χ_Q).

Referring to Section 2.6, beginning on page 67, we see that the Riemann integral is developed by first defining the integral of a step function $h = \sum_{k=1}^{n} a_k \chi_{I_k}$ on $[a, b]$ to be

$$\int_a^b h(x)\, dx = \sum_{k=1}^{n} a_k \ell(I_k),$$

where $\ell(I)$ denotes the length of an interval I. Therefore, the definition of the Riemann integral ultimately depends on the concept of length, which applies only to intervals of real numbers.

To obtain an integral that applies to all Borel measurable functions, we proceed by analogy with the development of the Riemann integral. Specifically, we must first define the integral of a Borel measurable function of the form $s = \sum a_k \chi_{B_k}$, where the B_ks are Borel sets. If the B_ks are intervals, then s is a step function and we simply define the integral to equal the Riemann integral $\sum a_k \ell(B_k)$. If the B_ks are not intervals, then what? It seems that we need to generalize the concept of length so that it applies to arbitrary Borel sets.

The Definition of Lebesgue Outer Measure

The concept of length will be extended and replaced by that of *measure*. As we will see, this is by no means a simple procedure. Let us denote the required measure by the Greek letter μ, and the collection of subsets of $\mathcal{R}$ to which it applies by the letter $\mathcal{A}$. Subsets of $\mathcal{R}$ that belong to $\mathcal{A}$ are called *measurable sets*. We will now list some properties that μ and $\mathcal{A}$ should satisfy.

Since measure is to be a generalization of length, we require that the measure of an interval be its length; that is, $\mu(I) = \ell(I)$ for all intervals I. Also, for purely mathematical reasons, we require that $\mathcal{A}$ be a σ-algebra; and as we want all Borel sets to be measurable, we require that $\mathcal{A} \supset \mathcal{B}$.

Now, clearly, the measure of the union of two disjoint intervals should be the sum of their lengths (measures). More generally, then, we require that the measure of the union of two disjoint measurable sets be the sum of their measures. That is, if A, $B \in \mathcal{A}$ and $A \cap B = \emptyset$, then

$$\mu(A \cup B) = \mu(A) + \mu(B). \tag{3.3}$$

Using mathematical induction, we can show that the previous condition implies that if A_1, A_2, ..., $A_n \in \mathcal{A}$ and $A_i \cap A_j = \emptyset$ for $i \neq j$, then

$$\mu\left(\bigcup_{k=1}^{n} A_k\right) = \sum_{k=1}^{n} \mu(A_k). \tag{3.4}$$

This condition on μ is called **finite additivity.**

For purposes of modern mathematical analysis, we need to impose a somewhat stronger condition on our measure than finite additivity; namely, that (3.4) hold not only for finite collections of pairwise disjoint measurable sets but also for countably infinite collections of pairwise disjoint measurable sets. This condition is called **countable additivity.**

In summary, if μ is the required generalization of length and $\mathcal{A}$ is the collection of subsets of $\mathcal{R}$ that have a length in this extended sense, then the following conditions should be satisfied:

(M1) $\mathcal{A}$ is a σ-algebra and $\mathcal{A} \supset \mathcal{B}$.

(M2) $\mu(I) = \ell(I)$, for all intervals I.

(M3)   If $A_1, A_2, \ldots$ are in $\mathcal{A}$, with $A_i \cap A_j = \emptyset$ for $i \neq j$, then

$$\mu\left(\bigcup_n A_n\right) = \sum_n \mu(A_n).$$

Conditions (M1)–(M3) provide us with the means for extending the notion of length to all open sets. First, since every open set is a Borel set, Condition (M1) implies that every open set should be measurable. Now, let O be an open set. Then O is a countable union of disjoint open intervals, say $O = \bigcup_n I_n$. Now applying, in turn, Conditions (M3) and (M2), we infer that

$$\mu(O) = \mu\left(\bigcup_n I_n\right) = \sum_n \mu(I_n) = \sum_n \ell(I_n).$$

So, we now see how to extend the notion of length to all open sets.

For sets that are more complicated than open sets, however, it is not at all obvious what to do. In fact, defining a suitable measure for subsets of $\mathcal{R}$ constituted a major problem for mathematicians until the beginning of the twentieth century, when Henri Lebesgue found the key. His idea was as follows: For a subset $A \subset \mathcal{R}$, consider all open sets that contain A as a subset. Then define the measure of A to be the greatest lower bound of the measures of all those open sets:

$$\inf\{\,\mu(O) : O \text{ open}, O \supset A\,\}. \tag{3.5}$$

With this definition, we "close down on A" or "come at A from the outside," so we call this measure of A its **outer measure.** Outer measure is defined for all subsets of $\mathcal{R}$. But, as we will see, it is countably additive (i.e., satisfies Condition (M3)) only when restricted to a proper subcollection of subsets of $\mathcal{R}$. Consequently, we will denote outer measure not by μ, but instead by λ^*.

Below we give a formal definition of outer measure. The definition that we present does not use (3.5) but is equivalent to it.

DEFINITION 3.6   Lebesgue Outer Measure

For each subset $A \subset \mathcal{R}$, the **Lebesgue outer measure of A,** denoted by $\boldsymbol{\lambda^*(A)}$, is defined by

$$\lambda^*(A) = \inf\left\{\sum_n \ell(I_n) : \{I_n\}_n \text{ open intervals}, \bigcup_n I_n \supset A\right\}.$$

Note: A sequence of open intervals $\{I_n\}_n$ appearing in Definition 3.6 can be either a finite or infinite sequence.

Basic Properties of Lebesgue Outer Measure

Some basic properties of Lebesgue outer measure are proved in the next two propositions.

☐ ☐ ☐ **PROPOSITION 3.1**

Lebesgue outer measure λ^* has the following properties:

a) $\lambda^*(A) \geq 0$, for all $A \subset \mathcal{R}$. **(nonnegativity)**

b) $\lambda^*(\emptyset) = 0$.

c) $A \subset B \Rightarrow \lambda^*(A) \leq \lambda^*(B)$. **(monotonicity)**

d) $\lambda^*(x + A) = \lambda^*(A)$ for $x \in \mathcal{R}$, $A \subset \mathcal{R}$, where $x + A = \{ x + y : y \in A \}$. **(translation invariance)**

e) If $\{A_n\}_n$ is a sequence of subsets of $\mathcal{R}$, then

$$\lambda^*\left(\bigcup_n A_n\right) \leq \sum_n \lambda^*(A_n). \tag{3.6}$$

In particular if A, $B \subset \mathcal{R}$, then $\lambda^*(A \cup B) \leq \lambda^*(A) + \lambda^*(B)$. The relation in (3.6) is called **countable subadditivity.**

PROOF For each $E \subset \mathcal{R}$, let

$$S_E = \left\{ \sum_n \ell(I_n) : \{I_n\}_n \text{ open intervals}, \bigcup_n I_n \supset E \right\}.$$

Then, by definition, $\lambda^*(E) = \inf\{\, x : x \in S_E \,\}$.

a) If $A \subset \mathcal{R}$, then $S_A \subset [0, \infty]$ so that $\lambda^*(A) = \inf\{\, x : x \in S_A \,\} \geq 0$.

b) For $\epsilon > 0$, the interval $I_\epsilon = (-\epsilon/2, \epsilon/2)$ contains $\emptyset$; so, $\epsilon = \ell(I_\epsilon) \in S_\emptyset$. Hence, $\lambda^*(\emptyset) = \inf\{\, x : x \in S_\emptyset \,\} \leq \epsilon$ for all $\epsilon > 0$. This implies that $\lambda^*(\emptyset) = 0$.

c) Let $u \in S_B$. Then there is a sequence $\{I_n\}$ of open intervals such that $B \subset \bigcup I_n$ and $u = \sum \ell(I_n)$. But $B \subset \bigcup I_n \Rightarrow A \subset \bigcup I_n \Rightarrow u \in S_A$. Therefore, $S_B \subset S_A$ and, consequently,

$$\lambda^*(A) = \inf\{\, x : x \in S_A \,\} \leq \inf\{\, x : x \in S_B \,\} = \lambda^*(B).$$

d) The proof of this part is left to the reader as an exercise.

e) If $\lambda^*(A_n) = \infty$ for some n, then, by part (c), $\lambda^*(\bigcup A_n) = \infty$; consequently, (3.6) holds. So, assume that $\lambda^*(A_n) < \infty$ for all n. Let $\epsilon > 0$ be given. For each n, choose a sequence $\{I_{nk}\}_k$ of open intervals such that $\bigcup_k I_{nk} \supset A_n$ and $\sum_k \ell(I_{nk}) < \lambda^*(A_n) + \epsilon/2^n$. The collection of intervals $\{I_{nk}\}_{n,k}$ is countable (because $\mathcal{N} \times \mathcal{N}$ is countable) and $\bigcup_{n,k} I_{nk} = \bigcup_n (\bigcup_k I_{nk}) \supset \bigcup_n A_n$. Hence,

$$\lambda^*\left(\bigcup_n A_n\right) \leq \sum_{n,k} \ell(I_{nk}) = \sum_n \sum_k \ell(I_{nk})$$

$$\leq \sum_n \left(\lambda^*(A_n) + \frac{\epsilon}{2^n}\right) \leq \sum_n \lambda^*(A_n) + \epsilon.$$

As $\epsilon > 0$ was arbitrary, this proves that $\lambda^*(\bigcup_n A_n) \leq \sum_n \lambda^*(A_n)$. ■

As we have noted, the domain of λ^* is $\mathcal{P}(\mathcal{R})$; that is, every subset of $\mathcal{R}$ has an outer measure. Our question now is whether λ^* is the desired extension of length. That is, do Conditions (M1)–(M3) hold with $\mu = \lambda^*$ and $\mathcal{A} = \mathcal{P}(\mathcal{R})$? Certainly, Condition (M1) holds; and the next proposition shows that Condition (M2) holds also.

□ □ □ **PROPOSITION 3.2**

The outer measure of an interval is its length. That is, $\lambda^*(I) = \ell(I)$ for every interval I.

PROOF First assume $I = [a, b]$, that is, that I is a bounded and closed interval. If $\epsilon > 0$, then $(a - \epsilon/2, b + \epsilon/2) \supset [a, b]$ and so

$$\lambda^*([a, b]) \leq \ell\left(\left(a - \frac{\epsilon}{2}, b + \frac{\epsilon}{2}\right)\right) = b - a + \epsilon.$$

Thus, for any $\epsilon > 0$, $\lambda^*([a, b]) \leq b - a + \epsilon$ and, hence, $\lambda^*([a, b]) \leq b - a$.

Consequently, it remains to establish that $\lambda^*([a, b]) \geq b - a$. Let $\{I_n\}$ be a sequence of open intervals such that $\bigcup I_n \supset [a, b]$. We claim that $\sum \ell(I_n) > b - a$. Since $\{I_n\}$ is an open cover for $[a, b]$, the Heine-Borel theorem implies that there is a finite subcover, say $\{I_k\}_{k=1}^N$. Now, clearly, $\sum_{k=1}^N \ell(I_k) \leq \sum_n \ell(I_n)$. So, we need only show that $\sum_{k=1}^N \ell(I_k) > b - a$.

Because $a \in [a, b]$, there must be an interval, say $J_1 = (a_1, b_1)$, in the collection $\{I_k\}_{k=1}^N$ with $a_1 < a < b_1$. If $b < b_1$, then

$$\sum_{k=1}^N \ell(I_k) \geq \ell(J_1) = b_1 - a_1 > b - a.$$

Otherwise, $b_1 \in [a, b]$, so there must be an interval, say $J_2 = (a_2, b_2)$, in the collection $\{I_k\}_{k=1}^N$ with $a_2 < b_1 < b_2$. Note that, necessarily, $J_2 \neq J_1$. If $b < b_2$, then

$$\sum_{k=1}^N \ell(I_k) \geq \ell(J_1) + \ell(J_2) = (b_1 - a_1) + (b_2 - a_2)$$

$$= (b_2 - a_1) + (b_1 - a_2) > b_2 - a_1 > b - a.$$

Otherwise, $b_2 \in [a, b]$, so there must be an interval, say $J_3 = (a_3, b_3)$, in the collection $\{I_k\}_{k=1}^N$ such that $a_3 < b_2 < b_3$ and, necessarily, $J_3 \neq J_2$ and $J_3 \neq J_1$.

This process can continue at most N times. Consequently, there is an $m \in \mathcal{N}$ with $m \leq N$ such that $J_i = (a_i, b_i) \in \{I_k\}_{k=1}^N$ for $1 \leq i \leq m$ and

$$a_1 < a, \ a_2 < b_1 < b_2, \ \ldots, \ a_m < b_{m-1} < b_m, \ b < b_m.$$

Therefore,

$$\sum_{k=1}^N \ell(I_k) \geq \sum_{i=1}^m \ell(J_i) = (b_1 - a_1) + (b_2 - a_2) + \cdots + (b_m - a_m)$$

$$= (b_m - a_1) + (b_1 - a_2) + (b_2 - a_3) + \cdots + (b_{m-1} - a_m)$$

$$> b_m - a_1 > b - a.$$

So, if $\{I_n\}$ is a sequence of open intervals with $\bigcup I_n \supset [a,b]$, then $\sum \ell(I_n) > b - a$. But then, by definition, $\lambda^*([a,b]) \geq b - a$. This last fact and the previously established reverse inequality show $\lambda^*([a,b]) = b - a$.

Now, let I be any finite (i.e., bounded) interval. Then for each $\epsilon > 0$, there is a closed interval J with $J \subset I$ and $\ell(I) < \ell(J) + \epsilon$ (why?). Thus,

$$\ell(I) - \epsilon < \ell(J) = \lambda^*(J) \leq \lambda^*(I).$$

Since $\epsilon > 0$ was arbitrary, it follows that $\lambda^*(I) \geq \ell(I)$. But, on the other hand, $\lambda^*(I) \leq \lambda^*\left(\overline{I}\right) = \ell\left(\overline{I}\right) = \ell(I)$, so that $\lambda^*(I) \leq \ell(I)$.

Finally, if I is an infinite (i.e., unbounded) interval, then for each real number M, there is a closed interval J of length M with $J \subset I$. It follows that $\lambda^*(I) \geq \lambda^*(J) = \ell(J) = M$. Hence, $\lambda^*(I) = \infty$. ∎

We have observed that Conditions (M1) and (M2) are satisfied with $\mu = \lambda^*$ and $\mathcal{A} = \mathcal{P}(\mathcal{R})$. Therefore, our question now is: Does Condition (M3) hold with $\mu = \lambda^*$ and $\mathcal{A} = \mathcal{P}(\mathcal{R})$? If the answer to this question were yes, then λ^* would be the desired extension of length and every subset of $\mathcal{R}$ would be measurable. The answer, however, is no! In fact, as we will discover in the next section, λ^* is not even finitely additive.

Exercises for Section 3.2

3.14 Let I be any finite interval. Show that for each $\epsilon > 0$, there is a closed interval J with $J \subset I$ and $\ell(I) < \ell(J) + \epsilon$.

3.15 Prove part (d) of Proposition 3.1. That is, show that $\lambda^*(x + A) = \lambda^*(A)$ for $x \in \mathcal{R}$, $A \subset \mathcal{R}$, where $x + A = \{x + y : y \in A\}$.

3.16 Suppose that A is a set with $\lambda^*(A) < \infty$. Show that the function g defined by $g(x) = \lambda^*\left(A \cap (-\infty, x]\right)$ is uniformly continuous on $\mathcal{R}$.

3.17 Show that the Cantor set P has Lebesgue outer measure zero.

3.18 Show that, for each $E \subset \mathcal{R}$, there is a sequence of open sets $\{O_n\}_{n=1}^{\infty}$ such that $O_1 \supset O_2 \supset \cdots \supset E$ and

$$\lambda^*(E) = \lambda^*\left(\bigcap_{n=1}^{\infty} O_n\right) = \lim_{n \to \infty} \lambda^*(O_n).$$

3.19 For $A \subset \mathcal{R}$ and $b \in \mathcal{R}$, define $bA = \{ba : a \in A\}$. Show that $\lambda^*(bA) = |b|\lambda^*(A)$.

3.20 Suppose that $f : \mathcal{R} \to \mathcal{R}$ is differentiable at each point of $\mathcal{R}$.
  a) If $|f'(x)| \leq 1$ for each $x \in \mathcal{R}$, prove that for each $A \subset \mathcal{R}$, $\lambda^*\left(f(A)\right) \leq \lambda^*(A)$. *Hint:* Use the mean-value theorem.
  b) Provide an example to show that the preceding inequality may fail to hold if $|f'(x)| > 1$ for some $x \in \mathcal{R}$.

3.3 FURTHER PROPERTIES OF LEBESGUE OUTER MEASURE

Recall that we are trying to extend the notion of length so that it applies to all Borel sets. Specifically, we are searching for a function μ defined on some collection $\mathcal{A}$ of subsets of $\mathcal{R}$ such that

(M1) $\mathcal{A}$ is a σ-algebra and $\mathcal{A} \supset \mathcal{B}$.

(M2) $\mu(I) = \ell(I)$, for all intervals I.

(M3)   If $A_1, A_2, \ldots$ are in $\mathcal{A}$, with $A_i \cap A_j = \emptyset$ for $i \neq j$, then

$$\mu\left(\bigcup_n A_n\right) = \sum_n \mu(A_n).$$

In Section 3.2, we discovered that Conditions (M1) and (M2) are satisfied with $\mu = \lambda^*$ (Lebesgue outer measure) and $\mathcal{A} = \mathcal{P}(\mathcal{R})$. We will prove in this section that Condition (M3) does not hold with $\mu = \lambda^*$ and $\mathcal{A} = \mathcal{P}(\mathcal{R})$.

In fact, we will show that even finite additivity does not obtain. That is, it is possible to find disjoint subsets A and B of $\mathcal{R}$ such that the equation

$$\lambda^*(A \cup B) = \lambda^*(A) + \lambda^*(B) \tag{3.7}$$

fails to hold. The idea is to choose A and B so that they are disjoint but "sufficiently intermingled."

Finite Additivity Properties of λ^*

It is best to begin by determining conditions on disjoint sets A and B that imply that (3.7) holds. Our first result is that if A and B are not only disjoint but are a positive distance apart, then (3.7) is true. Before proving that fact, we need some preliminary definitions and lemmas.

DEFINITION 3.7   Distance Between a Point and a Set or Two Sets

If $x \in \mathcal{R}$ and $E \subset \mathcal{R}$, then the **distance from x to E**, denoted by $d(x, E)$, is defined to be

$$d(x, E) = \inf\{\, |y - x| : y \in E \,\}.$$

If E and F are subsets of $\mathcal{R}$, then the **distance from E to F**, denoted by $d(E, F)$, is defined to be

$$d(E, F) = \inf\{\, |y - x| : y \in E, x \in F \,\}.$$

It is left as an exercise for the reader to show that (1) for fixed $E \subset \mathcal{R}$, the function $d(x, E)$ is continuous, (2) $d(E, F) = \inf\{\, d(y, F) : y \in E \,\}$, and (3) if $A \subset E$ and $B \subset F$, then $d(E, F) \leq d(A, B)$. The proof of the following lemma is also left to the reader as an exercise.

□ □ □   **LEMMA 3.8**

Suppose that I is a finite open interval and let $\epsilon, \delta > 0$ be given. Then there are a finite number of open intervals, say $J_1, \ldots, J_n$, such that $\ell(J_k) < \delta$, $1 \leq k \leq n$, $I \subset \bigcup_{k=1}^n J_k$, and $\sum_{k=1}^n \ell(J_k) < \ell(I) + \epsilon$.

□ □ □ **LEMMA 3.9**

Suppose that A is a subset of $\mathcal{R}$ with $\lambda^*(A) < \infty$. Then for each ϵ, $\delta > 0$, there is a sequence $\{I_n\}$ of open intervals such that $\ell(I_n) < \delta$ for all n, $\bigcup I_n \supset A$, and $\sum \ell(I_n) < \lambda^*(A) + \epsilon$.

PROOF Given $\epsilon > 0$, there is a sequence $\{J_n\}$ of open intervals such that $\bigcup J_n \supset A$ and $\sum \ell(J_n) < \lambda^*(A) + \epsilon/2$. By Lemma 3.8, for each J_n, there are a finite number of open intervals, say $J_{n1}, J_{n2}, \ldots, J_{nk_n}$, with $\ell(J_{nj}) < \delta$, $1 \le j \le k_n$, $\bigcup_{j=1}^{k_n} J_{nj} \supset J_n$, and $\sum_{j=1}^{k_n} \ell(J_{nj}) < \ell(J_n) + \epsilon/2^{n+1}$.

Now, the collection

$$\bigcup_n \{J_{nj}\}_{j=1}^{k_n} = \{J_{11}, J_{12}, \ldots, J_{1k_1}, J_{21}, J_{22}, \ldots, J_{2k_2}, \ldots\}$$

is countable, being a countable union of finite collections. We have $\ell(J_{nj}) < \delta$, for each n and j, and

$$\sum_{n,j} \ell(J_{nj}) = \sum_n \sum_{j=1}^{k_n} \ell(J_{nj}) \le \sum_n \left(\ell(J_n) + \frac{\epsilon}{2^{n+1}} \right)$$

$$< \lambda^*(A) + \frac{\epsilon}{2} + \sum_n \frac{\epsilon}{2^{n+1}} \le \lambda^*(A) + \epsilon.$$

Also, $\bigcup_{n,j} J_{nj} = \bigcup_n \left(\bigcup_{j=1}^{k_n} J_{nj} \right) \supset \bigcup_n J_n \supset A$. If we now reindex the collection $\{J_{11}, J_{12}, \ldots, J_{1k_1}, J_{21}, J_{22}, \ldots, J_{2k_2}, \ldots\}$ by using a single subscript and obtain $\{I_n\}_n$, then this sequence satisfies the conclusions of the lemma. ∎

□ □ □ **THEOREM 3.8**

Suppose that A and B are subsets of $\mathcal{R}$ that are a positive distance apart; that is, $d(A, B) > 0$. Then

$$\lambda^*(A \cup B) = \lambda^*(A) + \lambda^*(B).$$

PROOF Let $\delta = d(A, B)$. If $\lambda^*(A \cup B) = \infty$, then it follows from Proposition 3.1(e) that the conclusion of the theorem holds. So, assume that $\lambda^*(A \cup B) < \infty$. Let $\epsilon > 0$ be given. By Lemma 3.9, there is a sequence $\{I_n\}$ of open intervals such that $\ell(I_n) < \delta$ for all n, $\bigcup I_n \supset A \cup B$, and $\sum \ell(I_n) < \lambda^*(A \cup B) + \epsilon$.

Now, let $\{J_n\}$ denote the members of $\{I_n\}$ that contain a point of A and let $\{K_n\}$ denote the ones that do not contain a point of A. As $A \subset A \cup B \subset \bigcup I_n$, it follows that $A \subset \bigcup J_n$. Also, because $d(A, B) = \delta$ and $\ell(I_n) < \delta$ for all n, there can be no points of B in any J_n. Therefore, because $B \subset A \cup B \subset \bigcup I_n$, it must be that $B \subset \bigcup K_n$.

Using the definition of outer measure, we conclude that

$$\lambda^*(A) + \lambda^*(B) \le \sum \ell(J_n) + \sum \ell(K_n) = \sum \ell(I_n) < \lambda^*(A \cup B) + \epsilon.$$

Because $\epsilon > 0$ was arbitrary, $\lambda^*(A) + \lambda^*(B) \le \lambda^*(A \cup B)$. The reverse inequality is true by Proposition 3.1(e). ∎

We can, in fact, improve on Theorem 3.8. Indeed, suppose that A and B are two subsets of $\mathcal{R}$ with the property that there is an open set O with $A \subset O$ and $B \subset O^c$. Then the conclusion of Theorem 3.8 obtains.

Roughly speaking, the reason is as follows. Since O is open, it can be written as a countable union of disjoint open intervals. Because the points of A must lie within these open intervals and the points of B must lie outside of them, the sets A and B cannot be too intermingled. Before we can provide a rigorous proof of the improvement of Theorem 3.8, we need two more lemmas.

□ □ □ LEMMA 3.10

Let O be a proper open subset of $\mathcal{R}$ (i.e., O is open, nonempty, and not equal to $\mathcal{R}$). For each $n \in \mathcal{N}$, let

$$O_n = \left\{ x : d(x, O^c) > \frac{1}{n} \right\}.$$

Then,
a) O_n is open and $O_n \subset O$ for all $n \in \mathcal{N}$.
b) $O_1 \subset O_2 \subset \cdots$ and $\bigcup_n O_n = O$.
c) If $O_n \neq \emptyset$, then $d(O_n, O^c) = 1/n$.
d) If $O_n \neq \emptyset$, then $d(O_n, O_{n+1}^c) = 1/n(n+1)$.

PROOF The proofs of parts (a) and (b) are left to the reader.

c) Since $d(O_n, O^c) = \inf\{ d(x, O^c) : x \in O_n \}$, we see that $d(O_n, O^c) \geq 1/n$. To prove the reverse inequality, we first note that because O is open, it can be expressed as a countable union of disjoint open intervals, say the intervals $\{I_j\}_j$.

Now, by assumption, $O_n \neq \emptyset$. This means that there is a $y \subset O$ such that $d(y, O^c) > 1/n$. Since $y \in O$, there is a k such that $y \in I_k$. Clearly, the distance from y to O^c is the same as the distance from y to the nearest endpoint of I_k. Therefore, if we write $I_k = (a_k, b_k)$, then $y \in (a_k + 1/n, b_k - 1/n)$. It follows that $\emptyset \neq (a_k + 1/n, b_k - 1/n) \subset O_n$. Note that at least one of the two numbers a_k and b_k must be finite. We will assume that a_k is finite. (If a_k is infinite, a similar argument holds.)

Since $(a_k + 1/n, b_k - 1/n) \subset O_n$ and $a_k \in O^c$, we conclude by applying Exercise 3.21(c) that

$$d(O_n, O^c) \leq d\left(\left(a_k + \frac{1}{n}, b_k - \frac{1}{n} \right), \{a_k\} \right) = \frac{1}{n}.$$

This completes the proof of part (c).

d) We first show that $d(O_n, O_{n+1}^c) \geq 1/n(n+1)$. Suppose $y \in O_n$ and $z \in O_{n+1}^c$. By definition, $d(y, O^c) > 1/n$ and $d(z, O^c) \leq 1/(n+1)$. Let $\epsilon > 0$ be given. Then there is a $w \in O^c$ with $|w - z| < 1/(n+1) + \epsilon$. Also, $w \in O^c$ implies that $|w - y| > 1/n$. Thus,

$$|z - y| \geq |w - y| - |w - z| > \frac{1}{n} - \frac{1}{n+1} - \epsilon = \frac{1}{n(n+1)} - \epsilon.$$

As z and y do not depend on ϵ, we conclude that $|z - y| \geq 1/n(n+1)$. Consequently, because $y \in O_n$, $z \in O_{n+1}^c$ were arbitrary, $d(O_n, O_{n+1}^c) \geq 1/n(n+1)$.

To prove the reverse inequality, let $I_k = (a_k, b_k)$ be as in the proof of part (c), and assume as before that a_k is finite. Because $a_k \in O^c$, we have $a_k + 1/(n+1) \in O^c_{n+1}$. Therefore, by Exercise 3.21(c),

$$d(O_n, O^c_{n+1}) \leq d\left(\left(a_k + \frac{1}{n}, b_k - \frac{1}{n}\right), \left\{a_k + \frac{1}{n+1}\right\}\right) = \frac{1}{n(n+1)}.$$

This completes the proof of part (d). ∎

□ □ □ **LEMMA 3.11**

Suppose that $A \subset \mathcal{R}$ and $\lambda^*(A) < \infty$. Assume there is a proper open subset O of $\mathcal{R}$ with $A \subset O$. Let $O_n = \{x : d(x, O^c) > 1/n\}$. Then

$$\lambda^*(A) = \lim_{n \to \infty} \lambda^*(A \cap O_n).$$

PROOF Let $A_n = A \cap O_n$. Then, by Lemma 3.10(b), $A_1 \subset A_2 \subset \cdots$ and, consequently, $\lambda^*(A_1) \leq \lambda^*(A_2) \leq \cdots$. Also, since $A_n \subset A$ for all n, $\lambda^*(A_n) \leq \lambda^*(A)$ for all n. By assumption, $\lambda^*(A) < \infty$. Thus, $\{\lambda^*(A_n)\}_n$ is a monotone nondecreasing, bounded sequence of real numbers; and, hence, converges to a real number, say α. Clearly, $\alpha \leq \lambda^*(A)$.

Now, let $B_n = A \setminus A_n$ and $C_n = A_{n+1} \setminus A_n$. Then we have $A = A_n \cup B_n$ and $B_n = C_n \cup C_{n+1} \cup \cdots$. Thus,

$$\lambda^*(A) \leq \lambda^*(A_n) + \lambda^*(B_n) \tag{3.8}$$

and

$$\lambda^*(B_n) \leq \lambda^*(C_n) + \lambda^*(C_{n+1}) + \cdots. \tag{3.9}$$

Now, for $n \geq 2$, $A_{n+1} = A_n \cup C_n \supset A_{n-1} \cup C_n$, so that

$$\lambda^*(A_{n-1} \cup C_n) \leq \lambda^*(A_{n+1}). \tag{3.10}$$

Also, $A_{n-1} \subset O_{n-1}$ and $C_n \subset O^c_n$. So, by Lemma 3.10(d),

$$d(A_{n-1}, C_n) \geq d(O_{n-1}, O^c_n) = 1/n(n-1) > 0.$$

Therefore, Theorem 3.8 implies that

$$\lambda^*(A_{n-1} \cup C_n) = \lambda^*(A_{n-1}) + \lambda^*(C_n).$$

Using (3.10) and this last equation, we conclude that, for $n \geq 2$,

$$\lambda^*(C_n) \leq \lambda^*(A_{n+1}) - \lambda^*(A_{n-1}).$$

Then (3.9) implies

$$\lambda^*(B_n) \leq \sum_{k=1}^{\infty} [\lambda^*(A_{n+k}) - \lambda^*(A_{n+k-2})]$$

so that by (3.8)

$$\lambda^*(A) \leq \lambda^*(A_n) + \lambda^*(B_n) \leq \lambda^*(A_n) + \sum_{k=1}^{\infty}[\lambda^*(A_{n+k}) - \lambda^*(A_{n+k-2})].$$

But,

$$\lambda^*(A_n) + \sum_{k=1}^{\infty}[\lambda^*(A_{n+k}) - \lambda^*(A_{n+k-2})]$$

$$= \lim_{m\to\infty}\left\{\lambda^*(A_n) + \sum_{k=1}^{m}[\lambda^*(A_{n+k}) - \lambda^*(A_{n+k-2})]\right\}$$

$$= \lim_{m\to\infty}\{-\lambda^*(A_{n-1}) + \lambda^*(A_{n+m-1}) + \lambda^*(A_{n+m})\}$$

$$= -\lambda^*(A_{n-1}) + 2\alpha.$$

Consequently, we have shown that $\lambda^*(A) \leq -\lambda^*(A_{n-1}) + 2\alpha$ for all n. Letting $n \to \infty$, reveals that $\lambda^*(A) \leq \alpha$. As we know, however, $\alpha \leq \lambda^*(A)$. Thus, we have $\lambda^*(A) = \alpha = \lim_{n\to\infty}\lambda^*(A_n)$, as required. ■

□ □ □ **THEOREM 3.9**

Suppose that A and B are subsets of $\mathcal{R}$ with the property that there is an open set O with $A \subset O$ and $B \subset O^c$. Then

$$\lambda^*(A \cup B) = \lambda^*(A) + \lambda^*(B).$$

PROOF If either $\lambda^*(A)$ or $\lambda^*(B)$ is infinite, then the result is trivial. So, assume that both are finite. If $O = \emptyset$, then $A = \emptyset$; and if $O = \mathcal{R}$, then $B = \emptyset$. In either of those cases, the result is also trivial.

Consequently, we can assume that O is a proper open subset of $\mathcal{R}$. As before, let $O_n = \{x : d(x, O^c) > 1/n\}$ and $A_n = A \cap O_n$. Because $A_n \subset O_n$ and $B \subset O^c$, Lemma 3.10(c) implies that $d(A_n, B) \geq d(O_n, O^c) = 1/n$ and, thus, by Theorem 3.8, $\lambda^*(A_n \cup B) = \lambda^*(A_n) + \lambda^*(B)$. Since $A_n \cup B \subset A \cup B$, we have $\lambda^*(A_n \cup B) \leq \lambda^*(A \cup B)$. Therefore, for each $n \in \mathcal{N}$,

$$\lambda^*(A \cup B) \geq \lambda^*(A_n \cup B) = \lambda^*(A_n) + \lambda^*(B).$$

Letting $n \to \infty$ and applying Lemma 3.11, we get that

$$\lambda^*(A \cup B) \geq \lambda^*(A) + \lambda^*(B).$$

Proposition 3.1(e) shows that the reverse inequality holds. ■

Lebesgue Outer Measure Is Not Finitely Additive

We have now seen that, under certain conditions,

$$\lambda^*(A \cup B) = \lambda^*(A) + \lambda^*(B) \tag{3.11}$$

for disjoint subsets A and B of $\mathcal{R}$. Our next theorem, which we will state and prove shortly, shows that (3.11) does not hold for every pair of disjoint subsets A and B of $\mathcal{R}$, that is, that λ^* is not finitely additive.

In view of Theorem 3.9, it is clear that if (3.11) fails to hold for disjoint subsets A and B, then those sets must be considerably intermingled. To obtain this intermingling, we proceed as follows.

□ □ □ **LEMMA 3.12**

For x, $y \in \mathcal{R}$, define $x \sim y$ if and only if $y - x \in Q$. Then $\sim$ is an equivalence relation and, hence, partitions $\mathcal{R}$ into disjoint equivalence classes. Moreover, there is a set $S \subset [0, 1)$ containing exactly one element from each equivalence class.

PROOF   That $\sim$ is an equivalence relation is left as an exercise for the reader. By the axiom of choice (see page 14), we can select a set $T \subset \mathcal{R}$ that contains exactly one element from each equivalence class. Let us set $S = \{ x - [x] : x \in T \}$ where $[x]$ denotes the greatest integer in x. Because for each x, $x - [x] \in [0, 1)$ and $x - [x] \sim x$, the proof is complete. ■

□ □ □ **LEMMA 3.13**

Let S be the set defined in Lemma 3.12 and $W = (-1, 1) \cap Q$. Then
a) $\{S + r\}_{r \in Q}$ forms a collection of pairwise disjoint sets.
b) $(0, 1) \subset \bigcup_{r \in W}(S + r) \subset (-1, 2)$.

PROOF

a) Suppose q, $r \in Q$ and $(S + q) \cap (S + r) \neq \emptyset$. Let $y \in (S + q) \cap (S + r)$. Then there exist u, $v \in S$ such that $y = u + q = v + r$. Hence, $u \sim v$. Since S contains only one element from each equivalence class, we must have $u = v$, which, in turn, implies $q = r$.

b) Let $x \in (0, 1)$. Then there is a $u \in S$ such that $x \sim u$. Put $r = x - u$. Then $r \in Q$ and $x \in S + r$. Moreover, since $x \in (0, 1)$ and $S \subset [0, 1)$, $-1 < r < 1$. Thus, $(0, 1) \subset \bigcup_{r \in W}(S + r)$. That $\bigcup_{r \in W}(S + r) \subset (-1, 2)$ follows immediately from the fact that $S \subset [0, 1)$. ■

Note: Lemma 3.13(a) shows that the sets $\{S + r\}_{r \in Q}$ are pairwise disjoint. They are also considerably intermingled as is shown in Exercise 3.27.

□ □ □ **THEOREM 3.10**

Lebesgue outer measure λ^* is not finitely additive.

PROOF   Suppose to the contrary that λ^* is finitely additive. Let $\{q_n\}_{n=1}^{\infty}$ be an enumeration of the rationals in $(-1, 1)$ and set $E_n = S + q_n$. Using the assumed finite additivity of λ^*, Proposition 3.2, Lemma 3.13, and Proposition 3.1(c) and (e), we conclude that

$$1 = \lambda^*((0, 1)) \leq \lambda^*\left(\bigcup_{n=1}^{\infty} E_n\right) \leq \sum_{n=1}^{\infty} \lambda^*(E_n)$$

$$= \lim_{n \to \infty} \sum_{k=1}^{n} \lambda^*(E_k) = \lim_{n \to \infty} \lambda^*\left(\bigcup_{k=1}^{n} E_k\right) \leq \lambda^*((-1, 2)) = 3. \tag{3.12}$$

This shows $1 \leq \lim_{n \to \infty} \sum_{k=1}^{n} \lambda^*(E_k) \leq 3$. But, by Proposition 3.1(d),

$$\sum_{k=1}^{n} \lambda^*(E_k) = \sum_{k=1}^{n} \lambda^*(S + q_k) = \sum_{k=1}^{n} \lambda^*(S) = n\lambda^*(S).$$

Consequently, $1 \leq \lim_{n \to \infty} n\lambda^*(S) \leq 3$, which is impossible (why?). Hence, λ^* is not finitely additive. ■

□ □ □ COROLLARY 3.2

Lebesgue outer measure λ^* is not countably additive. That is, Condition (M3) does not hold with $\mu = \lambda^*$ and $\mathcal{A} = \mathcal{P}(\mathcal{R})$.

Exercises for Section 3.3

3.21 Prove the following facts:
a) For fixed $E \subset \mathcal{R}$, the function $d(x, E)$ is continuous.
b) If E and F are subsets of $\mathcal{R}$, then $d(E, F) = \inf\{\, d(y, F) : y \in E \,\}$.
c) If $A \subset E$ and $B \subset F$, then $d(E, F) \leq d(A, B)$.
d) $d(\overline{E}, \overline{F}) = d(E, F)$.

3.22 Prove the following facts:
a) Suppose that F is a closed subset of $\mathcal{R}$, K is a compact subset of $\mathcal{R}$, and $F \cap K = \emptyset$. Then $d(F, K) > 0$.
b) Show that part (a) is not true if it is assumed only that K is closed.

3.23 Prove Lemma 3.8 on page 95.

3.24 Verify parts (a) and (b) of Lemma 3.10 on page 97.

3.25 Suppose that O is open. Prove that

$$\lambda^*(W) = \lambda^*(W \cap O) + \lambda^*(W \cap O^c)$$

for all subsets W of $\mathcal{R}$.

3.26 Define $x \sim y$ if and only if $x - y \in Q$. Show that $\sim$ is an equivalence relation.

3.27 Let N be a positive integer, $\{r_n\}_{n=1}^{\infty}$ an enumeration of Q, and S as in Lemma 3.12. For each $n \in \mathcal{N}$, define $S_n = S + r_n$. Prove that there is no open set O with the property that $\bigcup_{n=1}^{N} S_n \subset O$ and $\bigcup_{n=N+1}^{\infty} S_n \subset O^c$.

3.28 Suppose that $0 \leq a < b \leq 1$. Prove that it is possible to select the elements of the set S in Lemma 3.12 so that $S \subset (a, b)$.

3.29 Provide a detailed justification for each step in (3.12).

3.4 LEBESGUE MEASURE

For ease in reference, we repeat once more that we are searching for a function μ defined on some collection $\mathcal{A}$ of subsets of $\mathcal{R}$ such that

(M1) $\mathcal{A}$ is a σ-algebra and $\mathcal{A} \supset \mathcal{B}$.

(M2) $\mu(I) = \ell(I)$, for all intervals I.

(M3) If $A_1, A_2, \ldots$ are in $\mathcal{A}$, with $A_i \cap A_j = \emptyset$ for $i \neq j$, then

$$\mu\left(\bigcup_n A_n\right) = \sum_n \mu(A_n).$$

As we have seen, Conditions (M1) and (M2) hold with $\mu = \lambda^*$ and $\mathcal{A} = \mathcal{P}(\mathcal{R})$, but Condition (M3) does not (Corollary 3.2). Note, however, that we do not need to have our measure μ defined for all subsets of $\mathcal{R}$; Condition (M1) requires only that it be defined on a σ-algebra $\mathcal{A}$ of subsets of $\mathcal{R}$ that contains the Borel sets.

Thus, one way to get Condition (M3) to hold might be to restrict λ^* to some proper subcollection of subsets of $\mathcal{R}$; that is, select $\mathcal{A}$ to be a proper subset of $\mathcal{P}(\mathcal{R})$. And, to do that, we need to identify a criterion for deciding whether a subset of $\mathcal{R}$ is measurable, that is, is a member of $\mathcal{A}$. By Condition (M1), we must have $\mathcal{B} \subset \mathcal{A}$; so, in particular, $\mathcal{A}$ must contain all open sets. Hence, the criterion we select must be satisfied by all open sets.

The Carathéodory Criterion

Theorem 3.9 states that if A and B are subsets of $\mathcal{R}$ with the property that there is an open set O with $A \subset O$ and $B \subset O^c$, then

$$\lambda^*(A \cup B) = \lambda^*(A) + \lambda^*(B).$$

As a consequence of Theorem 3.9, we obtain the following proposition.

□ □ □ **PROPOSITION 3.3**

Let O be an open set. Then

$$\lambda^*(W) = \lambda^*(W \cap O) + \lambda^*(W \cap O^c) \tag{3.13}$$

for every subset W of $\mathcal{R}$.

PROOF   For every subset W of $\mathcal{R}$, we have $W = (W \cap O) \cup (W \cap O^c)$. Since $W \cap O \subset O$ and $W \cap O^c \subset O^c$, we see that (3.13) is a simple consequence of Theorem 3.9.   ∎

Equation (3.13) provides an additivity relation for Lebesgue outer measure that is satisfied by all open sets. That relation shows the way to the required criterion for deciding whether a subset of $\mathcal{R}$ is measurable.

DEFINITION 3.8   **Carathéodory Criterion**

A set $E \subset \mathcal{R}$ is said to satisfy the **Carathéodory criterion** if

$$\lambda^*(W) = \lambda^*(W \cap E) + \lambda^*(W \cap E^c) \tag{3.14}$$

for all subsets W of $\mathcal{R}$. We denote by $\mathcal{M}$ the collection of all subsets of $\mathcal{R}$ that satisfy the Carathéodory criterion.

Note: By Proposition 3.1(e), the inequality

$$\lambda^*(W) \leq \lambda^*(W \cap E) + \lambda^*(W \cap E^c)$$

always holds. Consequently, to prove that a subset E of $\mathcal{R}$ is a member of $\mathcal{M}$, it suffices to establish the inequality

$$\lambda^*(W) \geq \lambda^*(W \cap E) + \lambda^*(W \cap E^c) \tag{3.15}$$

for all subsets W of $\mathcal{R}$.

The next theorem demonstrates that Condition (M1) holds for the collection $\mathcal{M}$ of subsets of $\mathcal{R}$ that satisfy the Carathéodory criterion.

□ □ □ THEOREM 3.11

$\mathcal{M}$ is a σ-algebra and $\mathcal{M} \supset \mathcal{B}$.

PROOF That $\mathcal{M}$ is closed under complementation is clear. First we prove that $\mathcal{M}$ is closed under finite unions. So, assume $A, B \in \mathcal{M}$. We claim that $A \cup B \in \mathcal{M}$. Let $W \subset \mathcal{R}$. Then, we must show that

$$\lambda^*(W) \geq \lambda^*\big(W \cap (A \cup B)\big) + \lambda^*\big(W \cap (A \cup B)^c\big). \tag{3.16}$$

(See the note following Definition 3.8.)

Now, we can write $W \cap (A \cup B) = (W \cap A) \cup (W \cap A^c \cap B)$ and, hence, by the subadditivity of λ^*,

$$\lambda^*\big(W \cap (A \cup B)\big) \leq \lambda^*(W \cap A) + \lambda^*(W \cap A^c \cap B).$$

Consequently,

$$\lambda^*\big(W \cap (A \cup B)\big) + \lambda^*\big(W \cap (A \cup B)^c\big)$$
$$\leq \lambda^*(W \cap A) + \lambda^*(W \cap A^c \cap B) + \lambda^*\big(W \cap (A \cup B)^c\big)$$
$$= \lambda^*(W \cap A) + \big[\lambda^*\big((W \cap A^c) \cap B\big) + \lambda^*\big((W \cap A^c) \cap B^c\big)\big].$$

Because $B \in \mathcal{M}$, the quantity between the square brackets in the previous expression equals $\lambda^*(W \cap A^c)$. Thus,

$$\lambda^*\big(W \cap (A \cup B)\big) + \lambda^*\big(W \cap (A \cup B)^c\big) \leq \lambda^*(W \cap A) + \lambda^*(W \cap A^c).$$

This last sum equals $\lambda^*(W)$ because $A \in \mathcal{M}$. Hence, (3.16) holds. We have now established that $\mathcal{M}$ is an algebra of sets.

Next, we show that $\mathcal{M}$ is closed under countable unions. To that end, let $\{E_n\}_{n=1}^{\infty} \subset \mathcal{M}$. We must prove that $\bigcup_{n=1}^{\infty} E_n \in \mathcal{M}$. To begin, we disjointize the sets E_n, $n = 1, 2, \ldots$. Let $A_1 = E_1$, $A_2 = E_2 \setminus E_1$, $A_3 = E_3 \setminus (E_1 \cup E_2)$, and, in general, $A_n = E_n \setminus \big(\bigcup_{k=1}^{n-1} E_k\big)$. Then, see Exercise 3.30, $A_i \cap A_j = \emptyset$, for $i \neq j$, and $\bigcup_{n=1}^{\infty} A_n = \bigcup_{n=1}^{\infty} E_n$. Moreover, because $\mathcal{M}$ is an algebra of sets and $E_n \in \mathcal{M}$ for $n \in \mathcal{N}$, it follows that $A_n \in \mathcal{M}$ for $n \in \mathcal{N}$.

Now, let W be any subset of $\mathcal{R}$ and set $E = \bigcup_{n=1}^{\infty} E_n = \bigcup_{n=1}^{\infty} A_n$. We must show that $\lambda^*(W) \geq \lambda^*(W \cap E) + \lambda^*(W \cap E^c)$. By the subadditivity of λ^*,

$$\lambda^*(W \cap E) = \lambda^*\left(W \cap \left(\bigcup_{n=1}^{\infty} A_n\right)\right)$$
$$= \lambda^*\left(\bigcup_{n=1}^{\infty}(W \cap A_n)\right) \leq \sum_{n=1}^{\infty} \lambda^*(W \cap A_n). \tag{3.17}$$

For each $n \in \mathcal{N}$, set $B_n = \bigcup_{k=1}^{n} A_k$. Then, because $\mathcal{M}$ is an algebra, $B_n \in \mathcal{M}$ for all $n \in \mathcal{N}$. Consequently, for all n,

$$\lambda^*(W) = \lambda^*(W \cap B_n) + \lambda^*(W \cap B_n^c). \tag{3.18}$$

Because $B_n \subset \bigcup_{m=1}^{\infty} A_m = E$, it follows that $E^c \subset B_n^c$. This last fact and (3.18) imply that

$$\lambda^*(W) \geq \lambda^*(W \cap B_n) + \lambda^*(W \cap E^c). \tag{3.19}$$

We will now prove by induction that for all $n \in \mathcal{N}$,

$$\lambda^*(W \cap B_n) = \sum_{k=1}^{n} \lambda^*(W \cap A_k). \tag{3.20}$$

The equation holds trivially when $n = 1$. So, assume that it holds for n. Since $A_{n+1} \in \mathcal{M}$, we have

$$\lambda^*(W \cap B_{n+1}) = \lambda^*\big((W \cap B_{n+1}) \cap A_{n+1}\big) \\ + \lambda^*\big((W \cap B_{n+1}) \cap A_{n+1}^c\big). \tag{3.21}$$

Because the A_ks are pairwise disjoint, we have $W \cap B_{n+1} \cap A_{n+1} = W \cap A_{n+1}$ and $W \cap B_{n+1} \cap A_{n+1}^c = W \cap B_n$. Thus, by (3.21) and the induction hypothesis,

$$\lambda^*(W \cap B_{n+1}) = \lambda^*(W \cap A_{n+1}) + \lambda^*(W \cap B_n)$$
$$= \lambda^*(W \cap A_{n+1}) + \sum_{k=1}^{n} \lambda^*(W \cap A_k) = \sum_{k=1}^{n+1} \lambda^*(W \cap A_k),$$

as required.

Employing (3.19) and (3.20), we conclude that

$$\lambda^*(W) \geq \sum_{k=1}^{n} \lambda^*(W \cap A_k) + \lambda^*(W \cap E^c)$$

for all $n \in \mathcal{N}$ and, consequently,

$$\lambda^*(W) \geq \sum_{n=1}^{\infty} \lambda^*(W \cap A_n) + \lambda^*(W \cap E^c).$$

Applying (3.17) to the previous inequality, we deduce that

$$\lambda^*(W) \geq \lambda^*(W \cap E) + \lambda^*(W \cap E^c).$$

This shows $E \in \mathcal{M}$. We have now established that $\mathcal{M}$ is a σ-algebra.

It remains to prove that $\mathcal{M} \supset \mathcal{B}$. By Proposition 3.3, $\mathcal{M}$ contains all open sets and, as we have just seen, $\mathcal{M}$ is a σ-algebra. Consequently, since $\mathcal{B}$ is the smallest σ-algebra that contains all open sets, it must be that $\mathcal{M} \supset \mathcal{B}$. ∎

Our next theorem shows that Condition (M3) is satisfied when Lebesgue outer measure λ^* is restricted to $\mathcal{M}$. We denote by λ the restriction of Lebesgue outer measure to $\mathcal{M}$; that is, $\lambda \colon \mathcal{M} \to \mathcal{R}$ is defined by $\lambda(E) = \lambda^*(E)$.

□ □ □ **THEOREM 3.12**

If A_1, A_2, ... are in $\mathcal{M}$, with $A_i \cap A_j = \emptyset$ for $i \neq j$, then

$$\lambda\left(\bigcup_n A_n\right) = \sum_n \lambda(A_n).$$

PROOF We first prove that λ is finitely additive on $\mathcal{M}$. So, let A, $B \in \mathcal{M}$ with $A \cap B = \emptyset$. Set $W = A \cup B$. Then $W \cap A = A$ and $W \cap A^c = B$. Consequently, as $A \in \mathcal{M}$, we have by (3.14) that

$$\lambda(A \cup B) = \lambda(W) = \lambda^*(W) = \lambda^*(W \cap A) + \lambda^*(W \cap A^c)$$
$$= \lambda^*(A) + \lambda^*(B) = \lambda(A) + \lambda(B).$$

This shows that λ is finitely additive.

Suppose now that $\{A_n\}_{n=1}^\infty \subset \mathcal{M}$ with $A_i \cap A_j = \emptyset$ for $i \neq j$. Using the fact that λ is finitely additive on $\mathcal{M}$ and the monotonicity of Lebesgue outer measure, we conclude that

$$\sum_{k=1}^m \lambda(A_k) = \lambda\left(\bigcup_{k=1}^m A_k\right) \leq \lambda\left(\bigcup_{n=1}^\infty A_n\right)$$

for all $m \in \mathcal{N}$. Letting $m \to \infty$ gives $\sum_{n=1}^\infty \lambda(A_n) \leq \lambda\left(\bigcup_{n=1}^\infty A_n\right)$. The reverse inequality obtains because of the countable subadditivity of Lebesgue outer measure. ∎

Lebesgue Measurable Sets and Lebesgue Measure

From Proposition 3.2 on page 93 and Theorems 3.11 and 3.12, we see that Conditions (M1)–(M3) are satisfied with $\mu = \lambda$ and $\mathcal{A} = \mathcal{M}$; that is,

(L1) $\mathcal{M}$ is a σ-algebra and $\mathcal{M} \supset \mathcal{B}$.

(L2) $\lambda(I) = \ell(I)$, for all intervals I.

(L3)   If A_1, A_2, ... are in $\mathcal{M}$, with $A_i \cap A_j = \emptyset$ for $i \neq j$, then

$$\lambda\left(\bigcup_n A_n\right) = \sum_n \lambda(A_n).$$

Consequently, *the set function* $\lambda \colon \mathcal{M} \to \mathcal{R}$ *is the required extension of length.* We will employ the following terminology:

DEFINITION 3.9   Lebesgue Measurable Sets and Lebesgue Measure

The members of $\mathcal{M}$ are called **Lebesgue measurable sets.** That is, E is a Lebesgue measurable set if and only if for every subset W of $\mathcal{R}$,

$$\lambda^*(W) = \lambda^*(W \cap E) + \lambda^*(W \cap E^c).$$

The restriction of Lebesgue outer measure to $\mathcal{M}$ is denoted by $\boldsymbol{\lambda}$ and is called **Lebesgue measure.**

In the next three propositions, we establish some additional properties of Lebesgue measure and Lebesgue measurable sets.

□ □ □ **PROPOSITION 3.4**

A subset of $\mathcal{R}$ with Lebesgue outer measure zero is a Lebesgue measurable set; that is, $\lambda^*(E) = 0 \Rightarrow E \in \mathcal{M}$.

PROOF   Suppose that $\lambda^*(E) = 0$. Let W be an arbitrary subset of $\mathcal{R}$. As $W \cap E \subset E$, the monotonicity of Lebesgue outer measure implies that $\lambda^*(W \cap E) \leq \lambda^*(E) = 0$. Using the fact that $W \cap E^c \subset W$, we now conclude that

$$\lambda^*(W) \geq \lambda^*(W \cap E^c) = \lambda^*(W \cap E) + \lambda^*(W \cap E^c).$$

This last inequality shows that $E \in \mathcal{M}$. ■

□ □ □ **PROPOSITION 3.5**

Every countable subset of $\mathcal{R}$ has Lebesgue measure zero.

PROOF   Let $E \subset \mathcal{R}$ be countable, say $E = \{x_n\}_{n=1}^{\infty}$. Then we can write $E = \bigcup_{n=1}^{\infty} \{x_n\}$. Note that if $a \in \mathcal{R}$, then, by (L2), $\lambda(\{a\}) = \lambda([a,a]) = a - a = 0$. Therefore, applying (L3), we conclude that

$$\lambda(E) = \lambda\left(\bigcup_{n=1}^{\infty} \{x_n\} \right) = \sum_{n=1}^{\infty} \lambda(\{x_n\}) = 0,$$

as required. ■

The next proposition shows that the converse of Proposition 3.5 does not hold.

□ □ □ **PROPOSITION 3.6**

The Cantor set P has Lebesgue measure zero.

PROOF   Let $G = [0,1] \setminus P$. From Chapter 2 (page 62), we know that G can be written as a countable union of disjoint open intervals $\{I_n\}_{n=1}^{\infty}$ with the property that $\sum_{n=1}^{\infty} \ell(I_n) = 1$. Hence, by (L3) and (L2),

$$\lambda(G) = \sum_{n=1}^{\infty} \lambda(I_n) = \sum_{n=1}^{\infty} \ell(I_n) = 1.$$

Clearly, P and G are disjoint and $P \cup G = [0,1]$. Therefore,

$$1 = \lambda(P) + \lambda(G) = \lambda(P) + 1,$$

which shows that $\lambda(P) = 0$. ■

Another useful result is the following.

□ □ □ **THEOREM 3.13**

If $\{E_n\}_{n=1}^{\infty}$ is a sequence of Lebesgue measurable sets with $E_1 \subset E_2 \subset \cdots$, then

$$\lambda\left(\bigcup_{n=1}^{\infty} E_n \right) = \lim_{n \to \infty} \lambda(E_n).$$

PROOF If $\lambda(E_n) = \infty$ for some n, then both sides of the previous equation equal ∞. So, assume $\lambda(E_n) < \infty$ for all n.

We begin by disjointizing the E_ns as follows. Let $A_1 = E_1$ and, for $n \geq 2$, let $A_n = E_n \setminus E_{n-1}$. Then it is easy to see that $\{A_n\}_{n=1}^{\infty} \subset \mathcal{M}$, $A_i \cap A_j = \emptyset$ for $i \neq j$, and $\bigcup_{n=1}^{\infty} A_n = \bigcup_{n=1}^{\infty} E_n$. Therefore, by countable additivity,

$$\lambda\left(\bigcup_{n=1}^{\infty} E_n\right) = \lambda\left(\bigcup_{n=1}^{\infty} A_n\right) = \sum_{n=1}^{\infty} \lambda(A_n).$$

As $E_{n-1} \subset E_n$, we have $\lambda(A_n) = \lambda(E_n \setminus E_{n-1}) = \lambda(E_n) - \lambda(E_{n-1})$ for $n \geq 2$. Consequently,

$$\lambda\left(\bigcup_{n=1}^{\infty} E_n\right) = \sum_{n=1}^{\infty} \lambda(A_n) = \lim_{n\to\infty} \sum_{k=1}^{n} \lambda(A_k)$$

$$= \lim_{n\to\infty}\left(\lambda(E_1) + \sum_{k=2}^{n}[\lambda(E_k) - \lambda(E_{k-1})]\right) = \lim_{n\to\infty} \lambda(E_n),$$

as required. ∎

The Relation Between $\mathcal{B}$ and $\mathcal{M}$

We close this section by discussing the relationship between the collection of Borel sets $\mathcal{B}$ and the collection of Lebesgue measurable sets $\mathcal{M}$. By Theorem 3.11, $\mathcal{B} \subset \mathcal{M}$. The question now is: Does $\mathcal{B} = \mathcal{M}$? In other words, is every Lebesgue measurable set a Borel set or are there Lebesgue measurable sets that are not Borel sets?

It is not easy to answer that question. In fact, Lebesgue and Borel argued the question without finding the answer. It turns out that the answer to the question is no—there are Lebesgue measurable sets that are not Borel sets. In other words, we have the following theorem:

□ □ □ **THEOREM 3.14**

The σ-algebra of Borel sets $\mathcal{B}$ is a proper subcollection of the σ-algebra of Lebesgue measurable sets $\mathcal{M}$.

PROOF See Exercise 3.50. ∎

Exercises for Section 3.4

3.30 Let $\{E_n\}_{n=1}^{\infty}$ be a sequence of subsets of $\mathcal{R}$. Define $A_1 = E_1$ and $A_n = E_n \setminus \left(\bigcup_{k=1}^{n-1} E_k\right)$ for $n \geq 2$. Prove that $A_i \cap A_j = \emptyset$, for $i \neq j$, and that $\bigcup_{n=1}^{\infty} A_n = \bigcup_{n=1}^{\infty} E_n$.

3.31 In Chapter 2, we introduced the concept of measure zero. Prove that this concept is equivalent to that of Lebesgue measure zero. In other words, show that a subset $E \subset \mathcal{R}$ has measure zero in the sense of Definition 2.19 on page 70 if and only if $\lambda(E) = 0$.

★**3.32** Verify that if $A \in \mathcal{M}$, $\lambda(A) = 0$, and $B \subset A$, then $B \in \mathcal{M}$ and $\lambda(B) = 0$.

3.33 Let $A, B \in \mathcal{M}$ with $A \subset B$ and $\lambda(A) < \infty$. Show that $\lambda(B \setminus A) = \lambda(B) - \lambda(A)$.

3.34 Use properties of Lebesgue measure to supply a simple proof that any (nondegenerate) interval of $\mathcal{R}$ is uncountable.

3.35 Suppose that $\{E_n\}_{n=1}^{\infty} \subset \mathcal{M}$ and that $E_1 \supset E_2 \supset \cdots$. Also suppose that $\lambda(E_1) < \infty$. Prove that

$$\lambda\left(\bigcap_{n=1}^{\infty} E_n\right) = \lim_{n \to \infty} \lambda(E_n).$$

Can the assumption that $\lambda(E_1) < \infty$ be dropped? Why?

3.36 Show that if $A, B \in \mathcal{M}$ and $\lambda(A \cap B) < \infty$, then

$$\lambda(A \cup B) = \lambda(A) + \lambda(B) - \lambda(A \cap B).$$

3.37 Suppose $\lambda^*(A) = 0$.
a) Show that for any set B, $\lambda^*(A \cup B) = \lambda^*(B)$.
b) Show that if $A \cup B \in \mathcal{M}$, then $B \in \mathcal{M}$.

3.38 Find a sequence of pairwise disjoint sets $\{A_n\}_{n=1}^{\infty}$ such that strict inequality holds in the relation

$$\lambda^*\left(\bigcup_{n=1}^{\infty} A_n\right) \leq \sum_{n=1}^{\infty} \lambda^*(A_n).$$

Hint: Is $\{A_n\}_{n=1}^{\infty} \subset \mathcal{M}$ possible?

★**3.39** If $0 < \alpha < 1$, construct a set P_α in a manner similar to that in which the Cantor set is constructed, except that at the nth step remove open intervals of length $\alpha/3^n$ instead of $1/3^n$. Show that P_α is closed and that $\lambda(P_\alpha) = 1 - \alpha > 0$.

3.40 Prove that there is a sequence of continuous functions $\{f_n\}_{n=1}^{\infty}$ on $[0, 1]$ that converge pointwise to a function $f \notin R([0, 1])$. *Hint:* Use Exercise 3.39.

3.41 Prove that there is a Riemann integrable function on $[0, 1]$ that is not a Borel measurable function. *Hint:* The proof of Theorem 3.14, which is carried out in Exercise 3.50, shows there is a subset of the Cantor set that is not a Borel set.

3.42 Suppose that $E \in \mathcal{M}$. Show that for each $\epsilon > 0$, there is an open set O with $O \supset E$ and $\lambda(O \setminus E) < \epsilon$. *Hint:* First consider the case where $\lambda(E) < \infty$ and use the definition of Lebesgue outer measure.

★**3.43** Suppose that $E \in \mathcal{M}$. Show that for each $\epsilon > 0$, there is a closed set F with $F \subset E$ and $\lambda(E \setminus F) < \epsilon$.

★**3.44** A set is called a G_δ-**set** if it is the intersection of a countable number of open sets; and a set is called an F_σ-**set** if it is the union of a countable number of closed sets. Note that G_δ-sets and F_σ-sets are Borel sets. Now suppose that $E \in \mathcal{M}$.
a) Establish that there is a G_δ-set G and an F_σ-set F such that $F \subset E \subset G$ and $\lambda(E \setminus F) = \lambda(G \setminus E) = 0$.
b) Referring to part (a), deduce that $\lambda(F) = \lambda(E) = \lambda(G)$.

3.45 Let $E \in \mathcal{M}$. Prove that $\lambda(E) = \inf\{\lambda(O) : O \supset E, O \text{ open}\}$.

3.46 Let $E \in \mathcal{M}$. Prove that $\lambda(E) = \sup\{\lambda(K) : K \subset E, K \text{ compact}\}$.

3.47 Let $E \subset \mathcal{R}$.
a) Suppose that there is a Borel set B such that $B \subset E$ and $\lambda^*(E \setminus B) = 0$. Show that $E \in \mathcal{M}$.
b) Suppose that $\lambda^*(E) < \infty$ and that

$$\lambda^*(E) = \sup\{\lambda(F) : F \subset E, F \text{ closed}\} = \inf\{\lambda(O) : O \supset E, O \text{ open}\}.$$

Show that $E \in \mathcal{M}$.

3.48　Suppose that $\{E_n\}_{n=1}^{\infty}$ is a sequence of pairwise disjoint Lebesgue measurable sets. Prove that

$$\lambda^*\left(A\cap\left(\bigcup_{n=1}^{\infty}E_n\right)\right)=\sum_{n=1}^{\infty}\lambda^*(A\cap E_n)$$

for all subsets A of $\mathcal{R}$.

3.49　Suppose that $E\in\mathcal{M}$ and that $\lambda(E)<\infty$. Show that for each $\epsilon>0$, there are a finite number of pairwise disjoint intervals $I_1,I_2,\ldots,I_n$ such that

$$\lambda\left(E\bigtriangleup\left(\bigcup_{k=1}^{n}I_k\right)\right)<\epsilon.$$

★3.50　Prove Theorem 3.14 on page 107. Proceed by establishing each of the following facts:
  a) If $C\in\mathcal{M}$ and $x\in\mathcal{R}$, then $C+x\in\mathcal{M}$ and $\lambda(C+x)=\lambda(C)$.
  b) Let S be the set defined in Lemma 3.12 on page 100. If $C\in\mathcal{M}$ and $C\subset S$, then $\lambda(C)=0$. *Hint:* Consider $\{C+r:r\in(-1,1)\cap Q\}$.
  c) If $D\subset\mathcal{R}$ and $\lambda^*(D)>0$, then there is a nonmeasurable subset of D. *Hint:* For each $r\in Q$, let $D_r=D\cap(S+r)$. Use parts (a) and (b) to show that, if $D_r\in\mathcal{M}$, then $\lambda(D_r)=0$.
  d) Define $f:[0,1]\to\mathcal{R}$ by $f(x)=x+\psi(x)$, where ψ denotes the Cantor function (see page 64). Then f is a strictly increasing function and maps $[0,1]$ onto $[0,2]$.
  e) The function $g=f^{-1}$ is continuous and, hence, Borel measurable.
  f) f maps the Cantor set onto a set A with $\lambda(A)=1$.
  g) Let $E\subset A$ with $E\notin\mathcal{M}$. [Such an E exists by parts (f) and (c).] Then $f^{-1}(E)\in\mathcal{M}$ but $f^{-1}(E)\notin\mathcal{B}$.

3.51　Prove that the set S defined in Lemma 3.12 on page 100 is not a Lebesgue measurable set.

Henri Léon Lebesgue
(1875–1941)

Henri Lebesgue was born at Beauvais, France, on June 28, 1875. His father died of tuberculosis shortly after his birth, as did his two sisters. Lebesgue himself also contracted the disease and, as a consequence, his health remained fragile throughout the remainder of his life.

Lebesgue's mother was an untiring worker who gladly supported the education of her son. Lebesgue was a brilliant student in primary school. Later, from 1894 through 1897, he attended the École Normale Supérieure in Paris, where he was a student of Émile Borel (see the Chapter 3 biography). Lebesgue worked on his doctoral dissertation—titled *Intégrale, longueur, aire*—between 1899 and 1902 while teaching mathematical science at the lycée in Nancy, and received his doctorate from the Sorbonne in 1902.

Lebesgue's interest in Riemannian integration and its associated problems led to his creation of the Lebesgue integral in 1902. Not only has the Lebesgue integral been important to the amplification of the theory of trigonometric series, curve rectification, and calculus, but it has also proved central to the development of measure theory.

Lebesgue did research in many different areas of mathematics, among which were function theory, set theory, the calculus of variation, complex analysis, and topology. His and Émile Borel's work provided the foundation for the modern theory of functions of a real variable.

Many honors were bestowed upon Lebesgue. Among these were the Prix Houllevique in 1912, the Prix Poncelet in 1914, the Prix Saintour in 1917, and the Prix Petit d'Ormoy in 1919. Lebesgue was also awarded honorary doctorates from several universities. Additionally, he was elected to the French Academy of Sciences in 1922 and to the Royal Society in 1934.

Lebesgue taught at the University of Rennes from 1902–1906; at the University of Poitiers from 1906–1910; at the Sorbonne from 1910–1921; and, finally, at the Collége de France. He died in Paris on July 26, 1941.

The Lebesgue Integral on the Real Line

In Chapter 3, we took the first steps for obtaining a more general and sophisticated framework for analysis than that provided by open sets, continuous functions, and the Riemann integral. We expanded the collection of continuous functions to the collection of Borel measurable functions, the smallest algebra of functions that contains the continuous functions and is closed under pointwise limits. In doing so, we were led to consider the collection of Borel sets, the smallest σ algebra of subsets of $\mathcal{R}$ that contains the open sets.

With the ultimate goal of extending the Riemann integral to an integral that applies to all Borel measurable functions, we then generalized the concept of length so that it applies to all Borel sets. We, in fact, generalized that concept to all Lebesgue measurable sets. Consequently, we will be able to extend the Riemann integral to an integral that applies to a much larger collection of functions than the Borel measurable functions. We call that larger collection of functions the *Lebesgue measurable functions.*

4.1 THE LEBESGUE INTEGRAL FOR NONNEGATIVE FUNCTIONS

There are two ways that we can approach the definition of Lebesgue measurable functions. Here is the first approach: Taking our cue from the development of the Riemann integral, we begin by defining the integral of a function of the form $s = \sum a_k \chi_{E_k}$. If the E_ks are intervals, then s is a step function and we simply define the integral to equal the Riemann integral $\sum a_k \ell(E_k)$. But now that we have generalized the concept of length, we can do much better. Provided only that the E_ks are Lebesgue measurable sets, we define the integral to be $\sum a_k \lambda(E_k)$.

In particular, we see that every function of the form χ_E, where $E \in \mathcal{M}$, should be a Lebesgue measurable function; that is, should be integrable in the extended sense. Since we want the collection of all Lebesgue measurable functions to constitute an algebra and be closed under pointwise limits, we make the following definition.

DEFINITION 4.1 **Lebesgue Measurable Functions**

We denote by $\mathcal{L}$ the smallest algebra of real-valued functions on $\mathcal{R}$ that contains all functions of the form χ_E, where $E \in \mathcal{M}$, and is closed under pointwise limits. The members of $\mathcal{L}$ are called **Lebesgue measurable functions.**

Our second approach to obtain the definition of Lebesgue measurable functions is by analogy with a characterization of Borel measurable functions. Specifically, as we know from Theorem 3.3 (page 86), a function f is Borel measurable if and only if the inverse image of each open set under f is a Borel set; that is, if and only if $f^{-1}(O) \in \mathcal{B}$ for all open sets O. This leads to the following definition of Lebesgue measurable functions:

DEFINITION 4.2 **Lebesgue Measurable Function**

A real-valued function f on $\mathcal{R}$ is said to be a **Lebesgue measurable function** if the inverse image of each open set under f is a Lebesgue measurable set; that is, if $f^{-1}(O) \in \mathcal{M}$ for all open sets O.

Note: For brevity, we will often indicate that a function is a Lebesgue measurable function by saying that it is an **$\mathcal{M}$-measurable function.**

It really doesn't matter whether we use Definition 4.1 or Definition 4.2 because the two definitions are equivalent (see Exercise 4.2). But to be specific, we will take Definition 4.2 as our definition of Lebesgue measurable functions.

Our next proposition, whose proof is similar to that of Lemma 3.5 on page 85 and is left to the reader, provides some useful equivalent conditions for a function to be Lebesgue measurable.

□ □ □ **PROPOSITION 4.1**

Let f be a real-valued function on $\mathcal{R}$. Then the following statements are equivalent:

a) f *is $\mathcal{M}$-measurable.*
b) For each $a \in \mathcal{R}$, $f^{-1}\big((-\infty, a)\big) \in \mathcal{M}$.
c) For each $a \in \mathcal{R}$, $f^{-1}\big((a, \infty)\big) \in \mathcal{M}$.
d) For each $a \in \mathcal{R}$, $f^{-1}\big((-\infty, a]\big) \in \mathcal{M}$.
e) For each $a \in \mathcal{R}$, $f^{-1}\big([a, \infty)\big) \in \mathcal{M}$.

Several important properties of Lebesgue measurable functions are given in the next two theorems. We postpone the proofs of those theorems until Chapter 5, where more general results will be established.

□ □ □ **THEOREM 4.1**

The collection of Lebesgue measurable functions forms an algebra. That is, if f and g are $\mathcal{M}$-measurable and $\alpha \in \mathcal{R}$, then
a) $f + g$ is $\mathcal{M}$-measurable.
b) $f \cdot g$ is $\mathcal{M}$-measurable.
c) αf is $\mathcal{M}$-measurable.

□ □ □ **THEOREM 4.2**

Suppose that f and g are $\mathcal{M}$-measurable functions and that $\{f_n\}_{n=1}^{\infty}$ is a sequence of $\mathcal{M}$-measurable functions that converges pointwise to a real-valued function. Then
a) $f \vee g$ is $\mathcal{M}$-measurable.
b) $f \wedge g$ is $\mathcal{M}$-measurable.
c) $\lim_{n \to \infty} f_n$ is $\mathcal{M}$-measurable.

The Lebesgue Integral of a Nonnegative Simple Function

We now begin our extension of the Riemann integral to an integral that applies to all Lebesgue measurable functions. First we introduce a special type of Lebesgue measurable function that generalizes the notion of step functions.

DEFINITION 4.3 **Simple Function and Canonical Representation**

An $\mathcal{M}$-measurable function s is said to be a **simple function** if it takes on only finitely many values; that is, if its range is a finite set. Let $a_1, a_2, \ldots, a_n$ denote the distinct nonzero values of s and, for $1 \le k \le n$, set $A_k = \{ x : s(x) = a_k \}$. Then we can write

$$s = \sum_{k=1}^{n} a_k \chi_{A_k}. \tag{4.1}$$

This is called the **canonical representation** of s.

It is easy to see that every step function is a simple function, but not every simple function is a step function. Also, we leave it as an exercise for the reader to show that the sets $A_1, A_2, \ldots, A_n$, appearing in the canonical representation of a simple function, are Lebesgue measurable and pairwise disjoint.

EXAMPLE 4.1 *Illustrates Definition 4.3*

The function $s = 3\chi_{(0,2)} + 2\chi_{(1,3]} - 6\chi_{\mathcal{N}}$ is a simple function. However, the given representation in not canonical. In fact, the canonical representation of s is

$$s = 3\chi_{(0,1)} - 3\chi_{\{1\}} + 5\chi_{(1,2)} - 4\chi_{\{2,3\}} + 2\chi_{(2,3)} - 6\chi_{\mathcal{N}\setminus\{1,2,3\}},$$

as is easily verified.

□

Here is the definition of the Lebesgue integral for a nonnegative simple function. As already noted, this definition is a natural generalization of the Riemann integral of a step function.

DEFINITION 4.4   Lebesgue Integral of a Nonnegative Simple Function

Let s be a nonnegative simple function with canonical representation given by $s = \sum_{k=1}^{n} a_k \chi_{A_k}$. Then the **Lebesgue integral of s over $\mathcal{R}$** is defined by

$$\int_{\mathcal{R}} s(x)\, d\lambda(x) = \sum_{k=1}^{n} a_k \lambda(A_k).$$

If $E \in \mathcal{M}$, then the **Lebesgue integral of s over E** is defined by

$$\int_{E} s(x)\, d\lambda(x) = \int_{\mathcal{R}} \chi_E(x) s(x)\, d\lambda(x).$$

The next proposition shows how we can obtain the Lebesgue integral of a nonnegative simple function from a possibly noncanonical representation.

□ □ □   PROPOSITION 4.2

Let s be a nonnegative simple function that can be expressed in the form $s = \sum_{k=1}^{m} b_k \chi_{B_k}$, where this representation is not necessarily canonical but where $B_k \in \mathcal{M}$ for $1 \le k \le m$ and $B_i \cap B_j = \emptyset$ for $i \ne j$. Then

$$\int_{\mathcal{R}} s(x)\, d\lambda(x) = \sum_{k=1}^{m} b_k \lambda(B_k). \qquad (4.2)$$

More generally, we have

$$\int_{E} s(x)\, d\lambda(x) = \sum_{k=1}^{m} b_k \lambda(B_k \cap E) \qquad (4.3)$$

for each $E \in \mathcal{M}$.

PROOF   Let $s = \sum_{i=1}^{n} a_i \chi_{A_i}$ be the canonical representation of s. Also, set $a_0 = 0$ and $A_0 = \{\, x : s(x) = 0 \,\}$. Because the B_ks are pairwise disjoint, we know that for each $k = 1, 2, \ldots, m$, there is an i $(0 \le i \le n)$ such that $b_k = a_i$. Let $D_i = \{\, k : b_k = a_i \,\}$. Then the D_is are pairwise disjoint, $\bigcup_{i=0}^{n} D_i = \{1, 2, \ldots, m\}$, and $A_i = \bigcup_{k \in D_i} B_k$ for $1 \le i \le n$. Consequently,

$$\int_{\mathcal{R}} s(x)\, d\lambda(x) = \sum_{i=1}^{n} a_i \lambda(A_i) = \sum_{i=1}^{n} a_i \sum_{k \in D_i} \lambda(B_k)$$

$$= \sum_{i=0}^{n} \sum_{k \in D_i} a_i \lambda(B_k) = \sum_{i=0}^{n} \sum_{k \in D_i} b_k \lambda(B_k) = \sum_{k=1}^{m} b_k \lambda(B_k).$$

Thus, (4.2) holds.

To establish (4.3), we first observe that

$$\chi_E s = \sum_{k=1}^{m} b_k \chi_E \chi_{B_k} = \sum_{k=1}^{m} b_k \chi_{B_k \cap E}.$$

Applying (4.2), we now conclude that

$$\int_E s(x)\,d\lambda(x) = \int_{\mathcal{R}} \chi_E(x)s(x)\,d\lambda(x)$$

$$= \int_{\mathcal{R}} (\chi_E s)(x)\,d\lambda(x) = \sum_{k=1}^{m} b_k \lambda(B_k \cap E).$$

This completes the proof of Proposition 4.2. ∎

We should point out that Definition 4.4 really does provide a generalization of the Riemann integral of a step function; that is, the Lebesgue integral of a step function equals its Riemann integral. Indeed, suppose that g is a step function on $[a, b]$, say $g = \sum_{k=1}^{n} a_k \chi_{I_k}$, where the I_ks are pairwise disjoint subintervals of $[a, b]$. Then, by Proposition 4.2,

$$\int_a^b g(x)\,dx = \sum_{k=1}^{n} a_k \ell(I_k) = \sum_{k=1}^{n} a_k \lambda(I_k)$$

$$= \sum_{k=1}^{n} a_k \lambda(I_k \cap [a, b]) = \int_{[a,b]} g(x)\,d\lambda(x).$$

A technicality: We have defined the Lebesgue integral only for functions whose domain is all of $\mathcal{R}$; but, the domain of the step function g is only $[a, b]$. To remedy this difficulty, define $g(x) = 0$ for $x \in [a, b]^c$.

EXAMPLE 4.2 *Illustrates the Integral of a Nonnegative Simple Function*

a) Let $s = 3\chi_{(-2,-1]} + 4\chi_{(-1,1)} + 8\chi_{\mathcal{N}}$. Then

$$\int_{\mathcal{R}} s(x)\,d\lambda(x) = 3\lambda\big((-2, -1]\big) + 4\lambda\big((-1, 1)\big) + 8\lambda(\mathcal{N})$$

$$= 3 \cdot 1 + 4 \cdot 2 + 8 \cdot 0 = 11.$$

b) Let $s = \chi_{Q^c} = 1\chi_{Q^c}$. Then, by Proposition 4.2,

$$\int_{[0,2]} s(x)\,d\lambda(x) = 1\lambda(Q^c \cap [0, 2]) = 2.$$

c) Let $f(x) \equiv 1$; that is, $f = \chi_{\mathcal{R}}$. Then $\int_{\mathcal{R}} f(x)\,d\lambda(x) = \lambda(\mathcal{R}) = \infty$. Thus, the Lebesgue integral of a nonnegative simple function can be ∞. $\quad\square$

◻ ◻ ◻ **LEMMA 4.1**

Suppose that s and t are nonnegative simple functions and that α, $\beta \geq 0$. Then $\alpha s + \beta t$ is a nonnegative simple function and

$$\int_E [\alpha s(x) + \beta t(x)]\, d\lambda(x) = \alpha \int_E s(x)\, d\lambda(x) + \beta \int_E t(x)\, d\lambda(x)$$

for each $E \in \mathcal{M}$.

PROOF Let $s = \sum_{k=1}^n a_k \chi_{A_k}$ and $t = \sum_{j=1}^m b_j \chi_{B_j}$ be the canonical representations of s and t, respectively. Moreover, let $a_0 = 0$, $A_0 = \{x : s(x) = 0\}$, $b_0 = 0$, and $B_0 = \{x : t(x) = 0\}$. For each k $(0 \leq k \leq n)$ and for each j $(0 \leq j \leq m)$, let $C_{kj} = A_k \cap B_j$. Then the C_{kj}s are pairwise disjoint, $s = \sum_{k=0}^n \sum_{j=0}^m a_k \chi_{C_{kj}}$, and $t = \sum_{k=0}^n \sum_{j=0}^m b_j \chi_{C_{kj}}$. Hence,

$$\alpha s + \beta t = \sum_{k=0}^n \sum_{j=0}^m (\alpha a_k + \beta b_j)\chi_{C_{kj}}.$$

This last equation shows that $\alpha s + \beta t$ is a nonnegative simple function. Moreover, by applying Proposition 4.2, we can deduce that

$$\int_E [\alpha s(x) + \beta t(x)]\, d\lambda(x) = \int_E (\alpha s + \beta t)(x)\, d\lambda(x)$$

$$= \sum_{k=0}^n \sum_{j=0}^m (\alpha a_k + \beta b_j)\lambda(C_{kj} \cap E)$$

$$= \alpha \sum_{k=0}^n \sum_{j=0}^m a_k \lambda(C_{kj} \cap E) + \beta \sum_{k=0}^n \sum_{j=0}^m b_j \lambda(C_{kj} \cap E)$$

$$= \alpha \int_E s(x)\, d\lambda(x) + \beta \int_E t(x)\, d\lambda(x),$$

as required. ∎

The Lebesgue Integral of a Nonnegative $\mathcal{M}$-measurable Function

Next we will define the Lebesgue integral of a nonnegative Lebesgue measurable function. Before doing so, however, it is useful for motivational purposes to prove the following proposition:

◻ ◻ ◻ **PROPOSITION 4.3**

a) *Let f be a nonnegative $\mathcal{M}$-measurable function. Then there is a nondecreasing sequence of nonnegative simple functions that converges pointwise to f. In other words, there is a sequence $\{s_n\}_{n=1}^\infty$ of nonnegative simple functions such that, for all $x \in \mathcal{R}$, $s_1(x) \leq s_2(x) \leq \cdots$ and $\lim_{n \to \infty} s_n(x) = f(x)$.*

b) *If $\{s_n\}_{n=1}^\infty$ is a sequence of nonnegative simple functions that converges pointwise to a real-valued function f, then f is a nonnegative $\mathcal{M}$-measurable function.*

PROOF   a) For each $n \in \mathcal{N}$, set

$$E_{nm} = \left\{ x : \frac{m-1}{2^n} \leq f(x) < \frac{m}{2^n} \right\},$$

for $m = 1, 2, \ldots, n2^n$, and $E_n = \{ x : f(x) \geq n \}$. As f is Lebesgue measurable, the sets E_{nm}, E_n are Lebesgue measurable sets. Let

$$s_n = \sum_{m=1}^{n2^n} \frac{m-1}{2^n} \chi_{E_{nm}} + n \chi_{E_n}.$$

Then $\{s_n\}_{n=1}^{\infty}$ is a sequence of nonnegative simple functions and, clearly, $s_n \leq f$ for all $n \in \mathcal{N}$. Also, by construction, $\left| f(x) - s_n(x) \right| < 2^{-n}$ as soon as n is large enough so that $f(x) < n$. Thus, $s_n \to f$ pointwise.

Next we show that $s_n \leq s_{n+1}$ for all $n \in \mathcal{N}$. Let $x \in \mathcal{R}$. If $x \in E_{nm}$ for some $m = 1, 2, \ldots, n2^n$, then $(m-1)2^{-n} \leq f(x) < m2^{-n}$. Therefore, either

$$\frac{m-1}{2^n} \leq f(x) < \frac{2m-1}{2^{n+1}} \qquad \text{or} \qquad \frac{2m-1}{2^{n+1}} \leq f(x) < \frac{m}{2^n}.$$

In the former case, $s_n(x) = s_{n+1}(x) = (m-1)2^{-n}$ and, in the latter case, $s_n(x) = (m-1)2^{-n} < (2m-1)2^{-(n+1)} = s_{n+1}(x)$. Consequently, in either case, $s_n(x) \leq s_{n+1}(x)$. If $x \in E_n$, then $f(x) \geq n = (n2^{n+1})/2^{n+1}$. This implies that $s_{n+1}(x) \geq (n2^{n+1})/2^{n+1} = n = s_n(x)$. This completes the proof of part (a).

b) This part follows immediately from Theorem 4.2(c) on page 113.   ■

Proposition 4.3 shows that the functions that can be approximated by nonnegative simple functions are precisely the nonnegative Lebesgue measurable functions. With that proposition in mind, we now define the Lebesgue integrable of an arbitrary nonnegative $\mathcal{M}$-measurable function.

DEFINITION 4.5   Lebesgue Integral of a Nonnegative Function

Let f be a nonnegative $\mathcal{M}$-measurable function. Then the **Lebesgue integral of f over $\mathcal{R}$** is defined by

$$\int_{\mathcal{R}} f(x) \, d\lambda(x) = \sup_s \int_{\mathcal{R}} s(x) \, d\lambda(x), \tag{4.4}$$

where the supremum is taken over all nonnegative simple functions that are dominated by f. If $E \in \mathcal{M}$, then the **Lebesgue integral of f over E** is defined by

$$\int_E f(x) \, d\lambda(x) = \int_{\mathcal{R}} \chi_E(x) f(x) \, d\lambda(x). \tag{4.5}$$

Note that (4.4) makes sense for any nonnegative function, Lebesgue measurable or not. Thus, we might ask: Why define the Lebesgue integral only for nonnegative Lebesgue measurable functions; why not define it for any nonnegative function? The reason lies in the previous proposition, Proposition 4.3. For, if f is not Lebesgue measurable, then it cannot be approximated by a sequence of nonnegative simple functions. Hence, the quantity on the right-hand side of (4.4) will generally not reflect the behavior of f.

We mention that there are several widely used notations for the Lebesgue integral:

$$\int_E f(x)\, d\lambda(x), \quad \int_E f(x)\, dx, \quad \int_E f(x)\lambda(dx), \quad \text{and} \quad \int_E f\, d\lambda$$

all denote the Lebesgue integral of f over E. By the way, we will refer to the Lebesgue integral simply as "the integral" when there is no possibility of confusion.

◻ ◻ ◻ **PROPOSITION 4.4**

Let f and g be nonnegative Lebesgue measurable functions, $\alpha \geq 0$, and $E \in \mathcal{M}$. Then

a) *$f \leq g \Rightarrow \int_E f\, d\lambda \leq \int_E g\, d\lambda$.*
b) *$A \subset E$ and $A \in \mathcal{M} \Rightarrow \int_A f\, d\lambda \leq \int_E f\, d\lambda$.*
c) *$f(x) = 0$ for all $x \in E \Rightarrow \int_E f\, d\lambda = 0$.*
d) *$\lambda(E) = 0 \Rightarrow \int_E f\, d\lambda = 0$.*
e) *$\int_E \alpha f\, d\lambda = \alpha \int_E f\, d\lambda$.*

PROOF a) Suppose that s is a nonnegative simple function that is dominated by $\chi_E f$. As $f \leq g$, it follows that s is also dominated by $\chi_E g$. Therefore,

$$\int_{\mathcal{R}} \chi_E f\, d\lambda = \sup_{\substack{0 \leq s \leq \chi_E f \\ s \text{ simple}}} \int_{\mathcal{R}} s\, d\lambda \leq \sup_{\substack{0 \leq s \leq \chi_E g \\ s \text{ simple}}} \int_{\mathcal{R}} s\, d\lambda = \int_{\mathcal{R}} \chi_E g\, d\lambda,$$

as required.

b) If $A \subset E$, then $\chi_A f \leq \chi_E f$. Thus, by part (a),

$$\int_A f\, d\lambda = \int_{\mathcal{R}} \chi_A f\, d\lambda \leq \int_{\mathcal{R}} \chi_E f\, d\lambda = \int_E f\, d\lambda,$$

as required.

c) If $f(x) = 0$ for all $x \in E$, then $\chi_E f \equiv 0$. Thus, the only nonnegative simple function dominated by $\chi_E f$ is 0. Consequently, $\int_{\mathcal{R}} \chi_E f\, d\lambda = 0$; that is, $\int_E f\, d\lambda = 0$.

d) The function $\chi_E f$ is zero off of E. Therefore, any nonnegative simple function s that is dominated by $\chi_E f$ must also be zero off of E. In other words, $s(x) = 0$ for $x \in E^c$. Since $\lambda(E) = 0$, we have $\lambda(A) = 0$ for all subsets A of E. It now follows that if s is a nonnegative simple function that is dominated by $\chi_E f$, then $\int_{\mathcal{R}} s\, d\lambda = 0$. Hence,

$$\int_E f\, d\lambda = \sup_{\substack{0 \leq s \leq \chi_E f \\ s \text{ simple}}} \int_{\mathcal{R}} s\, d\lambda = 0.$$

e) If $\alpha = 0$, then there is nothing to prove. So, assume that $\alpha > 0$. Clearly, the required result holds for simple functions. Now, let s be a nonnegative simple function that is dominated by $\chi_E \cdot (\alpha f)$. Then we have $0 \leq \alpha^{-1} s \leq \chi_E f$ and, hence, by part (a),

$$\alpha^{-1} \int_{\mathcal{R}} s \, d\lambda = \int_{\mathcal{R}} \alpha^{-1} s \, d\lambda \leq \int_{\mathcal{R}} \chi_E f \, d\lambda = \int_E f \, d\lambda.$$

Thus, $\int_{\mathcal{R}} s \, d\lambda \leq \alpha \int_E f \, d\lambda$ for each nonnegative simple function s that is dominated by $\chi_E \cdot (\alpha f)$. This last fact implies that

$$\int_E \alpha f \, d\lambda = \sup_{\substack{0 \leq s \leq \chi_E \cdot (\alpha f) \\ s \text{ simple}}} \int_{\mathcal{R}} s \, d\lambda \leq \alpha \int_E f \, d\lambda.$$

On the other hand, let s be a nonnegative simple function that is dominated by $\chi_E f$. Then we have $0 \leq \alpha s \leq \chi_E \cdot (\alpha f)$. Therefore, by part (a),

$$\alpha \int_{\mathcal{R}} s \, d\lambda = \int_{\mathcal{R}} \alpha s \, d\lambda \leq \int_{\mathcal{R}} \chi_E \cdot (\alpha f) \, d\lambda = \int_E \alpha f \, d\lambda.$$

Thus, $\alpha \int_{\mathcal{R}} s \, d\lambda \leq \int_E \alpha f \, d\lambda$ for each nonnegative simple function s that is dominated by $\chi_E f$. Consequently,

$$\alpha \int_E f \, d\lambda = \alpha \cdot \sup_{\substack{0 \leq s \leq \chi_E f \\ s \text{ simple}}} \int_{\mathcal{R}} s \, d\lambda \leq \int_E \alpha f \, d\lambda.$$

This completes the proof of part (e). ∎

Exercises for Section 4.1

Note: A ★ denotes an exercise that will be subsequently referenced.

4.1 Let $\mathcal{G} = \{ \chi_E : E \in \mathcal{M} \}$. Show that $\mathcal{G}$ is closed under pointwise limits.

4.2 Prove that Definitions 4.1 and 4.2 (page 112) are equivalent by proceeding as follows: Let

$$\mathcal{F} = \{ f : f^{-1}(O) \in \mathcal{M} \text{ for all open sets } O \}.$$

We must prove that $\mathcal{F} = \mathcal{L}$.
a) Why do we know that $\mathcal{F}$ is an algebra of functions and is closed under pointwise limits? *Hint:* See Theorems 4.1 and 4.2 on page 113.
b) Show that if $E \in \mathcal{M}$, then $\chi_E \in \mathcal{F}$.
c) Deduce from parts (a) and (b) that $\mathcal{F} \supset \mathcal{L}$.
d) Show that $\mathcal{F} \subset \mathcal{L}$ by using a suitable modification of the proof given in Lemma 3.7 on page 86.

4.3 Explain why every Borel measurable function is a Lebesgue measurable function. Is the converse true? Why?

4.4 Prove Proposition 4.1 on page 112. *Hint:* Refer to the proof of Lemma 3.5 on page 85.

4.5 Suppose that f is $\mathcal{M}$-measurable.
a) Show that $f^{-1}(B) \in \mathcal{M}$ for all $B \in \mathcal{B}$.

b) *True* or *False:* $f^{-1}(E) \in \mathcal{M}$ for all $E \in \mathcal{M}$. *Hint:* Refer to Exercise 3.50(g) on page 109.

4.6 Verify that, if f is $\mathcal{M}$-measurable, then $\{\, x : f(x) = \alpha \,\} \in \mathcal{M}$ for each $\alpha \in \mathcal{R}$. Show that the converse is not true. *Hint:* Let K be a nonmeasurable set (i.e., $K \notin \mathcal{M}$). Construct a function h such that $\{\, x : h(x) = \alpha \,\} \in \mathcal{M}$ for each $\alpha \in \mathcal{R}$ and $h^{-1}\big((0,\infty)\big) = K$.

4.7 Show that if f is $\mathcal{M}$-measurable, then so is $|f|$.

4.8 Show that every step function is a simple function but not every simple function is a step function.

4.9 Suppose that the sets $A_1, A_2, \ldots, A_n$ are the ones appearing in the canonical representation of a simple function s. Prove that those sets are Lebesgue measurable and pairwise disjoint.

4.10 Theorem 4.2(c) on page 113 indicates that if $\{f_n\}_{n=1}^{\infty}$ is a sequence of $\mathcal{M}$-measurable functions converging pointwise to f, then f is $\mathcal{M}$-measurable. What can be said if the family of functions is indexed by an uncountable set? Specifically, suppose that $\{f_t\}_{t \in (0,\infty)}$ is a family of $\mathcal{M}$-measurable functions that converges pointwise to f; that is, $\lim_{t \to \infty} f_t(x) = f(x)$ for all $x \in \mathcal{R}$. Is f necessarily $\mathcal{M}$-measurable?

4.11 Let E be a Lebesgue measurable set with $\lambda(E) < \infty$. Suppose that $\{f_n\}_{n=1}^{\infty}$ is a sequence of Lebesgue measurable functions that converges pointwise on E to a function f. Prove that for each pair of positive numbers ϵ and δ, there is an $N \in \mathcal{N}$ and a Lebesgue measurable set $A \subset E$ such that $\lambda(A) < \delta$ and $|f(x) - f_n(x)| < \epsilon$ for $x \in E \setminus A$ and $n \geq N$. *Hint:* Let $E_m = \{\, x \in E : |f(x) - f_n(x)| \geq \epsilon \text{ for some } n \geq m \,\}$ and apply Exercise 3.35.

★4.12 **Egorov's theorem:** The following result shows that, in a certain sense, pointwise convergence of measurable functions is close to being uniform convergence: Let E be a Lebesgue measurable set with $\lambda(E) < \infty$. Suppose that $\{f_n\}_{n=1}^{\infty}$ is a sequence of Lebesgue measurable functions that converges pointwise on E to a real-valued function f. Prove that for each $\delta > 0$, there is a Lebesgue measurable set $B \subset E$ with $\lambda(B) < \delta$ such that $f_n \to f$ uniformly on $E \setminus B$. *Hint:* Apply Exercise 4.11 with ϵ replaced by $1/k$ and δ replaced by $\delta/2^k$.

★4.13 Prove the following facts:

a) Suppose that F is a nonempty closed subset of $\mathcal{R}$ and O is a proper open subset of $\mathcal{R}$. Further suppose that $F \subset O$. Then there is a continuous function f such that $f(R) \subset [0,1]$, $f(F) = \{1\}$, and $f(O^c) = \{0\}$. *Hint:* f can be constructed from the functions $d(\cdot, F)$ and $d(\cdot, O^c)$.

b) Let $E \in \mathcal{M}$. Then there is a sequence of open sets $\{O_n\}_{n=1}^{\infty}$ and a sequence of closed sets $\{F_n\}_{n=1}^{\infty}$ such that for $n \in \mathcal{N}$, $F_n \subset E \subset O_n$, $F_n \subset F_{n+1}$, $O_n \supset O_{n+1}$, and $\lambda\big((\bigcap_{n=1}^{\infty} O_n) \setminus (\bigcup_{n=1}^{\infty} F_n)\big) = 0$.

c) Let $E \in \mathcal{M}$. Then there is a Lebesgue measurable set B with $\lambda(B) = 0$ and a sequence of continuous functions $\{g_n\}_{n=1}^{\infty}$ with $0 \leq g_n \leq 1$ for all $n \in \mathcal{N}$ such that $\lim_{n \to \infty} g_n(x) = \chi_E(x)$ for each $x \in B^c$.

d) Let s be a simple function with $|s(x)| \leq M$ for all $x \in \mathcal{R}$, where M is a real number. Then there is a sequence of continuous functions $\{g_n\}_{n=1}^{\infty}$ and a Lebesgue measurable set B such that $\lambda(B) = 0$, $|g_n(x)| \leq M$ for $x \in \mathcal{R}$ and $n \in \mathcal{N}$, and $\lim_{n \to \infty} g_n(x) = s(x)$ for $x \in B^c$.

e) Let f be a nonnegative $\mathcal{M}$-measurable function that is bounded by the real number M and vanishes outside of a finite interval, say, $[-L, L]$. Then there exists a Lebesgue measurable set $B \subset [-L, L]$ with $\lambda(B) = 0$ and a sequence of continuous functions $\{g_n\}_{n=1}^{\infty}$ such that $0 \leq g_n(x) \leq M$ for $x \in \mathcal{R}$ and $n \in \mathcal{N}$, $g_n(x) = 0$ for $x \notin [-L - 1/n, L + 1/n]$ and $n \in \mathcal{N}$, and $\lim_{n \to \infty} g_n(x) = f(x)$ for $x \in B^c$.

Hint: Define

$$E_{jk} = \left\{ x \in [-L, L] : \frac{jM}{k} \le f(x) < \frac{(j+1)M}{k} \right\}$$

and set $s_k = \sum_{j=0}^{k}(jM/k)\chi_{E_{jk}}$. Then $\{s_k\}_k$ is a sequence of simple functions with $0 \le s_k \le f$ and $f - s_k \le M/k$. By part (d), there is a sequence of continuous functions $\{g_{nk}\}_{n=1}^{\infty}$ and a Lebesgue measurable set C_k such that $\lambda(C_k) = 0$, $|g_{nk}| \le M$, and $\lim_{n \to \infty} g_{nk}(x) = s_k(x)$ for $x \notin C_k$. Furthermore, the g_{nk}s can be chosen so that they vanish outside of $[-L-1, L+1]$. Now apply Exercises 4.11 and 4.12 to the sequence $\{g_{nk}\}_{n=1}^{\infty}$.

f) Let f be a nonnegative $\mathcal{M}$-measurable function that is bounded by the real number M. Then there is a Lebesgue measurable set B with $\lambda(B) = 0$ and a sequence of continuous functions $\{g_n\}_{n=1}^{\infty}$ such that each g_n vanishes outside a finite interval, $0 \le g_n(x) \le M$ for $x \in \mathcal{R}$ and $n \in \mathcal{N}$, and $\lim_{n \to \infty} g_n(x) = f(x)$ for $x \in B^c$. *Hint:* Apply part (e) and Exercise 4.11 to the function $f_n = \chi_{[-n,n]}f$.

g) Let f be a nonnegative $\mathcal{M}$-measurable function. Then there is a sequence of nonnegative continuous functions $\{g_n\}_{n=1}^{\infty}$ and a Lebesgue measurable set B with $\lambda(B) = 0$ such that $\lim_{n \to \infty} g_n(x) = f(x)$ for $x \in B^c$. *Hint:* Apply part (f) to the function $F = f/(1+f)$.

4.14 Suppose that f is an $\mathcal{M}$-measurable function. Prove that there is a sequence of continuous functions $\{g_n\}_{n=1}^{\infty}$ and a Lebesgue measurable set B with $\lambda(B) = 0$ such that $\lim_{n \to \infty} g_n(x) = f(x)$ for $x \in B^c$. *Hint:* Use the fact that $f = f^+ - f^-$, where $f^+ = f \vee 0$ and $f^- = -(f \wedge 0)$, and apply Exercise 4.13(g).

4.15 **Lusin's theorem:** The following result shows that, in a certain sense, a measurable function is close to being a continuous function: Let f be a Lebesgue measurable function and E a Lebesgue measurable set with $\lambda(E) < \infty$. Prove that for each $\epsilon > 0$, there is a Lebesgue measurable set $A \subseteq E$ with $\lambda(A) < \epsilon$ such that f is continuous on $E \setminus A$. *Hint:* Employ Exercises 4.14 and 4.12.

4.16 Suppose that f is a nonnegative $\mathcal{M}$-measurable function and that $E \in \mathcal{M}$.
a) Let $c > 0$ and set $A_c = \{ x \in E : f(x) \ge c \}$. Prove that

$$\lambda(A_c) \le \frac{1}{c} \int_E f \, d\lambda.$$

b) Let $A = \{ x \in E : f(x) > 0 \}$. Show that if $\int_E f \, d\lambda = 0$, then $\lambda(A) = 0$.

4.17 Let f be a nonnegative $\mathcal{M}$-measurable function with $\int_{\mathcal{R}} f \, d\lambda < \infty$. Show that for each $\epsilon > 0$, there is a $\delta > 0$ such that $\lambda(E) < \delta$ implies $\int_E f \, d\lambda < \epsilon$.

4.2 CONVERGENCE PROPERTIES OF THE LEBESGUE INTEGRAL FOR NONNEGATIVE FUNCTIONS

An important problem in mathematics is to determine when it is permissible to interchange a limit and an integral. For example, suppose that $\{f_n\}_{n=1}^{\infty}$ is a sequence of functions that converges pointwise. Under what conditions can we conclude that

$$\int \lim_{n \to \infty} f_n = \lim_{n \to \infty} \int f_n ?$$

As we noted at the end of Chapter 2, one significant advantage of the Lebesgue integral over the Riemann integral is that the interchange of limit and integral can be justified under less restrictive conditions. In this section and the next, we will develop theorems that provide sufficient conditions for the interchange of those two operations.

Monotone Convergence Theorem

The first theorem that we will discuss is called the *monotone convergence theorem,* or MCT for short. We begin with the following lemma.

□ □ □ **LEMMA 4.2**

Suppose that s is a nonnegative simple function and that $\{E_n\}_{n=1}^{\infty}$ is a sequence of Lebesgue measurable sets with $E_1 \subset E_2 \subset \cdots$. Then,

$$\int_{\bigcup_{n=1}^{\infty} E_n} s \, d\lambda = \lim_{n \to \infty} \int_{E_n} s \, d\lambda.$$

PROOF For convenience, set $E = \bigcup_{n=1}^{\infty} E_n$. Since s is a simple function, we can write $s = \sum_{k=1}^{m} a_k \chi_{A_k}$. Then, by Proposition 4.2 on page 114, we have for each $n \in \mathcal{N}$,

$$\int_{E_n} s \, d\lambda = \sum_{k=1}^{m} a_k \lambda(A_k \cap E_n).$$

Now, for each $k = 1, 2, \ldots, m$, consider the sequence $\{A_k \cap E_n\}_{n=1}^{\infty}$ of Lebesgue measurable sets. Since $E_1 \subset E_2 \subset \cdots$ and $E = \bigcup_{n=1}^{\infty} E_n$, it follows that

$$A_k \cap E_1 \subset A_k \cap E_2 \subset \cdots \qquad \text{and} \qquad \bigcup_{n=1}^{\infty} (A_k \cap E_n) = A_k \cap E.$$

Therefore, by Theorem 3.13 on page 106, $\lim_{n \to \infty} \lambda(A_k \cap E_n) = \lambda(A_k \cap E)$, for each k ($1 \leq k \leq m$). Consequently,

$$\lim_{n \to \infty} \int_{E_n} s \, d\lambda = \lim_{n \to \infty} \sum_{k=1}^{m} a_k \lambda(A_k \cap E_n) = \sum_{k=1}^{m} a_k \lim_{n \to \infty} \lambda(A_k \cap E_n)$$

$$= \sum_{k=1}^{m} a_k \lambda(A_k \cap E) = \int_E s \, d\lambda.$$

This completes the proof of the lemma. ∎

Before we state and prove the monotone convergence theorem (MCT), it will be useful to introduce two common conventions. First, if the integral of a function f is over all of $\mathcal{R}$, then the $\mathcal{R}$ is often omitted; in other words, by convention, we have $\int f \, d\lambda = \int_{\mathcal{R}} f \, d\lambda$.

Second, we sometimes write $f_n \uparrow f$ to indicate that $\{f_n\}_{n=1}^{\infty}$ is a monotone nondecreasing sequence of functions that converges pointwise to the function f. And, likewise, we sometimes write $f_n \downarrow f$ to indicate that $\{f_n\}_{n=1}^{\infty}$ is a monotone nonincreasing sequence of functions that converges pointwise to the function f.

□ □ □ **THEOREM 4.3 Monotone Convergence Theorem (MCT)**

Suppose that $\{f_n\}_{n=1}^{\infty}$ is a monotone nondecreasing sequence of nonnegative Lebesgue measurable functions that converges pointwise to a real-valued function; in other words, for each $x \in \mathcal{R}$,

$$0 \leq f_1(x) \leq f_2(x) \leq \cdots \leq f_n(x) \leq \cdots$$

and $\lim_{n \to \infty} f_n(x) < \infty.$[†] Then

$$\int_E \lim_{n \to \infty} f_n \, d\lambda = \lim_{n \to \infty} \int_E f_n \, d\lambda$$

for each $E \in \mathcal{M}$.

PROOF Set $f = \lim_{n \to \infty} f_n$. For each $E \in \mathcal{M}$, we have $0 \leq \chi_E f_n \uparrow \chi_E f$. Hence, it suffices to prove the theorem for $E = \mathcal{R}$. Because $f_n \leq f_{n+1}$ for all $n \in \mathcal{N}$, Proposition 4.4(a) on page 118 implies that $\int f_n \, d\lambda \leq \int f_{n+1} \, d\lambda$ for all $n \in \mathcal{N}$. Thus, $\lim_{n \to \infty} \int f_n \, d\lambda$ exists (possibly infinite). Let $L = \lim_{n \to \infty} \int f_n \, d\lambda$.

We must show that $L = \int f \, d\lambda$. First, $f_n \leq f$ for all $n \in \mathcal{N}$, so it follows immediately that

$$L = \lim_{n \to \infty} \int f_n \, d\lambda \leq \int f \, d\lambda.$$

To establish the reverse inequality, let $0 < \alpha < 1$ and s be a nonnegative simple function dominated by f. Set $E_n = \{ x : f_n(x) \geq \alpha s(x) \}$ for each $n \in \mathcal{N}$. Since $f_1 \leq f_2 \leq \cdots$, it is clear that $E_1 \subset E_2 \subset \cdots$. Also, because $0 < \alpha < 1$, $f_n \uparrow f$, and $0 \leq s \leq f$, it follows that $\bigcup_{n=1}^{\infty} E_n = \mathcal{R}$. Applying Proposition 4.4(e) and Lemma 4.2, we conclude that

$$\alpha \int_{\mathcal{R}} s \, d\lambda = \alpha \lim_{n \to \infty} \int_{E_n} s \, d\lambda = \lim_{n \to \infty} \int_{E_n} \alpha s \, d\lambda$$

$$\leq \limsup_{n \to \infty} \int_{E_n} f_n \, d\lambda \leq \lim_{n \to \infty} \int_{\mathcal{R}} f_n \, d\lambda = L.$$

Consequently, $\int s \, d\lambda \leq \alpha^{-1} L$ for each nonnegative simple function s that is dominated by f. This result implies that

$$\int_{\mathcal{R}} f \, d\lambda = \sup_{\substack{0 \leq s \leq f \\ s \text{ simple}}} \int_{\mathcal{R}} s \, d\lambda \leq \alpha^{-1} L,$$

for each $0 < \alpha < 1$. Letting $\alpha \uparrow 1$ yields $\int f \, d\lambda \leq L$. This completes the proof of the theorem. ∎

Note: For a fixed $E \in \mathcal{M}$, the conclusion of the MCT remains valid if the hypotheses are satisfied only on E. (See Exercise 4.21.)

Proposition 4.4 on page 118 lists several properties of the Lebesgue integral for nonnegative functions. Conspicuous by its absence is the additivity property. By employing the MCT and Proposition 4.3 on page 116, that important property can now be established.

[†] Since, for each $x \in \mathcal{R}$, $\{f_n(x)\}_{n=1}^{\infty}$ is monotone nondecreasing, $\lim_{n \to \infty} f_n(x)$ exists but it may be ∞. We assume here that the limit is finite for each $x \in \mathcal{R}$ although, as we will learn in Chapter 5, the theorem is also true without that restriction.

□ □ □ **PROPOSITION 4.5**

Let f and g be nonnegative Lebesgue measurable functions. Then

$$\int_E (f + g)\, d\lambda = \int_E f\, d\lambda + \int_E g\, d\lambda$$

for each $E \in \mathcal{M}$.

PROOF We first observe that, by Lemma 4.1 (page 116), the additivity property holds for simple functions. Next, we use Proposition 4.3 to select sequences of nonnegative simple functions $\{s_n\}_{n=1}^{\infty}$ and $\{t_n\}_{n=1}^{\infty}$ such that $s_n \uparrow f$ and $t_n \uparrow g$. Noting that $s_n + t_n \uparrow f + g$, we can apply the MCT and Lemma 4.1 to conclude that

$$\int_E (f + g)\, d\lambda = \lim_{n \to \infty} \int_E (s_n + t_n)\, d\lambda$$

$$= \lim_{n \to \infty} \int_E s_n\, d\lambda + \lim_{n \to \infty} \int_E t_n\, d\lambda = \int_E f\, d\lambda + \int_E g\, d\lambda,$$

as required. ∎

By induction, it follows immediately from Proposition 4.5 that if $\{f_k\}_{k=1}^n$ is a finite sequence of nonnegative $\mathcal{M}$-measurable functions, then

$$\int_E \sum_{k=1}^n f_k\, d\lambda = \sum_{k=1}^n \int_E f_k\, d\lambda \tag{4.6}$$

for each $E \in \mathcal{M}$. However, with the aid of the MCT, we can prove the following stronger result.

□ □ □ **THEOREM 4.4**

Suppose that $\{f_n\}_{n=1}^{\infty}$ is a sequence of nonnegative Lebesgue measurable functions such that $\sum_{n=1}^{\infty} f_n$ converges to a real-valued function.[†] Then we have, for each $E \in \mathcal{M}$,

$$\int_E \sum_{n=1}^{\infty} f_n\, d\lambda = \sum_{n=1}^{\infty} \int_E f_n\, d\lambda$$

PROOF For convenience, set $f = \sum_{n=1}^{\infty} f_n$ and let $g_n = \sum_{k=1}^n f_k$ for each $n \in \mathcal{N}$. Then $\{g_n\}_{n=1}^{\infty}$ is a monotone nondecreasing sequence of nonnegative Lebesgue measurable functions and $g_n \uparrow f$. Thus, by the MCT and (4.6),

$$\int_E f\, d\lambda = \lim_{n \to \infty} \int_E g_n\, d\lambda = \lim_{n \to \infty} \int_E \sum_{k=1}^n f_k\, d\lambda$$

$$= \lim_{n \to \infty} \sum_{k=1}^n \int_E f_k\, d\lambda = \sum_{n=1}^{\infty} \int_E f_n\, d\lambda,$$

as required. ∎

[†] See the footnote on page 123.

□ □ □   COROLLARY 4.1

Let f be a nonnegative Lebesgue measurable function and $\{E_n\}_n$ be a sequence of pairwise disjoint Lebesgue measurable sets. Then

$$\int_{\bigcup_n E_n} f \, d\lambda = \sum_n \int_{E_n} f \, d\lambda.$$

In particular, if $A, B \in \mathcal{M}$ and $A \cap B = \emptyset$, then

$$\int_{A \cup B} f \, d\lambda = \int_A f \, d\lambda + \int_B f \, d\lambda. \tag{4.7}$$

PROOF   Because the E_ns are pairwise disjoint, we have $\chi_{\bigcup_n E_n} = \sum_n \chi_{E_n}$ and, hence, $\chi_{\bigcup_n E_n} \cdot f = \sum_n (\chi_{E_n} f)$. Therefore, by Theorem 4.4,

$$\int_{\bigcup_n E_n} f \, d\lambda = \int \chi_{\bigcup_n E_n} \cdot f \, d\lambda = \int \sum_n (\chi_{E_n} f) \, d\lambda$$
$$= \sum_n \int \chi_{E_n} f \, d\lambda = \sum_n \int_{E_n} f \, d\lambda.$$

The proof of Corollary 4.1 is now complete.   ∎

Remark: Equation (4.7) generalizes the property of Riemann integrals that we presented in Theorem 2.6(a) on page 70.

Further Convergence Properties

The MCT shows that it is permissable to interchange limit and integral for monotone nondecreasing sequences of nonnegative Lebesgue measurable functions. Two additional questions concerning integrals and sequences of nonnegative functions come to mind.

Question 1: Suppose that $\{f_n\}_{n=1}^{\infty}$ is a monotone nonincreasing sequence of nonnegative Lebesgue measurable functions; in other words, for each $x \in \mathcal{R}$,

$$f_1(x) \geq f_2(x) \geq \cdots \geq f_n(x) \geq \cdots \geq 0.$$

Is it true that the limit and integral can be interchanged, that is, does

$$\int_E \lim_{n \to \infty} f_n \, d\lambda = \lim_{n \to \infty} \int_E f_n \, d\lambda \, ? \tag{4.8}$$

The answer to Question 1 is no — in general, the limit and the integral cannot be interchanged! For example, define $f_n(x) = |x|/n$ for each $x \in \mathcal{R}$ and $n \in \mathcal{N}$. Then $f_n \downarrow 0$ pointwise; thus, $\int_{\mathcal{R}} \lim_{n \to \infty} f_n \, d\lambda = 0$. But, it is easy to see that $\int_{\mathcal{R}} f_n \, d\lambda = \infty$ for all $n \in \mathcal{N}$ and, so, $\lim_{n \to \infty} \int_{\mathcal{R}} f_n \, d\lambda = \infty$. Consequently, (4.8) fails in this case.

With an additional condition, however, we can answer Question 1 in the affirmative. Specifically, we have the following theorem.

☐ ☐ ☐ **THEOREM 4.5**

Suppose that $\{f_n\}_{n=1}^{\infty}$ is a monotone nonincreasing sequence of nonnegative Lebesgue measurable functions. Further suppose that $\int f_1 \, d\lambda < \infty$. Then

$$\int_E \lim_{n \to \infty} f_n \, d\lambda = \lim_{n \to \infty} \int_E f_n \, d\lambda$$

for each $E \in \mathcal{M}$.

PROOF   For convenience, set $f = \lim_{n \to \infty} f_n$. As $\{f_n\}_{n=1}^{\infty}$ is monotone nonincreasing, $\{f_1 - f_n\}_{n=1}^{\infty}$ is a monotone nondecreasing sequence of nonnegative Lebesgue measurable functions. Therefore, by the MCT,

$$\int_E (f_1 - f) \, d\lambda = \lim_{n \to \infty} \int_E (f_1 - f_n) \, d\lambda.$$

Because $\int f_1 \, d\lambda < \infty$ and $f_n \le f_1$ for all $n \in \mathcal{N}$, Proposition 4.4 on page 118 implies that $\int_E f_n \, d\lambda < \infty$ for $n \in \mathcal{N}$ and $E \in \mathcal{M}$. Also, by Proposition 4.5,

$$\int_E f_1 \, d\lambda = \int_E \big((f_1 - f_n) + f_n\big) \, d\lambda = \int_E (f_1 - f_n) \, d\lambda + \int_E f_n \, d\lambda.$$

Consequently, we see that $\int_E (f_1 - f_n) \, d\lambda = \int_E f_1 \, d\lambda - \int_E f_n \, d\lambda$. This last equality also holds when f_n is replaced by f.

It now follows that

$$\int_E f_1 \, d\lambda - \int_E f \, d\lambda = \int_E f_1 \, d\lambda - \lim_{n \to \infty} \int_E f_n \, d\lambda.$$

Since all integrals in the previous equation are finite, the proof of the theorem is now complete. ∎

Question 2: Suppose that $\{f_n\}_{n=1}^{\infty}$ is a sequence of nonnegative Lebesgue measurable functions that converges pointwise to a real-valued function. Does a general relationship hold between the sequence $\left\{\int_E f_n \, d\lambda\right\}_{n=1}^{\infty}$ and the number $\int_E \lim_{n \to \infty} f_n \, d\lambda$? Of course, $\lim_{n \to \infty} \int_E f_n \, d\lambda$ need not exist and, so, (4.8) may not even make sense. The most that one can say in general is related by the following theorem.

☐ ☐ ☐ **THEOREM 4.6 Fatou's Lemma**

Suppose that $\{f_n\}_{n=1}^{\infty}$ is a sequence of nonnegative Lebesgue measurable functions that converges pointwise to a real-valued function. Then

$$\int_E \lim_{n \to \infty} f_n \, d\lambda \le \liminf_{n \to \infty} \int_E f_n \, d\lambda \tag{4.9}$$

for each $E \in \mathcal{M}$.

PROOF   For convenience, set $f = \lim_{n \to \infty} f_n$ and let $g_n = \inf_{k \geq n} f_k$ for each $n \in \mathcal{N}$. Then $\{g_n\}_{n=1}^{\infty}$ is a monotone nondecreasing sequence of nonnegative Lebesgue measurable functions and $g_n \uparrow f$ pointwise (why?). Thus, by the MCT,

$$\int_E f \, d\lambda = \lim_{n \to \infty} \int_E g_n \, d\lambda.$$

However, since $g_n \leq f_n$ for each $n \in \mathcal{N}$, it follows from Proposition 4.4(a) that

$$\lim_{n \to \infty} \int_E g_n \, d\lambda \leq \liminf_{n \to \infty} \int_E f_n \, d\lambda.$$

The proof of Fatou's lemma is now complete.   ∎

EXAMPLE 4.3   *Illustrates Strict Inequality in Fatou's Lemma*

This example shows that the inequality in (4.9) cannot be replaced by an equality. For each $n \in \mathcal{N}$, define

$$f_n = \begin{cases} \chi_{[n,n+1]}, & n \text{ odd}; \\ \chi_{[n,n+2]}, & n \text{ even}. \end{cases}$$

Then $f_n \to 0$ pointwise and, hence, in particular, $\int_{\mathcal{R}} \lim_{n \to \infty} f_n \, d\lambda = 0$. But,

$$\int_{\mathcal{R}} f_n \, d\lambda = \begin{cases} 1, & n \text{ odd}; \\ 2, & n \text{ even}. \end{cases}$$

and, so, $\liminf_{n \to \infty} \int_{\mathcal{R}} f_n \, d\lambda = 1$. Thus, we see that

$$\int_{\mathcal{R}} \lim_{n \to \infty} f_n \, d\lambda < \liminf_{n \to \infty} \int_{\mathcal{R}} f_n \, d\lambda.$$

So, the inequality in Fatou's lemma cannot be replaced by an equality.   □

Exercises for Section 4.2

4.18   Let f be a nonnegative Lebesgue measurable function. Show that

$$\lim_{n \to \infty} \int_{[-n,n]} f \, d\lambda = \int_{\mathcal{R}} f \, d\lambda.$$

4.19   Let f be a nonnegative Lebesgue measurable function. For each $n \in \mathcal{N}$, let $f_n = f \wedge n$. Prove that $\lim_{n \to \infty} \int_E f_n \, d\lambda = \int_E f \, d\lambda$ for each $E \in \mathcal{M}$.

★4.20   Prove that Lemma 4.2 holds for all nonnegative $\mathcal{M}$-measurable functions. That is, if f is a nonnegative Lebesgue measurable function and $\{E_n\}_{n=1}^{\infty}$ is a sequence of Lebesgue measurable sets with $E_1 \subset E_2 \subset \cdots$, then

$$\int_{\bigcup_{n=1}^{\infty} E_n} f \, d\lambda = \lim_{n \to \infty} \int_{E_n} f \, d\lambda.$$

4.21 Show that for a fixed $E \in \mathcal{M}$, the conclusion of the MCT remains valid if the hypotheses are satisfied only on E. In other words, let E be a Lebesgue measurable set and $\{f_n\}_{n=1}^{\infty}$ a sequence of nonnegative Lebesgue measurable functions that is monotone nondecreasing on E and converges pointwise to a real-valued function on E; that is, for each $x \in E$,

$$0 \leq f_1(x) \leq f_2(x) \leq \cdots \leq f_n(x) \leq \cdots$$

and $\lim_{n \to \infty} f_n(x) < \infty$. Prove that

$$\int_E \lim_{n \to \infty} f_n \, d\lambda = \lim_{n \to \infty} \int_E f_n \, d\lambda.$$

4.22 Give an example where strict inequality holds in Fatou's lemma and $\lim_{n \to \infty} \int_E f_n \, d\lambda$ exists.

4.23 Let $\{f_n\}_{n=1}^{\infty}$ be a sequence of nonnegative $\mathcal{M}$-measurable functions such that $f_n \to f$ pointwise and $\int f_n \, d\lambda \to \int f \, d\lambda < \infty$. Show that for each $E \in \mathcal{M}$, $\int_E f_n \, d\lambda \to \int_E f \, d\lambda$. *Hint:* Use Fatou's lemma and $\limsup_{n \to \infty}(a_n + b_n) \geq \limsup_{n \to \infty} a_n + \liminf_{n \to \infty} b_n$.

4.24 Suppose that f is a nonnegative $\mathcal{M}$-measurable function and that $\{E_n\}_{n=1}^{\infty} \subset \mathcal{M}$ with $E_1 \supset E_2 \supset \cdots$. Further suppose that $\int_{\mathcal{R}} f \, d\lambda < \infty$. Prove that

$$\int_{\bigcap_{n=1}^{\infty} E_n} f \, d\lambda = \lim_{n \to \infty} \int_{E_n} f \, d\lambda.$$

Hint: Apply Theorem 4.5.

4.25 Supply a proof for the following improved version of Fatou's lemma: Suppose that $\{f_n\}_{n=1}^{\infty}$ is a sequence of nonnegative Lebesgue measurable functions. Then

$$\int_E \liminf_{n \to \infty} f_n \, d\lambda \leq \liminf_{n \to \infty} \int_E f_n \, d\lambda$$

for each $E \in \mathcal{M}$.

★4.26 Suppose f is a nonnegative $\mathcal{M}$-measurable function with $\int_{[0,\infty)} f \, d\lambda < \infty$. Then we define the **Laplace transform of f**, denoted, F, by

$$F(t) = \int_{[0,\infty)} e^{-tx} f(x) \, d\lambda(x), \qquad t \geq 0.$$

Show that
a) F is real valued.
b) F is continuous on $[0, \infty)$. *Hint:* First establish that F is nonincreasing.
c) $\lim_{t \to \infty} F(t) = 0$.

4.27 Establish the following results:
a) If O is an open set, then

$$\lambda(O) = \sup \left\{ \int f \, d\lambda : 0 \leq f \leq \chi_O \text{ and } f \text{ continuous} \right\}.$$

Hint: Consider the functions $f_n(x) = \left(d(x, O^c)/[1 + d(x, O^c)] \right)^{\frac{1}{n}}$ for $n \in \mathcal{N}$.
b) If F is a closed set, then

$$\lambda(F) = \inf \left\{ \int f \, d\lambda : f \geq \chi_F \text{ and } f \text{ continuous} \right\}.$$

Hint: If $\lambda(F) < \infty$, select an appropriate open set $O \supset F$ and consider the functions $f_n(x) = \left(d(x, O^c)/[d(x, O^c) + d(x, F)] \right)^n$ for $n \in \mathcal{N}$.

4.28 In the next section, we will see how to define the Lebesgue integral for Lebesgue measurable functions that are not necessarily nonnegative. Assuming that can be done, construct a sequence of Lebesgue measurable functions for which the conclusion of Fatou's lemma fails. *Hint:* A sequence consisting of characteristic functions and negatives of characteristic functions will do the trick.

4.3 THE GENERAL LEBESGUE INTEGRAL

Up to this point, we have defined the Lebesgue integral only for nonnegative Lebesgue measurable functions. In this section, we will define the Lebesgue integral for arbitrary Lebesgue measurable functions and present some of its most important properties.

Definition of the General Lebesgue Integral

Basically, the Lebesgue integral of an arbitrary $\mathcal{M}$-measurable function f is obtained as follows: (1) express f as the difference of two nonnegative functions and (2) define the Lebesgue integral of f to be the difference of the Lebesgue integrals of the two nonnegative functions. To make this idea precise, we begin by defining the positive and negative parts of a function.

DEFINITION 4.6 **Positive and Negative Parts of a Function**

Suppose that f is a real-valued function. Then the **positive part of f**, denoted by f^+, is defined by

$$f^+ = f \vee 0 = \max\{f, 0\}$$

and the **negative part of f**, denoted by f^-, is defined by

$$f^- = -(f \wedge 0) = -\min\{f, 0\}.$$

Note that both f^+ and f^- are nonnegative functions. Proposition 4.6 states some other basic properties of the positive and negative parts of a function. The proof of the proposition is left as an exercise for the reader.

□ □ □ **PROPOSITION 4.6**

Suppose that f is a real-valued function on $\mathcal{R}$. Then
a) $f = f^+ - f^-$.
b) $|f| = f^+ + f^-$.
c) If f is Lebesgue measurable, then so are f^+ and f^-.

We now see that if f is a Lebesgue measurable function, then it can be expressed as the difference of two nonnegative Lebesgue measurable functions; namely, $f = f^+ - f^-$. Consequently, it is quite natural to define the Lebesgue integral of an arbitrary Lebesgue measurable function in the following way.

DEFINITION 4.7   Lebesgue Integral; Lebesgue Integrable

Let f be a Lebesgue measurable function and $E \in \mathcal{M}$. Then the **Lebesgue integral of f over E** is defined by

$$\int_E f(x)\, d\lambda(x) = \int_E f^+(x)\, d\lambda(x) - \int_E f^-(x)\, d\lambda(x) \qquad (4.10)$$

provided that the right-hand side makes sense; that is, at least one of the integrals on the right-hand side of (4.10) is finite. In addition, we say that **f is Lebesgue integrable over E** if both integrals on the right-hand side of (4.10) are finite or, equivalently, if

$$\int_E |f(x)|\, d\lambda(x) = \int_E f^+(x)\, d\lambda(x) + \int_E f^-(x)\, d\lambda(x) < \infty. \qquad (4.11)$$

If f is Lebesgue integrable over $\mathcal{R}$, then we say that **f is Lebesgue integrable.**

EXAMPLE 4.4   *Illustrates Definition 4.7*

a) Let

$$f(x) = \begin{cases} 1, & x \geq 0; \\ -1, & x < 0. \end{cases}$$

Then $\int_{\mathcal{R}} f\, d\lambda$ is not defined. Indeed, $f^+ = \chi_{[0,\infty)}$ and $f^- = \chi_{(-\infty,0)}$, so that both $\int_{\mathcal{R}} f^+\, d\lambda$ and $\int_{\mathcal{R}} f^-\, d\lambda$ are infinite. But, the Lebesgue integral of f is defined (and, in fact, f is Lebesgue integrable) over any Lebesgue measurable set with finite measure. For instance, if $E = [-3, 4]$, then $\int_E f^+\, d\lambda = 4$ and $\int_E f^-\, d\lambda = 3$, so that $\int_E f\, d\lambda = 4 - 3 = 1$ and $\int_E |f|\, d\lambda = 4 + 3 = 7 < \infty$.

b) We can generalize part (a): If f is a bounded Lebesgue measurable function, then f is Lebesgue integrable over any measurable set E with $\lambda(E) < \infty$. For, if $|f| \leq L$, then by Proposition 4.4(a),

$$\int_E |f|\, d\lambda \leq \int_E L\, d\lambda = \int_{\mathcal{R}} L\chi_E\, d\lambda = L\lambda(E) < \infty.$$

c) Let

$$f(x) = \begin{cases} 2, & 0 < x < 1; \\ -3, & x \geq 1; \\ 0, & \text{elsewhere.} \end{cases}$$

Then $f^+ = 2\chi_{(0,1)}$ and $f^- = 3\chi_{[1,\infty)}$ and, consequently, $\int_{\mathcal{R}} f^+\, d\lambda = 2$ and $\int_{\mathcal{R}} f^-\, d\lambda = \infty$. This implies that f is not Lebesgue integrable over $\mathcal{R}$, although the Lebesgue integral is defined and $\int_{\mathcal{R}} f\, d\lambda = 2 - \infty = -\infty$.   □

Properties of the General Lebesgue Integral

The next theorem provides some important properties of Lebesgue integrable functions. In proving this theorem, we will employ the following lemma whose proof we leave as an exercise for the reader.

□ □ □   LEMMA 4.3

Suppose that f is a Lebesgue measurable function and that $E \in \mathcal{M}$. Further suppose that $f = f_1 - f_2$, where f_1 and f_2 are nonnegative and Lebesgue integrable over E. Then f is Lebesgue integrable over E and

$$\int_E f\, d\lambda = \int_E f_1\, d\lambda - \int_E f_2\, d\lambda.$$

□ □ □   THEOREM 4.7

Suppose that f and g are Lebesgue integrable over $E \in \mathcal{M}$ and that $\alpha \in \mathcal{R}$. Then

a) $f + g$ is Lebesgue integrable over E and

$$\int_E (f + g)\, d\lambda = \int_E f\, d\lambda + \int_E g\, d\lambda.$$

b) αf is Lebesgue integrable over E and

$$\int_E \alpha f\, d\lambda = \alpha \int_E f\, d\lambda.$$

c) $f \le g \Rightarrow \int_E f\, d\lambda \le \int_E g\, d\lambda$.

d) $\left| \int_E f\, d\lambda \right| \le \int_E |f|\, d\lambda$.

e) If A and B are measurable subsets of E with $A \cap B = \emptyset$, then

$$\int_{A \cup B} f\, d\lambda = \int_A f\, d\lambda + \int_B f\, d\lambda.$$

PROOF   a) Since $|f + g| \le |f| + |g|$, it follows from Proposition 4.4(a) on page 118 and Proposition 4.5 on page 124 that $f + g$ is Lebesgue integrable over E. Now, we have $f + g = (f^+ + g^+) - (f^- + g^-)$. Hence, by Lemma 4.3 and Proposition 4.5, we conclude that

$$\int_E (f + g)\, d\lambda = \int_E (f^+ + g^+)\, d\lambda - \int_E (f^- + g^-)\, d\lambda$$

$$= \int_E f^+\, d\lambda + \int_E g^+\, d\lambda - \int_E f^-\, d\lambda - \int_E g^-\, d\lambda$$

$$= \int_E f^+\, d\lambda - \int_E f^-\, d\lambda + \int_E g^+\, d\lambda - \int_E g^-\, d\lambda$$

$$= \int_E f\, d\lambda + \int_E g\, d\lambda.$$

b) Since $|\alpha f| = |\alpha||f|$, Proposition 4.4(e) implies that αf is Lebesgue integrable over E. If $\alpha \ge 0$, then $(\alpha f)^+ = \alpha f^+$ and $(\alpha f)^- = \alpha f^-$. Thus, by Proposition 4.4(e) again,

$$\int_E \alpha f\, d\lambda = \int_E \alpha f^+\, d\lambda - \int_E \alpha f^-\, d\lambda$$

$$= \alpha \int_E f^+\, d\lambda - \alpha \int_E f^-\, d\lambda = \alpha \int_E f\, d\lambda.$$

If $\alpha < 0$, then $(\alpha f)^+ = -\alpha f^-$ and $(\alpha f)^- = -\alpha f^+$. Consequently, by Proposition 4.4(e),

$$\int_E \alpha f \, d\lambda = \int_E (-\alpha f^-) \, d\lambda - \int_E (-\alpha f^+) \, d\lambda$$

$$= \alpha \left(\int_E f^+ \, d\lambda - \int_E f^- \, d\lambda \right) = \alpha \int_E f \, d\lambda.$$

c) $f \le g \Rightarrow g - f \ge 0 \Rightarrow \int_E (g - f) \, d\lambda \ge 0$ by Proposition 4.4(a). Now applying parts (a) and (b), we deduce that

$$\int_E g \, d\lambda - \int_E f \, d\lambda = \int_E (g - f) \, d\lambda \ge 0.$$

In other words, $\int_E f \, d\lambda \le \int_E g \, d\lambda$.

d) Because $f \le |f|$ and $-f \le |f|$, we can use parts (b) and (c) to conclude that

$$\int_E f \, d\lambda \le \int_E |f| \, d\lambda \qquad \text{and} \qquad - \int_E f \, d\lambda \le \int_E |f| \, d\lambda.$$

These last two relations imply that $\left| \int_E f \, d\lambda \right| \le \int_E |f| \, d\lambda$.

e) Since $A \subset E$, $|\chi_A f| \le |\chi_E f|$ and, consequently, $\int_A |f| \, d\lambda \le \int_E |f| \, d\lambda$. Thus, f is integrable over A or, equivalently, $\chi_A f$ is integrable over E. Similarly, $\chi_B f$ is integrable over E. However, because $A \cap B = \emptyset$, we have that $\chi_{A \cup B} f = \chi_A f + \chi_B f$. Therefore, by part (a),

$$\int_E \chi_{A \cup B} f \, d\lambda = \int_E \chi_A f \, d\lambda + \int_E \chi_B f \, d\lambda.$$

Since A and B are subsets of E, the previous equation is equivalent to

$$\int_{A \cup B} f \, d\lambda = \int_A f \, d\lambda + \int_B f \, d\lambda.$$

This completes the proof of the theorem. ■

Remark: Parts (a) and (b) of Theorem 4.7 together imply that if α, $\beta \in \mathcal{R}$ and f and g are Lebesgue integrable over E, then

$$\int_E (\alpha f + \beta g) \, d\lambda = \alpha \int_E f \, d\lambda + \beta \int_E g \, d\lambda.$$

This is called the **linearity property** of the Lebesgue integral.

The next theorem, called the *dominated convergence theorem*, or DCT for short, is one of the most important theorems in analysis. Like the monotone convergence theorem, it gives sufficient conditions for the interchange of limit and integral.

□ □ □ **THEOREM 4.8 Dominated Convergence Theorem (DCT)**

Suppose that $\{f_n\}_{n=1}^{\infty}$ is a sequence of Lebesgue measurable functions that converges pointwise to a real-valued function. Further suppose that there is a nonnegative Lebesgue integrable function g such that $|f_n| \leq g$ for all $n \in \mathcal{N}$. Then

$$\int_E \lim_{n \to \infty} f_n \, d\lambda = \lim_{n \to \infty} \int_E f_n \, d\lambda$$

for each $E \in \mathcal{M}$.

PROOF For convenience, set $f = \lim_{n \to \infty} f_n$. Because $|f_n| \leq g$ and g is Lebesgue integrable, it follows that f, f_1, f_2, ... are Lebesgue integrable. Now, $g - f_n \geq 0$ for all $n \in \mathcal{N}$ and $g - f_n \to g - f$ pointwise. Thus, by Fatou's lemma (page 126) and the linearity of the integral,

$$\int_E g \, d\lambda - \int_E f \, d\lambda = \int_E (g - f) \, d\lambda \leq \liminf_{n \to \infty} \int_E (g - f_n) \, d\lambda$$

$$= \liminf_{n \to \infty} \left(\int_E g \, d\lambda - \int_E f_n \, d\lambda \right)$$

$$= \int_E g \, d\lambda - \limsup_{n \to \infty} \int_E f_n \, d\lambda.$$

Since the previous integrals are all finite, we conclude that

$$\limsup_{n \to \infty} \int_E f_n \, d\lambda \leq \int_E f \, d\lambda. \tag{4.12}$$

However, we also have that $g + f_n \geq 0$ for all $n \in \mathcal{N}$ and $g + f_n \to g + f$ pointwise. Applying Fatou's lemma again, we obtain the relations

$$\int_E g \, d\lambda + \int_E f \, d\lambda = \int_E (g + f) \, d\lambda$$

$$\leq \liminf_{n \to \infty} \int_E (g + f_n) \, d\lambda = \int_E g \, d\lambda + \liminf_{n \to \infty} \int_E f_n \, d\lambda$$

or, in other words,

$$\int_E f \, d\lambda \leq \liminf_{n \to \infty} \int_E f_n \, d\lambda. \tag{4.13}$$

From (4.12) and (4.13), we see that

$$\limsup_{n \to \infty} \int_E f_n \, d\lambda = \liminf_{n \to \infty} \int_E f_n \, d\lambda = \int_E f \, d\lambda.$$

This last fact implies that $\lim_{n \to \infty} \int_E f_n \, d\lambda$ exists and

$$\int_E f \, d\lambda = \lim_{n \to \infty} \int_E f_n \, d\lambda,$$

as required. ■

Note: For a fixed $E \in \mathcal{M}$, the conclusion of the DCT remains valid if the hypotheses are satisfied only on E. (See Exercise 4.36.)

EXAMPLE 4.5   *Illustrates the DCT*

a) In general, the conclusion of the DCT may fail if there is no dominating integrable function g. For instance, let $f_n = n\chi_{(0,\frac{1}{n})}$. Then $f_n \to 0$ pointwise. Moreover, $\int f_n \, d\lambda = 1$ for all $n \in \mathcal{N}$. Thus,

$$\int \lim_{n\to\infty} f_n \, d\lambda = 0 \neq 1 = \lim_{n\to\infty} \int f_n \, d\lambda.$$

The problem here is that there is no integrable function that dominates the sequence $\{f_n\}_{n=1}^{\infty}$.

b) Let $f_n(x) = x^n \chi_{[0,1]}(x)$ for $x \in \mathcal{R}$, $n \in \mathcal{N}$. Then $f_n \to \chi_{\{1\}}$ pointwise. Now, $|f_n| \leq \chi_{[0,1]}$ for all $n \in \mathcal{N}$ and, clearly, $\chi_{[0,1]}$ is Lebesgue integrable. Thus, by the DCT,

$$\lim_{n\to\infty} \int_{[0,1]} x^n \, d\lambda(x) = \int_{[0,1]} \chi_{\{1\}}(x) \, d\lambda(x) = \lambda(\{1\}) = 0.$$

Note: Theorem 4.9 provides a simpler way to obtain this result.   □

There are many corollaries of the DCT. Two of the most important are stated next. The proofs of these two corollaries are left as exercises for the reader. (See Exercises 4.33 and 4.34.)

□ □ □   **COROLLARY 4.2**

Suppose that $\{f_n\}_{n=1}^{\infty}$ is a sequence of Lebesgue measurable functions such that $\sum_{n=1}^{\infty} |f_n|$ converges to a Lebesgue integrable function. Then $\sum_{n=1}^{\infty} f_n$ is Lebesgue integrable and

$$\int_E \sum_{n=1}^{\infty} f_n \, d\lambda = \sum_{n=1}^{\infty} \int_E f_n \, d\lambda$$

for each $E \in \mathcal{M}$.

□ □ □   **COROLLARY 4.3**

Let f be a Lebesgue integrable function and $\{E_n\}_n$ a sequence of pairwise disjoint Lebesgue measurable sets. Then

$$\int_{\bigcup_n E_n} f \, d\lambda = \sum_n \int_{E_n} f \, d\lambda.$$

The Lebesgue Integral is an Extension of the Riemann Integral

We will now establish that the Lebesgue integral is indeed an extension of the Riemann integral. In other words, we will show that a Riemann integrable function is also Lebesgue integrable and that the two integrals are equal. First, we need the following lemma.

□ □ □ **LEMMA 4.4**

Let f be a bounded Lebesgue measurable function on $[a, b]$. Then f is Lebesgue integrable over $[a, b]$ and, moreover,

$$\int_{[a,b]} f(x)\, d\lambda(x) = \sup_{\substack{s \leq f \\ s \text{ simple}}} \int_{[a,b]} s(x)\, d\lambda(x) \tag{4.14}$$

and

$$\int_{[a,b]} f(x)\, d\lambda(x) = \inf_{\substack{t \geq f \\ t \text{ simple}}} \int_{[a,b]} t(x)\, d\lambda(x). \tag{4.15}$$

PROOF   Example 4.4(b) on page 130 shows that f is Lebesgue integrable over $[a, b]$. We will prove (4.14). The proof of (4.15) is similar and is left as an exercise.

First note that if s is a simple function with $s \leq f$, then, by Theorem 4.7(c) on page 131, $\int_{[a,b]} s\, d\lambda \leq \int_{[a,b]} f\, d\lambda$. Consequently,

$$\sup_{\substack{s \leq f \\ s \text{ simple}}} \int_{[a,b]} s(x)\, d\lambda(x) \leq \int_{[a,b]} f(x)\, d\lambda(x).$$

It remains to prove the reverse inequality. To accomplish that, we will construct a sequence $\{s_n\}_{n=1}^{\infty}$ of simple functions with $s_n \leq f$ for all $n \in \mathcal{N}$ and

$$\int_{[a,b]} f\, d\lambda = \lim_{n \to \infty} \int_{[a,b]} s_n\, d\lambda. \tag{4.16}$$

Set $L = \sup\{\, |f(x)| : x \in [a, b]\,\}$. Then $f + L$ is nonnegative and Lebesgue measurable on $[a, b]$. Applying Proposition 4.3 (page 116), we get a sequence $\{u_n\}_{n=1}^{\infty}$ of nonnegative simple functions such that $u_n \uparrow f + L$. Setting $s_n = u_n - L$, we see that $\{s_n\}_{n=1}^{\infty}$ is a sequence of simple functions such that $s_n \leq f$ for all $n \in \mathcal{N}$ and $s_n \to f$ pointwise on $[a, b]$. Furthermore, because $|s_n| \leq L$ on $[a, b]$, the DCT implies that (4.16) holds.   ∎

□ □ □ **THEOREM 4.9**

Suppose that f is Riemann integrable on $[a, b]$. Then f is Lebesgue integrable on $[a, b]$ and

$$\int_{[a,b]} f(x)\, d\lambda(x) = \int_a^b f(x)\, dx.$$

PROOF   To begin, we extend the domain of f to all of $\mathcal{R}$ by defining $f(x) = 0$ for $x \in [a, b]^c$. Now, since f is Riemann integrable on $[a, b]$, it is bounded thereon. So, to prove that f is Lebesgue integrable on $[a, b]$, it suffices to show that f is Lebesgue measurable (why?).

Let O be an open subset of $\mathcal{R}$. We must verify that $f^{-1}(O) \in \mathcal{M}$. Set $E = \{ x \in \mathcal{R} : f \text{ is discontinuous at } x \}$. We have

$$f^{-1}(O) = (f^{-1}(O) \cap E) \cup (f^{-1}(O) \cap E^c). \tag{4.17}$$

Clearly, $E \subset [a, b]$ and, consequently, by Theorem 2.7 on page 72, $\lambda(E) = 0$. But, every subset of a set with Lebesgue measure zero is Lebesgue measurable (Exercise 3.32). Hence, the first intersection on the right of (4.17) is a Lebesgue measurable set.

Next we show that the second intersection on the right of (4.17) is a Lebesgue measurable set. To begin, note that $f^{-1}(O) \cap E^c = f_{|E^c}^{-1}(O)$. Now, by the definition of E, the function $f_{|E^c}$ is continuous. Therefore, by Theorem 2.5 on page 55, $f_{|E^c}^{-1}(O)$ is an open subset of E^c. In view of Theorem 2.3 on page 51, there is an open subset U of $\mathcal{R}$ such that $f_{|E^c}^{-1}(O) = U \cap E^c$. Both sets in this last intersection are Lebesgue measurable (why?). Thus, $f^{-1}(O) \cap E^c \in \mathcal{M}$.

To complete the proof of the theorem, we must prove that the Riemann and Lebesgue integrals of f over $[a, b]$ are equal. First recall that every step function is a simple function and that the Riemann and Lebesgue integrals agree for step functions (because the Lebesgue measure of an interval is the length of the interval). Applying Lemma 4.4 and the definition of the Riemann integral, we now obtain that

$$\int_a^b f(x)\, dx = \sup_{\substack{g \leq f \\ g \text{ step function}}} \int_a^b g(x)\, dx \leq \sup_{\substack{s \leq f \\ s \text{ simple}}} \int_{[a,b]} s(x)\, d\lambda(x)$$

$$= \int_{[a,b]} f(x)\, d\lambda(x) = \inf_{\substack{t \geq f \\ t \text{ simple}}} \int_{[a,b]} t(x)\, d\lambda(x)$$

$$\leq \inf_{\substack{h \geq f \\ h \text{ step function}}} \int_a^b h(x)\, dx = \int_a^b f(x)\, dx.$$

These relations imply that

$$\int_a^b f(x)\, dx = \int_{[a,b]} f(x)\, d\lambda(x),$$

as required. ∎

We have now verified that the Lebesgue integral is indeed a generalization of the Riemann integral. Consequently, we will frequently denote the Lebesgue integral of f over $[a, b]$ by

$$\int_a^b f(x)\, dx$$

regardless of whether f is Riemann integrable over $[a, b]$. In other words, the notation for the Riemann integral is also used for the Lebesgue integral. Moreover, as previously mentioned, we will often write $\int_E f(x)\, dx$ instead of $\int_E f(x)\, d\lambda(x)$.

EXAMPLE 4.6   *Illustrates the Lebesgue and Riemann Integrals*

a) By Theorem 4.9,

$$\int_{[a,b]} x^n \, d\lambda(x) = \int_a^b x^n \, dx = \frac{b^{n+1} - a^{n+1}}{n+1}.$$

b) Clearly, χ_Q is Lebesgue integrable over $[0,1]$. However, it is not Riemann integrable on $[0,1]$ because it is discontinuous everywhere.

c) Define $f(x) = 1/\sqrt{x}$, for $0 < x \leq 1$, and zero otherwise. Note that f has only two discontinuities and, hence, the set of points of discontinuity of f has measure zero. But, f is not Riemann integrable on $[0,1]$ because it is not bounded. It is, however, Lebesgue integrable on $[0,1]$, as we will now show. For each $n \in \mathcal{N}$, let $f_n = \chi_{[1/n,1]} f$. Then f_n is Riemann integrable on $[0,1]$ and, so, by Theorem 4.9,

$$\int_{[0,1]} f_n(x) \, d\lambda(x) = \int_0^1 f_n(x) \, dx = \int_{\frac{1}{n}}^1 \frac{dx}{\sqrt{x}} = 2 - 2\sqrt{n^{-1}}.$$

Now, $\{f_n\}_{n=1}^\infty$ is a monotone nondecreasing sequence of nonnegative Lebesgue measurable functions and $f_n \to f$ pointwise. Applying the MCT, we conclude that

$$\int_{[0,1]} f(x) \, d\lambda(x) = \lim_{n \to \infty} \int_{[0,1]} f_n(x) \, d\lambda(x) = 2 < \infty.$$

Hence, f is Lebesgue integrable over $[0,1]$.   □

Exercises for Section 4.3

4.29 Prove Proposition 4.6 on page 129.

4.30 Determine the positive and negative parts of the following functions:
a) $\sin x$.   b) $x^2 - 4$.   c) $|x|$.

4.31 Prove Lemma 4.3 on page 131.

4.32 Show that if f is Lebesgue integrable (over $\mathcal{R}$), then it is Lebesgue integrable over E for each $E \in \mathcal{M}$.

4.33 Prove Corollary 4.2 on page 134.

4.34 Prove Corollary 4.3 on page 134.

4.35 Suppose that f is Lebesgue integrable over E and that $\{E_n\}_{n=1}^\infty$ is a sequence of Lebesgue measurable sets with $E_1 \subset E_2 \subset \cdots$ and $\bigcup_{n=1}^\infty E_n = E$. Prove that

$$\int_E f \, d\lambda = \lim_{n \to \infty} \int_{E_n} f \, d\lambda.$$

4.36 Let E be a Lebesgue measurable set and $\{f_n\}_{n=1}^\infty$ a sequence of Lebesgue measurable functions that converges pointwise on E to a real-valued function. Suppose that g is Lebesgue integrable over E and that $|f_n(x)| \leq g(x)$ for $n \in \mathcal{N}$, $x \in E$. Prove that

$$\int_E \lim_{n \to \infty} f_n \, d\lambda = \lim_{n \to \infty} \int_E f_n \, d\lambda.$$

4.37 **Bounded convergence theorem (BCT):** Let $E \in \mathcal{M}$ with $\lambda(E) < \infty$ and $\{f_n\}_{n=1}^{\infty}$ a sequence of Lebesgue measurable functions that converges pointwise on E to a real-valued function. Further suppose that there is an $M \in \mathcal{R}$ such that $|f_n(x)| \le M$ for $n \in \mathcal{N}$ and $x \in E$. Show that

$$\int_E \lim_{n \to \infty} f_n \, d\lambda = \lim_{n \to \infty} \int_E f_n \, d\lambda.$$

4.38 Find a sequence $\{f_n\}_{n=1}^{\infty}$ of Lebesgue measurable functions such that $|f_n| \le M$ for all $n \in \mathcal{N}$, $f_n \to f$ pointwise, but $\int f_n d\lambda \not\to \int f d\lambda$. Why doesn't this situation contradict the BCT?

4.39 Complete the proof of Lemma 4.4 on page 135 by establishing (4.15).

★**4.40** Theorem 4.9 on page 135 shows that every Riemann integrable function is also a Lebesgue integrable function. This result refers only to the proper Riemann integral. In this exercise, we will exhibit a function that has an improper Riemann integral but is not Lebesgue integrable. Let

$$f(x) = \begin{cases} \dfrac{\sin x}{x}, & x \ne 0; \\[2mm] 1, & x = 0. \end{cases}$$

Show that
a) f has an improper Riemann integral over $\mathcal{R}$ equal to π.
b) f is Lebesgue measurable.
c) f is not Lebesgue integrable over $\mathcal{R}$.

4.41 Show that, if f is Lebesgue integrable (over $\mathcal{R}$) and the improper Riemann integral exists, then $\int_{\mathcal{R}} f(x) \, d\lambda(x) = \int_{-\infty}^{\infty} f(x) \, dx$.

4.42 Prove that the results of Exercise 4.26 on page 128 remain valid if f is Lebesgue integrable over $[0, \infty)$.

4.43 Let $\{f_n\}_{n=1}^{\infty}$ and $\{g_n\}_{n=1}^{\infty}$ be two sequences of Lebesgue measurable functions. Suppose that $E \in \mathcal{M}$ and that on E, $|f_n| \le g_n$, $f_n \to f$, and $g_n \to g$. Further suppose that g, g_1, g_2, ... are Lebesgue integrable over E and that $\int_E g_n \, d\lambda \to \int_E g \, d\lambda$. Prove that $\int_E f_n \, d\lambda \to \int_E f \, d\lambda$.

4.44 For $E \subset \mathcal{R}$ and $a \in \mathcal{R}$, let $E + a = \{x + a : x \in E\}$ and $aE = \{ax : x \in E\}$. Suppose that f is a Lebesgue integrable function.
a) Show that

$$\int_{\mathcal{R}} f(x + a) \, d\lambda(x) = \int_{\mathcal{R}} f(x) \, d\lambda(x)$$

and, if $a \ne 0$,

$$\int_{\mathcal{R}} f(ax) \, d\lambda(x) = \frac{1}{|a|} \int_{\mathcal{R}} f(x) \, d\lambda(x).$$

Hint: Start with the case $f = \chi_A$, where $A \in \mathcal{M}$.
b) Show that, for $E \in \mathcal{M}$,

$$\int_E f(x + a) \, d\lambda(x) = \int_{E+a} f(x) \, d\lambda(x)$$

and, if $a \ne 0$,

$$\int_E f(ax) \, d\lambda(x) = \frac{1}{|a|} \int_{aE} f(x) \, d\lambda(x).$$

4.45 Consider a function $F: \mathcal{R} \times I \to \mathcal{R}$, where I is a nonempty open interval. Suppose that $\partial F / \partial t$ exists at each point of $\mathcal{R} \times I$, $F(\cdot, t)$ is Lebesgue measurable for each $t \in I$, $F(\cdot, t_0)$ is Lebesgue integrable for some $t_0 \in I$, and there is a Lebesgue integrable function G such that $|(\partial F / \partial t)(x, t)| \leq G(x)$ for $x \in \mathcal{R}$ and $t \in I$. Prove that $F(\cdot, t)$ is Lebesgue integrable for each $t \in I$ and that

$$\frac{d}{dt} \int F(x, t) \, d\lambda(x) = \int \frac{\partial F}{\partial t}(x, t) \, d\lambda(x).$$

4.46 Consider a function $F: \mathcal{R} \times T \to \mathcal{R}$, where $T \subset \mathcal{R}$. Suppose that $F(\cdot, t)$ is Lebesgue measurable for each $t \in T$ and that there is a Lebesgue integrable function g such that $|F(x, t)| \leq g(x)$ for $x \in \mathcal{R}$ and $t \in T$. Establish the following facts.
a) If $F(x, \cdot)$ is continuous on T for each $x \in \mathcal{R}$, then the function defined on T by $f(t) = \int F(x, t) \, d\lambda(x)$ is continuous.
b) If T is an interval of the form (b, ∞) and if $\lim_{t \to \infty} F(x, t)$ exists for each $x \in \mathcal{R}$, then

$$\lim_{t \to \infty} \int F(x, t) \, d\lambda(x) = \int \lim_{t \to \infty} F(x, t) \, d\lambda(x).$$

4.47 Provide an example to show that the conditions given in the DCT are not necessary for the interchange of limit and integral.

4.4 LEBESGUE ALMOST EVERYWHERE

Frequently, we are not concerned whether a certain property holds everywhere as long as it holds "most places." For example, in order for a bounded function f to be Riemann integrable on $[a, b]$, it does not have to be continuous everywhere on $[a, b]$ — all that is required is that the set of points at which f is discontinuous have measure zero.

Consider, also, the sequence of functions $f_n(x) = \chi_{[-1,1]}(x) x^n$ for $n \in \mathcal{N}$. That sequence of functions does not converge pointwise on $\mathcal{R}$, but it almost does. Indeed, $f_n(x) \to \chi_{\{1\}}(x)$ except when $x = -1$. As the Lebesgue (or Riemann) integral is not affected by the value of a function at a single point, the lack of convergence of $\{f_n\}_{n=1}^{\infty}$ at $x = -1$ should really not disturb any convergence results involving the integral.

In this section, we will define the concept of a property holding *almost everywhere* and show that our previous results for the Lebesgue integral remain valid when "everywhere" is replaced by "almost everywhere."

DEFINITION 4.8   Lebesgue Almost Everywhere

A property is said to hold **Lebesgue almost everywhere,** or **λ-ae** for short, if it holds except on a set of Lebesgue measure zero, that is, except on a set N with $\lambda(N) = 0$.

EXAMPLE 4.7   *Illustrates Definition 4.8*

a) Two measurable functions, f and g, are *equal Lebesgue almost everywhere*, written $f = g$ λ-ae, if $\lambda(\{x : g(x) \neq f(x)\}) = 0$.

b) A sequence of measurable functions $\{f_n\}_{n=1}^{\infty}$ converges *Lebesgue almost everywhere to* f, written $f_n \to f$ λ-ae, if $\lim_{n \to \infty} f_n(x) = f(x)$ except on a set of Lebesgue measure zero. In other words, $f_n \to f$ λ-ae if and only if $\lambda(\{x : \lim_{n \to \infty} f_n(x) \neq f(x)\}) = 0$.   □

Out first proposition demonstrates that a function equal almost everywhere to a Lebesgue measurable function is itself Lebesgue measurable.

□ □ □   **PROPOSITION 4.7**

Suppose that f is a Lebesgue measurable function and that $g = f$ λ-ae. Then g is Lebesgue measurable.

PROOF   Set $B = \{x : g(x) = f(x)\}$. We claim that $g^{-1}(O) \in \mathcal{M}$ for each open set O. To begin, we write

$$g^{-1}(O) = \left(g^{-1}(O) \cap B\right) \cup \left(g^{-1}(O) \cap B^c\right). \tag{4.18}$$

We will show that both intersections on the right of (4.18) are Lebesgue measurable sets. As $g = f$ on B, it follows that $g^{-1}(O) \cap B = f^{-1}(O) \cap B$. However, this last intersection is a Lebesgue measurable set because $B \in \mathcal{M}$ (why?) and f is Lebesgue measurable. Hence, the first intersection on the right of (4.18) is a Lebesgue measurable set.

Now, by assumption, $\lambda(B^c) = 0$. Therefore, the second intersection on the right of (4.18) is a Lebesgue measurable set because it is a subset of a set that has Lebesgue measure zero.   ∎

Our next result shows that the collection of Lebesgue measurable functions is closed under almost-everywhere limits. More precisely, we have the following proposition.

□ □ □   **PROPOSITION 4.8**

Suppose that $\{f_n\}_{n=1}^{\infty}$ is a sequence of Lebesgue measurable functions and that $f_n \to f$ λ-ae. Then f is a Lebesgue measurable function.

PROOF   Set $B = \{x : \lim_{n \to \infty} f_n(x) = f(x)\}$. Then, by assumption, we have $\lambda(B^c) = 0$. Let $g_n = \chi_B f_n$ and $g = \chi_B f$. Then $\{g_n\}_{n=1}^{\infty}$ is a sequence of $\mathcal{M}$-measurable functions and $g_n \to g$ pointwise. Hence, by Theorem 4.2(c) on page 113, g is $\mathcal{M}$-measurable. But $f = g$ λ-ae and, consequently, f is $\mathcal{M}$-measurable by Proposition 4.7.   ∎

Remark: We should point out that Propositions 4.7 and 4.8 are not valid for Borel measurable functions. This is because subsets of Borel sets of measure zero are not necessarily Borel sets. (See Exercise 4.48.)

Next, we will prove that the Lebesgue integral of a function is not affected by changing its values on a set of measure zero.

□ □ □ **PROPOSITION 4.9**

Let f and g be Lebesgue measurable functions with $f = g$ λ-ae. If f is Lebesgue integrable, then so is g and, moreover,

$$\int_E g \, d\lambda = \int_E f \, d\lambda$$

for each $E \in \mathcal{M}$.

PROOF   Set $B = \{\, x : g(x) = f(x) \,\}$. Then, by assumption, $\lambda(B^c) = 0$. Applying Corollary 4.1 on page 125 and Proposition 4.4(d) on page 118, we find that

$$\int_{\mathcal{R}} |g| \, d\lambda = \int_B |g| \, d\lambda + \int_{B^c} |g| \, d\lambda$$
$$= \int_B |f| \, d\lambda + \int_{B^c} |g| \, d\lambda$$
$$= \int_B |f| \, d\lambda < \infty.$$

Therefore, g is Lebesgue integrable.

Now, let $E \in \mathcal{M}$. Then, by Theorem 4.7(e) on page 131,

$$\int_E g \, d\lambda = \int_{E \cap B} g \, d\lambda + \int_{E \cap B^c} g \, d\lambda$$
$$= \int_{E \cap B} f \, d\lambda + \int_{E \cap B^c} g \, d\lambda \qquad (4.19)$$
$$= \int_E f \, d\lambda - \int_{E \cap B^c} f \, d\lambda + \int_{E \cap B^c} g \, d\lambda.$$

We will complete the proof of the proposition by showing that the last two integrals in (4.19) are zero. Employing Theorem 4.7(d) and Proposition 4.4(d), we deduce that

$$\left| \int_{E \cap B^c} f \, d\lambda \right| \le \int_{E \cap B^c} |f| \, d\lambda = 0.$$

Similarly, $\int_{E \cap B^c} g \, d\lambda = 0$. ∎

We often encounter functions that are only defined Lebesgue almost everywhere. Since the integral of a Lebesgue measurable function is not affected by its values on a set of measure zero, it is reasonable to make the following definition.

DEFINITION 4.9   Integral of a Function Defined Almost Everywhere

Suppose that f is a function defined Lebesgue almost everywhere; that is, if D is the domain of f, then $\lambda(D^c) = 0$. Further suppose that there is a Lebesgue measurable function g such that $g(x) = f(x)$ for $x \in D$. Then, for $E \in \mathcal{M}$, we define the **Lebesgue integral of f over E** by

$$\int_E f \, d\lambda = \int_E g \, d\lambda,$$

provided that the integral on the right-hand side exists (i.e., the integrals of the positive and negative parts of g over E are not both infinite).

Finally, we should point out that Fatou's lemma and the DCT remain valid if the hypothesis of pointwise convergence is replaced by convergence λ-ae. The proofs are left as exercises for the reader.

Exercises for Section 4.4

★4.48   Show that a subset of a Borel set of Lebesgue measure zero is not necessarily a Borel set. *Hint:* Refer to Exercise 3.50 on page 109.

★4.49   Show that Proposition 4.7 (page 140) fails for Borel measurable functions.

4.50   For Lebesgue measurable functions f and g, define $f \sim g$ if and only if $f = g$ λ-ae. Prove that $\sim$ is an equivalence relation.

4.51   Respond *True* or *False* to each of the following statements. Justify your answer.
a) If f is continuous λ-ae, then f is equal to a continuous function λ-ae.
b) If f is equal to a continuous function λ-ae, then f is continuous λ-ae.

4.52   Let $\{f_n\}_{n=1}^{\infty}$ be a sequence of Lebesgue measurable functions such that $\lim_{n \to \infty} f_n(x)$ exists λ-ae. Define

$$f(x) = \begin{cases} \lim\limits_{n \to \infty} f_n(x), & \text{if } \lim\limits_{n \to \infty} f_n(x) \text{ exists;} \\ 0, & \text{otherwise.} \end{cases}$$

Prove that f is Lebesgue measurable.

4.53   Verify that Definition 4.9 is well posed. That is, assume that g and h are Lebesgue measurable functions that equal f on its domain, D. Show that, for $E \in \mathcal{M}$, either $\int_E h \, d\lambda = \int_E g \, d\lambda$ or neither integral exists.

4.54   Show that the DCT (page 133) remains valid if convergence pointwise is replaced by convergence λ-ae. In other words, suppose that $\{f_n\}_{n=1}^{\infty}$ is a sequence of Lebesgue measurable functions that converges λ-ae to a real-valued function. Further suppose that there is a nonnegative Lebesgue integrable function g such that $|f_n| \leq g$ for all $n \in \mathcal{N}$. Prove that

$$\int_E \lim_{n \to \infty} f_n \, d\lambda = \lim_{n \to \infty} \int_E f_n \, d\lambda$$

for each $E \in \mathcal{M}$. *Note:* $\lim_{n \to \infty} f_n$ is not defined on all of $\mathcal{R}$ unless, of course, $\{f_n\}_{n=1}^{\infty}$ converges everywhere.

4.55 Show that Fatou's lemma (page 126) remains valid if convergence pointwise is replaced by convergence λ-ae.

4.56 Verify that Egorov's theorem, Exercise 4.12 on page 120, remains valid if $f_n \to f$ λ-ae on E.

4.57 Let f and g be $\mathcal{M}$-measurable functions with $\int |f - g| \, d\lambda = 0$. Prove that $f = g$ λ-ae.

4.58 Show that, if f is Lebesgue integrable and $\int_E f \, d\lambda = 0$ for each $E \in \mathcal{M}$, then $f = 0$ λ-ae.

4.59 Let $\{f_n\}_{n=1}^\infty$ be a sequence of Lebesgue measurable functions and set

$$E = \left\{ x : \lim_{n \to \infty} f_n(x) \text{ does not exist} \right\}.$$

Show that $E \in \mathcal{M}$.

*Constantin Carathéodory
(1873–1950)*

Constantin Carathéodory was born on September 13, 1873, in Berlin, Germany, to Greek parents. His father, Stephanos, was an Ottoman Greek who had studied law in Berlin; his mother, Despina, came from a Greek family of businessmen. At the time of Carathéodory's birth, his family was in Berlin because his father had been appointed there as First Secretary to the Ottoman Legation.

Carathéodory was a student at the Belgian Military Academy from 1891–1895 and then worked as an assistant engineer with the British Asyūt Dam project in Egypt. Returning to Germany, Carathéodory entered the University of Berlin in 1900 to begin his study of mathematics. In 1902, he transferred to the University of Göttingen, where he received his doctorate in 1904 under Hermann Minkowski.

The works of Carathéodory encompassed many disciplines. Among them were a simplified proof of a central theorem in conformal representation, extensions of findings by Picard on function theory, and the amplification of measure theory begun by Émile Borel and Henri Lebesgue. Regarding the latter discipline, two particularly important results in measure theory are the Carathéodory criterion for measurability and the Carathéodory extension theorem.

Carathéodory wrote many outstanding books. Two of particular relevance here are *Vorlesungen Über Reelle Funktionen* (Lectures on Real Functions) and *Mass und Integral und Ihre Algebraisierung* (Measure and Integral and their Algebraisation).

Carathéodory taught at the Universities of Hannover (1909), Breslau (1910–1913), Göttingen (1913–1918), Berlin (1918–1920), all in Germany. In 1920, he went to Anatolia, Greece, to establish the University of Smyrna. After the Turks burned Smyrna in 1922, he went to the University of Athens. In 1924, he returned to the University of Munich where he remained, except for one year (1936-1937) at the University of Wisconsin, until his death on February 2, 1950.

□ **5** □

Elements of Measure Theory

In Chapter 3, we expanded the collection of continuous functions to the collection of Borel measurable functions, the smallest algebra that contains the continuous functions and is closed under pointwise limits. Subsequently, in Chapter 4, we extended the Riemann integral so that it applies to all Borel measurable functions and, in doing so, we encountered Lebesgue measure, the collection of Lebesgue measurable functions, and the Lebesgue integral.

We will discover, in this chapter, that the concepts and methods of Chapters 3 and 4 lend themselves to considerable generalization with relatively little effort and huge rewards. This generalized theory has extensive applications throughout mathematics and, as well, to a large variety of fields outside of mathematics.

5.1 MEASURE SPACES

When we examine the definition of the Lebesgue integral carefully, we find that it depends ultimately on the concept of measure. More precisely, the mathematical framework requires a set, a σ-algebra of subsets, and a set function that assigns to each set in the σ-algebra a nonnegative number (its measure). In Chapter 3, this consisted, respectively, of $\mathcal{R}$, $\mathcal{M}$, and λ. But we can abstract the mathematical framework to provide a broader setting for the integral. We begin by considering the general concept of measure.

In developing Lebesgue measure, we imposed three conditions; namely, Conditions (M1)–(M3) on page 91. The first two conditions are specific to the generalization of length; but the third is not. In fact, Condition (M3), the countable-additivity condition, is the primary property of an abstract measure.

DEFINITION 5.1 **Measure, Measurable Space, Measure Space**

Let Ω be a set and $\mathcal{A}$ a σ-algebra of subsets of Ω. A **measure** μ on $\mathcal{A}$ is an extended real-valued function satisfying the following conditions:

a) $\mu(A) \geq 0$ for all $A \in \mathcal{A}$.

b) $\mu(\emptyset) = 0$.

c) If A_1, A_2, ... are in $\mathcal{A}$, with $A_i \cap A_j = \emptyset$ for $i \neq j$, then

$$\mu\left(\bigcup_n A_n\right) = \sum_n \mu(A_n).$$

The pair $(\Omega, \mathcal{A})$ is called a **measurable space** and the triple $(\Omega, \mathcal{A}, \mu)$ is called a **measure space.**

Note: We will often refer to members of $\mathcal{A}$ as **$\mathcal{A}$-measurable sets.**

We should point out the following fact: If μ satisfies (a) and (c) of Definition 5.1, then it is a measure (i.e., also satisfies (b)) if and only if there is an $A \in \mathcal{A}$ such that $\mu(A) < \infty$. We leave the proof of this fact to the reader.

EXAMPLE 5.1 *Illustrates Definition 5.1*

a) $(\mathcal{R}, \mathcal{M}, \lambda)$ is a measure space, the one that we studied in Chapter 3.

b) $(\mathcal{R}, \mathcal{B}, \lambda_{|\mathcal{B}})$ is a measure space.

c) Let $(\Omega, \mathcal{A}, \mu)$ be a measure space. For $D \in \mathcal{A}$, define $\mathcal{A}_D = \{D \cap A : A \in \mathcal{A}\}$ and $\mu_D = \mu_{|\mathcal{A}_D}$. Then $\mathcal{A}_D$ is a σ-algebra of subsets of D, μ_D is a measure on $\mathcal{A}_D$, and, hence, $(D, \mathcal{A}_D, \mu_D)$ is a measure space.

d) Referring to part (c), let $\Omega = \mathcal{R}$, $\mathcal{A} = \mathcal{M}$, $\mu = \lambda$, and $D = [0, 1]$. Then $([0,1], \mathcal{M}_{[0,1]}, \lambda_{[0,1]})$ is a measure space. $\lambda_{[0,1]}$ is called *Lebesgue measure on* $[0, 1]$. More generally, if D is a Lebesgue measurable set, then $(D, \mathcal{M}_D, \lambda_D)$ is a measure space and λ_D is called **Lebesgue measure on D.**

e) Refer to part (c). By Theorem 3.7 on page 88, if $D \in \mathcal{B}$, then $\mathcal{B}_D = \mathcal{B}(D)$.

f) Let Ω be a nonempty set and $\mathcal{A} = \mathcal{P}(\Omega)$. Define μ on $\mathcal{A}$ by

$$\mu(E) = \begin{cases} N(E), & \text{if } E \text{ is finite;} \\ \infty, & \text{if } E \text{ is infinite.} \end{cases}$$

where $N(E)$ denotes the number of elements of E. Then μ is a measure on $\mathcal{A}$ and is called **counting measure.**

g) Let $\Omega = \mathcal{N}$, $\mathcal{A} = \mathcal{P}(\mathcal{N})$, and μ be counting measure on $\mathcal{A}$, as defined in part (f). Then, for instance, $\mu(\mathcal{N}) = \infty$ and $\mu(\{1, 3\}) = 2$. We will see later that $(\mathcal{N}, \mathcal{P}(\mathcal{N}), \mu)$ is the appropriate measure space for the analysis of infinite series.

h) Suppose that $(\Omega, \mathcal{A}, \mu)$ is a measure space. If $\mu(\Omega) = 1$, then $(\Omega, \mathcal{A}, \mu)$ is called a **probability space** and μ a **probability measure.** Furthermore, μ is usually replaced by a P (for probability). Two simple examples are as follows:

(i) $([0,1], \mathcal{M}_{[0,1]}, \lambda_{[0,1]})$ is a probability space since $\lambda([0,1]) = 1$. It is an appropriate measure space for analyzing the experiment of selecting a number at random from the unit interval.

(ii) Consider the experiment of tossing a coin twice. The set of possible outcomes for that experiment is $\Omega = \{\text{HH}, \text{HT}, \text{TH}, \text{TT}\}$ where, for instance, HT denotes the outcome of a head on the first toss and a tail on the second toss. Set $\mathcal{A} = \mathcal{P}(\Omega)$ and, for $E \in \mathcal{A}$, define $P(E) = N(E)/4$ where, as before, $N(E)$ denotes the number of elements of E. Then $(\Omega, \mathcal{A}, P)$ is a probability space — the appropriate measure space to use when the coin is balanced (i.e., equally likely to come up heads or tails). To illustrate: The probability of getting at least one head in two tosses of a balanced coin is $P(\{\text{HH}, \text{HT}, \text{TH}\}) = 3/4$.

i) Let Ω be a nonempty set, $\{x_n\}_n$ a sequence of distinct elements of Ω, and $\{a_n\}_n$ a sequence of nonnegative numbers. For $E \subset \Omega$, define

$$\mu(E) = \sum_{x_n \in E} a_n, \tag{5.1}$$

where the notation $\sum_{x_n \in E}$ means the sum over all indices n such that $x_n \in E$. Then μ is a measure on $\mathcal{P}(\Omega)$ and, consequently, $(\Omega, \mathcal{P}(\Omega), \mu)$ is a measure space. Here are two special cases:

(i) If Ω is countable, $\{x_n\}_n$ is an enumeration of Ω, and $a_n = 1$ for all n, then the measure μ defined in (5.1) is counting measure.
(ii) If the sequence $\{x_n\}_n$ consists of only one element, say x_0, and if $a_0 = 1$, then the measure μ defined in (5.1) takes the form

$$\mu(E) = \begin{cases} 1, & \text{if } x_0 \in E; \\ 0, & \text{if } x_0 \notin E. \end{cases}$$

This measure is denoted by δ_{x_0} and is called the **unit point mass** or **Dirac measure concentrated at x_0**. Note that δ_{x_0} is a probability measure.

j) Let $(\Omega, \mathcal{A})$ be a measurable space such that $\{x\} \in \mathcal{A}$ for each $x \in \Omega$. A measure μ on $\mathcal{A}$ is called **discrete** if there is a countable set $K \subset \Omega$ such that $\mu(K^c) = 0$. It is not too difficult to show that if μ is a discrete measure, then we can write $\mu = \sum_{x \in K} \mu(\{x\})\delta_x$. See Exercises 5.6 and 5.19 for more on discrete measures. □

The following theorem provides some important properties of measures. We leave the proof as an exercise for the reader.

□ □ □ **THEOREM 5.1**

Suppose that $(\Omega, \mathcal{A}, \mu)$ is a measure space and that A and B are $\mathcal{A}$-measurable sets. Then the following properties hold:

a) *If $\mu(A) < \infty$ and $A \subset B$, then $\mu(B \setminus A) = \mu(B) - \mu(A)$.*
b) *$A \subset B \Rightarrow \mu(A) \leq \mu(B)$.* **(monotonicity)**
c) *If $\{E_n\}_{n=1}^{\infty} \subset \mathcal{A}$ with $E_1 \supset E_2 \supset \cdots$ and $\mu(E_1) < \infty$, then*

$$\mu\left(\bigcap_{n=1}^{\infty} E_n\right) = \lim_{n \to \infty} \mu(E_n).$$

d) If $\{E_n\}_{n=1}^\infty \subset \mathcal{A}$ with $E_1 \subset E_2 \subset \cdots$, then

$$\mu\left(\bigcup_{n=1}^\infty E_n\right) = \lim_{n \to \infty} \mu(E_n).$$

e) If $\{E_n\}_n \subset \mathcal{A}$, then

$$\mu\left(\bigcup_n E_n\right) \leq \sum_n \mu(E_n).$$

This property is called **countable subadditivity.**

Almost Everywhere and Complete Measure Spaces

Recall from Section 4.4 that a property holds Lebesgue almost everywhere (λ-ae) if it holds except on a set of Lebesgue measure zero. That concept can be generalized to apply to any measure space.

DEFINITION 5.2 **Almost Everywhere**

A property is said to hold μ **almost everywhere,** or μ-**ae** for short, if it holds except on a set of μ-measure zero, that is, except on a set N with $\mu(N) = 0$.

Note: Several terms are used synonymously for "almost everywhere." Here are a few: **almost always, for almost all** $x \in \Omega$, and, in probability theory, **almost surely, with probability one,** and **almost certainly.**

Proposition 3.4 on page 106 implies that subsets of Lebesgue measurable sets of Lebesgue measure zero are also Lebesgue measurable sets. On the other hand, Exercise 4.48 on page 142 indicates that there exist subsets of Borel sets of Lebesgue measure zero that are not Borel sets.

Those two facts have relevance to almost-everywhere (ae) properties of measurable functions. For instance, by Proposition 4.7 on page 140, if f is Lebesgue measurable and $g = f$ λ-ae, then g is Lebesgue measurable. However, as Exercise 4.49 on page 142 shows, that result is not true for Borel measurable functions.

We now see that it is important to know whether subsets of sets of measure zero are measurable sets. Hence, we make the following definition.

DEFINITION 5.3 **Complete Measure Space**

A measure space $(\Omega, \mathcal{A}, \mu)$ is said to be **complete** if all subsets of $\mathcal{A}$-measurable sets of μ-measure zero are also $\mathcal{A}$-measurable; in other words, if $A \in \mathcal{A}$ and $\mu(A) = 0$, then $B \in \mathcal{A}$ for all $B \subset A$.

Thus, $(\mathcal{R}, \mathcal{M}, \lambda)$ is a complete measure space, whereas $(\mathcal{R}, \mathcal{B}, \lambda_{|\mathcal{B}})$ is not a complete measure space.

The following theorem shows that any measure space can be extended to a complete measure space. We leave the proof of the theorem as an exercise for the reader.

□ □ □ **THEOREM 5.2**

Let $(\Omega, \mathcal{A}, \mu)$ be a measure space. Denote by $\overline{\mathcal{A}}$ the collection of all sets of the form $B \cup A$ where $B \in \mathcal{A}$ and $A \subset C$ for some $C \in \mathcal{A}$ with $\mu(C) = 0$. For such sets, define $\overline{\mu}(B \cup A) = \mu(B)$. Then $\overline{\mathcal{A}}$ is a σ-algebra, $\overline{\mu}$ is a measure on $\overline{\mathcal{A}}$, and $(\Omega, \overline{\mathcal{A}}, \overline{\mu})$ is a complete measure space. Furthermore, $\mathcal{A} \subset \overline{\mathcal{A}}$ and $\overline{\mu}_{|\mathcal{A}} = \mu$. $(\Omega, \overline{\mathcal{A}}, \overline{\mu})$ is called the **completion** of $(\Omega, \mathcal{A}, \mu)$.

It can be shown that the measure space $(\mathcal{R}, \mathcal{M}, \lambda)$ is the completion of the measure space $(\mathcal{R}, \mathcal{B}, \lambda_{|\mathcal{B}})$. See Exercise 5.16.

Exercises for Section 5.1

Note: A ★ denotes an exercise that will be subsequently referenced.

5.1 Suppose that $(\Omega, \mathcal{A}, \mu)$ is a measure space and that D is an $\mathcal{A}$-measurable set. Define $\mathcal{A}_D = \{ D \cap A : A \in \mathcal{A} \}$ and $\mu_D = \mu_{|\mathcal{A}_D}$. Show that $(D, \mathcal{A}_D, \mu_D)$ is a measure space.

5.2 Let Ω be a nonempty set and $\mathcal{A} = \mathcal{P}(\Omega)$. Define μ on $\mathcal{A}$ by

$$\mu(E) = \begin{cases} N(E), & \text{if } E \text{ is finite}; \\ \infty, & \text{if } E \text{ is infinite}. \end{cases}$$

where $N(E)$ denotes the number of elements of E. Prove that μ is a measure on $\mathcal{A}$.

5.3 Consider the experiment of selecting a number at random from the interval $[-1, 1]$.
a) Construct an appropriate probability space for this experiment.
b) Determine the probability that the number selected exceeds 0.5.
c) Determine the probability that the number selected is rational.

5.4 Let $(\Omega, \mathcal{A})$ be a measurable space, μ and ν measures on $\mathcal{A}$, and $\alpha > 0$. Define set functions $\mu + \nu$ and $\alpha\mu$ on $\mathcal{A}$ by

$$(\mu + \nu)(A) = \mu(A) + \nu(A), \qquad (\alpha\mu)(A) = \alpha\mu(A).$$

a) Show that $\mu + \nu$ is a measure on $\mathcal{A}$.
b) Show that $\alpha\mu$ is a measure on $\mathcal{A}$.

5.5 Let $(\Omega, \mathcal{A})$ be a measurable space, $\{\mu_n\}_{n=1}^{\infty}$ a sequence of measures on $\mathcal{A}$, and $\{\alpha_n\}_{n=1}^{\infty}$ a sequence of nonnegative real numbers. Define $\sum_{n=1}^{\infty} \alpha_n \mu_n$ on $\mathcal{A}$ by

$$\left(\sum_{n=1}^{\infty} \alpha_n \mu_n \right)(A) = \sum_{n=1}^{\infty} \alpha_n \mu_n(A).$$

Prove that $\sum_{n=1}^{\infty} \alpha_n \mu_n$ is a measure on $\mathcal{A}$.

★5.6 Refer to Example 5.1(j). Let $(\Omega, \mathcal{A})$ be a measurable space and suppose that $\{x\} \in \mathcal{A}$ for each $x \in \Omega$. Show that a measure μ on $\mathcal{A}$ is discrete if and only if there is a countable subset K of Ω such that $\mu = \sum_{x \in K} \mu(\{x\})\delta_x$.

5.7 Suppose that a balanced coin is tossed three times.
a) Construct a probability space for this experiment in which each possible outcome is equally likely.

    b) Determine the probability of obtaining exactly two heads.

    c) Express the probability measure P as a finite linear combination of Dirac measures.

5.8 Let Ω be a nonempty set, $\{x_n\}_n$ a sequence of distinct elements of Ω, and $\{a_n\}_n$ a sequence of nonnegative real numbers. For $E \subset \Omega$, define

$$\mu(E) = \sum_{x_n \in E} a_n.$$

    a) Show that μ is a measure on $\mathcal{P}(\Omega)$.

    b) Interpret the a_ns in terms of the measure μ.

    c) Express μ as a linear combination of Dirac measures.

5.9 Suppose that two balanced dice are thrown.

    a) Construct a probability space for this experiment in which each possible outcome is equally likely.

    b) Use part (a) to determine the probability that the sum of the dice is seven or 11.

    c) Construct a probability space for this experiment in which the outcomes consist of the possible sums of the two dice.

    d) Use part (c) to determine the probability that the sum of the dice is seven or 11.

5.10 Prove Theorem 5.1.

5.11 Let $(\Omega, \mathcal{A})$ be a measurable space. A measure μ on $\mathcal{A}$ is called a **finite measure** if $\mu(\Omega) < \infty$. A measure space $(\Omega, \mathcal{A}, \mu)$ is called a **finite measure space** if μ is a finite measure. For a finite measure space, prove the following:

    a) If A and B are $\mathcal{A}$-measurable sets, then

$$\mu(A \cup B) = \mu(A) + \mu(B) - \mu(A \cap B).$$

    b) Generalize part (a) to an arbitrary finite number of $\mathcal{A}$-measurable sets.

5.12 Let $\{E_n\}_{n=1}^{\infty}$ be a sequence of $\mathcal{A}$-measurable sets. Prove that

$$\mu\left(\bigcup_{n=1}^{\infty} \left(\bigcap_{k=n}^{\infty} E_k \right) \right) \leq \liminf_{n \to \infty} \mu(E_n).$$

5.13 Let $\{E_n\}_{n=1}^{\infty}$ be a sequence of $\mathcal{A}$-measurable sets with $\mu\left(\bigcup_{n=1}^{\infty} E_n \right) < \infty$. Prove that

$$\mu\left(\bigcap_{n=1}^{\infty} \left(\bigcup_{k=n}^{\infty} E_k \right) \right) \geq \limsup_{n \to \infty} \mu(E_n).$$

★5.14 Let $(\Omega, \mathcal{A}, \mu)$ be a measure space and $\{E_n\}_{n=1}^{\infty}$ a sequence of $\mathcal{A}$-measurable sets. Define $E = \{ x : x \in E_n \text{ for infinitely many } n \}$.

    a) Prove that $E = \bigcap_{n=1}^{\infty} \left(\bigcup_{k=n}^{\infty} E_k \right)$.

    b) Prove that $\sum_{n=1}^{\infty} \mu(E_n) < \infty \Rightarrow \mu(E) = 0$.

5.15 Prove Theorem 5.2.

5.16 Prove that $(\mathcal{R}, \mathcal{M}, \lambda)$ is the completion of $(\mathcal{R}, \mathcal{B}, \lambda_{|\mathcal{B}})$. Use the following steps:

    a) Verify that $\overline{\mathcal{B}} \subset \mathcal{M}$ by employing Exercise 3.32 on page 107.

    b) Show that $\overline{\mathcal{B}} \supset \mathcal{M}$ by applying Exercise 3.44 on page 108.

    c) Prove that $\lambda = \overline{\lambda}_{|\mathcal{B}}$. *Hint:* Use the fact established in parts (a) and (b) that $\mathcal{M} = \overline{\mathcal{B}}$.

★5.17 Let $(\Omega, \mathcal{A}, \mu)$ be a measure space. Suppose that $(\Omega, \mathcal{F}, \nu)$ is a complete measure space with $\mathcal{F} \supset \mathcal{A}$ and $\nu_{|\mathcal{A}} = \mu$. Prove that $\mathcal{F} \supset \overline{\mathcal{A}}$ and that $\nu_{|\overline{\mathcal{A}}} = \overline{\mu}$. Conclude that $(\Omega, \overline{\mathcal{A}}, \overline{\mu})$ is the smallest complete measure space that contains $(\Omega, \mathcal{A}, \mu)$.

5.18 Let f be a nonnegative $\mathcal{M}$-measurable function. Define μ_f on $\mathcal{M}$ by $\mu_f(E) = \int_E f\, d\lambda$. Prove that μ_f is a measure on $\mathcal{M}$.

5.19 Let $(\Omega, \mathcal{A}, \mu)$ be a measure space such that $\{x\} \in \mathcal{A}$ for each $x \in \Omega$. An element $x \in \Omega$ is said to be an **atom** of μ if $\mu(\{x\}) > 0$. Assume now that μ is a finite measure, that is, $\mu(\Omega) < \infty$. Prove the following facts.

   a) μ has only countably many atoms.

   b) μ can be expressed uniquely as the sum of two measures, μ_c and μ_d, where μ_c has no atoms and μ_d is discrete. Moreover, we have that $\mu_d = \sum_{x \in K} \mu(\{x\})\delta_x$, where K is the set of atoms of μ.

5.2 MEASURABLE FUNCTIONS

The next step in developing the abstract Lebesgue integral is to introduce the concept of measurability for functions defined on an abstract space. In addition to real-valued functions, we will also consider complex-valued and extended real-valued functions. We begin with real-valued functions.

Real-Valued Measurable Functions

Let $(\Omega, \mathcal{A})$ be a measurable space and $f : \Omega \to \mathcal{R}$. We want to specify when f is measurable. In previous chapters, we discussed two kinds of measurable functions: Borel measurable functions and Lebesgue measurable functions. Recall that a real-valued function f is Borel measurable if and only if $f^{-1}(O) \in \mathcal{B}$ for each open set $O \subset \mathcal{R}$ and it is Lebesgue measurable if and only if $f^{-1}(O) \in \mathcal{M}$ for each open set $O \subset \mathcal{R}$. Hence, it is quite natural to make the following definition.

DEFINITION 5.4   Real-Valued Measurable Function

Let $(\Omega, \mathcal{A})$ be a measurable space. A real-valued function f on Ω is said to be an **$\mathcal{A}$-measurable function** if the inverse image of each open subset of $\mathcal{R}$ under f is an $\mathcal{A}$-measurable set, that is, if $f^{-1}(O) \in \mathcal{A}$ for all open sets $O \subset \mathcal{R}$.

EXAMPLE 5.2   Illustrates Definition 5.4

   a) Let $\Omega = \mathcal{R}$. Then, as we know from Chapters 3 and 4, the Borel measurable functions are the $\mathcal{B}$-measurable functions, and the Lebesgue measurable functions are the $\mathcal{M}$-measurable functions.

   b) Let $(\Omega, \mathcal{A})$ be a measurable space, $D \in \mathcal{A}$, and $\mathcal{A}_D = \{D \cap A : A \in \mathcal{A}\}$. Then a function $f : D \to \mathcal{R}$ is $\mathcal{A}_D$-measurable if and only if for each open subset O of $\mathcal{R}$, $f^{-1}(O)$ is of the form $D \cap A$ for some $A \in \mathcal{A}$.

   c) Every real-valued function on a nonempty set Ω is $\mathcal{P}(\Omega)$-measurable. An important special case: If $\Omega = \mathcal{N}$, then $\mathcal{A}$ is usually taken to be $\mathcal{P}(\mathcal{N})$; hence, all functions $f : \mathcal{N} \to \mathcal{R}$ are $\mathcal{A}$-measurable. But functions on $\mathcal{N}$ are infinite sequences. Consequently, in this case, the $\mathcal{A}$-measurable functions are precisely the infinite sequences. □

The following proposition provides some useful equivalent conditions for a function to be $\mathcal{A}$-measurable. To prove the proposition, we proceed in a similar manner as we did in the proof of Lemma 3.5 on page 85.

□ □ □ **PROPOSITION 5.1**

Let $(\Omega, \mathcal{A})$ be a measurable space and f a real-valued function on Ω. Then the following statements are equivalent:

a) f is $\mathcal{A}$-measurable.
b) For each $a \in \mathcal{R}$, $f^{-1}((-\infty, a)) \in \mathcal{A}$.
c) For each $a \in \mathcal{R}$, $f^{-1}((a, \infty)) \in \mathcal{A}$.
d) For each $a \in \mathcal{R}$, $f^{-1}((-\infty, a]) \in \mathcal{A}$.
e) For each $a \in \mathcal{R}$, $f^{-1}([a, \infty)) \in \mathcal{A}$.

Theorem 5.3, which we prove next, gives several important properties of real-valued $\mathcal{A}$-measurable functions. Theorem 4.1 on page 113 is a special case.

□ □ □ **THEOREM 5.3**

Let $(\Omega, \mathcal{A})$ be a measurable space. The collection of all real-valued $\mathcal{A}$-measurable functions forms an algebra. In other words, if f and g are $\mathcal{A}$-measurable and $\alpha \in \mathcal{R}$, then

a) $f + g$ is $\mathcal{A}$-measurable.
b) αf is $\mathcal{A}$-measurable.
c) $f \cdot g$ is $\mathcal{A}$-measurable.

PROOF a) By Proposition 5.1, to prove that $f + g$ is $\mathcal{A}$-measurable, it suffices to show that $\{ x : f(x) + g(x) > a \} \in \mathcal{A}$ for each $a \in \mathcal{R}$. Now,

$$\{ x : f(x) + g(x) > a \} = \{ x : f(x) > a - g(x) \}$$
$$= \bigcup_{r \in Q} \{ x : f(x) > r > a - g(x) \}$$
$$= \bigcup_{r \in Q} \left(f^{-1}((r, \infty)) \cap g^{-1}((a - r, \infty)) \right).$$

This last union is an $\mathcal{A}$-measurable set since f and g are $\mathcal{A}$-measurable functions, $\mathcal{A}$ is a σ-algebra, and Q is countable. Consequently, $f + g$ is an $\mathcal{A}$-measurable function.

b) If $\alpha = 0$, then $\alpha f \equiv 0$, which is $\mathcal{A}$-measurable (why?). So, assume $\alpha \neq 0$ and let O be any open set in $\mathcal{R}$. Then $\alpha^{-1}O = \{ \alpha^{-1} y : y \in O \}$ is open. Therefore, because f is $\mathcal{A}$-measurable, $(\alpha f)^{-1}(O) = f^{-1}(\alpha^{-1}O) \in \mathcal{A}$. This fact proves that αf is $\mathcal{A}$-measurable.

c) First we show that if f is $\mathcal{A}$-measurable, then so is f^2. If $a < 0$, then $(f^2)^{-1}((a, \infty)) = \Omega \in \mathcal{A}$. If $a \geq 0$, then we have

$$(f^2)^{-1}((a, \infty)) = \{ x : f(x)^2 > a \} = \{ x : f(x) > \sqrt{a} \} \cup \{ x : f(x) < -\sqrt{a} \}$$
$$= f^{-1}((\sqrt{a}, \infty)) \cup f^{-1}((-\infty, -\sqrt{a})).$$

This last union is an $\mathcal{A}$-measurable set because f is $\mathcal{A}$-measurable. Hence, f^2 is an $\mathcal{A}$-measurable function whenever f is.

Now, for any two functions f and g, we can write

$$f \cdot g = \frac{1}{4}\left((f+g)^2 - (f-g)^2\right).$$

Applying parts (a) and (b) of this theorem and the fact that the square of an $\mathcal{A}$-measurable function is $\mathcal{A}$-measurable, we conclude that $f \cdot g$ is an $\mathcal{A}$-measurable function. ∎

We should emphasize that the measurability (or nonmeasurability) of a function depends only on the σ-algebra, $\mathcal{A}$, of subsets of Ω; that is, it has nothing to do with a measure. Nonetheless, if $(\Omega, \mathcal{A}, P)$ is a probability space, then the $\mathcal{A}$-measurable functions are called **random variables.**

Thus, an $\mathcal{A}$-measurable function is a random variable only when considered in the context of a probability space. By the way, in probability theory, random variables are usually denoted by uppercase italicized English-alphabet letters that are near the end of the alphabet (e.g., X, Y, and Z) instead of the more usual f, g, and h.

EXAMPLE 5.3   *Illustrates Random Variables*

Let $(\Omega, \mathcal{A}, P)$ be the probability space from subpart (ii) of Example 5.1(h) on page 147. Define $X(\text{HH}) = 2$, $X(\text{HT}) = X(\text{TH}) = 1$, and $X(\text{TT}) = 0$. Then $X \colon \Omega \to \mathcal{R}$ is a random variable. It indicates the number of heads obtained when a balanced coin is tossed twice. □

Our next result is a generalization of Proposition 4.7 on page 140 to an arbitrary complete measure space. Its proof is essentially identical to that of Proposition 4.7.

□ □ □   PROPOSITION 5.2

Suppose that $(\Omega, \mathcal{A}, \mu)$ is a complete measure space. If f is $\mathcal{A}$-measurable and $g = f$ μ-ae, then g is $\mathcal{A}$-measurable.

Complex-Valued Measurable Functions

In applying real analysis, we often encounter complex-valued functions. This occurs, for instance, in Fourier analysis. We will denote the set of all complex numbers by $\mathbb{C}$. Here now is the definition of measurability for complex-valued functions.

DEFINITION 5.5   Complex-Valued Measurable Function

Let $(\Omega, \mathcal{A})$ be a measurable space. A complex-valued function f on Ω is said to be an **$\mathcal{A}$-measurable function** if the inverse image of each open subset of $\mathbb{C}$ under f is an $\mathcal{A}$-measurable set, that is, if $f^{-1}(O) \in \mathcal{A}$ for all open sets $O \subset \mathbb{C}$.

The following theorem provides a useful characterization of measurability for complex-valued functions. We leave the proof of the theorem as an exercise for the reader.

□ □ □ **THEOREM 5.4**

A complex-valued function f on Ω is $\mathcal{A}$-measurable if and only if both its real part $\Re f$ and its imaginary part $\Im f$ are (real-valued) $\mathcal{A}$-measurable functions.

EXAMPLE 5.4 *Illustrates Complex-Valued Measurable Functions*

a) If f is a real-valued $\mathcal{A}$-measurable function on Ω, then it is also a complex-valued $\mathcal{A}$-measurable function.

b) Let $\Omega = \mathcal{R}$ and $\mathcal{A} = \mathcal{B}$. Define $f : \mathcal{R} \to \mathbb{C}$ by $f(x) = e^{ix}$. The real and imaginary parts of $f(x)$ are $\cos x$ and $\sin x$, respectively. Since those two functions are continuous, they are $\mathcal{B}$-measurable. Consequently, by Theorem 5.4, f is a complex-valued $\mathcal{B}$-measurable function.

c) If g and h are real-valued $\mathcal{A}$-measurable functions, then, by Theorem 5.4, the complex-valued function $f = g + ih$ is also $\mathcal{A}$-measurable.

d) Let $\{a_n\}_{n=1}^{\infty}$ be a sequence of complex numbers and define $f : \mathcal{N} \to \mathbb{C}$ by $f(n) = a_n$. Then f is a complex-valued $\mathcal{P}(\mathcal{N})$-measurable function. □

Theorem 5.3 holds also for complex-valued $\mathcal{A}$-measurable functions. That is, the collection of complex-valued $\mathcal{A}$-measurable functions forms a (complex) algebra. See Exercise 5.32.

Extended Real-Valued Measurable Functions

In addition to real- and complex-valued functions, we frequently must deal with **extended real-valued functions,** in other words, functions that take values in $\mathcal{R}^* = \mathcal{R} \cup \{-\infty, \infty\}$. This is especially so when considering suprema, infima, and limits. For instance, define

$$f_n(x) = \frac{n}{\sqrt{2\pi}} e^{-(nx)^2/2}$$

for $x \in \mathcal{R}$ and $n \in \mathcal{N}$. Then, as $n \to \infty$, $f_n(x) \to 0$ if $x \neq 0$ and $f_n(0) \to \infty$. Consequently, the sequence $\{f_n\}_{n=1}^{\infty}$ of real-valued functions converges pointwise to the extended real-valued function f, where

$$f(x) = \begin{cases} 0, & \text{if } x \neq 0; \\ \infty, & \text{if } x = 0. \end{cases}$$

Thus, we next consider measurability for extended real-valued functions. Recall that, by definition, a real-valued function f is $\mathcal{A}$-measurable if $f^{-1}(O) \in \mathcal{A}$ for all open sets $O \subset \mathcal{R}$. Also, by definition, a complex-valued function f is $\mathcal{A}$-measurable if $f^{-1}(O) \in \mathcal{A}$ for all open sets $O \subset \mathbb{C}$. Hence, once we identify the open sets of $\mathcal{R}^*$, we have a natural way to define extended real-valued $\mathcal{A}$-measurable functions.

DEFINITION 5.6 Open Subsets of the Extended Real Numbers

A subset of $\mathcal{R}^*$ is said to be **open** if it can be expressed as a union of intervals of the form (a, b), $[-\infty, b)$, and $(a, \infty]$, where $a, b \in \mathcal{R}$.

DEFINITION 5.7 **Extended Real-Valued Measurable Function**

Let $(\Omega, \mathcal{A})$ be a measurable space. An extended real-valued function f on Ω is said to be an **$\mathcal{A}$-measurable function** if the inverse image of each open subset of $\mathcal{R}^*$ under f is an $\mathcal{A}$-measurable set, that is, if $f^{-1}(O) \in \mathcal{A}$ for all open sets $O \subset \mathcal{R}^*$.

The next proposition provides the analogue of Proposition 5.1 for extended real-valued functions. Its proof is left as an exercise.

□ □ □ **PROPOSITION 5.3**

Let $(\Omega, \mathcal{A})$ be a measurable space and f an extended real-valued function on Ω. Then the following statements are equivalent:

a) f is $\mathcal{A}$-measurable.
b) For each $a \in \mathcal{R}$, $f^{-1}\big([-\infty, a)\big) \in \mathcal{A}$.
c) For each $a \in \mathcal{R}$, $f^{-1}\big((a, \infty]\big) \in \mathcal{A}$.
d) For each $a \in \mathcal{R}$, $f^{-1}\big([-\infty, a]\big) \in \mathcal{A}$.
e) For each $a \in \mathcal{R}$, $f^{-1}\big([a, \infty]\big) \in \mathcal{A}$.

Theorem 5.3 shows that the collection of real-valued $\mathcal{A}$-measurable functions forms an algebra. In the case of extended real-valued functions, if we adopt the convention that $\infty - \infty$ is some fixed extended real number, then the collection of extended real-valued $\mathcal{A}$-measurable functions is closed under addition, scalar multiplication, and multiplication. See Exercises 5.39 and 5.40.

The following theorem establishes that the collection of extended real-valued $\mathcal{A}$-measurable functions is closed under maxima, minima, suprema, infima, and pointwise limits. Note that Theorem 4.2 on page 113 is an immediate consequence.

□ □ □ **THEOREM 5.5**

Suppose that f and g are extended real-valued $\mathcal{A}$-measurable functions and that $\{f_n\}_{n=1}^{\infty}$ is a sequence of extended real-valued $\mathcal{A}$-measurable functions. Then

a) $f \vee g$ and $f \wedge g$ are $\mathcal{A}$-measurable.
b) $\sup_n f_n$ and $\inf_n f_n$ are $\mathcal{A}$-measurable.
c) $\limsup_{n \to \infty} f_n$ and $\liminf_{n \to \infty} f_n$ are $\mathcal{A}$-measurable.
d) If $\{f_n\}_{n=1}^{\infty}$ converges pointwise, then $\lim_{n \to \infty} f_n$ is $\mathcal{A}$-measurable.

PROOF a) Let $h = f \vee g$ and $a \in \mathcal{R}$. Then $h^{-1}\big((a, \infty]\big) = f^{-1}\big((a, \infty]\big) \cup g^{-1}\big((a, \infty]\big)$. This union is in $\mathcal{A}$ because f and g are $\mathcal{A}$-measurable functions. Thus, $f \vee g$ is $\mathcal{A}$-measurable. Similarly, $f \wedge g$ is $\mathcal{A}$-measurable.

b) Let $h = \sup_n f_n$ and $a \in \mathcal{R}$. Then $h^{-1}\big((a, \infty]\big) = \bigcup_{n=1}^{\infty} f_n^{-1}\big((a, \infty]\big)$. This union is in $\mathcal{A}$ because each f_n is an $\mathcal{A}$-measurable function. Hence, we see that $\sup_n f_n$ is $\mathcal{A}$-measurable. Similarly, $\inf_n f_n$ is $\mathcal{A}$-measurable.

c) Noting that $\limsup_{n\to\infty} f_n = \inf_n \sup_{k\geq n} f_k$, it follows from part (b) that $\limsup_{n\to\infty} f_n$ is $\mathcal{A}$-measurable. Using an entirely similar argument, we find that $\liminf_{n\to\infty} f_n$ is $\mathcal{A}$-measurable.

d) If $\{f_n\}_{n=1}^{\infty}$ converges pointwise, then we have $\lim_{n\to\infty} f_n = \limsup_{n\to\infty} f_n$. So, $\lim_{n\to\infty} f_n$ is $\mathcal{A}$-measurable by part (c). ∎

A common application of Theorem 5.5 occurs when $\{f_n\}_{n=1}^{\infty}$ is a sequence of real-valued $\mathcal{A}$-measurable functions, but where at least one of the functions $\inf_n f_n$, $\sup_n f_n$, $\liminf_{n\to\infty} f_n$, $\limsup_{n\to\infty} f_n$, and $\lim_{n\to\infty} f_n$ is an extended real-valued $\mathcal{A}$-measurable function.

EXAMPLE 5.5   *Illustrates Theorem 5.5*

a) Let $(\Omega, \mathcal{A}, \mu) = (\mathcal{R}, \mathcal{M}, \lambda)$. For $x \in \mathcal{R}$ and $n \in \mathcal{N}$, define

$$f_n(x) = \frac{n}{\sqrt{2\pi}} e^{-(nx)^2/2}.$$

Then $\{f_n\}_{n=1}^{\infty}$ is a sequence of real-valued $\mathcal{M}$-measurable functions. Moreover, $f_n \to f$ pointwise, where

$$f(x) = \begin{cases} 0, & \text{if } x \neq 0; \\ \infty, & \text{if } x = 0. \end{cases}$$

By Theorem 5.5(d), f is an extended real-valued $\mathcal{A}$-measurable function, a fact that we can easily verify directly.

b) Let f be an extended real-valued $\mathcal{A}$-measurable function. By Theorem 5.5(a), $|f|$ is $\mathcal{A}$-measurable since $|f| = f \vee -f$. □

Theorem 5.5(d) shows that if a sequence $\{f_n\}_{n=1}^{\infty}$ of $\mathcal{A}$-measurable functions converges pointwise to a function f, then f is an $\mathcal{A}$-measurable function. What if the convergence is only almost everywhere? In general, we cannot conclude that f is $\mathcal{A}$-measurable; however, for complete measure spaces we can.

□ □ □   **PROPOSITION 5.4**

Let $(\Omega, \mathcal{A}, \mu)$ be a complete measure space. Suppose that $\{f_n\}_{n=1}^{\infty}$ is a sequence of complex-valued or extended real-valued $\mathcal{A}$-measurable functions and that $f_n \to f$ μ-ae. Then f is an $\mathcal{A}$-measurable function.

PROOF   The proof is essentially identical to that of Proposition 4.8 on page 140 and is left to the reader. ∎

Exercises for Section 5.2

5.20 Prove Proposition 5.1 on page 152.

5.21 Let $(\Omega, \mathcal{A})$ be a measurable space and f a real-valued function on Ω. Prove that f is $\mathcal{A}$-measurable if and only if $f^{-1}(B) \in \mathcal{A}$ for each $B \in \mathcal{B}$.

5.22 Suppose that $(\Omega, \mathcal{A})$ is a measurable space and that $f \colon \Omega \to \mathcal{R}$ is an $\mathcal{A}$-measurable function. Further suppose that $g \colon \mathcal{R} \to \mathcal{R}$ is a Borel measurable function. Prove that $g \circ f$ is $\mathcal{A}$-measurable.

5.23 Let $D \in \mathcal{B}$. Show that $\widehat{C}(D)$, the collection of Borel measurable functions on D, is precisely the collection of $\mathcal{B}_D$-measurable functions.

5.24 Let $(\Omega, \mathcal{A})$ be a measurable space, $D \in \mathcal{A}$, and $\mathcal{A}_D = \{ D \cap A : A \in \mathcal{A} \}$.
   a) If $f : \Omega \to \mathcal{R}$ is $\mathcal{A}$-measurable, show that $f_{|D}$ is $\mathcal{A}_D$-measurable.
   b) Suppose that $g : D \to \mathcal{R}$ is $\mathcal{A}_D$-measurable. Define $f : \Omega \to \mathcal{R}$ by

$$f(x) = \begin{cases} g(x), & x \in D; \\ 0, & x \notin D. \end{cases}$$

   Prove that f is $\mathcal{A}$-measurable. (This result shows that every $\mathcal{A}_D$-measurable function can be extended to an $\mathcal{A}$-measurable function.)

5.25 Prove Proposition 5.2 on page 153.

5.26 Provide an example to show that the hypothesis of completeness cannot be omitted from Proposition 5.2.

5.27 If O is an open subset of $\mathcal{R}$ and α is a nonzero real number, show that $\alpha^{-1}O$ is an open subset of $\mathcal{R}$.

5.28 Prove Theorem 5.4 on page 154. *Hint:* Use the fact that each open set in $\mathbb{C}$ is a countable union of open rectangles. [An open rectangle in $\mathbb{C}$ is a set of the form $\{ u + iv \in \mathbb{C} : a < u < b,\ c < v < d \}$.]

5.29 Show that every real-valued $\mathcal{A}$-measurable function is a complex-valued $\mathcal{A}$-measurable function.

5.30 The collection $\mathcal{B}_2$ of Borel sets of $\mathbb{C}$ is defined to be the smallest σ-algebra of subsets of $\mathbb{C}$ that contains all the open subsets of $\mathbb{C}$. Show that $f : \Omega \to \mathbb{C}$ is $\mathcal{A}$-measurable if and only if $f^{-1}(B) \in \mathcal{A}$ for all $B \in \mathcal{B}_2$.

5.31 Let $(\Omega, \mathcal{A}, P)$ be a probability space, X a random variable on Ω, and t a fixed real number. Define $g : \Omega \to \mathbb{C}$ by $g = e^{itX}$; that is, for each $x \in \Omega$, $g(x) = e^{itX(x)}$. Prove that g is $\mathcal{A}$-measurable. Is g a random variable? Explain your answer.

★5.32 Prove that the collection of complex-valued $\mathcal{A}$-measurable functions forms a complex algebra. That is, if f and g are complex-valued $\mathcal{A}$-measurable functions and $\alpha \in \mathbb{C}$, show that $f + g$, αf, and $f \cdot g$ are complex-valued $\mathcal{A}$-measurable functions.

5.33 Show that each open subset of $\mathcal{R}$ is also an open subset of $\mathcal{R}^*$.

5.34 Prove Proposition 5.3 on page 155. *Hint:* Show that each open set in $\mathcal{R}^*$ can be written as a countable union of intervals of the form (a, b), $[-\infty, b)$, and $(a, \infty]$, where $a, b \in \mathcal{R}$.

5.35 Show that $f : \Omega \to \mathcal{R}^*$ is $\mathcal{A}$-measurable if and only if (i) $f^{-1}(\{-\infty\})$ and $f^{-1}(\{\infty\})$ are in $\mathcal{A}$ and (ii) $f^{-1}(B) \in \mathcal{A}$ for all $B \in \mathcal{B}$.

5.36 Show that every real-valued $\mathcal{A}$-measurable function is also an extended real-valued $\mathcal{A}$-measurable function.

5.37 Show that a set $O \subset \mathcal{R}$ is open in $\mathcal{R}$ if and only if there is an open subset U of $\mathcal{R}^*$ such that $O = \mathcal{R} \cap U$.

5.38 Suppose that f and g are extended real-valued $\mathcal{A}$-measurable functions. Prove that the following three sets are $\mathcal{A}$-measurable:
   a) $\{ x : f(x) > g(x) \}$.
   b) $\{ x : f(x) \geq g(x) \}$.
   c) $\{ x : f(x) = g(x) \}$.

5.39 Let f and g be extended real-valued $\mathcal{A}$-measurable functions and let $\beta \in \mathcal{R}^*$. Set

$$E = \{ x : f(x) = \infty,\ g(x) = -\infty \} \cup \{ x : f(x) = -\infty,\ g(x) = \infty \}.$$

For $x \in E$, define $(f + g)(x) = \beta$; otherwise, define $(f + g)(x) = f(x) + g(x)$, as usual. Prove that $f + g$ is $\mathcal{A}$-measurable.

5.40 With the convention established in the preceding exercise, prove that the collection of extended real-valued $\mathcal{A}$-measurable functions is closed under scalar multiplication and multiplication.

5.41 Suppose that $\{f_n\}_{n=1}^{\infty}$ is a sequence of extended real-valued $\mathcal{A}$-measurable functions. Verify that $\{\, x : \lim_{n \to \infty} f_n(x) \text{ exists} \,\}$ is an $\mathcal{A}$-measurable set.

5.42 Suppose $\{f_n\}_{n=1}^{\infty}$ is a sequence of complex-valued $\mathcal{A}$-measurable functions that converges pointwise to a complex-valued function f. Prove that f is $\mathcal{A}$-measurable.

5.43 Construct a sequence $\{f_n\}_{n=1}^{\infty}$ of $\mathcal{A}$-measurable functions that converges almost everywhere to a function f that is not $\mathcal{A}$-measurable. *Hint:* Take $(\Omega, \mathcal{A}, \mu) = (\mathcal{R}, \mathcal{B}, \lambda_{|\mathcal{B}})$ and do something with a non-Borel measurable subset of the Cantor set.

5.44 Prove Proposition 5.4 on page 156.

★5.45 Suppose that $\{f_n\}_{n=1}^{\infty}$ is a sequence of complex-valued $\mathcal{A}$-measurable functions. Define

$$f(x) = \begin{cases} \lim_{n \to \infty} f_n(x), & \text{if } \lim_{n \to \infty} f_n(x) \text{ exists;} \\ 0, & \text{otherwise.} \end{cases}$$

Prove that f is $\mathcal{A}$-measurable.

5.46 Suppose that $\{f_n\}_{n=1}^{\infty}$ is a sequence of complex-valued $\mathcal{A}$-measurable functions and that $f_n \to g$ μ-ae. Prove that there exists an $\mathcal{A}$-measurable function f such that $f_n \to f$ μ-ae. *Note:* g need not be $\mathcal{A}$-measurable unless, of course, $(\Omega, \mathcal{A}, \mu)$ is complete.

5.47 Suppose that E is an open subset of $\mathbb{C}$ and that g is a real-valued continuous function on E. Further suppose that f is a complex-valued $\mathcal{A}$-measurable function on Ω with the range of f being a subset of E. Prove that $g \circ f$ is a real-valued $\mathcal{A}$-measurable function on Ω. Repeat the proof if E is a closed subset of $\mathbb{C}$.

5.48 Suppose that $f \colon \Omega \to \mathbb{C}$ is $\mathcal{A}$-measurable. Verify that f can be written in the "polar" form, $f = Re^{i\Theta}$, where $R \colon \Omega \to [0, \infty)$ and $\Theta \colon \Omega \to \mathcal{R}$ are $\mathcal{A}$-measurable functions.

5.3 THE ABSTRACT LEBESGUE INTEGRAL FOR NONNEGATIVE FUNCTIONS

Now that we have discussed measure spaces and measurable functions, we can proceed to develop the abstract Lebesgue integral, that is, the Lebesgue integral on an arbitrary measure space $(\Omega, \mathcal{A}, \mu)$. As we will see, the development of the abstract Lebesgue integral is almost identical to that of the Lebesgue integral on the real line, that is, on $(\mathcal{R}, \mathcal{M}, \lambda)$, given in Chapter 4. Consequently, many of the proofs will be left to the reader.

Following the procedure used in Chapter 4, we will first define the abstract Lebesgue integral of a simple function, then of a nonnegative $\mathcal{A}$-measurable function, and then of a real-valued $\mathcal{A}$-measurable function. In addition, we will also define the abstract Lebesgue integral of extended real-valued and complex-valued $\mathcal{A}$-measurable functions. Nonnegative functions will be considered in this section and general functions in the next.

The Lebesgue Integral of a Nonnegative Simple Function

Let $(\Omega, \mathcal{A}, \mu)$ be a measure space. An $\mathcal{A}$-measurable function on Ω is called a **simple function** if it takes on only finitely many values. More precisely, we have the following definition.

DEFINITION 5.8   **Simple Function and Canonical Representation**

An $\mathcal{A}$-measurable function s is said to be a **simple function** if its range is a finite set. Let a_1, a_2, ..., a_n denote the distinct nonzero values of s and set $A_k = \{\, x : s(x) = a_k \,\}$, $1 \le k \le n$. Then

$$s = \sum_{k=1}^{n} a_k \chi_{A_k}.$$

This is called the **canonical representation** of s.

We leave it as an exercise for the reader to show that the sets $A_1, A_2, \ldots, A_n$, appearing in the canonical representation of an $\mathcal{A}$-measurable simple function, are $\mathcal{A}$-measurable and pairwise disjoint.

EXAMPLE 5.6   *Illustrates Definition 5.8*

a) The Lebesgue measurable simple functions introduced in Chapter 4 are $\mathcal{M}$-measurable simple functions in the sense of Definition 5.8.
b) If Ω is a finite set, then every $\mathcal{A}$-measurable function is simple.   □

We now give the definition of the abstract Lebesgue integral of a nonnegative $\mathcal{A}$-measurable simple function. It is a straightforward generalization of the definition presented in Chapter 4 for the Lebesgue integral of a nonnegative Lebesgue measurable simple function.

DEFINITION 5.9   **Integral of a Nonnegative Simple Function**

Let $(\Omega, \mathcal{A}, \mu)$ be a measure space and s a nonnegative $\mathcal{A}$-measurable simple function on Ω with canonical representation $s = \sum_{k=1}^{n} a_k \chi_{A_k}$. Then the (abstract) **Lebesgue integral of s over Ω with respect to μ** is defined by

$$\int_{\Omega} s(x)\, d\mu(x) = \sum_{k=1}^{n} a_k \mu(A_k).$$

If $E \in \mathcal{A}$, then the (abstract) **Lebesgue integral of s over E with respect to μ** is defined by

$$\int_{E} s(x)\, d\mu(x) = \int_{\Omega} \chi_E(x) s(x)\, d\mu(x).$$

Note: The notations $\int_E s\, d\mu$ and $\int_E s(x)\, \mu(dx)$ are commonly used in place of $\int_E s(x)\, d\mu(x)$.

    The next proposition shows how we can obtain the abstract Lebesgue integral of a nonnegative simple function from a possibly noncanonical representation. The proof is identical to that of Proposition 4.2 on page 114.

□ □ □ **PROPOSITION 5.5**

Let s be a nonnegative $\mathcal{A}$-measurable simple function that can be expressed in the form $s = \sum_{k=1}^{m} b_k \chi_{B_k}$, where this representation is not necessarily canonical but $B_k \in \mathcal{A}$ for $1 \leq k \leq m$ and $B_i \cap B_j = \emptyset$ for $i \neq j$. Then

$$\int_{\Omega} s(x) \, d\mu(x) = \sum_{k=1}^{m} b_k \mu(B_k).$$

More generally, we have

$$\int_{E} s(x) \, d\mu(x) = \sum_{k=1}^{m} b_k \mu(B_k \cap E)$$

for each $E \in \mathcal{A}$.

The following fact is proved in precisely the same way as Lemma 4.1 on page 116.

□ □ □ **PROPOSITION 5.6**

Suppose that s and t are nonnegative $\mathcal{A}$-measurable simple functions and that $\alpha, \beta \geq 0$. Then $\alpha s + \beta t$ is a nonnegative $\mathcal{A}$-measurable simple function and

$$\int_{E} (\alpha s + \beta t) \, d\mu = \alpha \int_{E} s \, d\mu + \beta \int_{E} t \, d\mu$$

for each $E \in \mathcal{A}$.

The Lebesgue Integral of a Nonnegative $\mathcal{A}$-measurable Function

The next thing on the agenda is the definition of the abstract Lebesgue integral for a nonnegative extended real-valued $\mathcal{A}$-measurable function. Proposition 5.7, whose proof is left to the reader as an exercise, provides the motivation for that definition.

□ □ □ **PROPOSITION 5.7**

a) Suppose that f is a nonnegative extended real-valued $\mathcal{A}$-measurable function on Ω. Then there is a nondecreasing sequence of nonnegative $\mathcal{A}$-measurable simple functions that converges pointwise to f. In other words, there is a sequence $\{s_n\}_{n=1}^{\infty}$ of nonnegative $\mathcal{A}$-measurable simple functions such that, for all $x \in \Omega$, $s_1(x) \leq s_2(x) \leq \cdots$ and $\lim_{n \to \infty} s_n(x) = f(x)$.

b) If $\{s_n\}_{n=1}^{\infty}$ is a sequence of nonnegative $\mathcal{A}$-measurable simple functions that converges pointwise on Ω to a function f, then f is a nonnegative extended real-valued $\mathcal{A}$-measurable function.

Proposition 5.7 shows that the functions that can be approximated by nonnegative $\mathcal{A}$-measurable simple functions are precisely the nonnegative extended real-valued $\mathcal{A}$-measurable functions. Thus, we make the following definition.

DEFINITION 5.10 Lebesgue Integral of a Nonnegative Function

Let f be a nonnegative extended real-valued $\mathcal{A}$-measurable function on Ω. Then the (abstract) **Lebesgue integral of f over Ω with respect to μ** is defined by

$$\int_{\Omega} f(x)\, d\mu(x) = \sup_{s} \int_{\Omega} s(x)\, d\mu(x),$$

where the supremum is taken over all nonnegative $\mathcal{A}$-measurable simple functions that are dominated by f. If $E \in \mathcal{A}$, then the (abstract) **Lebesgue integral of f over E with respect to μ** is defined by

$$\int_{E} f(x)\, d\mu(x) = \int_{\Omega} \chi_E(x) f(x)\, d\mu(x).$$

Note: The abstract Lebesgue integral of a nonnegative $\mathcal{M}$-measurable function with respect to λ is identical to its Lebesgue integral, as defined in Chapter 4.

Some of the more important properties of the abstract Lebesgue integral for nonnegative extended real-valued $\mathcal{A}$-measurable functions are provided in Proposition 5.8. The proof is left as an exercise for the reader.

□ □ □ **PROPOSITION 5.8**

Let f and g be nonnegative extended real-valued $\mathcal{A}$-measurable functions on Ω, $\alpha \geq 0$, and $E \in \mathcal{A}$. Then

a) $f \leq g$ μ-ae $\Rightarrow \int_E f\, d\mu \leq \int_E g\, d\mu$.

b) $B \subset E$ and $B \in \mathcal{A} \Rightarrow \int_B f\, d\mu \leq \int_E f\, d\mu$.

c) $f(x) = 0$ for all $x \in E \Rightarrow \int_E f\, d\mu = 0$.

d) $\mu(E) = 0 \Rightarrow \int_E f\, d\mu = 0$.

e) $\int_E \alpha f\, d\mu = \alpha \int_E f\, d\mu$.

Convergence Properties of the Abstract Lebesgue Integral for Nonnegative $\mathcal{A}$-measurable Functions

We now present two major convergence theorems for the abstract Lebesgue integral of nonnegative extended real-valued $\mathcal{A}$-measurable functions — the monotone convergence theorem (MCT) and Fatou's lemma. The proofs are similar to those given in Section 4.2 (page 121 onward).

The MCT is stated first. Observe that it applies to extended real-valued $\mathcal{A}$-measurable functions as well as to real-valued $\mathcal{A}$-measurable functions.

□ □ □ **THEOREM 5.6 Monotone Convergence Theorem (MCT)**

Suppose that $\{f_n\}_{n=1}^{\infty}$ is a monotone nondecreasing sequence of nonnegative extended real-valued $\mathcal{A}$-measurable functions. Then

$$\int_E \lim_{n \to \infty} f_n\, d\mu = \lim_{n \to \infty} \int_E f_n\, d\mu$$

for each $E \in \mathcal{A}$.

◻ ◻ ◻ **COROLLARY 5.1**

Let $f, g, f_1, f_2, \ldots$ be nonnegative extended real-valued $\mathcal{A}$-measurable functions and let $E \in \mathcal{A}$. Then

a) $\int_E (f + g)\, d\mu = \int_E f\, d\mu + \int_E g\, d\mu$.

b) $\int_E \sum_{n=1}^{\infty} f_n\, d\mu = \sum_{n=1}^{\infty} \int_E f_n\, d\mu$.

c) *If $\{E_n\}_n \subset \mathcal{A}$ are pairwise disjoint, then $\int_{\bigcup_n E_n} f\, d\mu = \sum_n \int_{E_n} f\, d\mu$.*

Proposition 5.8(e) and Corollary 5.1(a) together imply that if f and g are nonnegative extended real-valued $\mathcal{A}$-measurable functions and $\alpha, \beta \geq 0$, then

$$\int_\Omega (\alpha f + \beta g)\, d\mu = \alpha \int_\Omega f\, d\mu + \beta \int_\Omega g\, d\mu. \tag{5.2}$$

Equation (5.2), Proposition 5.7, and the MCT are frequently used together for "bootstrapping arguments." That is, suppose we want to prove that a certain Lebesgue-integral property holds for all nonnegative $\mathcal{A}$-measurable functions. To bootstrap, we employ three steps: First we show that the property holds for characteristic functions of $\mathcal{A}$-measurable sets; next we apply (5.2) to conclude that the property holds for nonnegative simple functions; and then we use Proposition 5.7(a) and the MCT to deduce that the property holds for all nonnegative $\mathcal{A}$-measurable functions. Exercises 5.60 and 5.61 provide illustrations of bootstrapping.

Next we state Fatou's lemma. This version of Fatou's lemma not only generalizes to arbitrary measure spaces the version presented in Theorem 4.6 on page 126 but its hypotheses are less restrictive. Specifically, it does not impose any convergence conditions on $\{f_n\}_{n=1}^{\infty}$.

◻ ◻ ◻ **THEOREM 5.7 Fatou's Lemma**

Let $\{f_n\}_{n=1}^{\infty}$ be a sequence of nonnegative extended real-valued $\mathcal{A}$-measurable functions. Then

$$\int_E \liminf_{n \to \infty} f_n\, d\mu \leq \liminf_{n \to \infty} \int_E f_n\, d\mu$$

for each $E \in \mathcal{A}$.

EXAMPLE 5.7 *Illustrates the Abstract Lebesgue Integral*

a) Let $(\Omega, \mathcal{A}, \mu)$ be a measure space and f a nonnegative extended real-valued $\mathcal{A}$-measurable function on Ω. Suppose that $x_0 \in \Omega$ and that $\{x_0\} \in \mathcal{A}$. We claim that

$$\int_{\{x_0\}} f\, d\mu = f(x_0)\mu(\{x_0\}). \tag{5.3}$$

To see this, note that $\chi_{\{x_0\}} f$ is the simple function $f(x_0)\chi_{\{x_0\}}$ and, hence, by Definition 5.9 on page 159,

$$\int_{\{x_0\}} f\, d\mu = \int_\Omega \chi_{\{x_0\}} f\, d\mu = \int_\Omega f(x_0)\chi_{\{x_0\}}\, d\mu = f(x_0)\mu(\{x_0\}).$$

More generally, let $C = \{x_n\}_n$ be a countable subset of Ω with $\{x_n\} \in \mathcal{A}$ for each n. Then, by Corollary 5.1(c) and (5.3),

$$\int_C f \, d\mu = \int_{\bigcup_n \{x_n\}} f \, d\mu = \sum_n \int_{\{x_n\}} f \, d\mu = \sum_n f(x_n)\mu(\{x_n\}). \qquad (5.4)$$

b) Let μ be counting measure on $\mathcal{P}(\mathcal{N})$. Then, as we learned in Example 5.2(c), a nonnegative real-valued $\mathcal{P}(\mathcal{N})$-measurable function f on $\mathcal{N}$ is a nonnegative infinite sequence $\{a_n\}_{n=1}^{\infty}$, where we have let $a_n = f(n)$. Thus, by (5.4),

$$\int_{\mathcal{N}} f \, d\mu = \sum_{n=1}^{\infty} f(n)\mu(\{n\}) = \sum_{n=1}^{\infty} a_n.$$

Hence, we can apply abstract measure theory to study infinite series.

c) Let $(\Omega, \mathcal{A}, P)$ be a probability space and X a nonnegative random variable. Then the abstract Lebesgue integral of X over Ω with respect to P is called the **mean (expectation, expected value)** of X. The mean of X is denoted by $\mathcal{E}(X)$. Thus,

$$\mathcal{E}(X) = \int_\Omega X \, dP.$$

For instance, consider the random experiment of tossing a balanced coin twice. An appropriate probability space for that experiment is $(\Omega, \mathcal{A}, P)$, where $\Omega = \{\text{HH, HT, TH, TT}\}$, $\mathcal{A} = \mathcal{P}(\Omega)$ and, for $E \in \mathcal{A}$, $P(E) = N(E)/4$. Let X denote the number of heads obtained. Then, by (5.4), the mean of X equals

$$\mathcal{E}(X) = \int_\Omega X \, dP = X(\text{HH})P(\{\text{HH}\}) + X(\text{HT})P(\{\text{HT}\})$$

$$+ X(\text{TH})P(\{\text{TH}\}) + X(\text{TT})P(\{\text{TT}\})$$

$$= 2 \cdot \frac{1}{4} + 1 \cdot \frac{1}{4} + 1 \cdot \frac{1}{4} + 0 \cdot \frac{1}{4} = 1,$$

which is intuitively what it should be.

d) Let Ω be a set, $\{x_n\}_n$ a sequence of distinct elements of Ω, and $\{b_n\}_n$ a sequence of nonnegative real numbers. For $E \subset \Omega$, define

$$\mu(E) = \sum_{x_n \in E} b_n.$$

Then μ is a measure on $\mathcal{P}(\Omega)$. Let f be a nonnegative function on Ω and set $C = \{x_n\}_n$. Then, by Corollary 5.1(c), Proposition 5.8(d) on page 161, and (5.4),

$$\int_\Omega f \, d\mu = \int_C f \, d\mu + \int_{C^c} f \, d\mu = \int_C f \, d\mu + 0$$

$$= \sum_n f(x_n)\mu(\{x_n\}) = \sum_n f(x_n)b_n. \qquad (5.5)$$

We will employ (5.5) frequently.

e) Let Ω be a set, $\mathcal{A} = \mathcal{P}(\Omega)$, and μ counting measure on $\mathcal{A}$. If f is a nonnegative function on Ω, then

$$\int_\Omega f \, d\mu = \sum_{x \in \Omega} f(x),$$

where $\sum_{x \in \Omega} f(x) = \sup \left\{ \sum_{x \in F} f(x) : F \text{ finite}, F \subset \Omega \right\}$. The verification of this fact is left to the reader. □

Exercises for Section 5.3

5.49 Establish that the sets appearing in the canonical representation of an $\mathcal{A}$-measurable simple function are $\mathcal{A}$-measurable and pairwise disjoint.

5.50 Prove Proposition 5.7 on page 160. *Hint:* Refer to Proposition 4.3 on page 116.

5.51 Suppose that f is a nonnegative extended real-valued $\mathcal{A}$-measurable function on Ω, $c > 0$, and $A_c = \{ x : f(x) \geq c \}$. Prove that

$$\mu(A_c) \leq \frac{1}{c} \int_\Omega f \, d\mu.$$

★5.52 Let f be a nonnegative extended real-valued $\mathcal{A}$-measurable function on Ω and $E \in \mathcal{A}$. Prove that $\int_E f \, d\mu = 0$ if and only if $f = 0$ μ-ae on E.

★5.53 Suppose that f is a nonnegative extended real-valued $\mathcal{A}$-measurable function on Ω and that $\int_\Omega f \, d\mu < \infty$. Show that f is finite μ-ae.

5.54 Prove Proposition 5.8 on page 161. *Hint:* Refer to Proposition 4.4 on page 118.

5.55 Prove the MCT, Theorem 5.6 on page 161.

5.56 Show that for a fixed $E \in \mathcal{A}$, the conclusion of the MCT remains valid if the hypotheses are satisfied only on E.

5.57 Prove Corollary 5.1 on page 162.

5.58 Suppose that f is a nonnegative extended real-valued $\mathcal{A}$-measurable function on Ω. Also, suppose that $\{E_n\}_{n=1}^\infty \subset \mathcal{A}$ with $E_1 \subset E_2 \subset \cdots$. Prove that

$$\int_{\bigcup_{n=1}^\infty E_n} f \, d\mu = \lim_{n \to \infty} \int_{E_n} f \, d\mu.$$

5.59 Prove Fatou's lemma, Theorem 5.7 on page 162.

5.60 Suppose that $(\Omega, \mathcal{A}, \mu)$ is a measure space, $D \in \mathcal{A}$, and f is a nonnegative extended real-valued $\mathcal{A}$-measurable function on Ω. Let $(D, \mathcal{A}_D, \mu_D)$ be as defined in Example 5.1(c) on page 146. Show that

$$\int_D f \, d\mu = \int_D f_{|D} \, d\mu_D.$$

Hint: Use bootstrapping.

★5.61 Let $(\Omega, \mathcal{A}, \mu)$ be a measure space and g a nonnegative $\mathcal{A}$-measurable function on Ω. For $E \in \mathcal{A}$, define

$$\nu(E) = \int_E g \, d\mu.$$

a) Show that ν is a measure on $\mathcal{A}$.

b) Show that

$$\int_\Omega f \, d\nu = \int_\Omega fg \, d\mu$$

for each nonnegative $\mathcal{A}$-measurable function f. *Hint:* Bootstrap.

5.62 Let $\{a_{mn}\}_{m,n=1}^\infty$ be a double sequence of nonnegative numbers. Prove that

$$\sum_{n=1}^\infty \sum_{m=1}^\infty a_{mn} = \sum_{m=1}^\infty \sum_{n=1}^\infty a_{mn}.$$

Hint: Refer to Example 5.7(b) on page 163.

5.63 Let $f\colon \Omega \to [0,1]$ be an $\mathcal{A}$-measurable function.

a) Prove that $\lim_{n\to\infty} \int_\Omega f^{\frac{1}{n}} \, d\mu = \mu\big(f^{-1}\big((0,1]\big)\big)$.

b) If $\mu(\Omega) < \infty$, prove that $\lim_{n\to\infty} \int_\Omega f^n \, d\mu = \mu\big(f^{-1}(\{1\})\big)$.

5.4 THE GENERAL ABSTRACT LEBESGUE INTEGRAL

In the previous section, we discussed the abstract Lebesgue integral for non-negative extended real-valued $\mathcal{A}$-measurable functions. We will now expand the definition of the abstract Lebesgue integral so that it applies to $\mathcal{A}$-measurable functions that are not necessarily nonnegative. We begin with extended real-valued functions.

Lebesgue Integral of an Extended Real-Valued Function

Let $(\Omega, \mathcal{A}, \mu)$ be a measure space. To define the abstract Lebesgue integral of an extended real-valued $\mathcal{A}$-measurable function f on Ω, we follow the procedure used in Section 4.3 for defining the Lebesgue integral of a real-valued Lebesgue measurable function on $\mathcal{R}$.

DEFINITION 5.11   Integral of an Extended Real-Valued Function

Let f be an extended real-valued $\mathcal{A}$-measurable function on Ω and $E \in \mathcal{A}$. Then the (abstract) **Lebesgue integral of f over E with respect to μ** is defined by

$$\int_E f \, d\mu = \int_E f^+ \, d\mu - \int_E f^- \, d\mu \tag{5.6}$$

provided that the right-hand side makes sense; that is, at least one of the integrals on the right-hand side of (5.6) is finite. Here $f^+ = f \vee 0$ and $f^- = -(f \wedge 0)$ denote the positive and negative parts of f, respectively. In addition, we say that **f is Lebesgue integrable over E** if both integrals on the right-hand side of (5.6) are finite or, equivalently, if

$$\int_E |f| \, d\mu = \int_E f^+ \, d\mu + \int_E f^- \, d\mu < \infty. \tag{5.7}$$

If f is Lebesgue integrable over Ω, then we say that **f is Lebesgue integrable.**

We should mention that if f is Lebesgue integrable (over Ω), then it is Lebesgue integrable over every $E \in \mathcal{A}$. Here are some examples.

EXAMPLE 5.8 *Illustrates Definition 5.11*

a) Let $(\Omega, \mathcal{A}, \mu) = (\mathcal{R}, \mathcal{M}, \lambda)$ and $f(x) = x$. Then

$$f^+(x) = \begin{cases} x, & x \geq 0; \\ 0, & x < 0. \end{cases} \quad \text{and} \quad f^-(x) = \begin{cases} 0, & x \geq 0; \\ -x, & x < 0. \end{cases}$$

(i) If $E = \mathcal{R}$, then $\int_{\mathcal{R}} f^+ \, d\lambda = \int_{\mathcal{R}} f^- \, d\lambda = \infty$. Hence, the integral $\int_{\mathcal{R}} f \, d\lambda$ is not defined.

(ii) If $E = [-1, 2]$, then $\int_E f^+ \, d\lambda = 2$ and $\int_E f^- \, d\lambda = 1/2$ and, consequently, $\int_E f \, d\lambda = 2 - 1/2 = 3/2$. And, because $\int_E |f| \, d\lambda = 2 + 1/2 = 5/2 < \infty$, we see that f is Lebesgue integrable over $[-1, 2]$.

(iii) If $E = (-\infty, 1)$, then $\int_E f^+ \, d\lambda = 1/2$ and $\int_E f^- \, d\lambda = \infty$ and, consequently, $\int_E f \, d\lambda = 1/2 - \infty = -\infty$. But, as $\int_E |f| \, d\lambda = 1/2 + \infty = \infty$, we see that f is not Lebesgue integrable over $(-\infty, 1)$.

b) Let $(\Omega, \mathcal{A}, \mu) = (\mathcal{N}, \mathcal{P}(\mathcal{N}), \mu)$, where μ is counting measure on $\mathcal{P}(\mathcal{N})$. Then real-valued $\mathcal{A}$-measurable functions are simply infinite sequences of real numbers. Referring to Example 5.7(b) on page 163, we see that a sequence of real numbers $\{a_n\}_{n=1}^{\infty}$ is Lebesgue integrable (over $\mathcal{N}$) if and only if

$$\sum_{n=1}^{\infty} |a_n| < \infty, \tag{5.8}$$

that is, if and only if the series is absolutely convergent. For instance, the sequence $\{(-1)^n / n^p\}_{n=1}^{\infty}$ is Lebesgue integrable if and only if $p > 1$. Note that, although $\sum_{n=1}^{\infty} (-1)^n / n$ converges, $\{(-1)^n / n\}_{n=1}^{\infty}$ is not Lebesgue integrable as the series is not absolutely convergent. □

Lebesgue Integral of a Complex-Valued Function

Next we define the abstract Lebesgue integral for complex-valued $\mathcal{A}$-measurable functions. First we present some preliminaries.

DEFINITION 5.12 **Modulus of a Complex-Valued Function**

Let f be a complex-valued function on Ω. Then the **modulus of f**, denoted by $|f|$, is defined to be the real-valued function

$$|f| = \sqrt{(\Re f)^2 + (\Im f)^2}.$$

In other words, $|f|(x) = |f(x)|$, where $|f(x)|$ denotes the modulus of the complex number $f(x)$.

The following two propositions will be required. We leave the proofs as exercises for the reader.

□ □ □   PROPOSITION 5.9

Let f be a complex-valued function on Ω. Then
a) $|f| \leq |\Re f| + |\Im f|$.
b) $|\Re f| \leq |f|$ and $|\Im f| \leq |f|$.
c) $|f|$ is $\mathcal{A}$-measurable if f is.

□ □ □   PROPOSITION 5.10

Let f be a complex-valued $\mathcal{A}$-measurable function on Ω and $E \in \mathcal{A}$. Then $|f|$ is Lebesgue integrable over E if and only if both $\Re f$ and $\Im f$ are.

In view of Proposition 5.10 and the fact that $f = \Re f + i\Im f$, it is reasonable to make the following definition.

DEFINITION 5.13   Integral of a Complex-Valued Function

Let f be a complex-valued $\mathcal{A}$-measurable function on Ω and $E \in \mathcal{A}$. We say that **f is Lebesgue integrable over E with respect to μ** if $|f|$ is Lebesgue integrable over E with respect to μ; that is,

$$\int_E |f| \, d\mu < \infty.$$

In that case, the (abstract) **Lebesgue integral of f over E with respect to μ** is defined by

$$\int_E f \, d\mu = \int_E (\Re f) \, d\mu + i \int_E (\Im f) \, d\mu.$$

If f is Lebesgue integrable over Ω, then we say that **f is Lebesgue integrable.**

For a measure space $(\Omega, \mathcal{A}, \mu)$, the collection of all complex-valued Lebesgue integrable functions is denoted by $\mathcal{L}^1(\Omega, \mathcal{A}, \mu)$. When no confusion will arise, we write $\mathcal{L}^1(\mu)$ for $\mathcal{L}^1(\Omega, \mathcal{A}, \mu)$.

EXAMPLE 5.9   *Illustrates Definition 5.13*

a) Let $(\Omega, \mathcal{A}, \mu) = (\mathcal{R}, \mathcal{M}, \lambda)$ and consider the complex-valued function f defined by $f(x) = e^{ix}/(1 + x^2)$. Then it is easy to see that f is measurable. Moreover, $\Re f(x) = \cos x/(1 + x^2)$, $\Im f(x) = \sin x/(1 + x^2)$, and $|f(x)| = 1/(1 + x^2)$. By Exercise 4.20 on page 127 and Theorem 4.9 on page 135,

$$\int_{\mathcal{R}} |f(x)| \, d\lambda(x) = \int_{\mathcal{R}} (1 + x^2)^{-1} \, d\lambda(x)$$

$$= \lim_{n \to \infty} \int_{[-n,n]} (1 + x^2)^{-1} \, d\lambda(x) = \lim_{n \to \infty} \int_{-n}^{n} \frac{dx}{(1 + x^2)}$$

$$= \lim_{n \to \infty} \left(\arctan(n) - \arctan(-n) \right) = \pi < \infty.$$

Therefore, $f \in \mathcal{L}^1(\lambda)$.

b) Let $(\Omega, \mathcal{A}, \mu) = (\mathcal{N}, \mathcal{P}(\mathcal{N}), \mu)$, where μ is counting measure on $\mathcal{P}(\mathcal{N})$. Then complex-valued $\mathcal{A}$-measurable functions are simply infinite sequences of complex numbers. Referring to Example 5.7(b) on page 163, we see that a sequence of complex numbers $\{a_n\}_{n=1}^{\infty}$ is in $\mathcal{L}^1(\mu)$ if and only if the series $\sum_{n=1}^{\infty} a_n$ converges absolutely. We point out here that the notations ℓ^1 or $\ell^1(\mathcal{N})$ are generally used in place of $\mathcal{L}^1(\mathcal{N}, \mathcal{P}(\mathcal{N}), \mu)$.

c) Let $(\Omega, \mathcal{A})$ be a measurable space. A measure μ on $\mathcal{A}$ is said to be a **finite measure** if $\mu(\Omega) < \infty$. If μ is a finite measure, then $(\Omega, \mathcal{A}, \mu)$ is called a **finite measure space**. For a finite measure space, each bounded complex-valued $\mathcal{A}$-measurable function f is in $\mathcal{L}^1(\mu)$. Indeed, if $|f| \leq M$, then by Proposition 5.8(a) on page 161,

$$\int_{\Omega} |f| \, d\mu \leq \int_{\Omega} M \, d\mu = M\mu(\Omega) < \infty.$$

Note that boundedness is a sufficient but not necessary condition for integrability. For instance, let $(\Omega, \mathcal{A}, \mu) = ((0, 1), \mathcal{M}_{(0,1)}, \lambda_{(0,1)})$ and $f(x) = x^{-1/2}$. Then f is not bounded on $(0, 1)$ but is in $\mathcal{L}^1(\lambda_{(0,1)})$.

d) If $(\Omega, \mathcal{A}, P)$ is a probability space, then the integrable functions, that is, members of $\mathcal{L}^1(P)$, are called random variables with **finite mean** or **finite expectation**. ◻

The following theorem, whose proof is left as an exercise, provides some important properties of Lebesgue integrable functions.

◻ ◻ ◻ **THEOREM 5.8**

Suppose that f and g are in $\mathcal{L}^1(\Omega, \mathcal{A}, \mu)$ and that $\alpha \in \mathbb{C}$. Then

a) $f + g \in \mathcal{L}^1(\mu)$ and

$$\int_{\Omega} (f + g) \, d\mu = \int_{\Omega} f \, d\mu + \int_{\Omega} g \, d\mu.$$

b) $\alpha f \in \mathcal{L}^1(\mu)$ and

$$\int_{\Omega} \alpha f \, d\mu = \alpha \int_{\Omega} f \, d\mu.$$

c) If f and g are real-valued and $f \leq g$ on Ω, then $\int_{\Omega} f \, d\mu \leq \int_{\Omega} g \, d\mu$.

d) $\left| \int_{\Omega} f \, d\mu \right| \leq \int_{\Omega} |f| \, d\mu$.

e) $\mu(E) = 0 \Rightarrow \int_{E} f \, d\mu = 0$.

f) If A and B are disjoint $\mathcal{A}$-measurable sets, then

$$\int_{A \cup B} f \, d\mu = \int_{A} f \, d\mu + \int_{B} f \, d\mu.$$

Remark: Parts (a) and (b) of Theorem 5.8 together imply that if α, $\beta \in \mathbb{C}$ and f, $g \in \mathcal{L}^1(\mu)$, then

$$\int_{\Omega} (\alpha f + \beta g) \, d\mu = \alpha \int_{\Omega} f \, d\mu + \beta \int_{\Omega} g \, d\mu.$$

This result is called the **linearity property** of the abstract Lebesgue integral.

As mentioned in Section 4.4, we often encounter functions that are only defined almost everywhere. Because the integral of an $\mathcal{A}$-measurable function is not affected by its values on a set of measure zero, it is reasonable to make the following definition.

DEFINITION 5.14   Integral of a Function Defined Almost Everywhere

Let $(\Omega, \mathcal{A}, \mu)$ be a measure space. Suppose that f is a function defined μ-ae on Ω; that is, if D is the domain of f, then $\mu(D^c) = 0$. Further suppose that there is an $\mathcal{A}$-measurable function g such that $g(x) = f(x)$ for $x \in D$. Then, for $E \in \mathcal{A}$, we define the (abstract) **Lebesgue integral of f over E** by

$$\int_E f \, d\mu = \int_E g \, d\mu,$$

provided that the integral on the right-hand side exists.

Dominated Convergence Theorem

Theorem 4.8 on page 133 gives the dominated convergence theorem (DCT) for real-valued functions on the measure space $(\mathcal{R}, \mathcal{M}, \lambda)$. Our next theorem generalizes that version of the DCT so that it applies to complex-valued functions on an arbitrary measure space $(\Omega, \mathcal{A}, \mu)$. Note that the version of the DCT given here has weaker hypotheses than the one presented in Theorem 4.8.

□ □ □   THEOREM 5.9 Dominated Convergence Theorem (DCT)

Let $(\Omega, \mathcal{A}, \mu)$ be a measure space. Suppose that $\{f_n\}_{n=1}^{\infty}$ is a sequence of complex-valued $\mathcal{A}$-measurable functions that converges μ-ae. Further suppose that there is a nonnegative Lebesgue integrable function g such that $|f_n| \leq g$ μ-ae for each $n \in \mathcal{N}$. Then

$$\int_E \lim_{n \to \infty} f_n \, d\mu = \lim_{n \to \infty} \int_E f_n \, d\mu \tag{5.9}$$

for each $E \in \mathcal{A}$.

PROOF   Without loss of generality, we can assume that, for each $n \in \mathcal{N}$, $|f_n| \leq g$ everywhere on Ω (why?). Define

$$f(x) = \begin{cases} \lim\limits_{n \to \infty} f_n(x), & \text{if } \lim\limits_{n \to \infty} f_n(x) \text{ exists;} \\ 0, & \text{otherwise.} \end{cases}$$

By Exercise 5.45 on page 158, f is $\mathcal{A}$-measurable. Moreover, since $\{f_n\}_{n=1}^{\infty}$ converges μ-ae, $f_n \to f$ μ-ae. From Definition 5.14, $\int_E \lim_{n \to \infty} f_n \, d\mu = \int_E f \, d\mu$ and, therefore, to prove (5.9) it suffices to prove

$$\int_E f \, d\mu = \lim_{n \to \infty} \int_E f_n \, d\mu \tag{5.10}$$

for each $E \in \mathcal{A}$.

First suppose that each f_n is real-valued. Then (5.10) can be proved by employing Fatou's lemma (Theorem 5.7 on page 162) and the same argument that was used in the proof of the DCT for the Lebesgue integral on the real line (Theorem 4.8).

Next, we remove the restriction that each f_n is real-valued. Observe that $\{|f - f_n|\}_{n=1}^{\infty}$ is a sequence of real-valued $\mathcal{A}$-measurable functions that converges to 0 μ-ae. Furthermore, for each $n \in \mathcal{N}$, we have $|f - f_n| \leq |f| + |f_n| \leq 2g$, an integrable function. Consequently, by Theorem 5.8 and the previous paragraph, as $n \to \infty$,

$$\left| \int_E f \, d\mu - \int_E f_n \, d\mu \right| \leq \int_E |f - f_n| \, d\mu \to \int_E 0 \, d\mu = 0$$

for each $E \in \mathcal{A}$. This completes the proof of the DCT. ∎

Three of the many important corollaries of the DCT are given next. Several other corollaries are considered in the exercises.

□ □ □ **COROLLARY 5.2**

Suppose that $\{f_n\}_{n=1}^{\infty}$ is a sequence of complex-valued $\mathcal{A}$-measurable functions such that

$$\sum_{n=1}^{\infty} \int_{\Omega} |f_n| \, d\mu < \infty.$$

Then $\sum_{n=1}^{\infty} f_n$ converges μ-ae and

$$\int_E \sum_{n=1}^{\infty} f_n \, d\mu = \sum_{n=1}^{\infty} \int_E f_n \, d\mu$$

for each $E \in \mathcal{A}$.

PROOF   From Corollary 5.1(b) on page 162, we know that

$$\int_{\Omega} \sum_{n=1}^{\infty} |f_n| \, d\mu = \sum_{n=1}^{\infty} \int_{\Omega} |f_n| \, d\mu.$$

By assumption, the sum on the right-hand side of the previous equation is finite and, hence, so is the integral on the left-hand side. In other words, if we set $g = \sum_{n=1}^{\infty} |f_n|$, then g is Lebesgue integrable. From Exercise 5.53 on page 164, we conclude that g is finite μ-ae which, in turn, implies that $\sum_{n=1}^{\infty} f_n$ converges μ-ae.

Set $g_n = \sum_{k=1}^{n} f_k$. Then, for each $n \in \mathcal{N}$, $|g_n| \leq g$ and, as we have just seen, $\{g_n\}_{n=1}^{\infty}$ converges μ-ae (to $\sum_{n=1}^{\infty} f_n$). Therefore, by the DCT and Theorem 5.8(a),

$$\int_E \sum_{n=1}^{\infty} f_n \, d\mu = \int_E \lim_{n \to \infty} g_n \, d\mu = \lim_{n \to \infty} \int_E g_n \, d\mu$$

$$= \lim_{n \to \infty} \int_E \sum_{k=1}^{n} f_k \, d\mu = \lim_{n \to \infty} \sum_{k=1}^{n} \int_E f_k \, d\mu = \sum_{n=1}^{\infty} \int_E f_n \, d\mu$$

for each $E \in \mathcal{A}$. ∎

□ □ □ **COROLLARY 5.3**

Let $(\Omega, \mathcal{A}, \mu)$ be a measure space. Suppose that $f \in \mathcal{L}^1(\mu)$ and that $\{E_n\}_{n=1}^{\infty}$ is a sequence of $\mathcal{A}$-measurable sets with $E_1 \subset E_2 \subset \cdots$. Then

$$\int_{\bigcup_{n=1}^{\infty} E_n} f \, d\mu = \lim_{n \to \infty} \int_{E_n} f \, d\mu.$$

PROOF For convenience, let $E = \bigcup_{n=1}^{\infty} E_n$. It is easy to see that $\chi_{E_n} f \to \chi_E f$ pointwise and that $|\chi_{E_n} f| \leq |f| \in \mathcal{L}^1(\mu)$ for each $n \in \mathcal{N}$. Thus, by the DCT,

$$\int_E f \, d\mu = \int_{\Omega} \chi_E f \, d\mu = \lim_{n \to \infty} \int_{\Omega} \chi_{E_n} f \, d\mu = \lim_{n \to \infty} \int_{E_n} f \, d\mu,$$

as required. ■

□ □ □ **COROLLARY 5.4 Bounded Convergence Theorem**

Let $(\Omega, \mathcal{A}, \mu)$ be a finite measure space. Suppose that $\{f_n\}_{n=1}^{\infty}$ is a sequence of uniformly bounded, complex-valued, $\mathcal{A}$-measurable functions that converges μ-ae. Then

$$\int_E \lim_{n \to \infty} f_n \, d\mu = \lim_{n \to \infty} \int_E f_n \, d\mu$$

for each $E \in \mathcal{A}$.

PROOF By assumption, there is a real number M such that $|f_n| \leq M$ for all $n \in \mathcal{N}$. Because $(\Omega, \mathcal{A}, \mu)$ is a finite measure space, the function $g(x) \equiv M$ is Lebesgue integrable (why?). Applying the DCT completes the proof. ■

EXAMPLE 5.10 Illustrates the DCT

a) Suppose that, for each $n \in \mathcal{N}$, $\{a_{nk}\}_{k=1}^{\infty}$ is a sequence of complex numbers and that $\lim_{n \to \infty} a_{nk} = a_k$ for each $k \in \mathcal{N}$. Further suppose that there is a sequence of nonnegative numbers $\{b_k\}_{k=1}^{\infty}$ such that $\sum_{k=1}^{\infty} b_k < \infty$ and $|a_{nk}| \leq b_k$ for $k, n \in \mathcal{N}$. We claim that

$$\lim_{n \to \infty} \sum_{k=1}^{\infty} a_{nk} = \sum_{k=1}^{\infty} a_k. \tag{5.11}$$

Indeed, consider the measure space $(\mathcal{N}, \mathcal{P}(\mathcal{N}), \mu)$, where μ is counting measure. Define $f_n(k) = a_{nk}$, $f(k) = a_k$, and $g(k) = b_k$. By assumption, g is integrable, $|f_n| \leq g$ for all $n \in \mathcal{N}$, and $f_n \to f$ pointwise on $\mathcal{N}$. Thus, by the DCT, $\int_{\mathcal{N}} f_n \, d\mu \to \int_{\mathcal{N}} f \, d\mu$ as $n \to \infty$. However, $\int_{\mathcal{N}} f_n \, d\mu = \sum_{k=1}^{\infty} a_{nk}$ and $\int_{\mathcal{N}} f \, d\mu = \sum_{k=1}^{\infty} a_k$ (see Exercise 5.73). Thus, (5.11) holds.

Without a dominating integrable sequence, (5.11) may fail. For instance, take $a_{nk} = \delta_{nk}$ and $a_k \equiv 0$. Then $\lim_{n \to \infty} a_{nk} = a_k$ for each $k \in \mathcal{N}$. But, as $\sum_{k=1}^{\infty} a_{nk} = 1$ for all $n \in \mathcal{N}$, we see that

$$\lim_{n \to \infty} \sum_{k=1}^{\infty} a_{nk} = 1 \neq 0 = \sum_{k=1}^{\infty} a_k.$$

Therefore, (5.11) fails to hold.

b) Let $(\Omega, \mathcal{A}, P)$ be a probability space and X a real-valued random variable having finite expectation, that is, $X \in \mathcal{L}^1(P)$. Define f on $\mathcal{R}$ by $f(t) = \mathcal{E}(e^{itX})$. Note that the definition of f makes sense because $|e^{itX}| \le 1$. We claim that $f'(0) = i\mathcal{E}(X)$. To prove this result, let $\{t_n\}_{n=1}^{\infty}$ be an arbitrary sequence of nonzero real numbers that converges to 0. For each $n \in \mathcal{N}$, define $Y_n = (e^{it_n X} - 1)/t_n$. Then (see Exercise 5.75)

$$\frac{f(t_n) - f(0)}{t_n - 0} = \int_{\Omega} \frac{e^{it_n X} - 1}{t_n} \, dP = \int_{\Omega} Y_n \, dP. \qquad (5.12)$$

Now, for $x \in \mathcal{R}$, we have $|e^{ix} - 1| \le |x|$ and, therefore, $|Y_n| \le |X|$ for each $n \in \mathcal{N}$. As $Y_n \to iX$ pointwise on Ω, we can apply the DCT to conclude that

$$\lim_{n \to \infty} \frac{f(t_n) - f(0)}{t_n - 0} = \lim_{n \to \infty} \int_{\Omega} Y_n \, dP = \int_{\Omega} iX \, dP = i\mathcal{E}(X).$$

Because $\{t_n\}_{n=1}^{\infty}$ is an arbitrary sequence of nonzero real numbers converging to 0, it follows that $f'(0)$ exists and equals $i\mathcal{E}(X)$. □

Exercises for Section 5.4

5.64 Let $(\Omega, \mathcal{A}, \mu) = (\mathcal{N}, \mathcal{P}(\mathcal{N}), \mu)$, where μ is counting measure on $\mathcal{P}(\mathcal{N})$. Define $f: \mathcal{N} \to \mathcal{R}$ by $f(n) = (-1)^n/n$ for $n \in \mathcal{N}$. Is $\int_{\mathcal{N}} f \, d\mu$ defined? Explain your answer.

5.65 Prove Proposition 5.9 on page 167.

5.66 Prove Proposition 5.10 on page 167.

5.67 Let $f(x) = x^{-1/2}$. Show that $f \in \mathcal{L}^1\big((0,1), \mathcal{M}_{(0,1)}, \lambda_{(0,1)}\big)$.

5.68 Prove Theorem 5.8 on page 168.

5.69 Prove that Definition 5.14 on page 169 is well-posed. In other words, assume that f is defined μ-ae on Ω and that g and h are $\mathcal{A}$-measurable functions that equal f on its domain. Show that for $E \in \mathcal{A}$, either $\int_E g \, d\mu = \int_E h \, d\mu$ or neither integral exists.

5.70 Show that for a fixed $E \in \mathcal{A}$, the conclusion of the DCT remains valid if the hypotheses are satisfied only on E.

5.71 State and prove a version of the DCT for extended real-valued $\mathcal{A}$-measurable functions.

★5.72 Suppose that $f \in \mathcal{L}^1(\Omega, \mathcal{A}, \mu)$. Further suppose that $\{E_n\}_{n=1}^{\infty}$ is a sequence of pairwise disjoint $\mathcal{A}$-measurable sets. Prove that

$$\int_{\bigcup_{n=1}^{\infty} E_n} f \, d\mu = \sum_{n=1}^{\infty} \int_{E_n} f \, d\mu.$$

★5.73 Let $f \in \mathcal{L}^1(\Omega, \mathcal{A}, \mu)$ and $C = \{x_n\}_n$ a countable subset of Ω such that $\{x_n\} \in \mathcal{A}$ for each n. Prove that

$$\int_C f \, d\mu = \sum_n f(x_n)\mu(\{x_n\}).$$

Deduce that if $\{a_n\}_{n=1}^{\infty} \in \ell^1$, then

$$\int_{\mathcal{N}} f \, d\mu = \sum_{n=1}^{\infty} a_n,$$

where $f(n) = a_n$.

5.74 Let $\sum_{k=1}^{\infty} a_k$ be a convergent series of nonnegative numbers and, for $n, k \in \mathcal{N}$, let b_{nk} be complex numbers with $|b_{nk}| \leq M < \infty$. Assume that $\lim_{n \to \infty} b_{nk} = b_k$ for each $k \in \mathcal{N}$. Prove that

$$\lim_{n \to \infty} \sum_{k=1}^{\infty} a_k b_{nk} = \sum_{k=1}^{\infty} a_k b_k.$$

5.75 Provide a detailed justification of (5.12).

5.76 Let $(\Omega, \mathcal{A}, P)$ be a probability space and Y a real-valued random variable taking on only finitely many values, say $y_1, y_2, \ldots, y_n$. Verify that

$$\mathcal{E}(Y) = \sum_{k=1}^{n} y_k P(Y = y_k) \tag{5.13}$$

where, by convention, $\{Y = y\} = \{x \in \Omega : Y(x) = y\}$. *Note:* Equation (5.13) shows that the mean of a random variable Y taking on only finitely many values is a weighted average of the values of Y, weighted according to their probabilities.

5.77 Let $(\Omega, \mathcal{A}, \mu)$ be a measure space, $f \in \mathcal{L}^1(\mu)$, and $\{E_n\}_{n=1}^{\infty}$ a sequence of $\mathcal{A}$-measurable sets with $E_1 \supset E_2 \supset \cdots$. Prove that

$$\int_{\bigcap_{n=1}^{\infty} E_n} f \, d\mu = \lim_{n \to \infty} \int_{E_n} f \, d\mu.$$

5.78 Suppose that $f : [0,1] \times (0,1) \to \mathcal{R}$ is such that, for each fixed $y \in (0,1)$, the function $f^{[y]}$ defined by $f^{[y]}(x) = f(x,y)$ is $\mathcal{M}_{[0,1]}$-measurable. Further suppose that $\partial f / \partial y$ exists and is bounded on $[0,1] \times (0,1)$. Show that

$$\frac{d}{dy} \int_0^1 f(x,y) \, dx = \int_0^1 \frac{\partial f}{\partial y}(x,y) \, dx.$$

5.79 Let $f \in \mathcal{L}^1(\Omega, \mathcal{A}, \mu)$. Show that for each $\epsilon > 0$, there is an $A \in \mathcal{A}$ such that $\mu(A) < \infty$ and $\int_{A^c} |f| \, d\mu < \epsilon$.

★5.80 Suppose that $f \in \mathcal{L}^1(\Omega, \mathcal{A}, \mu)$. Show that for each $\epsilon > 0$, there is a $\delta > 0$ such that $\mu(E) < \delta \Rightarrow \int_E |f| \, d\mu < \epsilon$.

★5.81 Let $f \in \mathcal{L}^1(\mathcal{R}, \mathcal{M}, \lambda)$. Then we define the **Fourier transform of f**, denoted $\widehat{f}$, by

$$\widehat{f}(t) = \int_{\mathcal{R}} e^{-itx} f(x) \, d\lambda(x), \qquad t \in \mathcal{R}.$$

a) Prove that $\widehat{f}$ is continuous on $\mathcal{R}$.

b) Prove that if $\int_{\mathcal{R}} |xf(x)| \, d\lambda(x) < \infty$, then $\widehat{f}$ is differentiable on $\mathcal{R}$ and

$$\widehat{f}'(t) = \int_{\mathcal{R}} (-ix) e^{-itx} f(x) \, d\lambda(x), \qquad t \in \mathcal{R}.$$

★5.82 Suppose that $f \in \mathcal{L}^1(\Omega, \mathcal{A}, \mu)$.

a) Prove that, for each $\epsilon > 0$, there exists a bounded $\mathcal{A}$-measurable function g such that $\int_{\Omega} |f - g| \, d\mu < \epsilon$.

b) Prove that, for each $\epsilon > 0$, there exists an $\mathcal{A}$-measurable simple function s such that $\int_{\Omega} |f - s| \, d\mu < \epsilon$.

5.5 CONVERGENCE IN MEASURE

Until now, we have discussed three types of convergence for functions: pointwise convergence, uniform convergence, and almost-everywhere convergence. Another kind of convergence, important especially in probability theory, is **convergence in measure.**[†] Here is the definition.

DEFINITION 5.15 Convergence in Measure

Let $(\Omega, \mathcal{A}, \mu)$ be a measure space and $\{f_n\}_{n=1}^{\infty}$ a sequence of complex-valued $\mathcal{A}$-measurable functions on Ω. Then $\{f_n\}_{n=1}^{\infty}$ is said to **converge in measure** to the $\mathcal{A}$-measurable function f, if for each $\epsilon > 0$,

$$\lim_{n \to \infty} \mu(\{ x : |f(x) - f_n(x)| \geq \epsilon \}) = 0.$$

We often write $f_n \xrightarrow{\mu} f$ to indicate convergence in measure. Thus, $f_n \xrightarrow{\mu} f$ if the measure of the set where f_n differs from f by more than any prescribed positive number tends to zero as $n \to \infty$.

Is there a relationship between almost-everywhere convergence and convergence in measure? The following example shows that, generally speaking, there is no relationship.

EXAMPLE 5.11 *Illustrates Definition 5.15*

a) Let $(\Omega, \mathcal{A}, \mu) = (\mathcal{R}, \mathcal{M}, \lambda)$. Set $f(x) \equiv 0$ and $f_n(x) = x/n$ for $x \in \mathcal{R}$ and $n \in \mathcal{N}$. Then $f_n \to f$ pointwise and, hence, λ-ae. But $f_n \not\to f$ in measure. Indeed, for $\epsilon > 0$,

$$\{ x : |f(x) - f_n(x)| \geq \epsilon \} = (-\infty, -n\epsilon) \cup (n\epsilon, \infty),$$

which has infinite Lebesgue measure for every $n \in \mathcal{N}$. Thus, we see that $\lambda(\{ x : |f(x) - f_n(x)| \geq \epsilon \}) \not\to 0$. Consequently, almost-everywhere convergence does not imply convergence in measure.

b) Let $(\Omega, \mathcal{A}, \mu) = ([0,1], \mathcal{M}_{[0,1]}, \lambda_{[0,1]})$. Consider the sequence of functions defined by $f_1 = \chi_{[0,1]}$, $f_2 = \chi_{[0,1/2]}$, $f_3 = \chi_{[1/2,1]}$ and, in general, if $n = k + 2^j$, where $0 \leq k < 2^j$, $f_n = \chi_{[k2^{-j},(k+1)2^{-j}]}$. Then, for $\epsilon > 0$,

$$\mu(\{ x : |f_n(x)| \geq \epsilon \}) < \frac{2}{n} \to 0$$

as $n \to \infty$. So, $f_n \xrightarrow{\mu} 0$. But, for each $x \in [0,1]$, the sequence $\{f_n(x)\}_{n=1}^{\infty}$ contains infinitely many 1s and infinitely many 0s. Thus, $\{f_n\}_{n=1}^{\infty}$ converges for no $x \in [0,1]$ and, in particular, $f_n \not\to 0$ μ-ae. Consequently, convergence in measure does not imply almost-everywhere convergence. □

Example 5.11(a) shows that, in general, convergence almost everywhere does not imply convergence in measure. For finite measure spaces, however, the implication is correct.

[†] In probability theory, the terminology **convergence in probability** is used in place of convergence in measure.

□ □ □ **PROPOSITION 5.11**

Suppose that $(\Omega, \mathcal{A}, \mu)$ is a finite measure space and that $\{f_n\}_{n=1}^{\infty}$ is a sequence of complex-valued $\mathcal{A}$-measurable functions that converges μ-ae to the $\mathcal{A}$-measurable function f. Then $f_n \xrightarrow{\mu} f$.

PROOF Let $B = \{\, x : f_n(x) \nrightarrow f(x) \,\}$. Then, by assumption, $\mu(B) = 0$. For $\epsilon > 0$, define $E_n = \{\, x : |f(x) - f_n(x)| \geq \epsilon \,\}$ and $E = \bigcap_{n=1}^{\infty} \left(\bigcup_{k=n}^{\infty} E_k \right)$. We must show that $\lim_{n\to\infty} \mu(E_n) = 0$.

Note that $x \in E$ if and only if $x \in E_n$ for infinitely many n. It follows easily that $E \subset B$ and, so, $\mu(E) = 0$. Because $\mu(\Omega) < \infty$ and $\bigcup_{k=n}^{\infty} E_k \supset \bigcup_{k=n+1}^{\infty} E_k$ for each $n \in \mathcal{N}$, we conclude from Theorem 5.1(c) on page 147 that

$$\limsup_{n\to\infty} \mu(E_n) \leq \lim_{n\to\infty} \mu\left(\bigcup_{k=n}^{\infty} E_k \right) = \mu(E) = 0.$$

Hence, $\lim_{n\to\infty} \mu(E_n) = 0$, as required. ∎

As we discovered in Example 5.11(b), convergence in measure does not imply almost-everywhere convergence. Nonetheless, we do have the following useful result.

□ □ □ **PROPOSITION 5.12**

Suppose that $\{f_n\}_{n=1}^{\infty}$ is a sequence of complex-valued $\mathcal{A}$-measurable functions that converges in measure to the $\mathcal{A}$-measurable function f. Then there is a subsequence $\{f_{n_k}\}_{k=1}^{\infty}$ of $\{f_n\}_{n=1}^{\infty}$ such that $f_{n_k} \to f$ μ-ae.

PROOF We can, for each $k \in \mathcal{N}$, choose an $n_k \in \mathcal{N}$ such that

$$\mu\left(\left\{ x : |f(x) - f_{n_k}(x)| \geq \frac{1}{k} \right\} \right) \leq 2^{-k}. \tag{5.14}$$

Furthermore, $\{n_k\}_{k=1}^{\infty}$ can be selected so that $n_1 < n_2 < \cdots$. Now let us define $E_k = \{\, x : |f(x) - f_{n_k}(x)| \geq k^{-1} \,\}$ and $E = \bigcap_{k=1}^{\infty} \left(\bigcup_{j=k}^{\infty} E_j \right)$. Note that $x \in E$ if and only if $|f(x) - f_{n_k}(x)| \geq k^{-1}$ for infinitely many k.

From (5.14), we see that $\sum_{k=1}^{\infty} \mu(E_k) < \infty$ and, consequently, by Exercise 5.14 on page 150, $\mu(E) = 0$. We claim that $f_{n_k} \to f$ on E^c. So, let $x \in E^c$ and let $\epsilon > 0$ be given. Choose $k_1 \in \mathcal{N}$ so that $k_1^{-1} \leq \epsilon$. Since $x \notin E$, it follows that there is a $k_2 \in \mathcal{N}$ such that $x \notin E_k$ for $k \geq k_2$. Let $K = \max\{k_1, k_2\}$. Then we have that $|f(x) - f_{n_k}(x)| < k^{-1} \leq \epsilon$ for all $k \geq K$. ∎

The DCT for Convergence in Measure

By employing Proposition 5.12, we can prove that the dominated convergence theorem remains valid when almost-everywhere convergence is replaced by convergence in measure. That is, we have the following result.

□ □ □ **THEOREM 5.10**

Let $(\Omega, \mathcal{A}, \mu)$ be a measure space. Suppose that $\{f_n\}_{n=1}^{\infty}$ is a sequence of complex-valued $\mathcal{A}$-measurable functions that converges in measure to the $\mathcal{A}$-measurable function f. Further suppose that there is a nonnegative Lebesgue integrable function g such that $|f_n| \leq g$ μ-ae for each $n \in \mathcal{N}$. Then

$$\int_E f \, d\mu = \lim_{n \to \infty} \int_E f_n \, d\mu \tag{5.15}$$

for each $E \in \mathcal{A}$.

PROOF Let $E \in \mathcal{A}$. To prove (5.15) it suffices, by Exercise 2.32 on page 45, to show that every subsequence of $\left\{\int_E f_n \, d\mu\right\}_{n=1}^{\infty}$ has a subsequence that converges to $\int_E f \, d\mu$. So, let $\{n_k\}_{k=1}^{\infty}$ be a subsequence of $\mathcal{N}$. As $f_n \xrightarrow{\mu} f$, it is clear that $f_{n_k} \xrightarrow{\mu} f$. Applying Proposition 5.12, we deduce that $\{f_{n_k}\}_{k=1}^{\infty}$ has a subsequence $\{f_{n_{k_j}}\}_{j=1}^{\infty}$ with $f_{n_{k_j}} \to f$ μ-ae. Clearly, we have $|f_{n_{k_j}}| \leq g$ μ-ae for each $j \in \mathcal{N}$ and, hence, by the DCT (Theorem 5.9 on page 169),

$$\int_E f \, d\mu = \lim_{j \to \infty} \int_E f_{n_{k_j}} \, d\mu.$$

This completes the proof. ∎

Exercises for Section 5.5

5.83 Show that if $f_n \xrightarrow{\mu} f$ and $f_n \xrightarrow{\mu} g$, then $f = g$ μ-ae.

★5.84 Suppose that $f, f_1, f_2, \ldots$ are in $\mathcal{L}^1(\Omega, \mathcal{A}, \mu)$ and that $\int_\Omega |f - f_n| \, d\mu \to 0$ as $n \to \infty$. Show that $f_n \to f$ in measure.

5.85 Let $(\Omega, \mathcal{A}, \mu)$ be a measure space. We say that a sequence $\{f_n\}_{n=1}^{\infty}$ of complex-valued $\mathcal{A}$-measurable functions on Ω converges **almost uniformly** to the complex-valued $\mathcal{A}$-measurable function f if for each $\epsilon > 0$, there is a set $A \in \mathcal{A}$ such that $\mu(A) < \epsilon$ and $f_n \to f$ uniformly on A^c.
a) Prove that almost-uniform convergence implies convergence in measure; that is, if $f_n \to f$ almost uniformly, then $f_n \to f$ in measure.
b) Prove that almost-uniform convergence implies almost-everywhere convergence; that is, if $f_n \to f$ almost uniformly, then $f_n \to f$ μ-ae.
c) Does almost-uniform convergence imply pointwise convergence? Why or why not?

5.86 Provide a detailed justification for all statements in Example 5.11(b) on page 174.

5.87 Suppose that $f, g, f_1, f_2, \ldots$ are complex-valued $\mathcal{A}$-measurable functions. Further suppose that $f_n \to g$ μ-ae and that $f_n \to f$ in measure. Prove that $f = g$ μ-ae and, hence, that $f_n \to f$ μ-ae.

5.88 **Fatou's lemma for convergence in measure:** Suppose that $\{f_n\}_{n=1}^{\infty}$ is a sequence of nonnegative $\mathcal{A}$-measurable functions that converges in measure to f. Prove that

$$\int_E f \, d\mu \leq \liminf_{n \to \infty} \int_E f_n \, d\mu$$

for each $E \in \mathcal{A}$. *Hint:* Select a subsequence of $\left\{\int_E f_n \, d\mu\right\}_{n=1}^{\infty}$ that converges to $\liminf_{n \to \infty} \int_E f_n \, d\mu$.

5.89 Establish the following fact: If $\{f_n\}_{n=1}^{\infty}$ converges in measure, then it is also Cauchy in measure, that is, for each $\epsilon > 0$, $\mu(\{x : |f_n(x) - f_m(x)| \geq \epsilon\}) \to 0$ as $m, n \to \infty$.

5.90 Prove the following strengthened version of Proposition 5.12. Suppose that $\{f_n\}_{n=1}^{\infty}$ is a sequence of complex-valued $\mathcal{A}$-measurable functions that converges in measure to f. Then there is a subsequence $\{f_{n_k}\}_{k=1}^{\infty}$ of $\{f_n\}_{n=1}^{\infty}$ such that $f_{n_k} \to f$ almost uniformly. *Hint:* Show that there is a subsequence $\{n_k\}_{k=1}^{\infty}$ of $\mathcal{N}$ such that

$$\mu\left(\left\{x : |f_{n_k}(x) - f_{n_{k+1}}(x)| \geq 2^{-k}\right\}\right) \leq 2^{-k}.$$

You will also need to apply the Weierstrass M-test.

5.91 Suppose that $(\Omega, \mathcal{A}, \mu)$ is a measure space and that f, f_1, f_2, ... are complex-valued $\mathcal{A}$-measurable functions on Ω. Show that

$$\{x : \lim_{n \to \infty} f_n(x) = f(x)\} = \bigcap_{m=1}^{\infty}\left(\bigcup_{n=1}^{\infty}\left(\bigcap_{k=n}^{\infty}\{x : |f(x) - f_k(x)| < 1/m\}\right)\right).$$

★5.92 Suppose that $(\Omega, \mathcal{A}, \mu)$ is a finite measure space and that f, f_1, f_2, ... are complex-valued $\mathcal{A}$-measurable functions on Ω. Prove that $f_n \to f$ μ-ae if and only if

$$\lim_{n \to \infty} \mu\left(\bigcup_{k=n}^{\infty}\{x : |f(x) - f_k(x)| \geq \epsilon\}\right) = 0$$

for each $\epsilon > 0$. Compare this equation with the definition of convergence in measure.

5.93 Suppose that $(\Omega, \mathcal{A}, \mu)$ is a finite measure space and that $\{f_n\}_{n=1}^{\infty}$ is a sequence of complex-valued $\mathcal{A}$-measurable functions that converges in measure to f. Further suppose that $g: \mathbb{C} \to \mathbb{C}$ is continuous. Prove that $g \circ f_n \to g \circ f$ in measure. *Hint:* For a given $\epsilon > 0$, let $a_n = \mu\left(\left\{x : |g(f(x)) - g(f_n(x))| \geq \epsilon\right\}\right)$. Show that each subsequence of $\{a_n\}_{n=1}^{\infty}$ has a subsequence that converges to 0.

5.94 **Egorov's theorem:** Suppose that $(\Omega, \mathcal{A}, \mu)$ is a finite measure space and that f, f_1, f_2, ... are complex-valued $\mathcal{A}$-measurable functions on Ω. Prove that, if $f_n \to f$ μ-ae, then $f_n \to f$ almost uniformly.

Guido Fubini
(1879–1943)

Guido Fubini was born on the 19th of January 1879, in Venice, Italy. His father, Lazzaro, taught mathematics at the Scuola Macchinisti in Venice and influenced his son toward mathematics at an early age.

Fubini attended secondary school in Venice where he already showed his mathematical brilliance. Subsequently, in 1896, he entered the Scuola Normale Superiore di Pisa where he studied with Ulisse Dini and Luigi Bianchi. Fubini received his doctorate in 1900; his thesis was titled *Clifford's Parallelism in Elliptic Spaces*.

After getting his doctorate, Fubini stayed at Pisa to qualify as a university teacher. Following that, in 1901, he began teaching at the University of Catania in Sicily; shortly thereafter, he went to the University of Genoa, where he was appointed chair. In 1908, Fubini moved to Turin, where he first taught at the Politecnico in Turin and then at the University of Turin. He remained there until 1938, when, as a Jew, he was forced to resign from his chair on account of Mussolini's newly adopted Manifesto of Fascist Racism.

During his tenure at the University of Turin, Fubini's research focused primarily on topics in mathematical analysis, especially differential equations, functional analysis, and complex analysis. However, he also studied the calculus of variations, group theory, non-Euclidean geometry, and projective geometry. With the outbreak of World War I, his research shifted toward more applied topics, studying the accuracy of artillery fire, electrical circuits, and acoustics.

Fubini is particularly known for Fubini's theorem (which we will study in Section 6.4), Fubini's theorem on differentiation, and the Fubini–Study metric. He also wrote several books.

In the late 1930s, Fubini feared for the safety of his family because of fascist racism. So, in 1939, he accepted an invitation from the Institute for Advanced Study in Princeton and immediately emigrated with his family to the United States. Although Fubini was in poor health by this time, he was able to teach for a few years. He died in New York City on June 9, 1943, of heart problems.

6

Extensions to Measures and Product Measure

In Chapter 5, we began our investigation of measure theory. There we generalized the real-line concepts of Lebesgue measure, Lebesgue measurable functions, and Lebesgue integral from Chapters 3 and 4 to general measure spaces.

This chapter continues our development of measure theory. Here we will see how to expand the concepts of outer measure and measurable sets to more general settings for the purpose of extending certain real-valued set functions to measures. As an application of these ideas, we will discuss the Lebesgue-Stieltjes integral and product measure.

6.1 EXTENSIONS TO MEASURES

In Chapter 3, the concept of length was extended and replaced by that of measure. Specifically, we began with the collection of intervals and the set function ℓ that assigns to each interval its length. The problem was to extend ℓ to a measure defined on a σ-algebra of subsets of $\mathcal{R}$ that contains all intervals. We proceeded as follows: First we extended the concept of length to all subsets of $\mathcal{R}$ by defining Lebesgue outer measure λ^*:

$$\lambda^*(A) = \inf \left\{ \sum_n \ell(I_n) : \{I_n\}_n \text{ open intervals}, \ \bigcup_n I_n \supset A \right\}. \quad (6.1)$$

Then we defined the Lebesgue measurable sets $\mathcal{M}$ to be the collection of subsets E of $\mathcal{R}$ that satisfy

$$\lambda^*(W) = \lambda^*(W \cap E) + \lambda^*(W \cap E^c), \qquad W \subset \mathcal{R}. \quad (6.2)$$

Finally, we proved that $\mathcal{M}$ is a σ-algebra containing all intervals and that the set function $\lambda = \lambda^*_{|\mathcal{M}}$ is a measure on $\mathcal{M}$ satisfying $\lambda(I) = \ell(I)$ for all intervals I. Thus, Lebesgue measure λ provided the required extension of length.

In this section, we will use our experience from Chapter 3 to handle more general situations. Suppose then that Ω is a set, $\mathcal{C}$ is a nonempty collection of subsets of Ω, and ι is a nonnegative extended real-valued set function on $\mathcal{C}$. Our two primary questions are:

Question 1: Can ι be extended to a measure on a σ-algebra containing $\mathcal{C}$?

Question 2: If such an extension exists, when is it unique?

We begin by considering Question 1.

Necessary Conditions; Semialgebras

First we will obtain some necessary conditions on ι for an affirmative answer to Question 1. So, assume that ι can be extended to a measure μ on a σ-algebra $\mathcal{A} \supset \mathcal{C}$. Then, by Definition 5.1 on page 146, Theorem 5.1 on page 147, and the fact that μ is an extension of ι, we must have

(E1)   If $\emptyset \in \mathcal{C}$, then $\iota(\emptyset) = 0$.

(E2)   If $\{C_k\}_{k=1}^n$ is a finite sequence of pairwise disjoint members of $\mathcal{C}$ whose union is in $\mathcal{C}$, then

$$\iota\left(\bigcup_{k=1}^n C_k\right) = \sum_{k=1}^n \iota(C_k).$$

(E3)   If $C, C_1, C_2, \ldots$ are in $\mathcal{C}$ and $C \subset \bigcup_n C_n$, then

$$\iota(C) \le \sum_n \iota(C_n).$$

Conditions (E1)–(E3) are necessary conditions for the extension of ι to a measure on a σ-algebra containing $\mathcal{C}$. In other words, unless those three conditions hold, such an extension is impossible. Remarkably, as we will see, if $\mathcal{C}$ is a semialgebra (defined below), then those three conditions are also sufficient for the extension.

DEFINITION 6.1   **Semialgebra of Subsets**

Let Ω be a set. A nonempty collection $\mathcal{C}$ of subsets of Ω is called a **semialgebra** if the following conditions hold:
a) If $A, B \in \mathcal{C}$, then $A \cap B \in \mathcal{C}$.
b) If $C \in \mathcal{C}$, then there is a pairwise disjoint finite (possibly empty) sequence of members of $\mathcal{C}$ whose union is C^c.

In words, $\mathcal{C}$ is a semialgebra if it is closed under intersection and the complement of each member of $\mathcal{C}$ is a finite (possibly empty) disjoint union of members of $\mathcal{C}$.

Note the following:

- Unless a semialgebra $\mathcal{C}$ consists of the single set Ω (i.e., $\mathcal{C} = \{\Omega\}$), it must contain $\emptyset$. Indeed, suppose that $\mathcal{C} \neq \{\Omega\}$. Select $A \in \mathcal{C}$ with $A \neq \Omega$. Then $A^c \neq \emptyset$ and, hence, A^c must be a nonempty finite disjoint union of members of $\mathcal{C}$, say, $A^c = \bigcup_{i=1}^n A_i$. Consequently,

$$\emptyset = A \cap A^c = A \cap \left(\bigcup_{i=1}^n A_i \right) = \bigcup_{i=1}^n (A \cap A_i).$$

  Therefore, $A \cap A_i = \emptyset$ for $1 \leq i \leq n$. Hence, $\emptyset = A \cap A_1 \in \mathcal{C}$ because $\mathcal{C}$ is closed under intersection.

- From the preceding fact and the definition of a semialgebra, we conclude that, if $\mathcal{C}$ is a semialgebra, then Ω is a finite disjoint union of members of $\mathcal{C}$.

We present a few examples of semialgebras in Example 6.1. The justifications are left as exercises for the reader.

EXAMPLE 6.1 *Illustrates Definition 6.1*

a) Any algebra and, hence, any σ-algebra is a semialgebra.
b) Suppose that Ω is a finite set. Let $\mathcal{C}$ denote the collection of sets consisting of the empty set and all singleton sets, that is, sets of the form $\{x\}$, where $x \in \Omega$. Then $\mathcal{C}$ is a semialgebra.
c) Let $\mathcal{I}$ denote the collection of all intervals of $\mathcal{R}$, including intervals of the form (a, a) and $[a, a]$. Then $\mathcal{I}$ is a semialgebra of subsets of $\mathcal{R}$.
d) Let $\mathcal{I}_n$ denote the collection of all n-dimensional intervals in $\mathcal{R}^n$; that is, all sets of the form $I_1 \times I_2 \times \cdots \times I_n$ where $I_j \in \mathcal{I}$ for $1 \leq j \leq n$. Then $\mathcal{I}_n$ is a semialgebra of subsets of $\mathcal{R}^n$. □

The following result shows that the collection of all Cartesian products of sets formed from a finite number of semialgebras is also a semialgebra.

□ □ □ **PROPOSITION 6.1**

Suppose that $\mathcal{C}$ and $\mathcal{D}$ are semialgebras of subsets of Γ and Λ, respectively. Let

$$\mathcal{E} = \{ C \times D : C \in \mathcal{C}, D \in \mathcal{D} \}.$$

Then $\mathcal{E}$ is a semialgebra of subsets of $\Gamma \times \Lambda$. More generally, if $\mathcal{C}_1, \mathcal{C}_2, \ldots, \mathcal{C}_n$ are semialgebras of subsets of $\Omega_1, \Omega_2, \ldots, \Omega_n$, respectively, then

$$\mathcal{E} = \left\{ \underset{k=1}{\overset{n}{\times}} C_k : C_k \in \mathcal{C}_k, 1 \leq k \leq n \right\}$$

is a semialgebra of subsets of $\times_{k=1}^n \Omega_k$.

PROOF Suppose that $A, B \in \mathcal{E}$. Then there are sets $C_1, C_2 \in \mathcal{C}$ and $D_1, D_2 \in \mathcal{D}$ such that $A = C_1 \times D_1$ and $B = C_2 \times D_2$. We have that

$$A \cap B = (C_1 \cap C_2) \times (D_1 \times D_2).$$

Because $\mathcal{C}$ and $\mathcal{D}$ are semialgebras, $C_1 \cap C_2 \in \mathcal{C}$ and $D_1 \times D_2 \in \mathcal{D}$. Consequently, $A \cap B \in \mathcal{E}$.

Now suppose that $E \in \mathcal{E}$, say, $E = C \times D$, where $C \in \mathcal{C}$ and $D \in \mathcal{D}$. Then

$$E^c = (\Gamma \times D^c) \cup (C^c \times D).$$

Note that $\Gamma \times D^c$ and $C^c \times D$ are disjoint. As $\mathcal{C}$ is a semialgebra, there exist disjoint $C_1, \ldots, C_n \in \mathcal{C}$ such that $\Gamma = \bigcup_{k=1}^n C_k$. Furthermore, there exist disjoint $D_1, \ldots, D_m \in \mathcal{D}$ such that $D^c = \bigcup_{j=1}^m D_j$ and disjoint $C_1', \ldots, C_p' \in \mathcal{C}$ such that $C^c = \bigcup_{l=1}^p C_l'$. Then

$$E^c = \left(\left(\bigcup_{k=1}^n C_k \right) \times \left(\bigcup_{j=1}^m D_j \right) \right) \cup \left(\left(\bigcup_{l=1}^p C_l' \right) \times D \right)$$

$$= \bigcup_{k=1}^n \bigcup_{j=1}^m (C_k \times D_j) \cup \bigcup_{l=1}^p (C_l' \times D),$$

which is a finite disjoint union of members of $\mathcal{E}$. Hence, we have shown that $\mathcal{E}$ is a semialgebra. The generalization to n semialgebras follows by mathematical induction. ∎

Existence of an Extension

Suppose now that Ω is a set, $\mathcal{C}$ is a semialgebra of subsets of Ω, and ι is a nonnegative extended real-valued set function on $\mathcal{C}$ satisfying Conditions (E1)–(E3). As we mentioned earlier, under those assumptions, there exists an extension of ι to a measure μ on a σ-algebra $\mathcal{A}$ containing $\mathcal{C}$. To obtain the extension, we will mimic the procedure used in Chapter 3 for extending the concept of length.

The first step is to extend ι to all subsets of Ω using (6.1) on page 179 as a guide. This is done in Definition 6.2.

DEFINITION 6.2   **Outer Measure**

Let Ω be a set, $\mathcal{C}$ a semialgebra of subsets of Ω, and ι a nonnegative extended real-valued set function on $\mathcal{C}$ satisfying Conditions (E1)–(E3). Then the set function μ^* defined on $\mathcal{P}(\Omega)$ by $\mu^*(\emptyset) = 0$ and

$$\mu^*(A) = \inf \left\{ \sum_n \iota(C_n) : \{C_n\}_n \subset \mathcal{C}, \ \bigcup_n C_n \supset A \right\},$$

for $A \neq \emptyset$, is called the **outer measure induced by ι and $\mathcal{C}$.**

The next example provides some illustrations of outer measure. The details of verification are left to the reader as exercises.

EXAMPLE 6.2 *Illustrates Definition 6.2*

    a) Suppose that $\Omega = \{x_1, x_2, \ldots, x_n\}$ is a finite set and $\{a_1, a_2, \ldots, a_n\}$ are non-negative real numbers. Let $\mathcal{C}$ denote the collection of sets consisting of the empty set and all singleton sets. Define ι on $\mathcal{C}$ by $\iota(\emptyset) = 0$ and $\iota(\{x_k\}) = a_k$ for $1 \le k \le n$. Then Conditions (E1)–(E3) hold and

$$\mu^*(A) = \sum_{x_k \in A} a_k$$

for each $A \subset \Omega$.

    b) Let $\mathcal{I}$ denote the collection of all intervals of $\mathcal{R}$, including degenerate intervals of the form (a, a) and $[a, a]$. Take $\Omega = \mathcal{R}$, $\mathcal{C} = \mathcal{I}$, and $\iota = \ell$ (= length). Then Conditions (E1)–(E3) hold and $\mu^* = \lambda^*$; that is, the outer measure induced by ℓ and $\mathcal{I}$ is Lebesgue outer measure.

    c) Let $\mathcal{I}_n$ denote the collection of all n-dimensional intervals in $\mathcal{R}^n$. Take $\Omega = \mathcal{R}^n$, $\mathcal{C} = \mathcal{I}_n$, and $\iota = \ell_n = $ volume; that is, for $I_1 \times I_2 \times \cdots \times I_n \in \mathcal{I}_n$, let $\ell_n(I_1 \times I_2 \times \cdots \times I_n) = \ell(I_1)\ell(I_2)\cdots\ell(I_n)$. Then Conditions (E1)–(E3) hold. The outer measure induced by ℓ_n and $\mathcal{I}_n$ is called **n-dimensional Lebesgue outer measure** and is denoted by λ_n^*.   □

Some basic properties of outer measure are provided by the following proposition. Observe that part (a) of the proposition shows that μ^* is indeed an extension of ι.

□ □ □   **PROPOSITION 6.2**

The outer measure μ^* induced by ι and $\mathcal{C}$ satisfies
a) $\mu^*|_{\mathcal{C}} = \iota$; that is, $\mu^*(C) = \iota(C)$ for $C \in \mathcal{C}$.
b) $\mu^*(A) \ge 0$, for all $A \subset \Omega$. **(nonnegativity)**
c) $A \subset B \Rightarrow \mu^*(A) \le \mu^*(B)$. **(monotonicity)**
d) $\mu^*(\bigcup_n A_n) \le \sum_n \mu^*(A_n)$. **(countable subadditivity)**

PROOF   We leave the proofs of parts (b) and (c) as exercises.

    a) Let $C \in \mathcal{C}$. If $C = \emptyset$, then, by Condition (E1) and Definition 6.2, we have $\iota(\emptyset) = 0 = \mu^*(\emptyset)$. So, assume that $C \ne \emptyset$. Because $\{C\} \subset \mathcal{C}$ and $C \supset C$, we have $\mu^*(C) \le \iota(C)$. Moreover, if $\{C_n\}_n \subset \mathcal{C}$ and $\bigcup_n C_n \supset C$, then, by Condition (E3), $\iota(C) \le \sum_n \iota(C_n)$; thus, $\iota(C) \le \mu^*(C)$. Consequently, we have shown that $\mu^*(C) = \iota(C)$.

    d) If $\mu^*(A_n) = \infty$ for some n, then, by part (b), the required inequality holds. So, we can assume that $\mu^*(A_n) < \infty$ for all n. Let $\epsilon > 0$ be given. For each n, choose $\{C_{nk}\}_k \subset \mathcal{C}$ such that $\bigcup_k C_{nk} \supset A_n$ and $\sum_k \iota(C_{nk}) < \mu^*(A_n) + \epsilon/2^n$. Then $\{C_{nk}\}_{n,k} \subset \mathcal{C}$, $\bigcup_{n,k} C_{nk} \supset \bigcup_n A_n$ and, therefore,

$$\mu^*\left(\bigcup_n A_n\right) \le \sum_{n,k} \iota(C_{nk}) = \sum_n \sum_k \iota(C_{nk})$$

$$\le \sum_n \left(\mu^*(A_n) + \epsilon/2^n\right) \le \sum_n \mu^*(A_n) + \epsilon.$$

Because $\epsilon > 0$ was arbitrarily chosen, $\mu^*(\bigcup_n A_n) \le \sum_n \mu^*(A_n)$.   ■

We have now completed the first step in obtaining the extension of ι to a measure on a σ-algebra containing $\mathcal{C}$; namely, the construction of the outer measure μ^* which is an extension of ι to all subsets of Ω. The second step is to restrict μ^* to an appropriate σ-algebra of subsets of Ω so as to ensure countable additivity. Thus, with (6.2) on page 179 in mind, we make the following definition.

DEFINITION 6.3   Measurable Sets

A set $E \subset \Omega$ is said to be μ^*-**measurable** if

$$\mu^*(W) = \mu^*(W \cap E) + \mu^*(W \cap E^c) \tag{6.3}$$

for all subsets W of Ω. The collection of all μ^*-measurable sets is denoted by $\mathcal{A}$.

EXAMPLE 6.3   *Illustrates Definition 6.3*

a) Suppose that $\Omega = \{x_1, x_2, \ldots, x_n\}$ is a finite set and $\{a_1, a_2, \ldots, a_n\}$ are nonnegative real numbers. Let '$\mathcal{C}$ denote the collection of sets consisting of the empty set and all singleton sets. Define ι on $\mathcal{C}$ by $\iota(\emptyset) = 0$ and $\iota(\{x_k\}) = a_k$ for $1 \le k \le n$. Referring to Example 6.2(a), it is easy to see that $\mathcal{A} = \mathcal{P}(\Omega)$. In other words, all subsets of Ω are μ^*-measurable.

b) Take $\Omega = \mathcal{R}$, $\mathcal{C} = \mathcal{I}$, and $\iota = \ell$. Then, by Example 6.2(b), $\mu^* = \lambda^*$. Hence, in this case, the μ^*-measurable sets are the Lebesgue measurable sets; that is, $\mathcal{A} = \mathcal{M}$.

c) Take $\Omega = \mathcal{R}^n$, $\mathcal{C} = \mathcal{I}_n$, and $\iota = \ell_n = $ volume. Then the μ^*-measurable (i.e., λ_n^*-measurable) sets are called n-**dimensional Lebesgue measurable sets** and the collection of all such sets is denoted by $\mathcal{M}_n$.   □

We claim that the set function $\mu = \mu^*_{|\mathcal{A}}$ is the required extension of ι. To verify this, we must now establish three facts: $\mathcal{A} \supset \mathcal{C}$, $\mathcal{A}$ is a σ-algebra, and μ is a measure on $\mathcal{A}$. The proofs of these facts are considered in the following three propositions.

□ □ □   **PROPOSITION 6.3**

Every set $C \in \mathcal{C}$ is μ^*-measurable. That is, $\mathcal{A} \supset \mathcal{C}$.

PROOF   Let $C \in \mathcal{C}$. We must show that (6.3) holds with $E = C$. Because of countable subadditivity (Proposition 6.2(d)), it suffices to prove that

$$\mu^*(W) \ge \mu^*(W \cap C) + \mu^*(W \cap C^c). \tag{6.4}$$

If $C = \emptyset$, it is trivial. So assume $C \ne \emptyset$. If $\mu^*(W) = \infty$, then clearly (6.4) holds. So, assume that $\mu^*(W) < \infty$. Let $\epsilon > 0$ be given. Choose $\{C_n\}_n \subset \mathcal{C}$ such that $W \subset \bigcup_n C_n$ and

$$\sum_n \iota(C_n) < \mu^*(W) + \epsilon. \tag{6.5}$$

Now, $W \cap C \subset \bigcup_n (C_n \cap C)$ and, hence, by Proposition 6.2,

$$\mu^*(W \cap C) \leq \sum_n \mu^*(C_n \cap C) = \sum_n \iota(C_n \cap C). \tag{6.6}$$

Also, we have $W \cap C^c \subset \bigcup_n (C_n \cap C^c)$ and, so, Proposition 6.2 implies that $\mu^*(W \cap C^c) \leq \sum_n \mu^*(C_n \cap C^c)$. Since $C \in \mathcal{C}$ and $\mathcal{C}$ is a semialgebra, there exist a finite number of pairwise disjoint members of $\mathcal{C}$, say $A_1, \ldots, A_m$, such that $C^c = \bigcup_{k=1}^m A_k$. Then, for each n, $C_n \cap C^c = \bigcup_{k=1}^m (C_n \cap A_k)$ and, therefore, $\mu^*(C_n \cap C^c) \leq \sum_{k=1}^m \mu^*(C_n \cap A_k) = \sum_{k=1}^m \iota(C_n \cap A_k)$. Consequently,

$$\mu^*(W \cap C^c) \leq \sum_n \sum_{k=1}^m \iota(C_n \cap A_k). \tag{6.7}$$

Because of (6.6) and (6.7), we can conclude that

$$\mu^*(W \cap C) + \mu^*(W \cap C^c) \leq \sum_n \iota(C_n \cap C) + \sum_n \sum_{k=1}^m \iota(C_n \cap A_k)$$
$$= \sum_n \left(\iota(C_n \cap C) + \sum_{k=1}^m \iota(C_n \cap A_k) \right). \tag{6.8}$$

But,

$$C_n = (C_n \cap C) \cup (C_n \cap C^c) = (C_n \cap C) \cup \left(\bigcup_{k=1}^m (C_n \cap A_k) \right),$$

which is a finite disjoint union of members of $\mathcal{C}$. Thus, by Condition (E2),

$$\iota(C_n) = \iota(C_n \cap C) + \sum_{k=1}^m \iota(C_n \cap A_k). \tag{6.9}$$

Substituting the left-hand side of (6.9) for the right-hand side in (6.8) and employing (6.5), we can conclude that

$$\mu^*(W \cap C) + \mu^*(W \cap C^c) \leq \sum_n \iota(C_n) < \mu^*(W) + \epsilon.$$

As $\epsilon > 0$ was arbitrary, we see that (6.4) holds. ∎

□ □ □ **PROPOSITION 6.4**

$\mathcal{A}$ is a σ-algebra of subsets of Ω.

PROOF   The proof is a duplication of the one given for Theorem 3.11 on page 103 with $\mathcal{M}$ replaced by $\mathcal{A}$ and λ^* replaced by μ^*. ∎

□ □ □ **PROPOSITION 6.5**

Let $\mu = \mu^*{}_{|\mathcal{A}}$. Then μ is a measure on $\mathcal{A}$.

PROOF Since, by definition, $\mu^*(\emptyset) = 0$, it follows that $\mu(\emptyset) = 0$. Also, by Proposition 6.2(b), $\mu^*(A) \geq 0$ for all $A \subset \Omega$ and, hence, $\mu(A) \geq 0$ for all $A \in \mathcal{A}$. To show that μ is countably additive, we duplicate the proof of Theorem 3.12 on page 105, replacing $\mathcal{M}$ by $\mathcal{A}$, λ^* by μ^*, and λ by μ. ∎

We have now established that μ is the required extension of ι. As an added bonus, it turns out that the measure space $(\Omega, \mathcal{A}, \mu)$ is complete. To see this, let $A \in \mathcal{A}$ with $\mu(A) = 0$. We must show that if $B \subset A$, then $B \in \mathcal{A}$. By the monotonicity of μ^*, we have $\mu^*(B) \leq \mu^*(A) = \mu(A) = 0$. Therefore, $\mu^*(B) = 0$. Now, let $W \subset \Omega$. Then $\mu^*(W \cap B) \leq \mu^*(B) = 0$ and $\mu^*(W \cap B^c) \leq \mu^*(W)$. Thus,

$$\mu^*(W) \geq \mu^*(W \cap B^c) = \mu^*(W \cap B) + \mu^*(W \cap B^c),$$

which implies that $B \in \mathcal{A}$.

The results that we have obtained so far are summarized in the following theorem.

□ □ □ **THEOREM 6.1 Extension Theorem**

Suppose that Ω is a set, $\mathcal{C}$ is a semialgebra of subsets of Ω, and ι is a nonnegative extended real-valued function on $\mathcal{C}$ satisfying Conditions (E1)–(E3) on page 180. Let μ^* denote the outer measure induced by ι and $\mathcal{C}$, $\mathcal{A}$ the collection of μ^*-measurable sets, and $\mu = \mu^*{}_{|\mathcal{A}}$. Then $\mathcal{A}$ is a σ-algebra, $\mathcal{A} \supset \mathcal{C}$, μ is a measure on $\mathcal{A}$, and $\mu_{|\mathcal{C}} = \iota$. Moreover, the measure space $(\Omega, \mathcal{A}, \mu)$ is complete.

An important application of Theorem 6.1 is to *n-dimensional Lebesgue measure:* Let $\Omega = \mathcal{R}^n$, $\mathcal{C} = \mathcal{I}_n$, and $\iota = \ell_n = $ volume. Then $\mu^* = \lambda_n^*$ and $\mathcal{A} = \mathcal{M}_n$. The restriction of λ_n^* to $\mathcal{M}_n$ is denoted by λ_n and is called **n-dimensional Lebesgue measure.**

Uniqueness of an Extension

Theorem 6.1 states, in particular, that ι has an extension to a measure on a σ-algebra containing $\mathcal{C}$, thus answering Question 1 on page 180. Now we will consider Question 2, the question of uniqueness: Under the assumptions of Theorem 6.1, is an extension of ι to a measure on a σ-algebra containing $\mathcal{C}$ unique? In general, the answer to the uniqueness question is no (see, for instance, Exercise 6.13).

However, under certain conditions, we can establish uniqueness results. To begin, we define two collections of subsets of Ω associated with $\mathcal{C}$: $\mathcal{C}_\sigma$ denotes the collection of all subsets of Ω that are countable unions of members of $\mathcal{C}$; in other words, $E \in \mathcal{C}_\sigma$ if and only if there exists $\{C_n\}_n \subset \mathcal{C}$ such that $E = \bigcup_n C_n$. $\mathcal{C}_{\sigma\delta}$ denotes the collection of all subsets of Ω that are countable intersections of members of $\mathcal{C}_\sigma$; in other words, $F \in \mathcal{C}_{\sigma\delta}$ if and only if there exists $\{E_n\}_n \subset \mathcal{C}_\sigma$ such that $F = \bigcap_n E_n$.

Next, we establish three lemmas that are required in order for us to prove a uniqueness theorem.

□ □ □ **LEMMA 6.1**

Let $A \subset \Omega$.
a) Given $\epsilon > 0$, there is an $E \in \mathcal{C}_\sigma$ with $E \supset A$ and $\mu^*(E) \leq \mu^*(A) + \epsilon$.
b) There is an $F \in \mathcal{C}_{\sigma\delta}$ such that $F \supset A$ and $\mu^*(F) = \mu^*(A)$.

PROOF   a) If $\mu^*(A) = \infty$, then also $\mu^*(\Omega) = \infty$. The required result now follows because $\Omega \in \mathcal{C}_\sigma$. So, assume that $\mu^*(A) < \infty$. Then there exists $\{C_n\} \subset \mathcal{C}$ such that $\bigcup_n C_n \supset A$ and $\sum_n \iota(C_n) < \mu^*(A) + \epsilon$. Let $E = \bigcup_n C_n$. Then $E \in \mathcal{C}_\sigma$, $E \supset A$, and

$$\mu^*(E) \leq \sum_n \mu^*(C_n) = \sum_n \iota(C_n) < \mu^*(A) + \epsilon.$$

b) If $\mu^*(A) = \infty$, we can take $F = \Omega$. So, let us assume that $\mu^*(A) < \infty$. By part (a), we can, for each $n \in \mathcal{N}$, choose $E_n \in \mathcal{C}_\sigma$ such that $E_n \supset A$ and $\mu^*(E_n) \leq \mu^*(A) + 1/n$. Let $F = \bigcap_{n=1}^{\infty} E_n$. Then $F \in \mathcal{C}_{\sigma\delta}$ and $F \supset A$. In particular, then, $\mu^*(F) \geq \mu^*(A)$. However, because $F \subset E_n$, we have $\mu^*(F) \leq \mu^*(E_n) \leq \mu^*(A) + 1/n$ for all $n \in \mathcal{N}$. Therefore, $\mu^*(F) \leq \mu^*(A)$. ∎

□ □ □ **LEMMA 6.2**

The algebra generated by $\mathcal{C}$, $\sigma_0(\mathcal{C})$, consists of the empty set and all finite disjoint unions of members of $\mathcal{C}$.

PROOF   Let $\mathcal{D}$ denote the collection of sets consisting of the empty set and all finite disjoint unions of members of $\mathcal{C}$. We must prove that $\mathcal{D} = \sigma_0(\mathcal{C})$. Clearly any algebra of sets containing $\mathcal{C}$ must contain $\mathcal{D}$; so, $\mathcal{D} \subset \sigma_0(\mathcal{C})$. To establish the reverse inequality, it suffices to prove that $\mathcal{D}$ is an algebra, because $\sigma_0(\mathcal{C})$ is the smallest algebra containing $\mathcal{C}$.

First we show $\mathcal{D}$ is closed under finite intersections. Let $A \in \mathcal{D}$ and $B \in \mathcal{D}$. We claim that $A \cap B \in \mathcal{D}$. If either A or B is empty, then $A \cap B = \emptyset \in \mathcal{D}$. So, assume neither A nor B is empty. Then there exists a pairwise disjoint sequence $\{A_i\}_{i=1}^m$ of members of $\mathcal{C}$ such that $A = \bigcup_{i=1}^m A_i$, and a pairwise disjoint sequence $\{B_j\}_{j=1}^n$ of members of $\mathcal{C}$ such that $B = \bigcup_{j=1}^n B_j$. Consequently,

$$A \cap B = \bigcup_{i=1}^m \left(A_i \cap \left(\bigcup_{j=1}^n B_j \right) \right) = \bigcup_{i=1}^m \bigcup_{j=1}^n (A_i \cap B_j).$$

Since $A_i, B_j \in \mathcal{C}$, we have $A_i \cap B_j \in \mathcal{C}$. Moreover, since the A_is and B_js are each pairwise disjoint, so are the $(A_i \cap B_j)$s. Hence, $A \cap B$ is a finite disjoint union of members of $\mathcal{C}$ and, consequently, is a member of $\mathcal{D}$.

Next, we show $\mathcal{D}$ is closed under complementation. Let $A \in \mathcal{D}$. If $A = \emptyset$, then $A^c = \Omega \in \mathcal{D}$. If $A \neq \emptyset$, then there exists a pairwise disjoint sequence $\{A_i\}_{i=1}^m$ of members of $\mathcal{C}$ such that $A = \bigcup_{i=1}^m A_i$. As $A_i \in \mathcal{C}$, A_i^c is a finite disjoint union of members of $\mathcal{C}$; hence, $A_i^c \in \mathcal{D}$. From the previous paragraph, we know $\mathcal{D}$ is closed under finite intersections. Thus, $A^c = \bigcap_{i=1}^m A_i^c \in \mathcal{D}$. ∎

□ □ □   **LEMMA 6.3**

Each $E \in \mathcal{C}_\sigma$ can be written as a countable disjoint union of members of $\mathcal{C}$.

PROOF By definition, there exists $\{C_n\}_n \subset \mathcal{C}$ such that $E = \bigcup_n C_n$. In particular, $\{C_n\}_n \subset \sigma_0(\mathcal{C})$. Let $D_1 = C_1$ and $D_n = C_n \setminus \bigcup_{k=1}^{n-1} C_k$ for $n \geq 2$. Then the D_ns are pairwise disjoint and $E = \bigcup_n D_n$. Moreover, $D_n \in \sigma_0(\mathcal{C})$ for each n. Without loss of generality, we can assume $D_n \neq \emptyset$ for all n. As $D_n \in \sigma_0(\mathcal{C})$, we know by Lemma 6.2 that there is a finite sequence $\{E_{nj}\}_{j=1}^{k_n}$ of pairwise disjoint members of $\mathcal{C}$ such that $D_n = \bigcup_{j=1}^{k_n} E_{nj}$. It follows that $\{E_{11}, \ldots, E_{1k_1}, E_{21}, \ldots, E_{2k_2}, \ldots\}$ is a countable collection of pairwise disjoint members of $\mathcal{C}$ whose union is E. ■

We are now in a position to prove a theorem that deals with the question of uniqueness for an extension of ι to a measure on a σ-algebra containing $\mathcal{C}$.

□ □ □   **THEOREM 6.2**

Let Ω be a set, $\mathcal{C}$ a semialgebra of subsets of Ω, and ι a nonnegative extended real-valued function on $\mathcal{C}$ satisfying Conditions (E1)–(E3) on page 180. Suppose there is a sequence $\{C_n\}_n$ of subsets of Ω such that

(E4) $\{C_n\}_n \subset \mathcal{C}$, $\bigcup_n C_n = \Omega$, and $\iota(C_n) < \infty$ for each n.

Then there exists a unique extension of ι to a measure on $\sigma(\mathcal{C})$, the σ-algebra generated by $\mathcal{C}$.

PROOF Let μ^* be the outer measure induced by ι and $\mathcal{C}$, $\mathcal{A}$ the collection of μ^*-measurable sets, and $\mu = \mu^*|_\mathcal{A}$. By Theorem 6.1, $\mathcal{A}$ is a σ-algebra, $\mathcal{A} \supset \mathcal{C}$, μ is a measure on $\mathcal{A}$, and $\mu_{|\mathcal{C}} = \iota$. It follows that $\mathcal{A} \supset \sigma(\mathcal{C})$ and that if we define $\nu = \mu_{|\mathcal{A}(\mathcal{C})}$, then ν is an extension of ι to $\sigma(\mathcal{C})$. Therefore, the existence portion of the theorem is established.

It remains to prove the uniqueness portion of the theorem, that ν is the only extension of ι to $\sigma(\mathcal{C})$. In other words, we must show that if τ is a measure on $\sigma(\mathcal{C})$ with $\tau(C) = \iota(C)$ for all $C \in \mathcal{C}$, then

$$\tau(A) = \nu(A), \qquad A \in \sigma(\mathcal{C}). \tag{6.10}$$

In establishing (6.10), we will use the fact that $\mathcal{C}_\sigma \subset \sigma(\mathcal{C})$, which follows because $\sigma(\mathcal{C})$ is a σ-algebra containing $\mathcal{C}$.

First, we will show that

$$\tau(E) = \nu(E), \qquad E \in \mathcal{C}_\sigma. \tag{6.11}$$

Let $E \in \mathcal{C}_\sigma$. By Lemma 6.3, there exists $\{C_n\}_n \subset \mathcal{C}$ with $C_i \cap C_j = \emptyset$, for $i \neq j$, such that $E = \bigcup_n C_n$. Consequently,

$$\tau(E) = \sum_n \tau(C_n) = \sum_n \iota(C_n) = \sum_n \nu(C_n) = \nu(E),$$

which establishes (6.11).

Next, we will show that

$$\tau(A) = \nu(A), \qquad A \in \sigma(\mathcal{C}), \ \nu(A) < \infty. \tag{6.12}$$

For a given $\epsilon > 0$, we can, by Lemma 6.1(a), select a set $E \in \mathcal{C}_\sigma$ such that $E \supset A$ and $\mu^*(E) \le \mu^*(A) + \epsilon$ which, in this case, is equivalent to $\nu(E) \le \nu(A) + \epsilon$. As $E \supset A$ and $E \in \mathcal{C}_\sigma$, we conclude from (6.11) that

$$\tau(A) \le \tau(E) = \nu(E) \le \nu(A) + \epsilon.$$

As $\epsilon > 0$ was arbitrary, we see that

$$\tau(A) \le \nu(A), \qquad A \in \sigma(\mathcal{C}), \ \nu(A) < \infty. \tag{6.13}$$

To prove the reverse inequality, we again select, for a given $\epsilon > 0$, a set $E \in \mathcal{C}_\sigma$ such that $E \supset A$ and $\nu(E) \le \nu(A) + \epsilon$. Because $\nu(A) < \infty$, it follows that $\nu(E \setminus A) = \nu(E) - \nu(A) \le \epsilon$. Applying (6.13) to $E \setminus A$, we obtain $\tau(E \setminus A) \le \epsilon$. Hence, by (6.11), we can now conclude that

$$\nu(A) \le \nu(E) = \tau(E) = \tau(A) + \tau(E \setminus A) \le \tau(A) + \epsilon.$$

As $\epsilon > 0$ was arbitrary, we see that $\nu(A) \le \tau(A)$. This result and (6.13) imply that (6.12) holds.

It remains to establish (6.10) when $\nu(A) = \infty$. Let $\{C_n\}_n$ be as in Condition (E4). By Exercise 6.12, we can assume that the C_ns are pairwise disjoint. Now, $A = A \cap \Omega = \bigcup_n (A \cap C_n)$. Because

$$\nu(A \cap C_n) \le \nu(C_n) = \iota(C_n) < \infty,$$

(6.12) implies that $\nu(A \cap C_n) = \tau(A \cap C_n)$. Consequently,

$$\nu(A) = \sum_n \nu(A \cap C_n) = \sum_n \tau(A \cap C_n) = \tau(A).$$

The proof of the theorem is now complete. ■

Three particularly important consequences of Theorem 6.2 are given here in Corollaries 6.1–6.3. We will refer to these corollaries frequently.

□ □ □ **COROLLARY 6.1**

Let $(\Omega, \mathcal{A}, \mu)$ be a measure space. Suppose that $\mathcal{C}$ is a semialgebra of subsets of Ω such that the σ-algebra generated by $\mathcal{C}$ is $\mathcal{A}$. Further suppose that there is a sequence $\{C_n\}_n \subset \mathcal{C}$ with $\bigcup_n C_n = \Omega$ and $\mu(C_n) < \infty$ for each n. If ν is a measure on $\mathcal{A}$ such that $\nu(C) = \mu(C)$ for all $C \in \mathcal{C}$, then $\nu = \mu$, that is, $\nu(A) = \mu(A)$ for all $A \in \mathcal{A}$.

PROOF Let $\iota = \mu_{|\mathcal{C}} (= \nu_{|\mathcal{C}})$. Since μ is a measure, it follows immediately that Conditions (E1)–(E3) are satisfied by ι and $\mathcal{C}$. Also, by assumption, Condition (E4) holds. Therefore, by Theorem 6.2, ι has a unique extension to the σ-algebra generated by $\mathcal{C}$, which, by hypothesis, is $\mathcal{A}$. Since both μ and ν are extensions of ι to $\mathcal{A}$, it must be that $\nu = \mu$. ■

□ □ □ **COROLLARY 6.2**

Suppose that μ and ν are two Borel measures (i.e., measures on $\mathcal{B}$) such that $\mu(I) = \nu(I) < \infty$ for all bounded intervals I. Then $\mu = \nu$.

PROOF   Note that $\mathcal{R} = \bigcup_{n=-\infty}^{\infty} (n, n+1]$. Thus, any interval I can be written as a countable union of disjoint bounded intervals, namely, $I = \bigcup_{n=-\infty}^{\infty} (I \cap (n, n+1])$. Hence, as $\mu(I) = \nu(I)$ for all bounded intervals I, it follows that $\mu(I) = \nu(I)$ for all $I \in \mathcal{I}$. By Exercise 6.4(b), $\mathcal{I}$ is a semialgebra and, by Exercise 6.14, the σ-algebra generated by $\mathcal{I}$ is $\mathcal{B}$. By assumption, $\mu((n, n+1]) < \infty$ for all $n \in \mathcal{Z}$. The required result now follows from Corollary 6.1.   ■

□ □ □ **COROLLARY 6.3**

Let $(\Omega, \mathcal{A})$ be a measurable space and $\mathcal{C}$ a semialgebra of subsets of Ω such that the σ-algebra generated by $\mathcal{C}$ is $\mathcal{A}$. If μ and ν are two finite measures on $\mathcal{A}$ such that $\mu(C) = \nu(C)$ for all $C \in \mathcal{C}$, then $\mu = \nu$.

PROOF   By Exercise 6.11(a), $\Omega \in \mathcal{C}_\sigma$. As μ is a finite measure, we see that all the assumptions of Corollary 6.1 are satisfied.   ■

Remark: In Corollary 6.3, we really need only assume that at least one of the measures, μ and ν, is finite (why?).

Exercises for Section 6.1

Note: A ★ denotes an exercise that will be subsequently referenced.

6.1   Let Ω be a (nonempty) set. Verify the following results.
   a) $\{\Omega\}$ is a semialgebra, but not an algebra.
   b) If $\mathcal{C}$ is a semialgebra that contains exactly one element, then $\mathcal{C} = \{\Omega\}$.
   c) $\{\emptyset, \Omega\}$ is a semialgebra.
   d) If $\mathcal{C}$ is a semialgebra that contains exactly two elements, then $\mathcal{C} = \{\emptyset, \Omega\}$.

★6.2   Let $\mathcal{J}$ denote the collection of intervals of $\mathcal{R}$ of the form $(a, b]$ and (c, ∞), where $-\infty \le a \le b < \infty$ and $-\infty \le c < \infty$. Prove that $\mathcal{J}$ is a semialgebra and that $\sigma(\mathcal{J}) = \mathcal{B}$.

6.3   Suppose that $\Omega = \{x_1, x_2, \ldots, x_n\}$ is a finite set and $\{a_1, a_2, \ldots, a_n\}$ are nonnegative real numbers. Let $\mathcal{C}$ denote the collection of sets consisting of the empty set and all singleton sets, that is, sets of the form $\{x\}$, where $x \in \Omega$. Define ι on $\mathcal{C}$ by $\iota(\emptyset) = 0$ and $\iota(\{x_k\}) = a_k$ for $1 \le k \le n$.
   a) Verify that Conditions (E1)–(E3) on page 180 hold.
   b) Show that $\mathcal{C}$ is a semialgebra of subsets of Ω.

6.4   Let $\mathcal{I}$ denote the collection of all intervals of $\mathcal{R}$, including degenerate intervals of the form (a, a) and $[a, a]$. Take $\Omega = \mathcal{R}$, $\mathcal{C} = \mathcal{I}$, and $\iota = \ell$ ($=$ length).
   a) Show that Conditions (E1)–(E3) on page 180 hold.
   b) Show that $\mathcal{I}$ is a semialgebra of subsets of $\mathcal{R}$.

6.5   Let $\mathcal{I}_2$ denote the collection of all two-dimensional intervals in $\mathcal{R}^2$; that is, all sets of the form $I_1 \times I_2$ where $I_j \in \mathcal{I}$ for $1 \le j \le 2$. Take $\Omega = \mathcal{R}^2$, $\mathcal{C} = \mathcal{I}_2$, and $\iota = \ell_2 = $ area; that is, for $I_1 \times I_2 \in \mathcal{I}_2$, $\ell_2(I_1 \times I_2) = \ell(I_1)\ell(I_2)$.
   a) Show that Conditions (E1)–(E3) on page 180 hold.
   b) Show that $\mathcal{I}_2$ is a semialgebra of subsets of $\mathcal{R}^2$.

6.6   Generalize Exercise 6.5 to n dimensions.

6.7 Refer to Exercise 6.3. Prove that $\mu^* = \sum_{k=1}^n a_k \delta_{x_k}$, that is, $\mu^*(A) = \sum_{x_k \in A} a_k$ for each $A \subset \Omega$.

6.8 Refer to Exercise 6.4. Prove that the outer measure μ^* induced by ℓ and $\mathcal{I}$ is Lebesgue outer measure.

6.9 Prove parts (b) and (c) of Proposition 6.2 on page 183.

6.10 Refer to Exercises 6.3 and 6.7. Establish that every subset of Ω is μ^*-measurable; that is, $\mathcal{A} = \mathcal{P}(\Omega)$.

6.11 Let $\mathcal{C}$ be a semialgebra of subsets of Ω.
   a) Why is $\Omega \in \mathcal{C}_\sigma$?
   b) Is it necessarily true that $\Omega \in \mathcal{C}$?

6.12 Suppose that Condition (E4) on page 188 holds. Prove there exists $\{E_n\}_n \subset \mathcal{C}$ with $\bigcup_n E_n = \Omega$, $E_i \cap E_j = \emptyset$, for $i \neq j$, and $\iota(E_n) < \infty$ for each n.

6.13 Prove that Condition (E4) cannot be omitted as a hypothesis in Theorem 6.2. *Hint:* Let $\mathcal{J}$ be as in Exercise 6.2, $\iota(\emptyset) = 0$, and $\iota(C) = \infty$ for $C \in \mathcal{J}$ and $C \neq \emptyset$.

6.14 Let $\mathcal{I}$ be as in Exercise 6.4. Show that $\sigma(\mathcal{I}) = \mathcal{B}$.

6.15 Let $\mathcal{I}$ be as in Exercise 6.4. Suppose that g is a nonnegative Lebesgue measurable function on $\mathcal{R}$ satisfying $\int_{(-\infty,n)} g \, d\lambda < \infty$ for each $n \in \mathcal{N}$. Define ι on $\mathcal{I}$ by

$$\iota(C) = \int_C g \, d\lambda.$$

   a) Verify that Conditions (E1)–(E3) are satisfied by ι and $\mathcal{I}$.
   b) Show there is a unique extension of ι to a measure μ on $\mathcal{B}$ and that $\mu(B) = \int_B g \, d\lambda$.

6.16 Let μ and ν be two finite Borel measures such that $\mu\big((-\infty, x]\big) = \nu\big((-\infty, x]\big)$ for all $x \in \mathcal{R}$. Prove that $\mu = \nu$.

6.17 Can the finiteness assumption be dropped in Exercise 6.16? Explain.

6.18 Let μ and ν be two Borel measures such that $\mu\big((-\infty, x]\big) = \nu\big((-\infty, x]\big) < \infty$ for all $x \in \mathcal{R}$. Prove that $\mu = \nu$.

★6.19 Let Ω be a set, $\mathcal{C}$ a semialgebra of subsets of Ω, and ι a nonnegative extended real-valued function on $\mathcal{C}$ satisfying Conditions (E1)–(E4). Also, let μ^* be the outer measure induced by ι and $\mathcal{C}$, $\mathcal{A}$ the collection of μ^*-measurable sets, and $\mu = \mu^*_{|\mathcal{A}}$. Suppose that $E \in \mathcal{A}$.
   a) Show that there is an $A \in \sigma(\mathcal{C})$ with $A \supset E$ and $\mu(A \setminus E) = 0$. *Hint:* First assume that $\mu(E) < \infty$ and employ Lemma 6.1.
   b) Show that there is a $B \in \sigma(\mathcal{C})$ with $B \subset E$ and $\mu(E \setminus B) = 0$.

★6.20 Let Ω be a set, $\mathcal{C}$ a semialgebra of subsets of Ω, and ι a nonnegative extended real-valued function on $\mathcal{C}$ satisfying Conditions (E1)–(E4). Also, let μ^* be the outer measure induced by ι and $\mathcal{C}$, $\mathcal{A}$ the collection of μ^*-measurable sets, $\mu = \mu^*_{|\mathcal{A}}$, and $\nu = \mu_{|\mathcal{A}(\mathcal{C})}$. Prove that $(\Omega, \mathcal{A}, \mu)$ is the completion of $(\Omega, \sigma(\mathcal{C}), \nu)$. *Hint:* Use Exercise 6.19 and Exercise 5.17 on page 150.

6.21 Consider the measure space $(\mathcal{R}, \mathcal{M}, \lambda)$.
   a) Can we deduce from Theorem 6.2 that λ is the unique extension of length to a measure on $\mathcal{M}$? Explain.
   b) Prove that λ is the unique extension of length to a measure on $\mathcal{M}$.

6.2 THE LEBESGUE-STIELTJES INTEGRAL

In the previous section, we developed existence and uniqueness theorems for extensions to measures. Specifically, suppose that Ω is a set, $\mathcal{C}$ is a semialgebra of subsets of Ω, and ι is a nonnegative extended real-valued function on $\mathcal{C}$. If Conditions (E1)–(E3) on page 180 hold, then there is an extension of ι to a measure on a σ-algebra containing $\mathcal{C}$; and if, in addition, Condition (E4) on page 188 holds, then an extension to the smallest σ-algebra containing $\mathcal{C}$ is unique.

Two important applications of this theory are to the Lebesgue-Stieltjes integral and to product measure spaces. We will discuss the former application in this section and the latter in the next.

Distribution Function of a Finite Borel Measure

Recall that a measure μ on the Borel sets $\mathcal{B}$ is called a **Borel measure** and that such a measure is called **finite** if $\mu(\mathcal{R}) < \infty$. With these conventions in mind, we make the following definition.

DEFINITION 6.4   Distribution Function of a Finite Borel Measure

Let μ be a finite Borel measure. Then the **distribution function of μ,** denoted F_μ, is the real-valued function defined on $\mathcal{R}$ by

$$F_\mu(x) = \mu\big((-\infty, x]\big).$$

Note: We will sometimes omit the subscript μ in F_μ, provided that no confusion will arise.

The following example gives some illustrations of distribution functions. The reader should supply the details of verification.

EXAMPLE 6.4   *Illustrates Definition 6.4*

a) Let $\mu = \lambda_{|\mathcal{B}}$. Then, although μ is a Borel measure, it is not a finite Borel measure because $\mu(\mathcal{R}) = \lambda(\mathcal{R}) = \infty$. Hence, we do not define the distribution function of μ.

b) For $B \in \mathcal{B}$, define $\mu(B) = \lambda\big(B \cap (0,1)\big)$. Then μ is a Borel measure and, as $\mu(\mathcal{R}) = \lambda\big((0,1)\big) = 1 < \infty$, it is a finite Borel measure. Its distribution function F_μ is easily seen to be

$$F_\mu(x) = \begin{cases} 0, & x < 0; \\ x, & 0 \le x < 1; \\ 1, & x \ge 1. \end{cases}$$

c) Recall that if $b \in \mathcal{R}$, then the set function,

$$\delta_b(E) = \begin{cases} 1, & b \in E; \\ 0, & b \notin E. \end{cases}$$

is a measure on $\mathcal{P}(\mathcal{R})$, called the Dirac measure concentrated at b. Let $\mu = \delta_b$ restricted to $\mathcal{B}$. Then μ is a finite Borel measure and

$$F_\mu(x) = \begin{cases} 0, & x < b; \\ 1, & x \ge b. \end{cases}$$

d) Let $\{a_n\}_{n=1}^{\infty}$ be a sequence of nonnegative real numbers with $\sum_{n=1}^{\infty} a_n < \infty$. Define μ on $\mathcal{B}$ by

$$\mu(B) = \sum_{n \in B} a_n.$$

Then μ is a finite Borel measure whose distribution function is

$$F_\mu(x) = \sum_{n=1}^{[x]} a_n,$$

where $[x]$ denotes the greatest integer in x. □

Some of the more important properties of distribution functions are presented in the next two propositions.

□ □ □ PROPOSITION 6.6

Let μ be a finite Borel measure and F its distribution function. Then
a) F is monotone nondecreasing.
b) F is right continuous.
c) F is bounded.
d) $\lim_{x \to -\infty} F(x) = 0$.

PROOF a) If $x \leq y$, then $(-\infty, x] \subset (-\infty, y]$ and, hence, by the monotonicity property of measures, $F(x) = \mu((-\infty, x]) \leq \mu((-\infty, y]) = F(y)$.
  b) Let $x \in \mathcal{R}$. Since F is nondecreasing, $\lim_{y \downarrow x} F(y) = F(x+)$ exists. Now, $(-\infty, x+1] \supset (-\infty, x+1/2] \supset \cdots$ and $\bigcap_{n=1}^{\infty}(-\infty, x+1/n] = (-\infty, x]$. So, because μ is a finite Borel measure, we have, by Theorem 5.1(c) on page 147,

$$F(x) = \mu((-\infty, x]) = \lim_{n \to \infty} \mu((-\infty, x+1/n])$$
$$= \lim_{n \to \infty} F(x+1/n) = F(x+).$$

Hence, $F(x+) = F(x)$, that is, F is right continuous at x. Because $x \in \mathcal{R}$ was arbitrarily chosen, we see that F is right continuous.
  c) Because μ is a finite measure, we have $F(x) = \mu((-\infty, x]) \leq \mu(\mathcal{R}) < \infty$, for each $x \in \mathcal{R}$. Hence F is bounded by $\mu(\mathcal{R})$.
  d) First note that, because F is monotone, $\lim_{x \to -\infty} F(x)$ exists. Also, we have $\emptyset = \bigcap_{n=1}^{\infty}(-\infty, -n]$ and $(-\infty, -1] \supset (-\infty, -2] \supset \cdots$. Thus, since μ is a finite measure,

$$0 = \mu(\emptyset) = \lim_{n \to \infty} \mu((-\infty, -n]) = \lim_{n \to \infty} F(-n) = \lim_{x \to -\infty} F(x).$$

The last equality holds because $\lim_{x \to -\infty} F(x)$ exists. ■

Proposition 6.6(a) shows that F_μ is monotone nondecreasing. Hence, $F_\mu(x)$ has a limit as both $x \to -\infty$ and $x \to \infty$. We denote those limits by $F_\mu(-\infty)$ and $F_\mu(\infty)$, respectively. By Proposition 6.6(d), $F_\mu(-\infty) = 0$; and it is easy to prove that $F_\mu(\infty) = \mu(\mathcal{R})$.

◻ ◻ ◻ **PROPOSITION 6.7**

Let μ be a finite Borel measure and let F be its distribution function. Then, for $-\infty \leq a \leq b < \infty$,

$$\mu\big((a,b]\big) = F(b) - F(a) \tag{6.14}$$

and, for $-\infty \leq c < \infty$,

$$\mu\big((c,\infty)\big) = F(\infty) - F(c). \tag{6.15}$$

PROOF If $a = -\infty$, then, by Definition 6.4 and Proposition 6.6(d),

$$\mu\big((a,b]\big) = F(b) = F(b) - F(-\infty) = F(b) - F(a).$$

If $-\infty < a < \infty$, then, since μ is a finite measure, we have

$$\mu\big((a,b]\big) = \mu\big((-\infty,b]\big) - \mu\big((-\infty,a]\big) = F(b) - F(a).$$

This result proves (6.14). To prove (6.15), note that

$$\mu\big((c,\infty)\big) = \mu(\mathcal{R}) - \mu\big((-\infty,c]\big) = F(\infty) - F(c),$$

as required. ∎

Lebesgue-Stieltjes Measure

We now consider the following two important questions concerning a real-valued function F on $\mathcal{R}$:

Question 1: Under what conditions is F the distribution function of some finite Borel measure?

Question 2: Can F be the distribution function for two different finite Borel measures?

As we have just seen, a necessary condition for a real-valued function F on $\mathcal{R}$ to be the distribution function of some finite Borel measure is that (a)–(d) of Proposition 6.6 hold. In other words, unless F satisfies the properties listed in Proposition 6.6, it cannot possibly be the distribution function of a finite Borel measure.

By employing Theorem 6.2 on page 188, we can show that the properties listed in Proposition 6.6 are not only necessary, but are also sufficient for F to be the distribution function of some finite Borel measure. Moreover, using that same theorem, we can prove that the answer to Question 2 is no.

So, assume that F satisfies (a)–(d) of Proposition 6.6. We will use F to define a nonnegative set function ι on a semialgebra $\mathcal{J}$ of subsets of $\mathcal{R}$. Then we will prove that Conditions (E1)–(E4) of Section 6.1 hold for ι and $\mathcal{J}$. Finally, we will show that the measure μ guaranteed by Theorem 6.2 is a finite Borel measure whose distribution function is F and that μ is the only such measure.

To begin, let $\mathcal{J}$ denote the collection of intervals of $\mathcal{R}$ of the form $(a,b]$ or (c,∞), where $-\infty \leq a \leq b < \infty$ and $-\infty \leq c < \infty$. Then, by Exercise 6.2 on page 190, $\mathcal{J}$ is a semialgebra and $\sigma(\mathcal{J}) = \mathcal{B}$.

Next, we want to use F to define a nonnegative set function ι on $\mathcal{J}$ in such a way that if μ is an extension of ι to a measure on $\mathcal{B}$, then F is the distribution function of μ. In view of (6.14) and (6.15), we see that ι should be defined on $\mathcal{J}$ as follows: For $-\infty \le a \le b < \infty$,

$$\iota\big((a,b]\big) = F(b) - F(a), \tag{6.16}$$

and, for $-\infty \le c < \infty$,

$$\iota\big((c,\infty)\big) = F(\infty) - F(c). \tag{6.17}$$

Note that ι is nonnegative because, by assumption, F is nondecreasing.

Now we will verify that Conditions (E1)–(E4) hold for ι and $\mathcal{J}$. Using (6.16) with $b = a$, we see that $\iota(\emptyset) = \iota\big((a,a]\big) = F(a) - F(a) = 0$. So, Condition (E1) on page 180 holds. To verify Condition (E4) on page 188, we can, for instance, take $\{C_n\}_n$ to consist of the single set $(-\infty, \infty)$.

It remains to show that Conditions (E2) and (E3) hold for ι and $\mathcal{J}$, which we establish in Lemmas 6.4 and 6.5, respectively. In proving those two lemmas, it is convenient to write (c, ∞) as $(c, \omega]$, with the conventions that $F(\omega) = F(\infty)$, $\iota\big((c,\omega]\big) = F(\infty) - F(c)$, and $(c, \omega]^c = (-\infty, c]$. Using this notation, $\mathcal{J}$ consists of all sets of the form $(a, b]$, where either $-\infty \le a \le b < \infty$ or $a \ge -\infty$ and $b = \omega$.

$\square\ \square\ \square$ **LEMMA 6.4**

Suppose that $\{C_k\}_{k=1}^n$ is a finite sequence of pairwise disjoint members of $\mathcal{J}$ whose union is in $\mathcal{J}$. Then $\iota\big(\bigcup_{k=1}^n C_k\big) = \sum_{k=1}^n \iota(C_k)$.

PROOF   Set $C = \bigcup_{k=1}^n C_k$. Then, by assumption, $C \in \mathcal{J}$. So we can write $C = (a, b]$ and $C_k = (a_k, b_k]$, $1 \le k \le n$. Without loss of generality, we can assume that $a_1 < a_2 < \cdots < a_n$. Because $\bigcup_{k=1}^n C_k = C$ and the C_ks are pairwise disjoint, $a = a_1 \le b_1 = a_2 \le b_2 = \cdots = a_{n-1} \le b_{n-1} = a_n$ and $b_n = b$. Hence,

$$\iota(C) = F(b) - F(a) = \sum_{k=1}^n \big(F(b_k) - F(a_k)\big) = \sum_{k=1}^n \iota(C_k),$$

as required.   ∎

$\square\ \square\ \square$ **LEMMA 6.5**

Assume $C, C_1, C_2, \ldots$ are in $\mathcal{J}$ and $C \subset \bigcup_n C_n$. Then $\iota(C) \le \sum_n \iota(C_n)$.

PROOF   We can write $C_n = (a_n, b_n]$, for each n, and $C = (a, b]$. Assume first that $a, b \in \mathcal{R}$; that is, C is a finite interval. Let $\epsilon > 0$ be given. Because F is right continuous, we can choose a $\delta > 0$ such that $F(a + \delta) < F(a) + \epsilon/2$ and, for each n, a $\delta_n > 0$ such that $F(b_n + \delta_n) \le F(b_n) + \epsilon/2^{n+1}$.

The interval $[a + \delta, b]$ is closed and bounded and, since $C \subset \bigcup_n C_n$, it follows that $[a + \delta, b] \subset \bigcup_n (a_n, b_n + \delta_n)$. Hence, by the Heine-Borel theorem, there is an $N \in \mathcal{N}$ such that $[a + \delta, b] \subset \bigcup_{n=1}^N (a_n, b_n + \delta_n)$. Set $I_n = (a_n, b_n + \delta_n)$. Arguing as in Proposition 3.2 on page 93, we find that there is an integer m, with $m \leq N$, and a sequence of intervals $\{J_i\}_{i=1}^m$ with $J_i = (c_i, d_i) \in \{I_n\}_{n=1}^N$, for $1 \leq i \leq m$, and

$$c_1 < a + \delta, c_2 < d_1 < d_2, \ldots, c_m < d_{m-1} < d_m, b < d_m.$$

Because F is nondecreasing and $\{J_i\}_{i=1}^m \subset \{I_n\}_{n=1}^N \subset \{(a_n, b_n + \delta_n)\}_n$, we conclude that

$$\begin{aligned}
F(b) - F(a + \delta) &\leq F(d_m) - F(c_1) \\
&\leq F(d_m) - F(c_1) + \big(F(d_1) - F(c_2)\big) \\
&\quad + \cdots + \big(F(d_{m-1}) - F(c_m)\big) \\
&= \sum_{i=1}^m \big(F(d_i) - F(c_i)\big) \leq \sum_n \big(F(b_n + \delta_n) - F(a_n)\big).
\end{aligned}$$

Consequently,

$$\begin{aligned}
F(b) - F(a) &\leq F(b) - F(a + \delta) + \epsilon/2 \\
&\leq \sum_n \big(F(b_n + \delta_n) - F(a_n)\big) + \epsilon/2 \\
&\leq \sum_n \big(F(b_n) + \epsilon/2^{n+1} - F(a_n)\big) + \epsilon/2 \\
&\leq \sum_n \big(F(b_n) - F(a_n)\big) + \epsilon.
\end{aligned}$$

In other words, $\iota(C) \leq \sum_n \iota(C_n) + \epsilon$. As $\epsilon > 0$ was arbitrary, we conclude that $\iota(C) \leq \sum_n \iota(C_n)$, as required.

The lemma has now been established when C is a finite interval. The proof for the case where C is an infinite interval is left as an exercise for the reader. ∎

We have now verified that Conditions (E1)–(E4) are satisfied by ι and $\mathcal{J}$. Using that fact, we can prove a theorem that answers Questions 1 and 2 on page 180.

□ □ □ **THEOREM 6.3**

Suppose that F is a real-valued function on $\mathcal{R}$ satisfying (a)–(d) of Proposition 6.6 on page 193. Then there is a unique finite Borel measure having F as its distribution function.

PROOF Let $\mathcal{J}$ and ι be as defined earlier. Since Conditions (E1)–(E4) are satisfied by $\mathcal{J}$ and ι, Theorem 6.2 implies that there is a unique extension of ι to a measure μ on $\sigma(\mathcal{J}) = \mathcal{B}$. Using the fact that μ is an extension of ι and the relation (6.16), we conclude that, for each $x \in \mathcal{R}$,

$$\mu\big((-\infty, x]\big) = \iota\big((-\infty, x]\big) = F(x) - F(-\infty) = F(x).$$

Thus, F is the distribution function of μ.

Suppose that ν is also a finite Borel measure having F as its distribution function. Then, by Proposition 6.7 and the definition of ι, we see that $\nu_{|\mathcal{J}} = \iota$. Therefore, by the uniqueness of the extension of ι, we must have $\nu = \mu$. ■

Theorem 6.3 reveals that the properties listed in Proposition 6.6 on page 193 are sufficient conditions for a real-valued function on $\mathcal{R}$ to be the distribution function of some finite Borel measure. Consequently, we make the following definition.

DEFINITION 6.5 Distribution Function; Lebesgue-Stieltjes Measure

A real-valued function F on $\mathcal{R}$ is called a **distribution function** provided that the following conditions hold:
a) F is monotone nondecreasing.
b) F is right continuous.
c) F is bounded.
d) $\lim_{x \to -\infty} F(x) = 0$.
For such a function, the unique finite Borel measure having F as its distribution function is called the **Lebesgue-Stieltjes measure corresponding to F**.

The next example provides some illustrations of Theorem 6.3 and Definition 6.5. The details of verification are left to the reader as exercises.

EXAMPLE 6.5 *Illustrates Theorem 6.3 and Definition 6.5*

a) Let F be defined by

$$F(x) = \begin{cases} 0, & x < 0; \\ x, & 0 \le x < 1; \\ 1, & x \ge 1. \end{cases}$$

Then F is bounded, nondecreasing, continuous, and $F(-\infty) = 0$. So, by Theorem 6.3, there is a unique finite Borel measure having F as its distribution function. Let μ be the Borel measure defined by $\mu(B) = \lambda\big(B \cap (0,1)\big)$. Then, as we discovered in Example 6.4(b) on page 192, μ has F as its distribution function. Hence, μ is the unique finite Borel measure having F as its distribution function; in other words, μ is the Lebesgue-Stieltjes measure corresponding to F.

b) Let g be a nonnegative Lebesgue integrable function on $\mathcal{R}$. Define F on $\mathcal{R}$ by

$$F(x) = \int_{(-\infty, x]} g(t) \, d\lambda(t). \tag{6.18}$$

Then F is nondecreasing, continuous, bounded, and $F(-\infty) = 0$. So, by Theorem 6.3, there is a unique finite Borel measure having F as its distribution function. For $B \in \mathcal{B}$, define $\mu(B) = \int_B g \, d\lambda$. Then μ is a finite Borel measure and, clearly, F is the distribution function of μ. Consequently, μ is the Lebesgue-Stieltjes measure corresponding to F. □

The Lebesgue-Stieltjes Integral

Assume F is a distribution function; that is, a real-valued function on $\mathcal{R}$ satisfying (a)–(d) of Proposition 6.6 on page 193. Then, as we know from Theorem 6.3, there is a unique finite Borel measure μ having F as its distribution function. Hence, it is natural to make the following definition.

DEFINITION 6.6 Lebesgue-Stieltjes Integral

Suppose that F is a distribution function and that μ is the Lebesgue-Stieltjes measure corresponding to F. Let f be a Borel measurable function and $B \in \mathcal{B}$. Then the **Lebesgue-Stieltjes integral of f over B with respect to F** is defined to be

$$\int_B f(x)\,dF(x) = \int_B f(x)\,d\mu(x),$$

provided the integral on the right-hand side makes sense.

EXAMPLE 6.6 *Illustrates Definition 6.6*

a) Let

$$F(x) = \begin{cases} 0, & x < 0; \\ x, & 0 \le x < 1; \\ 1, & x \ge 1. \end{cases}$$

By Example 6.5(a), F is a distribution function and the Lebesgue-Stieltjes measure corresponding to F is given by $\mu(B) = \lambda\big(B \cap (0,1)\big)$, $B \in \mathcal{B}$. Let f be a Borel measurable function and $B \in \mathcal{B}$. Then the Lebesgue-Stieltjes integral of f over B equals

$$\int_B f\,dF = \int_B f\,d\mu = \int_{B \cap (0,1)} f\,d\lambda, \tag{6.19}$$

provided the integral makes sense. To verify the last equality in (6.19), we apply the bootstrapping technique. The details are left to the reader as an exercise.

b) Let g be a nonnegative Lebesgue integrable function. Define F on $\mathcal{R}$ by

$$F(x) = \int_{(-\infty,x]} g(t)\,d\lambda(t).$$

By Example 6.5(b), F is a distribution function and the Lebesgue-Stieltjes measure corresponding to F is given by $\mu(B) = \int_B g\,d\lambda$, $B \in \mathcal{B}$. Let f be a Borel measurable function and $B \in \mathcal{B}$. Then the Lebesgue-Stieltjes integral of f over B equals

$$\int_B f\,dF = \int_B f\,d\mu = \int_B fg\,d\lambda, \tag{6.20}$$

provided the integral makes sense. To establish the last equality in (6.20), we proceed as follows. By Exercise 5.61 on page 164, the equality holds if f is a nonnegative Borel measurable function. If f is an extended real-valued Borel measurable function, write $f = f^+ - f^-$ and use the linearity of the abstract Lebesgue integral to conclude that (6.20) again holds. Finally, if f is a complex-valued Borel measurable function, write $f = \Re f + i\Im f$ and apply the linearity of the abstract Lebesgue integral to again conclude that (6.20) obtains.

Before leaving this example, we should point out that part (a) is a special case of part (b) with $g = \chi_{(0,1)}$. □

Exercises for Section 6.2

6.22 Provide the details for the illustrations given in parts (b)–(d) of Example 6.4 on pages 192–193.

6.23 Let $\{x_n\}_n$ be a sequence of distinct real numbers and $\{b_n\}_n$ a sequence of nonnegative real numbers with $\sum_n b_n < \infty$. Define μ on $\mathcal{B}$ by

$$\mu(B) = \sum_{x_n \in B} b_n.$$

a) Explain why μ is a finite Borel measure.
b) Determine the distribution function of μ.

6.24 Define μ on $\mathcal{B}$ by $\mu(B) = \int_B \chi_{[0,\infty)}(x) x e^{-x} \, d\lambda(x)$.
a) Explain why μ is a finite Borel measure.
b) Determine the distribution function of μ.

6.25 Let μ be a finite Borel measure. Prove that $F_\mu(\infty) = \mu(\mathcal{R})$.

6.26 Complete the proof of Lemma 6.5 on page 195. In other words, prove that Condition (E3) on page 180 is satisfied by ι and $\mathcal{J}$ when C is an infinite interval. *Hint:* First assume $C = (-\infty, b]$, where $b < \infty$, and note that $\iota(C) = \lim_{x \to -\infty} \iota((x, b])$.

6.27 Verify all statements made in Example 6.5(a) on page 197.

6.28 Verify all statements made in Example 6.5(b) on page 197.

6.29 Verify all statements made in Example 6.6(a).

6.30 Verify all statements made in Example 6.6(b).

6.31 Define F on $\mathcal{R}$ by

$$F(x) = \begin{cases} 0, & x < 0; \\ \frac{1}{8}, & 0 \le x < 1; \\ \frac{1}{2}, & 1 \le x < 2; \\ \frac{7}{8}, & 2 \le x < 3; \\ 1, & x \ge 3. \end{cases}$$

a) Show that F satisfies (a)–(d) of Proposition 6.6 on page 193.
b) Obtain the finite Borel measure μ whose distribution function is F.

6.32 Let $\{a_n\}_{n=1}^\infty$ be a sequence of nonnegative real numbers with $\sum_{n=1}^\infty a_n < \infty$. Define F on $\mathcal{R}$ by

$$F(x) = \sum_{n=1}^{[x]} a_n.$$

a) Show that F is a distribution function, that is, satisfies (a)–(d) of Proposition 6.6.
b) Determine the Lebesgue-Stieltjes measure corresponding to F.
c) If f is Borel measurable, determine $\int_{\mathcal{R}} f \, dF$.

6.33 Generalize the previous exercise as follows: Suppose that $\{x_n\}_n$ is a sequence of real numbers and that $\{a_n\}_n$ is a sequence of nonnegative real numbers with $\sum_n a_n < \infty$. Define F on $\mathcal{R}$ by

$$F(x) = \sum_{x_n \leq x} a_n.$$

a) Show that F is a distribution function, that is, satisfies (a)–(d) of Proposition 6.6.
b) Determine the Lebesgue-Stieltjes measure corresponding to F.
c) If f is Borel measurable, determine $\int_{\mathcal{R}} f \, dF$.

6.34 Let α be a positive constant and define $F(x) = 1 - e^{-\alpha x}$ for $x \geq 0$ and zero otherwise.
a) Show that F satisfies (a)–(d) of Proposition 6.6.
b) Find a nonnegative Borel measurable function g such that

$$F(x) = \int_{-\infty}^{x} g(t) \, dt$$

for all $x \in \mathcal{R}$.
c) Determine the unique finite Borel measure that has F as its distribution function.
d) Find $\int_{\mathcal{R}} x \, dF(x)$ and $\int_{\mathcal{R}} e^{itx} \, dF(x)$. *Hint:* Use Example 6.6(b) on page 198.

6.35 Let $F: \mathcal{R} \to \mathcal{R}$ be defined by

$$F(x) = \begin{cases} 0, & x < -2; \\ (x+2)/4, & -2 \leq x < 2; \\ 1, & x \geq 2. \end{cases}$$

a) Show that F is a distribution function.
b) Determine the Lebesgue-Stieltjes measure corresponding to F.
c) Find $\int_{\mathcal{R}} x \, dF(x)$, $\int_{\mathcal{R}} x^2 \, dF(x)$, and $\int_{\mathcal{R}} e^{itx} \, dF(x)$ for $t \in \mathcal{R}$.

6.36 Let $F: \mathcal{R} \to \mathcal{R}$ be defined by

$$F(x) = \sum_{n=0}^{[x]} e^{-\alpha} \frac{\alpha^n}{n!},$$

where α is a positive constant. Obtain $\int_{\mathcal{R}} x \, dF(x)$ and $\int_{\mathcal{R}} x^2 \, dF(x)$.

6.37 Suppose that F is a distribution function. Further suppose that F is differentiable on $\mathcal{R}$ and that $F' \in R([a,b])$ for all $a, b \in \mathcal{R}$. If f is Borel measurable, show that

$$\int_{\mathcal{R}} f(x) \, dF(x) = \int_{\mathcal{R}} f(x) F'(x) \, d\lambda(x)$$

whenever the integral on the right-hand side makes sense. *Hint:* Use the fundamental theorem of calculus and Example 6.6(b) on page 198.

6.38 Let ψ denote the Cantor function, as defined on page 64. Set

$$F(x) = \begin{cases} 0, & x < 0; \\ \psi(x), & 0 \leq x \leq 1; \\ 1, & x > 1. \end{cases}$$

a) Show that F is a distribution function.
b) Verify that $F' = 0$ λ-ae.
c) Prove that the conclusion of the previous exercise is not valid.

6.3 PRODUCT MEASURE SPACES

Our second application of the theory of extensions to measures, which we developed in Section 6.1, will be to product measure spaces. In this section, we will see how two measure spaces naturally give rise to a third measure space, called the *product measure space*. To help motivate product measure spaces, we consider the following example.

EXAMPLE 6.7 *Motivates Product Measure*

Note that in each of the illustrations below, a nonnegative set function is expressed in terms of a product of two measures.

a) Let λ denote Lebesgue measure on $\mathcal{R}$. If I and J are two intervals in $\mathcal{R}$, then the Cartesian product $I \times J$ is a rectangle in $\mathcal{R}^2$ ($= \mathcal{R} \times \mathcal{R}$) whose area can be expressed as
$$\text{area}(I \times J) = \ell(I)\ell(J) = \lambda(I)\lambda(J).$$

b) Let Γ and Λ be two finite sets and μ and ν counting measure on Γ and Λ, respectively. Also, as before, let $N(E)$ denote the number of elements of a finite set E. If $A \subset \Gamma$ and $B \subset \Lambda$, then the number of elements of $A \times B$ can be expressed as
$$N(A \times B) = \mu(A)\nu(B),$$
as we know from the fundamental principle of counting. □

Existence of a Product Measure

Suppose that $(\Gamma, \mathcal{S}, \mu)$ and $(\Lambda, \mathcal{T}, \nu)$ are two measure spaces. As usual, we let $\Gamma \times \Lambda$ denote the Cartesian product of Γ with Λ:
$$\Gamma \times \Lambda = \{ (x, y) : x \in \Gamma \text{ and } y \in \Lambda \}.$$

Our first task is to prove the existence of a σ-algebra $\mathcal{A}$ of subsets of $\Gamma \times \Lambda$ that contains all sets of the form $S \times T$, where $S \in \mathcal{S}$ and $T \in \mathcal{T}$, and a measure ω on $\mathcal{A}$ such that
$$\omega(S \times T) = \mu(S)\nu(T). \tag{6.21}$$

This task will be accomplished by applying the theory of extensions to measures. We begin with the following definition.

DEFINITION 6.7 Measurable Rectangles

Let $(\Gamma, \mathcal{S})$ and $(\Lambda, \mathcal{T})$ be measurable spaces. A subset of $\Gamma \times \Lambda$ of the form $S \times T$, where $S \in \mathcal{S}$ and $T \in \mathcal{T}$, is called a **measurable rectangle.** The collection of all measurable rectangles is denoted by $\mathcal{U}$.

As every σ-algebra is a semialgebra, it follows immediately from Proposition 6.1 on page 181 that the collection $\mathcal{U}$ of all measurable rectangles is a semialgebra of subsets of $\Gamma \times \Lambda$.

Next we define a nonnegative extended real-valued set function ι on $\mathcal{U}$. In view of (6.21), this should be done as follows: For $S \in \mathcal{S}$ and $T \in \mathcal{T}$, define

$$\iota(S \times T) = \mu(S)\nu(T). \tag{6.22}$$

If we can verify that Conditions (E1)–(E3) on page 180 hold for ι and $\mathcal{U}$, then Theorem 6.1 on page 186 will ensure the existence of a σ-algebra $\mathcal{A} \supset \mathcal{U}$ and a measure ω on $\mathcal{A}$ satisfying (6.21) — thereby completing our first task.

To verify Condition (E1), we note that $\emptyset \in \mathcal{U}$. In fact, $\emptyset = S \times T$ if and only if at least one of S and T are empty. But then $\iota(\emptyset) = \mu(S)\nu(T) = 0$, as required. The validity of Conditions (E2) and (E3) are established here in Lemmas 6.6 and 6.7, respectively.

□ □ □ **LEMMA 6.6**

Suppose that $\{C_k\}_{k=1}^{n}$ is a finite sequence of pairwise disjoint members of $\mathcal{U}$ whose union is in $\mathcal{U}$. Then $\iota\left(\bigcup_{k=1}^{n} C_k\right) = \sum_{k=1}^{n} \iota(C_k)$.

PROOF Set $C = \bigcup_{k=1}^{n} C_k$. By assumption, we can write $C = S \times T$ and $C_k = S_k \times T_k$, for $1 \le k \le n$, where $S, S_k \in \mathcal{S}$ and $T, T_k \in \mathcal{T}$. Let $x \in S$ and $N_x = \{k : x \in S_k\}$. If $y \in T$, then $(x, y) \in C$ and so there is a k such that $(x, y) \in S_k \times T_k$; thus, $y \in T_k$ for some $k \in N_x$. Conversely, if $y \in T_k$ for some $k \in N_x$, then we have that $(x, y) \in S_k \times T_k \subset S \times T$, so that $y \in T$. Hence, $T = \bigcup_{k \in N_x} T_k$ and, since the C_ks are pairwise disjoint, the sets T_k, $k \in N_x$, are also pairwise disjoint. Consequently, $\nu(T) = \sum_{k \in N_x} \nu(T_k)$. It follows (see Exercise 6.39) that

$$\nu(T)\chi_S(x) = \sum_{k=1}^{n} \nu(T_k)\chi_{S_k}(x) \tag{6.23}$$

for all $x \in \Gamma$. Therefore,

$$\iota(C) = \mu(S)\nu(T) = \int_\Gamma \nu(T)\chi_S(x)\,d\mu(x) = \int_\Gamma \sum_{k=1}^{n} \nu(T_k)\chi_{S_k}(x)\,d\mu(x)$$

$$= \sum_{k=1}^{n} \nu(T_k) \int_\Gamma \chi_{S_k}(x)\,d\mu(x) = \sum_{k=1}^{n} \mu(S_k)\nu(T_k) = \sum_{k=1}^{n} \iota(C_k),$$

as required. ∎

□ □ □ **LEMMA 6.7**

Assume $C, C_1, C_2, \dots$ are in $\mathcal{U}$ and $C \subset \bigcup_n C_n$. Then $\iota(C) \le \sum_n \iota(C_n)$.

PROOF We can write $C = S \times T$ and $C_n = S_n \times T_n$, where $S, S_n \in \mathcal{S}$ and $T, T_n \in \mathcal{T}$. Let $x \in S$ and set $N_x = \{n : x \in S_n\}$. If $y \in T$, then $(x, y) \in C$ and so there is an $n \in N_x$ such that $(x, y) \in S_n \times T_n$. Therefore, $T \subset \bigcup_{n \in N_x} T_n$ and hence $\nu(T) \le \sum_{n \in N_x} \nu(T_n)$. This result implies that $\nu(T)\chi_S(x) \le \sum_n \nu(T_n)\chi_{S_n}(x)$ for all $x \in \Gamma$. Thus,

$$\iota(C) = \mu(S)\nu(T) = \int_\Gamma \nu(T)\chi_S(x)\,d\mu(x) \le \int_\Gamma \sum_n \nu(T_n)\chi_{S_n}(x)\,d\mu(x)$$

$$= \sum_n \int_\Gamma \nu(T_n)\chi_{S_n}(x)\,d\mu(x) = \sum_n \mu(S_n)\nu(T_n) = \sum_n \iota(C_n),$$

as required. ∎

We have now verified that Conditions (E1)–(E3) are satisfied by ι and $\mathcal{U}$. Therefore, by Theorem 6.1 on page 186, we can deduce the following result, which completes our first task.

□ □ □ THEOREM 6.4

Suppose that $(\Gamma, \mathcal{S}, \mu)$ and $(\Lambda, \mathcal{T}, \nu)$ are measure spaces. Let

$$\mathcal{U} = \{\, S \times T : S \in \mathcal{S} \text{ and } T \in \mathcal{T} \,\}$$

and define ι on $\mathcal{U}$ by $\iota(S \times T) = \mu(S)\nu(T)$. Then there exists an extension of ι to a measure on a σ-algebra containing $\mathcal{U}$.

The Product Measure Space

In most of our work with product measure spaces, it will be necessary to impose a restriction on the factors $(\Gamma, \mathcal{S}, \mu)$ and $(\Lambda, \mathcal{T}, \nu)$, namely, that they are *σ-finite* measure spaces.

DEFINITION 6.8 σ-finite Measure Space

A measure space $(\Omega, \mathcal{A}, \mu)$ is called a **σ-finite measure space** if there is a sequence $\{A_n\}_n$ of $\mathcal{A}$-measurable sets such that $\bigcup_n A_n = \Omega$ and $\mu(A_n) < \infty$ for each n.

EXAMPLE 6.8   *Illustrates Definition 6.8*

a) $(\mathcal{R}, \mathcal{M}, \lambda)$ is a σ-finite measure space. Indeed, the sets $A_n = [-n, n]$, $n \in \mathcal{N}$, satisfy $\bigcup_{n=1}^{\infty} A_n = \mathcal{R}$ and $\lambda(A_n) < \infty$ for each $n \in \mathcal{N}$.

b) Let γ be counting measure on $\mathcal{P}(\mathcal{N})$. As $\mathcal{N} = \bigcup_{n=1}^{\infty} \{n\}$ and $\gamma(\{n\}) = 1 < \infty$ for each $n \in \mathcal{N}$, we see that $(\mathcal{N}, \mathcal{P}(\mathcal{N}), \gamma)$ is a σ-finite measure space.

c) Let Ω be a nonempty set and $\mathcal{A} = \{\Omega, \emptyset\}$. Define $\mu(\Omega) = \infty$ and $\mu(\emptyset) = 0$. Then $(\Omega, \mathcal{A}, \mu)$ is not a σ-finite measure space.

d) Clearly, any finite measure space is σ-finite. In particular, any probability space is a σ-finite measure space. □

As you probably noted, the condition of σ-finiteness is quite similar to Condition (E4) on page 188. In fact, the next proposition shows that, for product measure spaces, there is an important relationship between the two conditions.

□ □ □ PROPOSITION 6.8

Suppose that $(\Gamma, \mathcal{S}, \mu)$ and $(\Lambda, \mathcal{T}, \nu)$ are two σ-finite measure spaces. Let $\mathcal{U}$ be the semialgebra of measurable rectangles and ι the nonnegative extended real-valued set function on $\mathcal{U}$ as defined in (6.22). Then Condition (E4) is satisfied by ι and $\mathcal{U}$.

PROOF   By σ-finiteness, we can choose $\{S_n\}_n \subset \mathcal{S}$ and $\{T_n\}_n \subset \mathcal{T}$ such that $\mu(S_n) < \infty$ and $\nu(T_n) < \infty$, for all n, and $\Gamma = \bigcup_n S_n$ and $\Lambda = \bigcup_n T_n$. Let $A_n = \bigcup_{k=1}^n S_k$ and $B_n = \bigcup_{k=1}^n T_k$. Then $\{A_n\}_n \subset \mathcal{S}$ and $\{B_n\}_n \subset \mathcal{T}$; $\mu(A_n) < \infty$ and $\nu(B_n) < \infty$ for all n; and $\Gamma = \bigcup_n A_n$ and $\Lambda = \bigcup_n B_n$. Moreover, we have $A_1 \subset A_2 \subset \cdots$ and $B_1 \subset B_2 \subset \cdots$.

Let $C_n = A_n \times B_n$. We claim that $\{C_n\}_n$ is the required sequence of sets; that is, $\{C_n\}_n \subset \mathcal{U}$, $\bigcup_n C_n = \Gamma \times \Lambda$, and $\iota(C_n) < \infty$ for each n. The first and third properties of $\{C_n\}_n$ are obvious from the previous paragraph. To prove the second property, suppose that $(x, y) \in \Gamma \times \Lambda$. Since $x \in \Gamma$ and $y \in \Lambda$, there is an n_1 with $x \in A_{n_1}$, and an n_2 with $y \in B_{n_2}$. Let $n = \max\{n_1, n_2\}$. Then, because $\{A_n\}_n$ and $\{B_n\}_n$ are nondecreasing sequences of sets, we have $x \in A_n$ and $y \in B_n$ and, so, $(x, y) \in A_n \times B_n = C_n$. Thus, $\Gamma \times \Lambda = \bigcup_n C_n$. ∎

To summarize, we have now shown that Conditions (E1)–(E3) hold for ι and $\mathcal{U}$; and that, if $(\Gamma, \mathcal{S}, \mu)$ and $(\Lambda, \mathcal{T}, \nu)$ are both σ-finite, then Condition (E4) holds as well. Therefore, on account of Theorem 6.2 on page 188, we have the following result.

◻ ◻ ◻   **THEOREM 6.5**

Suppose that $(\Gamma, \mathcal{S}, \mu)$ and $(\Lambda, \mathcal{T}, \nu)$ are σ-finite measure spaces. Let

$$\mathcal{U} = \{\, S \times T : S \in \mathcal{S} \text{ and } T \in \mathcal{T} \,\}$$

and define ι on $\mathcal{U}$ by $\iota(S \times T) = \mu(S)\nu(T)$. Then there exists a unique extension of ι to a measure on the σ-algebra generated by $\mathcal{U}$.

Special notation and terminology are used for the extension of ι, the σ-algebra generated by $\mathcal{U}$, and the resulting measure space. We present this notation and terminology in the following definition.

DEFINITION 6.9   **Product Measure Space**

Suppose that $(\Gamma, \mathcal{S}, \mu)$ and $(\Lambda, \mathcal{T}, \nu)$ are σ-finite measure spaces and let $\mathcal{U}$ and ι be as in Theorem 6.5. The σ-algebra generated by $\mathcal{U}$, the smallest σ-algebra containing all measurable rectangles, is called the **product σ-algebra of $\mathcal{S}$ with $\mathcal{T}$** and is denoted by **$\mathcal{S} \times \mathcal{T}$**. The unique extension of ι to a measure on $\mathcal{S} \times \mathcal{T}$ is called the **product measure of μ with ν** and is denoted by **$\mu \times \nu$**. The measure space $(\Gamma \times \Lambda, \mathcal{S} \times \mathcal{T}, \mu \times \nu)$ is called the **product measure space** of $(\Gamma, \mathcal{S}, \mu)$ with $(\Lambda, \mathcal{T}, \nu)$.

Note: It is important to realize that $\mathcal{S} \times \mathcal{T}$ is a notation for the σ-algebra generated by $\mathcal{U}$ and is not the Cartesian product of the sets $\mathcal{S}$ and $\mathcal{T}$.

EXAMPLE 6.9   *Illustrates Definition 6.9*

a) Let $(\Gamma, \mathcal{S}, \mu) = (\Lambda, \mathcal{T}, \nu) = (\mathcal{R}, \mathcal{M}, \lambda)$. Since $\mathcal{M}$ contains all intervals, any rectangle $R \in \mathcal{R}^2$ is a measurable rectangle. If $R = I \times J$, where I and J are intervals, then $(\lambda \times \lambda)(R) = \lambda(I)\lambda(J) = \text{area}(R)$. So, $\lambda \times \lambda$ is a generalization of area to all $\mathcal{M} \times \mathcal{M}$-measurable sets.

b) Let γ be counting measure on $\mathcal{P}(\mathcal{N})$ and $(\Gamma, \mathcal{S}, \mu) = (\Lambda, \mathcal{T}, \nu) = (\mathcal{N}, \mathcal{P}(\mathcal{N}), \gamma)$. As we know from Example 6.8(b), $(\mathcal{N}, \mathcal{P}(\mathcal{N}), \gamma)$ is a σ-finite measure space. We leave it as an exercise for the reader to show that the product σ-algebra of $\mathcal{P}(\mathcal{N})$ with $\mathcal{P}(\mathcal{N})$ consists of all subsets of $\mathcal{N} \times \mathcal{N}$ and, furthermore, that the product measure of γ with γ is counting measure on $\mathcal{P}(\mathcal{N} \times \mathcal{N})$. In other words, the product measure space $(\mathcal{N} \times \mathcal{N}, \mathcal{P}(\mathcal{N}) \times \mathcal{P}(\mathcal{N}), \gamma \times \gamma)$ is the measure space $(\mathcal{N} \times \mathcal{N}, \mathcal{P}(\mathcal{N} \times \mathcal{N}), \kappa)$, where κ is counting measure on $\mathcal{P}(\mathcal{N} \times \mathcal{N})$.

c) If $(\Omega_1, \mathcal{A}_1, P_1)$ and $(\Omega_2, \mathcal{A}_2, P_2)$ are two probability spaces, then so is the product measure space $(\Omega_1 \times \Omega_2, \mathcal{A}_1 \times \mathcal{A}_2, P_1 \times P_2)$. As we will discover in Chapter 7, the product probability space is the appropriate mathematical model for the juxtaposition of two *independent* experiments. □

Sections of Sets and Functions in Product Spaces

We learned in calculus that a double (Riemann) integral can be evaluated as two iterated single integrals. Our next task is to prove a generalization of that result to product measure spaces; roughly speaking, a theorem of the following form: If $f: \Gamma \times \Lambda \to \mathcal{R}$ is $\mathcal{S} \times \mathcal{T}$-measurable, then

$$
\begin{aligned}
\int_{\Gamma \times \Lambda} f(x,y) \, d(\mu \times \nu)(x,y) &= \int_\Gamma \left[\int_\Lambda f(x,y) \, d\nu(y) \right] d\mu(x) \\
&= \int_\Lambda \left[\int_\Gamma f(x,y) \, d\mu(x) \right] d\nu(y).
\end{aligned}
\tag{6.24}
$$

In establishing (6.24), we must first show that it makes sense. For instance, we need to verify that if f is an $\mathcal{S} \times \mathcal{T}$-measurable function on $\Gamma \times \Lambda$, then the function $f_{[x]}$ defined on Λ by $f_{[x]}(y) = f(x,y)$ is $\mathcal{T}$-measurable; that the function g defined on Γ by $g(x) = \int_\Lambda f(x,y) \, d\nu(y)$ is $\mathcal{S}$-measurable; and so forth. To begin, we define the sections of a set.

DEFINITION 6.10   **Sections of a Set in a Product Space**

Suppose $A \subset \Gamma \times \Lambda$. Then the **$\Gamma$-sections of A** and the **Λ-sections of A** are defined, respectively, by

$$A_x = \{\, y \in \Lambda : (x,y) \in A \,\}, \qquad x \in \Gamma;$$

and

$$A^y = \{\, x \in \Gamma : (x,y) \in A \,\}, \qquad y \in \Lambda.$$

Note that each Γ-section is a subset of Λ and that each Λ-section is a subset of Γ. Figure 6.1 provides a visual representation of a Γ-section.

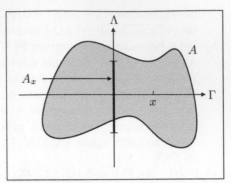

FIGURE 6.1 A Γ-section.

EXAMPLE 6.10 Illustrates Definition 6.10

a) Let $A = S \times T$, where $S \subset \Gamma$ and $T \subset \Lambda$. Then

$$A_x = \begin{cases} T, & \text{if } x \in S; \\ \emptyset, & \text{if } x \notin S. \end{cases} \quad \text{and} \quad A^y = \begin{cases} S, & \text{if } y \in T; \\ \emptyset, & \text{if } y \notin T. \end{cases}$$

b) Let $\Gamma = \Lambda = \mathcal{R}$ and $A = \{ (x, y) : x^2 + 4y^2 \leq 4 \}$. Then

$$A_x = \left[-\frac{1}{2}(4 - x^2)^{\frac{1}{2}}, \frac{1}{2}(4 - x^2)^{\frac{1}{2}} \right]$$

for $|x| \leq 2$, and $A_x = \emptyset$, otherwise. Similarly,

$$A^y = \left[-2(1 - y^2)^{\frac{1}{2}}, 2(1 - y^2)^{\frac{1}{2}} \right]$$

for $|y| \leq 1$, and $A^y = \emptyset$, otherwise. ◻

We next prove that sections of $\mathcal{S} \times \mathcal{T}$-measurable sets are themselves measurable. More precisely, we have the following proposition.

◻ ◻ ◻ **PROPOSITION 6.9**

Suppose that $(\Gamma, \mathcal{S})$ and $(\Lambda, \mathcal{T})$ are measurable spaces and that $A \in \mathcal{S} \times \mathcal{T}$. Then
a) $A_x \in \mathcal{T}$ for all $x \in \Gamma$.
b) $A^y \in \mathcal{S}$ for all $y \in \Lambda$.

PROOF We prove only (a). The proof of (b) is similar and is left as an exercise for the reader. Set

$$\mathcal{D} = \{ A \in \mathcal{S} \times \mathcal{T} : A_x \in \mathcal{T} \text{ for all } x \in \Gamma \}.$$

It follows immediately from Example 6.10(a) that $\mathcal{D}$ contains all measurable rectangles; that is, $\mathcal{D} \supset \mathcal{U}$. Because $\mathcal{S} \times \mathcal{T}$ is, by definition, the smallest σ-algebra containing $\mathcal{U}$, the proof will be complete once we show that $\mathcal{D}$ is a σ-algebra; because that will imply $\mathcal{D} = \mathcal{S} \times \mathcal{T}$ (why?).

So, assume that $A \in \mathcal{D}$. Then $A \in \mathcal{S} \times \mathcal{T}$ and $A_x \in \mathcal{T}$ for all $x \in \Gamma$. Therefore, $A^c \in \mathcal{S} \times \mathcal{T}$ and $(A_x)^c \in \mathcal{T}$ for all $x \in \Gamma$. But, $(A_x)^c = (A^c)_x$ and, hence, $A^c \in \mathcal{D}$.

Now assume that $\{A_n\}_n \subset \mathcal{D}$. Then $\{A_n\}_n \subset \mathcal{S} \times \mathcal{T}$ and $\{(A_n)_x\}_n \subset \mathcal{T}$ for all $x \in \Gamma$. Thus, $\bigcup_n A_n \in \mathcal{S} \times \mathcal{T}$ and $\bigcup_n (A_n)_x \in \mathcal{T}$ for all $x \in \Gamma$. However, $\bigcup_n (A_n)_x = (\bigcup_n A_n)_x$ and, so, $\bigcup_n A_n \in \mathcal{D}$. Consequently, $\mathcal{D}$ is a σ-algebra. ∎

Having discussed sections of subsets of $\Gamma \times \Lambda$, we now move on to the consideration of sections of functions on $\Gamma \times \Lambda$. The sections of such functions are obtained by holding one of the two variables fixed.

DEFINITION 6.11 **Sections of a Function on a Product Space**

Suppose that f is a function on $\Gamma \times \Lambda$. Then the **Γ-sections of f** and the **Λ-sections of f** are defined, respectively, by

$$f_{[x]}(y) = f(x,y), \qquad x \in \Gamma;$$

and

$$f^{[y]}(x) = f(x,y), \qquad y \in \Lambda.$$

Note that each Γ-section of f is a function on Λ and that each Λ-section of f is a function on Γ.

EXAMPLE 6.11 *Illustrates Definition 6.11*

Let $\Gamma = \mathcal{R}$ and $\Lambda = \mathcal{N}$. Define $f \colon \mathcal{R} \times \mathcal{N} \to \mathcal{R}$ by

$$f(x,y) = x^y + \frac{x^2}{y}.$$

Then $f_{[\frac{1}{2}]} \colon \mathcal{N} \to \mathcal{R}$ is given by $f_{[\frac{1}{2}]}(y) = 1/2^y + 1/4y$ and $f^{[2]} \colon \mathcal{R} \to \mathcal{R}$ is given by $f^{[2]}(x) = x^2 + x^2/2 = 3x^2/2$. □

Our next proposition shows that sections of $\mathcal{S} \times \mathcal{T}$-measurable functions are themselves measurable functions.

□ □ □ **PROPOSITION 6.10**

Let $(\Gamma, \mathcal{S})$ and $(\Lambda, \mathcal{T})$ be measurable spaces. *Suppose that f is an extended real-valued or complex-valued $\mathcal{S} \times \mathcal{T}$-measurable function on $\Gamma \times \Lambda$. Then*
a) $f_{[x]}$ *is $\mathcal{T}$-measurable for all $x \in \Gamma$.*
b) $f^{[y]}$ *is $\mathcal{S}$-measurable for all $y \in \Lambda$.*

PROOF To prove part (a), let $x \in \Gamma$. We will employ the bootstrapping technique to show that $f_{[x]}$ is $\mathcal{T}$-measurable. So, assume first that $f = \chi_A$, where $A \in \mathcal{S} \times \mathcal{T}$. Then,

$$f_{[x]}(y) = f(x,y) = \begin{cases} 1, & (x,y) \in A; \\ 0, & (x,y) \notin A; \end{cases} = \begin{cases} 1, & y \in A_x; \\ 0, & y \notin A_x. \end{cases} = \chi_{A_x}(y).$$

As $A \in \mathcal{S} \times \mathcal{T}$, we know by Proposition 6.9(a) that $A_x \in \mathcal{T}$ and, hence, that χ_{A_x} is $\mathcal{T}$-measurable. Next assume that f is a nonnegative simple function, say $f = \sum_{k=1}^{n} a_k \chi_{A_k}$, where $A_k \in \mathcal{S} \times \mathcal{T}$ for $1 \leq k \leq n$. Then $f_{[x]} = \sum_{k=1}^{n} a_k \chi_{(A_k)_x}$ which is $\mathcal{T}$-measurable, being a linear combination of $\mathcal{T}$-measurable functions.

Now assume f is a nonnegative extended real-valued $\mathcal{S} \times \mathcal{T}$-measurable function. Then, by Proposition 5.7(a) on page 160, there is a sequence $\{s_n\}_{n=1}^{\infty}$ of nonnegative $\mathcal{S} \times \mathcal{T}$-measurable simple functions that converges pointwise to f on $\Gamma \times \Lambda$. From the previous paragraph, we know that $\{(s_n)_{[x]}\}_{n=1}^{\infty}$ is a sequence of $\mathcal{T}$-measurable functions on Λ. Moreover, since $s_n \to f$ pointwise on $\Gamma \times \Lambda$, it is clear that $(s_n)_{[x]} \to f_{[x]}$ pointwise on Λ. Therefore, by Proposition 5.7(b), $f_{[x]}$ is $\mathcal{T}$-measurable.

Next assume that f is an extended real-valued $\mathcal{S} \times \mathcal{T}$-measurable function. We write $f = f^+ - f^-$ and note that $f_{[x]} = f_{[x]}^+ - f_{[x]}^-$. Using the result of the previous paragraph and the fact that the difference of two $\mathcal{T}$-measurable functions is $\mathcal{T}$-measurable, we conclude that $f_{[x]}$ is $\mathcal{T}$-measurable.

Finally, assume that f is a complex-valued $\mathcal{S} \times \mathcal{T}$-measurable function. We write $f = \Re f + i\Im f$ and note that $f_{[x]} = (\Re f)_{[x]} + i(\Im f)_{[x]}$. Then we apply the result of the previous paragraph and the fact that a linear combination of $\mathcal{T}$-measurable functions is $\mathcal{T}$-measurable to conclude that $f_{[x]}$ is $\mathcal{T}$-measurable. This completes the proof of part (a). The proof of part (b) is similar and is left as an exercise. ∎

In final preparation for our theorems on iterated integrals in product spaces, which we will consider in the next section, we prove the following two lemmas. Note the σ-finiteness assumptions in each lemma.

□ □ □ **LEMMA 6.8**

Suppose that $(\Gamma, \mathcal{S}, \mu)$ and $(\Lambda, \mathcal{T}, \nu)$ are two σ-finite measure spaces. Then, for each $A \in \mathcal{S} \times \mathcal{T}$,
a) the function g defined on Γ by $g(x) = \nu(A_x)$ is $\mathcal{S}$-measurable.
b) the function h defined on Λ by $h(y) = \mu(A^y)$ is $\mathcal{T}$-measurable.

PROOF Let
$$\mathcal{D} = \{\, A \in \mathcal{S} \times \mathcal{T} : \text{(a) and (b) hold}\,\}.$$

We will show that $\mathcal{D} \supset \sigma_0(\mathcal{U})$ and that $\mathcal{D}$ is closed under monotone limits. It will then follow from the monotone class theorem, Theorem 1.1 on page 25, that $\mathcal{D} \supset \mathcal{S} \times \mathcal{T}$. Since, by definition, $\mathcal{D} \subset \mathcal{S} \times \mathcal{T}$, we will have $\mathcal{D} = \mathcal{S} \times \mathcal{T}$, as required.

We first establish that $\mathcal{D} \supset \mathcal{U}$. So, let $S \times T$ be a measurable rectangle. Then $(S \times T)_x = T$, if $x \in S$, and is empty otherwise. Consequently, we have $g(x) = \nu((S \times T)_x) = \nu(T)\chi_S(x)$. Since $S \in \mathcal{S}$, g is $\mathcal{S}$-measurable. Similarly, we find that $h(y) = \mu((S \times T)^y) = \mu(S)\chi_T(y)$ is $\mathcal{T}$-measurable. So, $\mathcal{D} \supset \mathcal{U}$.

Next we show that $\mathcal{D} \supset \sigma_0(\mathcal{U})$. Let $A \in \sigma_0(\mathcal{U})$. Then, by Lemma 6.2 on page 187, A is a finite disjoint union of members of $\mathcal{U}$, say $A = \bigcup_{k=1}^{n} A_k$. Now, $A_x = \bigcup_{k=1}^{n} (A_k)_x$ and, because the A_ks are pairwise disjoint, so are the $(A_k)_x$s for each fixed $x \in \Gamma$. Consequently, $\nu(A_x) = \sum_{k=1}^{n} \nu((A_k)_x)$. Because $A_k \in \mathcal{U}$, the previous paragraph implies that $g_k(x) = \nu((A_k)_x)$ is $\mathcal{S}$-measurable. Hence,

$g(x) = \nu(A_x) = \sum_{k=1}^{n} g_k(x)$ is $\mathcal{S}$-measurable, being a sum of $\mathcal{S}$-measurable functions. Similarly, the function $h(y) = \mu(A^y)$ is $\mathcal{T}$-measurable. Thus, $\mathcal{D} \supset \sigma_0(\mathcal{U})$.

Next we prove that $\mathcal{D}$ is closed under nondecreasing limits: if $\{A_n\}_{n=1}^{\infty} \subset \mathcal{D}$ and $A_1 \subset A_2 \subset \cdots$, then $\bigcup_{n=1}^{\infty} A_n \in \mathcal{D}$. We have $(A_1)_x \subset (A_2)_x \subset \cdots$ and, hence, Theorem 5.1(d) on page 147 implies that

$$\nu\left(\left(\bigcup_{n=1}^{\infty} A_n\right)_x\right) = \nu\left(\bigcup_{n=1}^{\infty} (A_n)_x\right) = \lim_{n \to \infty} \nu((A_n)_x). \tag{6.25}$$

As $\{A_n\}_{n=1}^{\infty} \subset \mathcal{D}$, the function $g_n(x) = \nu((A_n)_x)$ is $\mathcal{S}$-measurable for each $n \in \mathcal{N}$ and, by (6.25), $g_n(x) \to \nu((\bigcup_{n=1}^{\infty} A_n)_x)$ pointwise on Γ. So, by Theorem 5.5(d) on page 155, $g(x) = \nu((\bigcup_{n=1}^{\infty} A_n)_x)$ is $\mathcal{S}$-measurable. A similar argument shows that the function $h(y) = \mu((\bigcup_{n=1}^{\infty} A_n)^y)$ is $\mathcal{T}$-measurable. Hence, $\bigcup_{n=1}^{\infty} A_n \in \mathcal{D}$.

Finally, we must verify that $\mathcal{D}$ is closed under nonincreasing limits; that is, if $\{A_n\}_{n=1}^{\infty} \subset \mathcal{D}$ and $A_1 \supset A_2 \supset \cdots$, then $\bigcap_{n=1}^{\infty} A_n \in \mathcal{D}$. Suppose first that there exist $S \in \mathcal{S}$ and $T \in \mathcal{T}$ with $\mu(S) < \infty$ and $\nu(T) < \infty$ such that $A_1 \subset S \times T$. Then $(A_1)_x \subset T$ and $(A_1)^y \subset S$ and, consequently, $\nu((A_1)_x)$ and $\mu((A_1)^y)$ are both finite. Applying Theorem 5.1(c) and an argument similar to the one used in the preceding paragraph, we find that $\bigcap_{n=1}^{\infty} A_n \in \mathcal{D}$.

To handle the general case—that is, no restriction on A_1—we must invoke the σ-finiteness assumption. We can select nondecreasing sequences $\{S_k\}_k \subset \mathcal{S}$ and $\{T_k\}_k \subset \mathcal{T}$ such that $\Gamma = \bigcup_k S_k$ and $\Lambda = \bigcup_k T_k$, and, for all k, $\mu(S_k) < \infty$ and $\nu(T_k) < \infty$. Define

$$\mathcal{E} = \{ E \in \mathcal{S} \times \mathcal{T} : E \cap (S_k \times T_k) \in \mathcal{D} \text{ for all } k \}. \tag{6.26}$$

We leave it as an exercise for the reader to prove that $\mathcal{E} = \mathcal{S} \times \mathcal{T}$. (See Exercise 6.50.)

Again, let $\{A_n\}_{n=1}^{\infty}$ be a nonincreasing sequence of members of $\mathcal{D}$, but this time with no restriction on A_1. For convenience, set $A = \bigcap_{n=1}^{\infty} A_n$. Then $A \in \mathcal{E}$ $(= \mathcal{S} \times \mathcal{T})$ and, thus, $A \cap (S_k \times T_k) \in \mathcal{D}$ for all k. The sequence $\{S_k \times T_k\}_k$ is nondecreasing because $\{S_k\}_k$ and $\{T_k\}_k$ are nondecreasing. This, in turn, implies that the sequence $\{A \cap (S_k \times T_k)\}_k$ is nondecreasing. Since we have already shown that $\mathcal{D}$ is closed under nondecreasing limits, we can conclude that $\bigcup_k (A \cap (S_k \times T_k)) \in \mathcal{D}$. But, $\bigcup_k (S_k \times T_k) = \Gamma \times \Lambda$ (why?) and, consequently,

$$A = A \cap (\Gamma \times \Lambda) = A \cap \left(\bigcup_k (S_k \times T_k)\right) = \bigcup_k (A \cap (S_k \times T_k)).$$

This proves that $A \in \mathcal{D}$.

We have now established that $\mathcal{D} \supset \sigma_0(\mathcal{U})$ and that $\mathcal{D}$ is closed under monotone limits. Therefore, by the monotone class theorem, $\mathcal{D}$ contains the σ-algebra generated by $\sigma_0(\mathcal{U})$, which is $\mathcal{S} \times \mathcal{T}$. Since, by definition, $\mathcal{D} \subset \mathcal{S} \times \mathcal{T}$, we deduce that $\mathcal{D} = \mathcal{S} \times \mathcal{T}$, as required. ∎

□ □ □ **LEMMA 6.9**

Suppose that $(\Gamma, \mathcal{S}, \mu)$ and $(\Lambda, \mathcal{T}, \nu)$ are two σ-finite measure spaces. Then, for each $A \in \mathcal{S} \times \mathcal{T}$,

a) $(\mu \times \nu)(A) = \int_{\Gamma} \nu(A_x) \, d\mu(x)$

b) $(\mu \times \nu)(A) = \int_{\Lambda} \mu(A^y) \, d\nu(y)$.

PROOF We will prove part (a). The proof of part (b) is similar and is left as an exercise for the reader. For $A \in \mathcal{S} \times \mathcal{T}$, define

$$\tau(A) = \int_\Gamma \nu(A_x) \, d\mu(x).$$

In view of Lemma 6.8, the integral exists because the function g defined on Γ by $g(x) = \nu(A_x)$ is a (nonnegative) $\mathcal{S}$-measurable function. We will show that τ is a measure on $\mathcal{S} \times \mathcal{T}$ and that $\tau = \mu \times \nu$ on $\mathcal{U}$. This will imply, by the uniqueness portion of Theorem 6.5 on page 204, that $\tau = \mu \times \nu$ on $\mathcal{S} \times \mathcal{T}$, as required.

Clearly, $\tau(A) \geq 0$, for all $A \in \mathcal{S} \times \mathcal{T}$, and $\tau(\emptyset) = 0$. Assume that $\{A_n\}_n$ is a sequence of pairwise disjoint members of $\mathcal{S} \times \mathcal{T}$. Then $\{(A_n)_x\}_n$ is a sequence of pairwise disjoint members of $\mathcal{T}$. Consequently,

$$\tau\left(\bigcup_n A_n\right) = \int_\Gamma \nu\left(\left(\bigcup_n A_n\right)_x\right) d\mu(x) = \int_\Gamma \nu\left(\bigcup_n (A_n)_x\right) d\mu(x)$$

$$= \int_\Gamma \sum_n \nu((A_n)_x) \, d\mu(x) = \sum_n \int_\Gamma \nu((A_n)_x) \, d\mu(x) = \sum_n \tau(A_n).$$

Hence, τ is a measure on $\mathcal{S} \times \mathcal{T}$.

Now suppose that $S \times T$ is a measurable rectangle. Then

$$\tau(S \times T) = \int_\Gamma \nu((S \times T)_x) \, d\mu(x) = \int_\Gamma \nu(T) \chi_S(x) \, d\mu(x)$$

$$= \mu(S)\nu(T) = (\mu \times \nu)(S \times T).$$

This result shows that τ agrees with $\mu \times \nu$ on $\mathcal{U}$. ∎

Exercises for Section 6.3

6.39 Verify (6.23) on page 202. *Hint:* Show that if $\nu(T_k)\chi_{S_k}(x) > 0$ for some k, then $x \in S$.

6.40 Let μ be counting measure on $\mathcal{P}(\mathcal{R})$. Show that $(\mathcal{R}, \mathcal{P}(\mathcal{R}), \mu)$ is not a σ-finite measure space.

6.41 Suppose that $(\Gamma, \mathcal{S}, \mu)$ and $(\Lambda, \mathcal{T}, \nu)$ are σ-finite measure spaces. Prove that the product measure space $(\Gamma \times \Lambda, \mathcal{S} \times \mathcal{T}, \mu \times \nu)$ is σ-finite.

6.42 Show that $\mathcal{M} \times \mathcal{M}$ contains all open and closed subsets of $\mathcal{R}^2$; that is, each open or closed subset of $\mathcal{R}^2$ is $\mathcal{M} \times \mathcal{M}$-measurable.

★6.43 Let γ be counting measure on $\mathcal{P}(\mathcal{N})$.
  a) Establish that the product σ-algebra of $\mathcal{P}(\mathcal{N})$ with $\mathcal{P}(\mathcal{N})$ consists of all subsets of $\mathcal{N} \times \mathcal{N}$; that is, $\mathcal{P}(\mathcal{N}) \times \mathcal{P}(\mathcal{N}) = \mathcal{P}(\mathcal{N} \times \mathcal{N})$. *Hint:* $\mathcal{N} \times \mathcal{N}$ is countable.
  b) Verify that the product measure of γ with γ is counting measure on $\mathcal{P}(\mathcal{N}) \times \mathcal{P}(\mathcal{N})$.

6.44 Suppose that $\Gamma = \{x_1, x_2, \ldots, x_m\}$ and $\Lambda = \{y_1, y_2, \ldots, y_n\}$ are finite sets and that $\{a_1, a_2, \ldots, a_m\}$ and $\{b_1, b_2, \ldots, b_n\}$ are two sets of nonnegative numbers. Define μ on $\mathcal{P}(\Gamma)$ and ν on $\mathcal{P}(\Lambda)$ by $\mu(A) = \sum_{x_j \in A} a_j$ and $\nu(B) = \sum_{y_k \in B} b_k$. Determine explicitly
  a) the product σ-algebra $\mathcal{P}(\Gamma) \times \mathcal{P}(\Lambda)$.
  b) the product measure $\mu \times \nu$.

6.45 Suppose that g is a complex-valued $\mathcal{S}$-measurable function on Γ and that h is a complex-valued $\mathcal{T}$-measurable function on Λ. Define f on $\Gamma \times \Lambda$ by $f(x, y) = g(x)h(y)$. Show that f is $\mathcal{S} \times \mathcal{T}$-measurable.

6.46 Let $\Gamma = \Lambda = \mathcal{R}$ and $A = \{(x, y) : 0 \leq y \leq x^2 \text{ and } x \geq 0\}$. Determine A_x and A^y.

6.47 Prove part (b) of Proposition 6.9 on page 206.

6.48 Prove part (b) of Proposition 6.10 on page 207.

6.49 *True* or *False:*
   a) If A, $B \subset \Gamma \times \Lambda$ are disjoint and $x \in \Gamma$, then A_x and B_x are disjoint.
   b) If $S_1 \times T_1$ and $S_2 \times T_2$ are disjoint rectangles in $\Gamma \times \Lambda$, then S_1 and S_2 are disjoint.
   c) If $S_1 \times T_1$ and $S_2 \times T_2$ are disjoint rectangles in $\Gamma \times \Lambda$, then either S_1 and S_2 are disjoint or T_1 and T_2 are disjoint.

6.50 Let $\mathcal{E}$ be defined as in (6.26) on page 209. Prove that $\mathcal{E} = \mathcal{S} \times \mathcal{T}$ by employing the following steps.
   a) Show that $\mathcal{E} \supset \sigma_0(\mathcal{U})$. *Hint:* $\sigma_0(\mathcal{U})$ is an algebra and $\sigma_0(\mathcal{U}) \subset \mathcal{D}$.
   b) Show that $\mathcal{E}$ is closed under nondecreasing limits.
   c) Show that $\mathcal{E}$ is closed under nonincreasing limits.
   d) Conclude that $\mathcal{E} = \mathcal{S} \times \mathcal{T}$ by employing the monotone class theorem.

6.51 This exercise shows that σ-finiteness cannot be omitted as a hypothesis in Lemma 6.8. Let S be the set defined in Lemma 3.12 on page 100. By Exercise 3.51, $S \notin \mathcal{M}$. Also, let $(\Gamma, \mathcal{S}, \mu) = (\mathcal{R}, \mathcal{M}, \lambda)$ and $(\Lambda, \mathcal{T}, \nu) = (\mathcal{R}, \mathcal{P}(\mathcal{R}), \nu)$, where $\nu(T) = \gamma(T \cap S)$ and γ is counting measure.
   a) Let $Q_0 = Q \setminus \{0\}$ and set $A = \{(x, x + r) : x \in \mathcal{R}, r \in Q_0\}$. Show $A \in \mathcal{M} \times \mathcal{P}(\mathcal{R})$. *Hint:* First show that the function $f(x, y) = y - x$ is $\mathcal{M} \times \mathcal{P}(\mathcal{R})$-measurable.
   b) Show that $\nu(A_x) = \chi_{S^c}(x)$ and conclude that $x \to \nu(A_x)$ is not $\mathcal{M}$-measurable.
   c) Why doesn't the result in part (b) contradict Lemma 6.8?

6.52 Prove part (b) of Lemma 6.9 on page 209.

Exercises 6.53–6.57 should be completed by all readers who plan to cover the probability material in Chapter 7.

★6.53 Denote by $\mathcal{B}_2$ the smallest σ-algebra of subsets of $\mathcal{R}^2$ that contains all open sets of $\mathcal{R}^2$. Members of $\mathcal{B}_2$ are called **two-dimensional Borel sets.**
   a) Show that $\mathcal{B}_2 = \mathcal{B} \times \mathcal{B}$.
   b) A measure on $\mathcal{B}_2$ is called a **two-dimensional Borel measure.** Suppose that μ and ν are finite two-dimensional Borel measures such that $\mu(A \times B) = \nu(A \times B)$ for all A, $B \in \mathcal{B}$. Prove that $\mu = \nu$.

6.54 Let $\mathcal{I}_2$ denote the collection of all two-dimensional intervals in $\mathcal{R}^2$, that is, all sets of the form $I \times J$ where $I, J \in \mathcal{I}$.
   a) Show that the σ-algebra generated by $\mathcal{I}_2$ is $\mathcal{B}_2$; that is, $\sigma(\mathcal{I}_2) = \mathcal{B}_2$. *Hint:* Use Exercise 6.53(a).
   b) Let μ and ν be two-dimensional Borel measures such that $\mu(K) = \nu(K) < \infty$ for all bounded two-dimensional intervals K. Prove that $\mu = \nu$.

6.55 Let $\mathcal{J}$ denote the collection of intervals of $\mathcal{R}$ of the form $(a, b]$ and (c, ∞), where $-\infty \leq a \leq b < \infty$ and $-\infty \leq c < \infty$. Also, let $\mathcal{J}_2$ denote the collection of all subsets of $\mathcal{R}^2$ of the form $I \times J$ where $I, J \in \mathcal{J}$. Prove that $\mathcal{J}_2$ is a semialgebra and that the σ-algebra generated by $\mathcal{J}_2$ is $\mathcal{B}_2$.

6.56 Suppose that μ and ν are finite two-dimensional Borel measures such that

$$\mu\big((-\infty, x] \times (-\infty, y]\big) = \nu\big((-\infty, x] \times (-\infty, y]\big)$$

for all x, $y \in \mathcal{R}$. Prove that $\mu = \nu$. *Hint:* It suffices to prove that $\mu = \nu$ on $\mathcal{J}_2$.

6.57 Suppose that μ and ν are finite Borel measures and that τ is a two-dimensional Borel measure. Further suppose that

$$\tau\big((-\infty, x] \times (-\infty, y]\big) = \mu\big((-\infty, x]\big)\nu\big((-\infty, y]\big)$$

for all $x, y \in \mathcal{R}$. Prove that $\tau = \mu \times \nu$.

6.4 ITERATION OF INTEGRALS IN PRODUCT MEASURE SPACES

In the preceding section, we discussed product measure and product measure spaces. Now we will learn how to evaluate integrals on product measure spaces by iteration; that is, by the evaluation of two integrals on the factor spaces. We will present several theorems of this type. The first theorem is known as Tonelli's theorem.[†]

□ □ □ **THEOREM 6.6 Tonelli's Theorem**

Suppose that $(\Gamma, \mathcal{S}, \mu)$ and $(\Lambda, \mathcal{T}, \nu)$ are σ-finite measure spaces. Let f be a nonnegative extended real-valued $\mathcal{S} \times \mathcal{T}$-measurable function on $\Gamma \times \Lambda$. Then

a) $f_{[x]}$ is $\mathcal{T}$-measurable for all $x \in \Gamma$.

b) $f^{[y]}$ is $\mathcal{S}$-measurable for all $y \in \Lambda$.

c) $g(x) = \int_\Lambda f(x, y)\, d\nu(y)$ is $\mathcal{S}$-measurable.

d) $h(y) = \int_\Gamma f(x, y)\, d\mu(x)$ is $\mathcal{T}$-measurable.

e) the equalities,

$$\int_{\Gamma \times \Lambda} f(x, y)\, d(\mu \times \nu)(x, y) = \int_\Gamma \left[\int_\Lambda f(x, y)\, d\nu(y) \right] d\mu(x)$$

$$= \int_\Lambda \left[\int_\Gamma f(x, y)\, d\mu(x) \right] d\nu(y),$$

hold.

PROOF Parts (a) and (b) are the contents of Proposition 6.10 on page 207. It remains to verify parts (c)–(e). To begin, we will show that if $f_1, f_2, \ldots, f_n$ are nonnegative $\mathcal{S} \times \mathcal{T}$-measurable functions on $\Gamma \times \Lambda$ that satisfy (c)–(e) and $c_1, c_2, \ldots, c_n$ are nonnegative real numbers, then $\sum_{k=1}^n c_k f_k$ satisfies (c)–(e). It suffices to verify this result for $n = 2$. Let $f = c_1 f_1 + c_2 f_2$. Then, by the linearity of the Lebesgue integral, we have

$$\int_\Lambda f(x, y)\, d\nu(y) = c_1 \int_\Lambda f_1(x, y)\, d\nu(y) + c_2 \int_\Lambda f_2(x, y)\, d\nu(y).$$

Since a linear combination of measurable functions is measurable and, by assumption, f_1 and f_2 satisfy (c), we conclude that $g(x) = \int_\Lambda f(x, y)\, d\nu(y)$ is $\mathcal{S}$-measurable. Similarly, $h(y) = \int_\Gamma f(x, y)\, d\mu(x)$ is $\mathcal{T}$-measurable. Now using

---

[†] Some mathematicians attribute this theorem to G. Fubini.

the linearity of the Lebesgue integral and the assumption that f_1 and f_2 satisfy (e), we get

$$\int_{\Gamma \times \Lambda} f(x,y)\, d(\mu \times \nu)(x,y) = c_1 \int_{\Gamma \times \Lambda} f_1\, d(\mu \times \nu) + c_2 \int_{\Gamma \times \Lambda} f_2\, d(\mu \times \nu)$$

$$= c_1 \int_{\Gamma} \left[\int_{\Lambda} f_1\, d\nu \right] d\mu + c_2 \int_{\Gamma} \left[\int_{\Lambda} f_2\, d\nu \right] d\mu$$

$$= \int_{\Gamma} \left[c_1 \int_{\Lambda} f_1\, d\nu + c_2 \int_{\Lambda} f_2\, d\nu \right] d\mu$$

$$= \int_{\Gamma} \left[\int_{\Lambda} (c_1 f_1 + c_2 f_2)\, d\nu \right] d\mu = \int_{\Gamma} \left[\int_{\Lambda} f(x,y)\, d\nu(y) \right] d\mu(x).$$

Hence, the first equation in (e) holds for f. A similar argument shows that the second equation in (e) holds for f.

We will now bootstrap to prove the theorem. If $f = \chi_A$, where $A \in \mathcal{S} \times \mathcal{T}$, then $g(x) = \nu(A_x)$ and $h(y) = \mu(A^y)$. Therefore, by Lemma 6.8 on page 208, (c) and (d) hold; and, by Lemma 6.9 on page 209, (e) holds. Hence, (c)–(e) are satisfied if f is the characteristic function of a set in $\mathcal{S} \times \mathcal{T}$. It now follows immediately from the previous paragraph that (c)–(e) hold for nonnegative simple functions.

If f is a nonnegative extended real-valued $\mathcal{S} \times \mathcal{T}$-measurable function, then, by Proposition 5.7(a) on page 160, we can choose a sequence $\{s_n\}_{n=1}^{\infty}$ of nonnegative simple functions such that $s_n \uparrow f$ pointwise on $\Gamma \times \Lambda$. It follows that $(s_n)_{[x]} \uparrow f_{[x]}$ pointwise on Λ and, so, by the MCT,

$$g(x) = \int_{\Lambda} f(x,y)\, d\nu(y) = \lim_{n \to \infty} \int_{\Lambda} s_n(x,y)\, d\nu(y).$$

This result shows that g is the pointwise limit of the $\mathcal{S}$-measurable functions $\int_{\Lambda} s_n(x,y)\, d\nu(y)$, $n \in \mathcal{N}$. Consequently, g is $\mathcal{S}$-measurable. Similarly, we find that $h(y) = \int_{\Gamma} f(x,y)\, d\mu(x)$ is $\mathcal{T}$-measurable. Finally, employing the MCT twice more yields

$$\int_{\Gamma \times \Lambda} f\, d(\mu \times \nu) = \lim_{n \to \infty} \int_{\Gamma \times \Lambda} s_n\, d(\mu \times \nu)$$

$$= \lim_{n \to \infty} \int_{\Gamma} \left[\int_{\Lambda} s_n\, d\nu \right] d\mu = \int_{\Gamma} \left[\int_{\Lambda} f\, d\nu \right] d\mu.$$

This result verifies the first equation in (e) and a similar argument establishes the second equation in (e). ∎

Tonelli's theorem deals with iterated integrals for nonnegative measurable functions. In that case, there is no issue of the existence of the integrals occurring in the theorem. Now we will consider the iteration of integrals for complex-valued measurable functions. To ensure the existence of the integrals involved, an integrability condition is imposed.

□ □ □ **THEOREM 6.7 Fubini's Theorem**

Suppose that $(\Gamma, \mathcal{S}, \mu)$ and $(\Lambda, \mathcal{T}, \nu)$ are σ-finite measure spaces. Let f be a complex-valued $\mathcal{S} \times \mathcal{T}$-measurable function on $\Gamma \times \Lambda$ such that at least one of the quantities,

$$\text{(i)} \quad \int_{\Gamma \times \Lambda} |f(x,y)| \, d(\mu \times \nu)(x,y),$$

$$\text{(ii)} \quad \int_{\Gamma} \left[\int_{\Lambda} |f(x,y)| \, d\nu(y) \right] d\mu(x), \quad \text{(iii)} \quad \int_{\Lambda} \left[\int_{\Gamma} |f(x,y)| \, d\mu(x) \right] d\nu(y),$$

is finite. Then

a) *$f_{[x]} \in \mathcal{L}^1(\nu)$ for μ-almost all $x \in \Gamma$.*

b) *$f^{[y]} \in \mathcal{L}^1(\mu)$ for ν-almost all $y \in \Lambda$.*

c) *$g(x) = \int_{\Lambda} f(x,y) \, d\nu(y)$ is defined μ-ae and is in $\mathcal{L}^1(\mu)$.*

d) *$h(y) = \int_{\Gamma} f(x,y) \, d\mu(x)$ is defined ν-ae and is in $\mathcal{L}^1(\nu)$.*

e) *the equalities*

$$\int_{\Gamma \times \Lambda} f(x,y) \, d(\mu \times \nu)(x,y) = \int_{\Gamma} \left[\int_{\Lambda} f(x,y) \, d\nu(y) \right] d\mu(x)$$

$$= \int_{\Lambda} \left[\int_{\Gamma} f(x,y) \, d\mu(x) \right] d\nu(y)$$

hold.

PROOF By Tonelli's theorem, the three integrals (i), (ii), and (iii) are equal. Since, by assumption, at least one of the integrals is finite, they all must be finite.

By Proposition 6.10 on page 207, $f_{[x]}$ is $\mathcal{T}$-measurable for all $x \in \Gamma$ and $f^{[y]}$ is $\mathcal{S}$-measurable for all $y \in \Lambda$. Assume f is real-valued and write $f = f^+ - f^-$. It will be convenient to let $f_1 = f^+$ and $f_2 = f^-$. Because $0 \leq f_j \leq |f|$, it follows from (ii) that $\int_{\Gamma} \left[\int_{\Lambda} f_j \, d\nu \right] d\mu < \infty$ for $j = 1, 2$. Consequently, by Exercise 5.53 on page 164,

$$\int_{\Lambda} f_j(x,y) \, d\nu(y) < \infty \quad \mu\text{-ae}, \qquad j = 1, 2. \tag{6.27}$$

Let $E = \{\, x \in \Gamma : \int_{\Lambda} f_j(x,y) \, d\nu(y) < \infty, \, j = 1, 2 \,\}$. Then, for $x \in E$, both $f_{[x]}^+$ and $f_{[x]}^-$ are in $\mathcal{L}^1(\nu)$; hence, so is $f_{[x]}$. Since, by (6.27), $\mu(E^c) = 0$, we see that (a) holds. A similar argument establishes (b).

Next, for $j = 1$ and 2, we define $g_j(x) = \chi_E(x) \int_{\Lambda} f_j(x,y) \, d\nu(y)$. Then g_j is real-valued and, by part (c) of Tonelli's theorem, is $\mathcal{S}$-measurable. Moreover, it follows immediately from (ii) that $\int_{\Gamma} g_j \, d\mu < \infty$. Hence, $g_j \in \mathcal{L}^1(\mu)$ for $j = 1$ and 2. But, then, Theorem 5.8 on page 168 implies that $g_1 - g_2 \in \mathcal{L}^1(\mu)$. However, if $x \in E$,

$$g_1(x) - g_2(x) = \int_{\Lambda} f_1(x,y) \, d\nu(y) - \int_{\Lambda} f_2(x,y) \, d\nu(y) = \int_{\Lambda} f(x,y) \, d\nu(y).$$

Therefore, the function $g(x) = \int_{\Lambda} f(x,y) \, d\nu(y)$ is defined μ-ae and is in $\mathcal{L}^1(\mu)$. This result proves that (c) holds and a similar argument verifies (d).

Employing part (e) of Tonelli's theorem, we deduce that

$$\int_{\Gamma \times \Lambda} f \, d(\mu \times \nu) = \int_{\Gamma \times \Lambda} f_1 \, d(\mu \times \nu) - \int_{\Gamma \times \Lambda} f_2 \, d(\mu \times \nu)$$

$$= \int_{\Gamma} \left[\int_{\Lambda} f_1 \, d\nu \right] d\mu - \int_{\Gamma} \left[\int_{\Lambda} f_2 \, d\nu \right] d\mu$$

$$= \int_{\Gamma} g_1 \, d\mu - \int_{\Gamma} g_2 \, d\mu = \int_{\Gamma} g \, d\mu = \int_{\Gamma} \left[\int_{\Lambda} f \, d\nu \right] d\mu.$$

This result establishes the first equation in (e) and a similar argument verifies the second equation in (e). We have now shown that Fubini's theorem holds if f is real-valued. The verification for complex-valued f is left to the reader as an exercise. ∎

The following example provides applications, illustrations, and remarks about the Tonelli and Fubini theorems.

EXAMPLE 6.12 *Illustrates the Tonelli and Fubini Theorems*

a) The following result is a theorem from calculus: Suppose that f is a real-valued function of two variables, defined and continuous on the rectangle

$$R = \{ (x, y) : a \le x \le b, \ c \le y \le d \}.$$

Then f is Riemann integrable on R and

$$\iint_R f(x, y) \, dx \, dy = \int_a^b \left[\int_c^d f(x, y) \, dy \right] dx. \qquad (6.28)$$

This result can be proved by employing Fubini's theorem, as outlined in Exercise 6.73.

b) Suppose that $\{a_{mn}\}_{m,n=1}^{\infty}$ is a double sequence of nonnegative real numbers. Then

$$\sum_m \left(\sum_n a_{mn} \right) = \sum_n \left(\sum_m a_{mn} \right). \qquad (6.29)$$

To prove (6.29), let $(\Gamma, \mathcal{S}, \mu) = (\Lambda, \mathcal{T}, \nu) = (\mathcal{N}, \mathcal{P}(\mathcal{N}), \gamma)$ where γ is counting measure. Then, by Exercise 6.43 on page 210, the product measure space is $(\mathcal{N} \times \mathcal{N}, \mathcal{P}(\mathcal{N} \times \mathcal{N}), \kappa)$, where κ is counting measure on $\mathcal{P}(\mathcal{N} \times \mathcal{N})$. Define $f{:}\mathcal{N} \times \mathcal{N} \to \mathcal{R}$ by $f(m, n) = a_{mn}$. Then, by Example 5.7(b) on page 163 and Tonelli's theorem,

$$\sum_m \left(\sum_n a_{mn} \right) = \sum_m \left[\int_{\mathcal{N}} f(m, n) \, d\gamma(n) \right]$$

$$= \int_{\mathcal{N}} \left[\int_{\mathcal{N}} f(m, n) \, d\gamma(n) \right] d\gamma(m)$$

$$= \int_{\mathcal{N}} \left[\int_{\mathcal{N}} f(m, n) \, d\gamma(m) \right] d\gamma(n) = \sum_n \left(\sum_m a_{mn} \right).$$

c) The σ-finiteness hypothesis cannot be dropped in Tonelli's theorem. Indeed, let γ be counting measure, $(\Gamma, \mathcal{S}, \mu) = (\mathcal{R}, \mathcal{M}, \lambda)$, $(\Lambda, \mathcal{T}, \nu) = (\mathcal{R}, \mathcal{P}(\mathcal{R}), \gamma)$, and $D = \{(x, y) : x = y\}$. Set $f = \chi_D$. We claim that f is $\mathcal{M} \times \mathcal{P}(\mathcal{R})$-measurable or, equivalently, $D \in \mathcal{M} \times \mathcal{P}(\mathcal{R})$. Since $\mathcal{M} \times \mathcal{P}(\mathcal{R})$ is a σ-algebra and contains all rectangles (in the geometry sense), it also contains all open sets in $\mathcal{R}^2$ and, hence, all closed sets in $\mathcal{R}^2$. Clearly D is a closed subset of $\mathcal{R}^2$. Now, it is easy to see that $f_{[x]}(y) = \chi_{\{x\}}(y)$ and $f^{[y]}(x) = \chi_{\{y\}}(x)$. Consequently, we have $\int_{\mathcal{R}} f(x, y) \, d\gamma(y) = \gamma(\{x\}) = 1$ for each $x \in \mathcal{R}$ and we have $\int_{\mathcal{R}} f(x, y) \, d\lambda(x) = \lambda(\{y\}) = 0$ for each $y \in \mathcal{R}$. Hence,

$$\int_{\mathcal{R}} \left[\int_{\mathcal{R}} f(x, y) \, d\gamma(y) \right] d\lambda(x) = \infty \neq 0 = \int_{\mathcal{R}} \left[\int_{\mathcal{R}} f(x, y) \, d\lambda(x) \right] d\gamma(y).$$

Therefore, the second equation in part (e) of Tonelli's theorem fails. Note that the measure space $(\mathcal{R}, \mathcal{P}(\mathcal{R}), \gamma)$ is not σ-finite.

d) In this part, we show that the integrability condition cannot be omitted from Fubini's theorem. Let $(\Gamma, \mathcal{S}, \mu) = (\Lambda, \mathcal{T}, \nu) = (\mathcal{Z}, \mathcal{P}(\mathcal{Z}), \gamma)$, where $\mathcal{Z}$ is the set of integers and γ is counting measure on $\mathcal{P}(\mathcal{Z})$. Clearly, $(\mathcal{Z}, \mathcal{P}(\mathcal{Z}), \gamma)$ is σ-finite. Let $f : \mathcal{Z} \times \mathcal{Z} \to \mathcal{R}$ be defined by

$$f(x, y) = \begin{cases} x, & y = x; \\ -x, & y = x + 1; \\ 0, & \text{elsewhere.} \end{cases}$$

Then, we have $\int_{\mathcal{Z}} f(x, y) \, d\gamma(y) = \sum_y f(x, y) = x + (-x) = 0$ for each $x \in \mathcal{Z}$, and, hence,

$$\int_{\mathcal{Z}} \left[\int_{\mathcal{Z}} f(x, y) \, d\gamma(y) \right] d\gamma(x) = \sum_x 0 = 0.$$

However, $\int_{\mathcal{Z}} f(x, y) \, d\gamma(x) = \sum_x f(x, y) = -(y - 1) + y = 1$ for each $y \in \mathcal{Z}$, and, consequently,

$$\int_{\mathcal{Z}} \left[\int_{\mathcal{Z}} f(x, y) \, d\gamma(x) \right] d\gamma(y) = \sum_y 1 = \infty.$$

Thus, the second equation in part (e) of Fubini's theorem fails. Note that here none of the integrals in (i)–(iii) of Fubini's theorem is finite. □

The Completion of the Product Measure Space

The product of two measure spaces may not be complete, even when each factor space is complete. For instance, we know the measure space $(\mathcal{R}, \mathcal{M}, \lambda)$ is complete. But the product of that measure space with itself, $(\mathcal{R}^2, \mathcal{M} \times \mathcal{M}, \lambda \times \lambda)$, is not complete. Indeed, let N be a non-Lebesgue measurable set, $A = N \times \{0\}$, and $B = \mathcal{R} \times \{0\}$. Then $(\lambda \times \lambda)(B) = 0$, $A \subset B$, but $A \notin \mathcal{M} \times \mathcal{M}$ because $A^0 = N \notin \mathcal{M}$ (see Proposition 6.9 on page 206). Consequently, we see that $(\mathcal{R}^2, \mathcal{M} \times \mathcal{M}, \lambda \times \lambda)$ is not a complete measure space.

Recall from Theorem 5.2 on page 149 that, given a measure space $(\Omega, \mathcal{A}, \mu)$, there is a complete measure space $(\Omega, \overline{\mathcal{A}}, \overline{\mu})$, called the completion of $(\Omega, \mathcal{A}, \mu)$, such that $\overline{\mathcal{A}} \supset \mathcal{A}$ and $\overline{\mu}_{|\mathcal{A}} = \mu$. It is often more appropriate to work with the completion of a product measure space than the product measure space itself. An important example of this occurs in classical analysis, as we now show.

We just discovered that the measure space $(\mathcal{R}^2, \mathcal{M} \times \mathcal{M}, \lambda \times \lambda)$ is not complete. This non-completeness can cause difficulties. For instance, as Exercise 6.68 reveals, a function can be Riemann integrable over a set $D \subset \mathcal{R}^2$ without being $\mathcal{M} \times \mathcal{M}$-measurable. However, this situation cannot happen with the completion $(\mathcal{R}^2, \overline{\mathcal{M} \times \mathcal{M}}, \overline{\lambda \times \lambda})$. In fact, we have the following two-dimensional analogue of Theorem 4.9 on page 135, whose proof is left as an exercise.

□ □ □ **THEOREM 6.8**

Suppose that f is Riemann integrable on $[a, b] \times [c, d]$. Then f is Lebesgue integrable on $[a, b] \times [c, d]$ with respect to $\overline{\lambda \times \lambda}$ and

$$\int_{[a,b] \times [c,d]} f(x, y) \, d\overline{\lambda \times \lambda}(x, y) = \int_a^b \int_c^d f(x, y) \, dx \, dy.$$

Note: Because of Theorem 6.8, we will often denote the integral on the left of the previous equation by the integral on the right, regardless of whether f is Riemann integrable.

We should point out that the measure space $(\mathcal{R}^2, \overline{\mathcal{M} \times \mathcal{M}}, \overline{\lambda \times \lambda})$ is identical to the measure space $(\mathcal{R}^2, \mathcal{M}_2, \lambda_2)$, discussed in Section 6.1 on page 186. In other words, $\overline{\lambda \times \lambda}$ is two-dimensional Lebesgue measure and $\overline{\mathcal{M} \times \mathcal{M}}$ is the collection of two-dimensional Lebesgue measurable sets. The verification of these facts is considered in Exercise 6.69.

We can derive analogues of Tonelli's theorem and Fubini's theorem for the completion of a product measure space provided that the factor spaces are complete and σ-finite. We begin with two lemmas.

□ □ □ **LEMMA 6.10**

Let $(\Omega, \mathcal{A}, \mu)$ be a measure space and f an $\overline{\mathcal{A}}$-measurable function on Ω. Then there exists an $\mathcal{A}$-measurable function ϕ on Ω such that $\phi = f$ $\overline{\mu}$-ae.

PROOF We employ the bootstrapping technique. Therefore, first suppose that $f = \chi_E$, where $E \in \overline{\mathcal{A}}$. By definition, we can select sets A, B, and C such that $E = B \cup A$, where $B, C \in \mathcal{A}$, $A \subset C$, and $\mu(C) = 0$. Now we define the function $\phi = \chi_B$ and note that, because $B \in \mathcal{A}$, ϕ is $\mathcal{A}$-measurable. Let $D = \{ x : \phi(x) \neq f(x) \}$. Then $D = E \setminus B \subset A \subset C$. Since $\overline{\mu}(C) = \mu(C) = 0$ and $D \subset C$, it follows by completeness that $D \in \overline{\mathcal{A}}$ and $\overline{\mu}(D) = 0$. Thus, $\phi = f$ $\overline{\mu}$-ae.

Suppose now that f is a simple function, say, $f = \sum_{k=1}^n a_k \chi_{A_k}$. As we have just seen, we can select $\mathcal{A}$-measurable functions ϕ_k such that $\phi_k = \chi_{A_k}$ $\overline{\mu}$-ae for $1 \leq k \leq n$. If $D_k = \{ x : \phi_k(x) \neq \chi_{A_k}(x) \}$, then the set $D = \bigcup_{k=1}^n D_k$ has $\overline{\mu}$-measure zero and the $\mathcal{A}$-measurable function $\phi = \sum_{k=1}^n a_k \phi_k$ equals f on D.

If f is nonnegative, choose a sequence $\{s_n\}_{n=1}^\infty$ of nonnegative $\overline{\mathcal{A}}$-measurable simple functions such that $s_n \uparrow f$ pointwise on Ω. Then, by using the result of the previous paragraph, we can select a sequence $\{t_n\}_{n=1}^\infty$ of $\mathcal{A}$-measurable functions such that, for each $n \in \mathcal{N}$, $t_n = s_n$ $\overline{\mu}$-ae. Define $\phi = \limsup_{n\to\infty} t_n$ and note that ϕ is $\mathcal{A}$-measurable. We claim that $\phi = f$ $\overline{\mu}$-ae. To prove this fact, let $A_n = \{\, x : t_n(x) \neq s_n(x) \,\}$ and $A = \bigcup_{n=1}^{\infty} A_n$. Then $\overline{\mu}(A) = 0$ and, if $x \notin A$,

$$\phi(x) = \limsup_{n\to\infty} t_n(x) = \lim_{n\to\infty} s_n(x) = f(x).$$

Consequently, $\phi = f$ $\overline{\mu}$-ae. We leave the remainder of the proof as an exercise for the reader. ∎

◻ ◻ ◻ **LEMMA 6.11**

Suppose that $(\Gamma, \mathcal{S}, \mu)$ and $(\Lambda, \mathcal{T}, \nu)$ are complete, σ-finite measure spaces. If ξ is an $\overline{\mathcal{S} \times \mathcal{T}}$-measurable function such that $\xi = 0$ $\overline{\mu \times \nu}$-ae, then

a) for μ-almost all $x \in \Gamma$, $\xi_{[x]} = 0$ ν-ae.

b) for ν-almost all $y \in \Lambda$, $\xi^{[y]} = 0$ μ-ae.

PROOF Let $E = \{\, (x, y) : \xi(x, y) \neq 0 \,\}$. Then $\overline{\mu \times \nu}(E) = 0$, by assumption. We can select sets A, B, and C such that $E = B \cup A$, where B, $C \in \mathcal{S} \times \mathcal{T}$, $A \subset C$, and $(\mu \times \nu)(C) = 0$. Because $\overline{\mu \times \nu}(E) = 0$, it is clear that $(\mu \times \nu)(B) = 0$. Let $D = B \cup C$. Then $D \in \mathcal{S} \times \mathcal{T}$ and $(\mu \times \nu)(D) = 0$. So, by Lemma 6.9(a) on page 209, $\int_\Gamma \nu(D_x)\, d\mu(x) = 0$. This result implies that $\nu(D_x) = 0$ μ-ae.

Set $N = \{\, x : \nu(D_x) \neq 0 \,\}$ and note that $\mu(N) = 0$. If $x \notin N$, then $\nu(D_x) = 0$ and, as $E_x \subset D_x$ and $(\Lambda, \mathcal{T}, \nu)$ is complete, it follows that $E_x \in \mathcal{T}$ and $\nu(E_x) = 0$. But $E_x = \{\, y : \xi_{[x]}(y) \neq 0 \,\}$ and, hence, we see that $\xi_{[x]} = 0$ ν-ae. Thus, part (a) holds and a similar argument establishes the validity of part (b). ∎

We are now in a position to prove the analogues of Tonelli's theorem and Fubini's theorem for the completion of a product measure space. The former theorem is presented as Theorem 6.9 and the latter theorem is left to the reader as an exercise.

◻ ◻ ◻ **THEOREM 6.9**

Suppose that $(\Gamma, \mathcal{S}, \mu)$ and $(\Lambda, \mathcal{T}, \nu)$ are complete, σ-finite measure spaces. Let f be a nonnegative extended real-valued $\overline{\mathcal{S} \times \mathcal{T}}$-measurable function on $\Gamma \times \Lambda$. Then

a) $f_{[x]}$ is $\mathcal{T}$-measurable for μ-almost all $x \in \Gamma$.

b) $f^{[y]}$ is $\mathcal{S}$-measurable for ν-almost all $y \in \Lambda$.

c) $g(x) = \int_\Lambda f(x, y)\, d\nu(y)$ is defined μ-ae and is equal to an $\mathcal{S}$-measurable function μ-ae.

d) $h(y) = \int_\Gamma f(x, y)\, d\mu(x)$ is defined ν-ae and is equal to a $\mathcal{T}$-measurable function ν-ae.

e) the equalities

$$\int_{\Gamma \times \Lambda} f(x, y)\, d\overline{\mu \times \nu}(x, y) = \int_\Gamma \left[\int_\Lambda f(x, y)\, d\nu(y) \right] d\mu(x)$$

$$= \int_\Lambda \left[\int_\Gamma f(x, y)\, d\mu(x) \right] d\nu(y)$$

hold.

PROOF Use Lemma 6.10 to choose an $\mathcal{S} \times \mathcal{T}$-measurable function ϕ such that $\phi = f$ $\overline{\mu \times \nu}$-ae. Now let $E = \{ (x, y) : \phi(x, y) \neq f(x, y) \}$. Arguing as in the proof of Lemma 6.11, we can select $D \in \mathcal{S} \times \mathcal{T}$ with $E \subset D$ and $(\mu \times \nu)(D) = 0$. If we define $h = \chi_{D^c} \phi$ and $\xi = \chi_D f$, then h is $\mathcal{S} \times \mathcal{T}$-measurable, $f = h + \xi$, and $\xi = 0$ $\overline{\mu \times \nu}$-ae.

By part (a) of Tonelli's theorem, $h_{[x]}$ is $\mathcal{T}$-measurable for $x \in \Gamma$. Since $f_{[x]} = h_{[x]} + \xi_{[x]}$ and $\xi = 0$ $\overline{\mu \times \nu}$-ae, we can conclude from Lemma 6.11(a) that, for μ-almost all $x \in \Gamma$, $f_{[x]} = h_{[x]}$ ν-ae. Consequently, because $(\Lambda, \mathcal{T}, \nu)$ is complete, $f_{[x]}$ is $\mathcal{T}$-measurable for μ-almost all $x \in \Gamma$. We have now established part (a), and a similar argument establishes part (b).

Referring to part (c) of Tonelli's theorem, we see that the function ρ defined on Γ by $\rho(x) = \int_\Lambda h(x, y) \, d\nu(y)$ is $\mathcal{S}$-measurable. Let $A = \{ x : f_{[x]} = h_{[x]} \ \nu\text{-ae} \}$. By the previous paragraph, $\mu(A^c) = 0$. If $x \in A$, then $f_{[x]}$ is $\mathcal{T}$-measurable and $\int_\Lambda f(x, y) \, d\nu(y) = \int_\Lambda h(x, y) \, d\nu(y)$. This result verifies part (c). Similarly, part (d) holds.

To establish the first equation in part (e), we apply Exercise 6.72, part (e) of Tonelli's theorem, and Definition 5.14 on page 169:

$$\int_{\Gamma \times \Lambda} f \, d\overline{\mu \times \nu} = \int_{\Gamma \times \Lambda} h \, d\overline{\mu \times \nu} = \int_{\Gamma \times \Lambda} h \, d(\mu \times \nu)$$

$$= \int_\Gamma \left[\int_\Lambda h \, d\nu \right] d\mu = \int_\Gamma \left[\int_\Lambda f \, d\nu \right] d\mu.$$

A similar argument verifies the second equation in part (e). ∎

The Product of More Than Two Measure Spaces

Up to this point, we have only considered product measure spaces in which there are two factors. Using similar techniques, we can develop the theory of product measure spaces in which there are a finite number of factors. We present only the highlights.

□ □ □ THEOREM 6.10

Suppose that $(\Omega_k, \mathcal{A}_k, \mu_k)$, $1 \leq k \leq n$, are σ-finite measure spaces. Let $\times_{k=1}^n \mathcal{A}_k$ denote the σ-algebra generated by the n-dimensional measurable rectangles. Then there is a unique measure $\times_{k=1}^n \mu_k$ on $\times_{k=1}^n \mathcal{A}_k$ such that

$$\left(\times_{k=1}^n \mu_k \right) \left(\times_{k=1}^n A_k \right) = \prod_{k=1}^n \mu_k(A_k)$$

for all n-dimensional measurable rectangles.

The σ-algebra $\times_{k=1}^n \mathcal{A}_k$ is called the **product σ-algebra** of $\mathcal{A}_1, \ldots, \mathcal{A}_n$; the measure $\times_{k=1}^n \mu_k$ is called the **product measure** of $\mu_1, \ldots, \mu_n$; and the measure space $\left(\times_{k=1}^n \Omega_k, \times_{k=1}^n \mathcal{A}_k, \times_{k=1}^n \mu_k \right)$ is called the **product measure space** of $(\Omega_1, \mathcal{A}_1, \mu_1), \ldots, (\Omega_n, \mathcal{A}_n, \mu_n)$.

Tonelli's theorem and Fubini's theorem generalize to n-dimensional product spaces. In particular, the integral of a nonnegative $\times_{k=1}^{n} \mathcal{A}_k$-measurable function or a function in $\mathcal{L}^1\left(\times_{k=1}^{n} \mu_k\right)$ can be evaluated by forming the iterated integrals in any order. For example, if $n = 3$ and $f \in \mathcal{L}^1(\mu_1 \times \mu_2 \times \mu_3)$, then

$$\int_{\Omega_1 \times \Omega_2 \times \Omega_3} f \, d(\mu_1 \times \mu_2 \times \mu_3) = \int_{\Omega_{i_1}} \left[\int_{\Omega_{i_2}} \left[\int_{\Omega_{i_3}} f \, d\mu_{i_3} \right] d\mu_{i_2} \right] d\mu_{i_1}$$

for each permutation i_1, i_2, i_3 of 1, 2, 3.

EXAMPLE 6.13 *Illustrates the Product of Finitely Many Measure Spaces*

a) Let $\mathcal{B}_n$ denote the σ-algebra generated by the open sets of $\mathcal{R}^n$. Members of $\mathcal{B}_n$ are called **n-dimensional Borel sets**. It is not too difficult to show that $\mathcal{B}_n = \mathcal{B} \times \cdots \times \mathcal{B}$.

b) It can be shown that $(\mathcal{R}^n, \overline{\mathcal{M} \times \cdots \times \mathcal{M}}, \overline{\lambda \times \cdots \times \lambda}) = (\mathcal{R}^n, \mathcal{M}_n, \lambda_n)$; that is, $\overline{\lambda \times \cdots \times \lambda}$ is n-dimensional Lebesgue measure and $\overline{\mathcal{M} \times \cdots \times \mathcal{M}}$ is the collection of n-dimensional Lebesgue measurable sets. We should also point out that Theorem 6.8 can be generalized to arbitrary dimensions. That is, if f is Riemann integrable on $\times_{k=1}^{n}[a_k, b_k]$, then f is Lebesgue integrable on $\times_{k=1}^{n}[a_k, b_k]$ with respect to λ_n and

$$\int_{\times_{k=1}^{n}[a_k, b_k]} f \, d\lambda_n = \int_{a_1}^{b_1} \cdots \int_{a_n}^{b_n} f(x_1, \ldots, x_n) \, dx_1 \cdots dx_n.$$

The proof of this fact is left to the reader as an exercise. □

Exercises for Section 6.4

Note: In some of the exercises, you will need the following two facts: (1) Let $D \subset \mathcal{R}^2$. A subset of D is open in D if and only if it can be expressed as the intersection of D with an open subset of $\mathcal{R}^2$. (2) A function $f: D \to \mathcal{R}$ is continuous if and only if $f^{-1}(O)$ is open in D for each open set $O \subset \mathcal{R}$.

6.58 Complete the proof of Fubini's theorem; that is, assuming that the theorem holds for real-valued functions, prove that it holds for complex-valued functions. *Hint:* Write $f = \Re f + i\Im f$ and use Proposition 5.9(b), found on page 167.

6.59 Suppose that f is a Lebesgue measurable function on $\mathcal{R}$ such that

$$\int_{-\infty}^{\infty} |f(x)| \, dx < \infty \qquad \text{and} \qquad \int_{-\infty}^{\infty} \frac{|f(x)|}{|x|} \, dx < \infty.$$

Define $G(x, y) = f(x)/(x^2 + y^2)$, if $(x, y) \neq (0, 0)$, and zero otherwise.
a) Show that G is $\mathcal{M} \times \mathcal{M}$-measurable.
b) Prove that $G \in \mathcal{L}^1(\lambda \times \lambda)$ and that

$$\int_{\mathcal{R}^2} G \, d(\lambda \times \lambda) = \pi \int_{-\infty}^{\infty} \frac{f(x)}{|x|} \, dx.$$

6.60 Suppose that $\{a_{mn}\}_{m,n=1}^{\infty}$ is a double sequence of complex numbers for which at least one of the quantities $\sum_m \left(\sum_n |a_{mn}| \right)$ and $\sum_n \left(\sum_m |a_{mn}| \right)$ is finite. Prove that both quantities are finite and that $\sum_m \left(\sum_n a_{mn} \right) = \sum_n \left(\sum_m a_{mn} \right)$. *Hint:* Use Exercise 5.73 on page 172.

6.61 Let $(\Gamma, \mathcal{S}, \mu) = (\Lambda, \mathcal{T}, \nu) = \left([0,1], \mathcal{M}_{[0,1]}, \lambda_{[0,1]} \right)$ and suppose $f \in \mathcal{L}^1(\mu \times \nu)$. Prove that

$$\int_0^1 \left[\int_0^x f(x,y)\,dy \right] dx = \int_0^1 \left[\int_y^1 f(x,y)\,dx \right] dy.$$

6.62 Let $(\Gamma, \mathcal{S}, \mu) = (\Lambda, \mathcal{T}, \nu) = \left([-1,1], \mathcal{M}_{[-1,1]}, \lambda_{[-1,1]} \right)$ and f be a continuous function on $\Gamma \times \Lambda$. Prove that $f \in \mathcal{L}^1\left(\lambda_{[-1,1]} \times \lambda_{[-1,1]} \right)$ and that, if $D = \{\, (x,y) : x^2 + y^2 \le 1 \,\}$, then

$$\iint_D f(x,y)\,dx\,dy = \int_{-1}^1 \left[\int_{-\sqrt{1-x^2}}^{\sqrt{1-x^2}} f(x,y)\,dy \right] dx.$$

★6.63 This exercise introduces the *convolution* of two Borel measurable functions. For convenience, we will write $\lambda_{|\mathcal{B}}$ simply as λ.
  a) Let f be a Borel measurable function on $\mathcal{R}$. Define F on $\mathcal{R}^2$ by $F(x,y) = f(x-y)$. Show that F is $\mathcal{B}_2$-measurable. *Hint:* $(x,y) \to x - y$ is continuous.
  b) Suppose that ξ is a Borel measurable function on $\mathcal{R}$. Prove that the function ϕ on $\mathcal{R}^2$ defined by $\phi(x,y) = \xi(y)$ is $\mathcal{B}_2$-measurable.
  c) Suppose that h is a nonnegative Borel measurable function. Show that

$$\int_{\mathcal{R}} h(x-y)\,d\lambda(x) = \int_{\mathcal{R}} h(x)\,d\lambda(x)$$

  for each $y \in \mathcal{R}$. *Hint:* Bootstrap.
  d) Suppose that $f, g \in \mathcal{L}^1(\mathcal{R}, \mathcal{B}, \lambda)$. Prove that the function $f * g$ defined on $\mathcal{R}$ by

$$(f * g)(x) = \int_{\mathcal{R}} f(x-y)g(y)\,d\lambda(y)$$

  exists for λ-almost all $x \in \mathcal{R}$ and is in $\mathcal{L}^1(\mathcal{R}, \mathcal{B}, \lambda)$. The function $f * g$ is called the **convolution** of f with g.

★6.64 This exercise introduces the *convolution* of two σ-finite Borel measures, μ and ν.
  a) If $E \in \mathcal{B}$, show that the set $A = \{\, (x,y) : x + y \in E \,\}$ is a two-dimensional Borel set, that is, is in $\mathcal{B}_2$.
  b) Show that the functions $g(x) = \nu(E - x)$ and $h(y) = \mu(E - y)$ are Borel measurable.
  c) Verify that

$$(\mu \times \nu)(A) = \int_{\mathcal{R}} \nu(E - x)\,d\mu(x) = \int_{\mathcal{R}} \mu(E - y)\,d\nu(y),$$

  where A is the set defined in part (a).
  d) For $E \in \mathcal{B}$, define

$$(\mu * \nu)(E) = \int_{\mathcal{R}} \mu(E - y)\,d\nu(y).$$

  Show that $\mu * \nu$ is a Borel measure. The measure $\mu * \nu$ is called the **convolution** of μ with ν. Part (c) shows that $\mu * \nu = \nu * \mu$.
  e) If $f \in \mathcal{L}^1(\mathcal{R}, \mathcal{B}, \mu * \nu)$, prove that $\int_{\mathcal{R}^2} f(x+y)\,d(\mu \times \nu)(x,y) = \int_{\mathcal{R}} f(t)\,d(\mu * \nu)(t)$. *Hint:* Bootstrap.

6.65 Let μ be a finite Borel measure. Define $\widehat{\mu} \colon \mathcal{R} \to \mathbb{C}$ by

$$\widehat{\mu}(s) = \int_{\mathcal{R}} e^{its} \, d\mu(t).$$

The function $\widehat{\mu}$ is called the **Fourier-Stieltjes transform** of μ.

a) Show that $\widehat{\mu}$ is well-defined, that is, the integral exists for each $s \in \mathcal{R}$.

b) Let ν be a finite Borel measure. Prove that $\widehat{\mu * \nu}(s) = \widehat{\mu}(s)\widehat{\nu}(s)$, where $\mu * \nu$ is the convolution of μ and ν as defined in Exercise 6.64. *Hint:* Use Exercise 6.64(e) and Fubini's theorem.

6.66 Suppose that $(\Gamma, \mathcal{S}, \mu)$ and $(\Lambda, \mathcal{T}, \nu)$ are two σ-finite measure spaces. Let $\mathcal{U}$ be the semialgebra of measurable rectangles and ι be defined on $\mathcal{U}$ by $\iota(S \times T) = \mu(S)\nu(T)$. Furthermore, let $(\Gamma \times \Lambda, \mathcal{A}, \tau)$ be the complete measure space induced by $\mathcal{U}$ and ι, as in Theorem 6.1. Prove that $\mathcal{A} = \overline{\mathcal{S} \times \mathcal{T}}$ and $\tau = \overline{\mu \times \nu}$.

6.67 Provide an example in which the measure space $(\Gamma \times \Lambda, \mathcal{S} \times \mathcal{T}, \mu \times \nu)$ is complete.

6.68 Construct a function f on $\mathcal{R}^2$ that is Riemann integrable on $[0, 1] \times [0, 1]$ but is not $\mathcal{M} \times \mathcal{M}$-measurable. *Hint:* Do something with a non-Lebesgue measurable set and use the fact that a function is Riemann integrable on $[0, 1] \times [0, 1]$ if and only if the set of its points of discontinuity has two-dimensional Lebesgue measure zero.

6.69 Prove that $(\mathcal{R}^2, \overline{\mathcal{M} \times \mathcal{M}}, \overline{\lambda \times \lambda}) = (\mathcal{R}^2, \mathcal{M}_2, \lambda_2)$. Use the following steps:

a) Let $\mathcal{I}_2$ denote the collection of all sets of the form $I \times J$, where I and J are intervals of $\mathcal{R}$. Show that $\overline{\mathcal{M} \times \mathcal{M}} \supset \sigma(\mathcal{I}_2)$ and that $\overline{\lambda \times \lambda}$ agrees with λ_2 on $\mathcal{I}_2$.

b) Use part (a) to conclude that $\overline{\mathcal{M} \times \mathcal{M}} \supset \mathcal{M}_2$ and $\overline{\lambda \times \lambda}_{|\mathcal{M}_2} = \lambda_2$. *Hint:* Employ Theorem 6.2, Exercise 6.20 on page 191, and Exercise 5.17 on page 150.

c) Show that $\mathcal{M} \times \mathcal{M} \subset \mathcal{M}_2$ and that λ_2 agrees with $\lambda \times \lambda$ on $\mathcal{M} \times \mathcal{M}$. *Hint:* First show that $B \times \mathcal{R} \in \mathcal{M}_2$ for all $B \in \mathcal{B}$ and then that $E \times \mathcal{R} \in \mathcal{M}_2$ for all $E \in \mathcal{M}$.

d) Use part (c) to conclude that $\mathcal{M}_2 \supset \overline{\mathcal{M} \times \mathcal{M}}$ and $\lambda_{2|\overline{\mathcal{M} \times \mathcal{M}}} = \overline{\lambda \times \lambda}$.

e) Deduce the required result.

6.70 Generalize the previous exercise to n-dimensions.

6.71 Complete the proof of Lemma 6.10 on page 217.

6.72 Suppose that $(\Omega, \mathcal{A}, \mu)$ is a measure space and that f is an $\mathcal{A}$-measurable function on Ω. Prove that $\int_{\Omega} f \, d\overline{\mu} = \int_{\Omega} f \, d\mu$. *Hint:* Bootstrap.

6.73 Establish the calculus theorem stated in Example 6.12(a) on page 215 by proceeding as follows:

a) Define h on $\mathcal{R}^2$ by $h(x, y) = f(x, y)$ if $(x, y) \in R$ and zero otherwise. Show that h is $\mathcal{M} \times \mathcal{M}$-measurable and is in $\mathcal{L}^1(\lambda \times \lambda)$.

b) Prove that $\iint_R f(x, y) \, dx \, dy = \int_R f \, d(\lambda \times \lambda)$, where the integral on the left is a double Riemann integral. *Hint:* Use Exercise 6.72.

c) Deduce (6.28) on page 215.

d) Does (6.28) remain valid in case that f is assumed only to be Riemann integrable on $[a, b] \times [c, d]$? Explain.

6.74 State and prove the analogue of Fubini's theorem for the completion of a product measure space.

6.75 **Integration by parts:** In this exercise, we will develop an integration by parts formula for Lebesgue-Stieltjes integrals. We proceed using the following steps:

a) Let μ be a finite Borel measure on $\mathcal{R}$ having distribution function F_μ. Define $F_\mu(x-) = \sup\{F_\mu(t) : t < x\}$. Show that $F_\mu(x-) = \mu\big((-\infty, x)\big)$ for each $x \in \mathcal{R}$.

b) Use part (a) to deduce that, for each $x \in \mathcal{R}$, $\mu(\{x\}) = F_\mu(x) - F_\mu(x-)$. [Thus, F_μ is continuous at x if and only if x is not an atom of μ.]

c) Let ν be a finite Borel measure having distribution function F_ν. Prove that, for $a, b \in \mathcal{R}$,

$$\int_{(a,b]} F_\mu(x)\, d\nu(x) + \int_{(a,b]} F_\nu(x)\, d\mu(x)$$

$$= F_\mu(b)F_\nu(b) - F_\mu(a)F_\nu(a) + \int_{(a,b]} \big(F_\mu(x) - F_\mu(x-)\big)\, d\nu(x).$$

Hint: Apply Tonelli's theorem to show that

$$\int_{(a,b]} F_\mu(y)\, d\nu(y) + \int_{(a,b]} F_\nu(x)\, d\mu(x) = \int_{\mathcal{R}^2} H(x,y)\, d(\mu \times \nu)(x,y),$$

where $H(x,y) = \chi_{(-\infty,y]}(x)\chi_{(a,b]}(y) + \chi_{(a,b]}(x)\chi_{(-\infty,x]}(y)$. Then show

$$H(x,y) = \chi_{\{y\}}(x)\chi_{(a,b]}(y) + \chi_{(-\infty,a]}(x)\chi_{(a,b]}(y)$$
$$+ \chi_{(a,b]}(x)\chi_{(-\infty,a]}(y) + \chi_{(a,b]}(x)\chi_{(a,b]}(y).$$

6.76 Let $(\Omega_k, \mathcal{A}_k)$, $1 \le k \le n$, be measurable spaces. Denote by $\mathcal{U}$ the collection of n-dimensional measurable rectangles; that is, $\mathcal{U} = \big\{ \underset{k=1}{\overset{n}{\times}} A_k : A_k \in \mathcal{A}_k,\ 1 \le k \le n \big\}$. Explain why $\mathcal{U}$ is a semialgebra.

Exercises 6.77–6.81 should be completed by all readers who plan to cover the probability material in Chapter 7.

★6.77 Denote by $\mathcal{B}_n$ the smallest σ-algebra of subsets of $\mathcal{R}^n$ that contains all open sets of $\mathcal{R}^n$.
  a) Show that $\mathcal{B}_n = \mathcal{B} \times \cdots \times \mathcal{B}$.
  b) A measure on $\mathcal{B}_n$ is called an **n-dimensional Borel measure**. Suppose μ and ν are two finite n-dimensional Borel measures such that $\mu\big(\underset{k=1}{\overset{n}{\times}} B_k\big) = \nu\big(\underset{k=1}{\overset{n}{\times}} B_k\big)$ for all $B_1, B_2, \ldots, B_n \in \mathcal{B}$. Prove that $\mu = \nu$; that is, $\mu(B) = \nu(B)$ for all $B \in \mathcal{B}_n$.

6.78 Let $\mathcal{I}_n$ denote the collection of all n-dimensional intervals in $\mathcal{R}^n$; that is, all sets of the form $I_1 \times I_2 \times \cdots \times I_n$ where $I_j \in \mathcal{I}$ for $1 \le j \le n$.
  a) Show that the σ-algebra generated by $\mathcal{I}_n$ is $\mathcal{B}_n$; that is, $\sigma(\mathcal{I}_n) = \mathcal{B}_n$. *Hint:* Use Exercise 6.77(a).
  b) Let μ and ν be two n-dimensional Borel measures such that $\mu(I) = \nu(I) < \infty$ for all bounded n-dimensional intervals I. Prove that $\mu = \nu$.

6.79 Let $\mathcal{J}$ denote the collection of intervals of $\mathcal{R}$ of the form $(a, b]$ and (c, ∞), where $-\infty \le a \le b < \infty$ and $-\infty \le c < \infty$. Also, let $\mathcal{J}_n$ denote the collection of all subsets of $\mathcal{R}^n$ of the form $J_1 \times J_2 \times \cdots \times J_n$ where $J_k \in \mathcal{J}$ for $1 \le k \le n$. Prove that $\mathcal{J}_n$ is a semialgebra and that the σ-algebra generated by $\mathcal{J}_n$ is $\mathcal{B}_n$.

★6.80 Suppose that μ and ν are two finite n-dimensional Borel measures such that

$$\mu\bigg(\overset{n}{\underset{k=1}{\times}} (-\infty, x_k]\bigg) = \nu\bigg(\overset{n}{\underset{k=1}{\times}} (-\infty, x_k]\bigg)$$

for all $x_1, x_2, \ldots, x_n \in \mathcal{R}$. Prove that $\mu = \nu$. *Hint:* It suffices to prove that $\mu = \nu$ on $\mathcal{J}_n$.

★6.81 Let $\mu_1, \ldots, \mu_n$ be finite Borel measures and μ an n-dimensional Borel measure. Suppose that

$$\mu\bigg(\overset{n}{\underset{k=1}{\times}} (-\infty, x_k]\bigg) = \prod_{k=1}^{n} \mu_k\big((-\infty, x_k]\big)$$

for all $x_1, x_2, \ldots, x_n \in \mathcal{R}$. Prove that $\mu = \underset{k=1}{\overset{n}{\times}} \mu_k$.

Andrei Nikolaevich Kolmogorov
(1903–1987)

Andrei Kolmogorov was born on April 25, 1903, in Tambov, Russia. His mother died in childbirth. His father, an agronomist, was deported from Saint Petersburg for participation in the revolutionary movement; he returned after the Revolution to head a department in the Ministry of Agriculture, but died in fighting in 1919.

Kolmogorov was raised by his mother's sister, Vera Yakovlena, in Tunoshna near Yaroslavl at the estate of his maternal grandfather who was a wealthy nobleman. He received his early education in his aunt's village school. At the age of 17, he entered Moscow State University, from which he graduated in 1925.

Kolmogorov's contributions to the world of mathematics encompass a formidable range of subjects. A partial listing includes the theory of functions of a real variable, trigonometric series, probability theory, theory of algorithms, functional analysis, topology, dynamical systems, information theory, and classical mechanics.

Kolmogorov revolutionized probability theory. He introduced the modern axiomatic approach to probability and proved many of the fundamental theorems that are a consequence of that approach. He also developed two systems of partial differential equations that play a crucial role in the theory of Markov processes.

In addition to his work in higher mathematics, Kolmogorov was interested in the mathematical education of schoolchildren. He was chairman of the Commission for Mathematical Education under the Presidium of the Academy of Sciences of the U.S.S.R. During that time, he was instrumental in the development of a new training program that was incorporated into Soviet schools.

Many articles and books were written by Kolmogorov. The book, *Introductory Real Analysis,* co-authored with S. V. Fomin, provides, in the bibliography, a listing of some of his publications. Kolmogorov was a member of the faculty at Moscow State University from his graduation in 1925 until his death in 1987.

7

Elements of Probability

Probability is the mathematical discipline dealing with the analysis of random phenomena. Intuitively, the probability of an event is a measure of the likelihood of its occurrence — a probability near 0 indicates that the event is unlikely to occur, whereas a probability near 1 suggests that the event is likely to occur.

The origins of the theory of probability are usually taken to be in the middle of the seventeenth century, although the basic concepts of probability date back to to well before the Common Era. With the development of the natural sciences in the early 1900s, it became increasingly important for probability to have a formal mathematical framework similar to that found in other branches of mathematics such as geometry and abstract algebra. Measure theory supplied the required framework.

In this chapter, we will introduce the elements of probability theory based on the axiomatic development by Andrei Nikolaevich Kolmogorov. The foundations will be presented in Sections 7.1–7.3. Then, as a first application, we will examine several theorems, known collectively as *laws of large numbers,* which comprise some of the most important results in probability. We will return to further explore probability theory in other chapters of the text.

7.1 THE MATHEMATICAL MODEL FOR PROBABILITY

In this section, we will develop the mathematical model for probability based on the theory of measure discussed in Chapters 5 and 6. However, before we begin with that development, it will be useful for motivational purposes to provide an interpretation of the meaning of probability.

To that end, let us think of an event as some specified result that may or may not occur when an experiment is performed; for example, a head comes up (the event) when a coin is tossed (the experiment). The usual interpretation of probability is the **relative-frequency interpretation,** which construes the probability of an event to be the relative frequency of its occurrence in a large number of repetitions of the experiment.

More formally, let E be an event and $P(E)$ its probability. For n repetitions of the experiment, let $n(E)$ denote the number of times that event E occurs. The relative-frequency interpretation is that, for large n, the proportion of times that event E occurs in the n repetitions of the experiment will be approximately equal to the probability that event E occurs on any particular trial:

$$\frac{n(E)}{n} \approx P(E), \qquad \text{for large } n. \tag{7.1}$$

To illustrate, consider the experiment of a single toss of a balanced coin. Because the coin is balanced, we reason that there is a 50-50 chance that the coin will come up heads (i.e., will land with heads facing up). Thus, we attribute probability 0.5 to that event. The relative-frequency interpretation is that in a large number of tosses of the coin, heads will come up about half the time. We used a computer to perform two simulations of tossing a balanced coin 100 times. The results are displayed in Figs. 7.1 and 7.2 and seem to corroborate the relative-frequency interpretation.

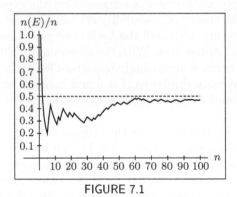

FIGURE 7.1

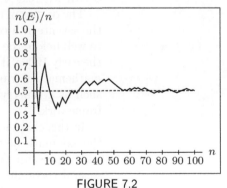

FIGURE 7.2

We should emphasize that all attempts to use (7.1) as a definition of probability have failed. Nonetheless, the relative-frequency interpretation is invaluable for motivational purposes in the axiomatic development. Furthermore, we shall see that once the axioms of probability are in place, a mathematically precise version of (7.1) can be proved as a theorem.

Probability Spaces

Consider now an experiment whose outcome cannot be predicted with certainty beforehand. Such an experiment is called a **random experiment.** The set of possible outcomes of the experiment is called the **sample space** and is usually denoted by the English letter S or the Greek letter Ω; we will use the latter notation. The possible outcomes themselves are denoted generically by the Greek letter ω.

Actually, we will permit as a sample space any set that contains all the possible outcomes of the experiment. This is because, *a priori,* it is sometimes difficult to know precisely the possible outcomes of an experiment. For instance, consider the experiment of rolling a die once and observing the number of dots on the face pointing up. The most natural choice for the sample space is $\Omega = \{1, 2, 3, 4, 5, 6\}$. However, it is conceivable, because of, say, an imperfection in the die, that four would never come up. The important factor in the choice of a sample space is that all possible outcomes are included as elements, not that all elements are possible outcomes.

Associated with a random experiment is a collection of events, usually called the **event class**, which we will denote by $\mathcal{A}$. The assumption is that any specified event will either occur or not occur when the experiment is performed. Each event $E \in \mathcal{A}$ can be considered a subset of the sample space; namely, the collection of outcomes that satisfy the conditions for the occurrence of E. Using this identification between events and sets, we see that an event E **occurs** if and only if the outcome ω of the experiment is a member of E, that is, $\omega \in E$.

We should point out that the empty set $\emptyset$ corresponds to an event that cannot occur and is called the **impossible event**. Two events A and B are called **mutually exclusive** if their joint occurrence is impossible; in other words, if A and B are disjoint. More generally, if each pair of events among a collection of events is mutually exclusive, then we say that the events in the collection are **pairwise mutually exclusive.**

EXAMPLE 7.1   *Illustrates Sample Spaces and Events*

a) Consider the experiment of tossing a coin three times. A sample space for the experiment is $\Omega = \{$HHH, HHT, HTH, HTT, THH, THT, TTH, TTT$\}$ where, for instance, HTT denotes the outcome of a head on the first toss and tails on the second and third tosses. Then, for instance, the event E that the first two tosses are heads consists of the two outcomes, HHH and HHT. In other words, $E = \{$HHH, HHT$\}$. Now, let F be the event that exactly two of the three tosses are tails. Clearly, it is not possible for both E and F to occur when the experiment is performed; hence, E and F are mutually exclusive. We can see this fact set theoretically by noting that $F = \{$HTT, THT, TTH$\}$ and, so, $E \cap F = \emptyset$.

b) Suppose that, starting at 6:00 PM, we observe the elapsed time, in hours, until the first patient arrives at a certain emergency room. For this experiment, we can take the sample space to be the nonnegative real numbers: $\Omega = [0, \infty)$. Then, for instance, the event E that the first patient arrives between 6:15 and 6:30 PM, inclusive, consists of all real numbers between 1/4 and 1/2, inclusive; that is, $E = [1/4, 1/2]$.   □

Next we need to decide on what properties an event class $\mathcal{A}$ must have. First of all, if $E \in \mathcal{A}$ (i.e., E is an event), then we can speak of the occurrence or nonoccurrence of E. However, the nonoccurrence of E is equivalent to the occurrence of the complement of E. Hence, if $E \in \mathcal{A}$, then we require that $E^c \in \mathcal{A}$.

Suppose that $A, B \in \mathcal{A}$. Then we can speak of the occurrence of each of the two events individually. Hence, it should be meaningful to speak of the occurrence of at least one of the two events. But, the occurrence of at least one of A and B is equivalent to the occurrence of the union of A and B. Thus, if $A, B \in \mathcal{A}$, then we require that $A \cup B \in \mathcal{A}$; that is, $\mathcal{A}$ should be closed under finite unions. For mathematical reasons, we will impose the stronger requirement that $\mathcal{A}$ be closed under countable unions.

To summarize, we see that the event class $\mathcal{A}$ should be closed under complementation and countable unions. In other words, $\mathcal{A}$ *should be a σ-algebra of subsets of* Ω.

We now turn our attention to probability. In the axiomatic treatment of probability, we assume that to each event E there corresponds a number $P(E)$ representing the probability that event E occurs. Thus, we can think of P as a set function defined on the collection $\mathcal{A}$ of events. We will employ the relative-frequency interpretation of probability in order to delineate the properties required of P.

So, assume that the experiment is repeated a large number n of times. Then, by (7.1), $P(E) \approx n(E)/n$ for each event E. Clearly, $n(E)/n \geq 0$ and, consequently, we require that $P(E) \geq 0$ for each event E. In other words, probabilities should be nonnegative numbers, an obvious restriction. Note also that because Ω contains all of the possible outcomes of the experiment, it must occur every time the experiment is performed. Hence, $n(\Omega)/n = 1$, which means that we should have $P(\Omega) = 1$, another obvious condition. Further, since $\emptyset$ represents an impossibility, $n(\emptyset)/n = 0$, which means that we should have $P(\emptyset) = 0$, again an obvious condition.

Finally, suppose that A and B are mutually exclusive (disjoint) events. Then, we have $n(A \cup B) = n(A) + n(B)$ and, consequently, by (7.1),

$$P(A \cup B) \approx \frac{n(A \cup B)}{n} = \frac{n(A) + n(B)}{n} = \frac{n(A)}{n} + \frac{n(B)}{n} \approx P(A) + P(B).$$

Hence, we require P to be finitely additive. Again, for mathematical reasons, we will impose the stronger condition of countable additivity. This condition and the previous paragraph indicate that P *should be a probability measure on the σ-algebra $\mathcal{A}$ of events.*

In summary, the mathematical model for a random experiment consists of a set Ω that contains the possible outcomes of the experiment, a σ-algebra $\mathcal{A}$ of subsets of Ω that represents the collection of events, and a probability measure P on $\mathcal{A}$, where, for each $E \in \mathcal{A}$, $P(E)$ is interpreted as the probability that event E occurs. As we learned, in Example 5.1(h) on pages 146–147, the triple $(\Omega, \mathcal{A}, P)$ is called a *probability space*.

DEFINITION 7.1   Probability Space

A **probability space** is a triple $(\Omega, \mathcal{A}, P)$, where Ω is a set, $\mathcal{A}$ is a σ-algebra of subsets of Ω, and P is a probability measure on $\mathcal{A}$.

The following examples illustrate the discussion of probability spaces. We leave any remaining details as exercises for the reader.

EXAMPLE 7.2 *Illustrates Definition 7.1*

a) Refer to Example 7.1(a) on page 227. In this case, we take $\mathcal{A} = \mathcal{P}(\Omega)$ so that every subset of Ω is an event. If the coin is balanced, then, by symmetry, each possible outcome should be equally likely, implying that each should have probability $1/8$. This requirement, in turn, implies that the appropriate probability measure is $P = \gamma/8$, where γ is counting measure on $\mathcal{P}(\Omega)$. In other words, for each $E \in \mathcal{A}$, $P(E) = N(E)/8$, where $N(E)$ denotes the number of elements of E.

b) Suppose that Ω is a countable sample space, that is, the experiment has either a finite or countably infinite number of possible outcomes, say ω_1, ω_2, $\ldots$. For a countable sample space, we always take $\mathcal{A} = \mathcal{P}(\Omega)$. Let $p_n = P(\{\omega_n\})$. Then $P = \sum_n p_n \delta_{\omega_n}$; that is, for $E \in \mathcal{A}$,

$$P(E) = \sum_{\omega_n \in E} p_n.$$

c) As a special case of part (b), suppose that Ω is finite and that each possible outcome is equally likely. Then we must have $p_n = 1/N(\Omega)$ for $n = 1, 2, \ldots, N(\Omega)$ and, moreover,

$$P(E) = \frac{N(E)}{N(\Omega)}$$

for each event E. This probability model is often referred to as the **discrete uniform model.** It can be used as the mathematical model for selecting a point at random from the finite set Ω.

d) Suppose that Ω is a bounded Lebesgue measurable subset of $\mathcal{R}^n$ with positive Lebesgue measure, and let $\mathcal{A} = \{\Omega \cap M : M \in \mathcal{M}_n\}$. For $E \in \mathcal{A}$, define $P(E) = \lambda_n(E)/\lambda_n(\Omega)$. Then $(\Omega, \mathcal{A}, P)$ is a probability space. This probability model is often referred to as the **continuous uniform model.** It can be used as the mathematical model for selecting a point at random from the set Ω. □

Because a probability space is, in particular, a finite measure space, we can immediately infer for probability spaces any properties of finite measure spaces. For future reference, we list some of the more important properties of probability measures in the following proposition.

□ □ □ **PROPOSITION 7.1**

Suppose that $(\Omega, \mathcal{A}, P)$ is a probability space and that A, B, and E are events, that is, $A, B, E \in \mathcal{A}$. Then the following properties hold:

a) If $A \subset B$, then $P(B \setminus A) = P(B) - P(A)$.

b) $P(E^c) = 1 - P(E)$.

c) $A \subset B \Rightarrow P(A) \leq P(B)$.

d) $0 \leq P(E) \leq 1$.

e) $P(A \cup B) = P(A) + P(B) - P(A \cap B)$.

f) If $\{E_n\}_{n=1}^{\infty} \subset \mathcal{A}$ with $E_1 \supset E_2 \supset \cdots$, then

$$P\left(\bigcap_{n=1}^{\infty} E_n\right) = \lim_{n \to \infty} P(E_n).$$

g) If $\{E_n\}_{n=1}^{\infty} \subset \mathcal{A}$ with $E_1 \subset E_2 \subset \cdots$, then

$$P\left(\bigcup_{n=1}^{\infty} E_n\right) = \lim_{n \to \infty} P(E_n).$$

h) If $\{E_n\}_n \subset \mathcal{A}$, then

$$P\left(\bigcup_n E_n\right) \leq \sum_n P(E_n).$$

In probability, this property is called **Boole's inequality**.

Conditional Probability

Frequently, we need to obtain the probability of an event B under the condition that another event A has occurred. For instance, consider the experiment of selecting an adult American at random. We might be interested in the probability that the person selected is a Democrat (event B). But we also might want to know the probability that the person selected is a Democrat, assuming that the person selected is a female (event A). The former probability, as we know, is denoted by $P(B)$. On the other hand, the latter probability is denoted by $P(B \mid A)$, read "the probability of B given A," and is called the *conditional probability* of event B given that event A has occurred.

More generally, we can refer to the relative-frequency interpretation of probability in order to obtain a formal definition of conditional probability; that is, a definition in terms of the original probability space $(\Omega, \mathcal{A}, P)$. So, assume that the experiment is repeated a large number n of times. Let E be an event with nonzero probability. Given that event E occurs, an event F will occur if and only if event $E \cap F$ occurs. Consequently, in the n repetitions of the experiment, the relative frequency of occurrence of event F among those times in which event E has occurred equals $n(E \cap F)/n(E)$. But, by (7.1),

$$\frac{n(E \cap F)}{n(E)} = \frac{n(E \cap F)/n}{n(E)/n} \approx \frac{P(E \cap F)}{P(E)}.$$

Therefore, we make the following definition:

DEFINITION 7.2 Conditional Probability

Let $(\Omega, \mathcal{A}, P)$ be a probability space and $E \in \mathcal{A}$ with $P(E) > 0$. Then, for $F \in \mathcal{A}$, the **conditional probability** of event F given that event E has occurred is defined by

$$P(F \mid E) = \frac{P(E \cap F)}{P(E)}.$$

EXAMPLE 7.3 *Illustrates Definition 7.2*

Refer to Examples 7.1(a) and 7.2(a). Suppose that a balanced coin is tossed three times. Let B denote the event that a total of two heads are tossed and A denote the event that the first toss is a head. We have $A = \{\text{HHH, HHT, HTH, HTT}\}$ and $B = \{\text{HHT, HTH, THH}\}$. Consequently, the conditional probability of event B given that event A has occurred is

$$P(B \mid A) = \frac{P(A \cap B)}{P(A)} = \frac{P(\{\text{HHT, HTH}\})}{P(\{\text{HHH, HHT, HTH, HTT}\})} = \frac{\frac{2}{8}}{\frac{4}{8}} = 0.5.$$

Observe that the (unconditional) probability of event B is

$$P(B) = \frac{3}{8} = 0.375.$$

Hence, the information that event A has occurred affects the probability that event B occurs. □

The next proposition, whose proof we leave as an exercise for the reader, shows that, for fixed E, the set function $P(\cdot \mid E)$ is a probability measure on $\mathcal{A}$. That probability measure provides the likelihood of events under the condition that event E has occurred.

□ □ □ **PROPOSITION 7.2**

Let $(\Omega, \mathcal{A}, P)$ be a probability space and $E \subset \mathcal{A}$ with $P(E) > 0$. Define P_E on $\mathcal{A}$ by $P_E(A) = P(A \mid E)$. Then P_E is a probability measure.

Independent Events

Next we will define *independence* for events. Intuitively, event F is independent of event E if the occurrence or nonoccurrence of event E does not affect the probability of F; that is, if $P(F \mid E) = P(F)$. In view of Definition 7.2, this is equivalent to the condition that $P(E \cap F)/P(E) = P(F)$. Clearing fractions yields the equation $P(E \cap F) = P(E)P(F)$. This last equation has the advantages of symmetry and not requiring the event E to have positive probability. Hence, we make the following definition.

DEFINITION 7.3 Independent Events

Two events E and F are said to be **independent**[†] if

$$P(E \cap F) = P(E)P(F).$$

If E and F are not independent, then they are called **dependent.**

[†] The terms **statistically independent, stochastically independent,** and **probabilistically independent** are also used.

EXAMPLE 7.4 *Illustrates Definition 7.3*

Refer to Example 7.3. Suppose that a balanced coin is tossed three times. Let A denote the event that the first toss is a head, let B denote the event that a total of two heads are tossed, and let C denote the event that the last two tosses are heads. We have $P(A) = 4/8 = 0.5$, $P(B) = 3/8 = 0.375$, $P(C) = 2/8 = 0.25$, $P(A \cap B) = 2/8 = 0.25$, and $P(A \cap C) = 1/8 = 0.125$. It follows that $P(A \cap B) \neq P(A)P(B)$ and $P(A \cap C) = P(A)P(C)$. Thus, events A and B are dependent, whereas events A and C are independent. □

We have defined independence for two events. For more than two events, we must be careful to distinguish between two types of independence, *pairwise independence* and *mutual independence*. Events A_1, A_2, ..., A_n are said to be **pairwise independent** if, for $i \neq j$, A_i and A_j are independent in the sense of Definition 7.3. In probability theory, however, the concept of mutual independence plays a more prominent role.

DEFINITION 7.4   Mutually Independent Events

Let $(\Omega, \mathcal{A}, P)$ be a probability space. Events A_1, A_2, ..., A_n are said to be **mutually independent** if for each subset $\{i_1, i_2, \ldots, i_m\}$ of $\{1, 2, \ldots, n\}$, we have

$$P(A_{i_1} \cap A_{i_2} \cap \cdots \cap A_{i_m}) = P(A_{i_1})P(A_{i_2}) \cdots P(A_{i_m}).$$

The events of an arbitrary (not necessarily finite) collection are called *mutually independent* if every finite number of them are mutually independent.

Note: Although mutually independent events are pairwise independent, the converse is not true. See Exercise 7.18(a).

One advantage of mutual independence over pairwise independence is that, with mutual independence, events formed by set operations on disjoint subcollections are also mutually independent. For example, if E, F, and G are mutually independent events, then $E \cup F$ and G are also independent events.

The following theorem plays a crucial role in many probabilistic arguments. In interpreting the theorem, observe that for a sequence of events $\{A_n\}_{n=1}^{\infty}$, the event $\bigcap_{n=1}^{\infty}(\bigcup_{k=n}^{\infty} A_k)$ occurs if and only if infinitely many of the A_ns occur. Note that, as a set, this event is the same as the limit superior of the A_ns, as defined in Exercise 1.8 on page 9. In probability, it is often denoted $\{A_n \text{ i.o.}\}$, where i.o. is an abbreviation of "infinitely often."

□ □ □   THEOREM 7.1 Borel-Cantelli Lemma

Suppose that $(\Omega, \mathcal{A}, P)$ is a probability space and that $\{A_n\}_{n=1}^{\infty} \subset \mathcal{A}$.
a) If $\sum_{n=1}^{\infty} P(A_n) < \infty$, then

$$P\left(\bigcap_{n=1}^{\infty}\left(\bigcup_{k=n}^{\infty} A_k\right)\right) = 0.$$

b) If A_1, A_2, ... are mutually independent and $\sum_{n=1}^{\infty} P(A_n) = \infty$, then

$$P\left(\bigcap_{n=1}^{\infty}\left(\bigcup_{k=n}^{\infty} A_k\right)\right) = 1.$$

PROOF For convenience, set $E_n = \bigcup_{k=n}^{\infty} A_k$.

a) We have $E_1 \supset E_2 \supset \cdots$ and $\bigcap_{n=1}^{\infty} E_n = \bigcap_{n=1}^{\infty}\left(\bigcup_{k=n}^{\infty} A_k\right)$. Applying Proposition 7.1(f) and Boole's inequality, we obtain that

$$P\left(\bigcap_{n=1}^{\infty} E_n\right) = \lim_{n\to\infty} P(E_n) \leq \lim_{n\to\infty} \sum_{k=n}^{\infty} P(A_k) = 0,$$

where the last equation holds because $\sum_{n=1}^{\infty} P(A_n) < \infty$.

b) Here we will use the fact that, for $x \geq 0$, $e^{-x} \geq 1 - x$. Let $n \in \mathcal{N}$ be fixed but arbitrary. Applying Proposition 7.1(f) to the sequence of events $\bigcap_{k=n}^{m} A_k^c$ for $m = n, n+1, \ldots$, and using Exercise 7.20(b), we get that

$$P(E_n^c) = P\left(\bigcap_{k=n}^{\infty} A_k^c\right) = \lim_{m\to\infty} P\left(\bigcap_{k=n}^{m} A_k^c\right)$$

$$= \lim_{m\to\infty} \prod_{k=n}^{m} P(A_k^c) = \lim_{m\to\infty} \prod_{k=n}^{m} \left[1 - P(A_k)\right]$$

$$\leq \lim_{m\to\infty} \prod_{k=n}^{m} e^{-P(A_k)} = \lim_{m\to\infty} \exp\left[-\sum_{k=n}^{m} P(A_k)\right] = 0,$$

where the last equality holds since $\sum_{k=n}^{\infty} P(A_k) = \infty$. Consequently, for each $n \in \mathcal{N}$, $P(E_n) = 1$. The required result now follows easily. ■

Exercises for Section 7.1

Note: A ★ denotes an exercise that will be subsequently referenced.

7.1 Marilyn vos Savant published a column in *Parade* magazine. A variation of the following problem appeared in her column and caused tremendous controversy among the mathematical community: On a game show, there are three doors behind which there is one prize each. Two of the prizes are worthless and one is valuable. A contestant selects one of the doors following which the game-show host, who knows where the valuable prize lies, opens one of the remaining two doors to reveal a worthless prize. The host then offers the contestant the opportunity to change his selection. Should he switch? *Hint:* Use the relative-frequency interpretation of probability.

7.2 Refer to Example 7.2 on page 229. Provide the details for parts (a)–(d) of that example.

7.3 Suppose that $(\Gamma, \mathcal{S}, \mu)$ is a measure space and that $\Omega \in \mathcal{S}$ is such that $0 < \mu(\Omega) < \infty$. Let $\mathcal{A} = \mathcal{S}_\Omega$ and, for $E \in \mathcal{A}$, define $P(E) = \mu(E)/\mu(\Omega)$. Show that $(\Omega, \mathcal{A}, P)$ is a probability space.

7.4 Refer to Example 7.1(b) on page 227, where $\Omega = [0, \infty)$ and $\mathcal{A} = \mathcal{M}_{[0,\infty)}$. Experience shows that the probability is $1 - e^{-7t}$ that the first patient arrives within t hours of 6:00 PM.

a) Prove that there exists a unique probability measure on $\mathcal{A}$ consistent with the previous sentence.

b) Determine explicitly the probability measure in part (a).

c) Determine the probability that the first patient arrives between 6:15 and 6:30 PM.

7.5 Provide the proof for Proposition 7.1 on pages 229–230. You may cite any theorems from Chapters 5 and 6.

7.6 Let $(\Omega, \mathcal{A}, P)$ be a probability space and $\{A_n\}_n$ a sequence of events with $P(A_n) = 1$ for each n. Prove that $P(\bigcap_n A_n) = 1$.

7.7 Use induction to prove the following generalization of Proposition 7.1(e): If E_1, E_2, ..., E_n are n events, then

$$P(E_1 \cup E_2 \cup \cdots \cup E_n) = \sum_{i=1}^{n} P(E_i) - \sum_{i_1 < i_2} P(E_{i_1} \cap E_{i_2}) + \cdots$$

$$+ (-1)^{k+1} \sum_{i_1 < i_2 < \cdots < i_k} P(E_{i_1} \cap E_{i_2} \cap \cdots \cap E_{i_k})$$

$$+ \cdots + (-1)^{n+1} P(E_1 \cap E_2 \cap \cdots \cap E_n).$$

7.8 Suppose that a coin has probability p of coming up heads, where $0 < p < 1$. Consider the experiment of tossing the coin until a head appears.

a) Determine a sample space for this experiment.

b) Assign probabilities to each of the possible outcomes.

c) Construct a probability space for the experiment.

d) Repeat parts (a)–(c) if $p = 1$.

e) Repeat parts (a)–(c) if $p = 0$.

7.9 Consider the experiment of rolling two balanced dice.

a) Construct a probability space for the experiment.

b) Determine the probability of rolling doubles, that is, of both dice coming up the same number.

c) Use Definition 7.2 to obtain the conditional probability of rolling doubles given that the sum of the dice is four.

d) Solve part (c) without using Definition 7.2 but instead by constructing a new sample space based upon the condition that the sum of the dice is four.

7.10 Suppose that two cards are selected at random from an ordinary deck of 52 playing cards, where the first card selected is not replaced prior to the drawing of the second card.

a) Employ counting techniques to determine the number of possible outcomes of the experiment.

b) Use Definition 7.2 and counting techniques to obtain the conditional probability that the second card selected is a heart given that the first card selected is a heart.

c) Solve part (b) without using Definition 7.2 but instead by constructing a new sample space based on the condition that the first card selected is a heart.

7.11 Refer to Exercise 7.4.

a) Determine the probability that the first patient arrives after 6:15 PM.

b) Determine the (conditional) probability that the first patient arrives after 6:15 PM given that the first arrival occurs after 6:10 PM.

7.12 Prove Proposition 7.2 on page 231.

★7.13 Let $(\Omega, \mathcal{A}, P)$ be a probability space. Suppose that $\{E_n\}_n$ is a sequence of pairwise mutually exclusive events with $\bigcup_n E_n = \Omega$.

a) Prove that, for each event A,

$$P(A) = \sum_n P(E_n \cap A).$$

b) Assuming also that $P(E_n) > 0$ for each n, prove the **law of total probability:** For each event A,

$$P(A) = \sum_n P(A \mid E_n)P(E_n) = \sum_n P_{E_n}(A)P(E_n).$$

c) Assuming also that $P(A) > 0$, prove **Bayes' rule** (named in honor of the 18th century clergyman, Thomas Bayes): For each k,

$$P(E_k \mid A) = \frac{P(A \mid E_k)P(E_k)}{\sum_n P(A \mid E_n)P(E_n)}.$$

7.14 This exercise considers some basic properties of independence.
   a) Show that if events E and F are both mutually exclusive and independent, then either $P(E) = 0$ or $P(F) = 0$. Equivalently, two events with positive probability cannot be both mutually exclusive and independent.
   b) Show that if event E and event F are independent and $E \subset F$, then either $P(E) = 0$ or $P(F) = 1$.

7.15 Refer to Example 7.2(d) on page 229. Take $n = 1$ and $\Omega = [0, 1]$. Suppose that $[a, b]$ is a nonempty, proper subinterval of $[0, 1]$. Determine all subintervals of $[0, 1]$ that are independent of $[a, b]$.

7.16 Suppose that a card is randomly selected from an ordinary deck of 52 playing cards. Let A denote the event that the card selected is a king, B the event that the card selected is a heart, and C the event that the card selected is a face card.
   a) Are events A and B independent?
   b) Are events A and C independent?

7.17 Refer to Example 7.2(d) on page 229.
   a) Let $\Omega = \{ (x, y) \in \mathcal{R}^2 : 0 \le x, y \le 2 \}$. Suppose that a point is selected at random from Ω. Let A denote the event that the x-coordinate of the point selected is at most one and let B denote the event that the y-coordinate of the point selected is at most 0.5. Determine whether A and B are independent events.
   b) Repeat part (a) if $\Omega = \{ (x, y) \in \mathcal{R}^2 : 0 \le y \le x \le 2 \}$.

7.18 Suppose that two balanced dice, one orange and the other black, are rolled. Let

$$
\begin{aligned}
A &= \text{event the orange die comes up even;} \\
B &= \text{event the black die comes up even;} \\
C &= \text{event the sum of the dice is even;} \\
D &= \text{event the orange die comes up 1, 2, or 3;} \\
E &= \text{event the orange die comes up 3, 4, or 5;} \\
F &= \text{event the sum of the dice is 5.}
\end{aligned}
$$

   a) Show that the events A, B, and C are pairwise independent but not mutually independent.
   b) Show that $A \cup B$ and C are dependent events.
   c) Show that $P(D \cap E \cap F) = P(D)P(E)P(F)$ but that D, E, and F are not pairwise independent (and, hence, not mutually independent).

7.19 Prove that if E and F are independent events, then so are E and F^c.

7.20 Suppose that $A_1, A_2, \ldots, A_n$ are mutually independent events.
   a) Prove that $A_1 \cup A_2 \cup \cdots \cup A_{n-1}$ and that A_n are independent events. *Hint:* Use induction.
   b) Prove that $P\left(\bigcap_{k=1}^n A_k^c\right) = \prod_{k=1}^n P(A_k^c)$. *Hint:* Use induction, part (a), and Exercise 7.19.

7.2 RANDOM VARIABLES

When a random experiment is performed, it is often some numerical quantity associated with the outcome that is of interest, rather than the outcome itself. For example, consider the classical (noncasino) game of craps in which two balanced dice are rolled. Each possible outcome of the experiment can be represented as an ordered pair of integers (i, j), where i and j are the number of dots showing on the two dice. But what is of concern here is the sum $i + j$, not the outcome (i, j) itself. Similarly, in studying the relationship between height and weight, we might sample individuals from the population. Here we would be interested in the heights and weights of the individuals selected, not the individuals themselves.

In the first example of the previous paragraph, we have a real-valued function, sum of the two dice, defined on a sample space; and, in the second example, a vector-valued function, (height, weight), defined on a sample space. Traditionally, in probability, real-valued functions on a sample space are called *random variables* and vector-valued functions on a sample space are called *random vectors*. It is also traditional to denote random variables and vectors by uppercase italicized English-alphabet letters near the end of the alphabet.

Random Variables and Their Distributions

For a rigorous development of random variables and random vectors, we need to be more precise. So, suppose that $(\Omega, \mathcal{A}, P)$ is a probability space and that X is a real-valued function on Ω. Usually, we are interested in the probability that X takes on various values (e.g., the probability that X equals two, that X exceeds 7.5). More generally, for each Borel set B, we want to know the probability that the value of X is a member of B; that is, $P(\{\omega : X(\omega) \in B\})$. But, for that probability to exist, $\{\omega : X(\omega) \in B\}$ must be an event. Hence, we make the following definition.

DEFINITION 7.5   Random Variable

Let $(\Omega, \mathcal{A}, P)$ be a probability space. A real-valued function X on Ω is called a **random variable** if $\{\omega : X(\omega) \in B\} \in \mathcal{A}$ for each $B \in \mathcal{B}$.

Remark: From Exercise 5.21 on page 156, we know that a real-valued function f on Ω is $\mathcal{A}$-measurable if and only if $f^{-1}(B) \in \mathcal{A}$ for each $B \in \mathcal{B}$. Thus, we see that random variables are just real-valued $\mathcal{A}$-measurable functions. However, as we mentioned in Section 5.2, the term "random variable" is used for measurable functions in the context of probability spaces, even though the measurability (or nonmeasurability) of a function has nothing at all to do with a measure.

In probability, we ordinarily employ the notation $\{X \in B\}$ in place of the more common notations $X^{-1}(B)$ or $\{\omega : X(\omega) \in B\}$. The reason is that the former notation is more suggestive. Also, for brevity, commas usually replace intersection symbols in probability expressions involving events defined in terms of random variables. For instance, we generally write $P(X \in A, Y \in B)$ instead of $P(\{X \in A\} \cap \{Y \in B\})$.

One of the most important quantities affiliated with a random variable is its *probability distribution*. Roughly speaking, the probability distribution of a random variable describes the probabilities associated with the various values of the random variable. More precisely, we have the following definition.

DEFINITION 7.6 **Probability Distribution**

Let X be a random variable on the probability space $(\Omega, \mathcal{A}, P)$. Then the **probability distribution** of X, denoted μ_X, is the set function on $\mathcal{B}$ defined by $\mu_X(B) = P(X \in B)$.

The proof of the next proposition is left to the reader as Exercise 7.21. The proposition shows that the probability distribution of a random variable is a probability measure.

□ □ □ **PROPOSITION 7.3**

Let X be a random variable on the probability space $(\Omega, \mathcal{A}, P)$. Then μ_X is a probability measure on $\mathcal{B}$.

In the following example, we will present some illustrations of random variables and their probability distributions. The reader should supply the required details of verification.

EXAMPLE 7.5 *Illustrates Definition 7.6*

a) A random variable X is said to be a **discrete random variable** if there is a countable set K such that $P(X \in K) = 1$. For such a random variable, write $K = \{x_n\}_n$. Then the probability distribution of X is given by $\mu_X = \sum_n p_n \delta_{x_n}$, where $p_n = P(X = x_n)$. For a discrete random variable, the function $p_X \colon \mathcal{R} \to [0, 1]$ defined by $p_X(x) = P(X = x)$ is called the **probability mass function (pmf)** of X. Note that p_X is zero on K^c and that $p_X(x_n) = p_n$.

b) Suppose that two balanced dice are rolled. An appropriate probability space is obtained by taking $\Omega = \{(i, j) : i, j = 1, 2, \ldots, 6\}$, $\mathcal{A} = \mathcal{P}(\Omega)$, and $P = \gamma/36$, where γ is counting measure. Let X denote the sum of the dice. Because $P(X \in \{2, 3, \ldots, 12\}) = 1$, we see that X is a discrete random variable. The pmf of X is

$$p_X(x) = \begin{cases} (x-1)/36, & x = 2, 3, \ldots, 7; \\ (13-x)/36, & x = 8, 9, \ldots, 12; \\ 0, & \text{otherwise.} \end{cases}$$

c) A random variable X is said to be an **absolutely continuous random variable** if there is a nonnegative Borel measurable function f such that $\mu_X(B) = \int_B f\,d\lambda$ for all $B \in \mathcal{B}$.[†] For such a random variable, we usually write $f = f_X$ and call f_X the **probability density function (pdf)** of X.

d) Suppose that a number is selected at random from the interval $[0, 1]$ and let X denote the number obtained. Then, for $B \in \mathcal{B}$, we have

$$\mu_X(B) = P(X \in B) = \lambda(B \cap [0, 1]) = \int_B \chi_{[0,1]}\,d\lambda.$$

Hence, X is an absolutely continuous random variable with pdf $f_X = \chi_{[0,1]}$. Such a random variable is said to have the **uniform distribution** on $[0, 1]$.

e) A random variable X is said to be a **continuous random variable** if $P(X = x) = 0$ for all $x \in \mathcal{R}$. Note that if X is a continuous random variable, then $P(X \in K) = 0$ for each countable subset $K \subset \mathcal{R}$; thus, a continuous random variable is not discrete and vice versa. Also, note that an absolutely continuous random variable is a continuous random variable. However, the converse is not true. See Exercise 7.28.

f) There are random variables that are neither discrete nor continuous. See, for instance, Exercise 7.32. □

Closely associated with the probability distribution of a random variable is the *probability distribution function*. We define this term next.

DEFINITION 7.7 **Probability Distribution Function**

Let X be a random variable on the probability space $(\Omega, \mathcal{A}, P)$. Then the **probability distribution function** of X, denoted F_X, is the real-valued function on $\mathcal{R}$ defined by $F_X(x) = P(X \leq x)$.

Remark: From Definitions 7.6 and 7.7, we see immediately that, for a random variable X, the probability distribution and probability distribution function are related by the equation $F_X(x) = \mu_X\big((-\infty, x]\big)$. In other words, the probability distribution function of X is also the distribution function of μ_X, in the sense of Definition 6.4 on page 192.

EXAMPLE 7.6 *Illustrates Definition 7.7*

a) For a discrete random variable, as described in Example 7.5(a), we have $F_X(x) = \sum_{x_n \leq x} p_n = \sum_{t \leq x} p_X(t)$.

b) For an absolutely continuous random variable, as described in Example 7.5(c), we have $F_X(x) = \int_{-\infty}^{x} f_X(t)\,dt$, where, in general, the integral is a Lebesgue integral. □

[†] In elementary probability courses, absolutely continuous random variables are usually referred to simply as continuous random variables. However, as we will see in part (e), to be precise, we need to include the adjective "absolutely."

Clearly, two random variables with the same probability distribution must also have the same probability distribution function. The converse is also true, as the next theorem shows.

□ □ □ **THEOREM 7.2**

Two random variables with the same probability distribution function have the same probability distribution; that is, $F_X = F_Y \Rightarrow \mu_X = \mu_Y$.

PROOF Let $F = F_X = F_Y$. By assumption, both of the finite Borel measures μ_X and μ_Y have F as their distribution function. Therefore, by the uniqueness portion of Theorem 6.3 on page 196, we must have $\mu_X = \mu_Y$. ∎

Given a probability measure μ on the Borel sets, or, equivalently, a distribution function with $F(\infty) = 1$, does there exist a probability space and a random variable defined thereon whose probability distribution is μ? The answer is yes! See Exercise 7.25.

Random Vectors and Their Distributions

Frequently, we are interested in two or more numerical quantities associated with the outcome of a random experiment, for example, the height and weight of a randomly selected individual. This leads to the notion of a *random vector* or, equivalently, two or more random variables considered simultaneously.

To begin our discussion of random vectors, we recall that $\mathcal{B}_n$ denotes the σ-algebra generated by the open sets of $\mathcal{R}^n$ and that the members of $\mathcal{B}_n$ are called **n-dimensional Borel sets.** Exercise 6.77(a) shows that $\mathcal{B}_n = \mathcal{B} \times \cdots \times \mathcal{B}$; in other words, $\mathcal{B}_n$ is also the σ-algebra generated by the n-dimensional Borel rectangles — sets of the form $B_1 \times \cdots \times B_n$, where $B_k \in \mathcal{B}$, $1 \leq k \leq n$. With these facts in mind, we now state and prove Proposition 7.4.

□ □ □ **PROPOSITION 7.4**

Let $X_1, \ldots, X_n$ be n random variables, all defined on the same probability space $(\Omega, \mathcal{A}, P)$. Then $\{ \omega : (X_1(\omega), \ldots, X_n(\omega)) \in B \} \in \mathcal{A}$ for all $B \in \mathcal{B}_n$.

PROOF Let $B_k \in \mathcal{B}$, $1 \leq k \leq n$. Because each X_k is a random variable, $\{X_k \in B_k\} \in \mathcal{A}$ for $1 \leq k \leq n$. Therefore, because $\mathcal{A}$ is a σ-algebra,

$$\{ \omega : (X_1(\omega), \ldots, X_n(\omega)) \in B_1 \times \cdots \times B_n \} = \bigcap_{k=1}^{n} \{X_k \in B_k\} \in \mathcal{A}. \qquad (7.2)$$

Now, let

$$\mathcal{F} = \Big\{ B \in \mathcal{B}_n : \{ \omega : (X_1(\omega), \ldots, X_n(\omega)) \in B \} \in \mathcal{A} \Big\}.$$

Since $\mathcal{B}_n$ and $\mathcal{A}$ are σ-algebras, so is $\mathcal{F}$. Furthermore, by (7.2), $\mathcal{F}$ contains all n-dimensional Borel rectangles. Thus, $\mathcal{F} = \mathcal{B}_n$. ∎

In view of Proposition 7.4, we now make the following definition.

DEFINITION 7.8 Joint Probability Distribution

Let $X_1, \ldots, X_n$ be n random variables, all defined on the same probability space $(\Omega, \mathcal{A}, P)$. Then the **joint probability distribution** of $X_1, \ldots, X_n$, denoted $\mu_{X_1,\ldots,X_n}$, is the set function on $\mathcal{B}_n$ defined by

$$\mu_{X_1,\ldots,X_n}(B) = P\big((X_1, \ldots, X_n) \in B\big), \qquad B \in \mathcal{B}_n.$$

The proof of the following proposition is left to the reader.

□ □ □ **PROPOSITION 7.5**

Let $X_1, \ldots, X_n$ be n random variables, all defined on the same probability space $(\Omega, \mathcal{A}, P)$. Then $\mu_{X_1,\ldots,X_n}$ is a probability measure on $\mathcal{B}_n$.

Here now are some examples of joint probability distributions. The details of verification should be supplied by the reader.

EXAMPLE 7.7 Illustrates Definition 7.8

a) Random variables $X_1, \ldots, X_n$, all defined on the same probability space $(\Omega, \mathcal{A}, P)$, are said to be **jointly discrete** if there is a countable set $K \subset \mathcal{R}^n$ such that $P\big((X_1, \ldots, X_n) \in K\big) = 1$. It is easy to see that if $X_1, \ldots, X_n$ are jointly discrete, then each X_k, $1 \le k \le n$, must be a discrete random variable. The function $p_{X_1,\ldots,X_n} \colon \mathcal{R}^n \to [0, 1]$, defined by

$$p_{X_1,\ldots,X_n}(x_1, \ldots, x_n) = P(X_1 = x_1, \ldots, X_n = x_n),$$

is called the **joint probability mass function (joint pmf)** of $X_1, \ldots, X_n$. In this context, each individual pmf p_{X_k}, $1 \le k \le n$, is called a **marginal probability mass function (marginal pmf)**.

b) Random variables $X_1, \ldots, X_n$, all defined on the same probability space $(\Omega, \mathcal{A}, P)$, are said to be **jointly absolutely continuous** if there is a non-negative $\mathcal{B}_n$-measurable function f on $\mathcal{R}^n$ such that

$$\mu_{X_1,\ldots,X_n}(B) = \int_B f \, d\lambda_n$$

for all $B \in \mathcal{B}_n$. For such random variables, we usually write $f = f_{X_1,\ldots,X_n}$ and call $f_{X_1,\ldots,X_n}$ the **joint probability density function (joint pdf)** of $X_1, \ldots, X_n$. It is not too difficult to show that if $X_1, \ldots, X_n$ are jointly absolutely continuous, then each X_k, $1 \le k \le n$, must be absolutely continuous. In this context, each individual pdf f_{X_k}, $1 \le k \le n$, is called a **marginal probability density function (marginal pdf)**. □

In analyzing jointly distributed random variables, it is useful to generalize the concept of a probability distribution function to apply to several random variables. This generalization is done in Definition 7.9.

DEFINITION 7.9 Joint Probability Distribution Function

Let $X_1, \ldots, X_n$ be n random variables, all defined on the same probability space $(\Omega, \mathcal{A}, P)$. Then their **joint probability distribution function,** denoted $F_{X_1,\ldots,X_n}$, is the real-valued function on $\mathcal{R}^n$ defined by

$$F_{X_1,\ldots,X_n}(x_1,\ldots,x_n) = P(X_1 \leq x_1,\ldots,X_n \leq x_n), \qquad (x_1,\ldots,x_n) \in \mathcal{R}^n.$$

Remark: From Definitions 7.8 and 7.9, we see that the joint probability distribution and joint probability distribution function are related by the equation

$$F_{X_1,\ldots,X_n}(x_1,\ldots,x_n) = \mu_{X_1,\ldots,X_n}\big((-\infty, x_1] \times \cdots \times (-\infty, x_n]\big).$$

By the previous remark, it is clear that if $X_1, \ldots, X_n$ and $Y_1, \ldots, Y_n$ have the same joint probability distribution, then they must also have the same joint probability distribution function. That the converse is also true is an immediate consequence of Exercise 6.80 on page 223.

□ □ □ THEOREM 7.3

Two random vectors with the same joint probability distribution function have the same joint probability distribution; that is,

$$F_{X_1,\ldots,X_n} = F_{Y_1,\ldots,Y_n} \quad \Rightarrow \quad \mu_{X_1,\ldots,X_n} = \mu_{Y_1,\ldots,Y_n}.$$

Given a probability measure μ on $\mathcal{B}_n$, does there exist a probability space and random variables defined thereon whose joint probability distribution is μ? The answer is yes! See Exercise 7.44.

Independent Random Variables

Next we will discuss independence for random variables. Let us begin by considering two random variables. Intuitively, two random variables are independent if knowing the value of one of the variables does not affect the probability distribution of the other random variable.

To be precise, two random variables X and Y, defined on the same probability space, are called **independent** if, for each pair of Borel sets A and B, the events $\{X \in A\}$ and $\{Y \in B\}$ are independent in the sense of Definition 7.3 on page 231; that is, if $P(X \in A, Y \in B) = P(X \in A)P(Y \in B)$. More generally, we have the following definition.

DEFINITION 7.10 Mutually Independent Random Variables

Random variables $X_1, \ldots, X_n$, defined on the same probability space $(\Omega, \mathcal{A}, P)$, are said to be **mutually independent** if

$$P(X_1 \in B_1, \ldots, X_n \in B_n) = P(X_1 \in B_1) \cdots P(X_n \in B_n),$$

for all Borel sets $B_1, \ldots, B_n$. The random variables of an infinite collection are called *mutually independent* if the random variables of each finite subcollection are mutually independent. In other words, if I is an infinite set, then the random variables $\{X_\iota\}_{\iota \in I}$ are *mutually independent* if, for each $n \in \mathcal{N}$ and subset $\{\iota_1, \ldots, \iota_n\} \subset I$, the n random variables $X_{\iota_1}, \ldots, X_{\iota_n}$ are mutually independent.

We can also define *pairwise independence* for random variables: Random variables $\{X_\iota\}_{\iota \in I}$, all defined on the same probability space, are said to be **pairwise independent** if, for each pair of distinct elements $\iota, \jmath \in I$, the random variables X_ι and $X_\jmath$ are independent. It is easy to see that mutually independent random variables are pairwise independent. However, the converse is not true. See Exercise 7.45(b).

EXAMPLE 7.8 *Illustrates Definition 7.10*

Consider the experiment of rolling three balanced dice, say, one orange, one green, and one black. Let X_1, X_2, and X_3 denote the number of dots facing up on the orange, green, and black dice, respectively, and let X_4 denote the sum of the three dice. Then it is clear intuitively that X_1, X_2, and X_3 are mutually independent but that X_1, X_2, X_3, and X_4 are not even pairwise independent. The reader should justify these statements mathematically. ◻

An important property of mutual independence is that functions of disjoint subcollections of mutually independent random variables are also mutually independent. That is, we have the following proposition:

◻ ◻ ◻ **PROPOSITION 7.6**

Suppose that $X_1, \ldots, X_n$ are mutually independent random variables and that $n_j \in \mathcal{N}$, $1 \le j \le k$, with $n_1 < n_2 < \cdots < n_k = n$. Further suppose that f_1 is $\mathcal{B}_{n_1}$-measurable, f_2 is $\mathcal{B}_{n_2-n_1}$-measurable, ..., and f_k is $\mathcal{B}_{n_k-n_{k-1}}$-measurable. Then the random variables

$$f_1(X_1, \ldots, X_{n_1}), \ f_2(X_{n_1+1}, \ldots, X_{n_2}), \ \ldots, \ f_k(X_{n_{k-1}+1}, \ldots, X_{n_k})$$

are mutually independent.

PROOF We will prove the proposition in the special case $n_j = j$, $1 \le j \le k = n$. The general case is left as an exercise for the reader. Let $B_j \in \mathcal{B}$ for $1 \le j \le n$. For the special case, we have

$$\begin{aligned} P\big(f_1(X_1) \in B_1, \ldots, f_n(X_n) \in B_n\big) &= P\big(X_1 \in f_1^{-1}(B_1), \ldots, X_n \in f_n^{-1}(B_n)\big) \\ &= P\big(X_1 \in f_1^{-1}(B_1)\big) \cdots P\big(X_n \in f_n^{-1}(B_n)\big) \\ &= P\big(f_1(X_1) \in B_1\big) \cdots P\big(f_n(X_n) \in B_n\big), \end{aligned}$$

as required. ∎

We will now obtain two equivalent conditions for the mutual independence of random variables. In doing so, we will refer to some of the exercises in Section 6.4.

□ □ □ **THEOREM 7.4**

Suppose that $X_1, \ldots, X_n$ are random variables, all defined on the same proba-bility space $(\Omega, \mathcal{A}, P)$. Then $X_1, \ldots, X_n$ are mutually independent if and only if

$$\mu_{X_1,\ldots,X_n} = \mu_{X_1} \times \cdots \times \mu_{X_n}; \tag{7.3}$$

that is, if and only if the joint probability distribution of $X_1, \ldots, X_n$ is equal to the product measure induced by the n marginal probability distributions.

PROOF Let B_k, for $1 \leq k \leq n$, be any n Borel sets. Suppose first that (7.3) holds. Then

$$P(X_1 \in B_1, \ldots, X_n \in B_n) = \mu_{X_1,\ldots,X_n}\left(\underset{k=1}{\overset{n}{\times}} B_k\right) = \left(\underset{k=1}{\overset{n}{\times}} \mu_{X_k}\right)\left(\underset{k=1}{\overset{n}{\times}} B_k\right)$$

$$= \prod_{k=1}^{n} \mu_{X_k}(B_k) = \prod_{k=1}^{n} P(X_k \in B_k).$$

Hence, $X_1, \ldots, X_n$ are mutually independent.

Conversely, suppose that $X_1, \ldots, X_n$ are mutually independent. Then we have

$$\mu_{X_1,\ldots,X_n}\left(\underset{k=1}{\overset{n}{\times}} B_k\right) = P(X_1 \in B_1, \ldots, X_n \in B_n) = \prod_{k=1}^{n} P(X_k \in B_k)$$

$$= \prod_{k=1}^{n} \mu_{X_k}(B_k) = \left(\underset{k=1}{\overset{n}{\times}} \mu_{X_k}\right)\left(\underset{k=1}{\overset{n}{\times}} B_k\right).$$

Thus, $\mu_{X_1,\ldots,X_n}$ agrees with $\times_{k=1}^{n} \mu_{X_k}$ on n-dimensional Borel rectangles. So, by Exercise 6.77(b) on page 223, $\mu_{X_1,\ldots,X_n} = \times_{k=1}^{n} \mu_{X_k}$. ∎

Our second equivalent condition for mutual independence is, in practice, easier to verify than the one given in Theorem 7.4.

□ □ □ **THEOREM 7.5**

Suppose that $X_1, \ldots, X_n$ are random variables, all defined on the same proba-bility space $(\Omega, \mathcal{A}, P)$. Then $X_1, \ldots, X_n$ are mutually independent if and only if for all $x_1, \ldots, x_n \in \mathcal{R}$,

$$F_{X_1,\ldots,X_n}(x_1,\ldots,x_n) = F_{X_1}(x_1) \cdots F_{X_n}(x_n); \tag{7.4}$$

in other words, if and only if the joint probability distribution function of X_1, $\ldots, X_n$ is equal to the product of the marginal probability distribution functions.

PROOF Let $x_1, \ldots, x_n$ be any n real numbers. Suppose first that $X_1, \ldots, X_n$ are mutually independent. Then

$$F_{X_1,\ldots,X_n}(x_1,\ldots,x_n) = P(X_1 \leq x_1, \ldots, X_n \leq x_n)$$

$$= \prod_{k=1}^{n} P(X_k \leq x_k) = \prod_{k=1}^{n} F_{X_k}(x_k).$$

Hence, (7.4) holds. Conversely, suppose that (7.4) holds. Then we have

$$\mu_{X_1,\ldots,X_n}\left(\underset{k=1}{\overset{n}{\times}} (-\infty, x_k]\right) = \prod_{k=1}^{n} \mu_{X_k}\big((-\infty, x_k]\big).$$

Consequently, by Exercise 6.81 on page 223, $\mu_{X_1,\ldots,X_n} = \underset{k=1}{\overset{n}{\times}}\,\mu_{X_k}$ and, therefore, because of Theorem 7.4, $X_1, \ldots, X_n$ are mutually independent. ∎

We should point out that special equivalent conditions for mutual independence exist for jointly discrete and jointly absolutely continuous random variables. See Exercises 7.51 and 7.53 for details.

Exercises for Section 7.2

7.21 Prove Proposition 7.3 on page 237.

7.22 Provide the details of verification for parts (a), (b), (d), and (e) of Example 7.5 on pages 237–238.

7.23 Let $(\Omega, \mathcal{A}, P)$ be a probability space and X a random variable defined thereon. Respond *True* or *False* to each of the following statements. Justify your answer.
a) If Ω is countable, then X is a discrete random variable.
b) If the range of X is countable, then X is a discrete random variable.
c) If X is a discrete random variable, then the range of X is countable.

7.24 Prove that X is a continuous random variable if and only if its probability distribution function F_X is a continuous function on $\mathcal{R}$. *Hint:* Refer to Exercise 6.75(b).

7.25 Let μ be a probability measure on $\mathcal{B}$. Show that there exists a probability space and a random variable defined thereon whose probability distribution is μ. *Hint:* Define an appropriate random variable on $(\mathcal{R}, \mathcal{B}, \mu)$.

7.26 Refer to Example 7.5(a).
a) Assume that X is a discrete random variable with pmf p_X. Let $\{x_n\}_n$ be a sequence of real numbers such that $P\big(X \in \{x_n\}_n\big) = 1$, and set $p_n = p_X(x_n)$. Prove that $\{p_n\}_n$ is a sequence of nonnegative real numbers whose sum is one.
b) Conversely, suppose that $\{x_n\}_n$ is a sequence of real numbers and that $\{p_n\}_n$ is a sequence of nonnegative real numbers whose sum is one. Define $p(x) = p_n$, if $x = x_n$ for some n, and zero otherwise. Prove that there is a discrete random variable X that has p as its pmf. *Hint:* Employ Exercise 7.25.

7.27 Refer to Example 7.5(c).
a) Assume that X is an absolutely continuous random variable with pdf f_X. Show that $\int_{\mathcal{R}} f_X \, d\lambda = 1$.
b) Conversely, suppose that f is a nonnegative Borel measurable function such that $\int_{\mathcal{R}} f \, d\lambda = 1$. Prove that there is an absolutely continuous random variable X that has f as its pdf. *Hint:* Employ Exercise 7.25.

★7.28 Let ψ be the Cantor function, and define F on $\mathcal{R}$ by

$$F(x) = \begin{cases} 0, & x < 0; \\ \psi(x), & 0 \le x < 1; \\ 1, & x \ge 1. \end{cases}$$

a) Show that F is the probability distribution function of a random variable X.
b) Prove that the random variable X in part (a) is continuous but not absolutely continuous.

★7.29 An absolutely continuous random variable with pdf $f(x) = (2\pi)^{-\frac{1}{2}} e^{-x^2/2}$ is said to have the **standard normal distribution.** Suppose that X has the standard normal distribution and let $Y = X^2$.
a) Obtain the probability distribution function of the random variable Y in terms of that of X.

      b) Show that Y is absolutely continuous and determine its pdf.

      c) Obtain the probability distribution of Y. (This probability distribution is called the **chi-square distribution** with one degree of freedom.)

★7.30   Suppose that a number is selected at random from the interval $[\alpha, \beta]$ and let X denote the number obtained.

      a) Find the probability distribution function of the random variable X.

      b) Show that X is absolutely continuous and determine its pdf.

      c) Determine the probability distribution of X. (This probability distribution is called the **uniform distribution** on $[\alpha, \beta]$.)

7.31   Let X have the uniform distribution on $[0, 1]$ and let $m \in \mathcal{N}$. Define $Y = 1 + [mX]$, where $[x]$ denotes the greatest integer in x. Obtain the pmf of Y.

7.32   Construct an example of a random variable X that is neither discrete nor continuous. *Hint:* Let Y have the uniform distribution on $[-1, 1]$ and set $X = Y^+$.

7.33   Suppose that a point is selected at random from the unit disk, that is, from the set $D = \{ (x, y) : x^2 + y^2 \le 1 \}$. Let R denote the distance from the origin to the point obtained.

      a) Find the probability distribution function of the random variable R.

      b) Show that R is absolutely continuous and determine its pdf.

      c) Determine the probability distribution of R.

★7.34   Refer to Exercise 7.32. Obtain the probability distribution function of X.

7.35   Prove Proposition 7.5 on page 240.

7.36   Refer to Example 7.7(a) on page 240. Write $K = \{\boldsymbol{x}_j\}_j$, where $\boldsymbol{x}_j \in \mathcal{R}^n$ for each j. Determine the joint probability distribution of $X_1, \ldots, X_n$.

7.37   Refer to Example 7.7 on page 240.

      a) Suppose that X and Y are jointly discrete random variables. Show that, individually, X and Y are discrete random variables and determine their (marginal) probability mass functions in terms of the joint pmf.

      b) Suppose that X and Y are jointly absolutely continuous random variables. Show that, individually, X and Y are absolutely continuous random variables and determine their (marginal) probability density functions in terms of the joint pdf.

7.38   Refer to Example 7.7 on page 240. This exercise generalizes the previous one from $n = 2$ to general n.

      a) In Example 7.7(a), show that each X_k must be a discrete random variable and obtain its (marginal) pmf in terms of the joint pmf.

      b) In Example 7.7(b), show that each X_k must be an absolutely continuous random variable and obtain its (marginal) pdf in terms of the joint pdf.

7.39   Respond *True* or *False* to each of the following statements. Justify your answers.

      a) If $X_1, \ldots, X_n$ are discrete random variables, all defined on the same probability space, then they are jointly discrete.

      b) If $X_1, \ldots, X_n$ are absolutely continuous random variables, all defined on the same probability space, then they are jointly absolutely continuous.

7.40   Suppose that $X_1, \ldots, X_n$ are mutually independent, absolutely continuous random variables. Prove that they are jointly absolutely continuous.

★7.41   Suppose that two balanced dice are rolled. Let X and Y be, respectively, the minimum and the maximum of the two numbers observed.

      a) Show that X and Y are jointly discrete.

      b) Determine the joint pmf of X and Y.

      c) Obtain the marginal pmf of X; of Y.

7.42 Suppose that a point is selected at random from the unit square, that is, from the set $S = \{(x, y) : 0 \leq x, y \leq 1\}$. Let X and Y denote, respectively, the x- and y-coordinates of the point obtained.
a) Show that X and Y are jointly absolutely continuous.
b) Determine the joint pdf of X and Y.
c) Obtain the marginal pdf of X; of Y.

★7.43 Repeat the previous exercise if S is replaced by the unit disk, D.

7.44 Let μ be a probability measure on $\mathcal{B}_n$. Show that there exists a probability space and random variables defined thereon whose joint probability distribution is μ.

7.45 This exercise examines the relationship between mutual independence and pairwise independence of random variables.
a) Suppose that $X_1, \ldots, X_n$ are mutually independent random variables. Prove that they are also pairwise independent.
b) Construct an example to show that pairwise independence does not imply mutual independence.

7.46 Provide a detailed verification for all statements made in Example 7.8 on page 242.

7.47 Supply the proof for Proposition 7.6 in the general case.

★7.48 Consider an experiment that has two possible outcomes, say, success s and failure f, with respective probabilities p and $q = 1 - p$. Suppose now that the experiment is repeated independently a finite number of times. Such repetitions are called **Bernoulli trials** in honor of James Bernoulli.
a) Construct a probability space for a sequence of n Bernoulli trials.
b) Let X denote the total number of successes in n Bernoulli trials. Obtain the pmf and probability distribution of the random variable X. (This probability distribution is called the **binomial distribution** with parameters n and p.)

★7.49 Refer to Exercise 7.48. Suppose that, for each $n \in \mathcal{N}$, X_n has a binomial distribution with parameters n and λ/n, where λ is a positive constant.
a) Prove that, for each nonnegative integer k,
$$\lim_{n \to \infty} P(X_n = k) = e^{-\lambda} \frac{\lambda^k}{k!}. \tag{7.5}$$
b) Let p_k denote the quantity on the right-hand side of (7.5). Show that the function defined on $\mathcal{R}$ by $p(x) = p_k$, if $x = k$ for some nonnegative integer k, and zero elsewhere, is the probability mass function of a random variable. (The probability distribution of such a random variable is called the **Poisson distribution** with parameter λ.)

7.50 Consider an experiment with a finite number r of possible outcomes, say, $o_1, \ldots, o_r$, with respective probabilities $p_1, \ldots, p_r$. Suppose now that the experiment is repeated independently a finite number of times. Such repetitions are called **multinomial trials.**
a) Construct a probability space for a sequence of n multinomial trials.
b) For each k, $1 \leq k \leq r$, let X_k denote the total number of times that outcome o_k occurs in the n multinomial trials. Determine the joint pmf and the joint probability distribution of the random variables $X_1, \ldots, X_r$. (This probability distribution is called the **multinomial distribution** with parameters n and $p_1, \ldots, p_r$.)
c) For each k, $1 \leq k \leq r$, determine the (marginal) probability distribution of X_k. *Hint:* Reformulate the model so that each trial has only two possible outcomes.

7.51 Suppose that $X_1, \ldots, X_n$ are jointly discrete random variables. Prove that they are mutually independent if and only if their joint probability mass function is equal to the product of the marginal probability mass functions; that is, if and only if $p_{X_1,\ldots,X_n}(x_1, \ldots, x_n) = p_{X_1}(x_1) \cdots p_{X_n}(x_n)$ for all $x_1, \ldots, x_n \in \mathcal{R}$.

7.52 Let X and Y be the random variables defined in Exercise 7.41. Apply Exercise 7.51 to determine whether X and Y are independent.

7.53 Suppose that $X_1, \ldots, X_n$ are jointly absolutely continuous random variables. Prove that they are mutually independent if and only if the function f defined on $\mathcal{R}^n$ by $f(x_1, \ldots, x_n) = f_{X_1}(x_1) \cdots f_{X_n}(x_n)$ is a joint probability density function for X_1, $\ldots, X_n$.

7.54 Apply Exercise 7.53 to determine whether the random variables X and Y are independent, where X and Y are as in
a) Exercise 7.42.
b) Exercise 7.43.

★7.55 Let X and Y be jointly absolutely continuous random variables with joint pdf given by

$$f_{X,Y}(x,y) = \frac{1}{2\pi\sqrt{1-\rho^2}}\, e^{-(x^2 - 2\rho x y + y^2)/2(1-\rho^2)},$$

where $0 \leq \rho < 1$.
a) Determine the marginal pdf of X and of Y.
b) Show that X and Y are independent if and only if $\rho = 0$.

★7.56 Suppose that X and Y are independent random variables.
a) Prove that $\mu_{X+Y} = \mu_X * \mu_Y$, where $*$ denotes convolution of measures, as defined in Exercise 6.64 on page 221.
b) If X is absolutely continuous, prove that $X + Y$ is absolutely continuous and has pdf $f_{X+Y}(z) = \int_{\mathcal{R}} f_X(z - y)\, d\mu_Y(y)$.
c) If both X and Y are absolutely continuous, prove that $f_{X+Y} = f_X * f_Y$, where $*$ denotes convolution of functions, as defined in Exercise 6.63 on page 221.

7.3 EXPECTATION OF RANDOM VARIABLES

In this section, we will discuss the *expectation of a random variable,* a concept that is central to the theory of probability and its applications. To motivate the formal definition of expectation, we will first provide an interpretation of its meaning. The most common interpretation is the **long-run-average interpretation,** which construes the expectation of a random variable to be the average value of the random variable in a large number of independent observations.

More formally, let X be a random variable on a probability space $(\Omega, \mathcal{A}, P)$, and let $\mathcal{E}(X)$ denote its expectation. For n independent repetitions of the experiment, let $X_1, \ldots, X_n$ represent the n values of the random variable X. The long run average interpretation is that for large n, the average value of X_1, $\ldots, X_n$ will be approximately equal to $\mathcal{E}(X)$:

$$\frac{X_1 + \cdots + X_n}{n} \approx \mathcal{E}(X), \qquad \text{for large } n. \tag{7.6}$$

We will now employ (7.6) to motivate the formal definition of expectation for a simple random variable, that is, a random variable that takes on only finitely many values, say, $x_1, \ldots, x_m$. So, assume that the experiment is repeated independently a large number n of times. Then, in view of the long-run-average interpretation of expectation and the relative-frequency interpretation of probability, we have

$$\mathcal{E}(X) \approx \frac{X_1 + \cdots + X_n}{n} = \frac{1}{n} \sum_{k=1}^{m} x_k \cdot n(\{X = x_k\})$$

$$= \sum_{k=1}^{m} x_k \cdot \frac{n(\{X = x_k\})}{n} \approx \sum_{k=1}^{m} x_k P(X = x_k), \tag{7.7}$$

where, as usual, $n(E)$ denotes the number of times that an event E occurs in n repetitions of the experiment.

Because of (7.7), we see that the expectation of a simple random variable X should be defined by

$$\mathcal{E}(X) = \sum_{k=1}^{m} x_k P(X = x_k), \tag{7.8}$$

where $x_1, \ldots, x_m$ are the possible values of X. But the quantity on the right-hand side of (7.8) is the abstract Lebesgue integral of the simple random variable X over Ω with respect to P. Generalizing now to arbitrary random variables, we make the following definition:

DEFINITION 7.11   **Expectation of a Random Variable**

Let X be a random variable on a probability space $(\Omega, \mathcal{A}, P)$. Then the **expectation** of X, denoted $\mathcal{E}(X)$, is defined by

$$\mathcal{E}(X) = \int_{\Omega} X(\omega) \, dP(\omega), \tag{7.9}$$

provided the integral on the right-hand side exists. If $X \in \mathcal{L}^1(\Omega, \mathcal{A}, P)$, that is, the integral on the right-hand side of (7.9) exists and is finite, then we say that X has **finite expectation**.

Remark: Terms used synonymously for expectation are **mean, expected value,** and **first moment.**

EXAMPLE 7.9   *Illustrates Definition 7.11*

a) Suppose that two balanced dice are rolled and let X denote the sum of the dice. Note that X is a simple random variable, taking on the values 2, 3, ..., 12. And, as we found in Example 7.5(b) on page 237,

$$P(X = k) = \begin{cases} (k-1)/36, & k = 2, 3, \ldots, 7; \\ (13-k)/36, & k = 8, 9, \ldots, 12. \end{cases}$$

Therefore, the expectation of X equals

$$\mathcal{E}(X) = \int_{\Omega} X \, dP = \sum_{k=2}^{12} k P(X = k) = 2 \cdot \frac{1}{36} + 3 \cdot \frac{2}{36} + \cdots + 12 \cdot \frac{1}{36} = 7.$$

b) Suppose that a point is selected at random from the unit disk, that is, from the set $D = \{(x, y) : x^2 + y^2 \le 1\}$. Let R denote the distance from the origin to the point obtained. Referring to Example 7.2(d) on page 229, we see that an appropriate probability space for the experiment is $(\Omega, \mathcal{A}, P)$, where $\Omega = D$, $\mathcal{A} = \{D \cap M : M \in \mathcal{M}_2\}$, and $P(A) = \pi^{-1}\lambda_2(A)$ for $A \in \mathcal{A}$. Referring to Theorem 6.8 on page 217, we have

$$\mathcal{E}(R) = \int_\Omega R\, dP = \frac{1}{\pi} \int_D \sqrt{x^2 + y^2}\, d\lambda_2(x, y) = \frac{1}{\pi} \iint_D \sqrt{x^2 + y^2}\, dx\, dy = \frac{2}{3},$$

where the last equality is easily obtained by using polar coordinates. $\qquad\square$

Since the expectation of a random variable is, by definition, its abstract Lebesgue integral, all properties of abstract Lebesgue integration apply immediately to expectation, for example, linearity, MCT, DCT. On the other hand, because a probability space is a finite measure space with total measure equal to one, expectation has properties that are not shared by abstract Lebesgue integrals on arbitrary measure spaces. For instance, a bounded random variable has finite expectation and the expectation of a constant random variable is equal to the constant.

Expectation in Terms of Probability Distributions

All of the probabilistic information about a random variable X is contained in its probability distribution μ_X. This fact indicates that we should be able to express $\mathcal{E}(X)$, the expectation of X, in terms of μ_X. As the next theorem shows, we can do considerably more.

$\square\ \square\ \square$ THEOREM 7.6

Let X be a random variable on the probability space $(\Omega, \mathcal{A}, P)$. Then, for each Borel-measurable function g on $\mathcal{R}$, we have

$$\mathcal{E}\big(g(X)\big) = \int_\mathcal{R} g(x)\, d\mu_X(x), \tag{7.10}$$

in the sense that if one side exists, then so does the other and they are equal.[†] In particular,

$$\mathcal{E}(X) = \int_\mathcal{R} x\, d\mu_X(x). \tag{7.11}$$

PROOF We employ the bootstrapping technique. Therefore, suppose first that $g = \chi_D$, where $B \in \mathcal{B}$. Then

$$\mathcal{E}\big(g(X)\big) = \mathcal{E}\big(\chi_{\{X \in B\}}\big) = \int_\Omega \chi_{\{X \in B\}}\, dP = P(X \in B)$$

$$= \mu_X(B) = \int_\mathcal{R} \chi_B(x)\, d\mu_X(x) = \int_\mathcal{R} g(x)\, d\mu_X(x).$$

Hence, (7.10) holds for characteristic functions.

[†] Recall that $g(X)$ is another notation for the composition $g \circ X$ of g with X.

Next suppose that g is a nonnegative $\mathcal{B}$-measurable simple function, say, $g = \sum_{k=1}^{m} b_k \chi_{B_k}$. Noting that $g(X) = \sum_{k=1}^{m} b_k \chi_{\{X \in B_k\}}$, we can apply the linearity property of abstract Lebesgue integrals and the result of the previous paragraph to conclude that (7.10) again holds.

Now assume that g is a nonnegative Borel measurable function. Using Proposition 5.7(a) on page 160, we select a nondecreasing sequence of nonnegative $\mathcal{B}$-measurable simple functions $\{s_n\}_{n=1}^{\infty}$ that converges pointwise on $\mathcal{R}$ to g. Then it is easy to see that $\{s_n(X)\}_{n=1}^{\infty}$ is a nondecreasing sequence of nonnegative random variables converging pointwise on Ω to $g(X)$. Consequently, by the MCT (applied twice) and the result of the previous paragraph, we have

$$\mathcal{E}(g(X)) = \lim_{n \to \infty} \mathcal{E}(s_n(X)) = \lim_{n \to \infty} \int_{\mathcal{R}} s_n(x) \, d\mu_X(x) = \int_{\mathcal{R}} g(x) \, d\mu_X(x).$$

The remainder of the proof proceeds in the usual way and is left as an exercise for the reader. ∎

It is important to note that Theorem 7.6 provides us with two methods for obtaining the expectation of a function of a random variable. Specifically, suppose that Y is a random variable and that h is a Borel measurable function. Applying (7.10) with $X = Y$ and $g = h$, we have the formula

$$\mathcal{E}(h(Y)) = \int_{\mathcal{R}} h(x) \, d\mu_Y(x).$$

On the other hand, by using (7.10) with $X = h(Y)$ and g the identity function, we get the formula

$$\mathcal{E}(h(Y)) = \int_{\mathcal{R}} x \, d\mu_{h(Y)}(x).$$

Generally speaking, the first formula is easier to use because it avoids having to determine the probability distribution of $h(Y)$. However, there are cases when the second formula is more efficient.

We should also point out that Theorem 7.6 implies that the expectation of a function of a random variable depends only on the probability distribution of the random variable. In other words, if X and Y have the same probability distribution, then $\mathcal{E}(g(X)) = \mathcal{E}(g(Y))$ for all Borel measurable functions g.

EXAMPLE 7.10   Illustrates Theorem 7.6

a) Refer to Example 7.5(a) on page 237. Suppose that X is a discrete random variable with probability mass function p_X. Let $\{x_n\}_n$ be such that $P(X \in \{x_n\}_n) = 1$ and set $p_n = P(X = x_n)$. Then $\mu_X = \sum_n p_n \delta_{x_n}$ and, hence, by (7.10),

$$\mathcal{E}(g(X)) = \int_{\mathcal{R}} g(x) \, d\mu_X(x) = \sum_n g(x_n) p_n = \sum_x g(x) p_X(x)$$

for each Borel-measurable function g.

b) Refer to Example 7.5(c) on page 238. Suppose that X is an absolutely continuous random variable with probability density function f_X. Then $\mu_X(B) = \int_B f_X \, d\lambda$ and, hence, by (7.10),

$$\mathcal{E}(g(X)) = \int_{\mathcal{R}} g(x) \, d\mu_X(x) = \int_{\mathcal{R}} g(x) f_X(x) \, d\lambda(x)$$

for each Borel-measurable function g.

c) Let X be a random variable on $(\Omega, \mathcal{A}, P)$ and $n \in \mathcal{N}$. If $X^n \in \mathcal{L}^1(P)$, then we say that X has a **finite nth moment** and we define the **nth moment** of X to be $\mathcal{E}(X^n)$. By Theorem 7.6, specifically, (7.10), we have $\mathcal{E}(X^n) = \int_{\mathcal{R}} x^n \, d\mu_X(x)$. It can be shown that if X has a finite nth moment, then it has a finite moment of each order less than n. □

Next we will discuss a generalization of Theorem 7.6 to random vectors. The proof of this generalization is essentially identical to that of Theorem 7.6 and is left as an exercise for the reader.

□ □ □ **THEOREM 7.7**

Let $X_1, \ldots, X_n$ be random variables, all defined on the same probability space $(\Omega, \mathcal{A}, P)$. Then, for each $\mathcal{B}_n$-measurable function g on $\mathcal{R}^n$,

$$\mathcal{E}(g(X_1, \ldots, X_n)) = \int_{\mathcal{R}^n} g(x_1, \ldots, x_n) \, d\mu_{X_1, \ldots, X_n}(x_1, \ldots, x_n),$$

in the sense that if one side exists, then so does the other and the two sides are equal.

We will apply Theorem 7.7 to obtain an important result concerning the expectation of the product of random variables. By the linearity property of the abstract Lebesgue integral, we know that the expectation of the sum of two random variables equals the sum of their expectations. Although, in general, the expectation of the product of two random variables does not equal the product of their expectations, we do have the following result.

□ □ □ **PROPOSITION 7.7**

Suppose that X and Y are independent random variables with finite expectations. Then XY has finite expectation and $\mathcal{E}(XY) = \mathcal{E}(X)\mathcal{E}(Y)$.

PROOF   First note that, on account of Theorem 7.6, we have

$$\int_{\mathcal{R}} \left[\int_{\mathcal{R}} |xy| \, d\mu_Y(y) \right] d\mu_X(x) = \int_{\mathcal{R}} |x| \, d\mu_X(x) \int_{\mathcal{R}} |y| \, d\mu_Y(y)$$
$$= \mathcal{E}(|X|)\mathcal{E}(|Y|) < \infty.$$

Because X and Y are independent, Theorem 7.4 on page 243 implies that $\mu_{X,Y} = \mu_X \times \mu_Y$. Consequently, applying Theorem 7.7, Fubini's theorem, and

Theorem 7.6, we get

$$\mathcal{E}(XY) = \int_{\mathcal{R}^2} xy \, d\mu_{X,Y} = \int_{\mathcal{R}} \left[\int_{\mathcal{R}} xy \, d\mu_Y(y) \right] d\mu_X(x)$$

$$= \int_{\mathcal{R}} x \, d\mu_X(x) \int_{\mathcal{R}} y \, d\mu_Y(y) = \mathcal{E}(X)\mathcal{E}(Y).$$

This completes the proof. ∎

We can generalize Proposition 7.7 to n mutually independent random variables. The proof can be accomplished either by employing the n-dimensional version of Fubini's theorem or by using induction, Proposition 7.6 on page 242, and Proposition 7.7.

□ □ □ COROLLARY 7.1

Suppose that $X_1, \ldots, X_n$ are mutually independent random variables with finite expectations. Then the random variable $\prod_{k=1}^{n} X_k$ also has finite expectation and $\mathcal{E}\left(\prod_{k=1}^{n} X_k\right) = \prod_{k=1}^{n} \mathcal{E}(X_k)$.

Variance of a Random Variable

If X is a random variable, then the expectation of $\left(X - \mathcal{E}(X)\right)^2$ is of particular importance. That quantity is called the *variance* of X.

DEFINITION 7.12   Variance of a Random Variable

Let X be a random variable with finite expectation. Then we define the **variance** of X, denoted Var(X), by

$$\text{Var}(X) = \mathcal{E}\left(\left(X - \mathcal{E}(X)\right)^2\right).$$

If Var$(X) < \infty$, then X is said to have **finite variance**. The square root of the variance of X is called the **standard deviation** of X.

Note: We leave it as an exercise for the reader to prove that

$$\text{Var}(X) = \mathcal{E}\left(X^2\right) - \left(\mathcal{E}(X)\right)^2.$$

It is often simpler to compute the variance of a random variable by using this latter formula. The formula also makes it clear that X has finite variance if and only if it has a finite second moment.

The variance of a random variable X is a measure of its dispersion relative to the mean because it is the expected value of the square of the distance from X to $\mathcal{E}(X)$. Thus, the smaller the variance, the less likely that X will take a value far from its mean. More precisely, we have the following fact.

□ □ □ PROPOSITION 7.8 Chebyshev's Inequality

Suppose that X is a random variable defined on the probability space $(\Omega, \mathcal{A}, P)$ and that it has finite variance. Then, for each $\epsilon > 0$,

$$P\big(|X - \mathcal{E}(X)| \geq \epsilon\big) \leq \frac{\operatorname{Var}(X)}{\epsilon^2}. \tag{7.12}$$

PROOF   We have

$$\operatorname{Var}(X) = \mathcal{E}\Big(\big(X - \mathcal{E}(X)\big)^2\Big) = \int_\Omega \big(X - \mathcal{E}(X)\big)^2 dP$$

$$\geq \int_{\{|X - \mathcal{E}(X)| \geq \epsilon\}} \big(X - \mathcal{E}(X)\big)^2 dP$$

$$\geq \int_{\{|X - \mathcal{E}(X)| \geq \epsilon\}} \epsilon^2 dP = \epsilon^2 P\big(|X - \mathcal{E}(X)| \geq \epsilon\big),$$

as required.   ∎

Note: It is trivial that Chebyshev's inequality also holds for random variables with finite expectation and infinite variance, but it is of little value in that case.

Although Chebyshev's inequality is quite easy to prove, it is, nonetheless, indispensable as a tool in probability theory. The importance of Chebyshev's inequality is due to its universality — it holds for every random variable with finite variance. And despite the fact that (7.12) will usually not be sharp, it is the best that can be said in general. See Exercise 7.73.

Variance of a Sum

Many probabilistic arguments require the analysis of the variance of a sum of random variables. We will see this fact, for instance, in the next section when we discuss laws of large numbers. To begin, it will be useful to make the following definition.

DEFINITION 7.13   Covariance of Two Random Variables

Suppose that X and Y have finite variances and are defined on the same probability space. Then the **covariance** of X and Y, denoted $\operatorname{Cov}(X, Y)$, is defined by

$$\operatorname{Cov}(X, Y) = \mathcal{E}\Big(\big(X - \mathcal{E}(X)\big)\big(Y - \mathcal{E}(Y)\big)\Big).$$

The finite-variance assumption, stated in Definition 7.13, assures the existence of $\mathcal{E}(XY)$. This fact is a consequence of Cauchy's inequality, which will be proved in a more general setting in Section 13.2 (see Theorem 13.1).

Note that $\operatorname{Cov}(X, X) = \operatorname{Var}(X)$. Also, it follows easily from properties of expectation that

$$\operatorname{Cov}(X, Y) = \mathcal{E}(XY) - \mathcal{E}(X)\mathcal{E}(Y).$$

We now present a formula for the variance of the sum of a finite number of random variables.

◻ ◻ ◻ **PROPOSITION 7.9**

Suppose that $X_1, \ldots, X_n$ have finite variances and are defined on the same probability space. Then $X_1 + \cdots + X_n$ has finite variance and

$$\operatorname{Var}\left(\sum_{k=1}^{n} X_k\right) = \sum_{k=1}^{n} \operatorname{Var}(X_k) + 2\sum_{i<j} \operatorname{Cov}(X_i, X_j). \qquad (7.13)$$

PROOF We have

$$\operatorname{Var}\left(\sum_{k=1}^{n} X_k\right) = \mathcal{E}\left(\left(\sum_{k=1}^{n} X_k - \mathcal{E}\left(\sum_{k=1}^{n} X_k\right)\right)^2\right) = \mathcal{E}\left(\left(\sum_{k=1}^{n}(X_k - \mathcal{E}(X_k))\right)^2\right)$$

$$= \mathcal{E}\left(\sum_{k=1}^{n}(X_k - \mathcal{E}(X_k))^2 + 2\sum_{i<j}(X_i - \mathcal{E}(X_i))(X_j - \mathcal{E}(X_j))\right)$$

$$= \sum_{k=1}^{n} \operatorname{Var}(X_k) + 2\sum_{i<j} \mathcal{E}\Big((X_i - \mathcal{E}(X_i))(X_j - \mathcal{E}(X_j))\Big),$$

as required. ∎

From (7.13), we see that a significant simplification will occur in the formula for the variance of the sum of random variables if the covariances are all zero. This leads to the following definition and corollaries:

DEFINITION 7.14 **Uncorrelated Random Variables**

Suppose X and Y have finite variances and are defined on the same probability space. Then they are said to be **uncorrelated** if $\operatorname{Cov}(X, Y) = 0$. Random variables $\{X_\iota\}_{\iota \in I}$ are called *uncorrelated* if, for each pair of distinct elements $\iota, \jmath \in I$, the two random variables X_ι and $X_\jmath$ are uncorrelated.

Note: It follows immediately from Proposition 7.7 on page 251 that two independent random variables with finite variances are uncorrelated. The converse, however, is not true. See Exercise 7.80.

◻ ◻ ◻ **COROLLARY 7.2**

If $X_1, \ldots, X_n$ are uncorrelated random variables, then

$$\operatorname{Var}\left(\sum_{k=1}^{n} X_k\right) = \sum_{k=1}^{n} \operatorname{Var}(X_k).$$

◻ ◻ ◻ **COROLLARY 7.3**

If $X_1, \ldots, X_n$ are pairwise independent random variables with finite variances, then

$$\operatorname{Var}\left(\sum_{k=1}^{n} X_k\right) = \sum_{k=1}^{n} \operatorname{Var}(X_k).$$

In particular, the previous equation holds for mutually independent random variables with finite variances.

Exercises for Section 7.3

7.57 Let Ω be a finite set, P a probability measure on $\mathcal{P}(\Omega)$, and X a random variable on $(\Omega, \mathcal{P}(\Omega), P)$. Show that $\mathcal{E}(X) = \sum_\omega X(\omega) P(\{\omega\})$, so that $\mathcal{E}(X)$ is a weighted average of the values of X, weighted by probabilities.

7.58 Suppose that a balanced coin is tossed three times. If X denotes the total number of times the coin comes up heads, determine the expectation of the random variable X.

7.59 Suppose that a point is selected at random from the unit square, that is, from the set $S = \{(x, y) : 0 \le x, y \le 1\}$. Let U denote the larger of the two coordinates of the point obtained. Compute the expectation of the random variable U.

7.60 Provide a detailed verification for parts (a) and (b) of Example 7.10.

7.61 Find the first two moments for a random variable that has the
a) uniform distribution on $[\alpha, \beta]$ (refer to Exercise 7.30 on page 245).
b) standard normal distribution (refer to Exercise 7.29 on page 244).

7.62 Let $n \in \mathcal{N}$. Construct a random variable having a finite nth moment but no finite moment of any higher order.

7.63 Suppose X is a random variable with finite nth moment. Prove that X has a finite mth moment for all nonnegative integers $m < n$.

★7.64 Suppose that X is a nonnegative random variable and that $n \in \mathcal{N}$.
a) Prove that

$$\mathcal{E}(X^n) = n \int_0^\infty x^{n-1} P(X > x) \, dx.$$

Hint: Express x^n as an integral and apply Tonelli's theorem.
b) If, in addition, X is nonnegative-integer valued, deduce from part (a) that

$$\mathcal{E}(X^n) = \sum_{k=1}^\infty \left(k^n - (k-1)^n \right) P(X \ge k).$$

7.65 Prove Theorem 7.7 on page 251.

7.66 Show that, in general, the expectation of the product of two random variables is not equal to the product of their expectations.

7.67 Prove Corollary 7.1 on page 252.

7.68 Suppose that a point is selected at random from the unit ball, that is, from the set $B = \{(x, y, z) : x^2 + y^2 + z^2 \le 1\}$. Let X, Y, Z, and R denote, respectively, the x-coordinate, y-coordinate, z-coordinate, and distance to the origin of the point obtained.
a) Determine $\mathcal{E}(R)$ by employing Theorem 7.7.
b) Determine $\mathcal{E}(R)$ by first finding the probability distribution of R and then applying (7.11).

★7.69 This exercise examines some basic properties of variance. Let $c \in \mathcal{R}$ and let X be a random variable with finite expectation. Prove that
a) $\mathrm{Var}(X) = \mathcal{E}(X^2) - \left(\mathcal{E}(X) \right)^2$.
b) $\mathrm{Var}(cX) = c^2 \mathrm{Var}(X)$.
c) $\mathrm{Var}(c + X) = \mathrm{Var}(X)$.
d) $\mathrm{Var}(X) = 0$ if and only if X is constant P-ae.

7.70 Let Y have the uniform distribution on $[-1, 1]$ and set $X = Y^+$. Obtain the mean and standard deviation of X. Refer to Exercise 7.34 on page 245.

7.71 Refer to Exercise 7.48 on page 246. Let X have the binomial distribution with parameters n and p. Determine the mean and variance of X.

7.72 Refer to Exercise 7.49 on page 246. Let X have the Poisson distribution with parameter λ. Determine the mean and variance of X.

7.73 Construct an example where equality holds in Chebyshev's inequality for some $\epsilon > 0$.

7.74 The following result, known as **Markov's inequality,** is due to the Russian mathematician Andrey Andreyevich Markov (1856–1922): Suppose that X is a nonnegative random variable on the probability space $(\Omega, \mathcal{A}, P)$. Then, for each $\epsilon > 0$, we have $P(X \geq \epsilon) \leq \mathcal{E}(X)/\epsilon$.
a) Prove Markov's inequality.
b) Deduce Chebyshev's inequality from Markov's inequality.

★7.75 Let X be a random variable on $(\Omega, \mathcal{A}, P)$ and suppose that ϕ is a function on $\mathcal{R}$ that is positive, increasing on $(0, \infty)$, and satisfies $\phi(-x) = \phi(x)$. Prove that, for each $\epsilon > 0$, $P(|X| \geq \epsilon) \leq \mathcal{E}\big(\phi(X)\big)/\phi(\epsilon)$.

7.76 This exercise investigates some basic properties of covariance. All random variables are assumed to have finite variance and to be defined on the same probability space.
a) Show that $\text{Cov}(X, Y) = \mathcal{E}(XY) - \mathcal{E}(X)\mathcal{E}(Y)$.
b) Let a_i, $1 \leq i \leq m$, and b_j, $1 \leq j \leq n$, be sequences of real numbers. Prove that

$$\text{Cov}\left(\sum_{i=1}^{m} a_i X_i, \sum_{j=1}^{n} b_j Y_j\right) = \sum_{i=1}^{m} \sum_{j=1}^{n} a_i b_j \text{Cov}(X_i, Y_j).$$

This result is called the **bilinearity** property of covariance.

7.77 Suppose that two balanced dice are rolled. Let X and Y denote, respectively, the minimum and maximum of the two numbers observed. Determine $\text{Cov}(X, Y)$. *Note:* Refer to Exercise 7.41 on page 245.

7.78 Obtain the covariance of the two random variables in each part that follows:
a) Suppose that a point is selected at random from the unit square, that is, from the set $S = \{(x, y) : 0 \leq x, y \leq 1\}$. Let X and Y denote, respectively, the x- and y-coordinates of the point obtained.
b) Repeat part (a) if S is replaced by the unit disk D. *Note:* Refer to Exercise 7.43 on page 246.

7.79 Refer to Exercise 7.55 on page 247. Determine $\text{Cov}(X, Y)$.

7.80 Let Θ be uniformly distributed on $[0, 2\pi]$. Define $X = \cos\Theta$ and $Y = \sin\Theta$. Show that X and Y are uncorrelated but not independent.

7.81 Redo Exercise 7.71 using the following steps: For $1 \leq k \leq n$, let X_k be 1 or 0 according as the kth trial results in success or failure.
a) Obtain $\mathcal{E}(X_k)$ and $\text{Var}(X_k)$.
b) Explain why $X_1, \ldots, X_n$ are independent. (No work is required here!)
c) Explain why $X = X_1 + \cdots + X_n$.
d) Use parts (a)–(c) to find the mean and variance of X. Compare the work done here with that in Exercise 7.71.

7.82 Suppose $X_1, \ldots, X_n$ are mutually independent random variables with identical probability distributions (such random variables are said to be **iid,** short for "independent and identically distributed"). Further suppose those random variables have finite variance and denote by μ and σ^2 their common mean and variance, respectively. Set

$$\overline{X}_n = (X_1 + \cdots + X_n)/n.$$

Prove that
a) $\mathcal{E}(\overline{X}_n) = \mu$. Did you use independence here?

b) $\mathrm{Var}(\overline{X}_n) = \sigma^2/n$.

c) $\mathcal{E}\left(\sum_{k=1}^n (X_k - \overline{X}_n)^2\right) = (n-1)\sigma^2$.

7.83 For a finite numerical population $\{a_1, \ldots, a_N\}$, the **population mean** μ and **population standard deviation** σ are defined by

$$\mu = \frac{1}{N}\sum_{i=1}^N a_i \quad \text{and} \quad \sigma = \sqrt{\frac{1}{N}\sum_{i=1}^N (a_i - \mu)^2}.$$

Suppose that n members of the population are selected at random, where we assume that $n \le N$ if the sampling is done without replacement. Denote by X_k, $1 \le k \le n$, the value of the kth member obtained. Set

$$\overline{X}_n = (X_1 + \cdots + X_n)/n.$$

Prove that

a) $\mathcal{E}(X_k) = \mu$ and $\mathrm{Var}(X_k) = \sigma^2$, $1 \le k \le n$. *Hint:* The value of X_k is equally likely to be any of the N population values.

b) $\mathcal{E}(\overline{X}_n) = \mu$.

c) $\mathrm{Var}(\overline{X}_n) = \sigma^2/n$, if the sampling is with replacement.

d) $\mathrm{Var}(\overline{X}_n) = \dfrac{N-n}{N-1} \cdot \sigma^2/n$, if the sampling is without replacement.

★**7.84** Let $(\Omega, \mathcal{A}, P)$ be a probability space and $E \in \mathcal{A}$ with $P(E) > 0$. For $A \in \mathcal{A}$, define $P_E(A) = P(A \,|\, E)$. By Proposition 7.2 (see page 231), P_E is a probability measure on $\mathcal{A}$. Hence, we can define expectation with respect to P_E. This is called the **conditional expectation** relative to E. Thus, by definition, the conditional expectation relative to E of a random variable X, denoted by $\mathcal{E}(X \,|\, E)$ or $\mathcal{E}_E(X)$, is

$$\mathcal{E}(X \,|\, E) = \mathcal{E}_E(X) = \int_\Omega X(\omega)\, dP_E(\omega),$$

provided the right-hand side exists.

a) Prove that $\mathcal{E}_E(X) = \int_E X(\omega)\, dP(\omega)/P(E)$.

b) Use part (a) to interpret conditional expectation.

c) Suppose that $\{E_n\}_n$ are pairwise mutually exclusive events with positive probability and that satisfy $\Omega = \bigcup_n E_n$. Further suppose that X is a random variable with finite expectation. Prove that $\mathcal{E}_{E_n}(X)$ exists and is finite for each n and that

$$\mathcal{E}(X) = \sum_n \mathcal{E}(X \,|\, E_n)P(E_n) = \sum_n \mathcal{E}_{E_n}(X)P(E_n).$$

This result is called the **law of total expectation.**

d) Interpret the previous equation in words.

e) Compare the law of total expectation with the law of total probability (Exercise 7.13(b) on pages 234–235). Precisely how are they related?

7.4 THE LAW OF LARGE NUMBERS

At the beginning of Section 7.3, we introduced the long-run-average interpretation of expectation, (7.6) on page 247, in order to motivate the formal definition of the expectation of a random variable. Now that we have made that formal definition and established some basic properties of expectation, it is natural to ask whether we can prove a mathematically precise version of (7.6) as a theorem.

So, let X be a random variable associated with some random experiment. Suppose that the experiment is repeated independently and let X_k represent the value of X on the kth trial. More precisely, we assume that X_1, X_2, ... are mutually independent random variables, all having the same probability distribution as X.[†] Then we want to prove that, in some sense,

$$\frac{X_1 + \cdots + X_n}{n} \to \mathcal{E}(X), \tag{7.14}$$

as $n \to \infty$. The question now is, in what sense do we take the convergence in (7.14)? Naively, we might want the convergence to be pointwise. But that is too much to expect, as the following example shows.

EXAMPLE 7.11 *Illustrates (7.14)*

Consider the experiment of tossing a balanced coin once. Let $X = 1$ or 0 according to whether the coin comes up a head or a tail. As the coin is balanced, $\mathcal{E}(X) = 0 \cdot (1/2) + 1 \cdot (1/2) = 1/2$. Here, repeating the experiment independently means tossing the coin over and over again. Also, $X_k = 1$ or 0 according to whether the kth toss is a head or a tail. In this context, (7.14) becomes

$$\frac{X_1 + \cdots + X_n}{n} \to \frac{1}{2}, \tag{7.15}$$

as $n \to \infty$, which says simply that, in the long run, the coin comes up heads half of the time. However, it is clear that (7.15) does not hold pointwise, that is, for every possible infinite sequence of heads and tails. For instance, if every toss comes up heads, then the limit is one, while if every toss comes up tails, then the limit is zero. ◻

Example 7.11 shows that it is unreasonable to expect the convergence in (7.14) to be pointwise. As a next best choice, we might try to prove almost-everywhere convergence — and that is exactly what we will do. Before proceeding, we recall from Section 5.1 that, in probability theory, the terms *almost surely, with probability one,* and *almost certainly* are used synonymously for "almost everywhere."

Preliminaries

Several preliminary results will be needed in order to prove (7.14). We begin with the following three lemmas. The proofs of the first two are left to the reader as exercises.

[†] The existence of such random variables is a consequence of the Kolmogorov extension theorem. See, for example, Robert B. Ash's *Real Analysis and Probability* (Cambridge, MA: Academic Press, 1972).

$\square$ $\square$ $\square$ **LEMMA 7.1**

Let $\{c_{mn}\}_{m,n=1}^{\infty}$ be a double sequence of real numbers such that

- for each $n \in \mathcal{N}$, $c_{mn} \to 0$ as $m \to \infty$, and
- the sequence $\left\{ \sum_{n=1}^{\infty} |c_{mn}| \right\}_{m=1}^{\infty}$ is bounded.

For a bounded sequence $\{y_n\}_{n=1}^{\infty}$ of real numbers, let $z_m = \sum_{n=1}^{\infty} c_{mn} y_n$ for each $m \in \mathcal{N}$. Then

a) $y_n \to 0$, as $n \to \infty$, $\Rightarrow z_m \to 0$, as $m \to \infty$.

b) $\sum_{n=1}^{\infty} c_{mn} \to 1$, as $m \to \infty$, and $y_n \to y$, as $n \to \infty$ $\Rightarrow z_m \to y$, as $m \to \infty$.

$\square$ $\square$ $\square$ **LEMMA 7.2 Toeplitz's Lemma**

Suppose that $\sum_{n=1}^{\infty} a_n$ is a divergent series of positive real numbers and that $\{s_n\}_{n=1}^{\infty}$ is a convergent sequence of real numbers. Then

$$\lim_{n \to \infty} \frac{\sum_{k=1}^{n} a_k s_k}{\sum_{k=1}^{n} a_k} = \lim_{n \to \infty} s_n.$$

$\square$ $\square$ $\square$ **LEMMA 7.3 Kronecker's Lemma**

Suppose that $\{b_n\}_{n=1}^{\infty}$ is an increasing sequence of positive real numbers such that $\lim_{n \to \infty} b_n = \infty$. Further suppose that $\sum_{n=1}^{\infty} x_n$ is a convergent series of real numbers. Then

$$\lim_{n \to \infty} \frac{1}{b_n} \sum_{k=1}^{n} b_k x_k = 0.$$

PROOF Define $x = \sum_{n=1}^{\infty} x_n$, $s_0 = 0$ and, for $n \in \mathcal{N}$, $s_n = \sum_{k=1}^{n} x_k$. Also define $b_0 = 0$ and, for $n \in \mathcal{N}$, $a_n = b_n - b_{n-1}$. Using summation by parts (see Exercise 7.87), we get that

$$\sum_{k=1}^{n} b_k x_k = b_n s_n - \sum_{k=1}^{n} s_{k-1} a_k. \tag{7.16}$$

Note that $b_n = \sum_{k=1}^{n} a_k$ and that $s_n \to x$ as $n \to \infty$. Hence, by (7.16) and Toeplitz's lemma,

$$\lim_{n \to \infty} \frac{1}{b_n} \sum_{k=1}^{n} b_k x_k = \lim_{n \to \infty} \left(s_n - \frac{\sum_{k=1}^{n} a_k s_{k-1}}{\sum_{k=1}^{n} a_k} \right) = x - x = 0,$$

as required. $\blacksquare$

Next we will prove two propositions, both due to Kolmogorov.

$\square$ $\square$ $\square$ **PROPOSITION 7.10 Kolmogorov's Inequality**

Let $X_1, \ldots, X_n$ be mutually independent random variables, each with finite variance. Set $S_j = X_1 + \cdots + X_j$, $1 \le j \le n$. Then, for each $\epsilon > 0$,

$$P\left(\max_{1 \le j \le n} |S_j - \mathcal{E}(S_j)| \ge \epsilon \right) \le \frac{\text{Var}(S_n)}{\epsilon^2}.$$

PROOF Without loss of generality, we can assume that $\mathcal{E}(X_j) = 0$ for $j = 1, \ldots, n$ (why?). Let $A = \{\max_{1 \le j \le n} |S_j| \ge \epsilon\}$ and, for $1 \le k \le n$,

$$A_k = \{|S_j| < \epsilon, \, j = 1, \ldots, k-1, \, |S_k| \ge \epsilon\}.$$

Note that the A_ks are mutually exclusive and that $A = \bigcup_{k=1}^n A_k$. Now,

$$\mathrm{Var}(S_n) = \mathcal{E}(S_n^2) = \int_\Omega S_n^2 \, dP \ge \int_A S_n^2 \, dP = \sum_{k=1}^n \int_{A_k} S_n^2 \, dP. \tag{7.17}$$

Let $Y_k = X_{k+1} + \cdots + X_n$. Then $S_n = S_k + Y_k$ and, hence,

$$\int_{A_k} S_n^2 \, dP = \int_{A_k} S_k^2 \, dP + 2 \int_{A_k} S_k Y_k \, dP + \int_{A_k} Y_k^2 \, dP. \tag{7.18}$$

Because $\chi_{A_k} S_k$ is a $\mathcal{B}_k$-measurable function of $X_1, \ldots, X_k$ and Y_k is a $\mathcal{B}_{n-k}$-measurable function of $X_{k+1}, \ldots, X_n$, Proposition 7.6 on page 242 implies that $\chi_{A_k} S_k$ and Y_k are independent random variables. Thus, by Proposition 7.7 on page 251 and the fact that $\mathcal{E}(Y_k) = 0$,

$$\int_{A_k} S_k Y_k \, dP = \int_\Omega \chi_{A_k} S_k Y_k \, dP = \mathcal{E}(\chi_{A_k} S_k \cdot Y_k) = \mathcal{E}(\chi_{A_k} S_k) \mathcal{E}(Y_k) = 0.$$

This last equation and (7.18) imply $\int_{A_k} S_n^2 \, dP \ge \int_{A_k} S_k^2 \, dP \ge \epsilon^2 P(A_k)$. Consequently, by (7.17),

$$\mathrm{Var}(S_n) \ge \epsilon^2 \sum_{k=1}^n P(A_k) = \epsilon^2 P(A).$$

This completes the proof of the proposition. ■

Note: If $n = 1$, Kolmogorov's inequality reduces to Chebyshev's inequality.

□ □ □ PROPOSITION 7.11

Suppose that $X_1, X_2, \ldots$ are mutually independent random variables and that $\sum_{n=1}^\infty \mathrm{Var}(X_n) < \infty$. Then $\sum_{n=1}^\infty \big(X_n - \mathcal{E}(X_n)\big)$ converges P-ae.

PROOF We can assume without loss of generality that $\mathcal{E}(X_n) = 0$ for all $n \in \mathcal{N}$. Set $S_n = \sum_{k=1}^n X_k$. We want to prove that, with probability one, $\{S_n\}_{n=1}^\infty$ converges. First we will show that, for each $\epsilon > 0$,

$$\lim_{m \to \infty} P\left(\bigcup_{k=1}^\infty \{|S_{m+k} - S_m| \ge \epsilon\}\right) = 0. \tag{7.19}$$

By Proposition 7.1(g) on page 230, we have, for each $m \in \mathcal{N}$,

$$P\left(\bigcup_{k=1}^\infty \{|S_{m+k} - S_m| \ge \epsilon\}\right) = \lim_{n \to \infty} P\left(\bigcup_{k=1}^n \{|S_{m+k} - S_m| \ge \epsilon\}\right)$$

$$= \lim_{n \to \infty} P\left(\max_{1 \le k \le n} |S_{m+k} - S_m| \ge \epsilon\right). \tag{7.20}$$

For $1 \leq k \leq n$, let $Y_k = X_{m+k}$ and

$$T_k = \sum_{j=1}^{k} Y_j = \sum_{j=m+1}^{m+k} X_j = S_{m+k} - S_m.$$

As $X_1, X_2, \ldots$ are mutually independent, so are $Y_1, \ldots, Y_n$. Applying Kolmogorov's inequality to the T_ks, we get

$$P\left(\max_{1 \leq k \leq n} |S_{m+k} - S_m| \geq \epsilon\right) \leq \frac{\mathrm{Var}(S_{m+n} - S_m)}{\epsilon^2} = \frac{1}{\epsilon^2} \sum_{k=m+1}^{m+n} \mathrm{Var}(X_k).$$

The previous relation and (7.20) imply that, for each $m \in \mathcal{N}$,

$$P\left(\bigcup_{k=1}^{\infty} \{|S_{m+k} - S_m| \geq \epsilon\}\right) \leq \frac{1}{\epsilon^2} \sum_{k=m+1}^{\infty} \mathrm{Var}(X_k).$$

Letting $m \to \infty$ in this last relation and using $\sum_{k=1}^{\infty} \mathrm{Var}(X_k) < \infty$, we see that (7.19) holds.

Now we can show that $\{S_n\}_{n=1}^{\infty}$ converges with probability one. Let

$$E = \left\{\omega : \{S_n(\omega)\}_{n=1}^{\infty} \text{ does not converge}\right\}.$$

Then $\omega \in E$ if and only if $\{S_n(\omega)\}_{n=1}^{\infty}$ is not a Cauchy sequence, which means that there exists an $r \in \mathcal{N}$ such that, for each $n \in \mathcal{N}$, there is a $k \in \mathcal{N}$ with $|S_{n+k}(\omega) - S_n(\omega)| \geq r^{-1}$. In other words,

$$E = \bigcup_{r=1}^{\infty} \left(\bigcap_{n=1}^{\infty} \left(\bigcup_{k=1}^{\infty} \left\{|S_{n+k} - S_n| \geq \frac{1}{r}\right\}\right)\right). \tag{7.21}$$

But, for each $r \in \mathcal{N}$, we have, by Proposition 7.1(f) and (7.19), that

$$P\left(\bigcap_{n=1}^{\infty}\left(\bigcup_{k=1}^{\infty}\left\{|S_{n+k} - S_n| \geq \frac{1}{r}\right\}\right)\right) = \lim_{m \to \infty} P\left(\bigcap_{n=1}^{m}\left(\bigcup_{k=1}^{\infty}\left\{|S_{n+k} - S_n| \geq \frac{1}{r}\right\}\right)\right)$$

$$\leq \lim_{m \to \infty} P\left(\bigcup_{k=1}^{\infty}\left\{|S_{m+k} - S_m| \geq \frac{1}{r}\right\}\right) = 0.$$

This last fact and (7.21) imply that $P(E) = 0$. ∎

The Strong Law of Large Numbers

Before proving our next theorem, we will introduce some additional terminology. We say that the random variables $X_1, X_2, \ldots$ obey the **strong law of large numbers** if there exists a sequence $\{a_n\}_{n=1}^{\infty}$ of real numbers and a sequence $\{b_n\}_{n=1}^{\infty}$ of positive real numbers tending to infinity such that, with probability one,

$$\lim_{n \to \infty} \frac{X_1 + \cdots + X_n - a_n}{b_n} = 0. \tag{7.22}$$

If the convergence in (7.22) is in probability (i.e., in P-measure), then we say that $X_1, X_2, \ldots$ obey the **weak law of large numbers.** Because a probability space is a finite measure space, Proposition 5.11 on page 175 implies that if a sequence of random variables obeys the strong law of large numbers, then it obeys the weak law of large numbers.

The next result, also due to Kolmogorov, provides a sufficient condition for a sequence of random variables to obey the strong law of large numbers.

□ □ □ THEOREM 7.8 Kolmogorov's Strong Law of Large Numbers

Let $X_1, X_2, \ldots$ be mutually independent random variables with finite variances and set $S_n = X_1 + \cdots + X_n$. Suppose that $\{b_n\}_{n=1}^{\infty}$ is an increasing sequence of positive real numbers satisfying $\lim_{n \to \infty} b_n = \infty$ and

$$\sum_{n=1}^{\infty} \frac{\mathrm{Var}(X_n)}{b_n^2} < \infty.$$

Then, with probability one,

$$\lim_{n \to \infty} \frac{S_n - \mathcal{E}(S_n)}{b_n} = 0.$$

That is, $X_1, X_2, \ldots$ obey the strong law of large numbers with $a_n = \mathcal{E}(S_n)$.

PROOF For $n \in \mathcal{N}$, let $Y_n = \big(X_n - \mathcal{E}(X_n)\big)/b_n$ and note that $\mathcal{E}(Y_n) = 0$. In view of Exercise 7.69 on page 255,

$$\sum_{n=1}^{\infty} \mathrm{Var}(Y_n) = \sum_{n=1}^{\infty} \mathrm{Var}\left(\frac{X_n - \mathcal{E}(X_n)}{b_n}\right) = \sum_{n=1}^{\infty} \frac{\mathrm{Var}(X_n)}{b_n^2} < \infty.$$

Therefore, by Proposition 7.11, $\sum_{n=1}^{\infty} Y_n$ converges with probability one. But, we have

$$\frac{S_n - \mathcal{E}(S_n)}{b_n} = \frac{1}{b_n} \sum_{k=1}^{n} b_k \frac{X_k - \mathcal{E}(X_k)}{b_k} = \frac{1}{b_n} \sum_{k=1}^{n} b_k Y_k.$$

The required result now follows from Kronecker's lemma. ∎

An immediate corollary of Kolmogorov's strong law of large numbers is the following result. Its proof is left to the reader as Exercise 7.88.

□ □ □ COROLLARY 7.4

Suppose that $X_1, X_2, \ldots$ are mutually independent random variables with common finite mean μ and variance σ^2. Then

$$\lim_{n \to \infty} \frac{X_1 + \cdots + X_n}{n} = \mu$$

with probability one.

In Kolmogorov's strong law of large numbers and the foregoing corollary, besides presuming that the random variables $X_1, X_2, \ldots$ are mutually independent, we impose a restriction on their variances. If we assume that $X_1, X_2, \ldots$ all have the same probability distribution, then the restriction on the variances can be eliminated. To prove this statement, we first establish the following lemma.

□ □ □ **LEMMA 7.4**

Suppose that X is a nonnegative random variable. Then

$$\sum_{n=1}^{\infty} P(X \geq n) \leq \mathcal{E}(X) \leq \sum_{n=0}^{\infty} P(X \geq n).$$

Thus, X has finite expectation if and only if $\sum_{n=1}^{\infty} P(X \geq n) < \infty$.

PROOF   For $n \leq x < n+1$, $P(X \geq n+1) \leq P(X > x) \leq P(X \geq n)$. Integrating these inequalities from n to $n+1$ and summing the results, we conclude that

$$\sum_{n=0}^{\infty} P(X \geq n+1) \leq \sum_{n=0}^{\infty} \int_{n}^{n+1} P(X > x)\, dx \leq \sum_{n=0}^{\infty} P(X \geq n).$$

However, by Corollary 4.1 on page 125, the integral in the previous relation equals $\int_{0}^{\infty} P(X > x)\, dx$. Applying Exercise 7.64(a) on page 255, we obtain the required result.   ∎

We will now prove the strong law of large numbers in the case where the random variables X_1, X_2, ... are mutually independent and have the same probability distributions. Such random variables are said to be **iid,** short for "independent and identically distributed." Note that there is no assumption made about the common variance of the X_ks; in particular, the common variance may be infinite.

□ □ □ **THEOREM 7.9 Strong Law of Large Numbers (iid Case)**

Suppose that X_1, X_2, ... are mutually independent and identically distributed random variables with finite mean μ. Then

$$\lim_{n \to \infty} \frac{X_1 + \cdots + X_n}{n} = \mu \tag{7.23}$$

with probability one.

PROOF   We can assume without loss of generality that $\mu = 0$. Because X_1, X_2, ... are identically distributed and have finite mean, Lemma 7.4 implies that

$$\sum_{n=1}^{\infty} P(|X_n| \geq n) = \sum_{n=1}^{\infty} P(|X_1| \geq n) < \infty. \tag{7.24}$$

The idea of the proof is to truncate the X_ks and then apply Theorem 7.8. Let $E = \bigcap_{n=1}^{\infty} \left(\bigcup_{k=n}^{\infty} \{|X_k| \geq k\} \right)$. Then (7.24) and part (a) of the Borel-Cantelli lemma (page 232) imply that $P(E) = 0$. Define the sequence of random variables Y_1, Y_2, ... by

$$Y_n = X_n \chi_{\{|X_n| < n\}} = \begin{cases} X_n, & \text{if } |X_n| < n; \\ 0, & \text{if } |X_n| \geq n; \end{cases}$$

and note that, if $\omega \in E^c$, then $Y_n(\omega) = X_n(\omega)$ for n sufficiently large. Therefore, to establish (7.23) with $\mu = 0$, it suffices to prove that, with probability one,

$$\lim_{n \to \infty} \frac{1}{n} \sum_{k=1}^{n} Y_k = 0. \tag{7.25}$$

Since X_1, X_2, ... have the same probability distribution, Theorem 7.6 on page 249 implies that $\mathcal{E}(Y_n) = \mathcal{E}(X_n \chi_{\{|X_n|<n\}}) = \mathcal{E}(X_1 \chi_{\{|X_1|<n\}})$. Hence, by Corollary 5.3 on page 171,

$$\mathcal{E}(Y_n) = \int_{\{|X_1|<n\}} X_1 \, dP \to \mathcal{E}(X_1) = \mu = 0,$$

as $n \to \infty$. This implies that $n^{-1} \sum_{k=1}^{n} \mathcal{E}(Y_k) \to 0$ as $n \to \infty$. Consequently, proving (7.25) is equivalent to proving

$$\lim_{n \to \infty} \frac{\sum_{k=1}^{n} Y_k - \mathcal{E}\left(\sum_{k=1}^{n} Y_k\right)}{n} = 0,$$

with probability one; and, to accomplish that, we will verify that the Y_ns satisfy the hypotheses of Theorem 7.8 with $b_n = n$.

As X_1, X_2, ... are mutually independent, so are Y_1, Y_2, ... (why?). Furthermore, we have

$$\sum_{n=1}^{\infty} \frac{\text{Var}(Y_n)}{n^2} \le \sum_{n=1}^{\infty} \frac{\mathcal{E}(Y_n^2)}{n^2} = \sum_{n=1}^{\infty} \frac{1}{n^2} \int_{\{|X_n|<n\}} X_n^2 \, dP$$

$$= \sum_{n=1}^{\infty} \frac{1}{n^2} \int_{\{|X_1|<n\}} X_1^2 \, dP = \sum_{n=1}^{\infty} \frac{1}{n^2} \sum_{m=1}^{n} \int_{\{m-1 \le |X_1|<m\}} X_1^2 \, dP$$

$$= \sum_{m=1}^{\infty} \int_{\{m-1 \le |X_1|<m\}} X_1^2 \, dP \sum_{n=m}^{\infty} \frac{1}{n^2}$$

$$\le \sum_{m=1}^{\infty} m \int_{\{m-1 \le |X_1|<m\}} |X_1| \, dP \sum_{n=m}^{\infty} \frac{1}{n^2}$$

$$\le 2 \sum_{m=1}^{\infty} \int_{\{m-1 \le |X_1|<m\}} |X_1| \, dP = 2\mathcal{E}(|X_1|) < \infty,$$

where, in the previous line, we have used the fact that, for each $m \in \mathcal{N}$, we have that $\sum_{n=m}^{\infty} n^{-2} \le 2/m$ (see Exercise 7.92). ∎

Theorem 7.9 indicates that the intuitive notion of expectation as the long-run-average value of a random variable in repeated, independent observations can be formulated and proved mathematically as a consequence of the axioms of probability. A simple corollary of that theorem shows that this is also true for the relative-frequency interpretation of probability.[†]

□ □ □ COROLLARY 7.5 Borel's Strong Law of Large Numbers

Suppose that E is an event associated with some random experiment and let p be its probability. Denote by $n(E)$ the number of times that event E occurs in n independent repetitions of the experiment. Then

$$\lim_{n \to \infty} \frac{n(E)}{n} = p$$

with probability one.

[†] Actually, the following corollary is also a corollary of Corollary 7.4.

PROOF For each $n \in \mathcal{N}$, define $X_n = 1$ or 0 according to whether event E occurs or does not occur on the nth repetition of the experiment. Then $n(E) = X_1 + \cdots + X_n$ and, as the repetitions of the experiment are independent of one another, the random variables $X_1, X_2, \ldots$ are iid. Their common mean is

$$\mu = 0 \cdot (1 - p) + 1 \cdot p = p.$$

The required result now follows from Theorem 7.9. ∎

We have concentrated our discussion in this section on the strong law of large numbers. As we know, if a sequence of random variables obeys the strong law of large numbers, then it must also obey the weak law of large numbers. Nonetheless, the weak law is important in its own right because, for example, it can be proved under weaker conditions than the strong law. Several versions of the weak law will be considered in the exercises.

Exercises for Section 7.4

Note: In the exercises below, we will use the notation $S_n = X_1 + \cdots + X_n$.

7.85 Prove Lemma 7.1 on page 259.

7.86 Prove Toeplitz's lemma, Lemma 7.2 on page 259.

7.87 Prove the summation by parts formula, (7.16) on page 259. *Hint:* Write $b_k = \sum_{j=1}^{k} a_j$ and interchange summations.

7.88 Prove Corollary 7.4 on page 262.

7.89 Describe, in words, the difference between the weak and strong laws of large numbers for iid random variables with finite mean. Refer to Definition 5.15 on page 174 and Exercise 5.92 on page 177.

7.90 The following result is known as **Cantelli's strong law of large numbers:** Suppose that $X_1, X_2, \ldots$ are mutually independent random variables with uniformly bounded fourth moments. Then, as $n \to \infty$, $(S_n - \mathcal{E}(S_n))/n \to 0$ with probability one.
   a) Deduce Cantelli's strong law from Kolmogorov's strong law.
   b) Prove Cantelli's strong law without reference to Kolmogorov's strong law. *Hint:* Use Exercise 7.75 on page 256 with $\phi(x) = x^4$, the Borel-Cantelli lemma (pages 232–233), and Exercise 5.92 (page 177).

7.91 Suppose that independent trials are performed in which an event E occurs on the kth trial with probability p_k. Let $n(E)$ denote the number of times that event E occurs in the first n trials. Show that, as $n \to \infty$,

$$\frac{n(E)}{n} - \frac{\sum_{k=1}^{n} p_k}{n} \to 0$$

with probability one.

7.92 Prove that for $m \in \mathcal{N}$, $\sum_{n=m}^{\infty} \frac{1}{n^2} \leq 2/m$.

7.93 Let $X_1, X_2, \ldots$ be iid with finite mean μ, and let f be a bounded continuous function on $\mathcal{R}$.
   a) Prove that

$$\lim_{n \to \infty} \mathcal{E}\left(f\left(\frac{X_1 + \cdots + X_n}{n} \right) \right) = f(\mu).$$

b) Deduce from part (a) that, for each $t \in [0, 1]$,

$$\lim_{n \to \infty} \sum_{k=1}^{n} f\left(\frac{k}{n}\right) \binom{n}{k} t^k (1 - t)^{n-k} = f(t).$$

Hint: Refer to Exercise 7.81 on page 256.

7.94 Each number in $[0, 1]$ has a decimal expansion, and the expansion is unique except for numbers of the form $m/10^n$. For definiteness, we will use the unique terminating expansion for numbers of the form $m/10^n$. Now, let $x \in [0, 1]$ have decimal expansion $.x_1 x_2 \ldots$; that is, $x = \sum_{n=1}^{\infty} x_n/10^n$. For each $n \in \mathcal{N}$ and $k \in \{0, 1, \ldots, 9\}$, denote by $n_k(x)$ the number of the first n decimal digits of x that equal k. Then x is said to be a **normal number** if $n_k(x)/n \to 1/10$ as $n \to \infty$ for all digits k; in other words, if the relative-frequency of occurrence of each decimal digit in x is $1/10$. In this exercise, we will prove the following result due to Borel: Except for a Borel set of Lebesgue measure zero, every number in $[0, 1]$ is normal.

a) Let $(\Omega, \mathcal{A}, P) = \left([0, 1], \mathcal{B}_{[0,1]}, \lambda_{[0,1]}\right)$. Define $Y_1, Y_2, \ldots$ on Ω by $Y_n(x) = x_n$, where x_n is the nth decimal digit of x. Prove that $Y_1, Y_2, \ldots$ are random variables, that is, are $\mathcal{B}_{[0,1]}$-measurable. *Hint:* Note that

$$\{Y_n = k\} = \bigcup_{k_1=0}^{9} \cdots \bigcup_{k_{n-1}=0}^{9} \{Y_1 = k_1, \ldots, Y_{n-1} = k_{n-1}, Y_n = k\},$$

and show that each set in the union is an interval.

b) Prove that the random variables $Y_1, Y_2, \ldots$ are iid.

c) Show that, for each decimal digit k, we have $\lim_{n \to \infty} n_k(x)/n = \frac{1}{10}$ λ-ae. *Hint:* Let $X_j^{(k)} = \chi_{\{Y_j = k\}}$, for $j = 1, 2, \ldots$.

d) Deduce that, except for a Borel set of Lebesgue measure zero, every number in $[0, 1]$ is normal.

7.95 Repeat Exercise 7.94 for binary instead of decimal expansions. Explain how this provides a model for the random experiment of tossing a balanced coin indefinitely.

7.96 Prove **Markov's weak law of large numbers:** Suppose that $X_1, X_2, \ldots$ are random variables, all defined on the same probability space and with finite variances. Further suppose that $\text{Var}(X_1 + \cdots + X_n) = o(n^2)$. Then $\lim_{n \to \infty} \left(S_n - \mathcal{E}(S_n)\right)/n = 0$, in probability. That is, $X_1, X_2, \ldots$ obey the weak law of large numbers with $a_n = \mathcal{E}(S_n)$ and $b_n = n$.

7.97 Prove **Chebyshev's weak law of large numbers:** Suppose that $X_1, X_2, \ldots$ are uncorrelated random variables and that their variances are (uniformly) bounded. Then $\lim_{n \to \infty} \left(S_n - \mathcal{E}(S_n)\right)/n = 0$, in probability.

7.98 Establish the following generalization of Chebyshev's weak law of large numbers: Suppose that $X_1, X_2, \ldots$ are random variables, all defined on the same probability space. Further suppose that they have uniformly bounded variances and that

$$\lim_{n \to \infty} \frac{1}{n} \sum_{k=1}^{n} \text{Cov}(X_k, X_n) = 0. \tag{7.26}$$

a) Prove that $\lim_{n \to \infty} \left(S_n - \mathcal{E}(S_n)\right)/n = 0$, in probability.

b) The random variables $X_1, X_2, \ldots$ are said to be **asymptotically uncorrelated** if $\text{Cov}(X_i, X_j) \to 0$ as $|i - j| \to \infty$. Prove that asymptotically uncorrelated random variables with uniformly bounded variances satisfy (7.26) and, hence, the weak law of large numbers.

7.99 A standard example of a series that is convergent but not absolutely convergent is

$$\sum_{n=1}^{\infty}(-1)^n\frac{1}{n}. \tag{7.27}$$

Suppose, instead of (7.27), we consider a similar series $\sum_{n=1}^{\infty}\pm n^{-1}$, where the signs are chosen at random. In other words, suppose we consider the series

$$\sum_{n=1}^{\infty}X_n\cdot\frac{1}{n} \tag{7.28}$$

where the X_ns are iid, taking the values ± 1 each with probability $1/2$.

a) Show that the series in (7.28) converges with probability one.

b) What can be said about convergence of the series $\sum_{n=1}^{\infty}a_nX_n$, when $\{a_n\}$ is a sequence of real numbers with $\sum_{n=1}^{\infty}a_n^2<\infty$?

7.100 Let $X_1, X_2, \ldots$ be a sequence of mutually independent random variables with

$$P\big(X_n=n^b\big)=P\big(X_n=-n^b\big)=\frac{1}{2},$$

where b is a positive real number. Prove the following facts.

a) If $b<1/2$, then $S_n/n\to 0$ with probability one.

b) If $b>1$, then $\limsup_{n\to\infty}|S_n|/n=\infty$ with probability one.

c) Conclude that Theorem 7.8 fails if the hypothesis $\sum\mathrm{Var}(X_n)/b_n^2<\infty$ is removed.

7.101 Let $X_1, X_2, \ldots$ be iid random variables such that $\mathcal{E}(X_1^+)=\infty$ and $\mathcal{E}(X_1^-)<\infty$. Prove that $S_n/n\to\infty$ with probability one.

Giuseppe Vitali
(1875–1932)

Giuseppe Vitali was born on August 26, 1875, in Ravenna, Italy. He graduated from the Scuola Normale Superiore in Pisa in 1899. Subsequently, he assisted the Italian mathematician and politician Ulisse Dini for 2 years before leaving mathematical research to teach in high schools due to financial problems.

Between 1904 and 1923, Vitali taught at the Liceo classico Cristoforo Colombo in Genoa, where he developed an interest in politics and became a Socialist councillor. In 1922, however, the Fascists came to power and dissolved the Socialist Party. Vitali then returned to mathematical research.

As a result of winning a competition for the chair of infinitesimal analysis in 1923, Vitali was appointed to a chair at the University of Modena and Reggio Emilia. The following year he was appointed as chair of mathematics at the University of Padua and, then, in 1930, as chair of mathematics at the University of Bologna.

Vitali was the first to give an example of a nonmeasurable subset of real numbers, as presented in Chapter 3. The Vitali covering theorem, which we will discuss in this chapter, is of fundamental importance in measure theory. Also due to Vitali are absolutely continuous functions (also discussed in this chapter), the Vitali convergence theorem (a generalization of the DCT), and the Vitali-Hahn-Saks theorem (jointly with Hans Hahn and Stanisław Saks). Another theorem bearing Vitali's name provides a sufficient condition for the uniform convergence of a sequence of uniformly bounded holomorphic functions to a holomorphic function.

In 1926, Vitali developed a serious illness. Nevertheless, about half of his research papers were written in the last 4 years of his life. During his last years, he also worked on a new absolute differential calculus and a geometry of Hilbert spaces. After his death, Vitali's text *Moderna teoria delle funzioni di variabile reale* (Modern theory of functions of a real variable) was completed and published in 1935. Vitali died on February 29, 1932, in Bologna, Italy.

□ 8 □

Differentiation and Absolute Continuity

Up to this point, we have been concentrating on the theory of integration. In this chapter, we will study the theory of differentiation in the classical sense of derivatives of functions. We will state and prove Lebesgue's remarkable theorem that any monotone function is differentiable almost everywhere. We will also introduce the concepts of bounded variation and absolute continuity, and use them to generalize the two fundamental theorems of calculus to Lebesgue integration.

8.1 DERIVATIVES AND DINI-DERIVATES

In this section, we will introduce derivatives and establish the fact that any monotone function has a (finite) derivative almost everywhere. *Note:* For brevity, we will use the phrase *almost everywhere* instead of *Lebesgue almost everywhere* and use the notation *ae* instead of λ-*ae*. To begin, we recall the following definition from elementary calculus.

DEFINITION 8.1   Derivative of a Real-Valued Function

A real-valued function f defined in some open interval about $x \in \mathcal{R}$ is said to be **differentiable at x** if

$$\lim_{h \to 0} \frac{f(x+h) - f(x)}{h}$$

exists and is finite. In that case the limit is called the **derivative of f at x** and is denoted by $f'(x)$.

Note: If $\lim_{h\to 0} (f(x+h) - f(x))/h = \infty$, we will write $f'(x) = \infty$ but will not say that f is differentiable at x and, similarly, if the limit is $-\infty$.

For our study of differentiation, we require the concept of the *Dini-derivates* of a function at a point. To begin, we present the following definition.

DEFINITION 8.2   Lower and Upper Limits

Let g be a real-valued function defined in a deleted interval about the point x, that is, a set of the form $(c, d) \setminus \{x\}$, where $c < x < d$. Then we define

$$\limsup_{y\to x^+} g(y) = \inf_{\delta > 0} \sup_{0 < y - x < \delta} g(y)$$

$$\liminf_{y\to x^+} g(y) = \sup_{\delta > 0} \inf_{0 < y - x < \delta} g(y)$$

$$\limsup_{y\to x^-} g(y) = \inf_{\delta > 0} \sup_{0 < x - y < \delta} g(y)$$

$$\liminf_{y\to x^-} g(y) = \sup_{\delta > 0} \inf_{0 < x - y < \delta} g(y).$$

These extended real numbers are called, respectively, the **upper right, lower right, upper left,** and **lower left limits of g at x.**

We introduce these lower and upper limits for the same reason that we introduce the limit inferior and limit superior of sequences; namely, although $\lim_{y\to x^+} g(y)$, etc., may not exist, $\limsup_{y\to x^+} g(y)$, etc., always exist (in $\mathcal{R}^*$). We leave it as an exercise for the reader to prove that the right-hand limit of g at x, $\lim_{y\to x^+} g(y)$, exists in $\mathcal{R}^*$ if and only if the lower and upper right limits of g at x are equal; in that case, we denote the right-hand limit by $g(x+)$. An analogous result holds for left-hand limits.

EXAMPLE 8.1   *Illustrates Definition 8.2*

Let $g(y) = \sin(1/y)$ for $y \neq 0$. It is easy to see that for each $\delta > 0$,

$$\sup_{0 < y < \delta} g(y) = 1 \qquad \text{and} \qquad \inf_{0 < y < \delta} g(y) = -1.$$

Consequently, we have $\limsup_{y\to 0^+} g(y) = 1$ and $\liminf_{y\to 0^+} g(y) = -1$. Similarly, $\limsup_{y\to 0^-} g(y) = 1$ and $\liminf_{y\to 0^-} g(y) = -1$.   ◻

Dini-Derivates

We now define the Dini-derivates of a real-valued function.

DEFINITION 8.3 Dini-Derivates

Let f be a real-valued function defined in an open interval about the point x. Set

$$D^+f(x) = \limsup_{h\to 0^+} \frac{f(x+h)-f(x)}{h}$$

$$D_+f(x) = \liminf_{h\to 0^+} \frac{f(x+h)-f(x)}{h}$$

$$D^-f(x) = \limsup_{h\to 0^-} \frac{f(x+h)-f(x)}{h}$$

$$D_-f(x) = \liminf_{h\to 0^-} \frac{f(x+h)-f(x)}{h}.$$

These four extended real numbers are called the **Dini-derivates of f at x**. They are, respectively, the **upper right, lower right, upper left,** and **lower left derivates**.

It follows that f is differentiable at x if and only if all four of the Dini-derivates are equal and finite. It also follows that

$$f'_+(x) = \lim_{h\to 0^+} \frac{f(x+h)-f(x)}{h}$$

exists in $\mathcal{R}^*$ if and only if $D_+f(x) = D^+f(x)$; and similarly for $f'_-(x)$.

EXAMPLE 8.2 Illustrates Definition 8.3

Let $f = \chi_Q$ and $x \in Q$. Then,

$$\frac{f(x+h)-f(x)}{h} = \begin{cases} -1/h, & \text{if } h \notin Q; \\ 0, & \text{if } h \in Q. \end{cases} \tag{8.1}$$

It follows easily from (8.1) that for each $\delta > 0$,

$$\sup_{0<h<\delta} \frac{f(x+h)-f(x)}{h} = 0 \quad \text{and} \quad \inf_{0<h<\delta} \frac{f(x+h)-f(x)}{h} = -\infty.$$

Thus, $D^+f(x) = 0$ and $D_+f(x) = -\infty$. Similarly, we find that for each $\delta > 0$,

$$\sup_{-\delta<h<0} \frac{f(x+h)-f(x)}{h} = \infty \quad \text{and} \quad \inf_{-\delta<h<0} \frac{f(x+h)-f(x)}{h} = 0,$$

so that $D^-f(x) = \infty$ and $D_-f(x) = 0$. □

An Everywhere-Continuous, Nowhere-Differentiable Function

It is an elementary fact proved in calculus that if f is differentiable at a point x, then it is continuous at x. The converse of this fact fails. For example, $f(x) = |x|$ is continuous but not differentiable at $x = 0$.

We now present a more striking example, namely, a function that is continuous at every point of $\mathcal{R}$ but differentiable at no points of $\mathcal{R}$. The idea is to construct a function that is everywhere continuous but oscillates so wildly as to be nowhere differentiable.

EXAMPLE 8.3   *A Continuous, Nowhere-Differentiable Function*

Define $\phi(x)$ on $[0, 1]$ by

$$\phi(x) = \begin{cases} x, & 0 \le x \le \frac{1}{2}; \\ 1 - x, & \frac{1}{2} < x \le 1. \end{cases}$$

Extend ϕ to all of $\mathcal{R}$ via $\phi(x + k) = \phi(x)$ for $k \in \mathcal{Z}$. See Fig. 8.1.

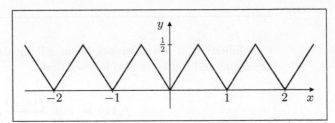

FIGURE 8.1 Graph of ϕ

Next, define the functions u_n, $n = 0,\ 1,\ 2,\ \ldots$, by $u_n(x) = \phi(4^n x)/4^n$, as portrayed in Fig. 8.2. Note that, for each n, u_n is continuous on $\mathcal{R}$.

Now consider the function f defined for all $x \in \mathcal{R}$ by

$$f(x) = \sum_{n=0}^{\infty} u_n(x).$$

For $x \in \mathcal{R}$, $|u_n(x)| \le 1/(2 \cdot 4^n)$ and so the series converges uniformly on $\mathcal{R}$. Hence, f is continuous on $\mathcal{R}$. Note also that for $k = 0,\ 1,\ 2,\ \ldots$ and $n \ge k$,

$$u_n(x \pm 4^{-k}) = \frac{\phi\big(4^n(x \pm 4^{-k})\big)}{4^n} = \frac{\phi\big(4^n x \pm 4^{n-k}\big)}{4^n} = \frac{\phi(4^n x)}{4^n} = u_n(x), \quad (8.2)$$

for all $x \in \mathcal{R}$.

To show that f is nowhere differentiable we consider two cases. Reference to Fig. 8.2 will prove helpful during the discussion.

Case 1: x is not of the form $m/4^k$, for some $m \in \mathcal{Z}$ and $k \in \mathcal{N}$.

We will find a sequence $\{h_n\}_{n=1}^{\infty}$ such that $h_n \neq 0$ for all $n \in \mathcal{N}$, $h_n \to 0$, but $\big(f(x + h_n) - f(x)\big)/h_n$ does not have a limit. This result will show that f is not differentiable at x.

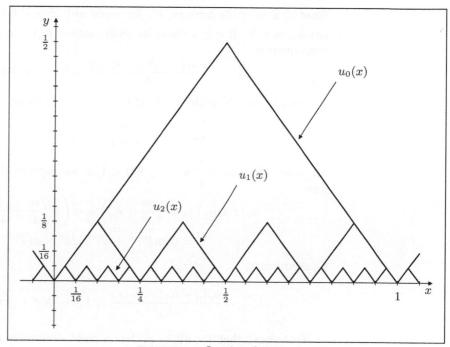

FIGURE 8.2 Graphs of the u_ns

We can assume without loss of generality that $0 < x < 1$. Then x lies in exactly one of the intervals $(0, 1/4)$, $(1/4, 1/2)$, $(1/2, 3/4)$, $(3/4, 1)$. Hence, we can choose h_1 so that $|h_1| = 1/4$ and that $x + h_1 \in (0, 1/2)$ if $x \subset (0, 1/2)$ and $x + h_1 \in (1/2, 1)$ if $x \in (1/2, 1)$. It then follows from (8.2) that

$$\left| \frac{f(x + h_1) - f(x)}{h_1} \right| = \left| \frac{\phi(x + h_1) - \phi(x)}{h_1} \right| = 1.$$

Next, x lies in one of the intervals $(0, 1/16)$, $(1/16, 1/8)$, ..., $(15/16, 1)$. Hence, we can choose h_2 so that $|h_2| = 1/4^2$ and that $x + h_2 \in (0, 1/8)$ if $x \in (0, 1/8)$, $x + h_2 \in (1/8, 1/4)$ if $x \in (1/8, 1/4)$, ..., $x + h_2 \in (7/8, 1)$ if $x \in (7/8, 1)$. It then follows from (8.2) that

$$\left| \frac{f(x + h_2) - f(x)}{h_2} \right| = \begin{cases} 2, & x \in (\frac{k-1}{8}, \frac{k}{8}), \ k = 1, 3, 6, 8; \\ 0, & x \in (\frac{k-1}{8}, \frac{k}{8}), \ k = 2, 4, 5, 7. \end{cases}$$

Continuing in this manner, we get a sequence $\{h_n\}_{n=1}^{\infty}$ such that $|h_n| = 1/4^n$ and

$$\frac{f(x + h_n) - f(x)}{h_n} = \begin{cases} \text{odd integer}, & n \text{ odd}; \\ \text{even integer}, & n \text{ even}. \end{cases}$$

Thus, $h_n \to 0$, but $\lim_{n \to \infty} (f(x + h_n) - f(x))/h_n$ does not exist. Hence, f is not differentiable at x.

Case 2: x is of the form $m/4^k$, for some $m \in \mathcal{Z}$ and $k \in \mathcal{N}$.

Let $h_n = 4^{-n}$. If $r \geq n$ then, by (8.2), $u_r(x + h_n) = u_r(x + 4^{-n}) = u_r(x)$ and, consequently,

$$\frac{u_r(x + h_n) - u_r(x)}{h_n} = 0, \qquad r \geq n. \tag{8.3}$$

Now, let $n \in \mathcal{N}$ with $n > k$. If $k \leq r \leq n - 1$, then

$$u_r(x) = u_r\left(\frac{m}{4^k}\right) = \phi(m4^{r-k})/4^r = 0.$$

Moreover, because $0 < 4^{r-n} \leq 1/4 < 1/2$, we have $\phi(4^{r-n}) = 4^{r-n}$, and therefore,

$$u_r(x + h_n) = u_r\left(\frac{m}{4^k} + h_n\right) = \phi\left(4^r\left(\frac{m}{4^k} + \frac{1}{4^n}\right)\right)/4^r$$

$$= \phi(m4^{r-k} + 4^{r-n})/4^r = \phi(4^{r-n})/4^r = 4^{-n}.$$

Consequently,

$$\frac{u_r(x + h_n) - u_r(x)}{h_n} = 1, \qquad k \leq r \leq n - 1. \tag{8.4}$$

Next, note that for all r, u_r has a right derivative at all points and so, in particular, at x. Hence, it follows that

$$\lim_{n \to \infty} \sum_{r=0}^{k-1} \left(\frac{u_r(x + h_n) - u_r(x)}{h_n}\right) \tag{8.5}$$

exists and is finite.

Now, for convenience, let $d_r = \left(u_r(x + h_n) - u_r(x)\right)/h_n$. For $n > k$, we have

$$\frac{f(x + h_n) - f(x)}{h_n} = \sum_{r=0}^{\infty}\left(\frac{u_r(x + h_n) - u_r(x)}{h_n}\right) = \sum_{r=0}^{\infty} d_r$$

$$= \sum_{r=0}^{k-1} d_r + \sum_{r=k}^{n-1} d_r + \sum_{r=n}^{\infty} d_r.$$

By (8.4), the second term equals $n - k$ and, by (8.3), the third term equals 0. Hence, for $n > k$,

$$\frac{f(x + h_n) - f(x)}{h_n} = \sum_{r=0}^{k-1}\left(\frac{u_r(x + h_n) - u_r(x)}{h_n}\right) + n - k.$$

Applying (8.5), we can now conclude that

$$\lim_{n \to \infty} \frac{f(x + h_n) - f(x)}{h_n} = \infty.$$

In particular, f is not differentiable at x. $\qquad\qquad\square$

Vitali Covers

Example 8.3 shows that continuity is by no means sufficient for differentiability. The function f, constructed in that example, is everywhere continuous but nowhere differentiable. Essentially, the reason that function is nowhere differentiable is because it "oscillates vigorously." In Section 8.2, we will show that functions that do not oscillate vigorously (a concept that will soon be defined precisely) are differentiable almost everywhere.

Our next goal is to prove that a monotone function is differentiable almost everywhere, a theorem due to Lebesgue. The proof we give uses the concept of Vitali covers. Roughly speaking, a family $\mathcal{V}$ of closed intervals is a Vitali cover of a set E of real numbers if every point of E is in arbitrarily small intervals of $\mathcal{V}$. More precisely, we have the following definition.

DEFINITION 8.4 **Vitali Cover**

Let $E \subset \mathcal{R}$. A family $\mathcal{V}$ of nondegenerate closed intervals is said to be a **Vitali cover** of E if for each $x \in E$ and each $\delta > 0$, there is an $I \in \mathcal{V}$ such that $x \in I$ and $\ell(I) < \delta$.

The following theorem, called the Vitali covering theorem, uses the concept of Lebesgue outer measure λ^* (Definition 3.6 on page 91).

□ □ □ **THEOREM 8.1 Vitali Covering Theorem**

Let $E \subset \mathcal{R}$ with $\lambda^*(E) < \infty$ and suppose $\mathcal{V}$ is a Vitali cover of E. Then for each $\epsilon > 0$ there is a finite disjoint collection $\{I_k\}_{k=1}^n \subset \mathcal{V}$ such that

$$\lambda^*\left(E \setminus \bigcup_{k=1}^n I_k\right) < \epsilon.$$

PROOF Since $\lambda^*(E) < \infty$, we can choose an open set O such that $O \supset E$ and $\lambda^*(O) < \infty$. Set $\mathcal{W} = \{I \in \mathcal{V} : I \subset O\}$. Then $\mathcal{W}$ is a Vitali cover for E. (See Exercise 8.12.)

The idea of the proof is this: Starting with some $I_1 \in \mathcal{W}$, select an $I_2 \in \mathcal{W}$ as large as possible but missing I_1; then select an $I_3 \in \mathcal{W}$ as large as possible but missing $I_1 \cup I_2$; and continue the process until $\lambda^*\left(E \setminus \bigcup_{k=1}^n I_k\right)$ becomes small.

So, let $I_1 \in \mathcal{W}$. If $E \subset I_1$, then we are done. Otherwise, let

$$\delta_1 = \sup\{\ell(I) : I \in \mathcal{W}, I \cap I_1 = \emptyset\}.$$

Because $\mathcal{W}$ is a Vitali cover of E and $E \setminus I_1 \neq \emptyset$, it follows that $\delta_1 > 0$. Also, because $I \subset O$ for all $I \in \mathcal{W}$, it follows that $\delta_1 \leq \lambda^*(O) < \infty$. Hence we can choose $I_2 \in \mathcal{W}$ with $I_2 \cap I_1 = \emptyset$ and $\ell(I_2) > \delta_1/2$. Again, if $E \subset I_1 \cup I_2$, then we are done.

We now proceed inductively. Suppose $I_1, \ldots, I_n \in \mathcal{W}$ are pairwise disjoint. If $E \subset \bigcup_{k=1}^n I_k$, then we are done. Otherwise, let

$$\delta_n = \sup\{\ell(I) : I \in \mathcal{W}, I \cap I_k = \emptyset, 1 \leq k \leq n\}.$$

Since $\mathcal{W}$ is a Vitali cover of E and $E \setminus \bigcup_{k=1}^{n} I_k \neq \emptyset$, it follows that $\delta_n > 0$. Also, because $I \subset O$ for all $I \in \mathcal{W}$ and $\lambda^*(O) < \infty$, it follows that $\delta_n < \infty$. Hence, there is a member of $\mathcal{W}$, say I_{n+1}, such that $I_{n+1} \cap I_k = \emptyset$, $1 \leq k \leq n$, and $\ell(I_{n+1}) > \delta_n/2$.

If this process terminates after a finite number of steps, then we are done. Otherwise, it yields a sequence $\{I_n\}_{n=1}^{\infty}$ of pairwise disjoint members of $\mathcal{W}$ such that $\ell(I_{n+1}) > \delta_n/2$ and $\sum \ell(I_n) < \infty$. Because $\sum \ell(I_n) < \infty$, there is an $N \in \mathcal{N}$ such that

$$\sum_{n=N+1}^{\infty} \ell(I_n) < \epsilon/5.$$

Set $A = E \setminus \bigcup_{n=1}^{N} I_n$. We claim that $\lambda^*(A) < \epsilon$.

Let $x \in A$. Then $x \notin \bigcup_{n=1}^{N} I_n$ and so $d(x, \bigcup_{n=1}^{N} I_n) = \delta > 0$. Because $x \in E$ and $\mathcal{W}$ is a Vitali cover for E, there is an $I \in \mathcal{W}$ with $x \in I$ and $\ell(I) < \delta$. It follows that $I \cap I_n = \emptyset$ for $n = 1, 2, \ldots, N$.

Now, there must be an $n \in \mathcal{N}$ with $I \cap I_n \neq \emptyset$. Suppose to the contrary. Then for each $n \in \mathcal{N}$, $I \cap I_k = \emptyset$, $1 \leq k \leq n$. Applying the definition of δ_n, we get that for each $n \in \mathcal{N}$, $\ell(I) \leq \delta_n < 2\ell(I_{n+1})$. But this is impossible because $\ell(I) > 0$ and $\ell(I_{n+1}) \to 0$ as $n \to \infty$. Let $m = \min\{n : I \cap I_n \neq \emptyset\}$; note that $m > N$. Let y_m be the midpoint of I_m. Then,

$$|x - y_m| \leq \ell(I) + \frac{1}{2}\ell(I_m) \leq \delta_{m-1} + \frac{1}{2}\ell(I_m) < 2\ell(I_m) + \frac{1}{2}\ell(I_m) = \frac{5}{2}\ell(I_m).$$

Consequently, $x \in [y_m - \frac{5}{2}\ell(I_m), y_m + \frac{5}{2}\ell(I_m)] = J_m$.

Hence, if $x \in A$, there is an $m > N$ such that $x \in J_m$; in other words, we have that $A \subset \bigcup_{m=N+1}^{\infty} J_m$. It follows that

$$\lambda^*(A) \leq \sum_{m=N+1}^{\infty} \ell(J_m) = 5 \sum_{m=N+1}^{\infty} \ell(I_m) < \epsilon.$$

This completes the proof. ∎

Differentiability of Monotone Functions

We are just about ready to prove Lebesgue's famous theorem on the almost-everywhere differentiability of monotone functions. First a lemma.

□ □ □ **LEMMA 8.1**

Let f be a real-valued function on (a, b). Then the set of points in (a, b) where $f'_+(x)$ and $f'_-(x)$ exist (possibly $\pm\infty$) but are unequal is countable and, hence, has Lebesgue measure zero.

PROOF In what follows, we show that

$$\{x \in (a, b) : f'_+(x) \text{ and } f'_-(x) \text{ exist and } f'_+(x) < f'_-(x)\},$$

which we denote by E, is countable. An analogous argument shows that

$$\{x \in (a, b) : f'_+(x) \text{ and } f'_-(x) \text{ exist and } f'_+(x) > f'_-(x)\}$$

is also countable.

We will set up a one-to-one correspondence between E and a countable set, thus establishing the countability of E. So, let $x \in E$. Choose $r_x \in Q$ such that $f'_+(x) < r_x < f'_-(x)$. By the definitions of $f'_+(x)$ and $f'_-(x)$, we can choose rational numbers s_x and t_x such that $a < t_x < x < s_x < b$ and

$$\frac{f(y) - f(x)}{y - x} < r_x, \quad x < y < s_x, \qquad \text{and} \qquad \frac{f(y) - f(x)}{y - x} > r_x, \quad t_x < y < x,$$

or, in other words,

$$f(y) - f(x) < r_x(y - x), \qquad t_x < y < s_x, \ y \neq x.$$

Now consider the mapping $\phi \colon E \to Q^3$ defined by $\phi(x) = (r_x, s_x, t_x)$. Because Q^3 is countable, it will follow that E is countable if we can prove that ϕ is one-to-one.

Assume to the contrary that there exist $x, z \in E$ with $x \neq z$ and $\phi(x) = \phi(z)$. Then $r_z = r_x$, $s_z = s_x$, and $t_z = t_x$. So,

$$f(y) - f(x) < r_x(y - x), \qquad t_x < y < s_x, \ y \neq x$$
$$f(y) - f(z) < r_x(y - z), \qquad t_x < y < s_x, \ y \neq z.$$

Since $t_x = t_z < z < s_z = s_x$ and $z \neq x$, and $t_x < x < s_x$ and $x \neq z$, we conclude that

$$f(z) - f(x) < r_x(z - x) \qquad \text{and} \qquad f(x) - f(z) < r_x(x - z),$$

which is impossible. ∎

□ □ □ **THEOREM 8.2**

Let f be a monotone function on $[a, b]$. Then f is differentiable almost everywhere on $[a, b]$.

PROOF We can assume without loss of generality that f is nondecreasing. Let

$$E = \{\, x \in (a, b) : D_+ f(x) < D^+ f(x) \,\}.$$

We will show that $\lambda^*(E) = 0$. For each $r, s \in Q$ with $0 < r < s$, let

$$E_{rs} = \{\, x \in (a, b) : D_+ f(x) < r < s < D^+ f(x) \,\}.$$

Then $E = \bigcup \{\, E_{rs} : r, s \in Q, \ 0 < r < s \,\}$, a countable union. If we can prove that $\lambda^*(E_{rs}) = 0$ for all $r, s \in Q$ with $0 < r < s$, then we will have established that $\lambda^*(E) = 0$.

So, let $r, s \in Q$ with $0 < r < s$, and set $\alpha = \lambda^*(E_{rs})$. Let $\epsilon > 0$ be given, and choose O open with $O \supset E_{rs}$ and $\lambda(O) < \alpha + \epsilon$. If $x \in E_{rs}$, then $D_+ f(x) < r$. Consequently, for each $\delta > 0$, $\inf_{0 < h < \delta}\big(f(x + h) - f(x)\big)/h < r$ and, hence, there is an h, with $0 < h < \delta$, and $f(x + h) - f(x) < rh$.

Now, let $\mathcal{V}$ be the collection of all closed intervals of the form $[x, x + h]$, where $h > 0$, $x \in E_{rs}$, $f(x + h) - f(x) < rh$, and $[x, x + h] \subset O \cap (a, b)$. Then it follows from the previous paragraph that $\mathcal{V}$ is a Vitali cover of E_{rs}. Thus, by the

Vitali covering theorem, Theorem 8.1, there is a finite sequence $I_1, I_2, \ldots, I_n$ of pairwise disjoint members of $\mathcal{V}$ such that

$$\lambda^* \left(E_{rs} \setminus \bigcup_{k=1}^{n} I_k \right) < \epsilon.$$

Let $I_k = [x_k, x_k + h_k]$. We need to work with the open intervals $(x_k, x_k + h_k)$. Set $U = \bigcup_{k=1}^{n} (x_k, x_k + h_k)$ and note that

$$\lambda^*(E_{rs} \setminus U) < \epsilon. \tag{8.6}$$

Also, because $U \subset O$, we have

$$\sum_{k=1}^{n} h_k = \lambda(U) \leq \lambda(O) < \alpha + \epsilon. \tag{8.7}$$

Then, by the definition of $\mathcal{V}$, we may conclude that

$$\sum_{k=1}^{n} \big(f(x_k + h_k) - f(x_k)\big) < r \sum_{k=1}^{n} h_k < r(\alpha + \epsilon). \tag{8.8}$$

Next, assume that $y \in E_{rs} \cap U$. Then $D^+ f(y) > s$ so that for each $\delta > 0$ there is a k, with $0 < k < \delta$, such that $f(y + k) - f(y) > sk$. As U is open, $[y, y + k] \subset U$ for sufficiently small k. Consequently, if we let $\mathcal{W}$ be the collection of all closed intervals of the form $[y, y + k]$, where $k > 0$, $y \in E_{rs} \cap U$, $f(y + k) - f(y) > sk$, and $[y, y + k] \subset U$, then $\mathcal{W}$ is a Vitali cover of $E_{rs} \cap U$. Using the Vitali covering theorem again, we can choose pairwise disjoint members of $\mathcal{W}$, say $J_1, J_2, \ldots, J_m$, such that

$$\lambda^* \left((E_{rs} \cap U) \setminus \bigcup_{j=1}^{m} J_j \right) < \epsilon. \tag{8.9}$$

From (8.6) and (8.9), we get

$$\alpha = \lambda^*(E_{rs}) \leq \lambda^*(E_{rs} \cap U) + \lambda^*(E_{rs} \setminus U) \leq \epsilon + \sum_{j=1}^{m} \ell(J_j) + \epsilon. \tag{8.10}$$

Setting $J_j = [y_j, y_j + k_j]$, we obtain from (8.10) and the definition of $\mathcal{W}$ that

$$\sum_{j=1}^{m} \big(f(y_j + k_j) - f(y_j)\big) > s \sum_{j=1}^{m} k_j = s \sum_{j=1}^{m} \ell(J_j) \geq s(\alpha - 2\epsilon). \tag{8.11}$$

Now, $\bigcup_{j=1}^{m} [y_j, y_j + k_j] \subset U \subset \bigcup_{k=1}^{n} [x_k, x_k + h_k]$ and, consequently, (see Exercise 8.14),

$$\sum_{j=1}^{m} \big(f(y_j + k_j) - f(y_j)\big) \leq \sum_{k=1}^{n} \big(f(x_k + h_k) - f(x_k)\big).$$

This result along with (8.8) and (8.11) imply that $s(\alpha - 2\epsilon) < r(\alpha + \epsilon)$. As $\epsilon > 0$ was arbitrary, it follows that $s\alpha \leq r\alpha$. Because $r < s$ and $\alpha \geq 0$, we must have $\alpha = 0$.

Thus, we have shown that $\{x \in (a, b) : D_+ f(x) < D^+ f(x)\}$ has measure zero. A similar argument shows that $\{x \in (a, b) : D_- f(x) < D^- f(x)\}$ also has measure zero. Hence, $\{x \in (a, b) : f'_+(x) \text{ and } f'_-(x) \text{ exist in } \mathcal{R}^*\}^c$ has measure zero. Applying Lemma 8.1, we conclude that $f'(x)$ exists for λ-almost all $x \in (a, b)$, although its value may be infinite at some points.

Set $A = \{x \in (a, b) : f'(x) = +\infty\}$. We want to show $\lambda(A) = 0$. Let $x \in A$ and let $N \in \mathcal{N}$. As $D^+ f(x) = \infty$, for each $\delta > 0$, there is an h with $0 < h < \delta$ such that

$$f(x + h) - f(x) > Nh. \tag{8.12}$$

So, if we let $\mathcal{U}$ be the collection of all closed intervals of the form $[x, x + h]$, where $h > 0$, $x \in A$, $f(x + h) - f(x) > Nh$, and $[x, x + h] \subset (a, b)$, then $\mathcal{U}$ is a Vitali cover of A. By the Vitali covering theorem, there exist pairwise disjoint members of $\mathcal{U}$, say, $I_1, I_2, \ldots, I_n$, such that

$$\lambda^*\left(A \setminus \bigcup_{k=1}^{n} I_k\right) < \frac{1}{N}. \tag{8.13}$$

Set $I_k = [x_k, x_k + h_k]$. Then, by (8.12) and (8.13),

$$N\lambda^*(A) \leq 1 + N\lambda^*\left(\bigcup_{k=1}^{n} I_k\right) = 1 + N\sum_{k=1}^{n} h_k$$

$$< 1 + \sum_{k-1}^{n} \left(f(x_k + h_k) - f(x_k)\right) \leq 1 + f(b) - f(a),$$

where the last inequality follows from the facts that f is nondecreasing and $I_k \subset (a, b)$, $1 \leq k \leq n$. (See Exercise 8.13.) But, $N\lambda^*(A) < 1 + f(b) - f(a)$ for each $N \in \mathcal{N}$ implies that $\lambda^*(A) = 0$. ∎

Derivatives of Complex-Valued Functions

We conclude this section by briefly discussing differentiation of complex-valued functions of a real variable.

DEFINITION 8.5   **Derivative of a Complex-Valued Function**

A complex-valued function f defined in an open interval about $x \in \mathcal{R}$ is said to be **differentiable at x** if

$$\lim_{h \to 0} \frac{f(x + h) - f(x)}{h}$$

exists. In that case the limit is called the **derivative of f at x** and is denoted by $\boldsymbol{f'(x)}$.

The proof of the following proposition is left as an exercise for the reader.

□ □ □ **PROPOSITION 8.1**

Suppose that f is a complex-valued function defined in an open interval about x. Let $u = \Re f$ and $v = \Im f$. Then f is differentiable at x if and only if u and v are differentiable at x and, in that case, $f'(x) = u'(x) + iv'(x)$.

Exercises for Section 8.1

8.1 Show that $\lim_{y \to x+} g(y) = L \in \mathcal{R}^*$ if and only if for each sequence $\{y_n\}_{n=1}^{\infty}$ with $y_n > x$ and $y_n \to x$, we have $\lim_{n \to \infty} g(y_n) = L$. Establish a similar result for left-hand limits.

8.2 Suppose that $\{y_n\}_{n=1}^{\infty}$ is a sequence with $y_n > x$ and $y_n \to x$. Show that

$$\liminf_{y \to x+} g(y) \leq \liminf_{n \to \infty} g(y_n) \leq \limsup_{n \to \infty} g(y_n) \leq \limsup_{y \to x+} g(y).$$

Establish a similar result for left-hand limits.

8.3 Show that $\lim_{y \to x+} g(y) = L \in \mathcal{R}^*$ if and only if

$$\liminf_{y \to x+} g(y) = \limsup_{y \to x+} g(y) = L.$$

Establish a similar result for left-hand limits.

8.4 Prove that $\lim_{y \to x} g(y) = L \in \mathcal{R}^*$ if and only if

$$\liminf_{y \to x-} g(y) = \limsup_{y \to x-} g(y) = \liminf_{y \to x+} g(y) = \limsup_{y \to x+} g(y) = L.$$

8.5 Find the Dini-derivates of $f = \chi_Q$ at x if $x \notin Q$.

8.6 Set $f(0) = 0$ and $f(x) = x \sin(1/x)$ for $x \neq 0$. Determine the Dini-derivates of f at each $x \in \mathcal{R}$.

8.7 Suppose f attains its minimum at a point x and that f is defined in an open interval about x. Show that $D_+ f(x) \geq 0 \geq D^- f(x)$.

Exercises 8.8–8.11 discuss differentiation of convex functions.

A real-valued function f on (a, b) is called **convex** if

$$f\big(cx + (1 - c)y\big) \leq cf(x) + (1 - c)f(y)$$

for all $x, y \in (a, b)$ and $0 \leq c \leq 1$.

8.8 Prove that a convex function on (a, b) is continuous thereon.

8.9 Let f be a convex function on (a, b). Prove that $f'_+(x)$ and $f'_-(x)$ exist for all $x \in (a, b)$.

8.10 Let f be a convex function on (a, b). Prove that f'_+ and f'_- are nondecreasing on (a, b).

8.11 Let f be a convex function on (a, b). Prove that f' exists almost everywhere on (a, b).

8.12 Let $E \subset \mathcal{R}$ and O an open set with $O \supset E$. Suppose $\mathcal{V}$ is a Vitali cover of E. Show that $\mathcal{W} = \{ I \in \mathcal{V} : I \subset O \}$ is also a Vitali cover of E.

8.13 Suppose f is nondecreasing on $[a, b]$. Let I_k, $1 \leq k \leq n$, be a sequence of pairwise disjoint subintervals of $[a, b]$ with endpoints a_k and b_k, $1 \leq k \leq n$. Prove that

$$\sum_{k=1}^{n} \big(f(b_k) - f(a_k) \big) \leq f(b) - f(a).$$

8.14 Suppose f is nondecreasing on $[a, b]$. Let $\{J_j\}_{j=1}^m$ and $\{I_k\}_{k=1}^n$ be two disjoint sequences of closed subintervals of $[a, b]$ such that $\bigcup_{j=1}^m J_j \subset \bigcup_{k=1}^n I_k$. Denote the left and right endpoints of J_j by a_j and b_j, respectively, and those of I_k by c_k and d_k, respectively. Prove that

$$\sum_{j=1}^m \big(f(b_j) - f(a_j)\big) \leq \sum_{k=1}^n \big(f(d_k) - f(c_k)\big).$$

8.15 Suppose f is nondecreasing on an interval I having a nonempty interior.
   a) Show that at each interior point x, both $f(x+)$ and $f(x-)$ exist and that

$$f(x+) = \inf\{\, f(y) : y \in I \text{ and } y > x \,\},$$
$$f(x-) = \sup\{\, f(y) : y \in I \text{ and } y < x \,\}.$$

   Furthermore, verify that $f(x-) \leq f(x) \leq f(x+)$. Conclude that f is continuous at x if and only if $f(x+) = f(x-)$.
   b) Formulate and prove appropriate analogues of the statements in part (a) in the cases where x is either a left or right endpoint of I.
   c) Suppose that $a, b \in I$ with $a < b$. Let $x_1, x_2, \ldots, x_n$ be points of (a, b). Prove that $\sum_{j=1}^n \big(f(x_j+) - f(x_j-)\big) \leq f(b) - f(a)$.
   d) Deduce from the previous parts that the set of points in I where f is discontinuous is countable.

8.16 Show that the derivative of the Cantor function equals zero almost everywhere on $[0, 1]$.

8.17 We know that the Cantor function ψ is nondecreasing on $[0, 1]$ and, from the preceding exercise, $\psi' = 0$ ae on $[0, 1]$. Although ψ is nondecreasing and maps $[0, 1]$ to $[0, 1]$, it is "usually constant" in the sense that it is constant on each subinterval of the complement of the Cantor set. In this exercise, the Cantor function is used to construct a *strictly* increasing continuous function f on $[0, 1]$ such that $f' = 0$ ae on $[0, 1]$. For each $n \in \mathcal{N}$ and nonnegative integer $k < 2^n$, let

$$f_{nk}(x) = \begin{cases} 0, & x < \frac{k}{2^n}; \\ \psi(2^n x - k), & \frac{k}{2^n} \leq x < \frac{k+1}{2^n}; \\ 1, & x \geq \frac{k+1}{2^n}. \end{cases}$$

Define the function f on $[0, 1]$ by

$$f(x) = \sum_{n=1}^\infty 4^{-n} \sum_{k=0}^{2^n - 1} f_{nk}(x).$$

   a) Show that f is well-defined and continuous.
   b) Show that f is strictly increasing on $[0, 1]$.
   c) Prove that $f' = 0$ ae on $[0, 1]$.

8.18 Let f be a real-valued function on $[a, b]$. Suppose that $E \subset [a, b]$ and that f' exists and is bounded on E, say, by M. Prove that $\lambda^*\big(f(E)\big) \leq M\lambda^*(E)$.

8.19 Prove Proposition 8.1.

8.2 FUNCTIONS OF BOUNDED VARIATION

In Section 8.1 we proved that every monotone function is differentiable almost everywhere (Theorem 8.2). Furthermore, we stated that not only monotone functions but any function that does not "oscillate vigorously" is differentiable almost everywhere. Definition 8.6 makes precise the notion of not oscillating vigorously.

DEFINITION 8.6   **Total Variation; Bounded Variation**

Let f be a complex-valued function on $[a, b]$. The **total variation** of f over $[a, b]$, denoted by $V_a^b f$, is defined by

$$V_a^b f = \sup \left\{ \sum_{k=1}^{n} |f(x_k) - f(x_{k-1})| : a = x_0 < x_1 < \cdots < x_n = b \right\}.$$

If $V_a^b f < \infty$, then f is said to be of **bounded variation** on $[a, b]$.

EXAMPLE 8.4   *Illustrates Definition 8.6*

a) Any monotone function on $[a, b]$ is of bounded variation. In fact, we have that $V_a^b f = f(b) - f(a)$ if f is nondecreasing and that $V_a^b f = f(a) - f(b)$ if f is nonincreasing.

b) Define f on $[0, 1]$ by $f(0) = 0$ and $f(x) = x \sin(1/x)$, for $x \neq 0$. Then f is not of bounded variation on $[0, 1]$. To see this, consider for each $n \in \mathcal{N}$ the partition of $[0, 1]$ given by

$$0 < \frac{2}{(4n+1)\pi} < \frac{2}{4n\pi} < \frac{2}{(4n-1)\pi} < \cdots < \frac{2}{2\pi} < \frac{2}{\pi} < 1.$$

That is, $x_0 = 0$; $x_k = 2/(4n + 2 - k)\pi$, $k = 1, 2, \ldots, 4n + 1$; and $x_{4n+2} = 1$. Then,

$$\sum_{k=1}^{4n+2} |f(x_k) - f(x_{k-1})| = \frac{2}{(4n+1)\pi} + \frac{2}{(4n+1)\pi}$$
$$+ \frac{2}{(4n-1)\pi} + \frac{2}{(4n-1)\pi}$$
$$+ \cdots + \frac{2}{3\pi} + \frac{2}{3\pi} + \frac{2}{\pi} + \left| \sin 1 - \frac{2}{\pi} \right|,$$

and so,

$$\sum_{k=1}^{4n+2} |f(x_k) - f(x_{k-1})| \geq \frac{2}{\pi} \sum_{k=2}^{4n+2} \frac{1}{k}.$$

Because $\sum_{k=2}^{4n+2} k^{-1} \to \infty$ as $n \to \infty$, it follows that $V_0^1 f = \infty$.   □

Our next goal is to prove that any function of bounded variation on $[a, b]$ is differentiable almost everywhere on $[a, b]$. This is an immediate consequence of two facts: that any monotone function is differentiable almost everywhere (Theorem 8.2) and that any real-valued function of bounded variation can be written as the difference of two nondecreasing functions. This latter fact is the content of Theorem 8.3.

□ □ □ **THEOREM 8.3**

Let f be a real-valued function of bounded variation on $[a, b]$. Then

$$V_a^x f + V_x^y f = V_a^y f, \qquad a \leq x < y \leq b. \tag{8.14}$$

Moreover, f can be written as the difference of two nondecreasing functions on $[a, b]$.

PROOF   Define $V_a^a f = 0$. First we prove (8.14). If $x = a$, there is nothing to prove. So, assume $x > a$. Let $a = x_0 < x_1 < \cdots < x_n = y$ be a partition of $[a, y]$ and set $k = \min\{ j : x \leq x_j \}$. Then, by the definition of total variation,

$$\sum_{j=1}^{k-1} |f(x_j) - f(x_{j-1})| + |f(x) - f(x_{k-1})| \leq V_a^x f$$

and

$$|f(x_k) - f(x)| + \sum_{j=k+1}^{n} |f(x_j) - f(x_{j-1})| \leq V_x^y f.$$

Consequently, by the triangle inequality,

$$\sum_{j=1}^{n} |f(x_j) - f(x_{j-1})| \leq \sum_{j=1}^{k-1} |f(x_j) - f(x_{j-1})| + |f(x) - f(x_{k-1})|$$

$$+ |f(x_k) - f(x)| + \sum_{j=k+1}^{n} |f(x_j) - f(x_{j-1})|$$

$$\leq V_a^x f + V_x^y f.$$

Because the preceding inequality has been established for any partition of $[a, y]$, we have $V_a^y f \leq V_a^x f + V_x^y f$.

In order to establish the reverse inequality, let $a = x_0 < x_1 < \cdots < x_m = x$ and $x = y_0 < y_1 < \cdots < y_n = y$ be partitions of $[a, x]$ and $[x, y]$, respectively. Then

$$a = x_0 < x_1 < \cdots < x_m = x = y_0 < y_1 < \cdots < y_n = y$$

is a partition of $[a, y]$. Therefore,

$$V_a^y f \geq \sum_{j=1}^{m} |f(x_j) - f(x_{j-1})| + \sum_{k=1}^{n} |f(y_k) - f(y_{k-1})|.$$

Taking the supremum over partitions of $[a, x]$, we obtain that, for any partition, $x = y_0 < y_1 < \cdots < y_n = y$ of $[x, y]$,

$$V_a^y f \geq V_a^x f + \sum_{k=1}^{n} |f(y_k) - f(y_{k-1})|.$$

Taking the supremum over all partitions of $[x, y]$ yields $V_a^y f \geq V_a^x f + V_x^y f$. Consequently, we have now verified (8.14).

Write $f(x) = V_a^x f - (V_a^x f - f(x))$. We claim that $V_a^x f$ and $V_a^x f - f(x)$ are nondecreasing functions on $[a, b]$. From (8.14) and the fact that $V_x^y f \geq 0$ for $x < y$, we deduce that, as a function of x, $V_a^x f$ is nondecreasing on $[a, b]$.

It remains to show $V_a^x f - f(x)$ is nondecreasing on $[a, b]$. Let $a \leq x < y \leq b$. By (8.14),

$$\left(V_a^y f - f(y)\right) - \left(V_a^x f - f(x)\right) = V_x^y f - \left(f(y) - f(x)\right)$$
$$\geq V_x^y f - |f(y) - f(x)| \geq 0,$$

where the last inequality holds since $x = x_0 < x_1 = y$ is a partition of $[x, y]$. ∎

□ □ □ **COROLLARY 8.1**

Suppose f is a complex-valued function of bounded variation on $[a, b]$. Then f can be written in the form

$$f = (f_1 - f_2) + i(f_3 - f_4),$$

where f_j, $1 \leq j \leq 4$, are nondecreasing functions on $[a, b]$.

PROOF Because f is of bounded variation, so are $\Re f$ and $\Im f$. (See Exercise 8.24.) Applying Theorem 8.3 to $\Re f$ and $\Im f$ completes the proof. ∎

Since any monotone function on $[a, b]$ is differentiable almost everywhere on $[a, b]$ (Theorem 8.2) and every function of bounded variation on $[a, b]$ can be written as a linear combination of nondecreasing functions (Corollary 8.1), we have the following theorem.

□ □ □ **THEOREM 8.4**

Any function of bounded variation on $[a, b]$ is differentiable almost everywhere on $[a, b]$.

Exercises for Section 8.2

Note: A ★ denotes an exercise that will be subsequently referenced.

8.20 Using only Definition 8.6, prove that if f is of bounded variation on $[a, b]$, then it is bounded thereon.

8.21 Define f on $[0, 1]$ by $f(0) = 0$ and $f(x) = x^2 \sin(1/x)$ for $x \neq 0$. Show that f is of bounded variation on $[0, 1]$.

8.22 Define f on $[0, 1]$ by $f(0) = 0$ and $f(x) = x^2 \sin(1/x^2)$ for $x \neq 0$. Show that f is not of bounded variation on $[0, 1]$.

8.23 Define f on $[0, 1]$ by $f(0) = 0$ and $f(x) = x^\alpha \sin(1/x)$ for $x \neq 0$. Show that f is of bounded variation on $[0, 1]$ if and only if $\alpha > 1$.

8.24 Prove that f is of bounded variation on $[a, b]$ if and only if $\Re f$ and $\Im f$ are of bounded variation on $[a, b]$.

★8.25 Let f and g be complex-valued functions on $[a, b]$ and $\alpha \in \mathbb{C}$. Prove that
a) $V_a^b(f + g) \leq V_a^b f + V_a^b g$.
b) $V_a^b(\alpha f) = |\alpha| V_a^b f$.

8.26 Suppose f is a function of bounded variation on $[a, b]$. Show that f has only a countable number of discontinuities.

8.27 Let f be of bounded variation on $[a, b]$ and D denote the set of points of (a, b) at which f is discontinuous. By Exercise 8.26, we can write $D = \{x_n\}_n$. For each n, set $d_n = f(x_n+) - f(x_n-)$. Show that $\sum_n |d_n| \leq V_a^b f$.

★8.28 Let f and g be of bounded variation on $[a, b]$. Show that

$$V_a^b(fg) \leq \left(\sup\{\,|f(x)| : x \in [a, b]\,\}\right) V_a^b g + \left(\sup\{\,|g(x)| : x \in [a, b]\,\}\right) V_a^b f.$$

Deduce that the product of two functions of bounded variation on $[a, b]$ is also of bounded variation on $[a, b]$.

8.29 Let $\{f_n\}_{n=1}^\infty$ be a sequence of real-valued functions on $[a, b]$ that converge pointwise to the function f. Prove that $V_a^b f \leq \liminf_{n \to \infty} V_a^b f_n$.

8.30 Suppose that $f : [a, b] \to [c, d]$ is monotone and that g is of bounded variation on $[c, d]$. Prove that $V_a^b(g \circ f) \leq V_c^d g$.

8.31 Let f be of bounded variation on $[a, b]$. If f is continuous at $x_0 \in [a, b]$, show that the function $g(x) = V_a^x f$ is also continuous at x_0.

8.32 Suppose that $\{f_n\}_{n=1}^\infty$ is a sequence of functions of bounded variation on $[a, b]$ such that $\sum_n f_n(a)$ converges absolutely and $\sum_n V_a^b f_n < \infty$. Prove that

a) $\sum_n f_n(x)$ converges absolutely for each $x \in [a, b]$.

b) $V_a^b \left(\sum_n f_n\right) \leq \sum_n V_a^b f_n$.

8.3 THE INDEFINITE LEBESGUE INTEGRAL

Recall that the two fundamental theorems of calculus for Riemann integration show that differentiation and integration are inverse operations. More precisely, we have the following two facts.

First Fundamental Theorem of Calculus: Suppose that f is Riemann integrable on $[a, b]$. Let

$$F(x) = \int_a^x f(t)\, dt, \qquad a \leq x \leq b.$$

Then F is differentiable at all points x at which f is continuous (hence, almost everywhere by Theorem 2.7) and at such points $F'(x) = f(x)$. In other words,

$$\frac{d}{dx} \int_a^x f(t)\, dt = f(x) \tag{8.15}$$

at all continuity points of f.

Second Fundamental Theorem of Calculus: Suppose that f is defined on $[a, b]$ and f' exists and is Riemann integrable on $[a, b]$. Then

$$\int_a^x f'(t)\, dt = f(x) - f(a), \qquad a \leq x \leq b. \tag{8.16}$$

In this section, we will prove a generalization of (8.15) to Lebesgue integration theory. Then, in the next section, we will characterize all functions f for which (8.16) holds in the Lebesgue sense.

To begin, we introduce some useful abbreviations. We write $\mathcal{L}^1([a,b])$ for $\mathcal{L}^1([a,b], \mathcal{M}_{[a,b]}, \lambda_{[a,b]})$ and we write $\mathcal{L}^1(\mathcal{R})$ for $\mathcal{L}^1(\mathcal{R}, \mathcal{M}, \lambda)$. Recalling our convention for using Riemann-integral notation for Lebesgue integrals, we have that $f \in \mathcal{L}^1([a,b])$ means f is $\mathcal{M}_{[a,b]}$-measurable and $\int_a^b |f(x)|\, dx < \infty$ and that $f \in \mathcal{L}^1(\mathcal{R})$ means f is $\mathcal{M}$-measurable and $\int_{-\infty}^{\infty} |f(x)|\, dx < \infty$. Moreover, we will continue to use the phrase *almost everywhere* instead of *Lebesgue almost everywhere* and the notation *ae* instead of λ-*ae*.

Our first goal is to prove that whenever $f \in \mathcal{L}^1([a,b])$, (8.15) holds almost everywhere on $[a,b]$. Several preliminary results are needed to establish that fact.

□ □ □ PROPOSITION 8.2

Suppose that $f \in \mathcal{L}^1([a,b])$ and set

$$F(x) = \int_a^x f(t)\, dt, \qquad a \le x \le b.$$

Then F is continuous and of bounded variation on $[a,b]$. Moreover,

$$V_a^b F = \int_a^b |f(x)|\, dx. \tag{8.17}$$

PROOF Let $x \in [a,b]$. We will show that F is continuous at x. Let $\{x_n\}_{n=1}^{\infty} \subset [a,b]$ be such that $x_n \to x$ as $n \to \infty$. Set $f_n = f\chi_{[a,x_n)}$. Note that we have $f_n \to f\chi_{[a,x)}$ except possibly at x, so that $f_n \to f\chi_{[a,x)}$ ae. Moreover, as $|f_n| \le |f| \in \mathcal{L}^1([a,b])$, the DCT implies that

$$\lim_{n \to \infty} F(x_n) = \lim_{n \to \infty} \int_a^{x_n} f(t)\, dt = \lim_{n \to \infty} \int_a^b f_n(t)\, dt$$

$$= \int_a^b f(t)\chi_{[a,x)}(t)\, dt = \int_a^x f(t)\, dt = F(x).$$

Consequently, F is continuous at x.

Now we show that F is of bounded variation on $[a,b]$. Consider an arbitrary partition $a = x_0 < x_1 < \cdots < x_n = b$ of $[a,b]$. Then

$$\sum_{k=1}^{n} |F(x_k) - F(x_{k-1})| = \sum_{k=1}^{n} \left| \int_a^{x_k} f(x)\, dx - \int_a^{x_{k-1}} f(x)\, dx \right|$$

$$= \sum_{k=1}^{n} \left| \int_{x_{k-1}}^{x_k} f(x)\, dx \right| \le \sum_{k=1}^{n} \int_{x_{k-1}}^{x_k} |f(x)|\, dx$$

$$= \int_a^b |f(x)|\, dx.$$

Hence,

$$V_a^b F \le \int_a^b |f(x)|\, dx. \tag{8.18}$$

To establish the reverse of (8.18), we first consider the case where f is a continuous complex-valued function on the interval $[a, b]$. Let $\epsilon > 0$ be given. By the uniform continuity of f, there is a $\delta > 0$ such that $x, y \in [a, b]$ and $|x - y| < \delta$ implies $|f(x) - f(y)| < \epsilon/2(b - a)$.

Now, let $a = x_0 < x_1 < \cdots < x_n = b$ be a partition such that $|x_{j+1} - x_j| < \delta$ for $j = 0, 1, \ldots, n - 1$. Then

$$
\begin{aligned}
|F(x_{j+1}) - F(x_j)| &= \left| \int_{x_j}^{x_{j+1}} f(x)\, dx \right| \\
&\geq \left| \int_{x_j}^{x_{j+1}} f(x_j)\, dx \right| - \int_{x_j}^{x_{j+1}} |f(x) - f(x_j)|\, dx \\
&\geq \int_{x_j}^{x_{j+1}} |f(x_j)|\, dx - \frac{\epsilon}{2(b - a)}(x_{j+1} - x_j) \\
&\geq \int_{x_j}^{x_{j+1}} |f(x)|\, dx - \int_{x_j}^{x_{j+1}} \big||f(x)| - |f(x_j)|\big|\, dx \\
&\quad - \frac{\epsilon}{2(b - a)}(x_{j+1} - x_j) \\
&\geq \int_{x_j}^{x_{j+1}} |f(x)|\, dx - \frac{\epsilon}{b - a}(x_{j+1} - x_j).
\end{aligned}
\tag{8.19}
$$

Summing on both sides of (8.19) we obtain

$$
V_a^b F \geq \sum_{j=0}^{n-1} |F(x_{j+1}) - F(x_j)| \geq \int_a^b |f(x)|\, dx - \epsilon.
$$

As ϵ is an arbitrary positive number, we obtain the reverse of (8.18) in the case where f is a continuous complex-valued function on $[a, b]$.

To prove the reverse of (8.18) in the general case, let $f \in \mathcal{L}^1([a, b])$ and let $\epsilon > 0$. Using Exercise 5.82 on page 173, we select a simple function s such that $\int_a^b |f(x) - s(x)|\, dx < \epsilon/2$. Then applying Exercise 4.13(d) on page 120 and the dominated convergence theorem, we choose a continuous function f_0 such that $\int_a^b |s(x) - f_0(x)|\, dx < \epsilon/2$. It follows that

$$
\int_a^b |f(x) - f_0(x)|\, dx < \epsilon.
\tag{8.20}
$$

Now let $F_0(x) = \int_a^x f_0(t)\, dt$. From Exercise 8.25 on page 284, we have

$$
V_a^b F_0 - V_a^b(F_0 - F) \leq V_a^b F.
$$

It follows from (8.18) and (8.20) that

$$
V_a^b F_0 - \epsilon \leq V_a^b F.
\tag{8.21}
$$

We have already proved the proposition in the case $F = F_0$. Thus, in view of (8.20), (8.21), and the previous inequality, we conclude that

$$\int_a^b |f(x)|\, dx - \epsilon \le \int_a^b |f(x)|\, dx - \int_a^b |f(x) - f_0(x)|\, dx$$

$$\le \int_a^b |f_0(x)|\, dx = V_a^b F_0 \le V_a^b F + \epsilon.$$

Because ϵ was arbitrarily chosen, we have proved the reverse of (8.18) in the general case. ∎

Remark: Since any continuous function on $[a, b]$ is uniformly continuous on $[a, b]$ we have, in fact, that the function F in Proposition 8.2 is uniformly continuous on $[a, b]$.

EXAMPLE 8.5   *Illustrates Proposition 8.2*

a) It is quite easy to prove directly that $\sin x$ is of bounded variation on $[0, 2\pi]$. However, it is even easier to prove that fact by employing Proposition 8.2. We just note that $\cos x \in \mathcal{L}^1([0, 2\pi])$ and that

$$\sin x = \int_0^x \cos t\, dt, \qquad 0 \le x \le 2\pi.$$

b) Define F on $[0, 1]$ by $F(0) = 0$ and $F(x) = x \sin(1/x)$ for $x \ne 0$. Clearly, F is continuous on $[0, 1]$. However, as we discovered in Example 8.4(b) on page 282, F is not of bounded variation on $[0, 1]$. Consequently, by Proposition 8.2, it is impossible to find a function $f \in \mathcal{L}^1([0, 1])$ such that $x \sin(1/x) = \int_0^x f(t)\, dt$ for $0 < x \le 1$. In words, F is not the indefinite integral of a Lebesgue integrable function on $[0, 1]$.

c) The Cantor function ψ is continuous and of bounded variation on $[0, 1]$ (because it is monotone). Exercise 7.28 on page 244 shows, and we will show again in Example 8.6(a), that ψ is not the indefinite integral of a Lebesgue integrable function on $[0, 1]$. This proves that the converse of Proposition 8.2 fails. □

A result analogous to Proposition 8.2 is valid for functions defined on $\mathcal{R}$. Specifically, we have the following fact whose proof is left as an exercise for the reader.

□ □ □ **PROPOSITION 8.3**

Suppose that $f \in \mathcal{L}^1(\mathcal{R})$ and set

$$F(x) = \int_{-\infty}^x f(t)\, dt, \qquad -\infty < x < \infty.$$

Then F is continuous on $\mathcal{R}$ and is of bounded variation on every finite closed interval. Moreover,

$$V_{-\infty}^\infty F = \int_{-\infty}^\infty |f(x)|\, dx, \tag{8.22}$$

where, by definition, $V_{-\infty}^\infty F = \lim_{n \to \infty} V_{-n}^n F$.

First Fundamental Theorem of Calculus

By Proposition 8.2 (page 286) and Theorem 8.4 (page 284), if $f \in \mathcal{L}^1([a,b])$ and F is the indefinite integral of f, then F is differentiable almost everywhere on $[a,b]$. So we will have established the generalization of (8.15) to Lebesgue integration theory once we show that $F'(x) = f(x)$ ae. To accomplish that, we need the following two lemmas, whose proofs are left to the reader. (See Exercises 8.38 and 8.39.)

□ □ □ **LEMMA 8.2**

If $f \in \mathcal{L}^1([a,b])$ and $\int_a^x f(t)\,dt = 0$ for all $x \in [a,b]$, then $f = 0$ almost everywhere on $[a,b]$.

□ □ □ **LEMMA 8.3**

Suppose that g is defined in some open interval about $x \in \mathcal{R}$ and g is continuous at x. Then

$$\lim_{h \to 0} \frac{1}{h} \int_x^{x+h} g(t)\,dt = g(x).$$

We are now in a position to establish the **first fundamental theorem of calculus for Lebesgue integration**.

□ □ □ **THEOREM 8.5 First Fundamental Theorem of Calculus**

Suppose that $f \in \mathcal{L}^1([a,b])$ and set

$$F(x) = \int_a^x f(t)\,dt, \qquad a \le x \le b.$$

Then F is differentiable almost everywhere on $[a,b]$ and, in fact,

$$F'(x) = f(x) \tag{8.23}$$

for almost all $x \in [a,b]$.

PROOF　We have already observed that F is differentiable almost everywhere on $[a,b]$. Hence, it remains to prove (8.23). We do this first for bounded, nonnegative f. So, assume $0 \le f \le M$ on $[a,b]$. Extend the domain of f (and, hence, of F) by setting $f(x) = 0$ for $x \notin [a,b]$. Let

$$f_n(t) = \frac{F(t + 1/n) - F(t)}{1/n}, \qquad a \le t \le b.$$

Note that $f_n \to F'$ ae. Because $f_n(t) = n \int_t^{t+1/n} f(s)\,ds$ and f is bounded by M, we have $|f_n| \le M$ for all n. Applying the dominated convergence theorem, we get that

$$\lim_{n \to \infty} \int_a^x f_n(t)\,dt = \int_a^x F'(t)\,dt, \qquad a \le x \le b.$$

On the other hand, by Lemma 8.3, we have for $a \leq x \leq b$,

$$\lim_{n \to \infty} \int_a^x f_n(t)\, dt = \lim_{n \to \infty} n \int_a^x [F(t + 1/n) - F(t)]\, dt$$

$$= \lim_{n \to \infty} n \left[\int_x^{x+1/n} F(t)\, dt - \int_a^{a+1/n} F(t)\, dt \right]$$

$$= F(x) - F(a) = \int_a^x f(t)\, dt.$$

Note that this result does not require f to be bounded.

Hence, we see that $\int_a^x f(t)\, dt = \int_a^x F'(t)\, dt$ for $a \leq x \leq b$ or, equivalently, that $\int_a^x \left(f(t) - F'(t) \right) dt = 0$ for $a \leq x \leq b$. Thus, by Lemma 8.2, $F' = f$ ae. Consequently, we have established (8.23) in case f is nonnegative and bounded.

Next we will establish (8.23) for nonnegative f without the boundedness condition. Let $f_n(t)$ be as before. Because f is nonnegative, so are the f_ns. Applying Fatou's lemma and the previous displayed equation, we get that

$$\int_a^x F'(t)\, dt \leq \liminf_{n \to \infty} \int_a^x f_n(t)\, dt = \int_a^x f(t)\, dt. \tag{8.24}$$

We use the method of truncation to reduce this case to the bounded case. For each $n \in \mathcal{N}$, let

$$g_n(t) = \begin{cases} f(t), & f(t) \leq n; \\ n, & f(t) > n. \end{cases}$$

Then each g_n is a nonnegative bounded measurable function. Consequently, as we have shown that (8.23) holds for such functions, we conclude that for almost all $x \in [a, b]$,

$$\frac{d}{dx} \int_a^x g_n(t)\, dt = g_n(x).$$

Now,

$$F(x) = \int_a^x f(t)\, dt = \int_a^x \left(f(t) - g_n(t) \right) dt + \int_a^x g_n(t)\, dt.$$

The first term on the right-hand side is nondecreasing because $f \geq g_n$ and, hence, is differentiable almost everywhere and, clearly, its derivative is nonnegative where it exists. Thus, for almost all $x \in [a, b]$,

$$F'(x) = \frac{d}{dx} \int_a^x \left(f(t) - g_n(t) \right) dt + g_n(x) \geq g_n(x).$$

Because $g_n \uparrow f$ pointwise on $[a, b]$, the monotone convergence theorem and the previous inequality give

$$\int_a^x F'(t)\, dt \geq \lim_{n \to \infty} \int_a^x g_n(t)\, dt = \int_a^x f(t)\, dt. \tag{8.25}$$

From (8.24) and (8.25), we conclude that $\int_a^x F'(t)\,dt = \int_a^x f(t)\,dt$ for $a \le x \le b$ and, so, $F' = f$ ae on $[a,b]$.

It remains to establish (8.23) without the nonnegativity assumption. For $f \in \mathcal{L}^1([a,b])$, write $f = f_1 - f_2 + if_3 - if_4$, where $f_j \ge 0$, $1 \le j \le 4$. Then,

$$F(x) = \int_a^x f_1(t)\,dt - \int_a^x f_2(t)\,dt + i\int_a^x f_3(t)\,dt - i\int_a^x f_4(t)\,dt.$$

From the preceding equation and the fact that (8.23) holds for nonnegative functions, we deduce that $F' = f_1 - f_2 + if_3 - if_4 = f$ ae on $[a,b]$.   ■

As a corollary of Theorem 8.5, we obtain the following result whose proof is left as an exercise for the reader. (See Exercise 8.40.)

□ □ □   **COROLLARY 8.2**

Suppose that $f \in \mathcal{L}^1(\mathcal{R})$ and set

$$F(x) = \int_{-\infty}^x f(t)\,dt, \qquad -\infty < x < \infty.$$

Then F is differentiable almost everywhere on $\mathcal{R}$ and, in fact, $F'(x) = f(x)$ for almost all $x \in \mathcal{R}$.

Exercises for Section 8.3

8.33   Let $\{f_n\}_{n=1}^\infty \subset \mathcal{L}^1([a,b])$. For each $n \in \mathcal{N}$, set

$$F_n(x) = \int_a^x f_n(t)\,dt, \qquad a \le x \le b.$$

Suppose that $F_n \to 0$ ae. Does it follow that
a) $f_n \to 0$ ae?
b) $f_n \to 0$ in measure?

8.34   Suppose F_1 and F_2 are disjoint bounded closed sets. Prove that there exist disjoint open sets O_1 and O_2 such that $O_1 \supset F_1$ and $O_2 \supset F_2$. *Hint:* Refer to Exercises 3.21(a) and 3.22(a) on page 101.

8.35   Prove Proposition 8.3 on page 288.

8.36   Give an example of a function that is of bounded variation on $[0,1]$ but that is not continuous on $[0,1]$. Can such a function be the indefinite integral of a Lebesgue integrable function on $[0,1]$? Explain your answer.

8.37   Give an example of a function that is continuous on $[0,1]$ but not of bounded variation on $[0,1]$. Can such a function be the indefinite integral of a Lebesgue integrable function on $[0,1]$? Explain your answer.

8.38   Prove Lemma 8.2 on page 289 by proceeding as follows.
a) Explain why we can, without loss of generality, assume that f is real-valued.
b) Show that if f is positive on a set of positive measure, then there is a closed subset K of (a,b) such that $\int_K f(x)\,dx > 0$. *Hint:* Refer to Exercise 5.52 on page 164 and Exercise 3.43 on page 108.

c) Use part (b) to deduce that if f is positive on a set of positive measure, then $\int_I f(x)\,dx \neq 0$ for some open interval $I \subset (a, b)$. *Hint:* Write $O = (a, b) \setminus K$ and note that $\int_O f(x)\,dx = -\int_K f(x)\,dx$.

d) Use part (c) to deduce that if f is positive on a set of positive measure, then $\int_a^x f(t)\,dt \neq 0$ for some $x \in [a, b]$, contradicting a hypothesis of the lemma. Conclude that the set where f is positive has measure zero.

e) Explain why the set where f is negative must have measure zero.

8.39 Prove Lemma 8.3 on page 289.

8.40 Prove Corollary 8.2 on page 291.

8.4 ABSOLUTELY CONTINUOUS FUNCTIONS

We have now extended the first fundamental theorem of calculus to the setting of Lebesgue integration theory. We might expect the generalization of the second fundamental theorem of calculus to be as follows: Suppose that f is defined on $[a, b]$ and that f' exists almost everywhere and is Lebesgue integrable on $[a, b]$. Then

$$\int_a^x f'(t)\,dt = f(x) - f(a), \qquad a \leq x \leq b.$$

But this equality is not true in general.† For example, let ψ be the Cantor function. Then $\psi' = 0$ ae and so $\psi' \in \mathcal{L}^1([0, 1])$. However,

$$\int_0^1 \psi'(t)\,dt = 0 \neq 1 = \psi(1) - \psi(0).$$

In fact, for all $x \in (0, 1]$, $\int_0^x \psi'(t)\,dt = 0 \neq \psi(x) - \psi(0)$.

Evidently then, a generalization of the second fundamental theorem of calculus to Lebesgue integration theory requires more restrictive hypotheses on f. In the next few pages, we will characterize the functions for which that generalization holds. To begin, we give such functions a special name, the rationale for which will become apparent once we characterize them.

DEFINITION 8.7   **Absolutely Continuous Function on** $[a, b]$

Suppose that f is defined on $[a, b]$, f' exists almost everywhere and is Lebesgue integrable on $[a, b]$, and

$$f(x) = f(a) + \int_a^x f'(t)\,dt, \qquad a \leq x \leq b.$$

Then f is said to be **absolutely continuous** on $[a, b]$.

---

† The equality is true, however, if in the preceding paragraph, "f' exists almost everywhere" is replaced by "f' exists everywhere." See, for instance, Theorem 5 of John F. Randolph's *Basic Real and Abstract Analysis* (New York: Academic Press, 1968), p. 424.

DEFINITION 8.8   Absolutely Continuous Function on $\mathcal{R}$

Suppose that f is defined on $\mathcal{R}$, f' exists almost everywhere and is Lebesgue integrable on $\mathcal{R}$, and

$$f(x) = \int_{-\infty}^{x} f'(t)\,dt, \qquad -\infty < x < \infty.$$

Then f is said to be **absolutely continuous** on $\mathcal{R}$.

EXAMPLE 8.6   *Illustrates Definitions 8.7 and 8.8*

a) We just saw that the Cantor function ψ is not absolutely continuous on $[0,1]$. As $\psi' = 0$ ae on $[0,1]$, Theorem 8.5 on page 289 shows the impossibility of representing ψ as an indefinite integral of an $\mathcal{L}^1([0,1])$ function. For if $\psi(x) = \int_0^x g(t)\,dt$, $0 \le x \le 1$, then we must have $0 = \psi' = g$ ae, implying that $\psi(x) = 0$, $0 \le x \le 1$, which is not true.

b) Theorem 8.5, in particular (8.23), shows that if a function F on $[a,b]$ can be represented as the indefinite integral of a function in $\mathcal{L}^1([a,b])$, then we have $F(x) = \int_a^x F'(t)\,dt$ for $a \le x \le b$, so that F is absolutely continuous. *Note:* $F(a) = 0$.

c) Corollary 8.2 (page 291) shows that if a function F on $\mathcal{R}$ can be represented as the indefinite integral of a function in $\mathcal{L}^1(\mathcal{R})$, then $F(x) = \int_{-\infty}^x F'(t)\,dt$ for $x \in \mathcal{R}$, so that F is absolutely continuous. *Note:* $F(-\infty) = 0$.   □

The results of Examples 8.6(b) and 8.6(c) are summarized in the following proposition.

□ □ □   PROPOSITION 8.4

a) *Let F be a function defined on $[a,b]$ and suppose that there is a function $f \in \mathcal{L}^1([a,b])$ such that $F(x) = \int_a^x f(t)\,dt$ for $a \le x \le b$. Then F is absolutely continuous on $[a,b]$.*

b) *Let F be a function defined on $\mathcal{R}$ and suppose that there is a function $f \in \mathcal{L}^1(\mathcal{R})$ such that $F(x) = \int_{-\infty}^x f(t)\,dt$ for $x \in \mathcal{R}$. Then F is absolutely continuous on $\mathcal{R}$.*

□ □ □   PROPOSITION 8.5

If f is absolutely continuous on $[a,b]$, then it is continuous and of bounded variation on $[a,b]$. Moreover,

$$V_a^b f = \int_a^b |f'(x)|\,dx. \qquad (8.26)$$

PROOF   By assumption, $f' \in \mathcal{L}^1([a,b])$ and

$$f(x) = f(a) + \int_a^x f'(t)\,dt, \qquad a \le x \le b.$$

Proposition 8.2 (page 286) shows that the function $\int_a^x f'(t)\, dt$ is continuous and of bounded variation on $[a, b]$. Obviously, constant functions are continuous and of bounded variation. Hence, f is continuous and of bounded variation on $[a, b]$. Equation (8.26) follows immediately from Proposition 8.2 once we note that for any function g and constant α, $V_a^b(\alpha + g) = V_a^b g$. ∎

EXAMPLE 8.7   *Illustrates Proposition 8.5*

Define f on $[0, 1]$ by $f(0) = 0$ and $f(x) = x \sin(1/x)$ for $x \neq 0$. It is easy to see that f is continuous on $[0, 1]$. In Example 8.4(b) on page 282, we learned that f is not of bounded variation on $[0, 1]$. Hence, by Proposition 8.5, it is not absolutely continuous on $[0, 1]$. □

An Equivalent Condition for Absolute Continuity

Although, as we know from Proposition 8.5, continuity and bounded variation are necessary conditions for absolute continuity, they are not sufficient. Indeed, the Cantor function is continuous and of bounded variation on $[0, 1]$ but is not absolutely continuous.

In the next few pages, we will discover a continuity-type condition that is equivalent to absolute continuity. As we have seen, any such continuity-type condition must be stronger than ordinary continuity and, in fact, than uniform continuity.

□ □ □   **PROPOSITION 8.6**

Suppose f is absolutely continuous on $[a, b]$. Then for each $\epsilon > 0$, there is a $\delta > 0$ such that if $\{(a_k, b_k)\}_{k=1}^n$ is any finite sequence of pairwise disjoint subintervals of $[a, b]$ with $\sum_{k=1}^n (b_k - a_k) < \delta$, then $\sum_{k=1}^n |f(b_k) - f(a_k)| < \epsilon$.

PROOF   By assumption, $f' \in \mathcal{L}^1([a, b])$ and $f(x) = f(a) + \int_a^x f'(t)\, dt$ for $a \leq x \leq b$. It follows that

$$f(d) - f(c) = \int_c^d f'(t)\, dt, \qquad a \leq c \leq d \leq b.$$

Now, let $\epsilon > 0$ be given. Applying Exercise 5.80 on page 173, we can choose a $\delta > 0$ such that if $E \subset [a, b]$ is measurable and $\lambda(E) < \delta$, then $\int_E |f'(t)|\, dt < \epsilon$.

Let $\{(a_k, b_k)\}_{k=1}^n$ be a finite sequence of pairwise disjoint subintervals of $[a, b]$ such that $\sum_{k=1}^n (b_k - a_k) < \delta$. Set $E = \bigcup_{k=1}^n (a_k, b_k)$. Then E is measurable, $E \subset [a, b]$, and $\lambda(E) < \delta$. Therefore,

$$\sum_{k=1}^n |f(b_k) - f(a_k)| = \sum_{k=1}^n \left| \int_{a_k}^{b_k} f'(t)\, dt \right| \leq \sum_{k=1}^n \int_{a_k}^{b_k} |f'(t)|\, dt = \int_E |f'(t)|\, dt < \epsilon,$$

as desired. ∎

Proposition 8.6 implies that an absolutely continuous function on a finite closed interval is necessarily uniformly continuous. However, the converse is not true. Indeed, we know that any continuous function on a finite closed interval is uniformly continuous, and we have already encountered several functions that are continuous but not absolutely continuous on a finite closed interval.

The question that now arises is whether the necessary condition for absolute continuity in Proposition 8.6 is also sufficient. The answer is yes, as Proposition 8.7 shows.

□ □ □ **PROPOSITION 8.7**

Suppose that f is defined on $[a,b]$ and that for each $\epsilon > 0$, there is a $\delta > 0$ such that if $\{(a_k, b_k)\}_{k=1}^{n}$ is any finite sequence of pairwise disjoint subintervals of $[a,b]$ with $\sum_{k=1}^{n}(b_k - a_k) < \delta$, then $\sum_{k=1}^{n}|f(b_k) - f(a_k)| < \epsilon$. Then f is absolutely continuous on $[a,b]$.

To establish Proposition 8.7, we first prove the following two lemmas.

□ □ □ **LEMMA 8.4**

Suppose f satisfies the hypotheses of Proposition 8.7. Then f is of bounded variation on $[a,b]$ and $f' \in \mathcal{L}^1([a,b])$.

PROOF By assumption, we can choose $\delta > 0$ so that if $\{(a_k, b_k)\}_{k=1}^{n}$ is any finite sequence of pairwise disjoint subintervals of $[a,b]$ such that $\sum_{k=1}^{n}(b_k - a_k) < \delta$, then $\sum_{k=1}^{n}|f(b_k) - f(a_k)| < 1$.

Choose $N \in \mathcal{N}$ so that $(b-a)/N < \delta$, and let $a = x_0 < x_1 < \cdots < x_N = b$ be the partition of the interval $[a,b]$ that divides it into N subintervals of equal length $(b-a)/N$. If $x_{k-1} = y_0 < y_1 < \cdots < y_m = x_k$ is any partition of $[x_{k-1}, x_k]$, then $\sum_{j=1}^{m}(y_j - y_{j-1}) = x_k - x_{k-1} < \delta$ and, consequently, we have that $\sum_{j=1}^{m}|f(y_j) - f(y_{j-1})| < 1$. Therefore, $V_{x_{k-1}}^{x_k} f \le 1$ for $k = 1, 2, \ldots, N$. Using (8.14) on page 283, we can now conclude that

$$V_a^b f = \sum_{k=1}^{N} V_{x_{k-1}}^{x_k} f \le N < \infty.$$

Hence, f is of bounded variation on $[a,b]$.

Next we must prove that $f' \in \mathcal{L}^1([a,b])$. As f is of bounded variation on $[a,b]$, it is differentiable almost everywhere on $[a,b]$. To show that f' is measurable, we first extend the domain of f by setting $f(x) = f(b)$ for $x > b$. Then we note that because f is measurable, so is the function $f_n(x) = n(f(x + 1/n) - f(x))$ for each $n \in \mathcal{N}$ (why?); and, because we have $f_n(x) \to f'(x)$ ae, it follows from Proposition 4.8 on page 140 that f' is measurable.

Finally, we show that f' is Lebesgue integrable. As f is of bounded variation, we can write $f = f_1 - f_2 + if_3 - if_4$, where f_j, $1 \le j \le 4$, are nondecreasing.

Now, let g be any nondecreasing function on $[a,b]$. Then g' exists ae on $[a,b]$ and so

$$\lim_{n \to \infty} \frac{g(x + 1/n) - g(x)}{1/n} = g'(x) \qquad \text{ae.}$$

Define $g(x) = g(b)$ for $x > b$. The functions $n(g(x + 1/n) - g(x))$ are nonnegative and hence, by Fatou's lemma,

$$\int_a^b g'(x)\,dx \le \liminf_{n\to\infty} n \int_a^b [g(x + 1/n) - g(x)]\,dx$$

$$= \liminf_{n\to\infty} n \left\{ \int_{a+1/n}^{b+1/n} g(t)\,dt - \int_a^b g(t)\,dt \right\}$$

$$= \liminf_{n\to\infty} n \left\{ \int_b^{b+1/n} g(t)\,dt - \int_a^{a+1/n} g(t)\,dt \right\}$$

$$= g(b) - \limsup_{n\to\infty} n \int_a^{a+1/n} g(t)\,dt.$$

Noting $n \int_a^{a+1/n} g(t)\,dt \ge g(a)$ for all $n \in \mathcal{N}$, we conclude that

$$\int_a^b g'(x)\,dx \le g(b) - g(a). \qquad (8.27)$$

Returning to f, we have $f' = f_1' - f_2' + if_3' - if_4'$ almost everywhere, so that $|f'| \le \sum_{j=1}^4 f_j'$ ae. Hence, by (8.27),

$$\int_a^b |f'(x)|\,dx \le \sum_{j=1}^4 \int_a^b f_j'(x)\,dx \le \sum_{j=1}^4 \big(f_j(b) - f_j(a)\big) < \infty.$$

This result shows that $f' \in \mathcal{L}^1([a,b])$ and completes the proof of the lemma. ∎

☐ ☐ ☐ **LEMMA 8.5**

Suppose that g satisfies the hypotheses of Proposition 8.7. If $g' = 0$ ae on $[a, b]$, then g is constant on $[a, b]$.

PROOF Let $x \in (a, b]$ be fixed but arbitrary. We will prove the lemma by establishing that $g(x) = g(a)$. By hypothesis, $g'(y) = 0$ for almost all $y \in (a, x)$. Now let $E = \{ y \in (a, x) : g'(y) = 0 \}$ and note that $\lambda(E) = x - a$.

Let $\epsilon > 0$. Then, by assumption, we can choose $\delta > 0$ such that if $\{(a_k, b_k)\}_{k=1}^m$ is any finite disjoint collection of subintervals of $[a, b]$ with the property that $\sum_{k=1}^m (b_k - a_k) < \delta$, then

$$\sum_{k=1}^m |g(b_k) - g(a_k)| < \frac{\epsilon}{2}. \qquad (8.28)$$

If $y \in E$, then $\lim_{h\to 0+} \big(g(y + h) - g(y)\big)/h = 0$ and, therefore, for h sufficiently small,

$$|g(y + h) - g(y)| \le \frac{h\epsilon}{2(x - a)}. \qquad (8.29)$$

Because $y \in E \subset (a, x)$, we also have $y + h \in (a, x)$ for h sufficiently small.

It follows that the collection $\mathcal{V}$ of all closed intervals of the form $[y, y+h]$, where $h > 0$, $y \in E$, $|g(y+h) - g(y)| \leq h\epsilon/2(x-a)$, and $[y, y+h] \subset (a, x)$ is a Vitali cover of E. So, by the Vitali covering theorem, there exist pairwise disjoint members of $\mathcal{V}$, say $[y_j, y_j + h_j]$, $1 \leq j \leq n$, such that

$$\lambda\left(E \setminus \bigcup_{j=1}^{n} [y_j, y_j + h_j] \right) < \delta. \tag{8.30}$$

By relabeling, we can assume that $a < y_1 < y_2 < \cdots < y_n < x$. Therefore, we have Fig. 8.3, to which it will be helpful to refer in the ensuing discussion.

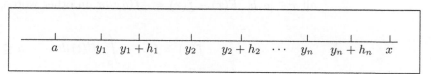

FIGURE 8.3

Now, from (8.30) and the fact that $\lambda(E) = x - a$, we can conclude that $\sum_{j=1}^{n} h_j > x - a - \delta$. It follows that

$$(y_1 - a) + \sum_{j=1}^{n-1} \left(y_{j+1} - (y_j + h_j) \right) + \left(x - (y_n + h_n) \right) < \delta.$$

So, the sum of the lengths of the pairwise disjoint intervals (a, y_1), $(y_1 + h_1, y_2)$, $\ldots$, $(y_n + h_n, x)$ is less than δ. Therefore, by (8.28),

$$|g(y_1) - g(a)| + \sum_{j=1}^{n-1} |g(y_{j+1}) - g(y_j + h_j)| + |g(x) - g(y_n + h_n)| < \frac{\epsilon}{2}.$$

On the other hand, by (8.29),

$$\sum_{k=1}^{n} |g(y_k + h_k) - g(y_k)| \leq \sum_{k=1}^{n} \frac{h_k \epsilon}{2(x-a)} \leq \frac{\epsilon}{2}.$$

Consequently, by the previous two relations and the triangle inequality,

$$\begin{aligned}
|g(x) - g(a)| \leq\; & |g(y_1) - g(a)| + |g(y_1 + h_1) - g(y_1)| \\
& + |g(y_2) - g(y_1 + h_1)| + |g(y_2 + h_2) - g(y_2)| \\
& + \cdots + |g(y_n + h_n) - g(y_n)| + |g(x) - g(y_n + h_n)| \\
< \;& \frac{\epsilon}{2} + \frac{\epsilon}{2} = \epsilon.
\end{aligned}$$

As $\epsilon > 0$ was chosen arbitrarily, the preceding display shows that $g(x) = g(a)$. ∎

We can now prove Proposition 8.7 (page 295). Let f satisfy the hypotheses of that proposition. From Lemma 8.4, we know $f' \in \mathcal{L}^1([a, b])$. Now, set

$$F(x) = \int_a^x f'(t) \, dt, \qquad a \leq x \leq b.$$

Then $F' = f'$ ae on $[a, b]$ (Theorem 8.5) and F is absolutely continuous on $[a, b]$ (Proposition 8.4(a)). The latter fact and Proposition 8.6 imply that F satisfies the hypotheses of Proposition 8.7 and, consequently, so does $F - f$.

But $(F - f)' = F' - f' = 0$ ae on $[a, b]$ and, consequently, by Lemma 8.5, $F - f$ is constant on $[a, b]$. Because $F(a) = 0$, we can now conclude that for all $x \in [a, b]$, $F(x) - f(x) = -f(a)$; or, in other words,

$$f(x) = f(a) + \int_a^x f'(t) \, dt, \qquad a \leq x \leq b.$$

This result shows that f is absolutely continuous and completes the proof of Proposition 8.7.

Second Fundamental Theorem of Calculus

We summarize Propositions 8.6 and 8.7 in the following theorem, which is often referred to as the **second fundamental theorem of calculus for Lebesgue integration**.

□ □ □ **THEOREM 8.6 Second Fundamental Theorem of Calculus**

Suppose that f is defined on $[a, b]$. A necessary and sufficient condition for f' to exist almost everywhere and be Lebesgue integrable on $[a, b]$ and for

$$\int_a^x f'(t) \, dt = f(x) - f(a), \qquad a \leq x \leq b,$$

is that for each $\epsilon > 0$ there is a $\delta > 0$ such that $\sum_{k=1}^n |f(b_k) - f(a_k)| < \epsilon$ whenever $\{(a_k, b_k)\}_{k=1}^n$ is a finite sequence of pairwise disjoint subintervals of $[a, b]$ with $\sum_{k=1}^n (b_k - a_k) < \delta$.

We conclude this section by giving necessary and sufficient conditions for a function to be absolutely continuous on $\mathcal{R}$. The proof is left as an exercise for the reader. (See Exercise 8.56.)

□ □ □ **THEOREM 8.7**

A function f is absolutely continuous on $\mathcal{R}$ if and only if it is absolutely continuous on every finite closed interval, $V_{-\infty}^\infty f < \infty$, and $\lim_{x \to -\infty} f(x) = 0$.

Exercises for Section 8.4

8.41 Establish the following facts.

a) Suppose f is defined on $[a, b]$ and f' exists and is Riemann integrable on $[a, b]$. Prove that f is absolutely continuous on $[a, b]$. Conclude that f is absolutely continuous on $[a, b]$ if f' is continuous on $[a, b]$.

b) Suppose f is continuous on $\mathcal{R}$, $V_{-\infty}^{\infty} f < \infty$, and $\lim_{x \to -\infty} f(x) = 0$. Further suppose there exists a finite number of points such that f is absolutely continuous on any finite closed interval that contains none of those points. Prove that f is absolutely continuous on $\mathcal{R}$.

c) Suppose F is a continuous distribution function such that F' exists and is continuous except at a finite number of points. Prove that F is absolutely continuous on $\mathcal{R}$.

8.42 Prove that $f(x) = \sqrt{x}$ is absolutely continuous on $[0, 1]$.

8.43 Show that if f and g are absolutely continuous on $[a, b]$, then $f + g$ is absolutely continuous on $[a, b]$.

8.44 Show that if f is absolutely continuous on $[a, b]$ and $\alpha \in \mathbb{C}$, then αf is absolutely continuous on $[a, b]$.

8.45 Let $\alpha > 0$ and define $f \colon \mathcal{R} \to \mathcal{R}$ by $f(x) = e^{-\alpha|x|}$. Show that f is absolutely continuous on $\mathcal{R}$.

8.46 Is $f(x) = 1$ absolutely continuous on $[0, 1]$? On $\mathcal{R}$?

8.47 Define f on $[0, 1]$ by $f(0) = 0$ and $f(x) = x \sin(1/x)$ for $x \neq 0$. From Example 8.7 on page 294, we know that f is not absolutely continuous on $[0, 1]$. Show that f is absolutely continuous on any interval $[a, b]$ not containing 0.

8.48 Define f on $[0, 1]$ by $f(0) = 0$ and $f(x) = x^{\alpha} \sin(1/x)$ for $x \neq 0$. Show that f is absolutely continuous on $[0, 1]$ if and only if $\alpha > 1$.

8.49 Let f be real-valued and absolutely continuous on $[a, b]$.

a) Prove that f takes sets of Lebesgue measure zero to sets of Lebesgue measure zero; that is, if $E \subset [a, b]$ and $\lambda(E) = 0$, then $\lambda(f(E)) = 0$.

b) Prove that f takes Lebesgue measurable sets to Lebesgue measurable sets; that is, if $A \subset [a, b]$ is Lebesgue measurable, then so is $f(A)$. *Hint:* Choose an increasing sequence of closed sets contained in A such that the set difference between A and the union of the closed sets has measure zero. Next apply part (a) to show that the set difference between the image of A and the union of the images of the closed sets has measure zero.

8.50 Suppose that $f \colon [a, b] \to [c, d]$ is absolutely continuous and monotone and that g is absolutely continuous on $[c, d]$. Prove that $g \circ f$ is absolutely continuous on $[a, b]$.

8.51 Proposition 8.5 on page 293 states in part that if f is absolutely continuous on $[a, b]$, then $V_a^b f = \int_a^b |f'(x)| \, dx$. Find a continuous function of bounded variation that does not satisfy that equation.

8.52 Give an example of a function f that is absolutely continuous on $[0, 1]$ but is such that f' is not Riemann integrable on $[0, 1]$.

8.53 Construct an absolutely continuous function on $[0, 1]$ that is strictly increasing but whose derivative vanishes on a set of positive measure. *Hint:* Let P_α be as in Exercise 3.39 on page 108 and set $f(x) = \int_0^x \chi_{P_\alpha^c}(t) \, dt$.

8.54 In establishing Lemma 8.4 on page 295, we proved that if f is of bounded variation on $[a, b]$, then $f' \in \mathcal{L}^1([a, b])$. Here is an alternate derivation of that result.

a) Show that f is Lebesgue measurable.

b) Show that f' is Lebesgue measurable.

c) Prove that $\int_a^b |f'(x)| \, dx \leq V_a^b f$ and, hence, that f' is Lebesgue integrable on $[a, b]$. *Hint:* Use (8.14) on page 283.

8.55 A function f is said to be **Lipschitzian** on $[a, b]$ if there is a constant M such that

$$|f(x) - f(y)| \leq M|x - y|, \qquad x, y \in [a, b].$$

a) Show that if f has a bounded derivative on $[a, b]$, then it is Lipschitzian on $[a, b]$.
b) Prove that if f is Lipschitzian on $[a, b]$, then it is absolutely continuous thereon.

8.56 Prove Theorem 8.7. *Hint:* Use (8.26).

8.57 **Integration by parts:** Suppose that f and g are absolutely continuous on $[a, b]$. Then fg' and $f'g$ are in $\mathcal{L}^1([a, b])$ and

$$\int_a^b f(x)g'(x)\, dx = f(b)g(b) - f(a)g(a) - \int_a^b f'(x)g(x)\, dx. \tag{8.31}$$

Establish this result by employing the following steps.
a) Show that fg' and $f'g$ are in $\mathcal{L}^1([a, b])$.
b) Prove that fg is absolutely continuous on $[a, b]$. *Hint:* Show that the hypotheses of Proposition 8.7 (page 295) are satisfied.
c) Prove that (8.31) holds. *Hint:* Recall the product rule from elementary calculus.

8.58 Give an example where the integration by parts formula fails in case f is absolutely continuous and g is uniformly continuous and of bounded variation but not absolutely continuous.

8.59 Let $h \in \mathcal{L}^1([a, b])$ and define

$$F(x) = \int_a^x h(t)(x - t)^n\, dt; \qquad a \leq x \leq b.$$

Show that F is n-times differentiable on $[a, b]$, $F^{(n)}$ is absolutely continuous on $[a, b]$, and $F^{(n+1)} = n!\, h$ ae on $[a, b]$. *Hint:* Use induction and integration by parts.

8.60 **Taylor's theorem:** Suppose f is defined on $[a, b]$, f is n-times differentiable on $[a, b]$, and $f^{(n)}$ is absolutely continuous on $[a, b]$. Then, for $a \leq x \leq b$,

$$f(x) = f(a) + f'(a)(x - a) + \cdots + \frac{f^{(n)}(a)}{n!}(x - a)^n$$
$$+ \frac{1}{n!}\int_a^x f^{(n+1)}(t)(x - t)^n\, dt. \tag{8.32}$$

Establish this result. *Hint:* Use induction on n, integration by parts, and Exercise 8.41.

8.61 Show that the hypothesis of absolute continuity cannot be removed in the version of Taylor's theorem given in the previous exercise.

8.62 Prove that the converse of Taylor's theorem is true: Suppose that f is defined on $[a, b]$ and that there are constants $a_0, a_1, \ldots, a_n$ and a function $h \in \mathcal{L}^1([a, b])$ such that for $a \leq x \leq b$,

$$f(x) = a_0 + a_1(x - a) + \cdots + a_n(x - a)^n + \int_a^x h(t)(x - t)^n\, dt.$$

Then, on $[a, b]$, f is n-times differentiable and $f^{(n)}$ is absolutely continuous. Moreover, $a_k = f^{(k)}(a)/k!$, $0 \leq k \leq n$, and $h = f^{(n+1)}/n!$ ae on $[a, b]$. *Hint:* Use Exercise 8.59.

★8.63 **Integration by substitution:** Suppose that g is monotone and absolutely continuous on $[a, b]$ with range $[c, d]$ and that $f \in \mathcal{L}^1([c, d])$. Then $(f \circ g)g' \in \mathcal{L}^1([a, b])$ and

$$\int_a^b f(g(x))|g'(x)|\, dx = \int_c^d f(y)\, dy.$$

Establish this result by employing the following steps.

a) Show that, without loss of generality, we can assume that g is nondecreasing.

b) Show that for each open set $O \subset [c, d]$,

$$\lambda(O) = \int_{g^{-1}(O)} g'(x)\, dx.$$

c) Show that if D is a G_δ-set, then D satisfies the equation in part (b).

d) Let $H = \{ x : g'(x) \neq 0 \}$ and let $E \subset [c, d]$. Prove that, if E has measure zero, then $\lambda^*\big(g^{-1}(E) \cap H\big) = 0$ and, hence, that $g^{-1}(E) \cap H$ is measurable and has measure zero.

e) Prove that, if $A \subset [c, d]$ is measurable, then so is $g^{-1}(A) \cap H$ and

$$\lambda(A) = \int_{g^{-1}(A) \cap H} g'(x)\, dx.$$

f) Prove that, if f is a nonnegative measurable function on $[c, d]$, then the function $(f \circ g)g'$ is measurable on $[a, b]$ and

$$\int_a^b f(g(x))g'(x)\, dx = \int_c^d f(y)\, dy.$$

g) Show that if $f \in \mathcal{L}^1\big([c, d]\big)$, then $(f \circ g)g' \in \mathcal{L}^1\big([a, b]\big)$ and the equation in part (f) holds.

8.64 Let f be the function in Exercise 8.53.

a) Show that there is a set E of measure zero such that $f^{-1}(E)$ is not measurable. *Hint:* Use the fact that any set of positive (outer) measure contains a nonmeasurable set. Also, refer to Exercise 8.63(e).

b) Show that the function $g = f^{-1}$ is not absolutely continuous. Hence, the inverse of an absolutely continuous function, when it exists, need not be absolutely continuous. *Hint:* Refer to Exercise 8.49(b).

Johann Radon
(1887–1956)

Johann Radon was born on December 16, 1887, in Tetschen, Bohemia, now Děčín, Czech Republic. He was the only child of Anton Radon, who worked at a local bank as the head bookkeeper, and Anna Schmiedeknecht.

Radon began his formal schooling at the age of 10 at the Gymnasium in Leitmaritz, Bohemia. Eight years later, in 1905, he enrolled at the University of Vienna to pursue the study of mathematics and physics. Radon presented his doctoral dissertation in 1910 on the calculus of variations.

Radon taught at several universities between 1910 and 1919; he spent a semester at the University of Göttingen, a year at the University of Brünn, and time at the Technische Hochschule of Vienna and at the University of Vienna. In 1919, he went to the University of Hamburg for 3 years, moved subsequently to Greifswald, Erlangen, and then to Breslau in 1928, where he remained until 1945.

In 1939, Radon became a corresponding member of the Austrian Academy of Sciences and, in 1947, a full member. From 1952 to 1956, Radon was this academy's Secretary of the Class of Mathematics and Science. He was also president of the Austrian Mathematical Society from 1948 to 1950. In 2003, the Austrian Academy of Sciences honored Radon by founding the Johann Radon Institute for Computational and Applied Mathematics (RICAM).

The calculus of variations continued to fascinate Radon because of its many applications to analysis, geometry, and physics. He applied it to differential geometry to discover Radon curves. Other work included the combination of Lebesgue's and Stieltjes's theories of integration (the development of the Radon integral), the Dirichlet problem of the logarithmic potential (application of the Radon-Nikodym theorem), and the development of the Radon transformation technique.

Radon spent the last 9 years of his life as a full professor at the University of Vienna in Vienna, Austria, where he died on May 25, 1956.

9

Signed and Complex Measures

In Chapter 8, we discussed differentiation of real- and complex-valued functions of a real variable. In doing so, we encountered the concept of absolutely continuous functions.

Now, in this chapter, we will extend the theory of differentiation and absolute continuity to measures and provide a relationship between absolutely continuous functions and absolutely continuous measures. Additionally, we will extend the notion of measure to include those that are real-valued and complex-valued, establish decomposition and representation theorems for measures, prove and apply the famous Radon-Nikodym theorem, and generalize the classical change-of-variable formula for integration.

9.1 SIGNED MEASURES

We discovered earlier (see, e.g., Exercise 5.61) that, if $(\Omega, \mathcal{A}, \mu)$ is a measure space and f is a nonnegative extended real-valued $\mathcal{A}$-measurable function on Ω, then the set function

$$\nu(A) = \int_A f \, d\mu, \qquad A \in \mathcal{A}, \tag{9.1}$$

is a measure on $\mathcal{A}$.

What about the converse: If $(\Omega, \mathcal{A}, \mu)$ is a measure space and ν is a measure on $\mathcal{A}$, can a nonnegative extended real-valued $\mathcal{A}$-measurable function f be found such that (9.1) holds? It is easy to see that the answer to this question is no!

For example, take $(\Omega, \mathcal{A}, \mu) = (\mathcal{R}, \mathcal{M}, \lambda)$ and $\nu = \delta_0$, the Dirac measure concentrated at 0. Then a representation of the form (9.1) is impossible. Indeed, if f is any nonnegative extended real-valued $\mathcal{M}$-measurable function, then

$$\delta_0(\{0\}) = 1 \neq 0 = \int_{\{0\}} f \, d\lambda.$$

What goes wrong here is the following: There exists a set A, namely, the set $\{0\}$, such that $\lambda(A) = 0$ but $\delta_0(A) \neq 0$, whereas, by Proposition 5.8(d), (9.1) forces $\nu(A) = 0$ whenever $\mu(A) = 0$. In other words,

$$\mu(A) = 0 \quad \Rightarrow \quad \nu(A) = 0 \tag{9.2}$$

is a *necessary* condition for a measure ν to be representable as in (9.1). Remarkably, in most important cases, (9.2) is also a *sufficient* condition for that representation. That fact is called the **Radon-Nikodym theorem** and will be proved in Section 9.2.

EXAMPLE 9.1   *Illustrates (9.1) and (9.2)*

a) Let $\mathcal{I}$ denote the collection of all intervals $I \subset \mathcal{R}$, including degenerate intervals of the form (a, a) and $[a, a]$. Also, let ν be the measure on $\mathcal{B}$ such that $\nu(I) = 0$ if $I \subset (-\infty, 0)$ and $\nu(I) = e^{-3a} - e^{-3b}$ if I has endpoints a and b, where $0 \leq a \leq b \leq \infty$; according to Corollary 6.2 on page 190, ν is determined by these conditions. If we let $\mu = \lambda_{|\mathcal{B}}$, then ν has a representation in the form (9.1), namely,

$$\nu(B) = \int_B f \, d\lambda, \qquad B \in \mathcal{B},$$

where $f(x) = 3e^{-3x}$ for $x > 0$, and $f(x) = 0$ otherwise. This is true because the measure $\omega(B) = \int_B f \, d\lambda$ agrees with ν on intervals and so, by Corollary 6.2, must equal ν.

b) Let $(\Omega, \mathcal{A}, \mu) = (\mathcal{R}, \mathcal{M}, \lambda)$ and $E \in \mathcal{M}$. Set $\nu(A) = \lambda(A \cap E)$ for $A \in \mathcal{M}$. Clearly (9.2) holds. Here it is obvious that ν can be represented in the form (9.1), namely, $\nu(A) = \int_A \chi_E \, d\lambda$.

c) Let $(\Omega, \mathcal{A}) = (\mathcal{R}, \mathcal{M})$ and $\nu = \delta_0$. As we have seen, ν cannot be represented in the form (9.1) if $\mu = \lambda$. However, if μ is counting measure on $(\mathcal{R}, \mathcal{M})$, then ν does have such a representation, namely, with $f = \chi_{\{0\}}$.

d) Let $(\Omega, \mathcal{A}) = (\mathcal{R}, \mathcal{B})$ and ψ be the Cantor function. Define

$$F(x) = \begin{cases} 0, & x < 0; \\ \psi(x), & 0 \leq x < 1; \\ 1, & x \geq 1. \end{cases}$$

Note that F is a distribution function, that is, it satisfies (a)–(d) of Definition 6.5 on page 197. Therefore, by Theorem 6.3 on page 196, there is a unique finite Borel measure ν with F as its distribution function. We claim that ν has no representation in the form (9.1) if $\mu = \lambda_{|\mathcal{B}}$. Suppose to the contrary that there is a nonnegative Borel measurable function f such that $\nu(A) = \int_A f \, d\lambda$ for $A \in \mathcal{B}$. Because $\nu(\mathcal{R}) = 1$, $f \in \mathcal{L}^1(\mathcal{R})$; hence, if we let $g = f_{|[0,1]}$, then $g \in \mathcal{L}^1([0,1])$. Moreover, for $0 \leq x \leq 1$,

$$\psi(x) = F(x) - F(0) = \nu((0, x]) = \int_{(0,x]} f \, d\lambda = \int_0^x g(t) \, dt.$$

This result implies that ψ is absolutely continuous on $[0, 1]$, which we know is not true. So ν cannot be represented in the form (9.1) if $\mu = \lambda_{|\mathcal{B}}$.   □

Signed Measures

In order to prove the Radon-Nikodym theorem, we need to introduce signed measures. This is a simple generalization of measures where the nonnegativity condition is dropped. Thus, any measure is a signed measure, but not conversely.

DEFINITION 9.1 **Signed Measure**

Let $(\Omega, \mathcal{A})$ be a measurable space. A **signed measure** ν on $\mathcal{A}$ is an extended real-valued function satisfying the following conditions:
a) $\nu(\emptyset) = 0$.
b) If $A_1, A_2, \ldots$ are in $\mathcal{A}$, with $A_i \cap A_j = \emptyset$ for $i \neq j$, then

$$\nu\left(\bigcup_n A_n\right) = \sum_n \nu(A_n).$$

Remark: The displayed equation in (b) of Definition 9.1 is taken in the extended sense; that is, any values in $\mathcal{R}^*$ are permitted. However, the right-hand side of the equation must make sense: it must converge or it must diverge to $\pm\infty$. In particular, ν cannot take on both ∞ and $-\infty$ as values; that is, if $\nu(E) = \infty$ for some $E \in \mathcal{A}$, then $\nu(A) > -\infty$ for all $A \in \mathcal{A}$, and if $\nu(E) = -\infty$ for some $E \in \mathcal{A}$, then $\nu(A) < \infty$ for all $A \in \mathcal{A}$.

EXAMPLE 9.2 **Illustrates Definition 9.1**

a) Let $(\Omega, \mathcal{A})$ be a measurable space, $\alpha_1, \alpha_2 \in \mathcal{R}$, and μ_1 and μ_2 measures on $\mathcal{A}$, at least one of which is finite. Define $\alpha_1\mu_1 + \alpha_2\mu_2$ on $\mathcal{A}$ by

$$(\alpha_1\mu_1 + \alpha_2\mu_2)(A) = \alpha_1\mu_1(A) + \alpha_2\mu_2(A).$$

It is easy to see that $\alpha_1\mu_1 + \alpha_2\mu_2$ is a signed measure on $\mathcal{A}$. Note that if $\alpha_1 \geq 0$, then $\alpha_1\mu_1$ is a measure on $\mathcal{A}$.
b) Let $(\Omega, \mathcal{A}, \mu)$ be a measure space and $f \in \mathcal{L}^1(\mu)$ be extended real-valued. Define

$$\nu(A) = \int_A f \, d\mu, \qquad A \in \mathcal{A}.$$

By Exercise 5.72 on page 172, ν is a signed measure on $\mathcal{A}$. If f is nonnegative, then ν is a measure. □

The Hahn Decomposition Theorem

From Example 9.2(a), we see that the difference of two measures on a σ-algebra, at least one of which is finite, is a signed measure. Our next goal is to show that the converse is also true: Any signed measure on a σ-algebra can be expressed as the difference of two measures on that σ-algebra, at least one of which is finite.

The idea is the following. Let ν be a signed measure on a σ-algebra $\mathcal{A}$. Suppose we can find a set $D \in \mathcal{A}$ such that $\nu(E) \geq 0$ for each $\mathcal{A}$-measurable subset of D and such that $\nu(E) \leq 0$ for each $\mathcal{A}$-measurable subset of D^c. Then we can define set functions ν^+ and ν^- on $\mathcal{A}$ by $\nu^+(A) = \nu(A \cap D)$ and $\nu^-(A) = -\nu(A \cap D^c)$. And it is easy to see that ν^+ and ν^- are measures on $\mathcal{A}$ and that $\nu = \nu^+ - \nu^-$.

The existence of such a set D is the substance of the **Hahn decomposition theorem**, and the pair (D, D^c) is called a **Hahn decomposition** for ν. We begin with the following definition.

DEFINITION 9.2   Positive and Negative Sets

Let $(\Omega, \mathcal{A})$ be a measurable space and ν a signed measure on $\mathcal{A}$. A set $P \in \mathcal{A}$ is called **positive** for ν if $\nu(E) \geq 0$ for all sets $E \in \mathcal{A}$ with $E \subset P$. A set $N \in \mathcal{A}$ is called **negative** for ν if $\nu(E) \leq 0$ for all sets $E \in \mathcal{A}$ with $E \subset N$.

EXAMPLE 9.3   Illustrates Definition 9.2

a) Let $(\Omega, \mathcal{A}) = (\mathcal{R}, \mathcal{M})$, $\mu = \delta_0 + \delta_1$, and $\nu = \lambda - \mu$. Note that for $A \in \mathcal{M}$,

$$\nu(A) = \begin{cases} \lambda(A), & \text{if } \{0,1\} \cap A = \emptyset; \\ \lambda(A) - 1, & \text{if exactly one of } 0, 1 \text{ are in } A; \\ \lambda(A) - 2, & \text{if } \{0,1\} \subset A. \end{cases}$$

The sets $\mathcal{R} \setminus \{0, 1\}$ and $(1, 2)$ are positive for ν; the sets $\{0, 1\}$ and Q are negative for ν; the set $\{2, 3, 4, \ldots\}$ is both positive and negative for ν; and the set $[0, 1]$ is neither positive nor negative for ν, although we have $\nu([0, 1]) = -1$. In fact, the positive sets for ν are the $\mathcal{M}$-measurable sets containing neither 0 nor 1; and the negative sets for ν are the $\mathcal{M}$-measurable sets with Lebesgue measure zero. Moreover, the pair consisting of $\mathcal{R} \setminus \{0, 1\}$ and its complement is a Hahn decomposition for ν, as is the pair consisting of $\mathcal{R} \setminus \{0, 1, 2\}$ and its complement, etc.

b) Let $(\Omega, \mathcal{A})$ be a measurable space and ν a measure on $\mathcal{A}$. Then every $\mathcal{A}$-measurable set is positive for ν; the negative sets for ν are the $\mathcal{A}$-measurable sets with ν-measure zero. A Hahn decomposition for ν is $(\Omega, \emptyset)$.

c) Any set that is both positive and negative for a signed measure ν must have ν-measure zero. However, there may be sets of ν-measure zero that are neither positive nor negative for ν. For instance, if ν is the signed measure defined in part (a), then the set $[0, 1)$ has ν-measure zero, but is neither positive nor negative for ν. □

Using the terminology introduced in Definition 9.2, we see that in a Hahn decomposition (D, D^c), the set D is positive for ν and its complement is negative for ν. Thus, intuitively, the set D should in some sense be a maximal positive set.

In proving the Hahn decomposition theorem, we will need several lemmas. The first lemma shows that signed measures share important properties with measures. Its proof is left as an exercise for the reader.

□ □ □   **LEMMA 9.1**

Let $(\Omega, \mathcal{A})$ be a measurable space and ν a signed measure on $\mathcal{A}$. Then the following properties hold:

a) If $A, B \in \mathcal{A}$, $A \subset B$, and $|\nu(B)| < \infty$, then $|\nu(A)| < \infty$.

b) If $\{E_n\}_{n=1}^{\infty} \subset \mathcal{A}$ with $E_1 \subset E_2 \subset \cdots$, then

$$\nu\left(\bigcup_{n=1}^{\infty} E_n \right) = \lim_{n \to \infty} \nu(E_n).$$

c) If $\{E_n\}_{n=1}^{\infty} \subset \mathcal{A}$ with $E_1 \supset E_2 \supset \cdots$ and $|\nu(E_1)| < \infty$, then

$$\nu\left(\bigcap_{n=1}^{\infty} E_n \right) = \lim_{n \to \infty} \nu(E_n).$$

□ □ □   **LEMMA 9.2**

A countable union of positive sets is positive.

PROOF   Let $(\Omega, \mathcal{A})$ be a measurable space and ν a signed measure on $\mathcal{A}$. Suppose that $\{P_n\} \subset \mathcal{A}$ is a sequence of positive sets for ν. We claim that $\bigcup_n P_n$ is positive for ν.

Let $E \subset \bigcup_n P_n$. We must show that $\nu(E) \geq 0$. To that end, we "disjointize" E as follows. Set $E_1 = E \cap P_1$ and, for $n \geq 2$, $E_n = E \cap P_n \setminus \bigcup_{k=1}^{n-1} P_k$. Then the E_ns are pairwise disjoint and $\bigcup_n E_n = E$. Since P_n is positive for ν and $E_n \subset P_n$, we have $\nu(E_n) \geq 0$. Hence, $\nu(E) = \sum_n \nu(E_n) \geq 0$.   ■

The next lemma shows that any set of finite positive ν-measure has a subset of positive ν-measure that is positive for ν.

□ □ □   **LEMMA 9.3**

Let $(\Omega, \mathcal{A})$ be a measurable space and ν a signed measure on $\mathcal{A}$. Suppose $A \in \mathcal{A}$ and $0 < \nu(A) < \infty$. Then there is an $\mathcal{A}$-measurable set $P \subset A$ that is positive for ν and satisfies $\nu(P) > 0$.

PROOF   Note that by Lemma 9.1(a), any subset of A has finite ν-measure. The idea of the proof is to keep extracting sets from A that have negative ν-measure of large magnitude.

If A is positive for ν, we are done. Otherwise, A contains a set with negative ν-measure. In that case, let $L_1 = \inf\{\nu(E) : E \subset A\}$. By assumption, $L_1 < 0$; so there is an $n \in \mathcal{N}$ such that $L_1 < -n^{-1}$. Let n_1 be the smallest such n. By definition of L_1, there is an $A_1 \subset A$ such that $\nu(A_1) < -n_1^{-1}$. Note that if $n < n_1$, then $L_1 \geq -n^{-1}$ and, hence, $\nu(E) \geq -n^{-1}$ for all $E \subset A$.

Extract A_1 from A; that is, consider $A \setminus A_1$. Note that

$$\nu(A \setminus A_1) = \nu(A) - \nu(A_1) > \nu(A) + \frac{1}{n_1} > 0.$$

If $A \setminus A_1$ is positive for ν, we are done. Otherwise, $A \setminus A_1$ contains a set with negative ν-measure. In that case, let $L_2 = \inf\{\nu(E) : E \subset A \setminus A_1\}$. By assumption, $L_2 < 0$; so there is an $n \in \mathcal{N}$ such that $L_2 < -n^{-1}$. Let n_2 be the smallest such n. By definition of L_2, there is an $A_2 \subset A \setminus A_1$ such that $\nu(A_2) < -n_2^{-1}$. Note that if $n < n_2$, then $L_2 \geq -n^{-1}$ and, so, $\nu(E) \geq -n^{-1}$ for all $E \subset A \setminus A_1$.

Extract A_2 from $A \setminus A_1$; that is, consider $A \setminus A_1 \setminus A_2 = A \setminus (A_1 \cup A_2)$. Note that

$$\nu(A \setminus (A_1 \cup A_2)) = \nu(A) - (\nu(A_1) + \nu(A_2)) > \nu(A) + \frac{1}{n_1} + \frac{1}{n_2} > 0.$$

If this process terminates after a finite number of steps, we are done. Otherwise, we obtain a sequence of pairwise disjoint subsets $\{A_k\}_{k=1}^\infty$ of A and a sequence of positive integers $\{n_k\}_{k=1}^\infty$ such that for each $k \in \mathcal{N}$, $A_k \subset A \setminus \bigcup_{j=1}^{k-1} A_j$, $\nu(A_k) < -n_k^{-1}$, and n_k is the smallest positive integer n for which there is a subset of $A \setminus \bigcup_{j=1}^{k-1} A_j$ with ν-measure less than $-n^{-1}$.

Let $P = A \setminus \bigcup_{k=1}^\infty A_k$, and note that

$$\nu(P) = \nu(A) - \sum_{k=1}^\infty \nu(A_k) > \nu(A) + \sum_{k=1}^\infty \frac{1}{n_k} > 0.$$

We claim that P is positive for ν. Suppose to the contrary that there is a set $B \subset P$ with $\nu(B) < 0$. As $\nu(A) < \infty$, Lemma 9.1(a) implies that $\nu(P) < \infty$. Thus, $\sum_{k=1}^\infty 1/n_k < \infty$ and so, in particular, $n_k \to \infty$ as $k \to \infty$. Since $n_k \to \infty$, there is a k_0 such that $(n_{k_0} - 1)^{-1} < -\nu(B)$; that is, $\nu(B) < -(n_{k_0} - 1)^{-1}$. But $B \subset P$ and so $B \subset A \setminus \bigcup_{j=1}^{k_0-1} A_j$. This contradicts the minimality of n_{k_0}. Hence P is positive for ν. ∎

□ □ □ **THEOREM 9.1 Hahn Decomposition Theorem**

Let $(\Omega, \mathcal{A})$ be a measurable space and ν a signed measure on $\mathcal{A}$. Then there is a set $D \in \mathcal{A}$ such that D is positive for ν and D^c is negative for ν. The pair (D, D^c) is called a Hahn decomposition for ν.

PROOF   We can assume without loss of generality that ν does not take on the value ∞. (Why can this assumption be made?)

As we mentioned earlier, D should in some sense be a maximal positive set for ν. With that idea in mind, let

$$p = \sup\{\,\nu(P) : P \text{ positive for } \nu\,\}.$$

Choose a sequence $\{P_n\}_{n=1}^\infty$ of positive sets for ν such that $\lim_{n\to\infty} \nu(P_n) = p$. Let $D_n = \bigcup_{k=1}^n P_k$ and $D = \bigcup_{n=1}^\infty D_n$. Applying Lemma 9.2 twice we see, in turn, that $D_1, D_2, \ldots,$ are positive for ν and D is positive for ν.

To show that (D, D^c) is a Hahn decomposition for ν, it remains to prove that D^c is negative for ν. To that end, we first note that because $D_1 \subset D_2 \subset \cdots$, Lemma 9.1(b) implies that $\nu(D) = \lim_{n\to\infty} \nu(D_n)$. Also, $D_n \supset P_n$ and, therefore, $D_n = P_n \cup (D_n \setminus P_n)$. Because $D_n \setminus P_n \subset D_n$ and D_n is positive for ν, $\nu(D_n \setminus P_n) \geq 0$; hence, $\nu(D_n) \geq \nu(P_n)$. Recalling that D is positive, we have

$$p \geq \nu(D) = \lim_{n\to\infty} \nu(D_n) \geq \lim_{n\to\infty} \nu(P_n) = p.$$

Therefore, $\nu(D) = p$. Since ν does not assume the value ∞, we must have $p < \infty$.

Now, suppose that D^c is not negative for ν. Then there is a set $A \subset D^c$ with $\nu(A) > 0$ and, by assumption, $\nu(A) < \infty$. Thus, by Lemma 9.3, A contains a set P that is positive for ν and has positive ν-measure. Since P and D are positive, so is $P \cup D$; and since $P \subset D^c$, $P \cap D = \emptyset$. Hence, $P \cup D$ is positive for ν and

$$\nu(P \cup D) = \nu(P) + \nu(D) > p.$$

This result contradicts the definition of p. Hence D^c is negative for ν. ■

Is the Hahn decomposition for a signed measure unique? In general, it is not. For example, consider the signed measure ν defined in Example 9.3(a) on page 306. Then $(\mathcal{R} \setminus \{0,1\}, \{0,1\})$ and $(\mathcal{R} \setminus \{0,1,2\}, \{0,1,2\})$ are both Hahn decompositions for ν. However, as Exercise 9.5 shows, if (D, D^c) and (E, E^c) are two Hahn decompositions for a signed measure ν, then D and E differ by a set that contains only sets of ν-measure zero, and likewise for D^c and E^c.

The Jordan Decomposition Theorem

Now that we have established the existence of a Hahn decomposition for a signed measure, we can easily prove that any such measure can be expressed as the difference of two measures, a result known as the **Jordan decomposition theorem.** Before doing so, we introduce the following terminology.

DEFINITION 9.3 Mutually Singular Measures

Let $(\Omega, \mathcal{A})$ be a measurable space. Two measures μ_1 and μ_2 on $\mathcal{A}$ are said to be **mutually singular,** denoted $\boldsymbol{\mu_1 \perp \mu_2}$, if there is a set $E \in \mathcal{A}$ such that $\mu_1(E^c) = 0$ and $\mu_2(E) = 0$.

Note that if $\mu_1 \perp \mu_2$, then μ_1 and μ_2 are *supported* by complementary sets: For each $A \in \mathcal{A}$, $\mu_1(A) = \mu_1(A \cap E)$ and $\mu_2(A) = \mu_2(A \cap E^c)$.

EXAMPLE 9.4 Illustrates Definition 9.3

a) Let λ be Lebesgue measure and μ any discrete measure on $\mathcal{M}$, that is, there is a countable set K such that $\mu(K^c) = 0$. Then $\mu \perp \lambda$.
b) Let λ be Lebesgue measure and μ be the Borel measure induced by the Cantor function. Then, considered as Borel measures, $\mu \perp \lambda$. Indeed, if P is the Cantor set, then $\mu(P^c) = 0$ and $\lambda(P) = 0$.
c) Let $D = \{ (x,y) \in \mathcal{R}^2 : y = x \}$. Define ν by

$$\nu(B) = \lambda(\{ x \in \mathcal{R} : (x,x) \in B \}), \qquad B \in \mathcal{B}_2.$$

Then $\lambda_2(D) = 0$ and $\nu(D^c) = 0$, so that $\nu \perp \lambda_2$. □

□ □ □ **THEOREM 9.2 Jordan Decomposition Theorem**

*Let $(\Omega, \mathcal{A})$ be a measurable space and ν a signed measure on $\mathcal{A}$. Then ν can be expressed uniquely as the difference of two mutually singular measures ν^+ and ν^- on $\mathcal{A}$. The representation $\nu = \nu^+ - \nu^-$ is referred to as the **Jordan decomposition** of ν.*

PROOF Let (D, D^c) be a Hahn decomposition for ν and, for $A \in \mathcal{A}$, define

$$\nu^+(A) = \nu(A \cap D) \quad \text{and} \quad \nu^-(A) = -\nu(A \cap D^c).$$

Clearly, $\nu = \nu^+ - \nu^-$ and, because (D, D^c) is a Hahn decomposition for ν, ν^+ and ν^- are nonnegative. Noting that $\nu^+(D^c) = 0$ and $\nu^-(D) = 0$, we see that $\nu^+ \perp \nu^-$.

Now suppose we can write $\nu = \mu_1 - \mu_2$, where μ_1 and μ_2 are mutually singular measures on $\mathcal{A}$. We must show that $\mu_1 = \nu^+$ and $\mu_2 = \nu^-$. Since $\mu_1 \perp \mu_2$, we can choose $E \in \mathcal{A}$ such that $\mu_1(E^c) = 0$ and $\mu_2(E) = 0$. We claim (E, E^c) is a Hahn decomposition for ν. Indeed, if $F \subset E$, then $\mu_2(F) = 0$, so that $\nu(F) = \mu_1(F) \geq 0$; hence, E is positive for ν. On the other hand, if $F \subset E^c$, then $\mu_1(F) = 0$, so that $\nu(F) = -\mu_2(F) \leq 0$; hence, E^c is negative for ν.

From Exercise 9.5, we have for all $A \in \mathcal{A}$ that $\nu(A \cap D) = \nu(A \cap E)$ and $\nu(A \cap D^c) = \nu(A \cap E^c)$. The former equality implies that for $A \in \mathcal{A}$,

$$\mu_1(A) = \mu_1(A \cap E) + \mu_1(A \cap E^c) = \mu_1(A \cap E) = \mu_1(A \cap E) - \mu_2(A \cap E)$$
$$= \nu(A \cap E) = \nu(A \cap D) = \nu^+(A).$$

Thus, $\mu_1 = \nu^+$. Similarly, $\mu_2 = \nu^-$. ∎

Exercises for Section 9.1

Note: A ★ denotes an exercise that will be subsequently referenced.

9.1 Refer to Example 9.1(d) on page 304, where ν denotes the Borel measure induced by the Cantor function. Show that there is no nonnegative Borel measurable function f such that $\nu(B) = \int_B f \, d\lambda$ for all $B \in \mathcal{B}$ by finding a set $A \in \mathcal{B}$ such that $\lambda(A) = 0$ and $\nu(A) \neq 0$.

9.2 Define $\nu(E) = \int_E x \, dx$.
   a) Show that, for each positive real number c, the preceding definition yields a signed measure on $\mathcal{M}_{[-c,c]}$.
   b) Is ν a signed measure on $\mathcal{M}$? Justify your answer.

9.3 Prove Lemma 9.1 on page 307.

9.4 Concerning the proof of the Hahn decomposition theorem:
   a) Why can we assume without loss of generality that ν does not take on the value ∞?
   b) What happens if $p = 0$?

9.5 Let $(\Omega, \mathcal{A})$ be a measurable space and ν a signed measure on $\mathcal{A}$. Suppose (D, D^c) and (E, E^c) are two Hahn decompositions for ν.
   a) Show that D and E differ by a set that contains only sets of ν-measure zero, that is, if $A \in \mathcal{A}$ and $A \subset (D \setminus E) \cup (E \setminus D)$, then $\nu(A) = 0$.
   b) Show that D^c and E^c differ by a set that contains only sets of ν-measure zero.
   c) Prove that $\nu(A \cap D) = \nu(A \cap E)$ and $\nu(A \cap D^c) = \nu(A \cap E^c)$ for all $A \in \mathcal{A}$.

9.6 Let $(\Omega, \mathcal{A}, \mu)$ be a measure space and ν and ω measures on $\mathcal{A}$. Show that if $\nu \perp \mu$ and $\omega \perp \mu$, then $\nu + \omega \perp \mu$.

9.7 Let $(\Omega, \mathcal{A})$ be a measurable space such that $\{x\} \in \mathcal{A}$ for each $x \in \Omega$. A measure μ on $\mathcal{A}$ is called **continuous** if it has no atoms, that is, $\mu(\{x\}) = 0$ for all $x \in \Omega$. If μ is a continuous measure on $\mathcal{A}$ and ν is a discrete measure on $\mathcal{A}$, show that $\mu \perp \nu$.

9.8 Refer to Example 9.4(c) on page 309. For $B \in \mathcal{B}_2$, let $A = \{x \in \mathcal{R} : (x, x) \in B\}$.
   a) Provide a geometric interpretation of the relation between A and B.
   b) Prove that $A \in \mathcal{B}$.

9.9 Let λ be Lebesgue measure and μ be the Borel measure induced by the Cantor function. Find a Borel measure ω such that $\omega \perp \lambda_{|B}$ and $\omega \perp \mu$.

9.10 Provide an example to show that the uniqueness condition in Theorem 9.2 fails without the requirement of mutual singularity.

★9.11 This exercise will be useful as motivation for the proof of the Radon-Nikodym theorem. Let $(\Omega, \mathcal{A}, \mu)$ be a measure space and $f \in \mathcal{L}^1(\mu)$ be extended real-valued. Define

$$\nu(A) = \int_A f \, d\mu, \qquad A \in \mathcal{A}.$$

As we have seen, ν is a signed measure. Let $D = \{ x : f(x) \geq 0 \}$.
a) Show that (D, D^c) is a Hahn decomposition for ν.
b) Prove that for $A \in \mathcal{A}$,

$$\nu^+(A) = \int_A f^+ \, d\mu \quad \text{and} \quad \nu^-(A) = \int_A f^- \, d\mu.$$

9.12 Let $(\Omega, \mathcal{A})$ be a measurable space and ν a signed measure on $\mathcal{A}$. Prove that for each $A \in \mathcal{A}$,

$$\nu^+(A) = \sup\{ \nu(E) : E \in \mathcal{A}, \ E \subset A \}$$

and

$$\nu^-(A) = -\inf\{ \nu(E) : E \in \mathcal{A}, \ E \subset A \}.$$

9.13 Let $(\Omega, \mathcal{A}) = (\mathcal{N}, \mathcal{P}(\mathcal{N}))$ and let $\{a_n\}_{n=1}^\infty$ be a sequence of real numbers such that $\sum_{n=1}^\infty |a_n| < \infty$. Define ν on $\mathcal{P}(\mathcal{N})$ by $\nu(A) = \sum_{n \in A} a_n$.
a) Prove that ν is a signed measure.
b) Determine ν^+ and ν^-.

9.14 Let $(\Omega, \mathcal{A})$ be a measurable space and ν_1 and ν_2 measures on $\mathcal{A}$, at least one of which is finite. Define $\nu = \nu_1 - \nu_2$. Prove that $\nu^+ \leq \nu_1$ and $\nu^- \leq \nu_2$.

9.2 THE RADON-NIKODYM THEOREM

Let $(\Omega, \mathcal{A}, \mu)$ be a measure space and ν a measure on $\mathcal{A}$. We want to determine when ν can be represented in the form

$$\nu(A) = \int_A f \, d\mu, \qquad A \in \mathcal{A},$$

for some nonnegative extended real-valued $\mathcal{A}$-measurable function f on Ω. As we have seen, a necessary condition for such a representation is that $\nu(A) = 0$ whenever $\mu(A) = 0$. The Radon-Nikodym theorem shows that, subject to σ-finiteness restrictions, that condition is also sufficient for such a representation. In the following definition, we give that condition a name.

DEFINITION 9.4 **Absolutely Continuous Measures**

Let $(\Omega, \mathcal{A})$ be a measurable space and μ and ν measures on $\mathcal{A}$. Then ν is said to be **absolutely continuous** with respect to μ, denoted $\boldsymbol{\nu \ll \mu}$, if $\nu(A) = 0$ whenever $\mu(A) = 0$.

Note: The reason for the term "absolutely continuous" will become clear soon.

EXAMPLE 9.5 Illustrates Definition 9.4

a) As we have noted, if f is a nonnegative extended real-valued $\mathcal{A}$-measurable function, then the measure

$$\nu(A) = \int_A f \, d\mu, \qquad A \in \mathcal{A},$$

is absolutely continuous with respect to μ.

b) Let ν be the Borel measure on $\mathcal{R}$ induced by the Cantor function. Then, as Borel measures, no one of the measures δ_0, ν, and λ, is absolutely continuous with respect to one of the others; that is, $\delta_0 \not\ll \lambda$, $\lambda \not\ll \nu$, and so forth.

c) Let $(\Omega, \mathcal{A}) = (\mathcal{R}, \mathcal{M})$. We have $\delta_0 \ll \delta_0 + \delta_1$, but $\delta_0 + \delta_1 \not\ll \delta_0$.

d) Let $(\Omega, \mathcal{A}) = (\mathcal{R}, \mathcal{M})$. Define $f(x) = 0$ for $x \le 0$, and $f(x) = e^{-x}$ for $x > 0$. Set $\nu(A) = \int_A f \, d\lambda$ for $A \in \mathcal{M}$. Then $\nu \ll \lambda$, but $\lambda \not\ll \nu$. □

□ □ □ **THEOREM 9.3 Radon-Nikodym Theorem**

Let $(\Omega, \mathcal{A}, \mu)$ be a σ-finite measure space and ν a σ-finite measure on $\mathcal{A}$. If $\nu \ll \mu$, then there is a nonnegative extended real-valued $\mathcal{A}$-measurable function f on Ω such that

$$\nu(A) = \int_A f \, d\mu, \qquad A \in \mathcal{A}. \tag{9.3}$$

Moreover, f is unique in the sense that if g is a nonnegative extended real-valued $\mathcal{A}$-measurable function with $\nu(A) = \int_A g \, d\mu$ for all $A \in \mathcal{A}$, then $g = f$ μ-ae.

Before proving the Radon-Nikodym theorem, let us consider the main idea behind the proof. Suppose, say, that μ is a finite measure on $(\Omega, \mathcal{A})$ and that $\nu \ll \mu$. We want to show that (9.3) holds for an appropriately chosen f. What would f have to look like?

Let $\alpha \ge 0$ and note that $\nu - \alpha\mu$ is a signed measure. If f is the required (but unknown) function, then

$$(\nu - \alpha\mu)(A) = \int_A f \, d\mu - \alpha\mu(A) = \int_A (f - \alpha) \, d\mu.$$

By Exercise 9.11 on page 311, it follows that if $D_\alpha = \{\, x : f(x) \ge \alpha \,\}$, then (D_α, D_α^c) is a Hahn decomposition for $\nu - \alpha\mu$. Thinking now of x as fixed and α as varying, we have $x \in D_\alpha$ if and only if $\alpha \le f(x)$ and, therefore,

$$f(x) = \sup\{\, \alpha : \alpha \le f(x) \,\} = \sup\{\, \alpha : x \in D_\alpha \,\}.$$

So, the procedure for finding f will be essentially as follows: For each $\alpha \ge 0$, let (D_α, D_α^c) be a Hahn decomposition for $\nu - \alpha\mu$. Define $f(x) = \sup\{\, \alpha : x \in D_\alpha \,\}$ for $x \in \Omega$. This should give a function f that satisfies (9.3). We now present a formal proof of the Radon-Nikodym theorem.

PROOF We first assume that μ is a finite measure. For each positive rational number r, let (D_r, D_r^c) be a Hahn decomposition for $\nu - r\mu$. Define f on Ω by

$$f(x) = \begin{cases} \sup\{\, r \in Q : x \in D_r \,\}, & \text{if } x \in D_r \text{ for some } r; \\ 0, & \text{otherwise.} \end{cases}$$

Clearly f is a nonnegative extended real-valued function on Ω. We assert that f is $\mathcal{A}$-measurable. To prove that, it suffices to show that $f^{-1}([\alpha, \infty]) \in \mathcal{A}$ for each $\alpha \in \mathcal{R}$. For $\alpha \leq 0$, the inverse image is Ω. Hence, we can assume $\alpha > 0$. We will show that for $\alpha > 0$,

$$f^{-1}([\alpha, \infty]) = \bigcap_{q < \alpha} \left(\bigcup_{r > q} D_r \right), \tag{9.4}$$

where here and until specified otherwise, r and q represent positive rational numbers.

From the definition of f, we have that $f^{-1}((q, \infty]) = \bigcup_{r > q} D_r$, for each q. Because $[\alpha, \infty] = \bigcap_{q < \alpha}(q, \infty]$, it follows that

$$f^{-1}([\alpha, \infty]) = \bigcap_{q < \alpha} f^{-1}((q, \infty]) = \bigcap_{q < \alpha} \left(\bigcup_{r > q} D_r \right),$$

and so (9.4) holds.

Because Q is countable and $D_r \in \mathcal{A}$ for each r, it follows from (9.4) that $f^{-1}([\alpha, \infty]) \in \mathcal{A}$ for each $\alpha > 0$. Thus, f is $\mathcal{A}$-measurable.

We next show that f satisfies (9.3). To that end, let $A \in \mathcal{A}$ and, for each pair of rational numbers α and β with $0 \leq \alpha < \beta$, define

$$E = \{ x \in A : \alpha \leq f(x) < \beta \}.$$

We claim that

$$\alpha \mu(E) \leq \nu(E) \leq \beta \mu(E). \tag{9.5}$$

We begin by establishing the first inequality in (9.5). If $\alpha = 0$, that inequality is trivial; so, we assume that $\alpha > 0$. By (9.4), $E \subset \{ x : f(x) \geq \alpha \} \subset \bigcup_{r > q} D_r$ for each $q < \alpha$. If we can show that, for each $q < \alpha$, the latter set is positive for $\nu - q\mu$, then we will have $(\nu - q\mu)(E) \geq 0$, that is, $\nu(E) \geq q\mu(E)$, for each $q < \alpha$, from which it follows that $\nu(E) \geq \alpha\mu(E)$.

So, suppose $r > q$ and let $F \subset D_r$. Because D_r is positive for $\nu - r\mu$, we have

$$0 \leq (\nu - r\mu)(F) = \nu(F) - r\mu(F) \leq \nu(F) - q\mu(F) = (\nu - q\mu)(F).$$

Hence, $(\nu - q\mu)(F) \geq 0$ for $F \subset D_r$ and, consequently, D_r is positive for $\nu - q\mu$. Lemma 9.2 now implies that $\bigcup_{r > q} D_r$ is positive for $\nu - q\mu$, as required.

To establish the second inequality in (9.5), we first note that by definition, if $f(x) < \beta$, then $x \notin D_\beta$ or, in other words, $\{ x : f(x) < \beta \} \subset D_\beta^c$. Because D_β^c is negative for $\nu - \beta\mu$ and $E \subset \{ x : f(x) < \beta \} \subset D_\beta^c$, it follows that $(\nu - \beta\mu)(E) \leq 0$. That is, $\nu(E) \leq \beta\mu(E)$. We have now shown that (9.5) holds.

To continue, we need to consider where f is infinite on A. To that end, let $H = \{ x : f(x) = \infty \}$. We will show $\mu(A \cap H) > 0$ implies $\nu(A \cap H) = \infty$. In doing so, we will use the already established fact that, for each q, $\bigcup_{r > q} D_r$ is positive for $\nu - q\mu$.

Using the definition of f, we see that if $f(x) = \infty$, then for each q, there is an $r > q$ such that $x \in D_r$. Thus $H \subset \bigcup_{r > q} D_r$ and, hence, H is positive for $\nu - q\mu$. So, for each q, $(\nu - q\mu)(A \cap H) \geq 0$, that is, $\nu(A \cap H) \geq q\mu(A \cap H)$. Hence, if $\mu(A \cap H) > 0$, then $\nu(A \cap H) = \infty$.

Now, assume that $\mu(A \cap H) > 0$. Then $\nu(A \cap H) = \infty$ and, hence, $\nu(A) = \infty$. On the other hand, because $\mu(A \cap H) > 0$ and $f = \infty$ on H,

$$\int_A f \, d\mu \geq \int_{A \cap H} f \, d\mu = \infty.$$

Thus, if $\mu(A \cap H) > 0$, then both sides of (9.3) equal ∞.

So, assume that $\mu(A \cap H) = 0$. Then, because $\nu \ll \mu$, we have $\nu(A \cap H) = 0$. For each $n \in \mathcal{N}$, set

$$A_{n,k} = \left\{ x \in A : \frac{k-1}{n} \leq f(x) < \frac{k}{n} \right\},$$

for $k = 1, 2, \ldots$. Then $A = \bigcup_{k=1}^{\infty} A_{n,k} \cup (A \cap H)$ and, since the $A_{n,k}$'s are pairwise disjoint and $\nu(A \cap H) = 0$, it follows that $\nu(A) = \sum_{k=1}^{\infty} \nu(A_{n,k})$.

By (9.5),

$$\frac{k-1}{n}\mu(A_{n,k}) \leq \nu(A_{n,k}) \leq \frac{k}{n}\mu(A_{n,k})$$

and, from the definition of $A_{n,k}$, we conclude that

$$\frac{k-1}{n}\mu(A_{n,k}) \leq \int_{A_{n,k}} f \, d\mu \leq \frac{k}{n}\mu(A_{n,k}).$$

Therefore, for $k \in \mathcal{N}$,

$$\int_{A_{n,k}} f \, d\mu - \frac{1}{n}\mu(A_{n,k}) \leq \nu(A_{n,k}) \leq \int_{A_{n,k}} f \, d\mu + \frac{1}{n}\mu(A_{n,k}).$$

Recalling that $\mu(A \cap H) = 0$, we get upon summing on k that, for each $n \in \mathcal{N}$,

$$\int_A f \, d\mu - \frac{1}{n}\mu(A) \leq \nu(A) \leq \int_A f \, d\mu + \frac{1}{n}\mu(A).$$

Since we are assuming μ is finite, $\mu(A) < \infty$, and therefore, letting $n \to \infty$ in the previous display, we get that

$$\int_A f \, d\mu \leq \nu(A) \leq \int_A f \, d\mu.$$

So again, (9.3) holds.

Suppose now that μ is a σ-finite measure. We can write Ω as a countable disjoint union of $\mathcal{A}$-measurable sets $\{E_n\}_n$, where $\mu(E_n) < \infty$ and $\nu(E_n) < \infty$ for each $n \in \mathcal{N}$. (See Exercise 9.19.) Let $(E_n, \mathcal{A}_{E_n}, \mu_{E_n})$ be as usual, and set $\mu_n = \mu_{|\mathcal{A}_{E_n}} = \mu_{E_n}$ and $\nu_n = \nu_{|\mathcal{A}_{E_n}}$.

We have $\mu_n(E_n) = \mu(E_n) < \infty$ and, since $\nu \ll \mu$, we have $\nu_n \ll \mu_n$. Hence, by what we have proved for finite measures, there is a nonnegative $\mathcal{A}_{E_n}$-measurable function g_n such that

$$\nu_n(B) = \int_B g_n \, d\mu_n, \qquad B \in \mathcal{A}_{E_n}.$$

For each $n \in \mathcal{N}$, define f_n on Ω by $f_n(x) = g_n(x)$ if $x \in E_n$, and $f_n(x) = 0$ otherwise. Then $f = \sum_{n=1}^{\infty} f_n$ is a nonnegative $\mathcal{A}$-measurable function. Moreover, if $A \in \mathcal{A}$,

$$\nu(A) = \sum_{n=1}^{\infty} \nu(A \cap E_n) = \sum_{n=1}^{\infty} \nu_n(A \cap E_n) = \sum_{n=1}^{\infty} \int_{A \cap E_n} g_n \, d\mu_n$$

$$= \sum_{n=1}^{\infty} \int_{A \cap E_n} f_n \, d\mu = \sum_{n=1}^{\infty} \int_{A \cap E_n} f \, d\mu = \int_A f \, d\mu.$$

Thus, (9.3) holds.

It remains to prove uniqueness. So suppose that g is also a nonnegative extended real-valued $\mathcal{A}$-measurable function on Ω such that

$$\nu(A) = \int_A g \, d\mu, \qquad A \in \mathcal{A}.$$

We must show that $g = f$ μ-ae.

Let $E = \{ x : f(x) > g(x) \}$. We claim that $\mu(E) = 0$. Let $\{E_n\}_n$ be the sequence of sets defined earlier in the proof. We have, for each $n \in \mathcal{N}$,

$$\int_{E \cap E_n} f \, d\mu = \int_{E \cap E_n} g \, d\mu = \nu(E \cap E_n) < \infty.$$

Thus, f and g are integrable over $E \cap E_n$ with respect to μ and, consequently, so is $f - g$. Moreover,

$$\int_{E \cap E_n} (f - g) \, d\mu = \int_{E \cap E_n} f \, d\mu - \int_{E \cap E_n} g \, d\mu = 0.$$

Since $f - g > 0$ on E and, hence, on $E \cap E_n$, we must have $\mu(E \cap E_n) = 0$. Thus, $\mu(E) = \sum_n \mu(E \cap E_n) = 0$. A similar argument shows that $\{ x : g(x) > f(x) \}$ has μ-measure zero. Therefore, $g = f$ μ-ae. ∎

DEFINITION 9.5 Radon-Nikodym Derivative

The function f in the statement of the Radon-Nikodym theorem is called the **Radon-Nikodym derivative** of ν with respect to μ and is denoted by $d\nu/d\mu$.

Remark: The Radon-Nikodym theorem shows that the Radon-Nikodym derivative is determined only up to sets of μ-measure zero.

EXAMPLE 9.6 *Illustrates Definition 9.5*

a) Let α be a positive constant. Define $F(x) = 1 - e^{-\alpha x}$ for $x \geq 0$, and $F(x) = 0$ otherwise; and let ν denote the unique Borel measure induced by F. Define the Borel measure ω by

$$\omega(B) = \int_B f \, d\lambda, \qquad B \in \mathcal{B},$$

where $f(x) = \alpha e^{-\alpha x}$ for $x > 0$, and $f(x) = 0$ otherwise. Then it is easy to see that ω has F for its distribution function and, so, $\omega = \nu$. Hence $\nu \ll \lambda$ and $d\nu/d\lambda = f$ λ-ae. Note that we also have, for example, $d\nu/d\lambda = g$ λ-ae, where $g(x) = \alpha e^{-\alpha x}$ for $x \geq 0$, and $g(x) = 0$ otherwise, because $g = f$ λ-ae.

b) Let μ be the measure on $(\mathcal{R}, \mathcal{P}(\mathcal{R}))$ defined by $\mu(A) = \gamma(A \cap \mathcal{N})$, where γ is counting measure on $\mathcal{P}(\mathcal{R})$. Consider the measure $\nu = \sum_{n=1}^{\infty} 2^{-n} \delta_n$ on $\mathcal{P}(\mathcal{R})$, that is,

$$\nu(A) = \sum_{n \in A \cap \mathcal{N}} \frac{1}{2^n}, \qquad A \in \mathcal{P}(\mathcal{R}).$$

If $\mu(A) = 0$, then $A \cap \mathcal{N} = \emptyset$; so, $\nu(A) = 0$. Hence, $\nu \ll \mu$. Let $f(x) = 1/2^x$ if $x \in \mathcal{N}$, and $f(x) = 0$ otherwise. We claim that $d\nu/d\mu = f$ μ-ae. Indeed, if $A \in \mathcal{P}(\mathcal{R})$, then

$$\int_A f \, d\mu = \int_{A \cap \mathcal{N}} f \, d\mu + \int_{A \cap \mathcal{N}^c} f \, d\mu = \int_{A \cap \mathcal{N}} f \, d\mu$$

$$= \int_{\mathcal{N}} f \chi_A \, d\mu = \sum_{n=1}^{\infty} \int_{\{n\}} f \chi_A \, d\mu = \sum_{n=1}^{\infty} f(n) \chi_A(n) \mu(\{n\})$$

$$= \sum_{n=1}^{\infty} \frac{1}{2^n} \chi_A(n) = \sum_{n \in A \cap \mathcal{N}} \frac{1}{2^n} = \nu(A).$$

Note that a nonnegative function g on $\mathcal{R}$ is a Radon-Nikodym derivative of ν with respect to μ if and only if $g = f$ on $\mathcal{N}$.

c) Let $(\Omega, \mathcal{A}, P)$ be a probability space and let $E \in \mathcal{A}$ be such that $P(E) > 0$. Recall that the conditional probability measure P_E corresponding to E is defined by $P_E(A) = P(A \,|\, E) = P(E \cap A)/P(E)$. Clearly, we have $P_E \ll P$. A Radon-Nikodym derivative for P_E with respect to P is $\chi_E/P(E)$ because, for each $A \in \mathcal{A}$,

$$\int_A \frac{\chi_E}{P(E)} \, dP = \frac{1}{P(E)} \int_{\Omega} \chi_{E \cap A} \, dP = \frac{1}{P(E)} \cdot P(E \cap A) = P_E(A).$$

d) Refer to Example 7.5(c) on page 238. Let X be an absolutely continuous random variable on $(\Omega, \mathcal{A}, P)$ with probability density function f_X. Then we have $\mu_X \ll \lambda$ (as Borel measures). Moreover, since by definition, f_X is a non-negative Borel measurable function such that $\mu_X(B) = \int_B f_X \, d\lambda$ for $B \in \mathcal{B}$, f_X is a Radon-Nikodym derivative of μ_X with respect to λ. □

□ □ □ **PROPOSITION 9.1**

Let $(\Omega, \mathcal{A}, \mu)$ be a σ-finite measure space and ν a σ-finite measure on $\mathcal{A}$ such that $\nu \ll \mu$. If $g \in \mathcal{L}^1(\nu)$, then $g \cdot d\nu/d\mu \in \mathcal{L}^1(\mu)$ and

$$\int_{\Omega} g \, d\nu = \int_{\Omega} g \frac{d\nu}{d\mu} \, d\mu.$$

PROOF   Exercise 5.61(b) on page 164 shows that the proposition is true if g is nonnegative. For real-valued g, write $g = g^+ - g^-$ and apply the result for nonnegative functions twice. For complex-valued g, write $g = \Re g + i\Im g$ and apply the result for real-valued functions twice. ■

A Relation Between Absolutely Continuous Functions and Absolutely Continuous Measures

In Section 8.4, we discussed absolutely continuous functions and, in this section, we discussed absolutely continuous measures. A relation between the two concepts is expressed in the following proposition.

□ □ □ **PROPOSITION 9.2**

Let ν be a finite Borel measure and F_ν its distribution function. Then ν is absolutely continuous with respect to Lebesgue measure if and only if F_ν is absolutely continuous on $\mathcal{R}$. In this case, $d\nu/d\lambda = F_\nu'$ λ-ae.

PROOF Suppose that $\nu \ll \lambda$. Then, for $B \in \mathcal{B}$, $\nu(B) = \int_B (d\nu/d\lambda)\, d\lambda$. In particular, for $B = (-\infty, x]$,

$$F_\nu(x) = \nu\big((-\infty, x]\big) = \int_{(-\infty, x]} \frac{d\nu}{d\lambda}\, d\lambda = \int_{-\infty}^{x} \frac{d\nu}{d\lambda}(t)\, dt.$$

Because $F_\nu(\infty) = \nu(\mathcal{R}) < \infty$, it follows that $d\nu/d\lambda \in \mathcal{L}^1(\mathcal{R})$. Consequently, by Proposition 8.4(b) on page 293, F_ν is absolutely continuous on $\mathcal{R}$ and, by Corollary 8.2 on page 291, $F_\nu' = d\nu/d\lambda$ λ-ae.

Conversely, suppose that F_ν is absolutely continuous on $\mathcal{R}$. Then, by definition, $F_\nu' \in \mathcal{L}^1(\mathcal{R})$ and $F_\nu(x) = \int_{-\infty}^{x} F_\nu'(t)\, dt$, $-\infty < x < \infty$. Define

$$\omega(B) = \int_B F_\nu'\, d\lambda, \qquad B \in \mathcal{B}.$$

Then ω has F_ν as its distribution function and, consequently, by Theorem 6.3 on page 196, $\omega = \nu$. It follows that $\nu \ll \lambda$ and $d\nu/d\lambda = F_\nu'$ λ-ae. ∎

Conditional Probability Given a σ-Algebra

Let us recall the definition of conditional probability from Section 7.1: Suppose that $(\Omega, \mathcal{A}, P)$ is a probability space and F is an event having positive probability, that is, $F \in \mathcal{A}$ and $P(F) > 0$. Then, for $E \in \mathcal{A}$, the *conditional probability* of event E given that event F has occurred is defined by

$$P(E \mid F) = \frac{P(F \cap E)}{P(F)}.$$

We can generalize the notion of conditional probability by conditioning on a σ-algebra instead of just an event. The idea is that if $\mathcal{G}$ is a σ-algebra with $\mathcal{G} \subset \mathcal{A}$, then conditioning on $\mathcal{G}$ means that we know whether or not each $G \in \mathcal{G}$ has occurred; and the conditional probability of an event E given $\mathcal{G}$, denoted $P(E \mid \mathcal{G})$, is the probability of E computed with that knowledge.

To see how to define $P(E \mid \mathcal{G})$, we consider the simplest nontrivial case. Suppose that F is an event with probability strictly between 0 and 1. Let $\mathcal{G}$ be the σ-algebra generated by F, that is, the smallest σ-algebra containing F;

clearly, $\mathcal{G} = \{\emptyset, F, F^c, \Omega\}$. Then, given $\mathcal{G}$, we know whether or not F has occurred, that is, whether F has occurred or F^c has occurred. In the former case, $P(E \mid \mathcal{G}) = P(E \mid F)$ and, in the latter case, $P(E \mid \mathcal{G}) = P(E \mid F^c)$. In other words,

$$P(E \mid \mathcal{G}) = P(E \mid F)\chi_F + P(E \mid F^c)\chi_{F^c}.$$

Note that $P(E \mid \mathcal{G})$ is not only a random variable (i.e., is $\mathcal{A}$-measurable), but is in fact $\mathcal{G}$-measurable. Furthermore, it is not too difficult to see that

$$P(G \cap E) = \int_G P(E \mid \mathcal{G}) \, dP, \qquad G \in \mathcal{G}. \tag{9.6}$$

We can use the Radon-Nikodym theorem to show the existence of conditional probability given a σ-algebra in the general case. Specifically, with (9.6) in mind, we have the following proposition.

□ □ □ **PROPOSITION 9.3 Existence of Conditional Probability**

Let $(\Omega, \mathcal{A}, P)$ be a probability space, $E \in \mathcal{A}$, and $\mathcal{G}$ a σ-algebra with $\mathcal{G} \subset \mathcal{A}$. Then there exists a nonnegative $\mathcal{G}$-measurable function $P(E \mid \mathcal{G})$ such that

$$P(G \cap E) = \int_G P(E \mid \mathcal{G}) \, dP, \qquad G \in \mathcal{G}. \tag{9.7}$$

*Moreover, such a function is unique P-ae and is called the **conditional probability of E given $\mathcal{G}$**.*

PROOF   Define $\mu_E(G) = P(G \cap E)$, for $G \in \mathcal{G}$. Then, clearly, μ_E is a finite measure on $\mathcal{G}$ and $\mu_E \ll P$. The result now follows from the Radon-Nikodym theorem. ■

Further properties of conditional probability given a σ-algebra will be considered in the exercises.

Exercises for Section 9.2

9.15   This exercise provides an alternative for the definition of absolute continuity of measures. Let $(\Omega, \mathcal{A})$ be a measurable space and μ and ν measures on $\mathcal{A}$ with ν finite.
   a) Prove that a necessary and sufficient condition for $\nu \ll \mu$ is that for each $\epsilon > 0$, there is a $\delta > 0$ such that $\nu(A) < \epsilon$, whenever $A \in \mathcal{A}$ and $\mu(A) < \delta$. *Hint:* For the necessity part, suppose to the contrary that there is an $\epsilon > 0$ such that for each $\delta > 0$, there is an $A \in \mathcal{A}$ with $\mu(A) < \delta$ and $\nu(A) \geq \epsilon$. For each $n \in \mathcal{N}$, let A_n correspond to $\delta = 2^{-n}$.
   b) Show that weakening the finiteness condition on ν to σ-finiteness invalidates the necessity portion of part (a).

9.16   Let ν be a finite Borel measure and F_ν its distribution function. Provide an alternate proof of Proposition 9.2 (page 317) without relying on the Radon-Nikodym theorem, but instead by using Theorem 8.6 (page 298), Theorem 8.7 (page 298), and Exercise 9.15.

9.17   Find two measures μ and ν such that $\nu \not\ll \mu$, $\mu \not\ll \nu$, and $\mu \not\perp \nu$.

9.18   Suppose that μ and ν are measures on $(\Omega, \mathcal{A})$ such that $\mu \perp \nu$ and $\mu \ll \nu$. What can you say about μ?

9.19 Let $(\Omega, \mathcal{A})$ be a measurable space and μ and ν σ-finite measures on $\mathcal{A}$. Show that Ω can be written as a countable disjoint union of $\mathcal{A}$-measurable sets $\{E_n\}_n$, where $\mu(E_n) < \infty$ and $\nu(E_n) < \infty$.

9.20 Let Ω be a nonempty set and μ counting measure on $\mathcal{P}(\Omega)$. Suppose ν is a discrete measure on $\mathcal{P}(\Omega)$, that is, there is a countable set $K \subset \Omega$ such that $\nu(K^c) = 0$.
   a) Show that $\nu \ll \mu$.
   b) Find $d\nu/d\mu$.
   c) Are the hypotheses of our version of the Radon-Nikodym theorem (Theorem 9.3 on page 312) necessarily satisfied in this problem?

9.21 Define ν on $(\mathcal{R}, \mathcal{M})$ by $\nu(A) = \lambda\big(A \cap [-2, 2]\big)$. Show that $\nu \ll \lambda$ and find $d\nu/d\lambda$.

9.22 Let μ be the measure on $\big(\mathcal{R}, \mathcal{P}(\mathcal{R})\big)$ defined by $\mu(A) = \gamma(A \cap \mathcal{N})$, where γ is counting measure on $\mathcal{P}(\mathcal{R})$. Let $\{a_n\}_{n=1}^{\infty}$ be a sequence of nonnegative real numbers and set $\nu = \sum_{n=1}^{\infty} a_n \delta_n$. Show that $\nu \ll \mu$ and find $d\nu/d\mu$.

9.23 Refer to Example 7.5(a) on page 237. Let X be a discrete random variable on $(\Omega, \mathcal{A}, P)$ with probability mass function p_X.
   a) Show that $\mu_X \ll \mu$, where μ is counting measure on $(\mathcal{R}, \mathcal{B})$.
   b) Prove that p_X is the unique Radon-Nikodym derivative of μ_X with respect to μ.

9.24 Provide an example showing that the σ-finiteness of μ cannot be dropped as an hypothesis in the Radon-Nikodym theorem.

9.25 Let $(\Omega, \mathcal{A}, \mu)$ be a σ-finite measure space and ν_1 and ν_2 σ-finite measures on $\mathcal{A}$. Assume that $\nu_1 \ll \mu$ and $\nu_2 \ll \mu$. Prove that $\nu_1 + \nu_2 \ll \mu$ and

$$\frac{d(\nu_1 + \nu_2)}{d\mu} = \frac{d\nu_1}{d\mu} + \frac{d\nu_2}{d\mu} \qquad \mu\text{-ae.}$$

9.26 Let $(\Omega, \mathcal{A})$ be a measurable space and let ω, ν, and μ be σ-finite measures on $\mathcal{A}$ such that $\omega \ll \nu \ll \mu$. Show that $\omega \ll \mu$ and

$$\frac{d\omega}{d\mu} = \frac{d\omega}{d\nu}\frac{d\nu}{d\mu} \qquad \mu\text{-ae.}$$

9.27 Let $(\Omega, \mathcal{A})$ be a measurable space and μ and ν σ-finite measures on $\mathcal{A}$ such that $\mu \ll \nu$ and $\nu \ll \mu$. Prove that

$$\frac{d\nu}{d\mu} = 1 \bigg/ \frac{d\mu}{d\nu} \qquad \mu\text{-ae.}$$

9.28 Suppose that μ and ν are two σ-finite Borel measures. Recall that the *convolution* of μ and ν is the measure defined by

$$(\mu * \nu)(B) = \int_{\mathcal{R}} \mu(B - y) \, d\nu(y), \qquad B \in \mathcal{B}.$$

   a) Show that if $\mu \ll \lambda$, then $\mu * \nu \ll \lambda$ and

$$\frac{d(\mu * \nu)}{d\lambda}(x) = \int_{\mathcal{R}} \frac{d\mu}{d\lambda}(x - y) \, d\nu(y), \qquad x \subset \mathcal{R}.$$

   Hint: Refer to Exercises 6.63 and 6.64 (page 221).
   b) Show that if both μ and ν are absolutely continuous with respect to Lebesgue measure, then

$$\frac{d(\mu * \nu)}{d\lambda}(x) = \int_{\mathcal{R}} \frac{d\mu}{d\lambda}(x - y)\frac{d\nu}{d\lambda}(y) \, d\lambda(y), \qquad x \in \mathcal{R}.$$

   In words, the Radon-Nikodym derivative of the convolution of two measures that are absolutely continuous with respect to Lebesgue measure is the convolution of their Radon-Nikodym derivatives.

In Exercises 9.29–9.36, $(\Omega, \mathcal{A}, P)$ is a probability space and $\mathcal{G}$ is a σ-algebra of subsets of Ω with $\mathcal{G} \subset \mathcal{A}$.

9.29 Let F be an event with probability strictly between 0 and 1. Let $\mathcal{G}$ be the σ-algebra generated by F, that is, $\mathcal{G} = \{\emptyset, F, F^c, \Omega\}$. Define $P(E \mid \mathcal{G}) = P(E \mid F)\chi_F + P(E \mid F^c)\chi_{F^c}$.
   a) Show that $P(E \mid \mathcal{G})$ is $\mathcal{G}$-measurable.
   b) Prove that for each $G \in \mathcal{G}$, $P(G \cap E) = \int_G P(E \mid \mathcal{G}) \, dP$.

9.30 Suppose that $\{F_n\}_n$ is a sequence of pairwise mutually exclusive events each with positive probability and such that $\bigcup_n F_n = \Omega$. Let $\mathcal{G}$ be the σ-algebra generated by $\{F_n\}_n$.
   a) Characterize the sets in $\mathcal{G}$.
   b) Prove that, with probability one, $P(E \mid \mathcal{G}) = \sum_n P(E \mid F_n)\chi_{F_n}$.

9.31 Suppose that $E \in \mathcal{G}$.
   a) What does your intuition tell you regarding $P(E \mid \mathcal{G})$?
   b) Prove your assertion in part (a).

9.32 Suppose that $\mathcal{G} = \{\Omega, \emptyset\}$.
   a) What does your intuition tell you regarding $P(E \mid \mathcal{G})$?
   b) Prove your assertion in part (a).

9.33 Establish that each of the following hold with probability one.
   a) $P(\Omega \mid \mathcal{G}) = 1$.
   b) For each $E \in \mathcal{A}$, $P(E \mid \mathcal{G}) \geq 0$.
   c) If $E_1, E_2, \ldots$ are in $\mathcal{A}$, with $E_i \cap E_j = \emptyset$ for $i \neq j$, then

$$P\left(\bigcup_n E_n \,\Big|\, \mathcal{G}\right) = \sum_n P(E_n \mid \mathcal{G}).$$

★9.34 **Conditional probability given a random variable:** An important case of conditional probability given a σ-algebra is when the σ-algebra is generated by a random variable, X. The **σ-algebra generated by X**, denoted $\sigma(X)$, is by definition the smallest σ-algebra of subsets of Ω for which X is measurable. We define the **conditional probability of E given X**, denoted $P(E \mid X)$, to be the conditional probability of E given $\sigma(X)$; that is, by definition, $P(E \mid X) = P(E \mid \sigma(X))$.
   a) Show that $\sigma(X) = \{\, \{X \in B\} : B \in \mathcal{B} \,\}$.
   b) Establish that there exists a nonnegative Borel measurable function ϕ such that $P(E \mid X) = \phi \circ X$, P-ae.
   c) Let ϕ be the function in part (b). For $x \in \mathcal{R}$, set $P(E \mid X = x) = \phi(x)$, called the **conditional probability of E given $X = x$.** Prove that

$$P(\{X \in B\} \cap E) = \int_B P(E \mid X = x) \, d\mu_X(x), \qquad B \in \mathcal{B},$$

   where μ_X is the probability distribution of X. *Hint:* Use Theorem 7.6 on page 249.
   d) Prove that if g is a nonnegative Borel measurable function such that

$$P(\{X \in B\} \cap E) = \int_B g(x) \, d\mu_X(x), \qquad B \in \mathcal{B},$$

   then, for μ_X-almost all x, $g(x) = P(E \mid X = x)$.

9.35 Refer to Example 7.7(a) on page 240. Suppose X and Y are jointly discrete random variables with joint probability mass function $p_{X,Y}$. Define

$$p_{Y \mid X}(y \mid x) = \begin{cases} \dfrac{p_{X,Y}(x, y)}{p_X(x)}, & p_X(x) > 0; \\ 0, & \text{otherwise.} \end{cases}$$

a) Prove that for each $y \in \mathcal{R}$,

$$P(Y = y \,|\, X = x) = p_{Y \,|\, X}(y \,|\, x)$$

for each possible value x of X. *Hint:* Use Exercise 9.34(d).

b) Determine $P(Y = y \,|\, X)$.

9.36   Refer to Example 7.7(b) on page 240. Suppose X and Y are jointly absolutely continuous random variables with joint probability density function $f_{X,Y}$. Define

$$f_{Y \,|\, X}(y \,|\, x) = \begin{cases} \dfrac{f_{X,Y}(x,y)}{f_X(x)}, & f_X(x) > 0; \\ 0, & \text{otherwise.} \end{cases}$$

a) Prove that for each $C \in \mathcal{B}$,

$$P(Y \in C \,|\, X = x) = \int_C f_{Y \,|\, X}(y \,|\, x) \, dy$$

for μ_X-almost all x. *Hint:* Use Exercise 9.34(d).

b) Determine $P(Y \in C \,|\, X)$.

9.3 SIGNED AND COMPLEX MEASURES

In Section 9.1, we introduced signed measures. Now, in this section, we will further investigate signed measures and also introduce the concept of complex measures. Recall that if $(\Omega, \mathcal{A})$ is a measurable space, then a *signed measure* ν on $\mathcal{A}$ is an extended real-valued function satisfying the following two conditions:

- $\nu(\emptyset) = 0$.

- If A_1, A_2, … are in $\mathcal{A}$, with $A_i \cap A_j = \emptyset$ for $i \neq j$, then

$$\nu\left(\bigcup_n A_n\right) = \sum_n \nu(A_n).$$

For this definition to make sense, ν cannot take on both ∞ and $-\infty$ as values.

We proved the Jordan decomposition theorem — that any signed measure ν can be expressed uniquely as the difference of two mutually singular measures ν^+ and ν^- on $\mathcal{A}$. The representation $\nu = \nu^+ - \nu^-$ is called the Jordan decomposition of ν. In fact, if (D, D^c) is a Hahn decomposition for ν, then

$$\nu^+(A) = \nu(A \cap D) \qquad \text{and} \qquad \nu^-(A) = -\nu(A \cap D^c).$$

Now we will give names to the measures ν^+ and ν^- and define yet another measure corresponding to a signed measure.

DEFINITION 9.6   Variations of a Signed Measure

Suppose that $(\Omega, \mathcal{A})$ is a measurable space and that ν is a signed measure on $\mathcal{A}$ with Jordan decomposition $\nu = \nu^+ - \nu^-$. Define

$$|\nu| = \nu^+ + \nu^-.$$

The measures ν^+, ν^-, and $|\nu|$ are called, respectively, the **positive variation, negative variation,** and **total variation** of ν.

Note: Observe that $|\nu|$ is a measure and that $|\nu| = \nu$ if ν is a measure.

Before proving our next result, we introduce the following terminology.

DEFINITION 9.7 **Measurable Partition**

Let $(\Omega, \mathcal{A})$ be a measurable space and $A \in \mathcal{A}$. A finite sequence $\{A_k\}_{k=1}^n$ of subsets of Ω is said to be a **measurable partition** of A if the A_ks are $\mathcal{A}$-measurable, pairwise disjoint, and their union is A. That is,

a) $A_k \in \mathcal{A}$, $k = 1, 2, \ldots, n$,

b) $A_i \cap A_j = \emptyset$ for $i \neq j$, and

c) $\bigcup_{k=1}^n A_k = A$.

The next proposition shows that the concept of the total variation of a signed measure is similar to that of the total variation of a function. Exercises 9.39 and 9.40 further explore the analogy.

□ □ □ **PROPOSITION 9.4**

Let $(\Omega, \mathcal{A})$ be a measurable space and ν a signed measure on $\mathcal{A}$. Then, for each $A \in \mathcal{A}$,

$$|\nu|(A) = \sup \left\{ \sum_{k=1}^n |\nu(A_k)| : \{A_k\}_{k=1}^n \text{ is a measurable partition of } A \right\}.$$

PROOF Let $\{A_k\}_{k=1}^n$ be a measurable partition of A. Then, because $|\nu|$ is a measure, $|\nu|(A) = \sum_{k=1}^n |\nu|(A_k)$. From Exercise 9.37(b), we know that $|\nu|(A_k) \geq |\nu(A_k)|$. Therefore, $|\nu|(A) \geq \sum_{k=1}^n |\nu(A_k)|$ and, so,

$$|\nu|(A) \geq \sup \left\{ \sum_{k=1}^n |\nu(A_k)| : \{A_k\}_{k=1}^n \text{ is a measurable partition of } A \right\}.$$

To prove the reverse inequality, let (D, D^c) be a Hahn decomposition for ν, and set $A_1 = A \cap D$ and $A_2 = A \cap D^c$. Then $\{A_1, A_2\}$ is a measurable partition of A and we have $|\nu(A_1)| + |\nu(A_2)| = \nu^+(A) + \nu^-(A) = |\nu|(A)$. Hence,

$$|\nu|(A) \leq \sup \left\{ \sum_{k=1}^n |\nu(A_k)| : \{A_k\}_{k=1}^n \text{ is a measurable partition of } A \right\}.$$

The proof of the proposition is now complete. ∎

Next we define the abstract Lebesgue integral of a measurable function with respect to a signed measure. It should be clear that the following definition is reasonable and natural.

DEFINITION 9.8   Integral with Respect to a Signed Measure

Suppose that $(\Omega, \mathcal{A})$ is a measurable space and that ν is a signed measure on $\mathcal{A}$ with Jordan decomposition $\nu = \nu^+ - \nu^-$. Let $E \in \mathcal{A}$ and f an extended real-valued or complex-valued $\mathcal{A}$-measurable function on Ω. Then the (abstract) **Lebesgue integral of f over E with respect to ν** is defined by

$$\int_E f \, d\nu = \int_E f \, d\nu^+ - \int_E f \, d\nu^-,$$

provided the right-hand side makes sense.

EXAMPLE 9.7   Illustrates Definition 9.8

Let $(\Omega, \mathcal{A}) = (\mathcal{R}, \mathcal{M})$ and $\nu = \lambda - \delta_0 - \delta_1$. Then $\nu^+ = \lambda$ and $\nu^- = \delta_0 + \delta_1$. Let $f(x) = x^3 + 2$. We have

$$\int_{[0,1)} f \, d\nu = \int_{[0,1)} f \, d\lambda - \int_{[0,1)} f \, d(\delta_0 + \delta_1)$$

$$= \int_0^1 (x^3 + 2) \, dx - \int_{[0,1)} (x^3 + 2) \, d(\delta_0 + \delta_1)(x)$$

$$= \frac{9}{4} - 2 = \frac{1}{4}$$

and

$$\int_{[0,1]} f \, d\nu = \int_0^1 (x^3 + 2) \, dx - \int_{[0,1]} (x^3 + 2) \, d(\delta_0 + \delta_1)(x)$$

$$= \frac{9}{4} - (2 + 3) = -\frac{11}{4},$$

as the reader should verify.   □

Complex Measures

Here now is the definition of a complex measure.

DEFINITION 9.9   Complex Measure

Let $(\Omega, \mathcal{A})$ be a measurable space. A **complex measure** ν on $\mathcal{A}$ is a complex-valued countably additive set function; that is,

a) $\nu(\Lambda) \subset \mathbb{C}$ for all $\Lambda \subset \mathcal{A}$.

b) If $A_1, A_2, \ldots$ are in $\mathcal{A}$, with $A_i \cap A_j = \emptyset$ for $i \neq j$, then

$$\nu\left(\bigcup_n A_n\right) = \sum_n \nu(A_n).$$

Remark: Because, for a complex measure, $\nu(A) \in \mathbb{C}$ for each $A \in \mathcal{A}$, we see that there are no sets of infinite ν measure. It follows easily from countable additivity that $\nu(\emptyset) = 0$.

EXAMPLE 9.8 *Illustrates Definition 9.9*

a) Any finite measure or any finite signed measure is a complex measure.

b) Let $(\Omega, \mathcal{A})$ be a measurable space and ν a complex measure on $\mathcal{A}$. Define, for $A \in \mathcal{A}$,

$$(\Re\nu)(A) = \Re\big(\nu(A)\big) \qquad \text{and} \qquad (\Im\nu)(A) = \Im\big(\nu(A)\big).$$

Then it is easy to see that $\Re\nu$ and $\Im\nu$ are finite signed measures on $\mathcal{A}$ and that $\nu = \Re\nu + i\Im\nu$.

c) Lebesgue measure is not a complex measure on $\mathcal{M}$. (Why?)

d) Let $(\Omega, \mathcal{A})$ be a measurable space and ν_k, $1 \leq k \leq 4$, finite measures on $\mathcal{A}$. Define

$$\nu = \nu_1 - \nu_2 + i\nu_2 - i\nu_4,$$

that is, $\nu(A) = \nu_1(A) - \nu_2(A) + i\nu_3(A) - i\nu_4(A)$ for $A \in \mathcal{A}$. Then ν is a complex measure on $\mathcal{A}$. As we will see shortly, all complex measures are of this form.

e) Let $(\Omega, \mathcal{A}, \mu)$ be a measure space and $f \in \mathcal{L}^1(\mu)$. Define

$$\nu(A) = \int_A f \, d\mu, \qquad A \in \mathcal{A},$$

Then, ν is a complex measure on $\mathcal{A}$. □

Using the Jordan decomposition theorem, we can easily establish the following theorem. Its verification is left to the reader as an exercise.

□ □ □ **THEOREM 9.4**

Let $(\Omega, \mathcal{A})$ be a measurable space and let ν be a complex measure on $\mathcal{A}$. Then there exist unique measures, ν_1^+, ν_1^-, ν_2^+, ν_2^-, such that $\nu_1^+ \perp \nu_1^-$, $\nu_2^+ \perp \nu_2^-$, and $\nu = \nu_1^+ - \nu_1^- + i\nu_2^+ - i\nu_2^-$.

Next we want to define the total variation $|\nu|$ of a complex measure ν. In view of Proposition 9.4 (page 322), we make the following definition.

DEFINITION 9.10 **Total Variation of a Complex Measure**

Let $(\Omega, \mathcal{A})$ be a measurable space and let ν be a complex measure on $\mathcal{A}$. For each $A \in \mathcal{A}$, define

$$|\nu|(A) = \sup\left\{ \sum_{k=1}^{n} |\nu(A_k)| : \{A_k\}_{k=1}^{n} \text{ is a measurable partition of } A \right\}.$$

The set function $|\nu|$ is called the **total variation** of ν.

Because of Proposition 9.4, Definition 9.10 is consistent with the definition of total variation of signed measures. The next proposition shows that, as is the case for signed measures, the total variation of a complex measure is a measure.

□ □ □   PROPOSITION 9.5

Let $(\Omega, \mathcal{A})$ be a measurable space and let ν be a complex measure on $\mathcal{A}$. Then $|\nu|$ is a finite measure on $\mathcal{A}$.

PROOF   It follows from Definition 9.10 and Theorem 9.4 that

$$|\nu|(A) \leq \nu_1^+(A) + \nu_1^-(A) + \nu_2^+(A) + \nu_2^-(A)$$

for all $A \in \mathcal{A}$. Since the right-hand side of the preceding inequality is finite, we see that $|\nu|(A) < \infty$ for all $A \in \mathcal{A}$.

Clearly, $|\nu|(\emptyset) = 0$ and $|\nu|(A) \geq 0$ for $A \in \mathcal{A}$. Now, let $\{A_n\}_n$ be a pairwise disjoint sequence of $\mathcal{A}$-measurable sets. We can, without loss of generality, assume that the sequence is infinite. Set $A = \bigcup_{n=1}^{\infty} A_n$. We claim that

$$|\nu|(A) = \sum_{n=1}^{\infty} |\nu|(A_n).$$

Let $\alpha < |\nu|(A)$. Then there is a measurable partition of A, say, $\{E_k\}_{k=1}^{m}$, such that $\alpha < \sum_{k=1}^{m} |\nu(E_k)|$. But

$$\sum_{k=1}^{m} |\nu(E_k)| = \sum_{k=1}^{m} \left| \sum_{n=1}^{\infty} \nu(E_k \cap A_n) \right| \leq \sum_{k=1}^{m} \sum_{n=1}^{\infty} |\nu(E_k \cap A_n)|$$

$$= \sum_{n=1}^{\infty} \sum_{k=1}^{m} |\nu(E_k \cap A_n)| \leq \sum_{n=1}^{\infty} |\nu|(A_n),$$

where the last inequality holds because, for each n, $\{E_k \cap A_n\}_{k=1}^{m}$ is a measurable partition of A_n. Consequently, we have $\sum_{n=1}^{\infty} |\nu|(A_n) > \alpha$ for each $\alpha < |\nu|(A)$. It now follows that $|\nu|(A) \leq \sum_{n=1}^{\infty} |\nu|(A_n)$.

Next we prove the reverse inequality. Let $\epsilon > 0$ be given. For each $n \in \mathcal{N}$, we can choose a measurable partition $\{E_{nk}\}_{k=1}^{k_n}$ of A_n such that

$$\sum_{k=1}^{k_n} |\nu(E_{nk})| > |\nu|(A_n) - \frac{\epsilon}{2^n}.$$

Let $N \in \mathcal{N}$ be fixed but arbitrary. Then $\{E_{nk} : 1 \leq k \leq k_n, \ 1 \leq n \leq N\}$ is a measurable partition of $\bigcup_{n=1}^{N} A_n$. Using Exercise 9.43, we obtain that

$$|\nu|(A) \geq |\nu| \left(\bigcup_{n=1}^{N} A_n \right) \geq \sum_{n=1}^{N} \sum_{k=1}^{k_n} |\nu(E_{nk})|$$

$$> \sum_{n=1}^{N} \left(|\nu|(A_n) - \frac{\epsilon}{2^n} \right) = \sum_{n=1}^{N} |\nu|(A_n) - \sum_{n=1}^{N} \frac{\epsilon}{2^n}.$$

Letting $N \to \infty$ gives $|\nu|(A) \geq \sum_{n=1}^{\infty} |\nu|(A_n) - \epsilon$. As $\epsilon > 0$ was chosen arbitrarily, we have $\sum_{n=1}^{\infty} |\nu|(A_n) \leq |\nu|(A)$.   ■

Radon-Nikodym Theorem for Complex Measures

Next, we will prove the Radon-Nikodym theorem for complex measures. We begin with the following obvious extension of the definition of absolute continuity.

DEFINITION 9.11 **Absolutely Continuous Complex Measures**

Let $(\Omega, \mathcal{A}, \mu)$ be a measure space and ν a complex measure on $\mathcal{A}$. Then ν is said to be **absolutely continuous** with respect to μ, denoted $\boldsymbol{\nu \ll \mu}$, if $\nu(A) = 0$ whenever $\mu(A) = 0$.

We have noted previously that if $f \in \mathcal{L}^1(\Omega, \mathcal{A}, \mu)$, then the set function

$$\nu(A) = \int_A f \, d\mu, \qquad A \in \mathcal{A},$$

is a complex measure and, clearly, $\nu \ll \mu$. The following generalization of the Radon-Nikodym theorem shows that, under σ-finite conditions, the converse is true.

□ □ □ **THEOREM 9.5 Radon-Nikodym Theorem for Complex Measures**

Let $(\Omega, \mathcal{A}, \mu)$ be a σ-finite measure space and ν a complex measure on $\mathcal{A}$. If $\nu \ll \mu$, then there is a function $f \in \mathcal{L}^1(\Omega, \mathcal{A}, \mu)$ such that

$$\nu(A) = \int_A f \, d\mu, \qquad A \in \mathcal{A}. \tag{9.8}$$

Moreover, f is unique in the sense that if g is an $\mathcal{A}$-measurable function with $\nu(A) = \int_A g \, d\mu$ for all $A \in \mathcal{A}$, then $g = f$ μ-ae.

PROOF First we assume that ν is a finite signed measure. Let (D, D^c) be a Hahn decomposition for ν, and let ν^+ and ν^- be the positive and negative variations of ν. Because ν is finite, so are ν^+ and ν^-.

Suppose $\mu(A) = 0$. Then $\mu(A \cap D) = 0$ and, therefore, because $\nu \ll \mu$, we have $\nu^+(A) = \nu(A \cap D) = 0$. Hence, $\nu^+ \ll \mu$. Similarly, $\nu^- \ll \mu$. Applying the Radon-Nikodym theorem (Theorem 9.3 on page 312), we conclude that there exist nonnegative $\mathcal{A}$-measurable functions f_1 and f_2 such that for each $A \in \mathcal{A}$,

$$\nu^+(A) = \int_A f_1 \, d\mu \quad \text{and} \quad \nu^-(A) = \int_A f_2 \, d\mu.$$

Because ν^+ and ν^- are finite measures, $f_1, f_2 \in \mathcal{L}^1(\mu)$. Letting $f = f_1 - f_2$, we see that $f \in \mathcal{L}^1(\mu)$ and that (9.8) holds.

To prove uniqueness, suppose that g is an $\mathcal{A}$-measurable function such that $\nu(A) = \int_A g \, d\mu$ for each $A \in \mathcal{A}$. Since ν is a finite measure, $g \in \mathcal{L}^1(\mu)$ and, hence, so is $f - g$.

Now, let $E = \{\, x : f(x) > g(x)\,\}$. We have

$$\int_E (f - g)\, d\mu = \int_E f\, d\mu - \int_E g\, d\mu = \nu(E) - \nu(E) = 0.$$

Because $f - g > 0$ on E, it must be that $\mu(E) = 0$. Hence, $f \le g$ μ-ae. Similarly, $f \ge g$ μ-ae. Consequently, $g = f$ μ-ae.

Now suppose that ν is a complex measure. Applying what was just proved to the finite signed measures $\Re\nu$ and $\Im\nu$, we find that there are functions $g_1, g_2 \in \mathcal{L}^1(\mu)$ such that for each $A \in \mathcal{A}$,

$$(\Re\nu)(A) = \int_A g_1\, d\mu \quad \text{and} \quad (\Im\nu)(A) = \int_A g_2\, d\mu.$$

Letting $f = g_1 + ig_2$ yields a function in $\mathcal{L}^1(\mu)$ such that (9.8) holds. Uniqueness follows easily from the already established uniqueness for finite signed measures. ∎

Conditional Expectation Given a σ-algebra

Let $(\Omega, \mathcal{A}, P)$ be a probability space and F an event with positive probability, that is, $F \in \mathcal{A}$ and $P(F) > 0$. Recall that the set function P_F, defined by

$$P_F(E) = P(E \mid F) = \frac{P(F \cap E)}{P(F)}, \qquad E \in \mathcal{A},$$

is a probability measure on $\mathcal{A}$. Hence, we can define expectation with respect to P_F. This is called conditional expectation relative to F.

Thus, if $Y \in \mathcal{L}^1(\Omega, \mathcal{A}, P)$, then the *conditional expectation* of Y given that event F has occurred is defined by

$$\mathcal{E}(Y \mid F) = \int_\Omega Y\, dP_F.$$

According to Exercise 7.84(a) on page 257,

$$\mathcal{E}(Y \mid F) = \frac{1}{P(F)} \int_F Y\, dP.$$

We can generalize the notion of conditional expectation by conditioning on a σ-algebra instead of just an event. The idea is that if $\mathcal{G}$ is a σ-algebra with $\mathcal{G} \subset \mathcal{A}$, then conditioning on $\mathcal{G}$ means that we know whether or not each $G \in \mathcal{G}$ has occurred; and the conditional expectation of a random variable Y given $\mathcal{G}$, denoted $\mathcal{E}(Y \mid \mathcal{G})$, is the expectation of Y computed with that knowledge.

To see how to define $\mathcal{E}(Y \mid \mathcal{G})$, we consider the simplest nontrivial case. Suppose that F is an event with probability strictly between 0 and 1. Let $\mathcal{G}$ be the σ-algebra generated by F, that is, the smallest σ-algebra containing F; clearly, $\mathcal{G} = \{\, \emptyset, F, F^c, \Omega\,\}$. Then, given $\mathcal{G}$, we know whether or not F has occurred, that is, whether F has occurred or F^c has occurred. In the former case, we have $\mathcal{E}(Y \mid \mathcal{G}) = \mathcal{E}(Y \mid F)$ and, in the latter case, $\mathcal{E}(Y \mid \mathcal{G}) = \mathcal{E}(Y \mid F^c)$. In other words,

$$\mathcal{E}(Y \mid \mathcal{G}) = \mathcal{E}(Y \mid F)\chi_F + \mathcal{E}(Y \mid F^c)\chi_{F^c}.$$

Note that $\mathcal{E}(Y \mid \mathcal{G})$ is not only a random variable (i.e., is $\mathcal{A}$-measurable), but is in fact $\mathcal{G}$-measurable. Furthermore, it is not too difficult to see that

$$\int_G Y \, dP = \int_G \mathcal{E}(Y \mid \mathcal{G}) \, dP, \qquad G \in \mathcal{G}. \tag{9.9}$$

We can use the complex version of the Radon-Nikodym theorem to show the existence of conditional expectation given a σ-algebra in the general case. Keeping (9.9) in mind, we have the following proposition.

□ □ □ **PROPOSITION 9.6 Existence of Conditional Expectation**

Let $(\Omega, \mathcal{A}, P)$ be a probability space, $Y \in \mathcal{L}^1(\Omega, \mathcal{A}, P)$, and $\mathcal{G}$ a σ-algebra with $\mathcal{G} \subset \mathcal{A}$. Then there exists a $\mathcal{G}$-measurable function $\mathcal{E}(Y \mid \mathcal{G})$ such that

$$\int_G Y \, dP = \int_G \mathcal{E}(Y \mid \mathcal{G}) \, dP, \qquad G \in \mathcal{G}. \tag{9.10}$$

*Moreover, such a function is unique P-ae and is called the **conditional expectation of Y given $\mathcal{G}$**.*

PROOF   Define $\mu_Y(G) = \int_G Y \, dP$, for $G \in \mathcal{G}$. Then μ_Y is a complex measure on $\mathcal{G}$ and $\mu_Y \ll P$. The result now follows from the complex version of the Radon-Nikodym theorem. ∎

It is important to note that $\mathcal{E}(\mathcal{E}(Y \mid \mathcal{G})) = \mathcal{E}Y$ even though that fact follows trivially from (9.10). We will investigate further properties of conditional expectation given a σ-algebra in the exercises.

Exercises for Section 9.3

9.37   Let ν be a signed measure on $(\Omega, \mathcal{A})$ and $\nu = \nu^+ - \nu^-$ its Jordan decomposition. Show that for $A \in \mathcal{A}$,
   a) $-\nu^-(A) \leq \nu(A) \leq \nu^+(A)$.
   b) $|\nu(A)| \leq |\nu|(A)$.

9.38   Let ν_1 and ν_2 be finite signed measures on $(\Omega, \mathcal{A})$. Prove that

$$|\nu_1 + \nu_2| \leq |\nu_1| + |\nu_2|;$$

that is, $|\nu_1 + \nu_2|(A) \leq |\nu_1|(A) + |\nu_2|(A)$ for each $A \in \mathcal{A}$.

9.39   Let f be an extended real-valued Borel-measurable function, integrable with respect to Lebesgue measure. Then $\nu(B) = \int_B f \, d\lambda$ is a finite signed measure on $\mathcal{B}$. Define $F_\nu(x) = \nu((-\infty, x]) = \int_{-\infty}^x f(t) \, dt$. Prove that

$$|\nu|((a, b]) = V_a^b F_\nu, \qquad -\infty < a < b < \infty.$$

9.40 Exercise 9.39 can be generalized as follows: Let ν be a finite signed measure on $\mathcal{B}$. Define $F_\nu(x) = \nu((-\infty, x])$ for $x \in \mathcal{R}$. Then, it can be proved that

$$|\nu|((a, b]) = V_a^b F_\nu, \qquad -\infty < a < b < \infty.$$

Show that the preceding equation does not necessarily hold for other types of intervals, even if ν is a measure.

9.41 Suppose that $(\Omega, \mathcal{A})$ is a measurable space and that ν is a signed measure on $\mathcal{A}$ with Jordan decomposition $\nu = \nu^+ - \nu^-$.
a) Show that $f \in \mathcal{L}^1(|\nu|)$ if and only if $f \in \mathcal{L}^1(\nu^+) \cap \mathcal{L}^1(\nu^-)$.
b) Suppose that f is $\mathcal{A}$-measurable and $|f| \leq M$ on Ω. Prove that for each $E \in \mathcal{A}$,

$$\left| \int_E f \, d\nu \right| \leq M \cdot |\nu|(E).$$

9.42 Prove Theorem 9.4 on page 324.

9.43 Let $(\Omega, \mathcal{A})$ be a measurable space and ν a complex measure on $\mathcal{A}$. Using only Definition 9.10 on page 324, prove that $|\nu|$ is monotone; that is, if $A, B \in \mathcal{A}$ and $A \subset B$, then $|\nu|(A) \leq |\nu|(B)$.

9.44 Let $(\Omega, \mathcal{A})$ be a measurable space and ν a complex measure on $\mathcal{A}$.
a) Prove that $|\nu(A)| \leq |\nu|(A)$ for each $A \in \mathcal{A}$.
b) From part (a), we have $|\nu(A)| \leq |\nu|(A)$ for each $A \in \mathcal{A}$. Prove that $|\nu|$ is the smallest measure dominating ν in that sense; that is, if τ is a measure on $\mathcal{A}$ such that $|\nu(A)| \leq \tau(A)$ for each $A \in \mathcal{A}$, then $|\nu| \leq \tau$.

9.45 Let $(\Omega, \mathcal{A}) = (\mathcal{R}, \mathcal{P}(\mathcal{R}))$ and $\nu = \delta_0 + i\delta_0$. Determine $|\nu|$.

9.46 Recall that the total variation of a signed measure ν is given by $|\nu| = \nu^+ + \nu^-$, where $\nu = \nu^+ - \nu^-$ is the Jordan decomposition of ν. According to Theorem 9.4, a complex measure can be uniquely decomposed into four measures as $\nu = \nu_1^+ - \nu_1^- + i\nu_2^+ - i\nu_2^-$, where $\nu_1^+ \perp \nu_1^-$ and $\nu_2^+ \perp \nu_2^-$. Establish that it is not generally the case that $|\nu|$ equals $\nu_1^+ + \nu_1^- + \nu_2^+ + \nu_2^-$.

9.47 Let $(\Omega, \mathcal{A})$ be a measurable space. Prove the following facts.
a) If ν is a complex measure on $\mathcal{A}$ and $\alpha \in \mathbb{C}$, then $|\alpha\nu| = |\alpha||\nu|$.
b) If ν_1 and ν_2 are complex measures on $\mathcal{A}$, then $|\nu_1 + \nu_2| \leq |\nu_1| + |\nu_2|$.

★9.48 Let $(\Omega, \mathcal{A})$ be a measurable space. If ν is a complex measure on $\mathcal{A}$, define $\|\nu\| = |\nu|(\Omega)$. Prove that
a) $\|\nu_1 + \nu_2\| \leq \|\nu_1\| + \|\nu_2\|$ for all complex measures ν_1 and ν_2 on $\mathcal{A}$.
b) $\|\alpha\nu\| = |\alpha|\|\nu\|$ for all complex numbers $\alpha \in \mathbb{C}$ and all complex measures ν on $\mathcal{A}$.
c) $\|\nu\| = 0$ implies that $\nu = 0$.

9.49 Let ν be a complex measure on $(\Omega, \mathcal{A})$ and $\nu = \nu_1^+ - \nu_1^- + i\nu_2^+ - i\nu_2^-$ the decomposition of ν given by Theorem 9.4 on page 324. Show that $f \in \mathcal{L}^1(|\nu|)$ if and only if $f \in \mathcal{L}^1(\nu_1^+) \cap \mathcal{L}^1(\nu_1^-) \cap \mathcal{L}^1(\nu_2^+) \cap \mathcal{L}^1(\nu_2^-)$.

9.50 Let ν be a complex measure on $(\Omega, \mathcal{A})$. If $f \in \mathcal{L}^1(|\nu|)$, define

$$\int_\Omega f \, d\nu = \int_\Omega f \, d\nu_1^+ - \int_\Omega f \, d\nu_1^- + i \int_\Omega f \, d\nu_2^+ - i \int_\Omega f \, d\nu_2^-.$$

(Exercise 9.49 shows that the right-hand side of the foregoing equation makes sense.)
Prove the following:
a) If $f, g \in \mathcal{L}^1(|\nu|)$, then

$$\int_\Omega (f + g) \, d\nu = \int_\Omega f \, d\nu + \int_\Omega g \, d\nu.$$

b) If $\alpha \in \mathbb{C}$ and $f \in \mathcal{L}^1(|\nu|)$, then

$$\int_\Omega \alpha f \, d\nu = \alpha \int_\Omega f \, d\nu.$$

★9.51 Refer to Exercise 9.50. Let $(\Omega, \mathcal{A}, \mu)$ be a measure space and $f \in \mathcal{L}^1(\mu)$. Define

$$\nu(A) = \int_A f \, d\mu, \qquad A \in \mathcal{A}.$$

Prove that for $g \in \mathcal{L}^1(|\nu|)$ and $A \in \mathcal{A}$,

$$\int_A g \, d\nu = \int_A gf \, d\mu.$$

9.52 Let $(\Omega, \mathcal{A}, \mu)$ be a measure space and $f \in \mathcal{L}^1(\mu)$. Define

$$\nu(A) = \int_A f \, d\mu, \qquad A \in \mathcal{A}.$$

Prove that

$$|\nu|(A) = \int_A |f| \, d\mu, \qquad A \in \mathcal{A}.$$

Hint: To prove that $|\nu|(A) \geq \int_A |f| \, d\mu$, choose a sequence $\{s_n\}_{n=1}^\infty$ of $\mathcal{A}$-measurable simple functions such that $s_n \to \chi_A \cdot \operatorname{sgn} \overline{f}$ pointwise on Ω, where $\bar{\ }$ denotes complex conjugation and

$$(\operatorname{sgn} \overline{f})(x) = \begin{cases} \overline{f(x)}/|f(x)|, & f(x) \neq 0; \\ 0, & f(x) = 0. \end{cases}$$

Show that each s_n can be chosen so that $|s_n| \leq 1$.

★9.53 Let $(\Omega, \mathcal{A})$ be a measurable space and ν a complex measure on $\mathcal{A}$.
a) Prove that there is an $\mathcal{A}$-measurable function ϕ on Ω such that $|\phi| = 1$ and

$$\nu(A) = \int_A \phi \, d|\nu|, \qquad A \in \mathcal{A}.$$

Hint: Use the Radon-Nikodym theorem and Exercise 9.52.
b) Suppose that f is $\mathcal{A}$-measurable and $|f| \leq M$ on Ω. Prove that for each $E \in \mathcal{A}$,

$$\left| \int_E f \, d\nu \right| \leq M \cdot |\nu|(E).$$

Hint: Refer to part (a) and Exercise 9.51.

In Exercises 9.54–9.65, $(\Omega, \mathcal{A}, P)$ is a probability space, $Y \in \mathcal{L}^1(\Omega, \mathcal{A}, P)$, and $\mathcal{G}$ is a σ-algebra of subsets of Ω with $\mathcal{G} \subset \mathcal{A}$.

9.54 Let F be an event with probability strictly between 0 and 1. Let $\mathcal{G}$ be the σ-algebra generated by F, that is, $\mathcal{G} = \{\emptyset, F, F^c, \Omega\}$. Define $\mathcal{E}(Y \mid \mathcal{G}) = \mathcal{E}(Y \mid F)\chi_F + \mathcal{E}(Y \mid F^c)\chi_{F^c}$.
a) Show that $\mathcal{E}(Y \mid \mathcal{G})$ is $\mathcal{G}$-measurable.
b) Prove that for each $G \in \mathcal{G}$, $\int_G Y \, dP = \int_G \mathcal{E}(Y \mid \mathcal{G}) \, dP$. *Hint:* Refer to Exercise 7.84 on page 257.

9.55 Suppose that $\{F_n\}_n$ is a sequence of pairwise mutually exclusive events each with positive probability and such that $\bigcup_n F_n = \Omega$. Let $\mathcal{G}$ be the σ-algebra generated by $\{F_n\}_n$.
   a) Characterize the sets in $\mathcal{G}$.
   b) Prove that, with probability one, $\mathcal{E}(Y \mid \mathcal{G}) = \sum_n \mathcal{E}(Y \mid F_n)\chi_{F_n}$.

9.56 Show that, for each $E \in \mathcal{A}$, $P(E \mid \mathcal{G}) = \mathcal{E}(\chi_E \mid \mathcal{G})$ P-ae. Thus, conditional probability is a special case of conditional expectation.

9.57 Suppose that Y is $\mathcal{G}$-measurable.
   a) What does your intuition tell you regarding $\mathcal{E}(Y \mid \mathcal{G})$?
   b) Prove your assertion in part (a).

9.58 Suppose that $\mathcal{G} = \{\Omega, \emptyset\}$.
   a) What does your intuition tell you regarding $\mathcal{E}(Y \mid \mathcal{G})$?
   b) Prove your assertion in part (a).

9.59 Let Y, Y_1, $Y_2 \in \mathcal{L}^1(\Omega, \mathcal{A}, P)$ and α, α_1, $\alpha_2 \in \mathbb{C}$. Establish that each of the following properties holds with probability one.
   a) If $Y = \alpha$, then $\mathcal{E}(Y \mid \mathcal{G}) = \alpha$.
   b) If $Y_1 \leq Y_2$, then $\mathcal{E}(Y_1 \mid \mathcal{G}) \leq \mathcal{E}(Y_2 \mid \mathcal{G})$.
   c) $\mathcal{E}(\alpha Y \mid \mathcal{G}) = \alpha \mathcal{E}(Y \mid \mathcal{G})$.
   d) $\mathcal{E}(Y_1 + Y_2 \mid \mathcal{G}) = \mathcal{E}(Y_1 \mid \mathcal{G}) + \mathcal{E}(Y_2 \mid \mathcal{G})$.
   e) $|\mathcal{E}(Y \mid \mathcal{G})| \leq \mathcal{E}(|Y| \mid \mathcal{G})$.

9.60 Let $\{Y_n\}_{n=1}^{\infty}$ be a nondecreasing sequence of nonnegative, integrable random variables converging to the integrable random variable Y. Prove that, with probability one, $\lim_{n \to \infty} \mathcal{E}(Y_n \mid \mathcal{G}) = \mathcal{E}(Y \mid \mathcal{G})$.

9.61 Suppose that $\mathcal{G}_1$ and $\mathcal{G}_2$ are σ-algebras such that $\mathcal{G}_1 \subset \mathcal{G}_2 \subset \mathcal{A}$. Prove that, with probability one,
   a) $\mathcal{E}(\mathcal{E}(Y \mid \mathcal{G}_1) \mid \mathcal{G}_2) = \mathcal{E}(Y \mid \mathcal{G}_1)$.
   b) $\mathcal{E}(\mathcal{E}(Y \mid \mathcal{G}_2) \mid \mathcal{G}_1) = \mathcal{E}(Y \mid \mathcal{G}_1)$.

9.62 Suppose that Z is $\mathcal{G}$-measurable and $ZY \in \mathcal{L}^1(\Omega, \mathcal{A}, P)$. Prove that, with probability one, $\mathcal{E}(ZY \mid \mathcal{G}) = Z\mathcal{E}(Y \mid \mathcal{G})$.

9.63 **Conditional expectation given a random variable:** A special case of conditional expectation given a σ-algebra is when the σ-algebra is generated by a random variable X. Recall from Exercise 9.34 on page 320 that the σ-algebra generated by X, denoted $\sigma(X)$, is the smallest σ-algebra of subsets of Ω for which X is measurable. Exercise 9.34(a) shows that $\sigma(X) = \{\{X \in B\} : B \in \mathcal{B}\}$. If $Y \in \mathcal{L}^1(\Omega, \mathcal{A}, P)$, then we define the **conditional expectation of Y given X,** denoted $\mathcal{E}(Y \mid X)$, to be the conditional expectation of Y given $\sigma(X)$; that is, by definition, $\mathcal{E}(Y \mid X) = \mathcal{E}(Y \mid \sigma(X))$.
   a) Prove that there is a Borel measurable function ϕ such that $\mathcal{E}(Y \mid X) = \phi \circ X$, P-ae. *Hint:* First assume $Y \geq 0$.
   b) Let ϕ be the function in part (a). For $x \in \mathcal{R}$, set $\mathcal{E}(Y \mid X = x) = \phi(x)$, called the **conditional expectation of Y given $X = x$.** Prove that

$$\int_{\{X \in B\}} Y \, dP = \int_B \mathcal{E}(Y \mid X = x) \, d\mu_X(x), \qquad B \in \mathcal{B},$$

where μ_X is the probability distribution of X. *Hint:* Use Theorem 7.6 on page 249.
   c) Prove that if g is a Borel measurable function such that

$$\int_{\{X \in B\}} Y \, dP = \int_B g(x) \, d\mu_X(x), \qquad B \in \mathcal{B},$$

then, for μ_X-almost all x, $g(x) = \mathcal{E}(Y \mid X = x)$.

9.64 Refer to Example 7.7(a) on page 240. Suppose X and Y are jointly discrete random variables with joint probability mass function $p_{X,Y}$. Define

$$p_{Y\mid X}(y\mid x) = \begin{cases} \dfrac{p_{X,Y}(x,y)}{p_X(x)}, & p_X(x) > 0; \\ 0, & \text{otherwise.} \end{cases}$$

a) Prove that

$$\mathcal{E}(Y\mid X = x) = \sum_y y p_{Y\mid X}(y\mid x)$$

for each possible value x of X. *Hint:* Use Theorem 7.7 on page 251 and Exercise 9.63(c).

b) Determine $\mathcal{E}(Y\mid X)$.

9.65 Refer to Example 7.7(b) on page 240. Suppose X and Y are jointly absolutely continuous random variables with joint probability density function $f_{X,Y}$. Define

$$f_{Y\mid X}(y\mid x) = \begin{cases} \dfrac{f_{X,Y}(x,y)}{f_X(x)}, & f_X(x) > 0; \\ 0, & \text{otherwise.} \end{cases}$$

a) Prove that

$$\mathcal{E}(Y\mid X = x) = \int y f_{Y\mid X}(y\mid x)\,dy$$

for μ_X-almost all x. *Hint:* Use Theorem 7.7 on page 251 and Exercise 9.63(c).

b) Determine $\mathcal{E}(Y\mid X)$.

9.4 DECOMPOSITION OF MEASURES

In this section, we will study several results regarding the decomposition of measures. Our first result, known as the Lebesgue decomposition theorem, is a consequence of the Radon-Nikodym theorem.

□ □ □ **THEOREM 9.6 Lebesgue Decomposition Theorem**

Let $(\Omega, \mathcal{A}, \mu)$ be a σ-finite measure space and ν a σ-finite measure on $\mathcal{A}$. Then there exist measures ν_1 and ν_2 on $\mathcal{A}$ such that $\nu_1 \ll \mu$, $\nu_2 \perp \mu$, and $\nu = \nu_1 + \nu_2$. Moreover, such a representation is unique. It is called the **Lebesgue decomposition** *of ν with respect to μ.*

PROOF Clearly $\mu \ll \mu + \nu$, and μ and $\mu + \nu$ are σ-finite. Therefore, the Radon-Nikodym theorem implies that there is a nonnegative $\mathcal{A}$-measurable function f on Ω such that

$$\mu(A) = \int_A f\,d(\mu + \nu), \qquad A \in \mathcal{A}.$$

Let $E = \{x : f(x) > 0\}$. Obviously, then, $\mu(E^c) = 0$.

Define measures ν_1 and ν_2 on $\mathcal{A}$ by

$$\nu_1(A) = \nu(A \cap E) \qquad \text{and} \qquad \nu_2(A) = \nu(A \cap E^c).$$

Clearly, $\nu = \nu_1 + \nu_2$. Moreover, as $\nu_2(E) = 0$ and $\mu(E^c) = 0$, we see that $\nu_2 \perp \mu$.

We claim that $\nu_1 \ll \mu$. So, suppose $\mu(A) = 0$. Then

$$\int_{A \cap E} f \, d(\mu + \nu) = \int_A f \, d(\mu + \nu) = \mu(A) = 0.$$

Because $f > 0$ on $A \cap E$, it must be that $(\mu + \nu)(A \cap E) = 0$ and, consequently, $\nu_1(A) = \nu(A \cap E)$ must equal zero.

It remains to prove uniqueness. Assume that $\nu = \omega_1 + \omega_2$, where $\omega_1 \ll \mu$ and $\omega_2 \perp \mu$. We must show that $\omega_1 = \nu_1$ and $\omega_2 = \nu_2$.

Since $\nu_2 \perp \mu$ and $\omega_2 \perp \mu$, there exist sets $B, C \in \mathcal{A}$ such that $\mu(B) = \mu(C) = 0$ and $\nu_2(B^c) = \omega_2(C^c) = 0$. In particular, any subset of $B \cup C$ has μ-measure zero and any subset of $B^c \cap C^c$ has ν_2- and ω_2-measure zero. Since $\nu_1 \ll \mu$ and $\omega_1 \ll \mu$, it follows that any subset of $B \cup C$ also has ν_1- and ω_1-measure zero. Thus, for $A \in \mathcal{A}$,

$$\omega_2(A) = \omega_2\big(A \cap (B \cup C)\big) + \omega_2\big(A \cap (B^c \cap C^c)\big) = \omega_2\big(A \cap (B \cup C)\big)$$

$$= \omega_1\big(A \cap (B \cup C)\big) + \omega_2\big(A \cap (B \cup C)\big) = \nu\big(A \cap (B \cup C)\big)$$

$$= \nu_1\big(A \cap (B \cup C)\big) + \nu_2\big(A \cap (B \cup C)\big) = \nu_2\big(A \cap (B \cup C)\big)$$

$$= \nu_2\big(A \cap (B \cup C)\big) + \nu_2\big(A \cap (B^c \cap C^c)\big) = \nu_2(A).$$

Hence $\omega_2 = \nu_2$. A similar argument shows that $\omega_1 = \nu_1$. ■

EXAMPLE 9.9 *Illustrates the Lebesgue Decomposition Theorem*

Let $F: \mathcal{R} \to \mathcal{R}$ be defined by

$$F(x) = \begin{cases} 0, & x < 0; \\ 3 - e^{-x}, & 0 \le x < 1; \\ 4 - e^{-x}, & x \ge 1. \end{cases}$$

Note that F is a distribution function; that is, F is nondecreasing, right continuous, bounded, and $F(x) \to 0$ as $x \to -\infty$. According to Theorem 6.3 on page 196, there is a unique finite Borel measure ν that has F as its distribution function. We will obtain the Lebesgue decomposition of ν with respect to λ (considered a Borel measure). Define $f(x) = e^{-x}$ for $x > 0$ and zero otherwise, and let

$$\nu_1(B) = \int_B f(t) \, dt, \qquad B \in \mathcal{B}.$$

Also, set $\nu_2 = 2\delta_0 + \delta_1$. Then $\nu_1 \ll \lambda$ and $\nu_2 \perp \lambda$. A simple calculation shows that $\nu_1 + \nu_2$ has F as its distribution function, which implies that $\nu = \nu_1 + \nu_2$. Therefore, we have found the unique Lebesgue decomposition of ν with respect to λ. Example 9.13 provides an alternate (and more straightforward) method for obtaining this decomposition. □

Further Decomposition of Measures

Suppose that $(\Omega, \mathcal{A}, \mu)$ is a σ-finite measure space such that $\{x\} \in \mathcal{A}$ for all $x \in \Omega$. We will show that if ν is a σ-finite measure on $\mathcal{A}$, then it can be decomposed into three mutually singular measures, of which one is absolutely continuous with respect to μ, one is singular with respect to μ and has no atoms, and one is singular with respect to μ and is discrete. To begin, we recall the following definitions.

DEFINITION 9.12 **Atoms**

Let $(\Omega, \mathcal{A})$ be a measurable space such that $\{x\} \in \mathcal{A}$ for all $x \in \Omega$ and let ν be a measure on $\mathcal{A}$. An element $x \in \Omega$ is said to be an **atom** of ν if $\nu(\{x\}) > 0$.

DEFINITION 9.13 **Continuous and Discrete Measures**

Let $(\Omega, \mathcal{A})$ be a measurable space such that $\{x\} \in \mathcal{A}$ for all $x \in \Omega$.
a) A measure ν on $\mathcal{A}$ is said to be **continuous** if it has no atoms, that is, $\nu(\{x\}) = 0$ for all $x \in \Omega$.
b) A measure ν on $\mathcal{A}$ is said to be **discrete** if there is a countable subset K of Ω such that $\nu(K^c) = 0$.

EXAMPLE 9.10 *Illustrates Definitions 9.12 and 9.13*

a) Let $(\Omega, \mathcal{A}) = (\mathcal{R}, \mathcal{M})$. Lebesgue measure λ is continuous; the measure $\delta_0 + \delta_1$ is discrete; and the measure $\lambda + \delta_0 + \delta_1$ is neither continuous nor discrete. For the latter two measures, the set of atoms is $\{0, 1\}$.
b) Refer to Example 7.5 on page 237. Let X be a random variable on $(\Omega, \mathcal{A}, P)$ and μ_X its probability distribution, that is,

$$\mu_X(B) = P(X \in B), \qquad B \in \mathcal{B}.$$

X is discrete if and only if μ_X is a discrete measure on $\mathcal{B}$; X is continuous if and only if μ_X is a continuous measure on $\mathcal{B}$. □

The following proposition was (essentially) proved in Exercise 5.6 on page 149, but we state it formally here for completeness. The proof is left to the reader.

□ □ □ **PROPOSITION 9.7**

Let $(\Omega, \mathcal{A})$ be a measurable space such that $\{x\} \in \mathcal{A}$ for all $x \in \Omega$, and let ν be a measure on $\mathcal{A}$. Then ν is discrete if and only if there is a countable subset K of Ω such that $\nu = \sum_{x \in K} \nu(\{x\}) \delta_x$. We can always take K to be the set of atoms of ν.

The next proposition shows that, under σ-finite conditions, we can decompose a measure as the sum of a continuous and discrete measure.

□ □ □ **PROPOSITION 9.8**

Let $(\Omega, \mathcal{A})$ be a measurable space such that $\{x\} \in \mathcal{A}$ for all $x \in \Omega$, and let ν be a σ-finite measure on $\mathcal{A}$. Then there exist mutually singular measures ν_c and ν_d on $\mathcal{A}$ such that ν_c is continuous, ν_d is discrete, and $\nu = \nu_c + \nu_d$. Moreover, such a representation is unique.

PROOF First assume ν is finite. We claim ν has countably many atoms. Let $F \subset \Omega$ be finite. Then $\sum_{x \in F} \nu(\{x\}) = \nu(F) \leq \nu(\Omega)$. Taking the supremum over all finite subsets of Ω, we deduce that $\sum_{x \in \Omega} \nu(\{x\}) \leq \nu(\Omega) < \infty$. Consequently, by Exercise 2.37 on page 46, only countably many of the $\nu(\{x\})$s are nonzero; that is, ν has only countably many atoms.

Let K denote the set of atoms of ν. As we have just seen, K is countable. Therefore, the measure $\nu_d = \sum_{x \in K} \nu(\{x\}) \delta_x$ is discrete. Let $\nu_c = \nu - \nu_d$. To show that ν_c is a measure on $\mathcal{A}$, it is enough to show that it is nonnegative, because the other two conditions for being a measure are clearly satisfied. Noting that $\nu(\{x\}) = \nu_d(\{x\})$ for each $x \in K$, we conclude that ν and ν_d agree on all subsets of K. Let $A \in \mathcal{A}$. Then

$$\nu(A) = \nu(A \cap K) + \nu(A \cap K^c) = \nu_d(A \cap K) + \nu(A \cap K^c)$$

and

$$\nu_d(A) = \nu_d(A \cap K) + \nu_d(A \cap K^c) = \nu_d(A \cap K).$$

Consequently,

$$\nu_c(A) = \nu(A) - \nu_d(A) = \nu(A \cap K^c) \geq 0.$$

Thus, ν_c is a measure.

Recalling that K denotes the set of atoms of ν, we can apply the previous equation with $A = \{x\}$ to conclude that ν_c is continuous. Indeed, if $x \in K$, then $\nu_c(\{x\}) = \nu(\emptyset) = 0$. On the other hand, if $x \notin K$, then it is not an atom of ν and we have $\nu_c(\{x\}) = \nu(\{x\}) = 0$.

Now assume that ν is σ-finite. Select a countable collection $\{E_n\}_n$ of disjoint $\mathcal{A}$-measurable sets of finite ν-measure whose union is Ω. For each $n \in \mathcal{N}$, define the measure ν_n on $\mathcal{A}$ by $\nu_n(A) = \nu(E_n \cap A)$. Then ν_n is a finite measure on $\mathcal{A}$. Consequently, by what we just proved for finite measures, we can write $\nu_n = \nu_{nc} + \nu_{nd}$, where ν_{nc} is continuous and ν_{nd} is discrete; moreover,

$$\nu_{nd} = \sum_{x \in K_n} \nu_n(\{x\}) \delta_x = \sum_{x \in K_n} \nu(\{x\}) \delta_x,$$

where K_n denotes the set of atoms of ν_n. Note that K_n consists of those atoms of ν that lie in E_n.

Let $K = \bigcup_n K_n$. As each K_n is countable, so is K; moreover, K is the set of atoms of ν. Let $\nu_c = \sum_n \nu_{nc}$ and

$$\nu_d = \sum_n \nu_{nd} = \sum_{x \in K} \nu(\{x\}) \delta_x. \tag{9.11}$$

Because each ν_{nc} is continuous, so is ν_c; and, because K is countable, ν_d is discrete. We have

$$\nu = \sum_n \nu_n = \sum_n (\nu_{nc} + \nu_{nd}) = \sum_n \nu_{nc} + \sum_n \nu_{nd} = \nu_c + \nu_d.$$

Moreover, since ν_c is continuous and ν_d is discrete, it follows easily that $\nu_c \perp \nu_d$. (See Exercise 9.71.)

It remains to prove the uniqueness of the decomposition. So, assume that we have $\nu = \tau_c + \tau_d$, where τ_c is continuous and τ_d is discrete. By Proposition 9.7, we can write

$$\tau_d = \sum_{x \in C} \tau_d(\{x\})\delta_x, \tag{9.12}$$

where C is the collection of atoms of τ_d. On the other hand, since τ_c is continuous, we have $\tau_d(\{x\}) = \nu(\{x\})$ for all x. Therefore, the set of atoms of τ_d is identical to that of ν, namely, K. Therefore, $C = K$. Since $C = K$ and $\tau_d(\{x\}) = \nu(\{x\})$ for all x, we see from (9.11) and (9.12) that $\tau_d = \nu_d$.

Next note that because τ_c and ν_c are continuous and K is countable, we have $\tau_c(A \cap K) = \nu_c(A \cap K) = 0$ for all $A \in \mathcal{A}$. Consequently, for $A \in \mathcal{A}$,

$$\begin{aligned}
\tau_c(A) &= \tau_c(A \cap K) + \tau_c(A \cap K^c) = \tau_c(A \cap K^c) \\
&= \tau_c(A \cap K^c) + \tau_d(A \cap K^c) = \nu(A \cap K^c) \\
&= \nu_c(A \cap K^c) + \nu_d(A \cap K^c) = \nu_c(A \cap K^c) \\
&= \nu_c(A \cap K) + \nu_c(A \cap K^c) = \nu_c(A).
\end{aligned}$$

In other words, $\tau_c = \nu_c$. ∎

□ □ □ **THEOREM 9.7**

Let $(\Omega, \mathcal{A}, \mu)$ be a σ-finite measure space such that $\{x\} \in \mathcal{A}$ for all $x \in \Omega$, and let ν be a σ-finite measure on $\mathcal{A}$. Then there exist measures ν_{ac}, ν_{sc}, and ν_d with the following properties:

a) $\nu_{ac} \ll \mu$, $\nu_{sc} \perp \mu$, and $\nu_d \perp \mu$
b) ν_{sc} is continuous and ν_d is discrete
c) $\nu = \nu_{ac} + \nu_{sc} + \nu_d$
Moreover, the representation in (c) is unique.

PROOF By the Lebesgue decomposition theorem, there exist unique measures ν_1 and ν_2 on $\mathcal{A}$ such that $\nu_1 \ll \mu$, $\nu_2 \perp \mu$, and $\nu = \nu_1 + \nu_2$. Set $\nu_{ac} = \nu_1$.

Because ν is σ-finite, so is ν_2. Consequently, by Proposition 9.8, there exist unique mutually singular measures ν_c and ν_d on $\mathcal{A}$ such that ν_c is continuous, ν_d is discrete, and $\nu_2 = \nu_c + \nu_d$. Setting $\nu_{sc} = \nu_c$, we have $\nu = \nu_{ac} + \nu_{sc} + \nu_d$. So, (b) and (c) hold. Because $\nu_2 \perp \mu$, it is easy to see that $\nu_{sc} \perp \mu$ and $\nu_d \perp \mu$. Thus, (a) holds.

It remains to prove uniqueness. So, suppose that τ_{ac}, τ_{sc}, and τ_d are measures on $\mathcal{A}$ that satisfy (a)–(c). Let $\tau_2 = \tau_{sc} + \tau_d$. Since $\tau_{sc} \perp \mu$ and $\tau_d \perp \mu$, we have $\tau_2 \perp \mu$. Therefore, by the uniqueness part of the Lebesgue decomposition theorem, $\tau_{ac} = \nu_{ac}$ and $\tau_2 = \nu_{sc} + \nu_d$. But then, by the uniqueness part of Proposition 9.8, $\tau_{sc} = \nu_{sc}$ and $\tau_d = \nu_d$. ∎

Remark: We leave it as an exercise for the reader to show that the measures ν_{ac}, ν_{sc}, and ν_d in Theorem 9.7 are mutually singular.

EXAMPLE 9.11 **Illustrates Theorem 9.7**

Let $(\Omega, \mathcal{A}, \mu) = (\mathcal{R}, \mathcal{M}, \mu)$, where μ is counting measure on $\mathcal{N}$ (i.e., $\mu = \sum_{n=1}^{\infty} \delta_n$). Let $\nu = \delta_0 + \delta_1 + \lambda$. Then $\nu_{ac} = \delta_1$, $\nu_{sc} = \lambda$, and $\nu_d = \delta_0$. Moreover, we have $d\nu_{ac}/d\mu = d\delta_1/d\mu = \chi_{\{1\}}$ μ-ae. This simple example shows that the absolutely continuous component of a measure need not be a continuous measure. □

Decomposition of Finite Borel Measures

An important application of Theorem 9.7 is to the decomposition of finite Borel measures on $\mathcal{R}$ with respect to Lebesgue measure (considered a Borel measure), that is, where $(\Omega, \mathcal{A}, \mu) = (\mathcal{R}, \mathcal{B}, \lambda_{|\mathcal{B}})$ and ν is a finite Borel measure. When there is no possibility of confusion, we will, as before, write λ in place of $\lambda_{|\mathcal{B}}$. First we have the following obvious corollary to Theorem 9.7.

□ □ □ **THEOREM 9.8 Decomposition Theorem for Finite Borel Measures**

Let ν be a finite Borel measure on $\mathcal{R}$. Then ν can be decomposed uniquely as the sum of three finite Borel measures ν_{ac}, ν_{sc}, and ν_{d}, where ν_{ac} is absolutely continuous with respect to λ, ν_{sc} is continuous and singular with respect to λ, and ν_{d} is discrete.

Our next task is to express the conclusions of Theorem 9.8 in terms of distribution functions. First we state the following proposition relating the atoms of a finite Borel measure to the discontinuities of its distribution function. The proof is left as an exercise for the reader.

□ □ □ **PROPOSITION 9.9**

Let ν be a finite Borel measure on $\mathcal{R}$ and F_ν its distribution function. Then

$$\nu(\{x\}) = F_\nu(x) - F_\nu(x-), \qquad x \in \mathcal{R}. \tag{9.13}$$

Consequently, the set of atoms of ν is precisely the set of discontinuities of F_ν. In particular, ν is a continuous measure if and only if F_ν is a continuous function.

EXAMPLE 9.12 *Illustrates Proposition 9.9*

a) Define

$$F(x) = \begin{cases} 0, & x < 0; \\ \psi(x), & 0 \leq x < 1; \\ 1, & x \geq 1, \end{cases}$$

where ψ denotes the Cantor function. Let ν be the unique Borel measure that has F as its distribution function. Because F is continuous on $\mathcal{R}$, Proposition 9.9 shows that ν is a continuous measure.

b) Let $\{x_n\}_n$ be a sequence of distinct real numbers and $\{a_n\}_n$ a sequence of positive real numbers such that $\sum_n a_n < \infty$. Set $\nu = \sum_n a_n \delta_{x_n}$. Then ν is a discrete measure and its set of atoms is $\{x_n\}_n$. The distribution function of ν is given by $F_\nu(x) = \sum_{x_n \leq x} a_n$. It follows from Proposition 9.9 that F is continuous except at x_n, $n = 1, 2, \ldots$, where it has, respectively, a "jump discontinuity" of magnitude a_n, $n = 1, 2, \ldots$. Note that if ν has only a finite number of atoms (i.e., $\{x_n\}_n$ and $\{a_n\}_n$ are finite sequences), then F_ν is a step function on $\mathcal{R}$. □

In view of Example 9.12(b), it is natural and reasonable to make the following definition.

DEFINITION 9.14   Discrete Distribution Function

A distribution function F is said to be **discrete** if it can be expressed in the form $F(x) = \sum_{x_n \le x} a_n$, where $\{x_n\}_n$ is a sequence of real numbers and $\{a_n\}_n$ a sequence of positive real numbers with $\sum_n a_n < \infty$.

The next two propositions will also be required for our decomposition of distribution functions. The proof of the first one is left to the reader as an exercise.

□ □ □   **PROPOSITION 9.10**

A distribution function F is discrete if and only if the Lebesgue-Stieltjes measure corresponding to F is a discrete measure.

□ □ □   **PROPOSITION 9.11**

Let ν be a finite Borel measure on $\mathcal{R}$ and F_ν its distribution function. Then ν is singular with respect to Lebesgue measure if and only if $F_\nu' = 0$ λ-ae.

PROOF   For convenience, set $F = F_\nu$. Suppose that $\nu \perp \lambda$. Select $A \in \mathcal{B}$ such that $\nu(A) = 0$ and $\lambda(A^c) = 0$. Let $E = \{ x : F'(x) > 0 \}$. We have

$$\int_E F' \, d\lambda = \int_{E \cap A} F' \, d\lambda + \int_{E \cap A^c} F' \, d\lambda = \int_{E \cap A} F' \, d\lambda \le \int_A F' \, d\lambda.$$

We will show that the last integral in the previous display is zero. This will imply that $\int_E F' \, d\lambda = 0$, which, in turn, implies that $\lambda(E) = 0$, as required.

Let $\mathcal{C}$ be the collection of intervals of $\mathcal{R}$ of the form $(a, b]$ and (c, ∞), where $-\infty \le a \le b < \infty$ and $-\infty \le c < \infty$. Then, by Exercise 6.2, $\mathcal{C}$ is a semialgebra and $\sigma(\mathcal{C}) = \mathcal{B}$.

Because F is nondecreasing, we have by (8.27) on page 296 that

$$\int_a^b F' \, d\lambda \le F(b) - F(a) = \nu((a, b]), \qquad -\infty < a < b < \infty.$$

From this result, it follows easily that

$$\int_C F' \, d\lambda \le \nu(C), \qquad C \in \mathcal{C}. \tag{9.14}$$

Let $\epsilon > 0$. By Lemma 6.1 (page 187) and Lemma 6.3 (page 188), we can select a sequence $\{C_n\}_n$ of pairwise disjoint members of $\mathcal{C}$ such that $\bigcup_n C_n \supset A$ and $\nu\left(\bigcup_n C_n\right) \le \nu(A) + \epsilon$. Because $\nu(A) = 0$ and the C_ns are pairwise disjoint, we thus have $\sum_n \nu(C_n) \le \epsilon$.

From Proposition 6.7 on page 194, we have

$$\nu((a, b]) = F(b) - F(a), \qquad -\infty \le a \le b \le \infty,$$

where we are using the convention that $(a, \infty] = (a, \infty)$. Writing $C_n = (a_n, b_n]$ and referring to (9.14), we conclude that

$$\int_A F' \, d\lambda \le \int_{\bigcup_n C_n} F' \, d\lambda = \sum_n \int_{C_n} F' \, d\lambda \le \sum_n \nu(C_n) \le \epsilon.$$

As $\epsilon > 0$ was chosen arbitrarily, it follows that $\int_A F' \, d\lambda = 0$.

Conversely, suppose that $F' = 0$ λ-ae. By the Lebesgue decomposition theorem, we can write $\nu = \nu_1 + \nu_2$, where $\nu_1 \ll \lambda$ and $\nu_2 \perp \lambda$. Because ν is finite, so are ν_1 and ν_2. We must show that $\nu_1 = 0$.

By the Radon-Nikodym theorem, there is a nonnegative Lebesgue measurable function f such that

$$\nu_1(B) = \int_B f \, d\lambda, \qquad B \in \mathcal{B}, \tag{9.15}$$

and, because ν_1 is finite, $f \in \mathcal{L}^1(\mathcal{R})$. By Corollary 8.2 on page 291, $f = F'_{\nu_1}$ λ-ae. Also, by the necessity part of this proposition (proved above), we know that $F'_{\nu_2} = 0$ λ-ae. Noting that $F_\nu = F_{\nu_1} + F_{\nu_2}$, we conclude that

$$0 = F'_\nu = F'_{\nu_1} + F'_{\nu_2} = F'_{\nu_1} \quad \lambda\text{-ae.}$$

Therefore, $f = F'_{\nu_1} = 0$ λ-ae, which, by (9.15), implies that $\nu_1 = 0$. ∎

□ □ □ **THEOREM 9.9 Decomposition Theorem for Distribution Functions**

Let F be a distribution function. Then F can be expressed uniquely in the form

$$F = F_{\mathrm{ac}} + F_{\mathrm{sc}} + F_{\mathrm{d}}, \tag{9.16}$$

where F_{ac} is absolutely continuous, F_{sc} is continuous and $F'_{\mathrm{sc}} = 0$ λ-ae, and F_{d} is discrete. Moreover,

$$F'_{\mathrm{ac}} = F' \quad \lambda\text{-ae.} \tag{9.17}$$

PROOF   Let ν be the Lebesgue-Stieltjes measure corresponding to F. Applying Theorem 9.8, we can write $\nu = \nu_{\mathrm{ac}} + \nu_{\mathrm{sc}} + \nu_{\mathrm{d}}$, where $\nu_{\mathrm{ac}} \ll \lambda$, $\nu_{\mathrm{sc}} \perp \lambda$ and is continuous, and ν_{d} is discrete. Let F_{ac}, F_{sc}, and F_{d} denote, respectively, the distribution functions of ν_{ac}, ν_{sc}, and ν_{d}. Then we have $F = F_{\mathrm{ac}} + F_{\mathrm{sc}} + F_{\mathrm{d}}$.

As $\nu_{\mathrm{ac}} \ll \lambda$, Proposition 9.2 on page 317 implies that F_{ac} is absolutely continuous; as $\nu_{\mathrm{sc}} \perp \lambda$ and $\nu_{\mathrm{d}} \perp \lambda$, Proposition 9.11 implies that $F'_{\mathrm{sc}} = F'_{\mathrm{d}} = 0$ λ-ae. Also, because ν_{sc} is continuous, Proposition 9.9 on page 337 implies that F_{sc} is continuous; and, because ν_{d} is discrete, Proposition 9.10 implies that F_{d} is discrete. Because $F = F_{\mathrm{ac}} + F_{\mathrm{sc}} + F_{\mathrm{d}}$ and $F'_{\mathrm{sc}} = F'_{\mathrm{d}} = 0$ λ-ae, it follows immediately that $F'_{\mathrm{ac}} = F'$ λ-ae.

It remains to establish uniqueness. So suppose $F = F_1 + F_2 + F_3$, where F_1 is absolutely continuous, F_2 is continuous and $F'_2 = 0$ λ-ae, and F_3 is discrete. For $1 \leq j \leq 3$, let τ_j denote the Lebesgue-Stieltjes measure corresponding to F_j.

Let $\mathcal{C}$ be the collection of intervals of $\mathcal{R}$ of the form $(a, b]$ and (c, ∞), where $-\infty \leq a \leq b < \infty$ and $-\infty \leq c < \infty$. Then, by Exercise 6.2 on page 190, $\mathcal{C}$ is a semialgebra and $\sigma(\mathcal{C}) = \mathcal{B}$. As $F = F_1 + F_2 + F_3$, the measure $\tau_1 + \tau_2 + \tau_3$ agrees with ν on $\mathcal{C}$ and, therefore, on $\mathcal{B}$; in other words, $\nu = \tau_1 + \tau_2 + \tau_3$.

Because F_1 is absolutely continuous, Proposition 9.2 on page 317 implies that $\tau_1 \ll \lambda$. Because F_2 is continuous, Proposition 9.9 (page 337) implies that τ_2 is a continuous measure and, because $F'_2 = 0$ λ-ae, Proposition 9.11 implies that $\tau_2 \perp \lambda$. And, because F_3 is a discrete distribution function, Proposition 9.10 implies that τ_3 is a discrete measure.

It now follows from the uniqueness portion of Theorem 9.8 that $\tau_1 = \nu_{ac}$, $\tau_2 = \nu_{sc}$, and $\tau_3 = \nu_d$. This implies that the corresponding distribution functions are equal: $F_1 = F_{ac}$, $F_2 = F_{sc}$, and $F_3 = F_d$. ∎

Theorem 9.9 provides a concrete method for determining the decomposition of a distribution function F and, hence, of its corresponding Lebesgue-Stieltjes measure ν. Specifically:

Step 1. Determine the derivative F' of F.

Step 2. F_{ac} is the indefinite integral of F', that is,

$$F_{ac}(x) = \int_{-\infty}^{x} F'(t)\, dt.$$

We have $\nu_{ac}(B) = \int_B F'\, d\lambda$.

Step 3. F_d is obtained from the discontinuities of F, that is,

$$F_d(x) = \sum_{x_n \le x} a_n,$$

where $\{x_n\}_n$ denotes the set of discontinuities of F and $\{a_n\}_n$ denotes the corresponding magnitudes of the jumps, that is, $a_n = F(x_n) - F(x_n-)$. (See Exercise 9.76.) We have $\nu_d = \sum_n a_n \delta_{x_n}$.

Step 4. $F_{sc} = F - F_{ac} - F_d$. We have $\nu_{sc} = \nu - \nu_{ac} - \nu_d$.

EXAMPLE 9.13 *Illustrates Decomposition*

Let $F: \mathcal{R} \to \mathcal{R}$ be defined by

$$F(x) = \begin{cases} 0, & x < 0; \\ 3 - e^{-x}, & 0 \le x < 1; \\ 4 - e^{-x}, & x \ge 1. \end{cases}$$

Note that F is a distribution function; that is, F is nondecreasing, right continuous, bounded, and $F(x) \to 0$ as $x \to -\infty$. We will apply the preceding Steps 1–4 to decompose F and its corresponding Lebesgue-Stieltjes measure.

Step 1. We have

$$F'(x) = \begin{cases} 0, & x < 0; \\ e^{-x}, & x > 0,\ x \ne 1. \end{cases}$$

Step 2. We have

$$F_{ac}(x) = \int_{-\infty}^{x} F'(t)\, dt = \begin{cases} 0, & x < 0; \\ 1 - e^{-x}, & x \ge 0. \end{cases}$$

Also, $\nu_{ac}(B) = \int_B F'\, d\lambda$.

Step 3. F is discontinuous at $x = 0$ and $x = 1$. The magnitudes of the jumps at those two points are, respectively,

$$F(0) - F(0-) = 2 - 0 = 2 \quad \text{and} \quad F(1) - F(1-) = (4 - e^{-1}) - (3 - e^{-1}) = 1.$$

Therefore,
$$F_{\mathrm{d}}(x) = \begin{cases} 0, & x < 0; \\ 2, & 0 \le x < 1; \\ 3, & x \ge 1. \end{cases}$$

Also, $\nu_{\mathrm{d}} = 2\delta_0 + \delta_1$.

Step 4. From Steps 2 and 3, we see that $F_{\mathrm{ac}} + F_{\mathrm{d}} = F$. Consequently,

$$F_{\mathrm{sc}} = F - F_{\mathrm{ac}} - F_{\mathrm{d}} = 0.$$

We have $\nu_{\mathrm{sc}} = \nu - \nu_{\mathrm{ac}} - \nu_{\mathrm{d}} = 0$. □

Exercises for Section 9.4

9.66 Let $(\Omega, \mathcal{A}, \mu) = (\mathcal{R}, \mathcal{B}, \lambda)$. For $E \in \mathcal{M}$, define $\tau(E) = \lambda(E \cap [0, 1])$. Also define

$$F(x) = \begin{cases} 0, & x < 0; \\ \psi(x), & 0 \le x < 1; \\ 1, & x \ge 1, \end{cases}$$

where ψ denotes the Cantor function, and let ω be the unique Borel measure that has F as its distribution function. Finally, set $\nu = \tau + \omega + \delta_0 + \delta_1$. Determine the Lebesgue decomposition of ν with respect to λ.

9.67 Let $(\Omega, \mathcal{A}, \mu) = (\mathcal{R}, \mathcal{M}, \mu)$, where μ is counting measure on $\mathcal{N}$, that is, $\mu = \sum_{n=1}^{\infty} \delta_n$. Set $\nu = \delta_0 + \delta_1 + \lambda$. Determine the Lebesgue decomposition of ν with respect to μ.

9.68 Let $(\Omega, \mathcal{A}, \mu)$ be a measure space and ν_1 and ν_2 measures on $\mathcal{A}$. If $\nu_1 \ll \mu$ and $\nu_2 \perp \mu$, show that $\nu_1 \perp \nu_2$. Hence, the two measures in the Lebesgue decomposition of a measure are mutually singular.

In each of Exercises 9.69–9.71, $(\Omega, \mathcal{A})$ denotes a measurable space such that $\{x\} \in \mathcal{A}$ for all $x \in \Omega$.

9.69 What can you say about a measure ν that is both discrete and continuous?

9.70 Find a measure ν on a measurable space $(\Omega, \mathcal{A})$ such that every element of Ω is an atom, but ν is not discrete.

9.71 Let μ and ν be measures on $(\Omega, \mathcal{A})$. Prove the following facts.
   a) If μ is discrete and ν is continuous, then $\mu \perp \nu$.
   b) If $\nu \ll \mu$ and μ is continuous, then so is ν.
   c) If $\nu \ll \mu$ and μ is discrete, then so is ν.

9.72 Show that the measures ν_{ac}, ν_{sc}, and ν_{d} in Theorem 9.7 are mutually singular.

9.73 Let X be a random variable on $(\Omega, \mathcal{A}, P)$ and let μ_X denote its probability distribution. Prove that μ_X can be expressed as a convex combination of probability measures μ_1, μ_2, and μ_3 on $\mathcal{B}$, where $\mu_1 \ll \lambda$, $\mu_2 \perp \lambda$ and is continuous, and μ_3 is discrete. *Convex combination* means we can write $\mu_X = \sum_{j=1}^{3} \alpha_j \mu_j$, where $\alpha_j \ge 0$, $1 \le j \le 3$, and $\sum_{j=1}^{3} \alpha_j = 1$.

9.74 Let ν be a finite Borel measure on $\mathcal{R}$ and F_ν its distribution function.
   a) Show that for each $x \in \mathcal{R}$, $\nu(\{x\}) = F_\nu(x) - F_\nu(x-)$.
   b) Prove that x is an atom of ν if and only if F_ν is discontinuous at x.

9.75 Prove Proposition 9.10 on page 338.

9.76 Suppose that F is a discrete distribution function, say, $F(x) = \sum_{x_n \leq x} a_n$.

a) Show that $\{x_n\}_n$ constitutes the set of discontinuities of F.

b) Prove that for each $n \in \mathcal{N}$, $F(x_n) - F(x_n-) = a_n$. This result shows that the magnitude of the discontinuity of F at the point x_n is a_n.

c) Show that F is constant on any interval not containing a discontinuity point of F.

9.77 Let $\{r_n\}_{n=1}^{\infty}$ be an enumeration of the rational numbers and let $\{a_n\}_{n=1}^{\infty}$ be a sequence of positive real numbers such that $\sum_{n=1}^{\infty} a_n < \infty$. Define the function F on $\mathcal{R}$ by $F(x) = \sum_{r_n \leq x} a_n$.

a) Prove that F is continuous at every irrational number and discontinuous at every rational number.

b) Show that F is strictly increasing.

c) Prove that $F' = 0$ λ-ae.

9.78 Let ν be a finite Borel measure on $\mathcal{R}$ and F_ν its distribution function. Let $\nu = \nu_1 + \nu_2$ be the Lebesgue decomposition of ν with respect to λ (considered a Borel measure). Prove that $d\nu_1/d\lambda = F_\nu'$ λ-ae.

9.79 Let ψ denote the Cantor function. Define

$$F(x) = \begin{cases} 0, & x < 0; \\ 2 + x + \psi(x), & 0 \leq x < 1; \\ 4 + x, & 1 \leq x < 2; \\ 9, & x \geq 2. \end{cases}$$

a) Decompose F into its absolutely continuous, singular continuous, and discrete parts.

b) Use part (a) to determine the decomposition of the Lebesgue-Stieltjes measure ν corresponding to F into its absolutely continuous, singular continuous, and discrete parts.

9.80 Refer to Exercise 7.30 on page 245. Let X be uniformly distributed on $[a, b]$, and let $a < c < b$.

a) Define $Y = \min\{c, X\}$. Determine the decomposition of μ_Y into its absolutely continuous, singular continuous, and discrete parts.

b) Define $Z = \max\{c, X\}$. Determine the decomposition of μ_Z into its absolutely continuous, singular continuous, and discrete parts.

c) Let X be an absolutely continuous random variable with probability density function f_X. Let M be a positive real number, and define

$$Y = \begin{cases} -M, & X \leq -M; \\ X, & |X| < M; \\ M, & X \geq M. \end{cases}$$

Determine the decomposition of μ_Y into its absolutely continuous, singular continuous, and discrete parts.

9.81 Let $(\Omega, \mathcal{A}, \mu) = (\mathcal{R}^2, \mathcal{B}_2, \lambda_2)$, and let $D = \{(x, y) \in \mathcal{R}^2 : x^2 + y^2 \leq 1\}$. Define μ, ω, and τ on $\mathcal{B}_2$ as follows: $\mu(B) = \lambda(\{x \in \mathcal{R} : (x, x) \in B\})$; $\omega(B) = 1$ if $(0, 0) \in B$, and 0 otherwise; and $\tau(B) = \lambda_2(B \cap D)$. Let $\nu = \mu + \omega + \tau$. Determine the decomposition of ν into its absolutely continuous, singular continuous, and discrete parts with respect to two-dimensional Lebesgue measure.

9.5 MEASURABLE TRANSFORMATIONS AND THE GENERAL CHANGE-OF-VARIABLE FORMULA

Recall the following change-of-variable formula from elementary calculus, often referred to as *integration by substitution*.

Change-of-Variable Formula for Riemann Integration: Suppose that g is a continuously differentiable monotone function on $[a, b]$ with range $[c, d]$ and that f is continuous on $[c, d]$. Then

$$\int_a^b f(g(x))|g'(x)|\, dx = \int_c^d f(y)\, dy.$$

In this section, we will generalize the change-of-variable formula by applying the theory of measurable transformations. We begin by presenting the following definition.

DEFINITION 9.15 Measurable Transformation

Let $(\Omega, \mathcal{A})$ and $(\Lambda, \mathcal{S})$ be measurable spaces. A mapping $T : \Omega \to \Lambda$ is called a **measurable transformation** if $T^{-1}(S) \in \mathcal{A}$ for each $S \in \mathcal{S}$.

EXAMPLE 9.14 *Illustrates Definition 9.15*

a) The measurable transformations from $(\mathcal{R}, \mathcal{B})$ to $(\mathcal{R}, \mathcal{B})$ are precisely the (real-valued) Borel measurable functions.

b) The measurable transformations from $(\mathcal{R}, \mathcal{M})$ to $(\mathcal{R}, \mathcal{B})$ are precisely the real-valued Lebesgue measurable functions.

c) More generally than in parts (a) and (b), let $(\Omega, \mathcal{A})$ be any measurable space and $(\Lambda, \mathcal{S}) = (\mathcal{R}, \mathcal{B})$. Then the measurable transformations from $(\Omega, \mathcal{A})$ to $(\mathcal{R}, \mathcal{B})$ coincide with the real-valued $\mathcal{A}$-measurable functions. □

Our next result shows that the composition of a measurable function with a measurable transformation is a measurable function.

□ □ □ **PROPOSITION 9.12**

Suppose that T is a measurable transformation from $(\Omega, \mathcal{A})$ to $(\Lambda, \mathcal{S})$ and that f is an $\mathcal{S}$-measurable function on Λ. Then $f \circ T$ is an $\mathcal{A}$-measurable function on Ω.

PROOF Let O be open in $\mathcal{R}$ (in $\mathbb{C}$ if f is complex-valued, in $\mathcal{R}^*$ if f is extended real-valued). Because f is $\mathcal{S}$-measurable, $f^{-1}(O) \in \mathcal{S}$, and because T is a measurable transformation from $(\Omega, \mathcal{A})$ to $(\Lambda, \mathcal{S})$, $T^{-1}(f^{-1}(O)) \in \mathcal{A}$. Consequently, we have that $(f \circ T)^{-1}(O) = T^{-1}(f^{-1}(O)) \in \mathcal{A}$ for each open set O. ■

From a measurable transformation and a measure on its domain space, we get in a natural way a measure on the range space. This fact is the content of the following proposition whose proof is left to the reader as an exercise.

□ □ □ **PROPOSITION 9.13**

Let T be a measurable transformation from $(\Omega, \mathcal{A})$ to $(\Lambda, \mathcal{S})$ and μ a measure on $\mathcal{A}$. Define

$$\mu \circ T^{-1}(S) = \mu(T^{-1}(S)), \qquad S \in \mathcal{S}.$$

*Then $\mu \circ T^{-1}$ is a measure on $\mathcal{S}$, called the **measure induced** by μ and T.*

EXAMPLE 9.15 *Illustrates Proposition 9.13*

Let $(\Omega, \mathcal{A}, P)$ be a probability space and X a random variable thereon. According to Definition 7.6, the set function

$$\mu_X(B) = P(X \in B), \qquad B \in \mathcal{B},$$

is the probability distribution of X. But note also that X is a measurable transformation from $(\Omega, \mathcal{A})$ to $(\mathcal{R}, \mathcal{B})$ and that the measure induced by P and X is

$$P \circ X^{-1}(B) = P(X^{-1}(B)) = P(X \in B) = \mu_X(B), \qquad B \in \mathcal{B}.$$

In other words, the measure induced by P and X is the probability distribution of X.   □

The General Change-of-Variable Formula

With Propositions 9.12 and 9.13 in mind, we now prove the general change-of-variable formula.

□ □ □   **THEOREM 9.10 General Change-of-Variable Formula**

Let $(\Omega, \mathcal{A}, \mu)$ be a measure space, $(\Lambda, \mathcal{S})$ a measurable space, and T a measurable transformation from $(\Omega, \mathcal{A})$ to $(\Lambda, \mathcal{S})$. Then, for any $\mathcal{S}$-measurable function f on Λ,

$$\int_\Omega f \circ T(x) \, d\mu(x) = \int_\Lambda f(y) \, d\mu \circ T^{-1}(y), \tag{9.18}$$

in the sense that if one of the integrals in (9.18) exists, then so does the other, and they are equal.

PROOF   Suppose first that f is the characteristic function of a set $S \in \mathcal{S}$. Then,

$$\int_\Omega f \circ T(x) \, d\mu(x) = \int_\Omega \chi_S(T(x)) \, d\mu(x) = \int_\Omega \chi_{T^{-1}(S)}(x) \, d\mu(x)$$

$$= \mu(T^{-1}(S)) = \mu \circ T^{-1}(S) = \int_\Lambda \chi_S(y) \, d\mu \circ T^{-1}(y)$$

$$= \int_\Lambda f(y) \, d\mu \circ T^{-1}(y).$$

Hence (9.18) holds if f is a characteristic function. It now follows easily that (9.18) holds if f is a nonnegative $\mathcal{S}$-measurable simple function.

If f is a nonnegative extended real-valued $\mathcal{S}$-measurable function, select a sequence $\{s_n\}_{n=1}^\infty$ of nonnegative $\mathcal{S}$-measurable simple functions such that $s_n \uparrow f$ on Λ. Then $s_n \circ T \uparrow f \circ T$ on Ω. Applying the monotone convergence theorem twice, we get that

$$\int_\Omega f \circ T(x) \, d\mu(x) = \lim_{n \to \infty} \int_\Omega s_n(T(x)) \, d\mu(x)$$

$$= \lim_{n \to \infty} \int_\Lambda s_n(y) \, d\mu \circ T^{-1}(y) = \int_\Lambda f(y) \, d\mu \circ T^{-1}(y).$$

If f is a complex-valued or extended real-valued $\mathcal{S}$-measurable function, we proceed in the usual manner. That is, we decompose f into a linear combination of nonnegative $\mathcal{S}$-measurable functions and apply the result of the previous paragraph to each component.   ■

As an immediate consequence of Theorem 9.10, we have the following corollaries. Their proofs are left to the reader as exercises.

□ □ □   **COROLLARY 9.1**

Let $(\Omega, \mathcal{A}, \mu)$ be a measure space, $(\Lambda, \mathcal{S})$ a measurable space, T a measurable transformation from $(\Omega, \mathcal{A})$ to $(\Lambda, \mathcal{S})$, and f an $\mathcal{S}$-measurable function on Λ. Then, for each $S \in \mathcal{S}$,

$$\int_{T^{-1}(S)} f \circ T(x) \, d\mu(x) = \int_S f(y) \, d\mu \circ T^{-1}(y),$$

in the sense that if one of the integrals exists, then so does the other, and they are equal.

□ □ □   **COROLLARY 9.2**

Let $(\Omega, \mathcal{A}, \mu)$ be a measure space and $(\Lambda, \mathcal{S}, \nu)$ a σ-finite measure space. Suppose that T is a measurable transformation from $(\Omega, \mathcal{A})$ to $(\Lambda, \mathcal{S})$ such that $\mu \circ T^{-1} \ll \nu$ and $\mu \circ T^{-1}$ is σ-finite. Then, for any $\mathcal{S}$-measurable function f on Λ,

$$\int_{\Omega} f \circ T(x) \, d\mu(x) = \int_{\Lambda} f(y) \frac{d\mu \circ T^{-1}}{d\nu}(y) \, d\nu(y),$$

in the sense that if one of the integrals exists, then so does the other, and they are equal.

Application to Euclidean n-Space

We will now apply the general change-of-variable formula (Theorem 9.10) to obtain a result that contains as a special case the classical change-of-variable formula in Euclidean n-space. To begin, we introduce some useful notation.

Let $E \subset \mathcal{R}^n$, $x \in \mathcal{R}^n$, and $a \in \mathcal{R}$. Then we define

$$x + E = \{\, x + y : y \in E \,\} \qquad \text{and} \qquad aE = \{\, ay : y \in E \,\}.$$

The set $x + E$ is called a **translate** of E, specifically, the translation of E by x. An important property of n-dimensional Lebesgue measure (defined on page 186) is that it is **translation invariant**:

$$\lambda_n(x + E) = \lambda_n(E), \qquad x \in \mathcal{R}^n, \, E \in \mathcal{M}_n.$$

The proof of this fact is left to the reader as Exercise 9.98.

By a **cell,** we mean a subset of $\mathcal{R}^n$ of the form

$$[a_1, b_1) \times [a_2, b_2) \times \cdots \times [a_n, b_n),$$

where, for $1 \leq j \leq n$, a_j and b_j are real numbers with $a_j < b_j$. We will be particularly interested in the following cells:

$$C_m = \{ (x_1, x_2, \ldots, x_n) : 0 \leq x_j < 2^{-m}, \ j = 1, \ldots, n \}$$

for $m = 0, 1, 2, \ldots$.

We denote by $\mathcal{B}_n$ the σ-algebra generated by the open sets of $\mathcal{R}^n$. Members of $\mathcal{B}_n$ are called **n-dimensional Borel sets.** Exercise 6.77(a) on page 223 shows that $\mathcal{B}_n = \mathcal{B} \times \cdots \times \mathcal{B}$; in other words, $\mathcal{B}_n$ is also the σ-algebra generated by the n-dimensional Borel rectangles—sets of the form $B_1 \times \cdots \times B_n$, where $B_k \in \mathcal{B}$, $1 \leq k \leq n$.

□ □ □ **LEMMA 9.4**

Let $A \colon \mathcal{R}^n \to \mathcal{R}^n$ be a nonsingular linear transformation. Then

$$\lambda_n(A(B)) = |\det A| \lambda_n(B) \tag{9.19}$$

for all $B \in \mathcal{B}_n$.

PROOF We first note that, if (9.19) holds for a $B \in \mathcal{B}_n$, then it holds for all translates of B. Indeed, by the linearity of A and the translation invariance of λ_n, we have

$$\lambda_n(A(x + B)) = \lambda_n(A(x) + A(B)) = \lambda_n(A(B))$$
$$= |\det A| \lambda_n(B) = |\det A| \lambda_n(x + B)$$

for all $x \in \mathcal{R}^n$.

Next we show that, if (9.19) holds for C_0, then it holds for all $B \in \mathcal{B}_n$. So, assume that $\lambda_n(A(C_0)) = |\det A| \lambda_n(C_0)$. Let $m \in \mathcal{N}$. We note that C_0 is a pairwise disjoint union of a collection of translates $\{ v + C_m : v \in S \}$, where S is a finite subset of $\mathcal{R}^n$ with, say, s elements. Using disjointness and translation invariance, we have that

$$\lambda_n(C_0) = \lambda_n\left(\bigcup_{v \in S} (v + C_m) \right) = \sum_{v \in S} \lambda_n(v + C_m) = \sum_{v \in S} \lambda_n(C_m) = s\lambda_n(C_m).$$

In addition, using the fact that A is one-to-one and linear, we get that

$$\lambda_n(A(C_0)) = \lambda_n\left(A\left(\bigcup_{v \in S} (v + C_m) \right) \right) = \lambda_n\left(\bigcup_{v \in S} A(v + C_m) \right)$$
$$= \sum_{v \in S} \lambda_n(A(v + C_m)) = \sum_{v \in S} \lambda_n(A(C_m)) = s\lambda_n(A(C_m)).$$

Consequently,

$$\lambda_n(A(C_m)) = s^{-1}\lambda_n(A(C_0)) = s^{-1}|\det A|\lambda_n(C_0)$$
$$= s^{-1}|\det A|s\lambda_n(C_m) = |\det A|\lambda_n(C_m).$$

Thus, (9.19) holds for C_m.

Next, let $D = [0, d_1) \times [0, d_2) \times \cdots \times [0, d_n)$, where the d_js are positive dyadic rationals; that is, each d_j is of the form $n_j/2^{m_j}$ with n_j a positive integer and m_j a nonnegative integer. By finding a common denominator for the d_js, we can assume that the m_js are all equal, say, to m. It follows that D is a pairwise disjoint union of a finite collection of translates of C_m. Arguing as we did above for C_0, we find that (9.19) holds for D.

Now let $E = [0, e_1) \times [0, e_2) \times \cdots \times [0, e_n)$, where the e_js are positive real numbers. Because the dyadic rationals are dense in $\mathcal{R}$, we can, for each j, choose a sequence of dyadic rationals increasing to e_j. Applying Theorem 5.1(d) on page 148 and the result of the preceding paragraph, it follows that (9.19) holds for E. Since any cell is a translate of a cell of the form E, we conclude that (9.19) holds for all cells.

We note that both sides of (9.19) define n-dimensional Borel measures, and we have just shown that, if (9.19) holds for C_0, then it holds for all cells. However, two n-dimensional Borel measures that agree and are finite on all cells must be equal, as the reader is asked to verify in Exercise 9.99. Consequently, if (9.19) holds for C_0, then it holds for all $B \in \mathcal{B}_n$.

Next we will verify that (9.19) holds for C_0 (and hence for all $B \in \mathcal{B}_n$) for three special types of nonsingular linear transformations. Let β be a nonzero real number, and let p and q be integers with $1 \le p < q \le n$. Define three types of linear transformations from $\mathcal{R}^n$ to $\mathcal{R}^n$ as follows:

(1) $A(x_1, \ldots, x_p, \ldots, x_q, \ldots, x_n) = (x_1, \ldots, x_q, \ldots, x_p, \ldots, x_n)$

(2) $A(x_1, \ldots, x_p, \ldots, x_n) = (x_1, \ldots, \beta x_p, \ldots, x_n)$

(3) $A(x_1, \ldots, x_p, \ldots, x_q, \ldots, x_n) = (x_1, \ldots, x_p + x_q, \ldots, x_q, \ldots, x_n)$

For a linear transformation of type (1), we have $\det A = -1$ and $A(C_0) = C_0$. Consequently,

$$\lambda_n(A(C_0)) = \lambda_n(C_0) = |\det A|\lambda_n(C_0),$$

and, hence, (9.19) holds.

For a linear transformation of type (2), we have $\det A = \beta$ and

$$A(C_0) = \begin{cases} \{(y_1, \ldots, y_n) : 0 \le y_j < 1 \text{ if } j \ne p, \ 0 \le y_p < \beta\}, & \beta > 0; \\ \{(y_1, \ldots, y_n) : 0 \le y_j < 1 \text{ if } j \ne p, \ \beta < y_p \le 0\}, & \beta < 0. \end{cases}$$

Consequently,

$$\lambda_n(A(C_0)) = |\beta| = |\det A| = |\det A| \cdot 1 = |\det A|\lambda_n(C_0),$$

and, hence, (9.19) holds.

For a linear transformation of type (3), we have $\det A = 1$ and

$$A(C_0) = \{(y_1, \ldots, y_n) : 0 \le y_j < 1 \text{ if } j \ne p, \ y_q \le y_p < 1 + y_q\}.$$

Applying Fubini's theorem, we find that

$$\lambda_n(A(C_0)) = \lambda_2(\{(y_p, y_q) : 0 \le y_q < 1, \ y_q \le y_p < 1 + y_q\}) = 1.$$

Consequently,
$$\lambda_n(A(C_0)) = 1 = 1 \cdot 1 = |\det A|\lambda_n(C_0),$$
and, hence, (9.19) holds.

In summary, we have established that (9.19) holds for all $B \in \mathcal{B}_n$ when A is a nonsingular linear transformation of type (1), (2), or (3). Now let $A: \mathcal{R}^n \to \mathcal{R}^n$ be any nonsingular linear transformation. From linear algebra, we know that we can write $A = A_1 \circ A_2 \circ \cdots \circ A_k$, where each A_j is a linear transformation of type (1), (2), or (3). For $B \in \mathcal{B}_n$, we have
$$\lambda_n((A_{k-1} \circ A_k)(B)) = \lambda_n(A_{k-1}(A_k(B))) = |\det A_{k-1}|\lambda_n(A_k(B))$$
$$= |\det A_{k-1}| \cdot |\det A_k|\lambda_n(B) = |\det(A_{k-1} \circ A_k)|\lambda_n(B).$$

Continuing in this way, we find that
$$\lambda_n(A(B)) = \lambda_n((A_1 \circ A_2 \circ \cdots \circ A_k)(B))$$
$$= |\det(A_1 \circ A_2 \circ \cdots \circ A_k)|\lambda_n(B) = |\det A|\lambda_n(B).$$

The proof of the lemma is now complete. ∎

For the remainder of this subsection, it is convenient to measure the "distance" between two points $x = (x_1, x_2, \ldots, x_n)$ and $y = (y_1, y_2, \ldots, y_n)$ of $\mathcal{R}^n$ by
$$\|x - y\|_\infty = \max\{ |x_j - y_j| : 1 \le j \le n \}.$$

We leave it to the reader as Exercise 9.100 to show that $\| \ \|_\infty$ satifies the following properties for all $\alpha \in \mathcal{R}$ and $x, y \in \mathcal{R}^n$:

(N1) $\|x\|_\infty \ge 0$, with equality holding if and only if $x = 0$.

(N2) $\|\alpha x\|_\infty = |\alpha|\|x\|_\infty$.

(N3) $\|x + y\|_\infty \le \|x\|_\infty + \|y\|_\infty$.

By a **neighborhood** of a point $a \in \mathcal{R}^n$, we mean an n-dimensional cube of the form $\{ x \in \mathcal{R}^n : \|x - a\|_\infty < \delta \}$, where δ is a positive real number.

Now let Ω and Λ be open subsets of $\mathcal{R}^n$ and $T: \Omega \to \Lambda$ a one-to-one onto transformation with inverse transformation denoted by U. We assume that T satisfies the following three conditions:

(T1) T is **differentiable** at each $a \in \Omega$; that is, for each $a \in \Omega$, there is a linear transformation $DT(a): \mathcal{R}^n \to \mathcal{R}^n$ such that in some neighborhood of a,
$$T(x) = T(a) + DT(a)(x - a) + e(x - a), \qquad (9.20)$$
where $\lim_{\|x-a\|_\infty \to 0} \|e(x - a)\|_\infty / \|x - a\|_\infty = 0$.

(T2) All first-order partial derivatives of T (i.e., $\partial T_i / \partial x_j$, $1 \le i, j \le n$) exist and are continuous on Ω; that is, the entries of the **Jacobian matrix** of T are continuous functions on Ω.

(T3) The **Jacobian** of T, defined by $J_T(a) = \det DT(a)$, is nonzero on Ω; that is, $J_T(a) \ne 0$ for all $a \in \Omega$.

We note that, if conditions (T1)–(T3) hold for T, then they also hold for the inverse transformation (i.e., with U in place of T and $b \in \Lambda$ in place of $a \in \Omega$). Furthermore, if $b = T(a)$, then $DU(b) = DT(a)^{-1}$ and $J_U(b) = 1/J_T(a)$.

Actually, (T2) implies (T1) but, for ease in reference, we have stated both conditions explicitly. A transformation that satisfies condition (T2), and hence also condition (T1), is said to be **continuously differentiable** on Ω.

In stating the next lemma, we use the following notation: For $D \subset \mathcal{R}^n$, we denote by $\mathcal{B}(D)$ the σ-algebra generated by the open sets of D. Members of $\mathcal{B}(D)$ are called Borel sets of D. It is easy to see that $\mathcal{B}(D) = \{ D \cap B : B \in \mathcal{B}_n \}$.

□ □ □   **LEMMA 9.5**

Let Ω and Λ be open subsets of $\mathcal{R}^n$ and let T be a one-to-one continuously differentiable transformation from Ω onto Λ with nonzero Jacobian. Define

$$\mu(E) = \int_E |J_T(x)| \, d\lambda_n(x), \qquad E \in \mathcal{B}(\Omega).$$

Then, as measures on $\mathcal{B}(\Lambda)$, we have $\lambda_n \leq \mu \circ T^{-1}$.

PROOF   Equivalently, we can prove that, as measures on $\mathcal{B}(\Omega)$, we have $\lambda_n \circ T \leq \mu$; in other words,

$$\lambda_n(T(E)) \leq \int_E |J_T(x)| \, d\lambda_n(x), \qquad E \in \mathcal{B}(\Omega). \tag{9.21}$$

To begin, we show that (9.21) holds for all cells of the form

$$C = [a_1, b_n) \times [a_2, b_2) \times \cdots [a_n, b_n),$$

where the a_js and b_js are dyadic rationals.

Let $\epsilon > 0$ be given. In view of (T1)–(T3), we can choose a positive integer m to obtain a cell of the form

$$B = [-\delta, \delta) \times [-\delta, \delta) \times \cdots \times [-\delta, \delta),$$

where $\delta = 2^{-m}$, and a finite set $S \subset \mathcal{R}^n$ such that the following hold:

(1) The cell C is the pairwise disjoint union of $\{ a + B : a \in S \}$.

(2) For each $a \in S$, (9.20) holds for all $x \in a + B$.

(3) For each $a \in S$, J_T varies by less than ϵ on $a + B$.

(4) For each $a \in S$,

$$\frac{\|DT(a)^{-1}(e(x - a))\|_\infty}{\|x - a\|_\infty} < \epsilon, \qquad x \in a + B.$$

(See Exercise 9.101.)

In view of (2) and the linearity of $DT(a)$, we have for each $a \in S$ that

$$T(x) = T(a) + DT(a)\left((x - a) + \frac{DT(a)^{-1}(e(x - a))}{\|x - a\|_\infty} \|x - a\|_\infty \right), \qquad x \in a + B.$$

Applying the triangle inequality for $\| \; \|_\infty$ and using (4), we observe that, for each $a \in S$,

$$\left\| (x - a) + \frac{DT^{-1}(a)(e(x - a))}{\|x - a\|_\infty} \|x - a\|_\infty \right\|_\infty \leq \delta + \epsilon\delta = (1 + \epsilon)\delta, \quad x \in a + B.$$

Hence, we deduce that, for each $a \in S$,

$$T(x) \in T(a) + DT(a)((1 + \epsilon)B), \quad x \in a + B.$$

Consequently,

$$T(a + B) \subset T(a) + DT(a)((1 + \epsilon)B), \quad a \in S.$$

Applying Lemma 9.4 on page 346 and the translation invariance of Lebesgue measure, we now obtain, for each $a \in S$, that

$$\lambda_n(T(a + B)) \leq \lambda_n\big(T(a) + DT(a)((1 + \epsilon)B)\big) = \lambda_n\big(DT(a)((1 + \epsilon)B)\big)$$

$$= |J_T(a)|\lambda_n((1 + \epsilon)B) = (1 + \epsilon)^n |J_T(a)|\lambda_n(B).$$

Using translation invariance again, we conclude that

$$\lambda_n(T(a + B)) \leq (1 + \epsilon)^n |J_T(a)|\lambda_n(a + B), \quad a \in S. \tag{9.22}$$

It follows from (9.22) and (3) that, for each $a \in S$,

$$\lambda_n(T(a + B)) \leq (1 + \epsilon)^n \int_{a+B} |J_T(a)| \, d\lambda_n(x)$$

$$\leq (1 + \epsilon)^n \left(\int_{a+B} |J_T(x)| \, d\lambda_n(x) + \int_{a+B} |J_T(a) - J_T(x)| \, d\lambda_n(x) \right)$$

$$\leq (1 + \epsilon)^n \left(\int_{a+B} |J_T(x)| \, d\lambda_n(x) + \epsilon\lambda_n(a + B) \right).$$

By (1), summing over $a \in S$ gives

$$\lambda_n(T(C)) \leq (1 + \epsilon)^n \left(\int_C |J_T(x)| \, d\lambda_n(x) + \epsilon\lambda_n(C) \right).$$

As $\epsilon > 0$ is arbitrary, we conclude that

$$\lambda_n(T(C)) \leq \int_C |J_T(x)| \, d\lambda_n(x). \tag{9.23}$$

It remains to establish (9.21) for any $E \in \mathcal{B}(\Omega)$. We leave the verification of that fact to the reader as Exercise 9.103. ∎

We are now in a position to prove a generalization of the classical change-of-variable formula in Euclidean n-space.

□ □ □ THEOREM 9.11

Let Ω and Λ be open subsets of $\mathcal{R}^n$ and let T be a one-to-one continuously differentiable transformation from Ω onto Λ with nonzero Jacobian. Then, for any $\mathcal{B}(\Lambda)$-measurable function f and any $E \in \mathcal{B}(\Omega)$,

$$\int_E f(T(x))|J_T(x)| \, d\lambda_n(x) = \int_{T(E)} f(y) \, d\lambda_n(y), \qquad (9.24)$$

in the sense that if one of the integrals exists, then so does the other, and they are equal.

PROOF It suffices to prove (9.24) when f is nonnegative. Define

$$\mu(E) = \int_E |J_T(x)| \, d\lambda_n(x), \qquad E \in \mathcal{B}(\Omega).$$

From Lemma 9.5, we have that $\lambda_n \leq \mu \circ T^{-1}$ as measures on $\mathcal{B}(\Lambda)$. Using that fact and Corollary 9.1 on page 345, we get that, for $E \in \mathcal{B}(\Omega)$,

$$\int_{T(E)} f(y) \, d\lambda_n(y) \leq \int_{T(E)} f(y) \, d\mu \circ T^{-1}(y) = \int_E f(T(x)) \, d\mu(x).$$

Thus,

$$\int_{T(E)} f(y) \, d\lambda_n(y) \leq \int_E f(T(x))|J_T(x)| \, d\lambda_n(x), \qquad E \in \mathcal{B}(\Omega). \qquad (9.25)$$

Because U satisfies the same hypotheses as T, it follows from (9.25) that

$$\int_{U(F)} g(x) \, d\lambda_n(x) \leq \int_F g(U(y))|J_U(y)| \, d\lambda_n(y), \qquad F \in \mathcal{B}(\Lambda), \qquad (9.26)$$

for all nonnegative $\mathcal{B}(\Omega)$-measurable functions g. By letting $g = (f \circ T)|J_T|$ and $F = T(E)$ in (9.26), we get that

$$\int_E f(T(x))|J_T(x)| \, d\lambda_n(x) = \int_E \big((f \circ T)|J_T|\big)(x) \, d\lambda_n(x)$$

$$\leq \int_{T(E)} \big((f \circ T)|J_T|\big)(U(y))|J_U(y)| \, d\lambda_n(y)$$

$$= \int_{T(E)} f(T(U(y)))|J_T(U(y))||J_U(y)| \, d\lambda_n(y)$$

$$= \int_{T(E)} f(y)|J_U(y)|^{-1}|J_U(y)| \, d\lambda_n(y).$$

Thus,

$$\int_E f(T(x))|J_T(x)| \, d\lambda_n(x) \leq \int_{T(E)} f(y) \, d\lambda_n(y), \qquad E \in \mathcal{B}(\Omega). \qquad (9.27)$$

From (9.25) and (9.27), we get (9.24). ■

Exercises for Section 9.5

9.82 *True* or *False:* Every real-valued Lebesgue measurable function is a measurable transformation from $(\mathcal{R}, \mathcal{M})$ to $(\mathcal{R}, \mathcal{M})$.

9.83 Prove Proposition 9.13 on page 343.

9.84 Let $(\Omega, \mathcal{A}, P)$ be a probability space and $X_1, \ldots, X_n$ random variables thereon. Define $X : \Omega \to \mathcal{R}^n$ by $X(\omega) = (X_1(\omega), \ldots, X_n(\omega))$.
a) Prove that X is a measurable transformation from $(\Omega, \mathcal{A})$ to $(\mathcal{R}^n, \mathcal{B}_n)$.
b) Identify the measure induced by P and X.

9.85 Prove Corollary 9.1.

9.86 Prove Corollary 9.2.

9.87 Suppose that g is an absolutely continuous and monotone function on $[a, b]$ with range $[c, d]$ and that f is Borel measurable and Lebesgue integrable on $[c, d]$. Use the general change-of-variable formula (Theorem 9.10 on page 344) to prove that

$$\int_a^b f(g(x))|g'(x)|\, dx = \int_c^d f(y)\, dy,$$

where both integrals are in the Lebesgue sense. *Hint:* Assume first that g is non-decreasing. Let $\mu(B) = \int_B g'\, d\lambda$ for $B \in \mathcal{B}_{[a,b]}$ and show that, as measures on $\mathcal{B}_{[c,d]}$, $\mu \circ g^{-1} = \lambda$.

9.88 Use Exercise 9.87 to establish the change-of-variable formula for Riemann integration given on page 343.

9.89 Let X be an absolutely continuous random variable on $(\Omega, \mathcal{A}, P)$ with probability density function f_X, and let ϕ be a strictly monotone function on $\mathcal{R}$ whose inverse is absolutely continuous on $\mathcal{R}$. Prove that $Y = \phi \circ X$ is an absolutely continuous random variable on $(\Omega, \mathcal{A}, P)$ with probability density function given by

$$f_Y(y) = f_X(\phi^{-1}(y)) \left| \frac{d}{dy} \phi^{-1}(y) \right|.$$

9.90 Let $(\Omega, \mathcal{A}, \mu)$ be a finite measure space and ϕ a nonnegative real-valued $\mathcal{A}$-measurable function on Ω. For $x \geq 0$, set $G(x) = \mu(\phi^{-1}((x, \infty)))$. Prove that

$$\int_\Omega \phi\, d\mu = \int_0^\infty G(x)\, dx.$$

9.91 Let $(\Omega, \mathcal{A}, P)$ be a probability space, and let X be a nonnegative random variable thereon. Use Exercise 9.90 to prove that

$$\mathcal{E}(X) = \int_0^\infty (1 - F_X(x))\, dx.$$

9.92 Suppose g is a real-valued Lebesgue measurable function such that if $B \in \mathcal{B}$ with $\lambda(B) = 0$, then $\lambda(g^{-1}(B)) = 0$; that is, the inverse image under g of any Borel set of Lebesgue measure zero has Lebesgue measure zero.
a) Prove that if $E \in \mathcal{M}$ with $\lambda(E) = 0$, then $g^{-1}(E)$ has Lebesgue (outer) measure zero, and hence is measurable; that is, the inverse image under g of any Lebesgue measurable set of Lebesgue measure zero is Lebesgue measurable and has Lebesgue measure zero.

b) Prove that $g^{-1}(E) \in \mathcal{M}$ for each $E \in \mathcal{M}$, so that g is a measurable transformation from $(\mathcal{R}, \mathcal{M})$ to $(\mathcal{R}, \mathcal{M})$.

9.93 Suppose that g is a real-valued Borel measurable function such that, as Borel measures, $\lambda \circ g^{-1} \ll \lambda$. Prove that g is a measurable transformation from $(\mathcal{R}, \mathcal{M})$ to $(\mathcal{R}, \mathcal{M})$. *Hint:* Exercise 9.92.

9.94 In each of the following parts, we have specified a real-valued Borel measurable function g. In each case, show that, as Borel measures, $\lambda \circ g^{-1} \ll \lambda$, and determine explicitly $d(\lambda \circ g^{-1})/d\lambda$.
a) $g(x) = x^2$
b) $g(x) = x^3$
c) $g(x) = e^x$

9.95 Let ψ denote the Cantor function. Show that, as Borel measures on $[0, 1]$, $\lambda \circ \psi^{-1} \perp \lambda$.

9.96 In Exercise 8.63 on pages 300–301, we proved the following generalization of the change-of-variable formula for Riemann integration: Suppose that g is an absolutely continuous and monotone function on $[a, b]$ with range $[c, d]$ and that $f \in \mathcal{L}^1\big([c, d]\big)$. Then we have that $(f \circ g)g' \in \mathcal{L}^1\big([a, b]\big)$ and

$$\int_a^b f(g(x))|g'(x)| \, dx = \int_c^d f(y) \, dy,$$

where both integrals are in the Lebesgue sense. Explain why this result does not follow directly from the general change-of-variable formula.

9.97 Use the general change-of-variable formula to provide a proof of Theorem 7.6 on page 249.

9.98 Prove that n-dimensional Lebesgue measure is translation invariant.

9.99 Let μ and ν be two n-dimensional Borel measures such that $\mu(C) = \nu(C) < \infty$ for all cells C. Prove that $\mu = \nu$.

9.100 Show that $\| \, \|_\infty$ satisfies properties (N1)–(N3) on page 348. That is, for all $\alpha \in \mathcal{R}$ and $x, y \in \mathcal{R}^n$,
a) $\|x\|_\infty \geq 0$, with equality holding if and only if $x = 0$.
b) $\|\alpha x\|_\infty = |\alpha| \|x\|_\infty$.
c) $\|x + y\|_\infty \leq \|x\|_\infty + \|y\|_\infty$.

9.101 Let $A: \mathcal{R}^n \to \mathcal{R}^m$ be a linear transformation. Prove that, for each $\epsilon > 0$, there is a $\delta > 0$ such that $\|x\|_\infty < \delta$ implies $\|A(x)\|_\infty < \epsilon$.

9.102 Let $\Omega \subset \mathcal{R}^n$. A measure τ on a σ-algebra $\mathcal{A} \supset \mathcal{B}(\Omega)$ is said to be *regular* if
(1) $\tau(K) < \infty$ for all compact sets $K \subset \Omega$.
(2) $\tau(A) = \inf\{\, \tau(O) : O \supset A, \, O \text{ open} \,\}$ for all $A \in \mathcal{A}$.
(3) $\tau(O) = \sup\{\, \tau(K) : K \subset O, \, K \text{ compact} \,\}$ for all open sets $O \subset \Omega$.

From the Heine-Borel theorem and Exercises 3.45 and 3.46 on page 108, we see that Lebesgue measure λ is regular; similarly, it can be shown that n-dimensional Lebesgue measure λ_n is regular. A special case of an important result concerning regular measures is as follows: *If Ω is an open subset of $\mathcal{R}^n$, then any measure on $\mathcal{B}(\Omega)$ for which all compact sets have finite measure is regular.* Use that fact to solve the following problems. *Note:* Recall that if $g: \mathcal{R}^n \to \mathcal{R}^m$ is continuous and $K \subset \mathcal{R}^n$ is compact, then $g(K)$ is compact.
a) Show that the measure μ defined in Lemma 9.5 on page 349 is regular.
b) Show that the measure $\lambda_n \circ T$ in Lemma 9.5 is regular.

9.103 In this exercise, you are asked to complete the proof of Lemma 9.5 on page 349. For $E \in \mathcal{B}(\Omega)$, set $\nu(E) = (\lambda_n \circ T)(E)$ and $\mu(E) = \int_E |J_T(x)| \, d\lambda_n(x)$. We want to prove that

$$\nu(E) \leq \mu(E) \tag{9.28}$$

for all $E \in \mathcal{B}(\Omega)$. Recall that we have already established (9.28) for cells with dyadic rational endpoints, as shown in (9.23) on page 350.

a) Use (9.23), the density of the dyadic rationals, and properties of measures to show that (9.28) holds for all cells.

b) Use part (a) and properties of measures to show that (9.28) holds for all closed cubes, that is, for all sets of the form

$$[a_1, b_1] \times [a_2, b_2] \times \cdots \times [a_n, b_n],$$

where the a_ks and b_ks are real numbers such that $b_1 - a_1 = b_2 - a_2 = \cdots = b_n - a_n$.

c) Use part (b) and the fact that any open subset of $\mathcal{R}^n$ can be written as a countable union of closed cubes whose interiors are pairwise disjoint to show that (9.28) holds for all open subsets of Ω. *Hint:* Observe that $\mu \ll \lambda_n$.

d) Use Exercises 9.102(a) and (b) and part (c) of this exercise to show that (9.28) holds for all $E \in \mathcal{B}(\Omega)$.

9.104 Use Theorem 9.11 on page 351 to express the change-of-variable formula in each of the following three cases.

a) Polar coordinates: $x = r \cos \theta$, $y = r \sin \theta$

b) Cylindrical coordinates: $x = r \cos \theta$, $y = r \sin \theta$, $z = z$

c) Spherical coordinates: $x = \rho \sin \phi \cos \theta$, $y = \rho \sin \phi \sin \theta$, $z = \rho \cos \phi$

PART THREE

□

Topological, Metric, and Normed Spaces

Felix Hausdorff
(1868–1942)

Felix Hausdorff was born in Breslau, Germany (now Wroclaw, Poland), on November 8, 1868, into a wealthy family in the textile business. When he was a young boy, Hausdorff's family moved from Breslau to Leipzig, where he spent the remainder of his youth.

At school, Hausdorff was interested in literature and music in addition to mathematics. Evidently, he wanted to pursue a career as a music composer but his parents dissuaded him. Subsequently, Hausdorff decided to study mathematics at the University of Leipzig, where he obtained his Ph.D. in 1891, specializing in applications of mathematics to astronomy. Hausdoff taught mathematics in Leipzig until 1910, when he became professor of mathematics at the University of Bonn. From 1913 to 1921, he was professor at the University of Greifswald, following which he returned to Bonn.

Considered one of the founders of modern topology, Hausdorff also contributed significantly to set theory, measure theory, function theory, and functional analysis. In his classic text, *Grundzüge der Mengenlehre* (Elements of Set Theory), he defined and examined partially ordered sets, axiomatized the topological concept of neighborhood, and introduced topological spaces that are now called Hausdorff spaces in his honor. Additionally, Hausdorff developed the concepts of Hausdorff measure and Hausdorff dimension, as they are currently called (see Chapter 17).

When the Nazis came to power, Hausdorff, who was Jewish, thought that he would be spared, given that he was a highly respected university professor. However, his abstract mathematics was denounced as Jewish, useless, and "un-German." Although Hausdorff was forced to relinquish his position in 1935 and could no longer publish in Germany, he continued as an active researcher.

In 1942, when it was certain that he would be sent to a concentration camp, Hausdorff committed suicide together with his wife, Charlotte Goldschmidt Hausdorff, and sister-in-law, Edith Goldschmidt Pappenheim, on January 26.

10

Topologies, Metrics, and Norms

In this chapter and the next, we will discuss topological, metric, and normed spaces. Here, in this chapter, we will begin with a section that introduces topological spaces, followed by one that does so for metric and normed spaces. In subsequent sections of this chapter, we will examine weak topologies, closed sets, convergence, completeness, nets and continuity, separation properties, and connected sets.

10.1 INTRODUCTION TO TOPOLOGICAL SPACES

Like most good ideas in mathematics, the concept of a topological space can be approached from several points of view. Because our perspective is from analysis, we will emphasize the connections between topological spaces and the concepts of limit and continuity.

In this section, we show how extending the limit concept leads naturally to the notion of a topological space. Suppose that we have a function $f \colon \Omega \to \Lambda$. What kinds of structures on Ω and Λ are needed to make sense of the formula

$$\lim_{x \to a} f(x) = b ? \tag{10.1}$$

In case $\Omega = \Lambda = \mathcal{R}$, (10.1) can be described verbally as follows: "$f(x)$ will be near b whenever x is sufficiently near (but not equal to) a." Of course, "$f(x)$ being near b" means that $f(x)$ lies in some prescribed open interval I centered at b, and "x is sufficiently near a" means that x lies in a certain interval J centered at a. Here, we seek to capture the idea of nearness in terms of intervals. In general, our approach to (10.1) is to consider collections of subsets of Ω that have properties mimicking those of collections of intervals with a common center.

Neighborhood Bases, Continuous Functions, and Open Sets

Consider an element $a \in \Omega$. We would like to define a collection $\mathfrak{N}_a$ of subsets of Ω that can be thought of as "neighborhoods" of a. Certainly, our collection should be nonempty and satisfy the condition:

$$a \in N \text{ for all } N \in \mathfrak{N}_a. \tag{10.2}$$

If we think of an element of Ω as being "near" a if it belongs to some member of the collection $\mathfrak{N}_a$, then at least some of the elements of the intersection of two members of $\mathfrak{N}_a$ should also be "near" a. Thus, it is reasonable to assume that the collection $\mathfrak{N}_a$ satisfies the condition:

$$\text{If } N_1, N_2 \in \mathfrak{N}_a, \text{ there exists } N_3 \in \mathfrak{N}_a \text{ such that } N_3 \subset N_1 \cap N_2. \tag{10.3}$$

A nonempty collection $\mathfrak{N}_a$ of subsets of Ω satisfying (10.2) and (10.3) is called a **neighborhood basis at the point a.** Members of $\mathfrak{N}_a$ are called **neighborhoods in $\mathfrak{N}_a$** or, simply, **neighborhoods.** Using the concept of a neighborhood basis at a point, we make the following definition.

DEFINITION 10.1   Neighborhood Basis on a Set

A collection $\mathfrak{N}$ of subsets of a set Ω is said to be a **neighborhood basis on Ω** if for each $a \in \Omega$, the collection $\{\, N \in \mathfrak{N} : a \in N \,\}$ is a neighborhood basis at the point a.

The next proposition provides an equivalent set of conditions for a collection of subsets of Ω to be a neighborhood basis on Ω. Its proof is left to the reader as an exercise.

◻ ◻ ◻   **PROPOSITION 10.1**

A collection $\mathfrak{N}$ of subsets of a set Ω is a neighborhood basis on Ω if and only if it satisfies the following two conditions:
a) $\Omega = \bigcup_{N \in \mathfrak{N}} N$.
b) *If $N_1, N_2 \in \mathfrak{N}$ and $x \in N_1 \cap N_2$, then there exists an $N_3 \in \mathfrak{N}$ such that*

$$x \in N_3 \subset N_1 \cap N_2.$$

EXAMPLE 10.1   *Illustrates Neighborhood Bases*

a) The collection $\mathcal{I}$ of all open intervals of $\mathcal{R}$ is a neighborhood basis on $\mathcal{R}$. And so is $\{\, (x - r, x + r) : x \in \mathcal{R},\ r > 0 \,\}$.
b) There are several natural neighborhood bases on $\mathcal{R}^2$. Two examples are

$$\mathcal{I}^2 = \{\, I \times J : I, J \in \mathcal{I} \,\}$$

and

$$\mathcal{D} = \{\, D_r(a, b) : (a, b) \in \mathcal{R}^2,\ r > 0 \,\},$$

where $D_r(a, b) = \{\, (x, y) : (x - a)^2 + (y - b)^2 < r^2 \,\}$. Additional examples of neighborhood bases on $\mathcal{R}^2$ can be found in the exercises at the end of this section. ◻

Using the notion of neighborhood basis, we can make sense of (10.1). Suppose that $\mathfrak{N}_a$ and $\mathfrak{M}_b$ are neighborhood bases at a and b, respectively. Then we will take (10.1) to mean that the following condition is satisfied: *For each $M \in \mathfrak{M}_b$, there exists $N \in \mathfrak{N}_a$ such that $f(N \setminus \{a\}) \subset M$.*

In cases where (10.1) holds with $b = f(a)$, the function f is said to be **continuous at a with respect to the neighborhood bases $\mathfrak{N}_a$ and $\mathfrak{M}_{f(a)}$.** If $\mathfrak{N}$ and $\mathfrak{M}$ are neighborhood bases on Ω and Λ, respectively, then f is said to be **continuous on Ω with respect to $\mathfrak{N}$ and $\mathfrak{M}$** if it is continuous at each $a \in \Omega$ with respect to the neighborhood bases $\mathfrak{N}_a$ and $\mathfrak{M}_{f(a)}$, where

$$\mathfrak{N}_a = \{N \in \mathfrak{N} : a \in N\}$$

and

$$\mathfrak{M}_{f(a)} = \{M \in \mathfrak{M} : f(a) \in M\}.$$

When we have a neighborhood basis $\mathfrak{N}$ on Ω, we can also generalize the idea of an open set. Referring to Definition 2.7 on page 47, we define a subset $O \subset \Omega$ to be **open with respect to $\mathfrak{N}$** if for each $x \in O$, there is an $N \in \mathfrak{N}$ such that $x \in N \subset O$. When it is clear from the context which neighborhood basis we are using, we will say simply that O is an **open set.**

We note that all of the sets belonging to $\mathfrak{N}$ are open with respect to $\mathfrak{N}$. More generally, as the reader should verify, a subset of Ω is open with respect to $\mathfrak{N}$ if and only if it is a union of members of $\mathfrak{N}$.

Theorem 2.2 on page 47 states three fundamental properties of the collection of open subsets of $\mathcal{R}$. The next proposition, whose proof is left to the reader, shows that those properties also hold for the collection of subsets of Ω that are open with respect to a neighborhood basis.

□ □ □ PROPOSITION 10.2

Let $\mathfrak{N}$ be a neighborhood basis on the set Ω. Then the open sets with respect to $\mathfrak{N}$ satisfy the following conditions:
a) The empty set and the set Ω are open.
b) The union of any collection of open sets is an open set.
c) The intersection of any finite collection of open sets is an open set.

Exercise 10.2 shows that the neighborhood bases $\mathcal{I}^2$ and $\mathcal{D}$, defined in Example 10.1(b), determine the same collection of open subsets of $\mathcal{R}^2$. It is easy to construct other examples where distinct neighborhood bases determine the same collection of open sets.

The following proposition shows, however, that the property of continuity for a function $f \colon \Omega \to \Lambda$, where Ω and Λ have neighborhood bases $\mathfrak{N}$ and $\mathfrak{M}$, respectively, depends only on the open sets determined by $\mathfrak{N}$ and $\mathfrak{M}$. The proof of the proposition is left to the reader as an exercise.

□ □ □ PROPOSITION 10.3

Let $\mathfrak{N}$ and $\mathfrak{M}$ be neighborhood bases on Ω and Λ, respectively. Then a function $f \colon \Omega \to \Lambda$ is continuous on Ω (with respect to $\mathfrak{N}$ and $\mathfrak{M}$) if and only if $f^{-1}(O)$ is open in Ω with respect to $\mathfrak{N}$ whenever O is open in Λ with respect to $\mathfrak{M}$.

Topological Spaces and Continuous Functions

We note that Proposition 10.3 generalizes and is motivated by Theorem 2.5 on page 55. It also shows that, with respect to the concept of continuity, the notion of open set is more fundamental than that of neighborhood basis since two distinct neighborhood bases on a set can determine the same collection of open sets and, hence, the same continuous functions. Thus, we are led to formalize the concept of open set via the following definition.

DEFINITION 10.2 **Topology, Topological Space**

Let Ω be a nonempty set. A collection $\mathcal{T}$ of subsets of Ω is said to be a **topology** on Ω if it satisfies the following conditions:
a) $\emptyset, \Omega \in \mathcal{T}$.
b) $\mathcal{S} \subset \mathcal{T}$ implies $\bigcup_{O \in \mathcal{S}} O \in \mathcal{T}$.
c) $O_1, O_2 \in \mathcal{T}$ implies $O_1 \cap O_2 \in \mathcal{T}$.

If $\mathcal{T}$ is a topology on Ω, then the pair $(\Omega, \mathcal{T})$ is called a **topological space;** the members of $\mathcal{T}$ are called $\mathcal{T}$-**open** or, if there is no danger of confusion, simply **open.**

Note: When the topology under consideration is clear from context, a topological space $(\Omega, \mathcal{T})$ will usually be referred to simply as Ω.

It follows from Proposition 10.2 that if $\mathfrak{N}$ is a neighborhood basis on a set Ω, then the subsets of Ω that are open with respect to $\mathfrak{N}$ constitute a topology on Ω, which we will call the **topology determined by** $\mathfrak{N}$. On the other hand, if $\mathcal{T}$ is a topology on the set Ω, then the collection $\{\, O \in \mathcal{T} : a \in O \,\}$ is a neighborhood basis at the point a for each $a \in \Omega$, $\mathcal{T}$ is a neighborhood basis on Ω, and the topology determined by $\mathcal{T}$ is $\mathcal{T}$. We also have the following definition.

DEFINITION 10.3 **Neighborhood Basis for a Topological Space**

Let $(\Omega, \mathcal{T})$ be a topological space. A collection $\mathfrak{N}$ of subsets of Ω is said to be a **neighborhood basis for $(\Omega, \mathcal{T})$** if the following two conditions are satisfied:
a) $\mathfrak{N}$ is a neighborhood basis on Ω.
b) The topology determined by $\mathfrak{N}$ is $\mathcal{T}$.

In such cases, we also say that the neighborhood basis $\mathfrak{N}$ **induces** or **determines** $\mathcal{T}$.

The reader should verify that each of the following conditions is necessary and sufficient for a collection $\mathfrak{N}$ of subsets of Ω to be a neighborhood basis for $(\Omega, \mathcal{T})$.

- $\mathfrak{N} \subset \mathcal{T}$ and each open set (i.e., member of $\mathcal{T}$) is a union of members of $\mathfrak{N}$.

- $\mathfrak{N} \subset \mathcal{T}$ and for each $O \in \mathcal{T}$ and $x \in O$, there is an $N \in \mathfrak{N}$ such that $x \in N \subset O$.

Motivated by Proposition 10.3 (page 359), we now extend the definition of continuity to functions $f : \Omega \to \Lambda$, where Ω and Λ are topological spaces.

DEFINITION 10.4   Continuous Functions on Topological Spaces

Let Ω and Λ be topological spaces. A function $f: \Omega \to \Lambda$ is said to be **continuous** if $f^{-1}(O)$ is open in Ω whenever O is open in Λ.

EXAMPLE 10.2   *Illustrates Topological Spaces and Continuous Functions*

a) Let $\Omega = \mathcal{R}$ and $\mathcal{T}$ consist of the usual open sets as given by Definition 2.7 on page 47. Then, according to Corollary 2.1 on page 55, a real-valued function on $\mathcal{R}$ is continuous in the sense of Definition 2.11 on page 54 if and only if it is continuous in the sense of Definition 10.4.

b) The neighborhood bases given in Example 10.1(b) determine the same topology $\mathcal{T}$ on $\mathcal{R}^2$. With respect to $\mathcal{T}$ and the usual topology on $\mathcal{R}$, the functions $f, g: \mathcal{R}^2 \to \mathcal{R}$ given by $f(x, y) = x$ and $g(x, y) = y$ are continuous.

c) Let Ω be any set and $\mathcal{T} = \{\emptyset, \Omega\}$. Then $\mathcal{T}$ is a topology on Ω, albeit not an interesting one. Nevertheless, this topology is sometimes useful as an illustrative example. It is not hard to show that a function $f: \Omega \to \mathcal{R}$ is continuous with respect to the topology $\mathcal{T}$ if and only if it is constant.

d) Let Ω be any set. Then the collection $\mathcal{P}(\Omega)$ of all subsets of Ω is a topology on Ω which is sometimes referred to as the **discrete topology.** It is not hard to see that $\mathcal{P}(\Omega)$ is determined by the neighborhood basis consisting of all the single-element subsets of Ω. Also, it is obvious that all functions from Ω to $\mathcal{R}$, or to any other topological space, are continuous with respect to the discrete topology on Ω.   □

Note: From now on, unless stated otherwise, we will assume that $\mathcal{R}$ is equipped with the topology determined by the neighborhood basis of all open intervals, which is the same topology as the one consisting of the open sets of $\mathcal{R}$ as given by Definition 2.7 on page 47.

Relative Topologies and Continuous Functions

Given a topological space, we can produce still others by considering subsets with topologies defined as follows. Let $(\Omega, \mathcal{T})$ be a topological space and $D \subset \Omega$. Then it is easy to check that the collection $\{D \cap O : O \in \mathcal{T}\}$ is a topology on D, that is, it satisfies (a)–(c) of Definition 10.2. This topology is given a special name.

DEFINITION 10.5   Relative Topology

Let $(\Omega, \mathcal{T})$ be a topological space and D a subset of Ω. The collection of sets $\mathcal{T}_D = \{D \cap O : O \in \mathcal{T}\}$ is a topology on D, called the **relative topology.** Sets in $\mathcal{T}_D$ are said to be **relatively open.**

Remark: The reader should compare the definition of relatively open set given here with that given for subsets of $\mathcal{R}$ in Chapter 2; specifically, see Definition 2.10 and Theorem 2.3, both on page 51.

Unless stated otherwise, when we say that a function is **continuous on a subset** of a topological space, we will mean that it is continuous with respect to the relative topology. For example, when we say a function is continuous on the interval $[0, 1]$, we are assuming that $[0, 1]$ is equipped with the relative topology inherited from $\mathcal{R}$.

We should note that, if $f: \Omega \to \Lambda$ is continuous and $D \subset \Omega$, then the function $f_{|D}: D \to \Lambda$, the restriction of f to D, is continuous with respect to the relative topology on D.

Homeomorphic Topological Spaces

We conclude this section by considering what it means for two topological spaces to be equivalent.

DEFINITION 10.6   Homeomorphic Spaces; Homeomorphism

Suppose $(\Omega, \mathcal{T})$ and $(\Lambda, \mathcal{U})$ are topological spaces and $h: \Omega \to \Lambda$ is a 1-1 correspondence. If $h^{-1}(U) \in \mathcal{T}$ for each $U \in \mathcal{U}$ and $h(O) \in \mathcal{U}$ for each $O \in \mathcal{T}$, then we say that $(\Omega, \mathcal{T})$ and $(\Lambda, \mathcal{U})$ are **homeomorphic** and call h a **homeomorphism.**

We note that, if h is a homeomorphism, then both h and h^{-1} are continuous and, moreover, $U \to h^{-1}(U)$ is a 1-1 correspondence from $\mathcal{U}$ to $\mathcal{T}$. Thus, homeomorphic spaces are equivalent as topological spaces.

EXAMPLE 10.3   *Illustrates Definition 10.6*

The function $f(x) = 2x$ is a homeomorphism of the interval $(0, 1)$ onto the interval $(0, 2)$. Indeed, it can be shown that any two open intervals of $\mathcal{R}$ are homeomorphic. On the other hand, $[0, 1]$ and $(0, 1)$ are not homeomorphic. See Exercises 10.14–10.16.   □

Exercises for Section 10.1

Note: A ★ denotes an exercise that will be subsequently referenced.

10.1 Let Ω be a nonempty set and suppose that for each $a \in \Omega$, $\mathfrak{N}_a$ is a neighborhood basis at the point a. *True* or *False*: The collection $\mathfrak{N} = \bigcup_{a \in \Omega} \mathfrak{N}_a$ is a neighborhood basis on Ω.

10.2 Refer to Example 10.1(b) on page 358. Show that the topologies determined by the neighborhood bases $\mathcal{I}^2$ and $\mathcal{D}$ are identical.

10.3 Show that each of the following collections are neighborhood bases on $\mathcal{R}^2$ and that each collection determines the same topology as that in Exercise 10.2.
   a) The collection
$$\mathfrak{L} = \{\, L_r(a, b) : (a, b) \in \mathcal{R}^2,\ r > 0 \,\},$$
   where $L_r(a, b) = \{\, (x, y) : |x - a| + |y - b| < r \,\}$.
   b) The collection
$$\mathfrak{M} = \{\, M_r(a, b) : (a, b) \in \mathcal{R}^2,\ r > 0 \,\},$$
   where $M_r(a, b) = \{\, (x, y) : |x - a|^{1/2} + |y - b|^{1/2} < r \,\}$.

10.4 Let $\mathcal{I}$ denote the neighborhood basis on $\mathcal{R}$ that consists of all open intervals and $\mathcal{T}$ the topology determined by $\mathcal{I}$. Show that $\mathcal{T}$ consists precisely of the open sets of $\mathcal{R}$ as given by Definition 2.7 on page 47.

10.5 Prove Proposition 10.1 on page 358.

10.6 Prove Proposition 10.2 on page 359.

10.7 Prove Proposition 10.3 on page 359.

10.8 Let $\mathfrak{N} = \{\,[a, b) : -\infty < a < b < \infty\,\}$
  a) Show that $\mathfrak{N}$ is a neighborhood basis on $\mathcal{R}$.
  b) Let $\mathcal{T}$ be the topology determined by $\mathfrak{N}$. Give an example of a real-valued function that is continuous with respect to $\mathcal{T}$ but not with respect to the usual topology on $\mathcal{R}$.

10.9 A collection $\mathcal{S}$ of subsets of a set Ω is called a **sub-basis** on Ω if the collection of finite intersections of members of $\mathcal{S}$ is a neighborhood basis on Ω. The topology determined by the neighborhood basis is also said to be determined by the sub-basis $\mathcal{S}$.
  a) Show that a collection $\mathcal{S}$ of subsets of Ω is a sub-basis on Ω if and only if $\bigcup_{S \in \mathcal{S}} S = \Omega$.
  b) Show that the topology determined by the basis $\mathcal{I}^2$ in Example 10.1(b) is also determined by the sub-basis $\{\, I \times \mathcal{R} : I \in \mathcal{I}\,\} \cup \{\, \mathcal{R} \times I : I \in \mathcal{I}\,\}$.

10.10 Verify the assertions made in parts (a)–(d) of Example 10.2 on page 361.

10.11 Let Ω be a set and $\mathcal{T} = \{\,\emptyset\,\} \cup \{\, U \subset \Omega : U^c \text{ is finite}\,\}$. Show that $\mathcal{T}$ is a topology.

★10.12 Refer to Exercise 1.33 on page 21. Suppose $(\Omega, \mathcal{T})$ is a topological space and $\equiv$ is an equivalence relation on Ω. Let $\Omega/\equiv$ denote the corresponding set of equivalence classes and, for each $x \in \Omega$, let $\langle x \rangle$ denote the equivalence class containing x. For any subset W of $\Omega/\equiv$, let $\widehat{W} = \bigcup_{\langle x \rangle \in W} \langle x \rangle$.
  a) Show that $\mathcal{T}_{\equiv} = \{\, W : \widehat{W} \in \mathcal{T}\,\}$ is a topology on $\Omega/\equiv$. The topology $\mathcal{T}_{\equiv}$ is often called the **quotient topology** determined by $\equiv$.
  b) Show that the function $p \colon \Omega \to \Omega/\equiv$ defined by $p(x) = \langle x \rangle$ is continuous with respect to $\mathcal{T}$ and $\mathcal{T}_{\equiv}$.

10.13 Prove that "homeomorphic to" is an equivalence relation on the collection of all topological spaces.

10.14 This exercise asks you to show that any two nonempty open intervals of $\mathcal{R}$ are homeomorphic.
  a) Prove that $(0, 1)$ and $\mathcal{R}$ are homeomorphic.
  b) Prove that any two nonempty open intervals are homeomorphic.

10.15 Show that if h is a homeomorphism from (a, b) onto (c, d), then h is either strictly increasing or strictly decreasing.

10.16 Show that no two of the intervals $(0, 1)$, $[0, 1)$, and $[0, 1]$ are homeomorphic.

★10.17 Let $\mathfrak{C}$ be a collection of topologies on a set Ω.
  a) Show that $\bigcap_{\mathcal{T} \in \mathfrak{C}} \mathcal{T}$ is also a topology on Ω.
  b) Does the result in part (a) hold if intersection is replaced by union?

10.2 METRICS AND NORMS

In the preceding section, we developed the notion of a neighborhood basis as a way of expressing the concept of "nearness." An alternative approach to the idea of nearness is through a generalized concept of distance.

In the case of the real line $\mathcal{R}$, we usually think of the distance $d(x, y)$ between the numbers x and y as being given by $d(x, y) = |x - y|$. Proofs of many of the fundamental theorems of analysis on $\mathcal{R}$ make use of three crucial properties of this distance function; namely, for all $x, y, z \in \mathcal{R}$,

(D1) $d(x, y) \geq 0$, with equality if and only if $x = y$.
(D2) $d(x, y) = d(y, x)$.
(D3) $d(x, z) \leq d(x, y) + d(y, z)$.

Properties (D1)–(D3) are the model for the general notion of a distance function or *metric*, which we will introduce in a moment. Of course, (D1)–(D3) are derived from the properties of the absolute value function (see Exercise 2.2 on page 34). The absolute value function is also the model for another idea that we will introduce later in this section, namely, that of a *norm*.

DEFINITION 10.7   **Metric, Metric Space**

Let Ω be a set. A function $\rho \colon \Omega \times \Omega \to \mathcal{R}$ is said to be a **metric** on Ω if it satisfies the following conditions for all $x, y, z \in \Omega$:
a) $\rho(x, y) \geq 0$, with equality if and only if $x = y$.
b) $\rho(x, y) = \rho(y, x)$.
c) $\rho(x, z) \leq \rho(x, y) + \rho(y, z)$.
If ρ is a metric on Ω, then the pair (Ω, ρ) is called a **metric space.**

Note: When it is clear which metric is defined on Ω, we will often suppress the ρ and simply write Ω for (Ω, ρ).

Normed Spaces

While the distance function on $\mathcal{R}$ given by $d(x, y) = |x - y|$ is the model for the concept of a metric, it has an algebraic aspect that is not present in Definition 10.7. We will combine algebraic and metric properties by adapting the notion of distance function to the setting of a *linear space*. To begin, we recall the definition of a linear space.

DEFINITION 10.8   **Linear Space**

A **linear space** (**vector space**) consists of a set Ω, a field $F,$[†] and two functions $+ \colon \Omega \times \Omega \to \Omega$ and $\cdot \colon F \times \Omega \to \Omega$, where we denote $+(x, y)$ by $x + y$ and $\cdot(\alpha, x)$ by αx, such that the following conditions are satisfied for all $x, y, z \in \Omega$ and $\alpha, \beta \in F$:
a) $x + y = y + x$.
b) $x + (y + z) = (x + y) + z$.
c) There exists a $0 \in \Omega$ such that $x + 0 = x$.
d) There exists $-x \in \Omega$ such that $x + (-x) = 0$.

[†] A **field** is a set along with two binary operations satisfying the field axioms (F1)–(F5) on page 30. In this book, F will always be either $\mathcal{R}$ or $\mathbb{C}$.

e) $\alpha(\beta x) = (\alpha\beta)x$.

f) $\alpha(x + y) = \alpha x + \alpha y$.

g) $(\alpha + \beta)x = \alpha x + \beta x$.

h) $1x = x$.

The field F is called the *field of scalars*, the function $+$ is called *vector addition*, and the function $\cdot$ is called *scalar multiplication*.

Note: On account of (b), sums of the form $x + y + \cdots + z$ are unambiguously defined. Also, it is conventional to write $x - y$ for $x + (-y)$.

The space $\mathcal{R}^n$ is a linear space that has $\mathcal{R}$ as its field of scalars, where $+$ and $\cdot$ are defined by

$$(x_1, x_2, \ldots, x_n) + (y_1, y_2, \ldots, y_n) = (x_1 + y_1, x_2 + y_2, \ldots, x_n + y_n)$$

and

$$\alpha(x_1, x_2, \ldots, x_n) = (\alpha x_1, \alpha x_2, \ldots, \alpha x_n).$$

Using analogous definitions, we can make $\mathbb{C}^n$ into a linear space that has $\mathbb{C}$ as its scalar field.

A nonempty subset D of Ω is called a **(linear) subspace** of Ω if (1) $x, y \in D$ implies $x + y \in D$ and (2) $\alpha \in F$, $x \in D$ imply $\alpha x \in D$. We observe that a subspace D of a linear space Ω is itself a linear space, where the operations of vector addition and scalar multiplication in D are the restrictions to D of those operations in Ω.

Often we will deal with linear spaces of real- or complex-valued functions on a set. When we do so, the operations of vector addition ($+$) and scalar multiplication ($\cdot$) will always be defined pointwise, as explained in Section 2.4 on page 54.

Similarly, we also define the following operations pointwise: multiplication of functions fg; maximum of two real-valued functions $f \vee g$; minimum of two real-valued functions $f \wedge g$; the real part of a complex-valued function $\Re f$; the imaginary part of a complex-valued function $\Im f$; the absolute value (modulus) of a complex-valued function $|f|$; and the complex conjugate of a complex-valued function $\overline{f} = \Re f - i\Im f$. Also, a function that is constantly equal to α is denoted simply by α.

Now that we have recalled the definition of a linear space, we can define a normed space as follows.

DEFINITION 10.9 Norm, Normed Space

Let Ω be a linear space that has as its scalar field F either $\mathcal{R}$ or $\mathbb{C}$. A function $\|\ \|: \Omega \to \mathcal{R}$, whose value at x is written as $\|x\|$, is said to be a **norm** on Ω if it satisfies the following conditions for all $x, y \in \Omega$ and $\alpha \in F$:

a) $\|x\| \geq 0$, with equality if and only if $x = 0$.

b) $\|\alpha x\| = |\alpha| \|x\|$.

c) $\|x + y\| \leq \|x\| + \|y\|$.

If $\|\ \|$ is a norm on Ω, then the pair $(\Omega, \|\ \|)$ is called a **normed space**.

Note: When it is clear from context which norm is being considered, the normed space $(\Omega, \| \ \|)$ will be indicated simply by Ω.

It is easy to check that if $(\Omega, \| \ \|)$ is a normed space, then

$$\rho(x, y) = \|x - y\|$$

defines a metric on Ω. We will call this the **metric induced by the norm** $\| \ \|$. Hence, any normed space can also be viewed as a metric space; indeed, the first examples of metric spaces that we consider arise from norms. However, as we will see, there is still a need for the more general theory of metric spaces to handle, among other cases, those metric spaces where there is no underlying linear-space structure.

EXAMPLE 10.4 Euclidean n-Space Equipped with Various Norms

The space $\mathcal{R}^n$ of n-tuples of real numbers is a linear space with respect to the operations of vector addition and multiplication by real scalars given on page 365. Here are three, naturally arising, norms defined on $\mathcal{R}^n$:

$$\|x\|_2 = (x_1^2 + x_2^2 + \cdots + x_n^2)^{1/2},$$
$$\|x\|_1 = |x_1| + |x_2| + \cdots + |x_n|,$$
$$\|x\|_\infty = \max\{|x_1|, |x_2|, \ldots, |x_n|\},$$

where $x = (x_1, x_2, \ldots, x_n)$. □

EXAMPLE 10.5 Unitary n-Space Equipped with Various Norms

The set of complex numbers $\mathbb{C} = \{ x + iy : x, y \in \mathcal{R} \}$ with the usual absolute value function (modulus) defined by $|x + iy| = (x^2 + y^2)^{1/2}$ is a normed space, where the scalar field is also $\mathbb{C}$.

The space $\mathbb{C}^n$ of n-tuples of complex numbers is a linear space with respect to the operations of vector addition and multiplication by complex scalars given on page 365. We will abuse notation slightly by also using $\| \ \|_2$, $\| \ \|_1$, and $\| \ \|_\infty$ to denote the norms defined, respectively, on $\mathbb{C}^n$ via:

$$\|z\|_2 = (|z_1|^2 + |z_2|^2 + \cdots + |z_n|^2)^{1/2},$$
$$\|z\|_1 = |z_1| + |z_2| + \cdots + |z_n|,$$
$$\|z\|_\infty = \max\{|z_1|, |z_2|, \ldots, |z_n|\},$$

where $z = (z_1, z_2, \ldots, z_n)$. □

EXAMPLE 10.6 Spaces of Measurable Functions

We will present three normed spaces of functions that are generalizations of those given in Example 10.5. Let $(\Omega, \mathcal{A}, \mu)$ be a measure space.

a) Recall from Section 5.4 that the set $\mathcal{L}^1(\mu)$ consists of all complex-valued $\mathcal{A}$-measurable functions satisfying $\int_\Omega |f| \, d\mu < \infty$. Parts (a) and (b) of Theorem 5.8 on page 168 show that $\mathcal{L}^1(\mu)$ is a linear space with scalar field $\mathbb{C}$. Furthermore, if we identify two functions in $\mathcal{L}^1(\mu)$ whenever they are equal μ-ae and define

$$\|f\|_1 = \int_\Omega |f| \, d\mu,$$

then $\| \ \|_1$ is a norm on $\mathcal{L}^1(\mu)$, called the $\mathcal{L}^1$-**norm.**

b) A somewhat more difficult task is to show that if we again identify two functions that are equal μ-ae, then

$$\|f\|_2 = \left(\int_\Omega |f|^2 \, d\mu \right)^{1/2}$$

defines a norm on the linear space $\mathcal{L}^2(\mu)$ consisting of all complex-valued $\mathcal{A}$-measurable functions such that $\int_\Omega |f|^2 \, d\mu < \infty$. The norm $\| \ \|_2$ is called the **$\mathcal{L}^2$-norm**.

c) Another important space of measurable functions is $\mathcal{L}^\infty(\mu)$. This space consists of all complex-valued $\mathcal{A}$-measurable functions that satisfy the following condition: There is a real number M such that $|f| \leq M$ μ-ae. Such functions are said to be **essentially bounded**. If we again identify two functions that agree μ-ae, then

$$\|f\|_\infty = \inf\{ M : |f| \leq M \ \mu\text{-ae} \}$$

defines a norm on the linear space $\mathcal{L}^\infty(\mu)$, called the **$\mathcal{L}^\infty$-norm** or **essential-supremum norm**.

d) In case μ is n-dimensional Lebesgue measure restricted to a measurable subset Ω of $\mathcal{R}^n$, we denote the spaces $\mathcal{L}^1(\mu)$, $\mathcal{L}^2(\mu)$, and $\mathcal{L}^\infty(\mu)$ by $\mathcal{L}^1(\Omega)$, $\mathcal{L}^2(\Omega)$, and $\mathcal{L}^\infty(\Omega)$, respectively. And when μ is counting measure on some set Ω, we denote the spaces $\mathcal{L}^1(\mu)$, $\mathcal{L}^2(\mu)$, and $\mathcal{L}^\infty(\mu)$ by $\ell^1(\Omega)$, $\ell^2(\Omega)$, and $\ell^\infty(\Omega)$, respectively. ☐

EXAMPLE 10.7 *Metric Spaces That Are Not Normed Spaces*

To see how metric spaces that are not normed spaces can arise naturally, we first consider a simple way of constructing new metric spaces from existing ones. Suppose that (Ω, ρ) is a metric and that D is a subset of Ω. Then we can define a metric ρ_D on D by restricting the function ρ to $D \times D$. When there is no danger of confusion, we will denote the metric space (D, ρ_D) by (D, ρ).

Now, suppose that D is a subset, but not a linear subspace, of a normed space and let ρ be the metric induced by the norm. Then (D, ρ) is a metric space that is not a normed space. ☐

Metric Spaces as Topological Spaces

We now show how metrics can be used to define topologies. Let (Ω, ρ) be a metric space. For $x \in \Omega$ and $r > 0$, let $B_r^\rho(x) = \{ y \in \Omega : \rho(x, y) < r \}$. We call $B_r^\rho(x)$ the **open ball of radius r centered at x**. When the metric with which we are dealing is given unambiguously, we write $B_r(x)$ for $B_r^\rho(x)$.

When $\Omega = \mathcal{R}$ and $\rho = d$ (the metric induced by absolute value), we have that $B_r(x) = (x - r, x + r)$. The next proposition shows that, just as the collection of open intervals $\{ (x - r, x + r) : x \in \mathcal{R}, \ r > 0 \}$ is a neighborhood basis on $\mathcal{R}$, the collection of open balls of a metric space Ω is a neighborhood basis on Ω. We leave the proof of the proposition as an exercise for the reader.

☐ ☐ ☐ **PROPOSITION 10.4**

Let (Ω, ρ) be a metric space. Then the collection $\{ B_r(x) : x \in \Omega, \ r > 0 \}$ of open balls of Ω is a neighborhood basis on Ω.

The neighborhood basis $\{\, B_r(x) : x \in \Omega, \ r > 0 \,\}$ determines a topology on Ω, denoted by $\mathcal{T}_\rho$, which we call the **topology induced by the metric** ρ. If the metric ρ is itself induced by a norm $\| \, \|$, we also say that $\mathcal{T}_\rho$ is the **topology induced by the norm** $\| \, \|$.

When we have a metric space (or a normed space), we assume, unless stated otherwise, that it has the topology induced by the metric (or norm). Thus, for example, suppose that Ω and Λ are each either a metric, normed, or topological space and that $f : \Omega \to \Lambda$. Then, when we say that f is *continuous* on Ω, we mean, unless stated otherwise, that it is continuous with respect to the induced topologies on Ω and Λ.

A topological space $(\Omega, \mathcal{T})$ is said to be **metrizable** if there is a metric ρ on Ω such that $\mathcal{T}_\rho = \mathcal{T}$. Later we will address the problem of determining when a topological space is metrizable. We will see that even in cases where a topological space is metrizable, the metric may not be defined by a simple usable formula.

When two metrics on the same set or two norms on the same linear space induce the same topology, we say that they are **equivalent.**

EXAMPLE 10.8 *Nonequivalent Norms*

Consider the space $C([a, b])$ of continuous complex-valued functions on the closed interval $[a, b]$.[†] $C([a, b])$ is a linear subspace of each of the spaces $\mathcal{L}^1([a, b])$, $\mathcal{L}^2([a, b])$, and $\mathcal{L}^\infty([a, b])$; hence, it can be given any of the norms $\| \, \|_1$, $\| \, \|_2$, and $\| \, \|_\infty$ defined in Example 10.6 on pages 366–367. It is left to the reader to show that no two of these norms on $C([a, b])$ are equivalent. □

The following proposition and its corollary provide useful equivalent conditions for two metrics or norms to be equivalent. We leave the proof of the corollary to the reader as an exercise.

□ □ □ **PROPOSITION 10.5**

Let ρ and σ be metrics defined on a set Ω. Then ρ and σ are equivalent if and only if the following condition is satisfied: for each $x \in \Omega$ and $\epsilon > 0$, there are positive numbers r and s such that $B_s^\sigma(x) \subset B_\epsilon^\rho(x)$ and $B_r^\rho(x) \subset B_\epsilon^\sigma(x)$.

PROOF Suppose that the condition specified in the statement of this proposition is satisfied. Let O be open with respect to the metric ρ and let $x \in O$. Then there is an $\epsilon > 0$ such that $B_\epsilon^\rho(x) \subset O$. So, by assumption, there is an $s > 0$ such that $B_s^\sigma(x) \subset B_\epsilon^\rho(x) \subset O$. Hence, O is also open with respect to σ. A similar argument shows that a set that is open with respect to σ is also open with respect to ρ.

Conversely, suppose that ρ and σ are equivalent. Let $x \in \Omega$ and $\epsilon > 0$. Then, since $B_\epsilon^\rho(x)$ is an open set containing x in the topology induced by ρ, it is also an open set containing x in the topology induced by σ. Thus, there is an $s > 0$ such that $B_s^\sigma(x) \subset B_\epsilon^\rho(x)$. A similar argument shows that there is an $r > 0$ such that $B_r^\rho(x) \subset B_\epsilon^\sigma(x)$. ■

[†] In the terminology of Definition 2.11 on page 54, $C([a, b])$ denotes the collection of continuous *real-valued* functions on $[a, b]$. But, as we said in a footnote to that definition, the notation introduced there was temporary.

□ □ □ **COROLLARY 10.1**

Let ‖ ‖ *and* ⫴ ⫴ *be norms on a linear space* Ω. *Then* ‖ ‖ *and* ⫴ ⫴ *are equivalent if and only if there are positive constants A and B such that*

$$A\|x\| \leq \|\|x\|\| \leq B\|x\|$$

for all $x \in \Omega$.

Exercise 10.25 shows that the three norms on $\mathcal{R}^n$ defined in Example 10.4 on page 366 are equivalent, that is, they induce the same topology $\mathcal{T}$. Unless otherwise stated, we assume that each subset D of $\mathcal{R}^n$ has the relative topology $\mathcal{T}_D$. Similar comments hold for $\mathbf{C}^n$.

We conclude this section with a construction showing that every metric is equivalent to a bounded metric.

□ □ □ **PROPOSITION 10.6**

Let (Ω, ρ) be a metric space. Then there is a bounded metric σ on Ω such that ρ and σ are equivalent.

PROOF It can be shown (see Exercise 10.30) that the function σ defined on $\Omega \times \Omega$ by

$$\sigma(x, y) = \frac{\rho(x, y)}{1 + \rho(x, y)}$$

is a metric. Clearly, $\sigma(x, y) \leq 1$ for all $x, y \in \Omega$ and, so, σ is bounded.

Now, because $\sigma(x, y) \leq \rho(x, y)$, it follows that for each $x \in \Omega$ and $\epsilon > 0$, we have $B_\epsilon^\rho(x) \subset B_\epsilon^\sigma(x)$. On the other hand, choosing $s = \epsilon/(1 + \epsilon)$ and using the fact that $\rho(x, y) = \sigma(x, y)/(1 - \sigma(x, y))$, we find that $B_s^\sigma(x) \subset B_\epsilon^\rho(x)$. Thus, the condition of Proposition 10.5 is satisfied by ρ and σ. ■

Exercises for Section 10.2

10.18 Let (Ω, ρ) be a metric space. Prove each of the following facts.
  a) For $x, y, z \in \Omega$, $|\rho(x, y) - \rho(z, y)| \leq \rho(x, z)$.
  b) For $x_1, x_2, \ldots, x_n \in \Omega$, $\rho(x_1, x_n) \leq \rho(x_1, x_2) + \rho(x_2, x_3) + \cdots + \rho(x_{n-1}, x_n)$.

10.19 Refer to Example 10.4 on page 366. Verify that each of ‖ ‖$_1$, ‖ ‖$_2$, and ‖ ‖$_\infty$ are norms.

10.20 Refer to Example 10.5 on page 366. Verify that each of ‖ ‖$_1$, ‖ ‖$_2$, and ‖ ‖$_\infty$ are norms.

10.21 Refer to Example 10.6 on pages 366–367. Verify that each of ‖ ‖$_1$, ‖ ‖$_2$, and ‖ ‖$_\infty$ are norms.

10.22 For each $x \in \mathcal{R}$, let $\langle x \rangle = |x|^{1/2}$.
  a) Show that $\langle \ \rangle$ satisfies conditions (a) and (c) of Definition 10.9 but not condition (b).
  b) Show that, nevertheless, $\rho(x, y) = \langle x - y \rangle$ defines a metric on $\mathcal{R}$ that is equivalent to the metric induced by the absolute value function.

10.23 Prove Proposition 10.4 on page 367.

10.24 Prove Corollary 10.1.

★**10.25** Prove that the three norms defined in Example 10.4 on page 366 are all equivalent.

10.26 Prove that no two of the norms in Example 10.8 are equivalent.

10.27 Let ρ and σ be metrics on Ω. Show that each of the following are also metrics on Ω.
  a) $\rho_1 = \rho + \sigma$.
  b) $\rho_2 = (\rho^2 + \sigma^2)^{1/2}$.
  c) $\rho_\infty = \max\{\rho, \sigma\}$.

10.28 Refer to Exercise 10.27. Show that any two of the three metrics ρ_1, ρ_2, and ρ_∞ are equivalent.

10.29 Refer to Example 10.2(d) on page 361. Let $\mathcal{T}$ be the discrete topology on a set Ω. Show that $(\Omega, \mathcal{T})$ is metrizable.

★10.30 Refer to the definition of a metric (Definition 10.7 on page 364).
a) Show that if ρ satisfies condition (c), then so does $\sigma = \rho/(1+\rho)$.
b) Deduce that the function σ in Proposition 10.6 on page 369 is a metric.

10.31 Provide an example of a topological space that is not metrizable.

★10.32 Suppose that (Ω, ρ) and (Λ, σ) are metric spaces and let $f: \Omega \to \Lambda$. Show that f is continuous on Ω if and only if for each $a \in \Omega$ and $\epsilon > 0$, there is a $\delta > 0$ such that $\rho(x, a) < \delta$ implies $\sigma(f(x), f(a)) < \epsilon$.

10.3 WEAK TOPOLOGIES

While metric spaces are ubiquitous in analysis, there are natural ways in which nonmetrizable spaces enter the subject. For example, we will see later that nonmetrizable spaces often arise in the context of weak topologies determined by families of functions. It is the concept of weak topology that we introduce in this section.

If $\mathcal{T}$ and $\mathcal{U}$ are two topologies on a set Ω and $\mathcal{T} \subset \mathcal{U}$, then we say that $\mathcal{T}$ is **weaker** than $\mathcal{U}$. If $\mathcal{T}$ is weaker than but not equal to $\mathcal{U}$, then $\mathcal{T}$ is said to be **strictly weaker** than $\mathcal{U}$.

Let Ω be a nonempty set. Consider a family of functions $\mathcal{F}$ such that for each $f \in \mathcal{F}$, $(\Lambda_f, \mathcal{T}_f)$ is a topological space and $f: \Omega \to \Lambda_f$. Can we find a topology $\mathcal{T}$ on Ω such that f is continuous with respect to $\mathcal{T}$ and $\mathcal{T}_f$ for each $f \in \mathcal{F}$? The answer to this question is "yes" because the discrete topology (Example 10.2(d) on page 361) on Ω will always do the trick.

However, the discrete topology is of little interest because, with respect to it, any function from Ω into a topological space is continuous. Therefore, it is better to ask the following question: Of all the topologies on Ω with respect to which each $f \in \mathcal{F}$ is continuous, is there a weakest one? The answer to this question is based on the observation that if $\mathfrak{T}$ is a nonempty collection of topologies on the set Ω, then the intersection, $\bigcap_{\mathcal{T} \in \mathfrak{T}} \mathcal{T}$ is also a topology on Ω. (See Exercise 10.17(a) on page 363.)

DEFINITION 10.10 **Weak Topology**

Let Ω be a nonempty set. Consider a family of functions $\mathcal{F}$ such that for each $f \in \mathcal{F}$, $(\Lambda_f, \mathcal{T}_f)$ is a topological space and $f: \Omega \to \Lambda_f$. Let $\mathfrak{T}$ denote the collection of topologies on Ω with respect to which all functions in $\mathcal{F}$ are continuous. Then the topology

$$\mathcal{T}_{\mathcal{F}} = \bigcap_{\mathcal{T} \in \mathfrak{T}} \mathcal{T}$$

is called the **weak topology** determined by the family $\mathcal{F}$.

We leave it to the reader as an exercise to prove that $\mathcal{T}_{\mathcal{F}}$ is the weakest topology on Ω for which all $f \in \mathcal{F}$ are continuous. (See Exercise 10.34.)

Usually, when there is no possibility of confusion, functions that are continuous with respect to $\mathcal{T}_{\mathcal{F}}$ are called **weakly continuous.** Furthermore, $\mathcal{T}_{\mathcal{F}}$-open sets are called **weakly open.**

The following proposition provides a useful alternative way of looking at the topology $\mathcal{T}_{\mathcal{F}}$.

□ □ □ **PROPOSITION 10.7**

Let Ω be a nonempty set. Consider a family of functions $\mathcal{F}$ such that for each $f \in \mathcal{F}$, $(\Lambda_f, \mathcal{T}_f)$ is a topological space and $f : \Omega \to \Lambda_f$. Suppose that the topology $\mathcal{T}_f$ is determined by a neighborhood basis $\mathfrak{N}_f$. Then sets of the form

$$\bigcap_{f \in \mathcal{D}} f^{-1}(W_f), \tag{10.4}$$

where $\mathcal{D}$ is a finite subset of $\mathcal{F}$ and, for each $f \in \mathcal{D}$, $W_f \in \mathfrak{N}_f$, form a neighborhood basis that induces $\mathcal{T}_{\mathcal{F}}$.

PROOF We first note that the collection of sets of the form (10.4) is a neighborhood basis on Ω. We need to show that the topology $\mathcal{T}$ determined by that neighborhood basis is $\mathcal{T}_{\mathcal{F}}$.

Let $f \in \mathcal{F}$ and $O \in \mathcal{T}_f$. Then

$$f^{-1}(O) = \bigcup_{W \subset O,\; W \in \mathfrak{N}_f} f^{-1}(W).$$

Because each $f^{-1}(W)$ belongs to $\mathcal{T}$, it follows that $f^{-1}(O) \in \mathcal{T}$. Thus, every function in $\mathcal{F}$ is continuous with respect to $\mathcal{T}$. It follows that $\mathcal{T}_{\mathcal{F}}$ is weaker than $\mathcal{T}$.

On the other hand, each set of the form (10.4) is weakly open, because it is the intersection of finitely many weakly open sets. Consequently, $\mathcal{T}$ is weaker than $\mathcal{T}_{\mathcal{F}}$. ∎

EXAMPLE 10.9 *Compares Weak and Metric Topologies*

The space $C([a,b])$ of continuous complex-valued functions on $[a,b]$ is a linear subspace of $\mathcal{L}^{\infty}([a,b])$. Thus, $C([a,b])$ is a normed space with norm $\|\ \|_{\infty}$. Let $\mathcal{T}_{\infty}$ denote the topology induced by this norm.

For each $x \in [a,b]$, the complex-valued function e_x on $C([a,b])$ defined by $e_x(f) = f(x)$ satisfies the inequality $|e_x(f) - e_x(g)| \leq \|f - g\|_{\infty}$. From this inequality, it is easy to show that each function e_x is continuous with respect to the topology $\mathcal{T}_{\infty}$. It follows that the weak topology $\mathcal{T}_{\mathcal{F}}$ determined by the family $\mathcal{F} = \{\, e_x : x \in [a,b] \,\}$ is weaker than $\mathcal{T}_{\infty}$.

Is it possible that $\mathcal{T}_{\mathcal{F}} = \mathcal{T}_{\infty}$? Suppose the answer is yes. Then, in particular, the open ball $B_1(0)$ is weakly open. Applying Proposition 10.7, we see that there exist $x_1, x_2, \ldots, x_n \in [a,b]$, $w_1, w_2, \ldots, w_n \in \mathbb{C}$, and positive numbers $\delta_1, \delta_2, \ldots, \delta_n$ such that

$$\{\, f : |f(x_j) - w_j| < \delta_j,\; j = 1,\, 2,\, \ldots,\, n \,\} \subset B_1(0).$$

However, it is easy to construct a function $g \in C([a,b])$ such that $g(x_j) = w_j$ for $j = 1, 2, \ldots, n$ and $g(c) = 2$ for some $c \in [a,b] \setminus \{x_1, x_2, \ldots, x_n\}$. Clearly, g is an element of the set on the left of the previous display but cannot belong to $B_1(0)$ since $\|g\|_\infty \geq 2$. Hence, we have a contradiction. Thus, $\mathcal{T}_\mathcal{F} \neq \mathcal{T}_\infty$ and we conclude that $\mathcal{T}_\mathcal{F}$ is strictly weaker than $\mathcal{T}_\infty$. □

Product Topologies

Suppose that $\{(\Omega_\iota, \mathcal{T}_\iota)\}_{\iota \in I}$ is an indexed family of topological spaces. The idea of a weak topology can be used to define a topology on the Cartesian product $\Omega = \times_{\iota \in I} \Omega_\iota$. We recall from Definition 1.11 on page 16 that each element $f \in \Omega$ is a function on I such that $f(\iota) \in \Omega_\iota$ for each $\iota \in I$. Furthermore, we know from the axiom of choice that $\Omega \neq \emptyset$ provided $\Omega_\iota \neq \emptyset$ for all $\iota \in I$.

DEFINITION 10.11   **Product Topology**

Let $\{(\Omega_\iota, \mathcal{T}_\iota)\}_{\iota \in I}$ be an indexed family of topological spaces and set $\Omega = \times_{\iota \in I} \Omega_\iota$. The function p_ι defined by $p_\iota(f) = f(\iota)$ is called the ιth **coordinate projection** on Ω. The weak topology on Ω determined by the family of coordinate projections $\{p_\iota : \iota \in I\}$ is called the **product topology**. Thus, the product topology is the weakest topology for which all coordinate projections are continuous.

Exercises for Section 10.3

10.33 Let $\mathfrak{T}$ be a collection of topologies on a set Ω. Show that if a function f is continuous with respect to every member of $\mathfrak{T}$, then it is continuous with respect to the intersection of $\mathfrak{T}$, that is, with respect to $\bigcap_{\mathcal{T} \in \mathfrak{T}} \mathcal{T}$.

10.34 Refer to Definition 10.10 on page 370. Prove that $\mathcal{T}_\mathcal{F}$ is the weakest topology on Ω for which all $f \in \mathcal{F}$ are continuous. That is, prove the following:
a) Each $f \in \mathcal{F}$ is continuous with respect to $\mathcal{T}_\mathcal{F}$.
b) If $\mathcal{U}$ is a topology on Ω such that each $f \in \mathcal{F}$ is continuous with respect to $\mathcal{U}$, then $\mathcal{T}_\mathcal{F}$ is weaker than $\mathcal{U}$.

10.35 Show that the function $L: C([a,b]) \to \mathbb{C}$ defined by $L(f) = \int_a^b f(x)\,dx$ is not continuous with respect to the weak topology defined in Example 10.9 on page 371.

10.36 Refer to Example 10.9 on page 371. For $A \subset [a,b]$, let $\mathcal{A} = \{e_x : x \in A\}$. Show that if A is a proper subset of B, then $\mathcal{T}_\mathcal{A}$ is strictly weaker than $\mathcal{T}_\mathcal{B}$.

10.37 In Example 10.9 on page 371, show that every $\mathcal{T}_\mathcal{F}$-open set that contains the constant function 0 must also contain a nonzero linear subspace of $C([a,b])$.

10.38 Let Ω and Λ be linear spaces that have the same scalar field F. A function $L: \Omega \to \Lambda$ is called a **linear mapping** or **linear operator** if for all $x, y \in \Omega$ and all scalars $\alpha \in F$,

$$L(x+y) = L(x) + L(y)$$

and

$$L(\alpha x) = \alpha L(x).$$

Suppose that $L: C([a,b]) \to \mathbb{C}$ is linear and continuous with respect to the weak topology $\mathcal{T}_\mathcal{F}$ defined in Example 10.9 on page 371. Show that there are finitely many points $x_1, x_2, \ldots, x_n \in [a, b]$ and constants $c_1, c_2, \ldots, c_n \in \mathbb{C}$ such that

$$L = c_1 e_{x_1} + c_2 e_{x_2} + \cdots + c_n e_{x_n},$$

where $e_x(f) = f(x)$. *Hint:* Find a finite set of points $\{x_1, \ldots, x_n\} \subset [a, b]$ such that if $g(x_j) = 0$ for $1 \leq j \leq n$, then $|L(g)| \leq 1$.

10.39 Refer to Exercise 10.12 on page 363. Show that if p is continuous with respect to some topology $\mathcal{U}$ on $\Omega/\equiv$, then $\mathcal{U}$ is weaker than $\mathcal{T}_\equiv$.

10.40 Show that the product topology on $\mathbb{C}^n$ is the same as the topology defined by any of the norms in Example 10.5 on page 366.

10.41 The space $\ell^2(\mathcal{N})$ is a subset of the Cartesian product $\mathbb{C}^\mathcal{N}$. Thus, $\ell^2(\mathcal{N})$ can be given the relative product topology $\mathcal{T}$. Show that $\mathcal{T}$ is strictly weaker than the topology induced by the norm $\| \ \|_2$.

10.42 Do Exercise 10.41 with $\ell^1(\mathcal{N})$ replacing $\ell^2(\mathcal{N})$ and $\| \ \|_1$ replacing $\| \ \|_2$.

10.43 Do Exercise 10.41 with $\ell^\infty(\mathcal{N})$ replacing $\ell^2(\mathcal{N})$ and $\| \ \|_\infty$ replacing $\| \ \|_2$.

10.44 Consider the Cantor set P as defined on page 62. Recall that each $x \in P$ has a unique ternary expansion of the form $x = \sum_{n=1}^\infty a_n(x) 3^{-n}$, where $a_n(x) \in \{0, 2\}$ for all $n \in \mathcal{N}$. Define $A: P \to \{0, 2\}^\mathcal{N}$ by $A(x)_n = a_n(x)$. Suppose that $\{0, 2\}$ is given the discrete topology and $\{0, 2\}^\mathcal{N}$ is given the corresponding product topology. Prove that A is continuous.

10.4 CLOSED SETS, CONVERGENCE, AND COMPLETENESS

In this section, we discuss closed sets and convergence in topological and metric spaces and some related topics as well. We assume throughout that $(\Omega, \mathcal{T})$ is a topological space.

DEFINITION 10.12 **Closed Set**

A subset F of a topological space is said to be **closed** if F^c is open.

Note: From Proposition 2.14 on page 50, we see that the closed subsets of $\mathcal{R}$, as given by Definition 2.9 on page 50, are also closed in the sense of Definition 10.12, and vice versa.

It follows immediately from Definition 10.12 and the definition of a topology (Definition 10.2 on page 360) that the collection $\mathcal{C}$ of closed sets satisfies the following conditions:

(C1) $\emptyset, \Omega \in \mathcal{C}$.

(C2) $\mathcal{E} \subset \mathcal{C}$ implies $\bigcap_{F \in \mathcal{E}} F \in \mathcal{C}$.

(C3) $F_1, F_2 \in \mathcal{C}$ implies $F_1 \cup F_2 \in \mathcal{C}$.

Conversely, if $\mathcal{C}$ is a collection of subsets of Ω that satisfies conditions (C1)–(C3), then $\{ F^c : F \in \mathcal{C} \}$ is a topology on Ω for which $\mathcal{C}$ is the collection of closed sets.

A simple example of a closed subset of $\mathcal{R}$ is $[a, b]$, where $a \leq b$, because $[a, b]^c = (-\infty, a) \cup (b, \infty)$. On the other hand, an interval of the form $[a, b)$ is not closed.

Limit Points, Closure, and Convergent Sequences

Next we define the limit points and closure of a set.

DEFINITION 10.13 **Limit Point, Closure**

Let E be a subset of a topological space Ω. A point $x \in \Omega$ is called a **limit point** of E if each open set that contains x intersects E; that is, if O is open and $x \in O$, then $O \cap E \neq \emptyset$. The set of all limit points of E, denoted $\overline{E}$, is called the **closure** of E.

Note: If $\overline{E} = \Omega$ (i.e., every point of Ω is a limit point of E), then we say that E is **dense** in Ω. Thus, we see that E is dense in Ω if and only if it has a nonempty intersection with every nonempty open set.

We leave it to the reader as an exercise to show that the following properties hold:

- $\overline{E}$ is the intersection of all closed sets that contain E and, hence, $\overline{E}$ is the smallest closed set that contains E.

- Let $\mathfrak{N}$ be a neighborhood basis that determines the topology. Then $x \in \overline{E}$ if and only if

$$x \in W \in \mathfrak{N} \text{ implies } W \cap E \neq \emptyset. \tag{10.5}$$

Condition (10.5) suggests that we can interpret $\overline{E}$ as the set of points of Ω that can be "approximated arbitrarily closely" by points of E. Thus, in the case of the real line $\mathcal{R}$, the rational numbers are dense since any real number can be approximated arbitrarily closely by rational numbers.

The next proposition, whose proof is left to the reader as an exercise, provides the basic properties of the closure operation.

□ □ □ **PROPOSITION 10.8**

Let $A, B \subset \Omega$. Then
a) $\overline{A} = A$ *if and only if A is closed.*
b) $\overline{\overline{A}} = \overline{A}$.
c) $\overline{A \cup B} = \overline{A} \cup \overline{B}$.
d) $A \subset B$ *implies* $\overline{A} \subset \overline{B}$.
e) $A \subset F$ *and F closed implies* $\overline{A} \subset F$.

Remark: It follows easily from Proposition 10.8(c) that $\overline{\bigcup_{k=1}^{n} A_k} = \bigcup_{k=1}^{n} \overline{A_k}$, whenever $A_1, A_2, \ldots, A_n$ are subsets of Ω. (See Exercise 10.49.)

When a topological space is metrizable, there is a useful characterization of limit point in terms of sequences. To give that characterization, we must first define convergence of a sequence in a topological space.

DEFINITION 10.14   **Convergent Sequence in a Topological Space**

Let $(\Omega, \mathcal{T})$ be a topological space. A sequence $\{x_n\}_{n=1}^{\infty}$ of points in Ω is said to **converge** to the point $x \in \Omega$ if for each open set O that contains x, there is an integer N such that $x_n \in O$ whenever $n \geq N$. Convergence of $\{x_n\}_{n=1}^{\infty}$ to x is denoted by

$$\lim_{n \to \infty} x_n = x \qquad \text{or} \qquad x_n \to x$$

or, in case it is important to indicate the topology in which convergence occurs, by $x_n \xrightarrow{\mathcal{T}} x$.

It is not hard to see that if the topology on Ω is induced by a metric ρ, then $x_n \to x$ if and only if $\rho(x_n, x) \to 0$. In case the topology is the weak topology determined by a family of functions $\mathcal{F}$, it can be shown that $x_n \to x$ if and only if $f(x_n) \to f(x)$ for each $f \in \mathcal{F}$. (See Exercise 10.54.)

□ □ □   **PROPOSITION 10.9**

Suppose that the set Ω has the topology induced by a metric ρ. Let $E \subset \Omega$. Then $x \in \overline{E}$ if and only if there exists a sequence $\{x_n\}_{n=1}^{\infty}$ of points of E such that $\lim_{n \to \infty} x_n = x$.

PROOF   Suppose that $x \in \overline{E}$. Then, by (10.5), for each positive integer n there exists an $x_n \in B_{1/n}(x) \cap E$. Because $\rho(x_n, x) < 1/n$, it follows that $\lim_{n \to \infty} x_n = x$.

Conversely, suppose that $\{x_n\}_{n=1}^{\infty}$ is a sequence of points of E such that $\lim_{n \to \infty} x_n = x$. Then, for each $\epsilon > 0$, there is a positive integer N such that $\rho(x_n, x) < \epsilon$ whenever $n \geq N$. It follows that $B_{\epsilon}(x) \cap E \neq \emptyset$. Thus, by the condition (10.5), we have that $x \in \overline{E}$. ∎

As a simple illustration of the preceding proposition, let $\Omega = \mathcal{R}$ and $E = Q$. Since every real number is the limit of a sequence of rational numbers, it follows that $\overline{Q} = \mathcal{R}$. A more elaborate application of Proposition 10.9 is discussed in the following example.

EXAMPLE 10.10   *Illustrates a Nonmetrizable Space*

Let Ω be the Cartesian product $\{0, 1\}^{\mathcal{R}}$. Let $\{0, 1\}$ have the discrete topology and Ω the corresponding product topology. Recall that Ω is the set consisting of all functions from $\mathcal{R}$ to $\{0, 1\}$ and the product topology on Ω is the weak topology determined by the family of functions $\{p_t : t \in \mathcal{R}\}$, where $p_t(f) = f(t)$. The product topology is determined by the neighborhood basis of sets of the form

$$\{f \in \Omega : f(t_k) = a_k, \; k = 1, \, 2, \, \ldots, \, n\}, \tag{10.6}$$

where $n \in \mathcal{N}$ and $a_k \in \{0, 1\}$, $1 \leq k \leq n$. Consider the set

$$U = \{f \in \Omega : f^{-1}(\{0\}) \text{ is countable}\}.$$

We claim that U is dense in Ω. Indeed, the intersection of U with each set of the form (10.6) contains the function g defined by $g(t_k) = a_k$ for $k = 1, 2, \ldots, n$ and $g(t) = 1$ for $t \in \mathcal{R} \setminus \{t_1, t_2, \ldots, t_n\}$. In particular, U has a nonempty intersection with every set in a neighborhood basis determining the topology of Ω. Hence, U is dense in Ω.

We claim that Ω is not metrizable. Suppose to the contrary. Then, by Proposition 10.9, there is a sequence $\{f_n\}_{n=1}^{\infty} \subset U$ converging to the function on $\mathcal{R}$ that is identically 0. Thus, from Exercise 10.54, $\lim_{n\to\infty} f_n(t) = 0$ for each $t \in \mathcal{R}$. But, for each t in the complement of the countable set $\bigcup_{n=1}^{\infty} f_n^{-1}(\{0\})$, we have $\lim_{n\to\infty} f_n(t) = 1$. This contradiction shows that Ω is not metrizable. □

Completeness

There is a powerful extension of the notion of closed set, namely, the idea of a *complete set*. Before we can give a formal definition of a complete set, however, we require the following:

DEFINITION 10.15 **Cauchy Sequence**

A sequence $\{x_n\}_{n=1}^{\infty}$ in a metric space (Ω, ρ) is said to be a **Cauchy sequence** if for each $\epsilon > 0$, there is an $N \in \mathcal{N}$ such that $\rho(x_n, x_m) < \epsilon$ whenever $n, m \geq N$.

In the space $\mathcal{R}$ with the usual metric, this definition of a Cauchy sequence is exactly the same as Definition 2.6 on page 43. Cauchy sequences in $\mathcal{R}$ always converge by Theorem 2.1 on page 43. In general, however, Cauchy sequences may fail to converge. Metric spaces for which all Cauchy sequences converge are called *complete*.

DEFINITION 10.16 **Complete Metric Space, Complete Set**

A metric space (Ω, ρ) is said to be **complete** if every Cauchy sequence converges; that is, if $\{x_n\}_{n=1}^{\infty}$ is a Cauchy sequence of elements of Ω, then there exists an $x \in \Omega$ such that $\lim_{n\to\infty} x_n = x$. A subset $E \subset \Omega$ is called **complete** if (E, ρ) is a complete metric space.

The real line $\mathcal{R}$ provides an example of a complete metric space. Many other examples will be encountered in the exercises in this section and in the text and exercises of future sections.

Our next proposition, whose proof is left to the reader as an exercise, relates the concepts of closed and complete.

□ □ □ **PROPOSITION 10.10**

Let Ω be a metric space and $E \subset \Omega$. Then the following properties hold.
a) If E is complete, then it is closed.
b) If Ω is complete and E is closed, then E is complete.

The converse of Proposition 10.10(a) fails. Indeed, the interval $(0, 1]$ is closed in the relative topology of the space $(0, 2)$; however, $(0, 1]$ is not complete because the sequence $\{1/n\}_{n=1}^{\infty}$ is a Cauchy sequence in $(0, 1]$ but not convergent in $(0, 1]$.

Interior of a Set

As we have seen, $\overline{E}$ is the smallest closed set that contains E. Similarly, there is a largest open set contained in E, defined as follows.

DEFINITION 10.17   Interior of a Set

Let E be a subset of a topological space Ω. A point $x \in \Omega$ is called an **interior point** of E if there is an open set O such that $x \in O \subset E$. The set of all interior points of E, denoted $\boldsymbol{E^\circ}$, is called the **interior** of E.

Remark: We note that the interior of a set may be empty. For example, if we take $\Omega = \mathcal{R}$, then $Q^\circ = \emptyset$.

We leave it to the reader as an exercise to show that each of the following properties hold:

- E° is the union of all open sets contained in E and, hence, E° is the largest open set contained in E.

- Let $\mathfrak{N}$ be a neighborhood basis that determines the topology. Then $x \in E^\circ$ if and only if there is a $W \in \mathfrak{N}$ such that $x \in W \subset E$.

The following proposition is the analogue of Proposition 10.8 on page 374 for the interior of a set. Its proof is left to the reader as an exercise. (See Exercise 10.63.)

□ □ □   **PROPOSITION 10.11**

Let $A, B \subset \Omega$. Then
a) $A^\circ = A$ if and only if A is open.
b) $(A^\circ)^\circ = A^\circ$.
c) $(A \cap B)^\circ = A^\circ \cap B^\circ$.
d) $A \subset B$ implies $A^\circ \subset B^\circ$.
e) $U \subset A$ and U open implies $U \subset A^\circ$.

Exercises for Section 10.4

10.45   Let E be a subset of a topological space Ω. Prove the following facts.
a) $\overline{E}$ is the intersection of all closed sets that contain E and, hence, $\overline{E}$ is the smallest closed set that contains E.
b) Let $\mathfrak{N}$ be a neighborhood basis that determines the topology. Then $x \in \overline{E}$ if and only if $x \in W \in \mathfrak{N}$ implies $W \cap E \neq \emptyset$.

10.46   Prove Proposition 10.8 on page 374.

★**10.47**   Let (Ω, ρ) be a metric space. For $x \in \Omega$ and $\emptyset \neq E \subset \Omega$, define

$$\rho(x, E) = \inf\{\, \rho(x, y) : y \in E \,\}.$$

$\rho(x, E)$ is called the **distance from x to E**. Prove the following facts:
a) There is a sequence $\{x_n\}_{n=1}^\infty \subset E$ such that $\lim_{n \to \infty} \rho(x, x_n) = \rho(x, E)$.
b) $\rho(x, E) = 0$ if and only if $x \in \overline{E}$.
c) $\rho(x, E) = \rho(x, \overline{E})$.
d) $|\rho(x_1, E) - \rho(x_2, E)| \leq \rho(x_1, x_2)$.

e) The function $f:\Omega \to \mathcal{R}$ defined by $f(x) = \rho(x, E)$ is continuous.

f) Let A and B be disjoint closed nonempty subsets of Ω. Define

$$f(x) = \frac{\rho(x, A)(1 + \rho(x, B))}{\rho(x, A) + (1 + \rho(x, A))\rho(x, B)}, \qquad x \in \Omega.$$

Prove that f is continuous, $f(\Omega) \subset [0, 1]$, $f(A) = \{0\}$, and $f(B) = \{1\}$.

★10.48 Consider an open ball $B_r(x)$ in a metric space (Ω, ρ).
a) Show that $\overline{B_r(x)} \subset \{y : \rho(x, y) \le r\}$.
b) Show that equality holds in part (a) for the case of a normed space.
c) Give an example where the containment in part (a) is strict.
d) The set $\overline{B}_r(x) = \{y : \rho(x, y) \le r\}$ is called the **closed ball of radius r centered at x**. Verify that $\overline{B}_r(x)$ is a closed set.

10.49 Verify the formula $\overline{\bigcup_{k=1}^n A_k} = \bigcup_{k=1}^n \overline{A_k}$.

10.50 Suppose that Ω and Λ are topological spaces and $f:\Omega \to \Lambda$. Show that f is continuous if and only if $f^{-1}(F)$ is closed in Ω whenever F is closed in Λ.

10.51 Suppose that Ω and Λ are topological spaces and $f:\Omega \to \Lambda$. Show that f is continuous if and only if $f(\overline{A}) \subset \overline{f(A)}$ for all $A \subset \Omega$.

10.52 Suppose Ω and Λ are topological spaces and $f:\Omega \to \Lambda$. If f is continuous, does it follow that $f(E)$ is closed (open) whenever E is a closed (open) subset of Ω? Explain your answer.

★10.53 Suppose Ω and Λ are topological spaces and $f:\Omega \to \Lambda$.
a) Show that the condition "$f(x_n) \to f(x)$ whenever $x_n \to x$" is necessary for f to be continuous.
b) Show that the condition in part (a) is sufficient when Ω is metrizable. *Hint:* Refer to Exercise 10.32 on page 370.

10.54 Suppose Ω has the weak topology determined by some set $\mathcal{F}$ of functions. Let $\{x_n\}_{n=1}^{\infty}$ be a sequence of points of Ω and $x \in \Omega$. Show that $x_n \to x$ if and only if $f(x_n) \to f(x)$ for all $f \in \mathcal{F}$. *Hint:* See Exercise 10.53.

★10.55 Show that a Cauchy sequence is convergent if it has a convergent subsequence.

10.56 Show that if a sequence in a metric space in convergent, then it is Cauchy.

10.57 Give an example of a nonconvergent Cauchy sequence.

10.58 Prove Proposition 10.10 on page 376.

★10.59 Show that $\mathcal{R}^n$ is complete in each of the norms defined in Example 10.4 on page 366.

★10.60 Show that $\mathbb{C}^n$ is complete in each of the norms defined in Example 10.5 on page 366.

10.61 Let Ω be a nonempty set. Show that each of the spaces $\ell^1(\Omega)$, $\ell^2(\Omega)$, and $\ell^{\infty}(\Omega)$, described in Example 10.6 on pages 366–367, is complete.

10.62 Let E be a subset of a topological space Ω. Prove the following facts.
a) E° is the union of all open sets contained in E and, hence, E° is the largest open set contained in E.
b) Let $\mathfrak{N}$ be a neighborhood basis that determines the topology. Then $x \in E^{\circ}$ if and only if there is a $W \in \mathfrak{N}$ such that $x \in W \subset E$.

10.63 Prove Proposition 10.11 on page 377.

★10.64 Let Ω be a topological space. For $E \subset \Omega$, define $\partial E = \overline{E} \setminus E^{\circ}$. The set ∂E is called the **boundary of E**. Prove the following facts:
a) ∂E is closed.
b) E is closed if and only if $\partial E \subset E$.
c) $(\partial E)^{\circ} = \emptyset$.
d) $\partial E = \overline{E} \cap \overline{E^c}$.

10.5 NETS AND CONTINUITY

Proposition 10.9 on page 375 describes the limit points of a subset of a metric space in terms of convergent sequences. Example 10.10 on pages 375–376, on the other hand, shows that a similar characterization fails to be correct for general topological spaces.

In this section, we present a generalization of sequences that is flexible enough to permit a version of Proposition 10.9 to hold for general topological spaces and also provides an alternative method for characterizing continuity. We first introduce the concept of a *directed set*.

DEFINITION 10.18   Directed Set

A **directed set** is a nonempty set I together with a relation $\preceq$ having the following properties:

a) $\iota \preceq \iota$ for each $\iota \in I$.

b) $\iota_1 \preceq \iota_2$ and $\iota_2 \preceq \iota_3$ imply $\iota_1 \preceq \iota_3$.

c) $\iota_1, \iota_2 \in I$ implies there exists $\iota_3 \in I$ such that $\iota_1 \preceq \iota_3$ and $\iota_2 \preceq \iota_3$.

An element of a directed set is called an **index**.

Remark: It follows easily from Definition 10.18 that for each finite subset J of I, there exists a $\kappa \in I$ such that $\iota \preceq \kappa$ for each $\iota \in J$.

EXAMPLE 10.11   *Illustrates Definition 10.18*

a) A nonempty subset of real numbers with the order relation $\leq$ is a directed set. In particular, the set of integers greater than or equal to some fixed integer is a directed set.

b) Let $\mathfrak{N}$ be a neighborhood basis and $\mathfrak{N}_x = \{\, N \in \mathfrak{N} : x \in N \,\}$. For $U, V \in \mathfrak{N}_x$, say that $U \preceq V$ if $U \supset V$. Then $\mathfrak{N}_x$ is a directed set with respect to the relation $\preceq$, that is, with respect to $\supset$.

c) Let $\mathcal{P}_{\mathcal{F}}(S)$ denote the collection of finite subsets of a set S. Then $\mathcal{P}_{\mathcal{F}}(S)$ is a directed set with respect to $\subset$. □

DEFINITION 10.19   Net, Convergence of Nets

A **net** of points in a set Ω is a function x from a directed set I into Ω. The set I is called the **index set** of the net. We write x_ι for $x(\iota)$ and denote the net by $\{x_\iota\}_{\iota \in I}$.

When $(\Omega, \mathcal{T})$ is a topological space, a net $\{x_\iota\}_{\iota \in I}$ of points in Ω is said to **converge** to the point $x \in \Omega$ if for each open set O containing x, there is an index $\iota_0 \in I$ such that $x_\iota \in O$ whenever $\iota_0 \preceq \iota$. Convergence of $\{x_\iota\}_{\iota \in I}$ to x is denoted by

$$\lim x_\iota = x \qquad \text{or} \qquad x_\iota \to x$$

or, in case it is important to indicate the topology in which convergence occurs, by $x_\iota \xrightarrow{\mathcal{T}} x$.

EXAMPLE 10.12 *Illustrates Definition 10.19*

a) A sequence is a net with $I = \mathcal{N}$ and $\preceq\ =\ \leq$. A sequence in a topological space that converges to a point x is also a net converging to x.

b) A somewhat more general situation than in part (a) occurs when a sequence of the form $\{x_n\}_{n=1}^{\infty}$ is replaced by a net indexed by the set of integers $\{j \in \mathcal{Z} : j \geq k\}$ for some integer k, where the relation is the usual $\leq$ ordering. Such nets are customarily also referred to as sequences.

c) Consider a function $f: [a, \infty) \to \mathcal{R}$. Since $[a, \infty)$ is a directed set with respect to the relation $\leq$, the function f can be viewed as a net $\{f_t\}_{t \in [a, \infty)}$ in $\mathcal{R}$, where $f_t = f(t)$. Furthermore, $\lim f_t = L$ if and only if for each $\epsilon > 0$, there is a number M such that $|f(t) - L| < \epsilon$ whenever $t \geq a \vee M$.

d) Refer to Example 10.11(b). Suppose that $x_U \in U$ for each $U \in \mathfrak{N}_x$. Then the net $\{x_U\}_{U \in \mathfrak{N}_x}$ converges to x.

e) Let f be a Riemann integrable function on $[a, b]$ and $\mathcal{S}$ the collection of step functions on $[a, b]$ that are dominated by f. Then $\mathcal{S}$ is a directed set with respect to the usual $\leq$ ordering of functions. For each $h \in \mathcal{S}$, let $y_h = \int_a^b h(x)\, dx$. Then $\{y_h\}_{h \in \mathcal{S}}$ is a net of real numbers and we have that $\lim y_h = \int_a^b f(x)\, dx$. (See Section 2.6 starting on page 67.) ◻

Infinite Series and Infinite Sums

Using nets, we can now discuss infinite series and infinite sums in normed linear spaces.

DEFINITION 10.20 **Infinite Series and Sums in Normed Spaces**

Let Ω be a normed space and S an infinite subset of $\mathcal{Z}$ with the usual $\leq$ ordering. Suppose that $x_j \in \Omega$ for each $j \in S$.

a) Assume that $S = \{j \in \mathcal{Z} : j \geq k\}$ for some integer k. Then the expression

$$\sum_{j=k}^{\infty} x_j$$

is called an **infinite series**. For each $n \in S$, let $s_n = \sum_{j=k}^{n} x_j$. If the net $\{s_n\}_{n \in S}$ converges in Ω to s, then we say that the infinite series $\sum_{j=k}^{\infty} x_j$ **converges** to s and write $s = \sum_{j=k}^{\infty} x_j$. Otherwise, we say that the infinite series fails to converge.

b) In general, the expression

$$\sum_{j \in S} x_j$$

is called an **infinite sum**. Denote by $\mathcal{P}_{\mathcal{F}}(S)$ the collection of all finite subsets of S with the $\subset$ ordering. For each $F \in \mathcal{P}_{\mathcal{F}}(S)$, let $s_F = \sum_{j \in F} x_j$. If the net $\{s_F\}_{F \in \mathcal{P}_{\mathcal{F}}(S)}$ converges in Ω to s, then we say that the infinite sum $\sum_{j \in S} x_j$ **converges** to s and write $s = \sum_{j \in S} x_j$. Otherwise, we say that the infinite sum fails to converge. In the special case that S is as in part (a), we denote the infinite sum $\sum_{j \in S} x_j$ by $\sum_{j \geq k} x_j$.

Remark: If $\Omega = \mathcal{R}$ and $x_j \geq 0$ for each $j \in S$, then infinite sums are a special case of generalized sums, discussed in Exercise 2.37 on page 46.

EXAMPLE 10.13   *Illustrates Definition 10.20*

a) Let Ω be a normed space, $S = \{j \in \mathcal{Z} : j \geq k\}$ for some integer k, and $x_j \in \Omega$ for each $j \in S$. We will show that convergence of the infinite sum to s implies convergence of the infinite series to s. Suppose that $\sum_{j \geq k} x_j$ converges to s. We must show that $\sum_{j=k}^{\infty} x_j$ also converges to s. Let $\epsilon > 0$ and choose a finite subset F_0 of S such that $\|s - s_F\| < \epsilon$ whenever $F_0 \subset F$. Let $N = \max F_0$. Then for $n \geq N$, $F_0 \subset \{j : k \leq j \leq n\}$; hence, $\|s - s_n\| < \epsilon$. Thus, $\sum_{j=k}^{\infty} x_j$ converges to s.

b) Let Ω be a normed space, S an infinite subset of $\mathcal{Z}$, and $x_j \in \Omega$ for each j. It is not difficult to show that the infinite series $\sum_{j=k}^{\infty} x_j$ converges if and only if the infinite series $\sum_{j=\ell}^{\infty} x_j$ converges, where $k < \ell$. Similarly, the infinite sum $\sum_{j \in S} x_j$ converges if and only if the infinite sum $\sum_{j \in S \setminus F} x_j$ converges, where F is a finite subset of S.

c) By Exercise 10.68(a), the series $\sum_{j=1}^{\infty} 1/j$ fails to converge. In fact, we have that $\lim_{n \to \infty} \sum_{j=1}^{n} 1/j = \infty$.

d) In part (a), we showed that if $S = \{j \in \mathcal{Z} : j \geq k\}$, then convergence of an infinite sum to s implies convergence of the corresponding infinite series to s. Here we show that the converse is false. By Exercise 10.68(b), the series $\sum_{j=1}^{\infty} (-1)^j/j$ converges. However, the infinite sum $\sum_{j \geq 1} (-1)^j/j$ fails to converge. Indeed, suppose that F_0 is a finite subset of positive integers. Let $N = \max F_0$ and set $F = F_0 \cup \{2j : N \leq j \leq n\}$. Then, by part (c) of this example,

$$\sum_{j \in F} (-1)^j/j = \sum_{j \in F_0} (-1)^j/j + (1/2) \sum_{j=N}^{n} 1/j$$

can be made arbitrarily large by choosing n sufficiently large. Hence, the infinite sum $\sum_{j \geq 1} (-1)^j/j$ fails to converge.

e) In part (a), we showed that if $S = \{j \in \mathcal{Z} : j \geq k\}$, then convergence of an infinite sum to s implies convergence of the corresponding infinite series to s and, in part (d), we showed that the converse of that statement is false. We will now show that in the special case of $\Omega = \mathcal{R}$ and $x_j \geq 0$ for each $j \in S$, the converse is true. Thus, assume that $x_j \in [0, \infty)$ for each $j \in S$ and that $\sum_{j=k}^{\infty} x_j$ converges to s. Let $\epsilon > 0$ be given. Choose $N \in \mathcal{N}$ so that $n \geq N$ implies that $|s - \sum_{j=k}^{n} x_j| < \epsilon$. Let $F_0 = \{j : k \leq j \leq N\}$. Then $F \in \mathcal{P}_{\mathcal{F}}(S)$ and $F_0 \subset F$ implies

$$s - \epsilon < \sum_{j=k}^{N} x_j \leq \sum_{j \in F} x_j \leq \sum_{j=k}^{M} x_j < s + \epsilon,$$

where $M = \max F$. Therefore, $\sum_{j \geq k} x_j$ converges to s.

f) Assume the x_js belong to a normed space. If the series $\sum_{j=k}^{\infty} \|x_j\|$ converges, then $\sum_{j \geq k} x_j$ converges to s if and only if $\sum_{j=k}^{\infty} x_j$ converges to s. □

Remark: When the x_js are nonnegative real numbers, $\sum_{j=k}^{\infty} x_j$ fails to converge if and only if the terms $\sum_{j=k}^{n} x_j$ become arbitrarily large as n increases, that is, $\lim_{n\to\infty} s_n = \infty$. Consequently, in this case, we often indicate convergence of $\sum_{j=k}^{\infty} x_j$ by $\sum_{j=k}^{\infty} x_j < \infty$ and lack of convergence by $\sum_{j=k}^{\infty} x_j = \infty$.

Nets and Topological Properties

Using nets, we can generalize Proposition 10.9 on page 375 to arbitrary topological spaces.

□ □ □ **PROPOSITION 10.12**

Let E be a subset of a topological space Ω. Then $x \in \overline{E}$ if and only if there is a net $\{x_\iota\}_{\iota \in I}$ of points in E such that $\lim x_\iota = x$.

PROOF   Suppose $x \in \overline{E}$. Let $\mathcal{T}_x$ be the collection of all open sets containing x. Then $\mathcal{T}_x$ is a directed set with the relation $\supset$. For each $O \in \mathcal{T}_x$, we have $O \cap E \neq \emptyset$; using the axiom of choice, we select $x_O \in O \cap E$. Then $\{x_O\}_{O \in \mathcal{T}_x}$ is a net of points in E such that $\lim x_O = x$.

Conversely, suppose there is a net $\{x_\iota\}_{\iota \in I}$ of points in E such that $\lim x_\iota = x$. Then, for each open set O containing x, we have $x_\iota \in O$ for some index ι. Because $x_\iota \in E$, it follows that $O \cap E \neq \emptyset$. Hence, $x \in \overline{E}$.   ∎

We can also use nets to characterize continuity of functions. Before we do so, however, it will be convenient to introduce the idea of a *subnet*. For motivation, we note that a subsequence $\{x_{n_k}\}_{k=1}^{\infty}$ of a sequence $\{x_n\}_{n=1}^{\infty}$ is really the composition of the sequence (i.e., the function x on $\mathcal{N}$) with the strictly increasing function $n: \mathcal{N} \to \mathcal{N}$ defined by $n(k) = n_k$. Thus, we have the following definition.

DEFINITION 10.21   **Subnet**

Let $\{x_\iota\}_{\iota \in I}$ be a net with order relation $\preceq$. A **subnet** of $\{x_\iota\}_{\iota \in I}$ is a composition of that net (i.e., the function x on I) with a function $h: K \to I$, where K is a directed set with order relation $\lhd$ such that the following conditions are satisfied:
a) If $\kappa_1 \lhd \kappa_2$, then $h(\kappa_1) \preceq h(\kappa_2)$.
b) For each $\iota \in I$, there is a $\kappa \in K$ such that $\iota \preceq h(\kappa)$.
Usually we write ι_κ instead of $h(\kappa)$ and, hence, $\{x_{\iota_\kappa}\}_{\kappa \in K}$ instead of $\{x_{h(\kappa)}\}_{\kappa \in K}$.

Of course, a subsequence is also a subnet. Other examples of subnets are considered in the exercises. We leave it to the reader as an exercise to show that if a net converges to x, then so does every subnet of that net.

□ □ □ **THEOREM 10.1**

Let Ω and Λ be topological spaces and $f: \Omega \to \Lambda$. Then the following conditions are equivalent:
a) *For each $x \in \Omega$ and each open set V containing $f(x)$, there is an open set U containing x such that $f(U) \subset V$.*

b) f is continuous, that is, $f^{-1}(O)$ is open in Ω whenever O is open in Λ.

c) $f^{-1}(F)$ is closed in Ω whenever F is closed in Λ.

d) If $\{x_\iota\}_{\iota\in I}$ is a net converging to x, then $\{f(x_\iota)\}_{\iota\in I}$ has a subnet converging to $f(x)$.

e) If $\{x_\iota\}_{\iota\in I}$ is a net converging to x, then $\{f(x_\iota)\}_{\iota\in I}$ converges to $f(x)$.

PROOF The equivalence of (a) and (b) is shown by Proposition 10.3 on page 359 and the observation that a topology is a neighborhood basis for itself. The equivalence of (b) and (c) follows at once from the set identity $f^{-1}(F^c) = (f^{-1}(F))^c$. To complete the proof, it suffices to establish the chain of implications (a) implies (e), (e) implies (d), and (d) implies (c).

Suppose (a) holds and that $\{x_\iota\}_{\iota\in I}$ is a net in Ω such that $\lim x_\iota = x$. Let V be an open set containing $f(x)$. Then, by the continuity of f, there is an open set U containing x such that $f(U) \subset V$. Since $\lim x_\iota = x$, there is an index ι_0 such that $x_\iota \in U$ whenever $\iota_0 \preceq \iota$. It follows that $f(x_\iota) \in f(U) \subset V$ whenever $\iota_0 \preceq \iota$. Thus, $\lim f(x_\iota) = f(x)$.

Next, suppose (e) holds and that $\{x_\iota\}_{\iota\in I}$ is a net in Ω such that $\lim x_\iota = x$. Then $\lim f(x_\iota) = f(x)$. Because $\{f(x_\iota)\}_{\iota\in I}$ is a subnet of itself, it follows that (d) holds.

Finally, suppose (d) holds and that F is a closed subset of Λ. We will show that $f^{-1}(F)$ is closed by proving that $\overline{f^{-1}(F)} = f^{-1}(F)$. Let $x \in \overline{f^{-1}(F)}$. By Proposition 10.12, there is a net $\{x_\iota\}_{\iota\in I}$ in $f^{-1}(F)$ that converges to x. It follows from (d) that $\{f(x_\iota)\}_{\iota\in I}$ has a subnet $\{f(x_{\iota_\kappa})\}_{\kappa\in K}$ that converges to $f(x)$. So, by Proposition 10.12 again, $f(x) \in \overline{F} = F$ and, hence, $x \in f^{-1}(F)$. We have shown that $\overline{f^{-1}(F)} \subset f^{-1}(F)$. Because the reverse containment is trivial, we have established that $\overline{f^{-1}(F)} = f^{-1}(F)$. Consequently, by Proposition 10.8(a) on page 374, $f^{-1}(F)$ is closed. ■

Motivated by Theorem 10.1(a), we can define continuity at a point for a function from one topological space to another. Let Ω and Λ be topological spaces and $f: \Omega \to \Lambda$. We say that **f is continuous at a point $x \in \Omega$** if for each open set V that contains $f(x)$, there is an open set U that contains x such that $f(U) \subset V$. We see from Theorem 10.1 that f is continuous if and only if it is continuous at each point of Ω.

Criteria for Convergence of Nets

The two main types of topological spaces we have studied thus far are metric spaces (including normed spaces) and spaces with weak topologies. For each of these two types of topological spaces, we have a simple characterization of convergence of nets.

□ □ □ **PROPOSITION 10.13**

Let $(\Omega, \mathcal{T})$ be a topological space and let $\{x_\iota\}_{\iota\in I}$ be a net in Ω.

a) If the topology $\mathcal{T}$ is induced by a metric ρ, then $\lim x_\iota = x$ if and only if $\lim \rho(x_\iota, x) = 0$.

b) If the topology $\mathcal{T}$ is the weak topology determined by a family of functions $\mathcal{F}$, then $\lim x_\iota = x$ if and only if $\lim f(x_\iota) = f(x)$ for each $f \in \mathcal{F}$.

PROOF a) Suppose that $\lim \rho(x_\iota, x) = 0$. If O is an open set that contains x, then there is an $\epsilon > 0$ such that $B_\epsilon(x) \subset O$. Moreover, there is an index ι_0 such that $\rho(x_\iota, x) < \epsilon$ for $\iota_0 \preceq \iota$. Thus, $x \in O$ for $\iota_0 \preceq \iota$. It follows that $\lim x_\iota = x$. Conversely, suppose that $\lim x_\iota = x$. Then for each $\epsilon > 0$, there is an index ι_0 such that $x_\iota \in B_\epsilon(x)$ for $\iota_0 \preceq \iota$. Hence, $\rho(x_\iota, x) < \epsilon$ whenever $\iota_0 \preceq \iota$. Thus, we have that $\lim \rho(x_\iota, x) = 0$.

b) Suppose that $\lim x_\iota = x$. Then, by Theorem 10.1(e), $\lim f(x_\iota) = f(x)$ for each $f \in \mathcal{F}$. Conversely, suppose that $\lim f(x_\iota) = f(x)$ for each $f \in \mathcal{F}$. Let O be an open set containing x. Then, by Proposition 10.7 on page 371, there exist $n \in \mathcal{N}$, $\{f_j\}_{j=1}^n \subset \mathcal{F}$, and $U_j \in \mathcal{T}_{f_j}$, $1 \leq j \leq n$, such that

$$x \in \bigcap_{j=1}^n f_j^{-1}(U_j) \subset O.$$

Now, for each $j = 1, 2, \ldots, n$, since $f_j(x) \in U_j$ and $\lim f_j(x_\iota) = f_j(x)$, there exists an index ι_j such that $f_j(x_\iota) \in U_j$ whenever $\iota_j \preceq \iota$. Because I is a directed set, there is an index ι_0 such that $\iota_j \preceq \iota_0$ for each j. Therefore, we have

$$x_\iota \in \bigcap_{j=1}^n f_j^{-1}(U_j) \subset O,$$

whenever $\iota_0 \preceq \iota$. Thus, $\lim x_\iota = x$. ∎

Exercises for Section 10.5

10.65 Verify the assertions made in Example 10.12 on page 380.

10.66 This exercise shows that the dominated convergence theorem (DCT), Theorem 5.9 on page 169, cannot be extended to nets. Consider the measure space $([0,1], \mathcal{M}_{[0,1]}, \lambda_{[0,1]})$. The collection I of all finite subsets of $[0,1]$ with the order relation $\subset$ is a directed set. Define the net of functions $\{f_\iota\}_{\iota \in I}$ on $[0,1]$ by $f_\iota = \chi_\iota$. Observe that each f_ι is $\mathcal{M}_{[0,1]}$-measurable.
a) Show that $\{f_\iota\}_{\iota \in I}$ converges pointwise to $\chi_{[0,1]}$ on $[0,1]$.
b) Show that there is a nonnegative Lebesgue integrable function g on $[0,1]$ such that $|f_\iota| \leq g$ for all $\iota \in I$.
c) Show that $\lim \int_0^1 f_\iota(x)\, dx \neq \int_0^1 \chi_{[0,1]}(x)\, dx$.
d) Conclude that the DCT cannot be extended to nets.

In Exercises 10.67–10.74, we are using the notation introduced in Definition 10.20 on page 380.

10.67 Refer to Definition 10.20.
a) Show that an infinite series $\sum_{j=k}^\infty x_j$ converges if and only if the series $\sum_{j=\ell}^\infty x_j$ converges, where $k < \ell$.
b) Show that an infinite sum $\sum_{j \in S} x_j$ converges if and only if the sum $\sum_{j \in S \setminus F} x_j$ converges, where F is a finite subset of S.

10.68 Show that
a) $\sum_{j=1}^\infty 1/j$ fails to converge.
b) $\sum_{j=1}^\infty (-1)^j/j$ converges.

10.69 Suppose $\sum_{j=k}^\infty \|x_j\| < \infty$. Show that $\sum_{j \geq k} x_j$ converges to s if and only if $\sum_{j=k}^\infty x_j$ converges to s.

★10.70 Assume the z_js are complex numbers.
  a) Prove that the infinite sum $\sum_{j \in S} z_j$ converges if there are nonnegative real numbers b_j, $j \in S$, such that $|z_j| \le b_j$ for all but finitely many $j \in S$ and $\sum_{j \in S} b_j < \infty$.
  b) Prove that $\sum_{j \in S} |z_j| < \infty$ implies $\sum_{j \in S} z_j$ converges.

10.71 Let α and β be scalars.
  a) Show that if the two infinite sums $\sum_{j \in S} x_j$ and $\sum_{j \in S} y_j$ converge, then so does the infinite sum $\sum_{j \in S} (\alpha x_j + \beta y_j)$ and, moreover,

$$\sum_{j \in S} (\alpha x_j + \beta y_j) = \alpha \sum_{j \in S} x_j + \beta \sum_{j \in S} y_j.$$

  b) Show that the results of part (a) remain valid for infinite series.

10.72 Let S and T be disjoint infinite subsets of integers. Suppose that two of the three infinite sums $\sum_{j \in S} x_j$, $\sum_{j \in T} x_j$, and $\sum_{j \in S \cup T} x_j$ are convergent. Prove that all three are convergent and that

$$\sum_{j \in S \cup T} x_j = \sum_{j \in S} x_j + \sum_{j \in T} x_j.$$

10.73 Let $\sum_{j=k}^{\infty} x_j$ be a convergent infinite series in a normed space.
  a) Prove that $\lim x_n = 0$.
  b) Prove that $\lim \sum_{j=n}^{\infty} x_j = 0$.

10.74 Let $\sum_{j \in S} x_j$ be a convergent infinite sum in a normed space. Show that the net $\left\{ \sum_{j \in S \setminus F} x_j \right\}_{F \in \mathcal{P}_{\mathcal{F}}(S)}$ converges to 0.

10.75 Consider a sequence $\{a_n\}_{n=1}^{\infty}$ of real numbers. Show that the function $f_t = a_{[t]}$, where $[\ ]$ denotes the greatest integer function and $\iota \in [1, \infty)$, is a subnet of $\{a_n\}_{n=1}^{\infty}$. The order relation on $[1, \infty)$ is understood to be $\le$.

10.76 Consider a function $f : [1, \infty) \to \mathcal{R}$. Let $f_t = f(t)$. Then $\{f_t\}_{t \in [1, \infty)}$ is a net in $\mathcal{R}$ with respect to the usual $\le$ ordering. Suppose that $\{t_n\}_{n=1}^{\infty}$ is a nondecreasing sequence in $[1, \infty)$. Show that $\{f_{t_n}\}_{n=1}^{\infty}$ is a subnet of $\{f_t\}_{t \in [1, \infty)}$ if and only if $\lim_{n \to \infty} t_n = \infty$.

10.77 Suppose $\{x_{\iota_\kappa}\}_{\kappa \in K}$ is a subnet of $\{x_\iota\}_{\iota \in I}$ and $\{x_{\iota_{\kappa_\mu}}\}_{\mu \in M}$ is a subnet of $\{x_{\iota_\kappa}\}_{\kappa \in K}$. Show that $\{x_{\iota_{\kappa_\mu}}\}_{\mu \in M}$ is a subnet of $\{x_\iota\}_{\iota \in I}$.

10.78 Let $\{x_\iota\}_{\iota \in I}$ be a net such that $\lim x_\iota = x$. Show that if $\{x_{\iota_\kappa}\}_{\kappa \in K}$ is a subnet of $\{x_\iota\}_{\iota \in I}$, then $\lim x_{\iota_\kappa} = x$.

★10.79 Prove that a Cauchy sequence converges if and only if it has a convergent subnet.

★10.80 Let $\mathcal{T}$ and $\mathcal{U}$ be topologies on a set Ω. Show that $\mathcal{T}$ is weaker than $\mathcal{U}$ if and only if $x_\iota \xrightarrow{\mathcal{U}} x$ implies $x_\iota \xrightarrow{\mathcal{T}} x$.

10.81 Let Ω be the Cartesian product of a family $\{\Omega_\kappa\}_{\kappa \in K}$ of topological spaces and suppose that Ω is given the product topology. Let $\{x_\iota\}_{\iota \in I}$ be a net in Ω. Show that $\lim x_\iota = x$ if and only if $\lim p_\kappa(x_\iota) = p_\kappa(x)$ for each $\kappa \in K$.

10.82 Prove that the sum and product of continuous complex-valued functions on a topological space are continuous.

10.83 Suppose that f is a complex-valued continuous function on a topological space Ω. Prove that the function g defined on $f^{-1}(\{0\}^c)$ by $g(x) = 1/f(x)$ is continuous.

10.84 Let Ω and Λ be topological spaces and $f : \Omega \to \Lambda$. Prove by using nets that f is continuous if and only if $f(\overline{E}) \subset \overline{f(E)}$ for each $E \subset \Omega$.

★10.85 Let $f : \Omega \to \Lambda$ and $g : \Lambda \to \Gamma$ be continuous. Show that $g \circ f : \Omega \to \Gamma$ is continuous. In words, show that the composition of continuous functions is continuous.

10.86 Let Ω be the Cartesian product of a family $\{\Omega_\kappa\}_{\kappa \in K}$ of topological spaces and suppose that Ω is given the product topology. Let Λ be a topological space and $f: \Lambda \to \Omega$. Show that f is continuous if and only if $p_\kappa \circ f$ is continuous for each $\kappa \in K$.

★**10.87** Let Ω be a topological space and $f: \Omega \to \mathcal{R}$. Prove that the following conditions are equivalent.
a) f is continuous.
b) $f^{-1}((-\infty, a))$ and $f^{-1}((a, \infty))$ are open for each $a \in \mathcal{R}$.
c) $f^{-1}((-\infty, a])$ and $f^{-1}([a, \infty))$ are closed for each $a \in \mathcal{R}$.

10.88 Let Ω be a set and (Λ, ρ) a metric space. A net $\{f_\iota\}_{\iota \in I}$ of functions from Ω to Λ is said to **converge uniformly** to the function f if for each $\epsilon > 0$, there is an index ι_0 such that $\rho(f_\iota(x), f(x)) < \epsilon$ for all $x \in \Omega$ whenever $\iota_0 \preceq \iota$. Show that if Ω is a topological space, f_ι is continuous for each $\iota \in I$, and $\{f_\iota\}_{\iota \in I}$ converges uniformly to f, then f is continuous.

★**10.89** Let Ω be a topological space, $(\Lambda, \| \ \|)$ a complete normed space, and S an infinite subset of integers. Suppose that for each $j \in S$, $f_j: \Omega \to \Lambda$ is continuous and that there is a $b_j \in \mathcal{R}$ such that $\|f_j(x)\| \le b_j$ for all $x \in \Omega$. Show that if $\sum_{j \in S} b_j < \infty$, then $f(x) = \sum_{j \in S} f_j(x)$ defines a continuous function from Ω into Λ. *Hint:* Refer to Exercise 10.88.

10.6 SEPARATION PROPERTIES

In this section, we take up the topic of *separation* in topological spaces. We begin with the following definition.

DEFINITION 10.22 Separated Sets

Two subsets A and B of a topological space Ω are said to be **separated** if there exist open sets U and V such that $A \subset U$, $B \subset V$, and $U \cap V = \emptyset$. When it is important to emphasize the role of the sets U and V, we will say that A and B are separated by U and V.

EXAMPLE 10.14 Illustrates Definition 10.22

a) Consider the normed space $(C([a, b]), \| \ \|_\infty)$ discussed in Example 10.8 on page 368. For $f \in C([a, b])$, we have $\|f\|_\infty = \sup\{|f(x)| : x \in [a, b]\}$. (Why?) Thus, when two functions f_1 and f_2 are "close" with respect to this norm, say, $\|f_1 - f_2\|_\infty < \delta$, for some small number δ, it means $|f_1(x) - f_2(x)| < \delta$ for all $x \in [a, b]$, that is, the two functions are uniformly close.

Suppose $E \subset C([a, b])$ and $f \in C([a, b])$. On the one hand, $f \in \overline{E}$ if and only if for each $\epsilon > 0$, there exists a function $g \in E$ such that $\|f - g\|_\infty < \epsilon$; in other words, $f \in \overline{E}$ if and only if it can be uniformly approximated arbitrarily closely by members of E. On the other hand, $f \notin \overline{E}$ if and only if there is an $\epsilon > 0$ such that $E \subset U_\epsilon = \{h : \|f - h\|_\infty > \epsilon\}$. Because we have $\{f\} \subset V_\epsilon = \{h : \|f - h\|_\infty < \epsilon\}$, it follows that f is not uniformly approximable by members of E if and only if $\{f\}$ and E are separated by the open sets U_ϵ and V_ϵ for some $\epsilon > 0$.

b) Suppose that A and B are disjoint closed disks in $\mathcal{R}^2$. Then there is a line $L = \{(x, y) : ax + by = c\}$ such that A and B are separated by the open *half-planes* $L_- = \{(x, y) : ax + by < c\}$ and $L_+ = \{(x, y) : ax + by > c\}$. The topic of separation by *half-spaces* is important in our subsequent study of normed linear spaces.

c) Let (Ω, ρ) be a metric space and A and B disjoint closed subsets of Ω. By Exercise 10.47 on pages 377–378, the function $f(x) = \rho(x, A) - \rho(x, B)$ is continuous on Ω. It follows that A and B are separated by the open sets $f^{-1}((-\infty, 0))$ and $f^{-1}((0, \infty))$. □

DEFINITION 10.23 **Hausdorff Space, Normal Space**

Let Ω be a topological space.

a) Ω is said to be a **Hausdorff space** if distinct points are separated; that is, $x \neq y$ implies that $\{x\}$ and $\{y\}$ are separated.

b) Ω is said to be a **normal space** if disjoint closed sets are separated; that is, A and B closed and $A \cap B = \emptyset$ implies that A and B are separated.

Whereas it is not true in general that a normal space is a Hausdorff space (see Exercise 10.91), it is obvious that a normal space is Hausdorff if all single element subsets are closed. A space with the property that all single element subsets are closed is called a T_1**-space.** Hausdorff spaces are always T_1-spaces. *From now on, whenever we consider a topological space, we will assume implicitly that it is a T_1-space.*

Example 10.14(c) shows that all metric spaces are normal. And it is an easy exercise to prove that all metric spaces are Hausdorff. Later we will see examples of normal and Hausdorff spaces that are not metric spaces.

Existence of Continuous Functions

Given an arbitrary topological space Ω, it is not at all clear that there exist nonconstant, real-valued continuous functions on Ω. However, as we will see momentarily, normal spaces always possess an abundance of such functions. First we need the following characterization of normal spaces.

□ □ □ **PROPOSITION 10.14**

A topological space Ω is normal if and only if for each closed set F and each open set O with $F \subset O$, there exists an open set W such that $F \subset W \subset \overline{W} \subset O$.

PROOF Suppose that Ω is normal. Let F be closed, O open, and $F \subset O$. Then F and O^c are disjoint closed sets. It follows that there are open sets W and U with $F \subset W$, $O^c \subset U$, and $U \cap W = \emptyset$. Because $W \subset U^c$ and U^c is closed, it follows that $\overline{W} \subset U^c \subset (O^c)^c = O$.

To prove the converse, let A and B be disjoint closed sets. Taking $F = A$ and $O = B^c$, there is, by assumption, an open set W with $A \subset W \subset \overline{W} \subset B^c$. But $\overline{W}^c$ is open and contains B, and $W \cap \overline{W}^c = \emptyset$. Thus, A and B are separated by W and $\overline{W}^c$. ■

The string of containments $F \subset W \subset \overline{W} \subset O$ in the statement of Proposition 10.14 invites iteration: We can find open sets U and V such that

$$F \subset U \subset \overline{U} \subset W \subset \overline{W} \subset V \subset \overline{V} \subset O. \qquad (10.7)$$

To iterate further, we need better notation. A natural and judicious choice is to use binary digits as follows: $W = W_{.10}$, $U = W_{.01}$, and $V = W_{.11}$. Then (10.7) becomes

$$F \subset W_{.01} \subset \overline{W}_{.01} \subset W_{.10} \subset \overline{W}_{.10} \subset W_{.11} \subset \overline{W}_{.11} \subset O.$$

This construction can now be carried on indefinitely to yield the following lemma. The details of its proof are left to the reader as an exercise.

□ □ □ **LEMMA 10.1**

Suppose that Ω is a normal space. Let F be closed, O open, and $F \subset O$. Furthermore, let T denote the set of numbers in the interval $(0,1)$ that have terminating binary expansions. Then there is a collection of open sets $\{W_t : t \in T\}$ such that $t, s \in T$ and $t < s$ implies $F \subset \overline{W}_t \subset W_s \subset O$.

Using Lemma 10.1, we can now construct nonconstant, continuous, real-valued functions on normal spaces.

□ □ □ **THEOREM 10.2 Urysohn's Lemma**

Let A and B be disjoint closed nonempty subsets of a normal space Ω. Then there is a continuous function $f : \Omega \to \mathcal{R}$ such that $f(\Omega) \subset [0,1]$, $f(A) = \{0\}$, and $f(B) = \{1\}$.

PROOF First we apply Lemma 10.1 with $F = A$ and $O = B^c$ to obtain a collection of open sets $\{W_t : t \in T\}$ such that $t, s \in T$ and $t < s$ implies $A \subset \overline{W}_t \subset W_s \subset B^c$. Also, we let $W_1 = \Omega$ and $T_0 = T \cup \{1\}$. Now we define a function f on Ω by

$$f(x) = \inf\{t \in T_0 : x \in W_t\}.$$

Clearly, f takes values only in $[0,1]$. If $x \in A$, then $x \in W_t$ for each $t \in T$. Because T is dense in $[0,1]$, it follows that $f(x) = 0$. On the other hand, if $x \in B$, then $\{t \in T_0 : x \in W_t\} = \{1\}$. Thus, $f(x) = 1$.

It remains to show that f is continuous on Ω. By Exercise 10.87 on page 386, it is enough to prove that for each real number s, $f^{-1}((-\infty, s))$ is open and $f^{-1}((-\infty, s])$ is closed. First note that

$$f^{-1}((-\infty, s)) = \begin{cases} \emptyset, & s \leq 0; \\ \Omega, & s > 1. \end{cases} \quad \text{and} \quad f^{-1}((-\infty, s]) = \begin{cases} \emptyset, & s < 0; \\ \Omega, & s \geq 1. \end{cases}$$

Again using the fact that T is dense in $[0,1]$, we have

$$f^{-1}((-\infty, s)) = \bigcup_{t < s} W_t, \qquad s \in (0,1], \qquad (10.8)$$

and

$$f^{-1}((-\infty, s]) = \bigcap_{s<t} \overline{W}_t, \qquad s \in [0,1). \tag{10.9}$$

Equation (10.8) shows that for $s \in (0, 1]$, $f^{-1}((-\infty, s))$ is open, being a union of open sets. And (10.9) shows that for $s \in [0, 1)$, $f^{-1}((-\infty, s])$ is closed, being an intersection of closed sets.

We have now shown that for all $s \in \mathcal{R}$, $f^{-1}((-\infty, s))$ is open and $f^{-1}((-\infty, s])$ is closed. Hence, f is continuous. $\blacksquare$

Remark: Exercise 10.47(f) on page 378 provides a quick elementary proof of the metric-space version of Urysohn's lemma.

Urysohn's lemma is frequently used to obtain continuous approximations to characteristic functions of closed sets. Typically, one has a closed subset F of some normal space Ω and an open set O containing F that is "close" to F in some sense. Applying Urysohn's lemma with $B = F$ and $A = O^c$, we obtain a continuous function f with values in $[0, 1]$ that agrees with the characteristic function of F everywhere except possibly on $O \setminus F$.

When $\Omega = \mathcal{R}$, $F = [a, b]$, and $O = (a - \epsilon, b + \epsilon)$, this approximation procedure is nicely illustrated by the continuous function that is 1 on $[a, b]$, 0 on O^c, and linear on each of the intervals $(a - \epsilon, a)$ and $(b, b + \epsilon)$. Later, when we study spaces of continuous functions, we will rely heavily on the approximation of characteristic functions. As a more immediate application of Urysohn's lemma, we present the following important result.

$\square\ \square\ \square$ **THEOREM 10.3 Tietze's Extension Theorem**

Let Ω be a normal space, F a closed subset of Ω, and $f: F \to \mathcal{R}$ a continuous function. Then there is a continuous function $\tilde{f}: \Omega \to \mathcal{R}$ such that $\tilde{f}(x) = f(x)$ for each $x \in F$. Moreover, if $M = \sup\{|f(x)| : x \in F\} < \infty$, then $\tilde{f}$ may be chosen such that $\sup\{|\tilde{f}(x)| : x \in \Omega\} = M$.

PROOF
If $M = 0$, the result is trivial. We next consider the case where M is finite and nonzero. Without loss of generality, we can assume $M = 1$. (Why is that so?)

Because f is a continuous function on F, the sets $A = f^{-1}((-\infty, -1/3])$ and $B = f^{-1}([1/3, \infty))$ are relatively closed in F and, because F is a closed subset of Ω, A and B are also closed in Ω. So, by Urysohn's lemma, there is a continuous function g_1 on Ω such that $g_1(\Omega) \subset [-1/3, 1/3]$, $g_1(x) = -1/3$ for all $x \in A$, and $g_1(x) = 1/3$ for all $x \in B$. It follows that the continuous function f_1 defined on F via $f_1 = f - g_1$ satisfies $|f_1(x)| \le 2/3$ for all $x \in F$.

Similarly, by applying Urysohn's lemma to the sets $A = f_1^{-1}((-\infty, -2/9])$ and $B = f_1^{-1}([2/9, \infty))$, we obtain a continuous function g_2 on Ω such that $g_2(\Omega) \subset [-2/9, 2/9]$, $g_2(x) = -2/9$ for all $x \in A$, and $g_2(x) = 2/9$ for all $x \in B$. It follows that the continuous function f_2 defined on F via

$$f_2 = f_1 - g_2 = f - (g_1 + g_2)$$

satisfies $|f_2(x)| \le 4/9$ for all $x \in F$.

We now proceed inductively to construct a sequence $\{g_n\}_{n=1}^{\infty}$ of continuous functions on Ω such that $|g_n(x)| \leq 2^{n-1}/3^n$ for all $x \in \Omega$ and

$$\left| f(x) - \sum_{j=1}^{n} g_j(x) \right| \leq (2/3)^n, \qquad x \in F.$$

It follows from Exercise 10.89 on page 386, that the function $\tilde{f} = \sum_{n=1}^{\infty} g_n$ is continuous on Ω. And the previous two inequalities show that $|\tilde{f}(x)| \leq 1$ for each $x \in \Omega$ and that $\tilde{f}(x) = f(x)$ for each $x \in F$.

It remains to consider the case where f is unbounded. To that end, define $f_0 = \arctan f$. Since $|f_0(x)| < \pi/2$ for all $x \in F$, we can apply the results just proved for bounded functions to obtain a continuous function $g_0 : \Omega \to \mathcal{R}$ such that $g_0(x) = f_0(x)$ for each $x \in F$. The function $\tilde{f} = \tan g_0$ is continuous on Ω and is such that $\tilde{f}(x) = f(x)$ for each $x \in F$. ∎

Exercises for Section 10.6

10.90   Let $A = \{(x, y) \in \mathcal{R}^2 : x > 0, \ y \geq 1/x\}$ and $B = \{(x, 0) \in \mathcal{R}^2 : x \geq 0\}$.

    a) Show that A and B are disjoint closed sets that cannot be separated by open half-planes in the sense of Example 10.14(b) on page 387.

    b) Find explicitly open sets U and V that separate A and B.

10.91   Provide an example of a normal space that is not Hausdorff. *Hint:* Refer to Example 10.2(c) on page 361.

10.92   Show that all metric spaces are Hausdorff.

10.93   Let $\mathcal{T} = \{\emptyset\} \cup \{W \subset \mathcal{N} : W^c \text{ is a finite set}\}$ where, as usual, $\mathcal{N}$ denotes the set of positive integers. Show that $\mathcal{T}$ is a topology on $\mathcal{N}$ and that $(N, \mathcal{T})$ is a T_1-space.

10.94   Refer to Exercise 10.93. Show that the topological space $(N, \mathcal{T})$ is neither a Hausdorff space nor a normal space.

10.95   Describe all continuous functions $f : \mathcal{N} \to \mathcal{R}$, where $\mathcal{N}$ is given the topology $\mathcal{T}$ defined in Exercise 10.93.

10.96   Describe all convergent sequences in $\mathcal{N}$, where $\mathcal{N}$ is given the topology $\mathcal{T}$ defined in Exercise 10.93.

10.97   Prove that a normal T_1-space is a Hausdorff space.

10.98   Prove that a Hausdorff space is a T_1-space.

10.99   Prove that a topological space is Hausdorff if and only if convergent nets have unique limits (i.e., $\lim x_\iota = x$ and $\lim x_\iota = y$ imply $x = y$).

10.100   Let Ω be a nonempty set and $\mathcal{T_F}$ the weak topology on Ω determined by a family of functions $\mathcal{F}$. Suppose that for each $f \in \mathcal{F}$, the space $f(\Omega)$ is Hausdorff. Show that $(\Omega, \mathcal{T})$ is a Hausdorff space if and only if $\mathcal{F}$ separates the points of Ω (i.e., $x, y \in \Omega$ and $x \neq y$ imply that there exists an $f \in \mathcal{F}$ such that $f(x) \neq f(y)$).

10.101   Provide the details of the proof of Lemma 10.1 on page 388.

10.102   Let S be a nonempty set. Formulate and prove a version of the Tietze extension theorem where $\mathcal{R}$ is replaced by the Cartesian product $\mathcal{R}^S$ with the product topology.

10.103   Show that Theorem 10.3 on page 389 is no longer valid if F is assumed to be open instead of closed.

10.104   Let F be a closed subset of $\mathcal{R}$ and $f : F \to \mathcal{R}$ be continuous. From Proposition 2.13 on page 48, we can write $F^c = \bigcup_{J \in \mathcal{S}} J$, where $\mathcal{S}$ is a countable collection of disjoint open intervals. Construct a continuous function $g : \mathcal{R} \to \mathcal{R}$ that agrees with f on F and is linear on each interval $J \in \mathcal{S}$.

10.7 CONNECTED SETS

If D is a subset of $\mathcal{R}$, then, except in trivial cases, the characteristic function χ_D is not continuous. There are, however, many topological spaces that have non-constant, continuous characteristic functions. For example, if $\Omega = [0,1] \cup [2,3]$ is given the relative topology from $\mathcal{R}$, then $\chi_{[0,1]}$ is a continuous function on Ω. Such topological spaces are called *disconnected*.

DEFINITION 10.24   **Disconnected and Connected Spaces**

A topological space that has at least one nonconstant, continuous characteristic function is said to be **disconnected**. A topological space that is not disconnected is said to be **connected**. A subset of a topological space is called connected or disconnected, respectively, if it is connected or disconnected, respectively, with respect to the relative topology.

If f is a nonconstant, continuous characteristic function on a topological space Ω, then $f^{-1}((-\infty, 1/2))$ and $f^{-1}((1/2, \infty))$ are disjoint nonempty open sets whose union is Ω. Thus, we see that each of the following conditions is equivalent to a topological space Ω being disconnected:

- Ω can be decomposed into two disjoint nonempty open sets.
- Ω contains a proper, nonempty subset that is both open and closed.

Here is another way of characterizing disconnected sets.

□ □ □   **PROPOSITION 10.15**

A subset D of a topological space Ω is disconnected if and only if there are nonempty sets A and B such that $D = A \cup B$, $A \cap \overline{B} = \emptyset$, and $\overline{A} \cap B = \emptyset$.

PROOF   Suppose D is a disconnected subset of Ω. Let f be a nonconstant characteristic function on D that is continuous with respect to the relative topology. Because $A = f^{-1}((1/2, 3/2)) = f^{-1}(\{1\})$, we have that A is nonempty and both open and closed in the relative topology. Similarly, $B = D \setminus A = f^{-1}(\{0\})$ is also relatively open, relatively closed, and nonempty. It follows that there is a closed subset F of Ω such that $A = F \cap D$. Because $F \cap B = \emptyset$ and $\overline{A} \subset F$, we have that $\overline{A} \cap B = \emptyset$. Similarly, we have that $A \cap \overline{B} = \emptyset$.

Conversely, suppose there are nonempty sets A and B such that $D = A \cup B$, $\overline{A} \cap B = \emptyset$, and $A \cap \overline{B} = \emptyset$. Then we have $A = D \cap \overline{A}$ and $B = D \cap \overline{B}$. Thus, A and B are relatively closed. Since $B = D \setminus A$, B is also relatively open and, similarly, A is relatively open. It follows easily that the characteristic function χ_A is nonconstant on D and continuous in the relative topology. Consequently, D is a disconnected subset of Ω. ■

EXAMPLE 10.15   *Illustrates Connected Sets*

Let Ω be a topological space and $x \in \Omega$. It follows from Proposition 10.15 that each singleton subset of Ω is connected. The Cantor set provides an example of a topological space in which the only connected subsets are singletons. □

EXAMPLE 10.16 *Illustrates Connected Subsets of* $\mathcal{R}$

In this example, we will establish the fact that the connected subsets of $\mathcal{R}$ are precisely the intervals (including degenerate intervals).

Let D be a connected subset of $\mathcal{R}$. If $D = \emptyset$, then it is also a degenerate interval (e.g., $(x, x]$). If D is a singleton set $\{x\}$, then it is a degenerate interval of the form $[x, x]$. So, assume that D contains more than one point. Let $a, b \in D$ with $a < b$ and let $c \in (a, b)$. If c does not lie in D, then the sets $A = D \cap (c, \infty)$ and $B = D \cap (-\infty, c)$ are relatively open, disjoint, and their union is D. Thus, D is disconnected, a contradiction. Hence, the interval (a, b) is contained in D whenever a and b are elements of D with $a < b$. It follows immediately that D is equal to $(\inf D, \sup D)$, $(\inf D, \sup D]$, $[\inf D, \sup D)$, or $[\inf D, \sup D]$.

Conversely, suppose D is an interval. We claim D is connected. Assume to the contrary. Then, by Proposition 10.15, there are nonempty sets A and B such that $D = A \cup B$, $A \cap \overline{B} = \emptyset$, and $\overline{A} \cap B = \emptyset$. Let $a \in A$ and $b \in B$, and assume without loss of generality that $a < b$. Consider the set $C = \{\, x : [a, x] \subset A \,\}$. We note that $C \neq \emptyset$ because $a \in C$. Because $b \notin A$, C is bounded above by b. Thus, $u = \sup C$ is a real number and $a \leq u \leq b$. There are three possibilities: $u \in A$, $u \in B$, or $u \notin D$. The last possibility can be eliminated because $a \leq u \leq b$ and both a and b lie in the interval D. Suppose $u \in A$. Since, for each $n \in \mathcal{N}$, $[a, u + 1/n]$ is not a subset of A, it follows that there is an element $b_n \in B \cap [u, u + 1/n]$. And, since $\lim_{n \to \infty} b_n = u$, we have $u \in A \cap \overline{B}$. But this result contradicts the assumption that $A \cap \overline{B} = \emptyset$. On the other hand, suppose $u \in B$. Then $u > a$ and, so, $u - 1/n \in A$ for sufficiently large n. Consequently, because $\lim_{n \to \infty} u - 1/n = u$, we have that $u \in \overline{A} \cap B$. But this result contradicts the assumption that $\overline{A} \cap B = \emptyset$. □

A useful property of connectedness is that the continuous image of a connected space is connected. More precisely, we have the following theorem.

□ □ □ **THEOREM 10.4**

Let Ω be a connected topological space and $f : \Omega \to \Lambda$ be a continuous function. Then $f(\Omega)$ is a connected subset of Λ.

PROOF Suppose to the contrary that $f(\Omega)$ is not connected. Then there is a nonconstant, continuous characteristic function g on $f(\Omega)$. It follows from Exercise 10.85 on page 385 that the nonconstant characteristic function $g \circ f$ is continuous on Ω. Thus, Ω is not connected, a contradiction. ■

From Theorem 10.4 and Example 10.16, we get the following two corollaries.

□ □ □ **COROLLARY 10.2**

Let f be a real-valued continuous function on a connected topological space Ω. Then $f(\Omega)$ is an interval, possibly degenerate.

□ □ □ **COROLLARY 10.3** Intermediate Value Theorem

Let f be a real-valued continuous function on $[a, b]$. Then for each number y between $f(a)$ and $f(b)$, there is an $x \in [a, b]$ such that $f(x) = y$.

Arcwise Connected Spaces

Let Ω be a topological space and p and q points of Ω. Then we say that **p is connected to q by an arc** if there exist $a, b \in \mathcal{R}$ and a continuous function $g: [a, b] \to \Omega$ such that $p = g(a)$ and $q = g(b)$. The set $A = g([a, b])$ is called an **arc** connecting p to q. It is easy to show that the following facts hold for all points $p, q, r \in \Omega$. (See Exercise 10.111.)

- p is connected to itself by an arc.

- If p is connected to q by an arc, then q is connected to p by an arc.

- If p is connected to q by an arc and q is connected to r by an arc, then p is connected to r by an arc.

Note: In view of the second bulleted item, we can unambiguously use phrases such as "p and q are connected by an arc" and "there is an arc connecting p and q."

The space Ω is called **arcwise connected** if for every pair of points $p, q \in \Omega$, there is an arc connecting p and q. The next proposition shows that arcwise connected spaces are always connected.

□ □ □ **PROPOSITION 10.16**

An arcwise connected topological space is connected.

PROOF   Suppose that Ω is arcwise connected but not connected. Let g be a nonconstant, continuous characteristic function on Ω. Let p and q be points of Ω such that $g(p) = 0$ and $g(q) = 1$. As Ω is arcwise connected, there is an interval $[a, b]$ and a continuous function $f: [a, b] \to \Omega$ such that $f(a) = p$ and $f(b) = q$. It follows that $g \circ f$ is a nonconstant, continuous characteristic function on $[a, b]$, implying that $[a, b]$ is disconnected. But, by Example 10.16, the interval $[a, b]$ is connected. Consequently, we have reached a contradiction. Hence, Ω must be connected. ■

The converse of Proposition 10.16 is false. (See Exercise 10.109.) There is, however, a converse for open subsets of a normed linear space.

□ □ □ **PROPOSITION 10.17**

A connected open subset of a normed space is arcwise connected.

PROOF   Suppose that D is a nonempty open subset of a normed space. Let $p \in D$ and W be the set of all points of D that are connected to p by an arc in D. Since $p \in W$, W is nonempty. We claim that W is open. Let $q \in W$. Then $q \in D$ and, hence, there is an $r > 0$ such that $B_r(q) \subset D$. If $x \in B_r(q)$, then the arc $\{q + t(x - q) : 0 \le t \le 1\}$ connects q to x and lies inside $B_r(q)$. It follows that p is connected to x by an arc in D. Thus, $B_r(q) \subset W$ and, hence, W is open.

We also claim that $D \setminus W$ is open. Let $q \in D \setminus W$. Then $q \in D$ and, so, as we discovered in the previous paragraph, there is an $r > 0$ such that $B_r(q) \subset D$ and any point of $B_r(q)$ is connected to q by an arc in D. If a point of $B_r(q)$ is connected to p by an arc in D, then there would be an arc in D connecting p

to q, contradicting the assumption that $q \in D \setminus W$. Thus, $B_r(q) \subset D \setminus W$ and, hence, $D \setminus W$ is open.

We have shown that W is both open and closed in D. Because D is connected and W is nonempty, it follows that $D = W$. As any point of D is connected to p by an arc in D, any two points of D must be connected to each other by an arc in D. Hence, D is arcwise connected. ∎

Remark: A normed space Ω is always arcwise connected. Indeed, if $x \in \Omega$, then the arc $\{tx : 0 \le t \le 1\}$ connects 0 to x. Hence, any point of Ω is connected to 0 by an arc and, so, any two points of Ω are connected to each other by an arc.

Connected Components, Totally Disconnected Spaces

We will now discover how a topological space can be decomposed as the union of a family of pairwise disjoint connected subsets. First we state two propositions, whose proofs are left to the reader as Exercises 10.113 and 10.114.

□ □ □ **PROPOSITION 10.18**

Suppose that S is a collection of connected subsets of a topological space Ω and that $D_1 \cap D_2 \ne \emptyset$ whenever $D_1, D_2 \in S$. Then $\bigcup_{D \in S} D$ is a connected subset of Ω.

□ □ □ **PROPOSITION 10.19**

Let Ω be a topological space and A a connected subset of Ω. Then $\overline{A}$ is also connected.

Given a point x in a topological space Ω, we can apply Proposition 10.18 with S equal to the collection of all connected subsets of Ω that contain x to obtain a connected set C_x. The set C_x is the largest connected subset of Ω that contains x and is called the **connected component of Ω that contains x.**

□ □ □ **THEOREM 10.5**

Let Ω be a topological space.
a) For each pair of elements $x, y \in \Omega$, either $C_x = C_y$ or $C_x \cap C_y = \emptyset$.
b) For each $x \in \Omega$, C_x is closed.
c) $\Omega = \bigcup_{x \in \Omega} C_x$.

PROOF   a) If $C_x \cap C_y \ne \emptyset$, then, by Proposition 10.18, $C_x \cup C_y$ is connected. It follows that $C_x \cup C_y \subset C_x$ and $C_x \cup C_y \subset C_y$. Hence, $C_x = C_y$.
b) By Proposition 10.19, $\overline{C_x}$ is connected for each $x \in \Omega$. Hence, $\overline{C_x} \subset C_x$ and, so, C_x is closed. Thus, (b) holds.
c) The proof of (c) is trivial because $x \in C_x$ for all $x \in \Omega$. ∎

A topological space Ω is said to be **totally disconnected** if all of its connected components are single element sets.

EXAMPLE 10.17 *Illustrates Totally Disconnected Spaces*

  a) A nonempty set is totally disconnected with respect to the discrete topology.

  b) The set Q of rational numbers, equipped with the relative topology inherited from $\mathcal{R}$, is totally disconnected.

  c) The Cantor set P, equipped with the relative topology inherited from $\mathcal{R}$, is totally disconnected.   □

Exercises for Section 10.7

10.105 Show that a topological space is disconnected if and only if it has a subset that is proper, nonempty, open, and closed.

10.106 Show that a continuous integer-valued function on a connected space must be constant.

10.107 Refer to Exercise 10.64 on page 378. Let Ω be a topological space and $A \subset \Omega$. Suppose $g : [0, 1] \to \Omega$ is a continuous function such that $g(0) \in A$ and $g(1) \in A^c$. Show that there exists an $s \in [0, 1]$ such that $g(s) \in \partial A$.

10.108 Provide the omitted details of Example 10.16 on page 392.

10.109 Give an example of a topological space that is connected but not arcwise connected. *Hint:* Consider the following subset of $\mathcal{R}^2$:

$$\{ (0, y) \in \mathcal{R}^2 : -1 < y < 1 \} \cup \{ (x, y) \in \mathcal{R}^2 : x > 0, \ y = \sin(1/x) \}.$$

10.110 Consider the normed linear space $(C([a, b]), \| \ \|_\infty)$ from Example 10.8 on page 368. Which of the following subsets of $C([a, b])$ are connected? Provide a proof in each case.

  a) $\{ g : g$ is real-valued and never $0 \}$,

  b) $\{ g : g(x) > 0$ for each $x \in [a, b] \}$,

  c) $\{ g : g$ is never 0 on $[a, b] \}$.

10.111 Let p, q, and r be points of a topological space Ω. Prove each of the following facts:

  a) p is connected to itself by an arc.

  b) If p is connected to q by an arc, then q is connected to p by an arc.

  c) If p is connected to q by an arc and q is connected to r by an arc, then p is connected to r by an arc.

10.112 Let Ω be a topological space. For $x \in \Omega$, define the **arcwise connected component** of x by $A_x = \{ y \in \Omega : y$ is connected to x by an arc $\}$.

  a) Prove analogues of parts (a) and (c) of Theorem 10.5 with arcwise connected components in place of connected components.

  b) Show that the analogue of part (b) of Theorem 10.5 is false in general, but is true if Ω is an open subset of a normed space and, so, in particular, is true if Ω is a normed space.

10.113 Prove Proposition 10.18.

10.114 Prove Proposition 10.19.

10.115 Let T denote the unit circle centered at 0 in the complex plane $\mathbb{C}$; let $C(T)$ denote the space of complex-valued continuous functions defined on T equipped with the norm $\|f\| = \sup\{ |f(z)| : z \in T \}$; and let G denote the set of non-vanishing functions in $C(T)$.

  a) Show that G is open.

  b) Describe the connected component of the constant function 1.

  c) Describe the connected components of G.

10.116 The Cantor function restricted to the Cantor set is an example of a continuous function that maps a totally disconnected space onto a connected space. Show that if Ω is a connected space, then there are no nonconstant continuous functions from Ω into the Cantor set.

Maurice Fréchet
(1878–1973)

Maurice Fréchet was born in Maligny, France, on September 2, 1878, to Jacques and Zoé Fréchet. He was the fourth of six children.

Fréchet attended the secondary school Lycée Buffon in Paris, where he learned mathematics from Jacques Hadamard. Recognizing Fréchet's mathematical aptitude, Hadamard decided to tutor him individually. Although Hadamard moved to the University of Bordeaux in 1894, he continuously sent Fréchet mathematical problems, harshly criticizing any errors made by Fréchet.

Upon his completion of high school, Fréchet enrolled in military service, as required. During this time, he was deciding whether to study mathematics or physics. He chose mathematics due to his dislike of chemistry classes, which were required of physics majors. In 1900, Fréchet enrolled in the École Normale Supérieure to study mathematics.

During his academic career, Fréchet served at many different institutions. From 1907 to 1908, he was a professor of mathematics at the Lycée in Besançon, moving in 1908 to the Lycée in Nantes for a year, following which he served at the University of Poitiers from 1910 to 1919. With the end of World War I, Fréchet was chosen to go to Strasbourg to help reestablish the university. There he was a professor of higher analysis and Director of the Mathematics Institute. In 1928, encouraged by Borel, Fréchet returned to Paris, where he held a variety of positions until his retirement in 1948.

Fréchet made major contributions to point-set topology and introduced the concept of metric spaces (although the name is due to Hausdorff). Additionally, he provided several significant contributions to statistics, probability, and calculus. In his 1906 dissertation, he opened the entire field of functionals on metric spaces and introduced the abstract notion of compactness.

Fréchet was elected to the Polish Academy of Sciences in 1929, to the Royal Society of Edinburgh in 1947, and, finally, in 1956, to the Academy of Sciences of the Institut de France. He died on June 4, 1973, in Paris, France.

□ 11 □

Separability and Compactness

In this chapter, we will continue our study of topological, metric, and normed spaces, which we began in Chapter 10. Here we will discuss separable, second countable, and metrizable topological spaces, compact metric spaces, compact topological spaces, locally compact spaces, and function spaces.

11.1 SEPARABILITY, SECOND COUNTABILITY, AND METRIZABILITY

In this section, we will discuss separable spaces and a related class of spaces known as second countable spaces. We will also prove a powerful theorem that gives a sufficient condition for a topological space to be metrizable.

Separable Spaces

Recall that a subset E of a topological space Ω is dense if $\overline{E} = \Omega$. A crucial property of the space $\mathcal{R}$ of real numbers is that it contains a countable subset that is dense; for example, the countable set Q of rational numbers is dense, as we know from Proposition 2.4 on page 32. Many of the topological spaces of interest in analysis share with $\mathcal{R}$ the property of having subsets that are both countable and dense. Such spaces are called *separable*.

DEFINITION 11.1   Separable Space

A topological space Ω is said to be **separable** if it contains a countable dense subset; that is, if there is a set $E \subset \Omega$ such that E is countable and $\overline{E} = \Omega$.

EXAMPLE 11.1 *Illustrates Definition 11.1*

In this example, we use the notation $Q + iQ$ for the set of complex numbers that have rational real and imaginary parts. We note that $Q + iQ$ is a countable set. (Why?)

a) Consider the space $\ell^1(\mathcal{N})$. For each $n \in \mathcal{N}$, the set

$$A_n = \{\, f \in \ell^1(\mathcal{N}) : f(j) \in Q + iQ,\ 1 \leq j \leq n,\ \text{and}\ f(j) = 0,\ j > n \,\}$$

is countable. Hence, by Proposition 1.10 on page 20, $A = \bigcup_{n=1}^{\infty} A_n$ is also countable. It is left for Exercise 11.4 to show that A is dense in $\ell^1(\mathcal{N})$. Thus, $\ell^1(\mathcal{N})$ is separable.

b) Consider the normed space $(C([a, b]), \|\ \|_{\infty})$ discussed in Example 10.8 on page 368. For each $n \in \mathcal{N}$, let PL_n denote the set of $f \in C([a, b])$ with the property that, for each $j = 0, 1, 2, \ldots, n - 1$, the restriction of f to the subinterval $[a + j(b - a)/n, a + (j + 1)(b - a)/n]$ is of the form $m_j x + b_j$, where $m_j, b_j \in Q + iQ$. Each function in PL_n is completely determined by a $2n$-tuple of numbers in $Q + iQ$. Hence, PL_n is a countable set and, so, $PL = \bigcup_{n=1}^{\infty} PL_n$ is also countable. It is left for Exercise 11.5 to show that PL is dense in $C([a, b])$. Thus, $C([a, b])$ is separable. □

EXAMPLE 11.2 *A Nonseparable Metric Space*

Consider the space $\ell^{\infty}([0, 1])$. The family $\{\, \chi_{[0,t]} : t \in [0, 1] \,\}$ of characteristic functions satisfies the condition

$$B_{1/2}(\chi_{[0,s]}) \cap B_{1/2}(\chi_{[0,t]}) = \emptyset, \qquad s \neq t. \tag{11.1}$$

Let E be a dense subset of $\ell^{\infty}([0, 1])$. Then, for each $t \in [0, 1]$, there is an $f_t \in E \cap B_{1/2}(\chi_{[0,t]})$. From (11.1), we see that no two f_ts can coincide. Because the collection $\{\, f_t : t \in [0, 1] \,\}$ is uncountable, it follows that E is not countable. Consequently, $\ell^{\infty}([0, 1])$ is not separable. □

Second Countable Spaces

We know that the collection $\mathcal{I}$ of open intervals forms a neighborhood basis that determines the topology of $\mathcal{R}$. By considering intervals in $\mathcal{I}$ with rational endpoints, we obtain a countable neighborhood basis that determines the topology of $\mathcal{R}$. There are many interesting spaces that, like $\mathcal{R}$, have countable neighborhood bases. Such spaces are called *second countable*.

DEFINITION 11.2 Second Countable Space

A topological space is said to be **second countable** if it has a countable neighborhood basis.

The following proposition relates the concepts of second countable and separable for topological spaces.

$\square\ \square\ \square$ **PROPOSITION 11.1**

a) *If a topological space is second countable, then it is separable.*
b) *If a metric space is separable, then it is second countable.*

PROOF a) Let $\mathfrak{N}$ be a countable neighborhood basis for a topological space Ω. For each nonempty $U \in \mathfrak{N}$, let $x_U \in U$. The set $\{\, x_U : U \in \mathfrak{N},\ U \neq \emptyset \,\}$ is dense because it has a nonempty intersection with each nonempty open set, and it is countable because $\mathfrak{N}$ is countable. Thus, Ω is separable.

b) Suppose that (Ω, ρ) is a metric space that contains a countable dense subset, say, $E = \{x_j\}_j$. Then the collection of open balls

$$\mathfrak{M} = \{\, B_{1/k}(x_j) : j,\ k = 1,\ 2,\ \dots \,\}$$

is countable. And it is easy to show that $\mathfrak{M}$ is a neighborhood basis on Ω. (See Exercise 11.6.)

We claim that $\mathfrak{M}$ is a neighborhood basis for the topology induced by ρ. Let O be open with respect to the topology induced by ρ and let $x \in O$. Choose $\epsilon > 0$ so that $B_\epsilon(x) \subset O$, and let $k \in \mathcal{N}$ be such that $2/k < \epsilon$. Since E is dense, there exists a j such that $x_j \in B_{1/k}(x)$. Then $x \in B_{1/k}(x_j)$ and

$$B_{1/k}(x_j) \subset B_{2/k}(x) \subset B_\epsilon(x) \subset O.$$

Thus, $\mathfrak{M}$ is a neighborhood basis for the topology induced by ρ and, so, (Ω, ρ) is second countable. $\blacksquare$

We next consider a consequence of second countability that will be useful later when we study compactness. Let E be a set. A collection $\mathcal{S}$ of sets such that $E \subset \bigcup_{S \in \mathcal{S}} S$ is called a **covering** of E. A subcollection of $\mathcal{S}$ that is also a covering of E is called a **subcovering.** If the members of $\mathcal{S}$ are open in some topology, then $\mathcal{S}$ is called an **open covering** of E. A topological space Ω is said to have the **Lindelöf property** if every open covering of Ω has a countable subcovering.

$\square\ \square\ \square$ **PROPOSITION 11.2**

A second countable topological space has the Lindelöf property.

PROOF Suppose Ω is a topological space with a countable neighborhood basis $\{U_n\}_n$, and let $\mathcal{S}$ be an open covering of Ω. For each $x \in \Omega$, we can choose an $O_x \in \mathcal{S}$ and an $n_x \in \mathcal{N}$ such that $x \in U_{n_x} \subset O_x$. The set of integers $B = \{\, n_x : x \in \Omega \,\}$ is countable, being a subset of a countable set. For each $m \in B$, we can choose an $O_m \in \mathcal{S}$ such that $U_m \subset O_m$. It follows that $\Omega \subset \bigcup_{m \in B} U_m \subset \bigcup_{m \in B} O_m$. Thus, $\{\, O_m : m \in B \,\}$ is a countable subcovering. $\blacksquare$

The converse of Proposition 11.2 is false in general, but it is true for metric spaces. See Exercises 11.7–11.8.

Metrization

We conclude this section by stating and proving a theorem that provides a simple pair of conditions that are sufficient for a topological space to be metrizable.

□ □ □ **THEOREM 11.1 Urysohn's Metrization Theorem**

A second countable, normal space is metrizable.

PROOF   Let $(\Omega, \mathcal{T})$ be a normal space with countable neighborhood basis $\mathfrak{N}$. Consider the countable set $\mathcal{W} = \{ (U, V) : U, V \in \mathfrak{N} \text{ and } \overline{U} \subset V \}$. We show first that, for each open set O and point $x \in O$, there is a pair $(U, V) \in \mathcal{W}$ such that

$$x \in U \subset \overline{U} \subset V \subset O. \qquad (11.2)$$

Indeed, as $\mathfrak{N}$ is a neighborhood basis, we can find a $V \in \mathfrak{N}$ such that $x \in V \subset O$. Applying Proposition 10.14 on page 387 with $F = \{x\}$, we obtain an open set W such that $x \in W \subset \overline{W} \subset V$.[†] Again using the assumption that $\mathfrak{N}$ is a neighborhood basis, we can find a $U \in \mathfrak{N}$ such that $x \in U \subset W$. It follows that the pair (U, V) belongs to $\mathcal{W}$ and satisfies (11.2).

Let $\{(U_n, V_n)\}_n$ be an enumeration of $\mathcal{W}$ and apply Urysohn's lemma (Theorem 10.2 on page 388) to obtain, for each n, a continuous function $f_n : \Omega \to [0, 1]$ that vanishes on $\overline{U_n}$ and is constantly 1 on V_n^c. Using the functions $\{f_n\}_n$, we define a function σ on $\Omega \times \Omega$ by

$$\sigma(x, y) = \sum_n 2^{-n} |f_n(x) - f_n(y)|.$$

We claim σ is a metric. It is easy to show that $\sigma(x, x) = 0$, $\sigma(x, y) = \sigma(y, x)$, and $\sigma(x, y) \leq \sigma(x, z) + \sigma(z, y)$. Thus, it remains only to show that $\sigma(x, y) > 0$ if $x \neq y$. Because $\{y\}$ is a closed set, (11.2) implies that there is a k such that

$$x \in U_k \subset \overline{U_k} \subset V_k \subset \Omega \setminus \{y\}.$$

Thus, $f_k(x) = 0$ and $f_k(y) = 1$. So, $\sigma(x, y) \geq 2^{-k} |f_k(x) - f_k(y)| = 2^{-k} > 0$.

The last step of the proof is to show that the topology $\mathcal{T}_\sigma$ induced by the metric σ is the same as $\mathcal{T}$. For fixed $y \in \Omega$, consider the function $g_y(x) = \sigma(x, y)$. It follows from Exercise 10.89 on page 386 that g_y is continuous with respect to the topology $\mathcal{T}$. Hence, for each $r > 0$, the ball $B_r^\sigma(y) = g_y^{-1}((-\infty, r))$ is $\mathcal{T}$-open. Consequently, the topology $\mathcal{T}_\sigma$ is weaker than $\mathcal{T}$.

To prove that $\mathcal{T}$ is weaker than $\mathcal{T}_\sigma$, it suffices to show that if O is $\mathcal{T}$-open and $x \in O$, then there exists an $s > 0$ such that $B_s^\sigma(x) \subset O$. Referring to (11.2), we see that we can find a positive integer m such that $x \in U_m \subset \overline{U_m} \subset V_m \subset O$. From $2^{-m} |f_m(x) - f_m(y)| \leq \sigma(x, y)$, we deduce that

$$f_m(y) = |f_m(x) - f_m(y)| \leq 2^m \sigma(x, y) < 1, \qquad y \in B_{2^{-m}}^\sigma(x).$$

Because f is constantly 1 on V_m^c, it follows that $B_{2^{-m}}^\sigma(x) \subset V_m \subset O$. ■

The following corollary of Urysohn's metrization theorem provides a sufficient condition for a space with a weak topology to be metrizable. Its proof is left to the reader as Exercise 11.11.

[†] See the paragraph following Definition 10.23 on page 387.

□ □ □ **COROLLARY 11.1**

Let Ω be a set equipped with the weak topology induced by a family of functions $\mathcal{F}$ that satisfies the following conditions:
a) $\mathcal{F}$ is countable.
b) If $x, y \in \Omega$ and $x \neq y$, then there is an $f \in \mathcal{F}$ such that $f(x) \neq f(y)$.
c) $f(\Omega)$ is metrizable for each $f \in \mathcal{F}$.
Then Ω is metrizable.

Exercises for Section 11.1

Note: A ★ denotes an exercise that will be subsequently referenced.

11.1 Show that the spaces $\mathcal{R}^n$ and $\mathbb{C}^n$ are separable.

11.2 Let E be a Lebesgue measurable subset of $\mathcal{R}$. Show that
a) $\mathcal{L}^1(E)$ and $\mathcal{L}^2(E)$ are separable.
b) $\mathcal{L}^\infty(E)$ is not separable except in the trivial case where E has Lebesgue measure zero.

11.3 Work Exercise 11.2 with $\mathcal{R}$ replaced by $\mathcal{R}^n$.

11.4 Show that the set A in Example 11.1(a) on page 398 is dense in $\ell^1(\mathcal{N})$.

11.5 Show that the set PL in Example 11.1(b) on page 398 is dense in $C([a, b])$.

11.6 Refer to the first paragraph in the proof of part (b) of Proposition 11.1 on page 399. Show that $\mathfrak{M}$ is a neighborhood basis on Ω.

11.7 Let $\mathcal{V}$ denote the topology on $\mathcal{R}$ determined by the neighborhood basis consisting of all intervals of the form $[a, b)$. Show that $(\mathcal{R}, \mathcal{V})$ is separable and has the Lindelöf property but is not second countable.

11.8 Show that a metric space with the Lindelöf property is second countable.

★11.9 A topological space is called **first countable** if at each point of the space, there is a countable neighborhood basis. Show that the space in Exercise 11.7 is first countable.

11.10 Show that the topological space in Example 11.2 on page 398 is not second countable.

11.11 Prove Corollary 11.1.

11.12 Show that the conclusion of Corollary 11.1 fails if the hypothesis that $\mathcal{F}$ is countable is omitted.

11.2 COMPACT METRIC SPACES

The idea of *compactness* of a set of real numbers can be formulated in several ways — for example, compactness as the Heine-Borel property (see next page) or compactness in terms of the Bolzano-Weierstrass condition (see Exercise 2.45 on page 53).

In this section, we will present a definition of compactness in the context of metric spaces that reduces to the Heine-Borel property in the case of the real line $\mathcal{R}$. We will also prove an important theorem that provides several alternative characterizations of compactness.

DEFINITION 11.3   Compact Set, Compact Metric Space

A subset E of a metric space Ω is called **compact** if every open covering of E has a finite subcovering. If Ω itself is compact, then it is said to be a **compact metric space.**

In practice, we often verify that a space is compact not directly from the definition but, rather, by using conditions equivalent to compactness. These conditions are generalizations of various formulations of compactness on the line. For example, the **Heine-Borel theorem** (page 52) asserts that a subset of $\mathcal{R}^n$ is compact if and only if it is closed and bounded. We will see that we can get an appropriate generalization of the Heine-Borel theorem by using the right analogues of the terms *closed* and *bounded*.

The following simple example shows that a condition more subtle than "closed and bounded" is needed to extend the Heine-Borel theorem to metric spaces. Let $\Omega = Q$, $\rho(x, y) = |x - y|$, and $E = [t, t + 1] \cap Q$, where t is any irrational number. We note that although E is a closed and bounded subset of Q, the collection $\{(t + 1/n, t + 1) \cap Q\}_{n=1}^{\infty}$ is an open covering of E without a finite subcovering.

It is also not clear what replacement for *bounded* is appropriate for general metric spaces. A naive approach would call a subset E of a metric space bounded if the set of distances between points of E is bounded. By Proposition 10.6 on page 369, however, any metric is equivalent to a bounded metric. Because a set that is compact with respect to a metric ρ is also compact with respect to an equivalent metric, it follows that imposing a boundedness condition on the distances between points of a set will be irrelevant to the problem of characterizing compactness.

We will show that one way to generalize the Heine-Borel theorem to arbitrary metric spaces is to replace the term *closed* by *complete* and the term *bounded* by what is called *totally bounded*.

DEFINITION 11.4   Totally Bounded Set

A subset E of a metric space Ω is said to be **totally bounded** if for each $\epsilon > 0$, there exist finitely many points $x_1, x_2, \ldots, x_n$ of E such that $E \subset \bigcup_{j=1}^{n} B_{\epsilon}(x_j)$.

We note that a compact subset E of a metric space Ω is totally bounded because, for each $\epsilon > 0$, the collection of balls $\{ B_{\epsilon}(x) : x \in E \}$ is an open covering of E. However, total boundedness is not by itself sufficient to guarantee compactness, as can be seen by considering again the example where $\Omega = Q$, $\rho(x, y) = |x - y|$, and $E = Q \cap [t, t + 1]$. The following theorem shows, among other things, that total boundedness and completeness together are equivalent to compactness.

□ □ □ **THEOREM 11.2**

For a nonempty subset E of a metric space Ω, the following conditions are equivalent:

a) E is compact.

b) If $\{F_n\}_{n=1}^{\infty}$ is a sequence of closed subsets of Ω such that for each $N \in \mathcal{N}$, we have $E \cap \left(\bigcap_{n=1}^{N} F_n\right) \neq \emptyset$, then $E \cap (\bigcap_{n=1}^{\infty} F_n) \neq \emptyset$.

c) Each sequence of points of E has a subsequence that converges to a point of E.

d) E is complete and totally bounded.

PROOF (a) $\Rightarrow$ (b): Suppose that E is compact. Let $\{F_n\}_{n=1}^{\infty}$ be a sequence of closed sets such that $E \cap \left(\bigcap_{n=1}^{N} F_n\right) \neq \emptyset$ for each $N \in \mathcal{N}$. We claim $E \cap (\bigcap_{n=1}^{\infty} F_n) \neq \emptyset$. Suppose to the contrary. Then

$$E \subset \left(\bigcap_{n=1}^{\infty} F_n\right)^c = \bigcup_{n=1}^{\infty} F_n^c.$$

Therefore, $\{F_n^c\}_{n=1}^{\infty}$ is an open covering of E. Because E is compact, we have that $E \subset \bigcup_{n=1}^{N} F_n^c$ for some $N \in \mathcal{N}$. Thus, $E \cap \left(\bigcap_{n=1}^{N} F_n\right) = \emptyset$. This contradiction shows that $E \cap (\bigcap_{n=1}^{\infty} F_n) \neq \emptyset$.

(b) $\Rightarrow$ (c): Suppose that E satisfies (b). Let $\{x_n\}_{n=1}^{\infty}$ be a sequence of points of E. The sets $F_n = \overline{\{x_k : k > n\}}$, $n \in \mathcal{N}$, satisfy the hypothesis of (b) and, hence, $\bigcap_{n=1}^{\infty} F_n$ contains some point $x \in E$. We will obtain a subsequence of $\{x_n\}_{n=1}^{\infty}$ that converges to x. Because $x \in F_1$, there exists an $n_1 > 1$ such that $\rho(x_{n_1}, x) < 1$. Suppose that integers $n_1 < n_2 < \cdots < n_k$ have been chosen such that $\rho(x_{n_j}, x) < 1/j$ for $j = 1, 2, \ldots, k$. Because $x \in F_{n_k}$, we can find an $n_{k+1} > n_k$ such that $\rho(x_{n_{k+1}}, x) < 1/(k+1)$. Thus, we have defined inductively a subsequence $\{x_{n_k}\}_{k=1}^{\infty}$ of $\{x_n\}_{n=1}^{\infty}$ that converges to the point x of E.

(c) $\Rightarrow$ (d): Suppose E satisfies (c). First we show that E is totally bounded. Let $\epsilon > 0$. If $x_1 \in E$, then either $E \subset B_\epsilon(x_1)$ or there is an $x_2 \in E \setminus B_\epsilon(x_1)$. In the former case, we have found an open ball of radius ϵ that covers E. In the latter case, we again have two possibilities—either $E \subset B_\epsilon(x_1) \cup B_\epsilon(x_2)$ or there is an $x_3 \in E \setminus (B_\epsilon(x_1) \cup B_\epsilon(x_2))$.

Clearly, we can continue with this line of reasoning to obtain either a finite collection of balls of radius ϵ that cover E or a sequence $\{x_n\}_{n=1}^{\infty} \subset E$ that satisfies $x_{n+1} \notin \bigcup_{j=1}^{n} B_\epsilon(x_j)$ for all $n \in \mathcal{N}$. The latter case, however, contradicts (c) because it implies $\rho(x_n, x_m) \geq \epsilon$ for $m \neq n$ which, in turn, implies that $\{x_n\}_{n=1}^{\infty}$ cannot have a convergent subsequence. Hence, E is totally bounded.

Next we show that E is complete. But this follows easily from Exercise 10.55 on page 378, which states that a Cauchy sequence with a convergent subsequence is convergent.

(d) $\Rightarrow$ (c): We will use a famous argument due to Georg Cantor. Let $\{x_n\}_{n=1}^{\infty}$ be a sequence of points of E. Since E is totally bounded, we can find a finite number of open balls of radius $1/2$ that cover E. It follows that one of those balls must contain x_n for infinitely many n. Hence, it is possible to find a subsequence of $\{x_n\}_{n=1}^{\infty}$ whose terms are all contained in a single ball. It is convenient to denote the nth term of this subsequence by $x_{[1,n]}$. Then we have $\rho(x_{[1,n]}, x_{[1,m]}) < 1$ for $n, m \in \mathcal{N}$. Similarly, by covering E with finitely many open balls of radius $1/4$, we can find a subsequence $\{x_{[2,n]}\}_{n=1}^{\infty}$ of $\{x_{[1,n]}\}_{n=1}^{\infty}$ such that $\rho(x_{[2,n]}, x_{[2,m]}) < 1/2$ for $n, m \in \mathcal{N}$.

By proceeding inductively, we obtain an infinite sequence of subsequences $\{\{x_{[k,n]}\}_{n=1}^{\infty}\}_{k=1}^{\infty}$ such that $\{x_{[k+1,n]}\}_{n=1}^{\infty}$ is a subsequence of $\{x_{[k,n]}\}_{n=1}^{\infty}$ and $\rho(x_{[k,n]}, x_{[k,m]}) < 1/k$ for $m, n \in \mathcal{N}$. It follows that $\{x_{[n,n]}\}_{n=1}^{\infty}$ is a subsequence of $\{x_n\}_{n=1}^{\infty}$ that satisfies $\rho(x_{[k,k]}, x_{[j,j]}) < \max\{1/j, 1/k\}$. Thus, $\{x_{[n,n]}\}_{n=1}^{\infty}$ is a Cauchy subsequence of $\{x_n\}_{n=1}^{\infty}$ and, so, by completeness, that subsequence converges to a point of E.

(d) $\Rightarrow$ (a): Let E satisfy (d). Suppose for the moment we can show that (E, ρ) is separable. Then, by Proposition 11.1 (page 399) and Proposition 11.2 (page 399), E has the Lindelöf property. Thus, if $\mathcal{O}$ is an open covering of E, then it has a countable subcovering $\{O_n\}_{n=1}^{\infty}$.

We claim that $\{O_n\}_{n=1}^{\infty}$ has a finite subcovering. Otherwise, we could choose an element $x_n \in E \setminus \bigcup_{j=1}^{n} O_j$ for $n = 1, 2, \ldots$. Since we have already shown that (c) and (d) are equivalent, it follows that the sequence $\{x_n\}_{n=1}^{\infty}$ has a subsequence $\{x_{n_k}\}_{k=1}^{\infty}$ that converges to a point $x \in E$. Because $\{O_n\}_{n=1}^{\infty}$ is a covering of E, we have $x \in O_m$ for some m. Because $\lim_{k\to\infty} x_{n_k} = x$ and, because O_m is open, it follows that there is a k such that $n_k > m$ and $x_{n_k} \in O_m$. But this is a contradiction as $x_{n_k} \in E \setminus \bigcup_{j=1}^{n_k} O_j \subset E \setminus O_m$.

To complete the proof of (d) $\Rightarrow$ (a), we need to show that (E, ρ) is separable. If $k \in \mathcal{N}$, then, because E is totally bounded, there are points $x_{j,k} \in E$, $j = 1, 2, \ldots, m_k$, such that $E \subset \bigcup_{j=1}^{m_k} B_{1/k}(x_{j,k})$. Let $A = \{x_{j,k} : 1 \le j \le m_k, k \in \mathcal{N}\}$. Then A is countable.

Now let $B_\epsilon(x)$ be an open ball centered at an element $x \in E$. Then, by choosing $1/k < \epsilon$, we can find a j with $1 \le j \le m_k$ such that $\rho(x, x_{j,k}) < 1/k$. It follows that $x_{j,k} \in B_\epsilon(x)$. Thus, every open ball around a point of E contains a point of A. Hence, A is a countable dense subset of E and, so, (E, ρ) is separable. ∎

In the last two paragraphs of the proof of Theorem 11.2, we established the following result.

□ □ □ **COROLLARY 11.2**

A totally bounded metric space is separable.

EXAMPLE 11.3   *Illustrates Theorem 11.2*

Let $\| \ \|$ denote one of the norms defined in Example 10.4 (or 10.5) on page 366. By Proposition 10.10 on page 376 and Exercise 10.59 (or 10.60) on page 378, a subset E of $\mathcal{R}^n$ (or $\mathbb{C}^n$) is complete if and only if it is closed. Exercise 11.13 shows that E is totally bounded if and only if E is bounded, that is, if and only if $\sup\{\|x\| : x \in E\} < \infty$. From Theorem 11.2, we can now deduce the classical Heine-Borel theorem: *A subset of $\mathcal{R}^n$ (or $\mathbb{C}^n$) is compact if and only if it is closed and bounded.* □

A set E in a normed space $(\Omega, \| \ \|)$ is called **bounded** if

$$\sup\{ \|x\| : x \in E \} < \infty.$$

Example 11.3 suggests that in a normed space, total boundedness might be equivalent to boundedness. That this statement is not correct is shown by the following example.

EXAMPLE 11.4 *A Noncompact, Closed and Bounded Set*

Refer to Exercise 10.48(d) on page 378. The closed unit ball $\overline{B}_1(0)$ in the space $\ell^2(\mathcal{N})$ is closed and bounded. For each $n \in \mathcal{N}$, let $e_n(k) = 1$ if $k = n$ and 0 if $k \neq n$. As $\|e_n - e_m\|_2 = \sqrt{2}$ for $n \neq m$, it follows that no ball of radius $1/2$ can contain more than one e_n. Thus, the sequence $\{e_n\}_{n=1}^{\infty}$ of elements of $\overline{B}_1(0)$ cannot be contained in a finite union of balls of radius $1/2$. Hence, $\overline{B}_1(0)$ is not totally bounded and, so, by Theorem 11.2(d), is not compact. □

Properties of Compact Metric Spaces

Next we discuss some useful properties of compact metric spaces. Proofs will be left for the exercises.

DEFINITION 11.5 **The Lebesgue Number of a Covering**

Let $\mathcal{O}$ be an open covering of a metric space (Ω, ρ). A number $\lambda > 0$ is called a **Lebesgue number** of $\mathcal{O}$ if for each $x \in \Omega$, the ball $B_\lambda(x)$ is entirely contained in some member of $\mathcal{O}$.

The following result is left to the reader as Exercise 11.21.

□ □ □ THEOREM 11.3

Let (Ω, ρ) be a compact metric space. Then every open covering of Ω has a Lebesgue number.

DEFINITION 11.6 **Uniformly Continuous Function**

Let (Ω, ρ) and (Λ, σ) be metric spaces. A function $f: \Omega \to \Lambda$ is called **uniformly continuous** if for each $\epsilon > 0$, there is a $\delta > 0$ such that $\sigma(f(x), f(y)) < \epsilon$ whenever $\rho(x, y) < \delta$.

Note: A crucial element of Definition 11.6 is that δ depends only on ϵ. It has no dependence on x and y.

The following result is left to the reader as Exercise 11.22.

□ □ □ THEOREM 11.4

Suppose (Ω, ρ) and (Λ, σ) are metric spaces, Ω is compact, and $f: \Omega \to \Lambda$ is continuous. Then f is uniformly continuous.

Exercises for Section 11.2

11.13 Consider $\mathcal{R}^n$ equipped with any one of the norms in Example 10.4 on page 366.
a) Show that a subset E of $\mathcal{R}^n$ is totally bounded if and only if it is bounded, that is, if and only if the set of norms of elements of E is bounded as a subset of $\mathcal{R}$.
b) Show that part (a) holds when $\mathcal{R}^n$ is replaced by $\mathbb{C}^n$. (Refer to Example 10.5 on page 366.)

11.14 In a metric space Ω, let $\{x_n\}_{n=1}^{\infty}$ be a sequence such that $\lim_{n\to\infty} x_n = x$. Show that the set $\{x_n : n = 1, 2, \ldots\} \cup \{x\}$ is compact.

11.15 Compactness can also be expressed in terms of the Bolzano-Weierstrass property. Let (Ω, ρ) be a metric space and $E \subset \Omega$. A point $x \in \Omega$ is called an **accumulation point** of E if for each $\epsilon > 0$, there is a $y \in E$ such that $0 < \rho(x, y) < \epsilon$. Prove that E is compact if and only if every infinite subset of E has an accumulation point that is a member of E. *Hint:* Show that this condition is equivalent to (c) of Theorem 11.2 on page 403.

11.16 Let $y \in \ell^2(\mathcal{N})$ and $K = \{x \in \ell^2(\mathcal{N}) : |x(j)| \le |y(j)| \text{ for each } j \in \mathcal{N}\}$. Show that K is a compact subset of $\ell^2(\mathcal{N})$.

11.17 Refer to Exercise 10.47 on pages 377–378. Let K be a compact subset of a metric space (Ω, ρ) and let $x \in \Omega$. Show there is an element $y \in K$ such that $\rho(x, y) = \rho(x, K)$.

11.18 Refer to Exercise 10.47 on pages 377–378. In a metric space (Ω, ρ), let F and K be, respectively, closed and compact subsets such that $F \cap K = \emptyset$. Show that $\rho(F, K) > 0$.

11.19 Consider the normed space $(C([a, b]), \|\ \|_\infty)$ discussed in Example 10.8 on page 368. Show that the closed unit ball $\overline{B}_1(0)$ is not compact.

11.20 Suppose (Ω, ρ) and (Λ, σ) are metric spaces. Let $\Omega \times \Lambda$ be given the product topology.
a) Show that $\Omega \times \Lambda$ is metrizable.
b) Show that if K and H are compact subsets of Ω and Λ, respectively, then $K \times H$ is a compact subset of $\Omega \times \Lambda$.

11.21 Prove Theorem 11.3 on page 405.

11.22 Prove Theorem 11.4 on page 405.

11.23 Let (Ω, ρ) and (Λ, σ) be metric spaces and let $f: \Omega \to \Lambda$ be continuous. Show that if K is a compact subset of Ω, then $f(K)$ is a compact subset of Λ. In words, the continuous image of a compact space is compact.

11.24 Prove that a continuous real-valued function on a compact metric space attains maximum and minimum values. *Hint:* See Exercise 11.23.

11.3 COMPACT TOPOLOGICAL SPACES

In Section 11.2, we examined compact metric spaces. We are now ready to discuss compactness in the setting of arbitrary topological spaces. Our main goal is to prove a generalization of Theorem 11.2 on page 403. Subsequently, we will derive some useful properties of compact topological spaces.

DEFINITION 11.7   Compact Set, Compact Topological Space

A subset E of a topological space Ω is called **compact** if every open covering of E has a finite subcovering. If Ω itself is compact, then it is said to be a **compact topological space.**

Remark: Certainly, any compact metric space satisfies Definition 11.7. Later, we will give examples of nonmetrizable compact topological spaces.

We note that E is a compact subset of Ω if and only if E equipped with the relative topology is a compact topological space. We observe also that the union of a finite collection of compact sets is compact.

By studying conditions (a)–(d) in Theorem 11.2, we find that only (d) involves the use of a metric in a crucial way — the conditions (a)–(c) have natural generalizations to the setting of any topological space.

We can generalize condition (c) by passing from sequences to nets. And we can generalize condition (b) by introducing the finite intersection property: A collection $\mathcal{C}$ of subsets of a set Ω is said to have the **finite intersection property** if the intersection of each finite subcollection of $\mathcal{C}$ is nonempty.

□ □ □ **THEOREM 11.5**

The following conditions on a topological space Ω are equivalent:
a) Ω *is compact.*
b) *If a collection $\mathcal{C}$ of closed subsets of Ω has the finite intersection property, then $\bigcap_{F \in \mathcal{C}} F \neq \emptyset$.*
c) *Every net in Ω has a convergent subnet.*

PROOF   The equivalence of (a) and (b) is left for Exercise 11.25.

(b) $\Rightarrow$ (c): Suppose (b) holds. Let $\{x_\iota\}_{\iota \in I}$ be a net in Ω that has index set I with relation $\preceq$. For each index ι, let $F_\iota = \overline{\{x_\eta : \iota \preceq \eta\}}$. We claim that the collection $\{F_\iota : \iota \in I\}$ of closed subsets of Ω has the finite intersection property. For, if $\iota_1, \iota_2, \ldots, \iota_n$ are indices, then, because I is directed, there is an index ι_0 such that $\iota_j \preceq \iota_0$ for each $j = 1, 2, \ldots, n$. It follows that $F_{\iota_0} \subset F_{\iota_j}$ for each j and, so, $\bigcap_{j=1}^n F_{\iota_j} \neq \emptyset$. Hence, by (b), $\bigcap_{\iota \in I} F_\iota$ contains an element x.

We will construct a subnet of $\{x_\iota\}_{\iota \in I}$ that converges to x. Let $\mathfrak{N}$ denote the collection of all open sets that contain x. For each $U \in \mathfrak{N}$ and $\iota \in I$, we have $\{x_\eta : \iota \preceq \eta\} \cap U \neq \emptyset$. Applying the axiom of choice, we obtain a function $f: \mathfrak{N} \times I \to I$ such that $\iota \preceq f(U, \iota)$ and $x_{f(U,\iota)} \in U$ for each pair (U, ι). We define a relation $\vartriangleleft$ on $\mathfrak{N} \times I$ as follows:

$$(U, \iota) \vartriangleleft (V, \eta) \text{ if } f(U, \iota) \preceq f(V, \eta) \text{ and } V \subset U.$$

It is not hard to show that $\mathfrak{N} \times I$ is a directed set with respect to the relation $\vartriangleleft$. (See Exercise 11.26.) Therefore, the net defined on $\mathfrak{N} \times I$ by $y_{(U,\iota)} = x_{f(U,\iota)}$ is a subnet of $\{x_\iota\}_{\iota \in I}$.

All that remains is to show that $\lim y_{(U,\iota)} = x$. Given $W \in \mathfrak{N}$, we choose any $\iota_0 \in I$. If $(W, \iota_0) \vartriangleleft (U, \iota)$, then $y_{(U,\iota)} = x_{f(U,\iota)} \in U \subset W$. It follows from the definition of convergence of nets that $\lim y_{(U,\iota)} = x$.

(c) $\Rightarrow$ (b): Suppose that (c) holds and that $\mathcal{C}$ is a collection of closed sets that has the finite intersection property. If $\mathcal{C}^*$ is the collection that consists of finite intersections of members of $\mathcal{C}$, then, clearly, $\bigcap_{C \in \mathcal{C}} C = \bigcap_{F \in \mathcal{C}^*} F$. Thus, to show that (c) implies (b), it is enough to show that $\bigcap_{F \in \mathcal{C}^*} F \neq \emptyset$.

The collection $\mathcal{C}^*$ is a directed set with respect to the relation $\preceq$ defined by $F_1 \preceq F_2$ if $F_2 \subset F_1$. Applying the axiom of choice, we obtain a net $\{x_F\}_{F \in \mathcal{C}^*}$, where $x_F \in F$ for each $F \in \mathcal{C}^*$.

From (c), we know that there is a subnet $\{x_{F_\kappa}\}_{\kappa \in K}$, with index set K and corresponding relation $\lhd$, that has a limit x. Given an $F \in \mathcal{C}^*$, there is a $\kappa \in K$ such that $F \preceq F_\kappa$. Thus, $x_{F_\eta} \in F$ when $\kappa \lhd \eta$. Because F is closed, Proposition 10.12 on page 382 implies that $x \in F$. As F was chosen arbitrarily from $\mathcal{C}^*$, we have that $x \in \bigcap_{F \in \mathcal{C}^*} F$. ∎

Properties of Compact Topological Spaces

From Theorem 11.5, we can derive one of the most useful properties of compact spaces. In words, it states that the continuous image of a compact space is compact.

□ □ □ **THEOREM 11.6**

Let Ω be a compact topological space and $f \colon \Omega \to \Lambda$ be a continuous function. Then $f(\Omega)$ is a compact subset of Λ.

PROOF　Suppose that $\{y_\iota\}_{\iota \in I}$ is a net in $f(\Omega)$. For each $\iota \in I$, we choose an $x_\iota \in \Omega$ such that $f(x_\iota) = y_\iota$, whereby we obtain a net $\{x_\iota\}_{\iota \in I}$ in Ω. By Theorem 11.5, there is a subnet $\{x_{\iota_\kappa}\}_{\kappa \in K}$ that has a limit $x \in \Omega$. It follows from Theorem 10.1 on page 382 that $\lim y_{\iota_\kappa} = \lim f(x_{\iota_\kappa}) = f(x)$. Noting that $f(x) \in f(\Omega)$, we conclude, by applying Theorem 11.5 again, that $f(\Omega)$ is compact. ∎

The following corollary of Theorem 11.6 is left to the reader as an exercise. (See Exercise 11.27.)

□ □ □ **COROLLARY 11.3**

If Ω is compact and f is a real-valued continuous function on Ω, then there exist points $x_1, x_2 \in \Omega$ such that $f(x_1) = \sup f(\Omega)$ and $f(x_2) = \inf f(\Omega)$.

Next, we discuss relationships between compactness and separation properties. The first result is left to the reader as Exercise 11.28.

□ □ □ **THEOREM 11.7**

a) A closed subset of a compact space is compact.
b) A compact subset of a Hausdorff space is closed.

□ □ □ **COROLLARY 11.4**

Let Ω be a compact space and Λ a Hausdorff space. Suppose that $f \colon \Omega \to \Lambda$ is continuous, one-to-one, and onto. Then f^{-1} is continuous and, so, f is a homeomorphism.

PROOF　From Theorem 10.1 on page 382, it suffices to prove that $(f^{-1})^{-1}(F) = f(F)$ is closed in Λ when F is closed in Ω. But, if F is closed in Ω, then, by Theorem 11.7(a), F is compact. Hence, $f(F)$ is compact by Theorem 11.6. Applying Theorem 11.7(b), we conclude that $f(F)$ is closed. ∎

The following corollary is also left to the reader as an exercise. (See Exercise 11.29.)

□ □ □ **COROLLARY 11.5**

Let $\mathcal{T}$ and $\mathcal{U}$ be topologies on a set Ω such that $\mathcal{T}$ is weaker than $\mathcal{U}$. If $(\Omega, \mathcal{T})$ is Hausdorff and $(\Omega, \mathcal{U})$ is compact, then $\mathcal{T} = \mathcal{U}$.

□ □ □ **THEOREM 11.8**

A compact Hausdorff space is a normal space.

PROOF   Suppose that Ω is a compact Hausdorff space and that A and B are disjoint closed subsets of Ω. Because Ω is compact, Theorem 11.7(a) implies that A and B are also compact.

We must find disjoint open sets U and V containing A and B, respectively. Let b be a fixed, but arbitrary, element of B. Since Ω is a Hausdorff space, we can, for each $a \in A$, find disjoint open sets O_a and P_a containing a and b, respectively. The collection $\{ O_a : a \in A \}$ is an open covering of A. As A is compact, there is a finite subcovering $\{ O_{a_j} : j = 1, 2, \ldots, m \}$.

Now, let $U_b = \bigcup_{j=1}^{m} O_{a_j}$ and $V_b = \bigcap_{j=1}^{m} P_{a_j}$. Then U_b is an open set that contains A, V_b is an open set that contains b, and $U_b \cap V_b = \emptyset$. The open covering $\{ V_b : b \in B \}$ of B has a finite subcovering $\{ V_{b_k} : k = 1, 2, \ldots, n \}$. Let $V = \bigcup_{k=1}^{n} V_{b_k}$ and $U = \bigcap_{k=1}^{n} U_{b_k}$. Then U and V are disjoint open sets that contain A and B, respectively. ■

The next corollary follows immediately from Theorem 11.8 and Theorem 11.1 on page 400.

□ □ □ **COROLLARY 11.6**

A second countable compact Hausdorff space is metrizable.

It is useful to note that Theorem 11.8 together with Urysohn's lemma (Theorem 10.2 on page 388) show that compact Hausdorff spaces carry an abundance of real-valued continuous functions.

Exercises for Section 11.3

11.25   Prove the equivalence of (a) and (b) in Theorem 11.5 on page 407.

11.26   Prove that the set $\mathfrak{N} \times I$, defined in the proof of (b) $\Rightarrow$ (c) in Theorem 11.5 on page 407, is directed with respect to the relation $\triangleleft$ defined there.

11.27   Prove Corollary 11.3.

11.28   Prove Theorem 11.7.

11.29   Prove Corollary 11.5.

11.30   Let Ω be a compact Hausdorff space. Suppose there is a sequence $\{f_n\}_n$ of continuous real-valued functions on Ω that has the following property: If $x \neq y$, then there is an n such that $f_n(x) \neq f_n(y)$. Prove that Ω is metrizable.

11.31   Refer to Exercise 11.9 on page 401. Show that, in a first countable compact Hausdorff space, every sequence has a convergent subsequence.

11.32   Suppose that Ω and Λ are compact spaces and that $\Omega \times \Lambda$ is given the product topology. Show that $\Omega \times \Lambda$ is compact.

★11.33 Let Ω be a topological space. A function $f:\Omega \to [-\infty, \infty)$ is said to be **upper semi-continuous** if $f^{-1}([-\infty, r))$ is open for each real number r; a function g is said to be **lower semicontinuous** if $-g$ is upper semicontinuous.
  a) Show that an upper semicontinuous function on a compact space is bounded above and attains the sup of its range.
  b) Show that a lower semicontinuous function on a compact space is bounded below and attains the inf of its range.

★11.34 Refer to Exercise 11.33. Suppose that f is an upper semicontinuous function on a compact Hausdorff space Ω.
  a) Prove that $f(x) = \inf\{ h(x) : h$ is continuous and $f < h\}$ for each $x \in \Omega$.
  b) State and prove an analogous result to part (a) for lower semicontinuous functions.

11.35 Refer to Exercise 11.33, Definition 8.6 (page 282), and Example 10.8 (page 368). Show that the mapping $f \to V_a^b f$ defines a lower semicontinuous function on the normed space $(C([a,b]), \|\,\|_\infty)$.

11.4 LOCALLY COMPACT SPACES

The space of real numbers $\mathcal{R}$ is not compact. We can see this directly by noting that the open covering $\{(-n, n) : n \in \mathcal{N}\}$ of $\mathcal{R}$ has no finite subcovering; or we can deduce it from the Heine-Borel theorem.

Although $\mathcal{R}$ is not compact, compactness plays an important role in its analysis because every element of $\mathcal{R}$ is contained in an open set that has compact closure. Many topological spaces share with $\mathcal{R}$ this important property, which is called *local compactness*.

DEFINITION 11.8 **Locally Compact Topological Space**

A topological space Ω is said to be **locally compact** if for each $x \in \Omega$ there is an open set W such that $x \in W$ and $\overline{W}$ is compact.

It is not hard to see that the spaces $\mathcal{R}^n$ and $\mathbb{C}^n$ in Examples 10.4 and 10.5, respectively, on page 366 are locally compact. The spaces $\mathcal{L}^1(\mu)$, $\mathcal{L}^2(\mu)$, and $\mathcal{L}^\infty(\mu)$ of Example 10.6 on pages 366–367 are not locally compact except in certain special instances. (See Exercise 11.36.)

In most cases of interest, the property of local compactness appears in conjunction with the Hausdorff property. The next several results provide some important properties of locally compact Hausdorff spaces.

□ □ □ **PROPOSITION 11.3**

Let O be an open subset of a locally compact Hausdorff space Ω.
  a) If $x \in O$, there is an open set V such that $\overline{V}$ is compact and $x \in V \subset \overline{V} \subset O$.
  b) If K is compact and $K \subset O$, there is an open set W such that $\overline{W}$ is compact and $K \subset W \subset \overline{W} \subset O$.

PROOF a) Let W be an open set containing x such that $\overline{W}$ is compact. By Theorem 11.8 on page 409, $\overline{W}$ equipped with the relative topology is a normal space.

We note that, in the relative topology of the compact space $\overline{W}$, $W \cap O$ is open and $\{x\}$ is closed. Hence, by Proposition 10.14 on page 387, there is a set V that has the following properties: $x \in V$, V is open in the relative topology of $\overline{W}$, and the closure of V in the relative topology of $\overline{W}$ is contained in $W \cap O$. Because $\overline{W}$ is closed in Ω, it follows that the closure of V in the relative topology of $\overline{W}$ coincides with its closure in Ω. Hence,

$$x \in V \subset \overline{V} \subset W \cap O \subset O.$$

The proof of (a) will be complete if we can show that V is open as a subset of Ω. By the definition of relative topology, there is an open subset $U \subset \Omega$ with $V = U \cap \overline{W}$. Then

$$V = V \cap W = U \cap \overline{W} \cap W = U \cap W.$$

Thus, V is open in Ω.

b) By part (a) we can, for each $x \in K$, find an open set V_x whose closure is compact and satisfies $x \in V_x \subset \overline{V_x} \subset O$. Because $K \subset \bigcup_{x \in K} V_x$ and K is compact, we can find finitely many points $x_1, x_2, \ldots, x_n$ of K such that $K \subset \bigcup_{j=1}^n V_{x_j}$. Letting $W = \bigcup_{j=1}^n V_{x_j}$, we obtain

$$K \subset W \subset \overline{W} = \bigcup_{j=1}^n \overline{V_{x_j}} \subset O.$$

As $\overline{W}$ is a finite union of compact sets, it is compact. ∎

Using Proposition 11.3, we can prove a version of Urysohn's lemma (Theorem 10.2 on page 388) for locally compact Hausdorff spaces.

□ □ □ **THEOREM 11.9**

Suppose that Ω is a locally compact Hausdorff space and that O and K are, respectively, open and compact subsets of Ω such that $K \subset O$. Then there is a continuous function $f \colon \Omega \to [0,1]$ such that $f(x) = 1$ for $x \in K$ and $f(x) = 0$ for $x \in O^c$.

PROOF By applying Proposition 11.3 twice, we obtain open sets W_1 and W_2 such that $\overline{W_2}$ is compact and

$$K \subset W_1 \subset \overline{W_1} \subset W_2 \subset \overline{W_2} \subset O.$$

By Theorem 11.7 on page 408, K is a closed subset of $\overline{W_2}$. Because the space $\overline{W_2}$ equipped with the relative topology is normal (Theorem 11.8 on page 409), it follows from Urysohn's lemma that there is a continuous function $g \colon \overline{W_2} \to [0,1]$ with g equal to 1 on K and 0 on $\overline{W_2} \setminus W_1$.

We now define a function $f \colon \Omega \to [0,1]$ by letting f be equal to g on $\overline{W_2}$ and equal to 0 on $\Omega \setminus \overline{W_2}$. It is left as an exercise for the reader to show that f is continuous on Ω. (See Exercise 11.38.) ∎

Theorem 11.9 is the basis of an important construction related to coverings of compact subsets of locally compact Hausdorff spaces. To describe this construction, it is helpful to introduce the following terminology. Let f be a complex-valued function on a topological space Ω. The closure of the set of points where f is not 0 is called the **support of f** and is denoted by **supp f**. Hence, supp $f = \overline{\{x \in \Omega : f(x) \neq 0\}}$.

◻ ◻ ◻ **THEOREM 11.10 Partition of Unity**

Let Ω be a *locally compact Hausdorff space*, K *a compact subset of* Ω, *and* $\mathcal{O}$ *an open covering of* K. *Then there are finitely many continuous real-valued functions* f_1, f_2, ..., f_n *on* Ω *such that:*

a) $f_j \geq 0$ *for each* j.

b) *For each* j, *there is an* $O_j \in \mathcal{O}$ *such that* $\operatorname{supp} f_j \subset O_j$.

c) $\sum_{j=1}^n f_j(x) = 1$ *for each* $x \in K$.

d) $\sum_{j=1}^n f_j(x) \leq 1$ *for each* $x \in \Omega$.

PROOF For each $x \in K$ we choose $O_x \in \mathcal{O}$ such that $x \in O_x$. By Proposition 11.3, there is an open set V_x such that $\overline{V_x}$ is compact and $x \in V_x \subset \overline{V_x} \subset O_x$. By Theorem 11.9, there is, for each $x \in K$, a continuous function g_x such that $0 \leq g_x \leq 1$, $g_x(x) = 1$, and $g_x(y) = 0$ for $y \in V_x^c$. We note that $\operatorname{supp} g_x \subset \overline{V_x} \subset O_x$.

Since $\{ g_x^{-1}((0, \infty)) : x \in K \}$ is an open covering of K, there are a finite number of points x_1, x_2, ..., x_n of K such that $K \subset \bigcup_{j=1}^n g_{x_j}^{-1}((0, \infty))$. Hence, the function $g = \sum_{j=1}^n g_{x_j}$ is strictly positive on K.

By Corollary 11.3 on page 408, we have $a = \inf g(K) > 0$. The closed set $F = g^{-1}((-\infty, a/2])$ is disjoint from K. Thus, again by Theorem 11.9, we find that there is a continuous function h such that $0 \leq h \leq 1$, $h(x) = 0$ for $x \in K$, and $h(x) = 1$ for $x \in F$. Because the function $g + h$ is positive everywhere on Ω, it follows that the functions

$$f_j = g_{x_j}(1 + h/n)/(g + h), \qquad j = 1, 2, \ldots, n,$$

are continuous. It is easy to check that the functions f_1, f_2, ..., f_n satisfy conditions (a)–(d). ∎

Theorem 11.9 can also be applied to extend Urysohn's metrization theorem (Theorem 11.1 on page 400) to locally compact Hausdorff spaces.

◻ ◻ ◻ **THEOREM 11.11**

If a topological space is locally compact, Hausdorff, and second countable, then it is metrizable.

PROOF Let Ω be a locally compact Hausdorff space with a countable neighborhood basis $\mathfrak{N}$. Let

$$\mathcal{W} = \{ (U, V) : U, V \in \mathfrak{N}, \ \overline{U} \text{ is compact, and } \overline{U} \subset V \}.$$

We will show that given an open set O and a point $x \in O$, there is a pair $(U, V) \in \mathcal{W}$ such that $x \in U$ and $V \subset O$. Since $\mathfrak{N}$ is a neighborhood basis, there is a $V \in \mathfrak{N}$ such that $x \in V \subset O$. By Proposition 11.3 on page 410, there is an open set W such that $\overline{W}$ is compact and $x \in W \subset \overline{W} \subset V$. We can now choose a $U \in \mathfrak{N}$ such that $x \in U \subset W$. It follows from Theorem 11.7 on page 408 that $\overline{U}$ is compact. Thus, $(U, V) \in \mathcal{W}$.

The remainder of the proof is the same as the proof of Urysohn's metrization theorem, where Theorem 11.9 on page 411 is used as a replacement for Urysohn's lemma. ∎

It is possible to extract from the proof of Theorem 11.11 the following corollary whose proof is left to the reader as Exercise 11.43.

□ □ □ **COROLLARY 11.7**

Let Ω be a locally compact space. Then there is a neighborhood basis $\mathfrak{N}$ such that $\overline{U}$ is compact for each $U \in \mathfrak{N}$. Furthermore, if Ω is second countable, then $\mathfrak{N}$ can be chosen to be countable.

Let Ω be a second countable topological space. Suppose that Ω has a countable neighborhood basis $\mathfrak{N} = \{U_n\}_{n=1}^{\infty}$ such that the closure of each U_n is compact. Let $W_1 = U_1$. The sets in $\mathfrak{N}$ cover $\overline{U_1}$ and, so, there is an integer $n_2 > 1$ such that $\overline{W_1} \subset \bigcup_{j=1}^{n_2} U_j$. Let $W_2 = \bigcup_{j=1}^{n_2} U_j$. Then $\overline{W_2}$ is compact and, hence, we can find an integer $n_3 > n_2$ such that $\overline{W_2} \subset \bigcup_{j=1}^{n_3} U_j$. Let $W_3 = \bigcup_{j=1}^{n_3} U_j$. Continuing in this fashion, we obtain an infinite sequence of sets that satisfies the conditions delineated in the next definition.

DEFINITION 11.9   Exhaustion

A sequence $\{W_n\}_{n=1}^{\infty}$ of subsets of a topological space Ω is called an **exhaustion** if it satisfies the following conditions:
a) Each W_n is open.
b) Each $\overline{W_n}$ is compact.
c) $\overline{W_n} \subset W_{n+1}$ for each n.
d) $\Omega = \bigcup_{n=1}^{\infty} W_n$.

Corollary 11.7 and the paragraph preceding Definition 11.9 show that a second countable locally compact space has an exhaustion. Here are some concrete examples of exhaustions.

EXAMPLE 11.5   Illustrates Definition 11.9

a) $\{(-n, n)\}_{n=1}^{\infty}$ is an exhaustion of $\mathcal{R}$.
b) $\{(1/n, 1 - 1/n)\}_{n=1}^{\infty}$ is an exhaustion of the interval $(0, 1)$.
c) $\{B_n(0)\}_{n=1}^{\infty}$ is an exhaustion of the normed space $(\mathcal{R}^n, \|\ \|_2)$ discussed in Example 10.4 on page 366. □

In the next section, exhaustions will be used to obtain metrization results for certain spaces of functions.

Compactification

Our next theorem shows that it is possible to turn any locally compact Hausdorff space into a compact space by the addition of a single point. To see how, first consider the set $\mathcal{N}$ of positive integers with the discrete topology. This space is locally compact, but not compact, and its compact subsets coincide with its finite subsets. We would like to add a "point at infinity," denoted ω, to this space to turn it into a compact space. The problem is to find the right topology for the set $\mathcal{N} \cup \{\omega\}$.

To see what to do, we pass from the set $\mathcal{N}$ to the subset of real numbers $E = \{1/n : n \in \mathcal{N}\}$ via the function $h(n) = 1/n$. Note that $\overline{E} = E \cup \{0\}$. E is a bounded set of real numbers but is not compact because it is not closed. However, $E \cup \{0\}$ is closed and bounded and, hence, compact with respect to the relative topology inherited from $\mathcal{R}$. We can easily describe the open sets (in the relative topology) of $E \cup \{0\}$: Each subset of E is open, and a subset D of $E \cup \{0\}$ that contains 0 is open if and only if its relative complement $(E \cup \{0\}) \setminus D$ is a finite set.

If we now extend the function h to $\mathcal{N} \cup \{\omega\}$ by $h(\omega) = 0$ and call a subset W of $\mathcal{N} \cup \{\omega\}$ open if $h(W)$ is open in $E \cup \{0\}$, we obtain a topology that makes $\mathcal{N} \cup \{\omega\}$ into a compact space. The open sets of this topology on $\mathcal{N} \cup \{\omega\}$ consist of all subsets of $\mathcal{N}$ as well as all complements of finite subsets of $\mathcal{N}$.

As the next theorem shows, the construction that we just performed can be generalized to arbitrary locally compact Hausdorff spaces. The proof of the theorem is left to the reader as Exercise 11.44.

□ □ □ **THEOREM 11.12 One-Point Compactification**

Suppose that $(\Omega, \mathcal{T})$ is a locally compact Hausdorff space. Let ω be an element not in Ω and set $\Omega^* = \Omega \cup \{\omega\}$. Let $\mathcal{T}^*$ denote the collection of subsets of Ω^* that are either members of $\mathcal{T}$ or whose complements are compact subsets of the space $(\Omega, \mathcal{T})$. Then $\mathcal{T}^*$ is a topology on Ω^* with the following properties:
a) $(\Omega^*, \mathcal{T}^*)$ is compact.
b) $\mathcal{T}$ coincides with the relative topology $\{\Omega \cap W : W \in \mathcal{T}^*\}$.
c) Ω is open in the topological space $(\Omega^*, \mathcal{T}^*)$.
d) Ω is dense in $(\Omega^*, \mathcal{T}^*)$ unless $(\Omega, \mathcal{T})$ is compact.

The space $(\Omega^*, \mathcal{T}^*)$, constructed in Theorem 11.12, is called the **one-point compactification** of Ω.

Exercises for Section 11.4

11.36 Show that the space $\ell^2(\Omega)$ of Example 10.6 on pages 366–367 is locally compact if and only if Ω is finite.

11.37 Suppose that Ω is a locally compact space, that $D \subset \Omega$, and that $\mathcal{T}_D$ is the relative topology on D. Prove that $(D, \mathcal{T}_D)$ is locally compact if D is closed in Ω.

11.38 Show that the function f defined in the last paragraph of the proof of Theorem 11.9 on page 411 is continuous and satisfies $f(K) = \{1\}$ and $f(O^c) = \{0\}$.

11.39 Show that the functions $f_1, f_2, \ldots, f_n$ defined in the last paragraph of the proof of Theorem 11.10 on page 412 satisfy conditions (a)–(d) of that theorem.

11.40 State and prove a version of Tietze's extension theorem (Theorem 10.3 on page 389) for locally compact Hausdorff spaces.

11.41 Let K be a compact subset of a locally compact Hausdorff space Ω. Show that there is a nonnegative continuous function f on Ω such that $K = f^{-1}(0)$ if and only if there is a sequence $\{G_n\}_{n=1}^{\infty}$ of open sets such that $K = \bigcap_{n=1}^{\infty} G_n$.

11.42 Let f be a complex-valued function on a set Ω. A point $x_0 \in \Omega$ is said to be a **peak point** of f if $|f(x)| < |f(x_0)|$ for all $x \neq x_0$. Show that in a metrizable locally compact space Ω, each point is a peak point for some complex-valued continuous function on Ω.

11.43 Prove Corollary 11.7 on page 413.

11.44 Prove Theorem 11.12.

★*11.45* Let f be a continuous function from a locally compact Hausdorff space Ω into a metric space (Λ, ρ). The collection $\mathcal{K}$ of compact subsets of Ω is a directed set with respect to the relation $\subset$. Now, let us define $\lim_{x \to \omega} f(x) = y$ to mean that the net $\left\{ \sup\{ \rho(f(x), y) : x \in K^c \} \right\}_{K \in \mathcal{K}}$ converges to 0. Show that f is the restriction of a continuous function on the one-point compactification of Ω if and only if $\lim_{x \to \omega} f(x)$ exists. *Note:* In case $\Omega = \mathcal{R}^n$, $\lim_{x \to \omega} f(x) = y$ if and only if $\lim_{\|x\| \to \infty} f(x) = y$, where $\| \ \|$ is any one of the norms defined in Example 10.4 on page 366.

11.46 Prove that the one-point compactification of $\mathcal{R}$ is homeomorphic to a circle in $\mathcal{R}^2$.

11.47 Define a "two-point compactification" of $\mathcal{R}$ that is homeomorphic to the interval $[-1, 1]$.

11.48 Prove that the one-point compactification of $\mathbb{C}$ is homeomorphic to the unit sphere $S = \{ x \in \mathcal{R}^3 : \|x\|_2 = 1 \}$. *Hint:* Let $h(z) = (1 + |z|^2)^{-1}(2\Re z, 2\Im z, |z|^2 - 1)$ for $z \in \mathbb{C}$, and $h(\omega) = (0, 0, 1)$.

11.5 FUNCTION SPACES

We will consider, in this section, what it means for a sequence of continuous functions to converge. In particular, we will construct a topology $\mathcal{T}(\Omega, \Lambda)$ for the collection of continuous functions from a topological space Ω to a metric space (Λ, ρ) such that convergence of a sequence with respect to $\mathcal{T}(\Omega, \Lambda)$ corresponds to uniform convergence on compact subsets. Related notions of pointwise and uniform convergence will also be discussed.

For a sequence $\{f_n\}_{n=1}^{\infty}$ of functions from a topological space Ω to a metric space (Λ, ρ), there are several meanings that can be attached to the expression

$$\lim_{n \to \infty} f_n = f. \tag{11.3}$$

One simple way to define (11.3) is the following.

DEFINITION 11.10 **Pointwise Convergence**

We say that a sequence of functions $\{f_n\}_{n=1}^{\infty}$ from a topological space Ω to a metric space (Λ, ρ) **converges pointwise** to the function f if for each $x \in \Omega$ and each $\epsilon > 0$, there is an $N \in \mathcal{N}$ such that $\rho(f_n(x), f(x)) < \epsilon$ whenever $n \geq N$.

Pointwise convergence of a sequence of functions $\{f_n\}_{n=1}^{\infty}$ to a function f requires that the sequence $\{f_n(x)\}_{n=1}^{\infty}$ of elements of Λ converges to $f(x)$ for each $x \in \Omega$. A much more demanding mode of convergence is as follows.

DEFINITION 11.11 **Uniform Convergence**

We say that a sequence of functions $\{f_n\}_{n=1}^{\infty}$ from a topological space Ω to a metric space (Λ, ρ) **converges uniformly** to the function f if for each $\epsilon > 0$, there is an $N \in \mathcal{N}$ such that $\rho(f_n(x), f(x)) < \epsilon$ for all $x \in \Omega$ whenever $n \geq N$.

The essential difference between Definitions 11.10 and 11.11 is that, in the latter, N may not depend on x whereas, in the former, it may.

For many applications, Definition 11.10 is too weak and Definition 11.11 is too strong. In this section, we will be concerned primarily with a mode of convergence that is intermediate between pointwise and uniform convergence. This mode of convergence is as follows.

DEFINITION 11.12 **Uniform Convergence on Compact Subsets**

We say that a sequence of functions $\{f_n\}_{n=1}^{\infty}$ from a topological space Ω to a metric space (Λ, ρ) **converges uniformly on compact subsets** to the function f if for each compact subset $K \subset \Omega$ and each $\epsilon > 0$, there is an $N \in \mathcal{N}$ such that $\rho(f_n(x), f(x)) < \epsilon$ for all $x \in K$ whenever $n \geq N$.

EXAMPLE 11.6 *Illustrates Definitions 11.10–11.12*

Let $\Omega = (0, 1)$, $\Lambda = \mathcal{R}$, and $f_n(x) = \sum_{j=0}^{n} x^j$. The sequence of functions $\{f_n\}_{n=1}^{\infty}$ converges both pointwise and uniformly on compact subsets to the function $f(x) = 1/(1-x)$. But, $\{f_n\}_{n=1}^{\infty}$ does not converge uniformly to f. ☐

Next we introduce some notation that will be used throughout the remainder of the text.

DEFINITION 11.13 **Collection of Continuous Functions**

Let Ω be a topological space and Λ a metric space. Then we denote by $C(\Omega, \Lambda)$ the collection of all continuous functions from Ω to Λ. In case $\Lambda = \mathbb{C}$, we write $C(\Omega)$ for $C(\Omega, \Lambda)$; thus, $C(\Omega)$ is the collection of all complex-valued continuous functions on Ω.

We will construct a topology for $C(\Omega, \Lambda)$ such that convergence of sequences in that topology is the same as uniform convergence on compact subsets. To aid our construction, it will be helpful to have the following notation. For $f, g \in C(\Omega, \Lambda)$ and $S \subset \Omega$, we let

$$\rho_S(f, g) = \sup\{\, \rho(f(x), g(x)) : x \in S \,\}. \tag{11.4}$$

Thus, ρ_S measures how (uniformly) close two functions are on S.

The following proposition, whose proof is left for Exercise 11.50, shows that as a function from $C(\Omega, \Lambda) \times C(\Omega, \Lambda)$ to $[0, \infty]$, ρ_S is almost a metric.

☐ ☐ ☐ **PROPOSITION 11.4**

Let Ω be a topological space and Λ a metric space. The function ρ_S, defined in (11.4), satisfies the following conditions for all $f, g, h \in C(\Omega, \Lambda)$:
a) $0 \leq \rho_S(f, g) \leq \infty$.

b) $\rho_S(f, f) = 0$.

c) $\rho_S(f, g) = \rho_S(g, f)$.

d) $\rho_S(f, g) \leq \rho_S(f, h) + \rho_S(h, g)$.

e) If S is compact, then $\rho_S(f, g) < \infty$.

f) If S is compact, then $|\rho_S(f, h) - \rho_S(g, h)| \leq \rho_S(f, g)$.

g) If Ω is compact, then ρ_Ω is a metric.

If $\rho_S(f, g) = 0$ for some $f \neq g$ or if $\rho_S(f, g) = \infty$ for some f and g, then ρ is not a metric. When S is a compact subset of Ω, then, as Proposition 11.4(e) shows, the latter obstacle cannot arise, but the former still remains. (See Exercise 11.52.) Nevertheless, by considering the entire family $\{\, \rho_K : K \text{ compact}\,\}$, we can produce a topology on $C(\Omega, \Lambda)$ that will be the correct one for studying uniform convergence on compact subsets.

In the next definition, $\rho_K(\cdot, g)$ denotes the function from $C(\Omega, \Lambda)$ to $[0, \infty)$ defined by $\rho_K(\cdot, g)(f) = \rho_K(f, g)$, where $K \subset \Omega$ is compact and $g \in C(\Omega, \Lambda)$.

DEFINITION 11.14   Topology of Uniform Convergence on Compacts

The weak topology on $C(\Omega, \Lambda)$ determined by the family of functions

$$\{\, \rho_K(\cdot, g) : K \text{ compact}, \ g \in C(\Omega, \Lambda)\,\}$$

is called the **topology of uniform convergence on compact subsets** and is denoted $\mathcal{T}(\Omega, \Lambda)$.

Note: Whenever we work with a function space of the form $C(\Omega, \Lambda)$, we will assume that it is equipped with the topology $\mathcal{T}(\Omega, \Lambda)$ unless explicitly stated otherwise.

The next proposition shows that convergence in the topology $\mathcal{T}(\Omega, \Lambda)$ is exactly the same as uniform convergence on compact subsets.

□ □ □   **PROPOSITION 11.5**

Let $\{f_\iota\}_{\iota \in I}$ be a net of functions in $C(\Omega, \Lambda)$. Then $\{f_\iota\}_{\iota \in I}$ converges to f if and only if

$$\lim \rho_K(f_\iota, f) = 0 \tag{11.5}$$

for each compact subset $K \subset \Omega$.

PROOF   By Proposition 10.13(b) on page 383, the net $\{f_\iota\}_{\iota \in I}$ converges to f if and only if

$$\lim \rho_K(f_\iota, g) = \rho_K(f, g) \tag{11.6}$$

for each compact set K and each function $g \in C(\Omega, \Lambda)$. The equivalence of (11.5) and (11.6) follows easily from parts (b) and (f) of Proposition 11.4. ∎

Remark: Because $\rho(f(x), g(x)) \leq \rho_K(f, g)$ for each $x \in K$, it is clear from Proposition 11.5 that convergence of a sequence with respect to the topology $\mathcal{T}(\Omega, \Lambda)$ corresponds to uniform convergence on compact subsets.

Although, in general, $\mathcal{T}(\Omega, \Lambda)$ is not metrizable, it is nevertheless possible to define analogues of Cauchy sequences and a notion of completeness for the space $C(\Omega, \Lambda)$.

DEFINITION 11.15 *k*-Cauchy Sequence; *k*-Complete

A sequence of functions $\{f_n\}_{n=1}^{\infty}$ in $C(\Omega, \Lambda)$ is said to be **k-Cauchy** if for each compact subset K and each $\epsilon > 0$, there is an $N \in \mathcal{N}$ such that $\rho_K(f_n, f_m) < \epsilon$ whenever $n, m \geq N$. If every k-Cauchy sequence in $C(\Omega, \Lambda)$ converges, then we say that $C(\Omega, \Lambda)$ is **k-complete.**

Remark: The concept of a k-Cauchy sequence expresses the idea of a sequence that is "uniformly Cauchy on compact subsets."

The most interesting examples of spaces of the type $C(\Omega, \Lambda)$ occur when Ω is a locally compact Hausdorff space. Theorem 11.14 describes some properties of $C(\Omega, \Lambda)$ in this case. Before stating and proving that theorem, however, we need some preliminary results.

◻ ◻ ◻ **THEOREM 11.13**

Let Ω be a topological space and (Λ, ρ) a metric space. If $\{f_n\}_{n=1}^{\infty}$ is a sequence in $C(\Omega, \Lambda)$ that converges uniformly to a function f, then $f \in C(\Omega, \Lambda)$.

PROOF Let $x_0 \in \Omega$ and $\epsilon > 0$ be given. To establish the continuity of f at x_0, we will show that there is an open set U such that $x_0 \in U$ and $\rho(f(x), f(x_0)) < \epsilon$ whenever $x \in U$. (See Theorem 10.1 on pages 382–383.)

By uniform convergence, there is an N such that $\rho(f_n(x), f(x)) < \epsilon/3$ for all $x \in \Omega$ whenever $n \geq N$. Because f_N is a continuous function, there is an open set U that contains x_0 such that $\rho(f_N(x), f_N(x_0)) < \epsilon/3$ whenever $x \in U$. It follows that for $x \in U$,

$$\rho(f(x), f(x_0)) \leq \rho(f(x), f_N(x)) + \rho(f_N(x), f_N(x_0)) + \rho(f_N(x_0), f(x_0))$$
$$< \epsilon/3 + \epsilon/3 + \epsilon/3 = \epsilon.$$

Hence, f is continuous at x_0. ∎

The proof of the following lemma is left to the reader as Exercise 11.54.

◻ ◻ ◻ **LEMMA 11.1**

Let Ω and Λ be topological spaces and $f: \Omega \to \Lambda$. Suppose that for each $x \in \Omega$, there is an open set U_x such that $x \in U_x$ and $f_{|U_x}$ is continuous. Then f is continuous.

◻ ◻ ◻ **THEOREM 11.14**

Suppose that Ω is a locally compact Hausdorff space and that (Λ, ρ) is a metric space.

a) If Λ is complete, then $C(\Omega, \Lambda)$ is k-complete.

b) If Ω is second countable, then $C(\Omega, \Lambda)$ is metrizable.

c) If Ω is second countable and Λ is complete, then $C(\Omega, \Lambda)$ is complete.

d) If Ω is compact, then the topology $\mathcal{T}(\Omega, \Lambda)$ is induced by the metric ρ_Ω.

PROOF a) Suppose that $\{f_n\}_{n=1}^\infty$ is a k-Cauchy sequence. Applying Definition 11.15 with $K = \{x\}$, we find that the sequence $\{f_n(x)\}_{n=1}^\infty$ is Cauchy in Λ for each $x \in \Omega$. Because Λ is complete, we conclude that, for each $x \in \Omega$, the sequence $\{f_n(x)\}_{n=1}^\infty$ converges in Λ.

Let the function $f: \Omega \to \Lambda$ be defined by $f(x) = \lim_{n\to\infty} f_n(x)$. We will show that f is continuous on Ω. Let $x_0 \in \Omega$ and let W be an open set such that $x_0 \in W$ and $\overline{W}$ is compact. Let $\epsilon > 0$. We can choose N such that $\rho_{\overline{W}}(f_n, f_m) < \epsilon$ for $m, n \geq N$. For each $x \in W$, we have

$$\rho(f_n(x), f(x)) = \lim_{m\to\infty} \rho(f_n(x), f_m(x)) \leq \limsup_{m\to\infty} \rho_{\overline{W}}(f_n, f_m) \leq \epsilon \qquad (11.7)$$

for $n \geq N$. It follows that the restrictions to W of the functions f_n converge uniformly to the restriction of f to W. Hence, by Theorem 11.13 and Lemma 11.1, f is continuous on all of Ω.

It remains to show that the sequence $\{f_n\}_{n=1}^\infty$ converges to f with respect to the topology $\mathcal{T}(\Omega, \Lambda)$. Replacing W by an arbitrary compact subset K in (11.7) and taking the supremum over $x \in K$, we get that $\rho_K(f_n, f) \leq \epsilon$ for $n \geq N$. Hence, by Proposition 11.5 on page 417, $\{f_n\}_{n=1}^\infty$ converges in $C(\Omega, \Lambda)$ to f. The proof of (a) is now complete.

b) If Ω is second countable, then it has an exhaustion $\{W_n\}_{n=1}^\infty$. (See Definition 11.9 on page 413 and the paragraph that follows that definition.) Let $\rho_n = \rho_{\overline{W_n}}$. By Proposition 11.4(e) on page 417, ρ_n is real-valued.

Let $\sigma = \sum_{n=1}^\infty 2^{-n} \rho_n / (1 + \rho_n)$. We claim that σ is a metric on $C(\Omega, \Lambda)$. That σ is a real-valued function follows from Exercise 10.70 on page 385. By Exercise 10.30 on page 370, σ satisfies Definition 10.7(c) on page 364. That σ satisfies Definition 10.7(b) is clear, as are the facts that σ is nonnegative and satisfies $\sigma(f, f) = 0$ for each f. Thus, to prove that σ is a metric, it remains only to show that $\sigma(f, g) = 0$ implies $f = g$. If $\sigma(f, g) = 0$, then $\rho_n(f, g)$ must vanish for each n. Hence, for each $n \in \mathcal{N}$, $f(x) = g(x)$ for all $x \in W_n$. Because $\{W_n\}_{n=1}^\infty$ is an exhaustion, it follows that $f = g$. Consequently, σ is a metric.

Let $\mathcal{T}$ denote the topology on $C(\Omega, \Lambda)$ induced by the metric σ. We will show that $\mathcal{T} = \mathcal{T}(\Omega, \Lambda)$. By the definition of the topology $\mathcal{T}(\Omega, \Lambda)$, for each fixed $g \in C(\Omega, \Lambda)$ and $n \in \mathcal{N}$, the function $\rho_n(\cdot, g)$ is continuous with respect to that topology. The sequence of sums $\sum_{j=1}^n 2^{-j} \rho_j(\cdot, g) / (1 + \rho_j(\cdot, g))$ converges uniformly on $C(\Omega, \Lambda)$ to the function $\sigma(\cdot, g)$. From Theorem 11.13, we conclude that $\sigma(\cdot, g)$ is continuous with respect to $\mathcal{T}(\Omega, \Lambda)$. Because $B_r^\sigma(g) = \sigma(\cdot, g)^{-1}(-\infty, r)$, it follows that every open ball $B_r^\sigma(g)$ is open with respect to $\mathcal{T}(\Omega, \Lambda)$. Hence, every $\mathcal{T}$-open set is $\mathcal{T}(\Omega, \Lambda)$-open, that is, $\mathcal{T}$ is weaker than $\mathcal{T}(\Omega, \Lambda)$.

To complete the proof of (b), we must show that $\mathcal{T}(\Omega, \Lambda)$ is weaker than $\mathcal{T}$. To do so, it suffices, by Exercise 10.80 on page 385, to show that if a net $\{f_\iota\}_{\iota \in I}$ converges to f with respect to the topology $\mathcal{T}$, then it converges to f with respect to the topology $\mathcal{T}(\Omega, \Lambda)$. Now, $\{f_\iota\}_{\iota \in I}$ converges to f with respect to $\mathcal{T}$ if and only if $\lim \sigma(f_\iota, f) = 0$. Let K be an arbitrary compact subset. Then, since the sets in the exhaustion $\{W_n\}_{n=1}^\infty$ are an open covering of K, there is an m such that $K \subset W_m$. Thus,

$$\rho_K(f_\iota, f) \leq \rho_{\overline{W_m}}(f_\iota, f) = \rho_m(f_\iota, f).$$

The inequality

$$2^{-m} \rho_m(f_\iota, f)/(1 + \rho_m(f_\iota, f)) \leq \sigma(f_\iota, f)$$

implies that $\lim \rho_m(f_\iota, f) = 0$. Hence, $\lim \rho_K(f_\iota, f) = 0$. It now follows from Proposition 11.5 on page 417 that $\{f_\iota\}$ converges to f with respect to the topology $\mathcal{T}(\Omega, \Lambda)$. This completes the proof of (b).

c) By (a) and the proof of (b), it suffices to establish that a sequence $\{f_n\}_{n=1}^\infty$ that is Cauchy with respect to the metric σ is k-Cauchy. So, let $\epsilon > 0$ and K a compact subset of Ω. As before, we can choose an m such that $K \subset W_m$. It follows from the definition of σ that

$$2^{-m} \frac{\rho_K(f_n, f_p)}{1 + \rho_K(f_n, f_p)} \leq 2^{-m} \frac{\rho_m(f_n, f_p)}{1 + \rho_m(f_n, f_p)} \leq \sigma(f_n, f_p).$$

If N is large enough so that $\sigma(f_n, f_p) < 2^{-m}\epsilon/(1 + \epsilon)$ for $n, p \geq N$, we obtain $\rho_K(f_n, f_p) < \epsilon$. Thus, $\{f_n\}_{n=1}^\infty$ is k-Cauchy. The proof of (c) is now complete.

d) The proof of part (d) is left for Exercise 11.55. ■

EXAMPLE 11.7   *The Case $\Lambda = \mathbb{C}$*

Let Ω be a locally compact Hausdorff space. Besides having a metric space structure derived from the usual distance function $\rho(z, w) = |z - w|$, the space $C(\Omega)$ has a linear-space structure, where addition and scalar multiplication are defined pointwise.

We now consider the relationship between the linear-space structure and the topology $\mathcal{T}(\Omega, \mathbb{C})$ which, for simplicity, we denote by $\mathcal{T}(\Omega)$. For $S \subset \Omega$, let

$$\|f\|_S = \rho_S(f, 0) = \sup\{|f(x)| : x \in S\}.$$

We note that $\rho_S(f, g) = \|f - g\|_S$ and that $\| \ \|_S$ has the defining properties of a norm except that $\|f\|_S$ can be ∞ for some functions and can be 0 for functions that do not vanish identically.

If $\{f_\iota\}_{\iota \in I}$ and $\{g_\iota\}_{\iota \in I}$ are nets in $C(\Omega)$ that converge to f and g, respectively, then, for each compact subset K of Ω,

$$\begin{aligned}
\rho_K(f_\iota + g_\iota, f + g) &= \|f_\iota + g_\iota - f - g\|_K \\
&\leq \|f_\iota - f\|_K + \|g_\iota - g\|_K \\
&= \rho_K(f_\iota, f) + \rho_K(g_\iota, g).
\end{aligned}$$

We now see from Proposition 11.5 on page 417 that $\{f_\iota + g_\iota\}_{\iota \in I}$ converges to $f + g$. If follows that the operation of addition is continuous as a function from $C(\Omega) \times C(\Omega)$ to $C(\Omega)$, where $C(\Omega) \times C(\Omega)$ is given the product topology.

By a similar argument, we find that the operation of scalar multiplication is continuous as a function from $\mathbb{C} \times C(\Omega)$ to $C(\Omega)$. But, actually, more is true. Scalar multiplication of a function by a complex number is a special case of pointwise multiplication of functions. If the product fg of two functions f and g in $C(\Omega)$ is defined pointwise, then $C(\Omega)$ becomes an algebra of functions.[†] Moreover, since the product operation on $C(\Omega)$ satisfies $\|fg\|_K \leq \|f\|_K \|g\|_K$, it follows by an argument similar to the one used to prove continuity of addition that the operation of multiplication is continuous as a function from $C(\Omega) \times C(\Omega)$ to $C(\Omega)$. □

EXAMPLE 11.8 *The Case Ω Compact and $\Lambda = \mathbb{C}$*

Refer to Example 11.7. If Ω is compact, then $\| \ \|_\Omega$ is a norm on $C(\Omega)$ called the **sup-norm,** also known as the **supremum norm** or **uniform norm.** Thus the sup-norm on $C(\Omega)$ is given by

$$\|f\|_\Omega = \sup\{ |f(x)| : x \in \Omega \}, \qquad f \in C(\Omega). \tag{11.8}$$

The sup-norm induces the topology $\mathcal{T}(\Omega)$ and, moreover, $(C(\Omega), \| \ \|_\Omega)$ is complete. Whenever we are considering a space of the form $C(\Omega)$ where Ω is compact, we will assume that it is equipped with the sup-norm unless explicitly stated otherwise. □

EXAMPLE 11.9 *The Case Ω Not Compact and $\Lambda = \mathbb{C}$*

Refer to Example 11.7. If Ω is not compact, $\| \ \|_\Omega$ is still a norm on some subspaces of $C(\Omega)$. Important instances are the following:

$$C_c(\Omega) = \{ f \in C(\Omega) : \text{supp } f \text{ is compact} \}$$
$$C_0(\Omega) = \{ f \in C(\Omega) : \lim_{x \to \omega} f(x) = 0 \}^{\ddagger}$$
$$C_b(\Omega) = \{ f \in C(\Omega) : \|f\|_\Omega < \infty \}.$$

The spaces $C_c(\Omega)$, $C_0(\Omega)$, and $C_b(\Omega)$ are called, respectively, the **continuous functions with compact support, continuous functions vanishing at infinity,** and **bounded continuous functions.**

$C_0(\Omega)$ is a closed subspace of $C_b(\Omega)$ with respect to the topology induced by $\| \ \|_\Omega$. $C_c(\Omega)$ is a linear subspace of $C_0(\Omega)$ but it is not closed. Indeed, it can be shown that $C_c(\Omega)$ is dense in $C_0(\Omega)$ with respect to the topology induced by $\| \ \|_\Omega$. See Exercise 11.57. □

For each $x \in \Omega$ define $e_x \colon C(\Omega, \Lambda) \to \Lambda$ by $e_x(f) = f(x)$. The weak topology $\mathcal{T}^p(\Omega, \Lambda)$ determined by the family of functions $\{ e_x : x \in \Omega \}$ is called the **topology of pointwise convergence.** In case $\Lambda = \mathbb{C}$, the topology $\mathcal{T}^p(\Omega, \mathbb{C})$ is denoted by $\mathcal{T}^p(\Omega)$. Whereas each function e_x is continuous with respect to $\mathcal{T}(\Omega, \Lambda)$, it follows that $\mathcal{T}^p(\Omega, \Lambda)$ is weaker than $\mathcal{T}(\Omega, \Lambda)$.

The space $C(\Omega, \Lambda)$ is a subset of the Cartesian product Λ^Ω. If Λ^Ω is equipped with the product topology, then $\mathcal{T}^p(\Omega, \Lambda)$ is the relative topology on $C(\Omega, \Lambda)$.

[†] A linear space L with a multiplication operation that satisfies the conditions $x(yz) = (xy)z$, $x(y + z) = xy + xz$, $(x + y)z = xz + yz$, and $\alpha(xy) = (\alpha x)y = x(\alpha y)$ for all $x, y, z \in L$ and all scalars α is called an **algebra.**

[‡] For the meaning of $\lim_{x \to \omega} f(x)$, see Exercise 11.45 on page 415.

EXAMPLE 11.10   *The Case* $\Omega = \mathcal{N}$ *and* $\Lambda = \mathbb{C}$

The set of positive integers $\mathcal{N}$ equipped with the discrete topology is a second countable, locally compact Hausdorff space. Thus, by Theorem 11.14 on pages 418–419, $(C(\mathcal{N}), \mathcal{T}(\mathcal{N}))$ is metrizable and complete. As the compact subsets of $\mathcal{N}$ are exactly the finite ones, it follows that $\mathcal{T}(\mathcal{N}) = \mathcal{T}^p(\mathcal{N})$.

The subspace $C_c(\mathcal{N})$ consists of all sequences of complex numbers that are zero except for finitely many indices; the subspace $C_0(\mathcal{N})$, often denoted in the literature by c_0, consists of all sequences of complex numbers that converge to 0; and the subspace $C_b(\mathcal{N})$ coincides with the space $\ell^\infty(\mathcal{N})$ of Example 10.6 on pages 366–367. Note also that on $C_b(\mathcal{N})$, $\|\ \|_{\mathcal{N}}$ is just $\|\ \|_\infty$. □

EXAMPLE 11.11   *Continuous Periodic Functions*

Consider the subspace P of $C(\mathcal{R})$ given by

$$P = \{\, h \in C(\mathcal{R}) : h(x + 2\pi) = h(x) \text{ for all } x \in \mathcal{R} \,\}.$$

Of course, P is just the space of continuous functions with period 2π. It is easy to see that P is a closed subspace of the normed space $C_b(\mathcal{R})$.

We will discuss a relationship between P and the space $C(T)$, where T is the unit circle centered at 0 in the complex plane $\mathbb{C}$. For $f \in C(T)$, define $J(f): \mathcal{R} \to \mathbb{C}$ by $J(f)(x) = f(\cos x + i \sin x)$. We note that the function J is a one-to-one linear function from $C(T)$ onto P that satisfies $J(fg) = J(f)J(g)$ and $\|J(f)\|_{\mathcal{R}} = \|f\|_T$ for all $f, g \in C(T)$. (See Exercise 11.63.) Thus, as normed spaces and as algebras, the spaces P and $C(T)$ are essentially copies of each other. It will be helpful later, when we study approximation by trigonometric polynomials, to identify the spaces P and $C(T)$ by means of the correspondence J. □

Exercises for Section 11.5

11.49   Show by examples that no two of the modes of convergence described by Definitions 11.10, 11.11, and 11.12 are equivalent.

11.50   Prove Proposition 11.4 on pages 416–417.

11.51   Give an example where $\rho_S(f, g) = \infty$.

11.52   Suppose that a Hausdorff space Ω is locally compact but not compact and that K is a compact subset of Ω. Prove that ρ_K cannot be a metric on $C(\Omega)$.

11.53   Prove that $(C(\Omega, \Lambda), \mathcal{T}(\Omega, \Lambda))$ is always a Hausdorff space.

11.54   Prove Lemma 11.1 on page 418.

11.55   Prove part (d) of Theorem 11.14 on page 419.

11.56   Prove that $\|\ \|_\Omega$ is a norm on each of the spaces $C_c(\Omega)$, $C_0(\Omega)$, and $C_b(\Omega)$ defined in Example 11.9 on page 421.

★11.57   Suppose that a Hausdorff space Ω is locally compact but not compact. Let $C_c(\Omega)$, $C_0(\Omega)$, and $C_b(\Omega)$ be given the norm $\|\ \|_\Omega$.
a) Show that $C_0(\Omega)$ is closed in but not equal to $C_b(\Omega)$.
b) Show that $C_c(\Omega)$ is dense in $C_0(\Omega)$.
c) Prove that $C_b(\Omega)$ and $C_0(\Omega)$ are complete.
d) Is $C_c(\Omega)$ complete? Justify your answer.

★*11.58* For $x_0 \in \Omega$, define $e_{x_0} \colon C(\Omega, \Lambda) \to \Lambda$ by $e_{x_0}(f) = f(x_0)$.
a) Show that e_{x_0} is continuous with respect to $\mathcal{T}(\Omega, \Lambda)$.
b) Conclude that $\mathcal{T}^p(\Omega, \Lambda)$ is weaker than $\mathcal{T}(\Omega, \Lambda)$.

11.59 Give an example where $\mathcal{T}^p(\Omega, \Lambda)$ is properly contained in $\mathcal{T}(\Omega, \Lambda)$.

11.60 Show that the relative topology on $C_0(\mathcal{N})$ determined by $\mathcal{T}(\mathcal{N})$ is strictly weaker than the topology induced by $\| \ \|_{\mathcal{N}}$.

11.61 Let $(\Omega, \mathcal{T})$ be a locally compact Hausdorff space. Show that the weak topology determined by the family of functions $C(\Omega)$ coincides with $\mathcal{T}$.

11.62 Let Ω be a compact Hausdorff space. Show that the following subsets of $C(\Omega)$ are open:
a) $\{\, f : gf = 1 \text{ for some } g \in C(\Omega) \,\}$.
b) $\{\, f : f = e^h \text{ for some } h \in C(\Omega) \,\}$, where e^h is defined by $e^h(x) = e^{h(x)}$.
Hint: Use the Taylor series expansions of $1/(1 - z)$ and $\log(1 - z)$.

11.63 Verify the asserted properties of the function J defined in Example 11.11.

Marshall Harvey Stone
(1903–1989)

Marshall Harvey Stone was born in New York City on April 8, 1903, to Agnes Harvey Stone and Harlan Fiske Stone. His father became the 12th Chief Justice of the United States Supreme Court in 1941.

Stone attended public schools in Englewood, New Jersey, following which he entered Harvard in 1919. Although it was presumed that he would become a lawyer like his father, Stone became fascinated with mathematics as a Harvard undergraduate.

In 1926, Stone completed his Ph.D. at the University of Alberta, Red Deer campus, with a dissertation on differential equations. He worked under the supervision of George D. Birkhoff.

Stone taught at Columbia, Yale, Harvard, and the University of Chicago. He retired from Chicago in 1968, but accepted a position at the University of Massachusetts where he worked full time until 1973.

Influenced by Birkhoff and John von Neumann, Stone obtained significant results in the spectral theory of unbounded operators, Boolean algebras, general topology, and rings of continuous functions. Perhaps most well known is his striking generalization of Weierstrass's theorem on polynomial approximation. Four theorems are named for him: Stone's representation theorems for one-parameter unitary groups and Boolean algebras, the Stone-Čech compactification theorem, and the Stone-Weierstrass approximation theorem.

Under Stone's leadership as chair, the University of Chicago's department of mathematics developed into a world center of mathematics research. In 1950, after modernizing and upgrading the University of Chicago's undergraduate and graduate mathematics programs, Stone devoted himself to improving the teaching of pre-university mathematics. He took a leading part in a series of international conferences on mathematical education and served as an active member of the governing board of the School Mathematics Study Group and on its panel on elementary school mathematics.

Stone was traveling in Madras, India, when he died on January 9, 1989.

$\cdot$ 12 $\cdot$

Complete and Compact Spaces

In this chapter, the ideas of Chapters 10 and 11 are used to prove some general theorems of analysis in metric and topological spaces.

Sections 12.1 and 12.2 deal with two important consequences of the completeness property, namely, the Baire category theorem and the contraction mapping principle. Sections 12.3 and 12.4 answer the following question for spaces of the form $C(\Omega, \Lambda)$ and for product spaces, respectively: "When is a set of functions compact?" Section 12.5 discusses generalizations of piecewise linear approximation of functions to spaces of the form $C(\Omega, \mathcal{R})$. And Section 12.6 considers generalizations of polynomial approximation of functions to spaces of the form $C(\Omega)$.

12.1 THE BAIRE CATEGORY THEOREM

Suppose that we think of a subset of a topological space as being "thin" if its closure has an empty interior. The Baire category theorem asserts that a complete metric space cannot be expressed as the union of countably many "thin" sets. Rather than use the term "thin," we adopt the following more widely used terminology.

DEFINITION 12.1 Nowhere Dense Set

A subset E of a topological space Ω is said to be **nowhere dense** if its closure has an empty interior, that is, if $(\overline{E})^{\circ} = \emptyset$.

EXAMPLE 12.1 *Illustrates Definition 12.1*

a) It is easy to see that a set is nowhere dense if and only if the complement of its closure is dense.

b) Clearly, any finite subset of $\mathcal{R}$ is nowhere dense; in fact, any subset of $\mathcal{R}$ whose closure is countable is nowhere dense. There are, of course, countable subsets of $\mathcal{R}$ that are not nowhere dense (e.g., Q). On the other hand, the Cantor set is an example of an uncountable set that is nowhere dense.

c) Any line segment in $\mathcal{R}^2$ is nowhere dense. ☐

We now state and prove the Baire category theorem.

☐ ☐ ☐ **THEOREM 12.1 Baire Category Theorem**

A complete metric space (Ω, ρ) is not the union of a countable collection of nowhere dense subsets.

PROOF Let $\{S_n\}_{n=1}^{\infty}$ be a sequence of nowhere dense subsets of Ω. We must show that

$$\left(\bigcup_{n=1}^{\infty} S_n\right)^c \neq \emptyset. \tag{12.1}$$

To prove (12.1) it suffices to verify that $\bigcap_{n=1}^{\infty} G_n \neq \emptyset$, where $G_n = (\overline{S_n})^c$. We will prove somewhat more, namely, that $U \cap (\bigcap_{n=1}^{\infty} G_n) \neq \emptyset$, whenever U is a nonempty open subset of Ω.

So, let U be open and nonempty. We first note that, for each $n \in \mathcal{N}$, G_n is dense because S_n is nowhere dense. Thus, there exists an element x_1 in the open set $U \cap G_1$. It follows that there is an $r_1 > 0$ such that $\overline{B}_{r_1}(x_1) \subset U \cap G_1$. Since G_2 is dense and open, we can choose an element $x_2 \in B_{r_1}(x_1) \cap G_2$ and an $r_2 > 0$ such that

$$r_2 < r_1/2 \quad \text{and} \quad \overline{B}_{r_2}(x_2) \subset B_{r_1}(x_1) \cap G_2.$$

Continuing inductively, we obtain a sequence $\{x_n\}_{n=1}^{\infty}$ in Ω and a sequence of positive numbers $\{r_n\}_{n=1}^{\infty}$ such that for all $n \in \mathcal{N}$,

$$r_{n+1} < r_n/2 \quad \text{and} \quad \overline{B}_{r_n}(x_n) \subset G_n \tag{12.2}$$

and

$$\overline{B}_{r_{n+1}}(x_{n+1}) \subset B_{r_n}(x_n). \tag{12.3}$$

Because $\rho(x_{n+1}, x_n) < r_n \leq 2^{-(n-1)} r_1$ for $n = 1, 2, \ldots$, it follows that

$$\rho(x_{n+k}, x_n) \leq \sum_{j=0}^{k-1} \rho(x_{n+j+1}, x_{n+j}) < \sum_{j=0}^{k-1} 2^{-(n+j-1)} r_1 < 2^{-n+2} r_1.$$

Hence, $\{x_n\}_{n=1}^{\infty}$ is a Cauchy sequence and, so, its limit exists; call it x. It follows from (12.2) and (12.3) that, for $n > m$,

$$x_n \in \overline{B}_{r_{m+1}}(x_{m+1}) \subset G_m.$$

Thus, $x \in G_m$ for $m = 1, 2, \ldots$. As we also have $x_n \in \overline{B}_{r_1}(x_1) \subset U$, we conclude that $x \in U$. Hence, $x \in U \cap (\bigcap_{n=1}^{\infty} G_n)$. ∎

The following corollary is an immediate consequence of the proof of Theorem 12.1 and is sometimes referred to as the Baire category theorem.

□ □ □ COROLLARY 12.1 Baire Category Theorem (Alternative Version)

Let $\{G_n\}_{n=1}^{\infty}$ be a sequence of dense open subsets of a complete metric space. Then $\bigcap_{n=1}^{\infty} G_n$ is dense.

A subset of a topological space is said to be of the **first category** if it can be expressed as a countable union of nowhere dense sets. A set that is not of the first category is said to be of the **second category.**

Using this terminology, we can restate the Baire category theorem as follows: *A complete metric space is of the second category.* It shows that sets of the first category in a complete metric space are in a sense much smaller than sets of the second category.

The Baire category theorem is frequently used as a tool for obtaining existence results. Specifically, the existence of an object of a certain type is established by showing that the complement of the collection of such objects is of the first category in an appropriate complete metric space. Examples 12.2 and 12.3 illustrate this technique.

EXAMPLE 12.2   *Illustrates the Baire Category Theorem*

We know that $\mathcal{R}$ is a complete metric space. Whereas every single-element set in $\mathcal{R}$ is nowhere dense and the set Q of rational numbers is countable, it follows that Q is a set of the first category. Hence, the Baire category theorem implies that Q^c is nonempty, that is, irrational numbers exist.   □

EXAMPLE 12.3   *Illustrates the Baire Category Theorem*

By Theorem 11.14 (pages 418–419), $C([0, 1], \mathcal{R})$, equipped with the norm $\| \ \|_{[0,1]}$, is a complete metric space, where

$$\|f\|_{[0,1]} = \sup\{ |f(x)| : x \in [0, 1] \}.$$

We will apply the Baire category theorem to show that "most" functions in $C([0, 1], \mathcal{R})$ vary erratically in the sense that they fail to be monotonic on any nonempty subinterval. A stronger result of this type is developed in Exercise 12.2.

Let $\mathcal{I}$ denote the collection of nonempty open subintervals of $[0, 1]$ with rational endpoints. For $I \in \mathcal{I}$, let U_I and D_I denote, respectively, the set of functions in $C([0, 1], \mathcal{R})$ that are nondecreasing and nonincreasing on I. If we let F denote the set of functions in $C([0, 1], \mathcal{R})$ that are monotonic on some nonempty subinterval of $[0, 1]$, then we have that

$$F = \bigcup_{I \in \mathcal{I}} (U_I \cup D_I). \tag{12.4}$$

It is not hard to see that for each $I \in \mathcal{I}$, U_I and D_I are closed in $C([0, 1], \mathcal{R})$. If we can show that, for each $I \in \mathcal{I}$, U_I and D_I have empty interiors, then it will follow from (12.4) that F is a set of the first category. In particular then, the Baire category theorem will imply that F^c is nonempty; that is, there are functions in $C([0, 1], \mathcal{R})$ that fail to be monotonic on any nonempty subinterval of $[0, 1]$.

We will prove that $(U_I)^\circ = \emptyset$. A similar proof shows that $(D_I)^\circ = \emptyset$. Let $\epsilon > 0$ and $f \in U_I$. We choose a point $t \in I$ and $\delta > 0$ small enough so that the function f varies by less than $\epsilon/2$ on the interval $[t - \delta, t + \delta] \subset I$. Define

$$g(x) = \begin{cases} f(x), & \text{if } x \in [0,1] \setminus [t - \delta, t + \delta]; \\ f(x) + \delta^{-2}\epsilon(\delta^2 - (x-t)^2), & \text{if } x \in [t - \delta, t + \delta]. \end{cases}$$

It is easy to see that g is continuous and that $\|f - g\|_{[0,1]} = \epsilon$. On the other hand, $g \notin U_I$ because $g(t)$ is greater than $g(t + \delta)$. This fact shows that every ball around f contains points outside U_I. □

We conclude this section with an important consequence of the Baire category theorem called the **uniform boundedness principle.** We will give some applications of the uniform boundedness principle in the exercises. It will also prove useful later when we return to our study of normed spaces.

□ □ □ **THEOREM 12.2 Uniform Boundedness Principle**

Let Ω be a complete metric space and $\mathcal{F}$ a family of continuous functions from Ω into a normed space $(\Lambda, \| \; \|)$. Suppose that for each $x \in \Omega$,

$$\sup\{ \|f(x)\| : f \in \mathcal{F} \} < \infty. \tag{12.5}$$

Then there is an $M \in \mathcal{R}$ and a nonempty open set O such that $\|f(u)\| \leq M$ for all $u \in O$ and $f \in \mathcal{F}$.

PROOF For each $n \in \mathcal{N}$ and $f \in \mathcal{F}$, the set $\{ x : \|f(x)\| \leq n \}$ is closed. Thus,

$$E_n = \bigcap_{f \in \mathcal{F}} \{ x : \|f(x)\| \leq n \} = \{ x : \|f(x)\| \leq n \text{ for all } f \in \mathcal{F} \}$$

is closed. By (12.5), $\Omega = \bigcup_{n=1}^{\infty} E_n$. The Baire category theorem now implies that there is an integer N such that $(E_N)^\circ \neq \emptyset$. Thus, the assertion of the theorem is verified with $M = N$ and $O = (E_N)^\circ$. ∎

Exercises for Section 12.1

12.1 Let $E = \{ f \in C([0,1]) : |f(x) - f(y)| \leq |x - y| \text{ for all } x, y \in [0,1] \}$. Show that E is nowhere dense in $C([0,1])$.

12.2 For this exercise, we need to extend the definition of differentiability given in Definition 8.1 on page 269. When we have a function f defined on a closed interval $[a, b]$, we say that f is differentiable at the left endpoint a if the limit

$$\lim_{h \to 0^+} \frac{f(a + h) - f(a)}{h}$$

exists and is finite. In that case the limit is called the derivative of f at a and is denoted by $f'(a)$. In a similar way, we define differentiability at the right endpoint b.

a) Let D denote the set of functions in $C([0,1])$ that are differentiable at some point of $[0,1]$. Show that D is of the first category. *Hint:* Consider the set of functions

$$D_n = \{\, f : |f(t) - f(x)| \le n|t - x| \text{ for all } t \in [0,1] \text{ for some } x \in [0,1] \,\}.$$

b) Deduce from part (a) that there are functions in $C([0,1])$ that are not differentiable at any point of $[0,1]$.

12.3 Show that, as a subset of $\big(\mathcal{L}^1([0,1]), \|\ \|_1\big)$, $\mathcal{L}^2([0,1])$ is of the first category. *Hint:* See Exercise 5.84 on page 176.

12.4 In this exercise, we will be considering functions from $\mathcal{R}$ to $\mathcal{R}$.

a) Give an example of a function that is continuous at each irrational number and discontinuous at each rational number. *Hint:* Consider the function that is 0 at each irrational number, 1 at 0, and $1/q$ at each rational number of the form p/q, where $q > 0$ and p and q have no common divisor except 1.

b) Is there a function that is continuous at each rational number and discontinuous at each irrational number? *Hint:* Associate with each function $f : \mathcal{R} \to \mathcal{R}$, functions u and ℓ defined by

$$u(x) = \inf\{\, \sup\{\, f(y) : |x - y| < \delta \,\} : \delta > 0 \,\}$$

and

$$\ell(x) = \sup\{\, \inf\{\, f(y) : |x - y| < \delta \,\} : \delta > 0 \,\}.$$

Consider the set where ℓ is strictly less than u.

12.5 Show that there is no sequence $\{f_n\}_{n=1}^{\infty}$ in $C([0,1])$ that satisfies

$$\lim_{n \to \infty} f_n(x) = \begin{cases} 1, & \text{if } x \text{ is a rational number;} \\ 0, & \text{if } x \text{ is an irrational number.} \end{cases}$$

*In Exercises 12.6 and 12.7, we will need the concept of a basis for a linear space. A subset S of a linear space Ω is said to be a **basis** for Ω if for each nonzero $x \in \Omega$, there is a unique subset $\{x_1, x_2, \ldots, x_n\}$ of S and a unique set of nonzero scalars $\{\alpha_1, \alpha_2, \ldots, \alpha_n\}$ such that $x = \alpha_1 x_1 + \alpha_2 x_2 + \cdots + \alpha_n x_n$. It follows from Zorn's lemma that every linear space has a basis. The number of elements in a basis is called the **dimension** of the linear space. A linear space is said to be **finite dimensional** if it has a basis containing finitely many elements; otherwise, it is said to be **infinite dimensional**.*

12.6 Let Ω be a normed space and D a proper, finite-dimensional, linear subspace of Ω.
a) Show that D is closed.
b) Show that D is nowhere dense.

12.7 Let Ω be a complete normed space and S a basis for Ω. Prove that S is either finite or uncountable. *Hint:* Refer to Exercise 12.6.

12.8 Let Ω be a normed space and D a proper closed linear subspace of Ω. Show that D is nowhere dense.

12.2 CONTRACTIONS OF COMPLETE METRIC SPACES

Let $f : \Omega \to \Omega$. An element $p \in \Omega$ is called a **fixed point** of f if $f(p) = p$. In this section, we will give a simple condition on f that guarantees the existence of a fixed point when Ω is complete. We will also give some applications of that condition to differential and integral equations.

DEFINITION 12.2 **Contraction**

Let (Ω, ρ) be a metric space. A mapping $f: \Omega \to \Omega$ is called a **contraction** if there is a constant $c \in [0, 1)$ such that

$$\rho\big(f(x), f(y)\big) \le c\rho(x, y)$$

for all $x, y \in \Omega$.

EXAMPLE 12.4 *Illustrates Definition 12.2*

a) It is obvious that contractions are always continuous functions.

b) Suppose α is a positive constant and define $f(x) = \alpha x$ for $x \in [0, 1]$. If $\alpha \le 1$, then $f: [0, 1] \to [0, 1]$. If $\alpha < 1$, then f is a contraction with $c = \alpha$. If $\alpha = 1$, then f is not a contraction, since

$$1 = |1 - 0| = |f(1) - f(0)| \le c|1 - 0| = c$$

implies $c \ge 1$. Note, however, that 0 is a fixed point of f for all α.

c) Suppose that $f: [0, 1] \to [0, 1]$ is continuous and has a derivative at each point of $(0, 1)$. Suppose further that

$$B = \sup\{\,|f'(x)| : x \in (0, 1)\,\} < \infty.$$

By the mean value theorem, for each $x, y \in [0, 1]$ with $x < y$, we have that $f(y) - f(x) = f'(t)(y - x)$ for some $t \in (x, y)$. So, $|f(y) - f(x)| \le B|y - x|$ for all $x, y \in [0, 1]$. Therefore, f is a contraction if $B < 1$.

d) An interesting special case of part (c) is the cosine function. Whereas the derivative of $\cos x$ is $-\sin x$ and

$$\sup\{\,|-\sin x| : x \in [0, 1]\,\} = \sin 1 < \sin(\pi/2) = 1,$$

it follows from part (c) that the cosine function is a contraction of $[0, 1]$.

e) Let $\Omega = \{\, z \in \mathbb{C} : |z| \le 1 \,\}$, the closed unit disk, equipped with the usual metric inherited from $\mathbb{C}$. Define $f: \Omega \to \Omega$ by $f(z) = \alpha z$, where α is a complex constant with $|\alpha| \le 1$. It is easy to see that f is a contraction if and only if $|\alpha| < 1$. Note, however, that 0 is a fixed point of f for all α.

f) Let $\Omega = \{\, z \in \mathbb{C} : |z| < 1 \,\}$, the open unit disk, equipped with the usual metric inherited from $\mathbb{C}$. Define $f: \Omega \to \Omega$ by $f(z) = (1 + z)/2$. Then f is a contraction with $c = 1/2$, but has no fixed point. □

□ □ □ **THEOREM 12.3 Contraction Mapping Principle**

Let (Ω, ρ) be a complete metric space and $f: \Omega \to \Omega$ a contraction. Then f has a unique fixed point. Moreover, if x is any point of Ω, then the sequence $\{x_n\}_{n=0}^{\infty}$ defined recursively by $x_0 = x$ and $x_{n+1} = f(x_n)$ converges to the unique fixed point of f.

PROOF   By Definition 12.2, there is a constant $c \in [0, 1)$ such that

$$\rho(x_{n+1}, x_n) = \rho\big(f(x_n), f(x_{n-1})\big) \leq c\rho(x_n, x_{n-1}),$$

for $n = 1, 2, \ldots$. It follows immediately that $\rho(x_{n+1}, x_n) \leq c^n \rho(x_1, x)$. Thus, if $1 \leq n < m$, we have

$$\rho(x_m, x_n) \leq \sum_{j=1}^{m-n} \rho(x_{n+j}, x_{n+j-1}) \leq \sum_{j=1}^{m-n} c^{n+j-1} \rho(x_1, x)$$

$$\leq c^n \rho(x_1, x) \sum_{j=1}^{\infty} c^{j-1} = c^n \rho(x_1, x)/(1 - c).$$

Because $0 \leq c < 1$, we have that $\{x_n\}_{n=0}^{\infty}$ is a Cauchy sequence. Hence, by completeness, $p = \lim_{n \to \infty} x_n$ exists. Using the continuity of f, we conclude that

$$f(p) = \lim_{n \to \infty} f(x_n) = \lim_{n \to \infty} x_{n+1} = p.$$

Thus, p is a fixed point of f.

It remains to establish uniqueness. Let q be a fixed point of f. Then

$$\rho(p, q) = \rho\big(f(p), f(q)\big) \leq c\rho(p, q).$$

Since $c < 1$, it follows that $\rho(p, q) = 0$. Hence, $p = q$.   ∎

We note from the proof of the contraction mapping principle that we can obtain not only the existence of a unique fixed point, but a method for approximating it and an error estimate. See Exercise 12.12.

EXAMPLE 12.5   *Illustrates the Contraction Mapping Principle*

We will illustrate the contraction mapping principle by using it to obtain an existence result for a certain class of integral equations. Let K be a real-valued Borel measurable function defined on the rectangle

$$I \times J = [x_0 - \alpha, x_0 + \alpha] \times [y_0 - \beta, y_0 + \beta]$$

where $\alpha, \beta > 0$. We make the following assumptions:

$$\big|K(x, y_1) - K(x, y_2)\big| \leq A|y_1 - y_2| \tag{12.6}$$

for all $x \in I$ and any pair $y_1, y_2 \in J$, where A is a constant, and

$$B = \sup\{\, K(x, y) : (x, y) \in I \times J \,\} < \infty. \tag{12.7}$$

We will show by using the contraction mapping principle that the integral equation given by

$$g(x) = y_0 + \int_{x_0}^{x} K\big(t, g(t)\big)\, dt \tag{12.8}$$

has a unique solution $g \in C(I, \mathcal{R})$ if $\alpha A < 1$ and $\alpha B \leq \beta$.

By Theorem 11.14 on pages 418–419, $C(I, \mathcal{R})$ is a complete metric space with respect to the norm $\| \ \|_I$. Hence, by Proposition 10.10 on page 376, the closed ball $\overline{B}_\beta(y_0)$, centered at the constant function y_0, is also complete. If $g \in \overline{B}_\beta(y_0)$, then the function $T(g)$ defined by

$$T(g)(x) = y_0 + \int_{x_0}^x K\big(t, g(t)\big) \, dt \qquad (12.9)$$

is continuous on I. (See Exercise 12.14.) Furthermore, since

$$\big|T(g)(x) - y_0\big| = \left| \int_{x_0}^x K\big(t, g(t)\big) \, dt \right| \leq \alpha B \leq \beta,$$

it follows that $T(g) \in \overline{B}_\beta(y_0)$. Consequently, the function T defined by (12.9) carries $\overline{B}_\beta(y_0)$ into itself.

We will show that (12.8) has a unique solution in $\overline{B}_\beta(y_0)$ by showing that T is a contraction. For $f, g \in \overline{B}_\beta(y_0)$ and $x \geq x_0$ we have, using (12.6),

$$\big|T(f)(x) - T(g)(x)\big| = \left| \int_{x_0}^x K\big(t, f(t)\big) - K\big(t, g(t)\big) \, dt \right|$$

$$\leq A \int_{x_0}^x \big|f(t) - g(t)\big| \, dt \leq \alpha A \|f - g\|_I.$$

Similar inequalities hold if $x < x_0$. Hence, $\big\|T(f) - T(g)\big\|_I \leq \alpha A \|f - g\|_I$ and, because $\alpha A < 1$, T is a contraction on $\overline{B}_\beta(y_0)$.   □

The conditions (12.6) and (12.7) appear rather restrictive. However, they can be used effectively to obtain a fundamental existence result for differential equations. See Exercise 12.16.

Exercises for Section 12.2

12.9   Give an example of a complete metric space (Ω, ρ) and a map $f \colon \Omega \to \Omega$ that satisfies $\rho\big(f(x), f(y)\big) \leq \rho(x, y)$ but has no fixed point.

12.10   Define $f \colon [0, 1] \to [0, 1]$ by $f(x) = x^2$. Show that f is not a contraction. Note, however, that f has two fixed points, 0 and 1.

12.11   Let $f \colon [0, 1] \to [0, 1]$ be continuous.
a) Show that f has a fixed point.
b) Must f have a unique fixed point?

12.12   Let (Ω, ρ), f, $\{x_n\}_{n=0}^\infty$, and p be as in the proof of the contraction mapping principle (Theorem 12.3). Establish the error estimate $\rho(x_n, p) \leq c^n \rho\big(x, f(x)\big) / (1 - c)$.

12.13   Recall Newton's method $x_{n+1} = x_n - G(x_n)/G'(x_n)$ for approximating the roots of a function G. Suppose that r is a root of G. Further suppose that δ is a positive number such that G' does not vanish on $[r - \delta, r + \delta]$ and

$$c = \sup \left\{ \left| G(x) G''(x) / \big(G'(x)\big)^2 \right| : x \in [r - \delta, r + \delta] \right\} < 1.$$

Show that if the initial guess x_0 in Newton's method is selected from the open interval $(r - \delta, r + \delta)$, then the sequence $\{x_n\}_{n=0}^{\infty}$ converges to r and satisfies

$$|x_n - r| \le \frac{c^n}{1 - c} |G(x_0)/G'(x_0)|.$$

Hint: See Exercise 12.12.

12.14 Refer to Example 12.5.
a) Verify that $K\big(\cdot, g(\cdot)\big)$ is Borel measurable if $g \in \overline{B}_\beta(y_0)$.
b) Use part (a) to prove that $T(g)$ is continuous on I.

12.15 Let $T\colon C([0,1]) \to C([0,1])$ be defined by $T(f)(x) = 1 + \int_0^x f(t)\, dt$.
a) Show that T is not a contraction.
b) Show that $T \circ T$ is a contraction.
c) Show by direct calculation that the sequence $\{f_n\}_{n=0}^{\infty}$ defined recursively by $f_0 = 1$ and $f_{n+1} = T(f_n)$ converges in $C([0,1])$ to the solution of $f(x) = 1 + \int_0^x f(t)\, dt$.

In Exercises 12.16 and 12.17, we use the following notation for partial derivatives. Suppose that f is a real- or complex-valued function defined on some open subset of $\mathcal{R}^n$ that contains the point $x = (x_1, x_2, \ldots, x_n)$. When

$$\lim_{h \to 0} \frac{f(x_1 + h, x_2, \ldots, x_n) - f(x_1, x_2, \ldots, x_n)}{h}$$

*exists and is finite, we denote it by $D_1 f(x)$ and call it the **partial derivative** of f at x with respect to x_1. Partial derivatives with respect to x_2, x_3, $\ldots$, x_n are defined similarly. We will make use of standard results about partial differentiation from multivariable calculus.*

12.16 Use Example 12.5 to establish the following existence theorem for differential equations: Suppose that f and $D_2 f$ are defined and continuous on some open set that contains the point $(x_0, y_0) \in \mathcal{R}^2$. Then there is a $\delta > 0$ and a unique continuously differentiable function g such that $g(x_0) = y_0$ and $g'(x) = f\big(x, g(x)\big)$ for $|x - x_0| < \delta$.

12.17 Show that the uniqueness part of Exercise 12.16 fails if the conditions on the derivative $D_2 f$ are removed. *Hint:* Consider $g' = 3g^{1/3}/2$.

12.3 COMPACTNESS IN THE SPACE $C(\Omega, \Lambda)$

We now take up the study of compactness in the space $\big(C(\Omega, \Lambda), \mathcal{T}(\Omega, \Lambda)\big)$, introduced in Section 11.5, beginning on page 415, where Ω is a topological space and (Λ, ρ) is a metric space. Under certain mild restrictions on Ω, we give useful necessary and sufficient conditions for a subset D of $C(\Omega, \Lambda)$ to be compact with respect to the topology $\mathcal{T}(\Omega, \Lambda)$ of uniform convergence on compact subsets. But first we present a simple example.

EXAMPLE 12.6 *Illustrates Compactness in Function Spaces*

Consider the space $C(\mathcal{R})$ equipped with the topology $\mathcal{T}(\mathcal{R})$ of uniform convergence on compact subsets.
a) As for any topological space, any finite subset of $C(\mathcal{R})$ is compact.

b) Let $f \in C(\mathcal{R})$ and define $g: [0,1] \to C(\mathcal{R})$ by $g(t)(x) = f(x+t)$. It is not difficult to show that g is continuous and, therefore, from Theorem 11.6 on page 408, $\{ f(\cdot + t) : t \in [0,1] \}$ is a compact subset of $C(\mathcal{R})$.

c) Let $f \in C(\mathcal{R})$. The set $\{ f(\cdot + t) : t \in \mathcal{R} \}$ may fail to be a compact subset of $C(\mathcal{R})$, as is the case when $f(x) = x$. □

The construction of more elaborate examples of compactness in function spaces requires some theory. First we develop a necessary condition for the compactness of $D \subset C(\Omega, \Lambda)$ by using the functions $\{ e_x : x \in \Omega \}$. We need a result from Exercise 11.58 on page 423, which we now state formally as a proposition.

□ □ □ **PROPOSITION 12.1**

For $x \in \Omega$, define $e_x : C(\Omega, \Lambda) \to \Lambda$ by $e_x(f) = f(x)$. Then e_x is a continuous function.

Applying Proposition 12.1 and Theorem 11.6 (page 408), we get the following necessary condition for compactness of a subset of $C(\Omega, \Lambda)$.

□ □ □ **PROPOSITION 12.2**

If $D \subset C(\Omega, \Lambda)$ is compact, then $\{ f(x) : f \in D \}$ is a compact subset of Λ for each $x \in \Omega$.

Next, using Proposition 12.2 and Theorem 11.2 on page 403, we derive another necessary condition for compactness of a subset of $C(\Omega, \Lambda)$.

□ □ □ **PROPOSITION 12.3**

Suppose that Ω is a locally compact Hausdorff space and D is a compact subset of $C(\Omega, \Lambda)$. Then, given $x \in \Omega$ and $\epsilon > 0$, there is an open set W containing x such that $\rho\big(f(x), f(y)\big) < \epsilon$ for all $y \in W$ and $f \in D$.

PROOF First recall that for $f, g \in C(\Omega, \Lambda)$ and $S \subset \Omega$,

$$\rho_S(f, g) = \sup \{ \rho\big(f(x), g(x)\big) : x \in S \}$$

and that $\mathcal{T}(\Omega, \Lambda)$ is the weak topology on $C(\Omega, \Lambda)$ determined by the family of functions $\{ \rho_K(\cdot, g) : K \text{ compact}, g \in C(\Omega, \Lambda) \}$.

Now let U be an open set containing x such that $\overline{U}$ is compact. Then, for each $h \in C(\Omega, \Lambda)$, the function $\rho_{\overline{U}}(\cdot, h)$ is continuous. It follows that the collection of sets $\{ J_f : f \in D \}$ where $J_f = \{ g : \rho_{\overline{U}}(g, f) < \epsilon/3 \}$ is an open covering of D. Hence, there are finitely many functions $f_1, f_2, \ldots, f_n \in D$ such that $D \subset \bigcup_{j=1}^n J_{f_j}$. Because there are only finitely many f_js, we can find an open set W such that $x \in W \subset U$ and $\rho\big(f_j(x), f_j(y)\big) < \epsilon/3$ for each $y \in W$ and $j = 1, 2, \ldots, n$.

If $f \in D$, we choose $k \in \{1, 2, \ldots, n\}$ such that $\rho_{\overline{U}}(f_k, f) < \epsilon/3$. Then for each $y \in W$, we have

$$\rho\big(f(x), f(y)\big) \leq \rho\big(f(x), f_k(x)\big) + \rho\big(f_k(x), f_k(y)\big) + \rho\big(f_k(y), f(y)\big)$$
$$\leq 2\rho_{\overline{U}}(f_k, f) + \rho\big(f_k(x), f_k(y)\big) < 2\epsilon/3 + \epsilon/3 = \epsilon,$$

as required. ■

The necessary condition derived in Proposition 12.3 is a natural extension of the notion of continuity. It is so important that it merits a formal definition, which we now give. Note that the definition does not require Ω to be locally compact or Hausdorff.

DEFINITION 12.3 **Equicontinuity**

A subset D of $C(\Omega, \Lambda)$ is said to be **equicontinuous** on Ω if for each $x \in \Omega$ and $\epsilon > 0$, there is an open set W containing x such that $\rho\big(f(x), f(y)\big) < \epsilon$ for all $y \in W$ and $f \in D$.

We will now show that under mild conditions on Ω, the necessary conditions derived in Propositions 12.2 and 12.3 are also sufficient.

□ □ □ THEOREM 12.4

Let Ω be a second countable locally compact Hausdorff space. A subset D of $C(\Omega, \Lambda)$ is compact if and only if it is closed and satisfies the following conditions:

a) $\overline{\{\, f(x) : f \in D \,\}}$ is a compact subset of Λ for each $x \in \Omega$.

b) D is equicontinuous.

PROOF Propositions 12.2 and 12.3 already show that conditions (a) and (b) are implied by the compactness of D.

Assume that (a) and (b) hold and that D is closed. From Theorem 11.14 on pages 418–419, we know that the space $C(\Omega, \Lambda)$ is metrizable. Hence, by Theorem 11.2 on page 403, the compactness of D will follow if we can show that every sequence in D has a subsequence that converges uniformly on compact subsets of Ω.

So, let $\{f_n\}_{n=1}^{\infty} \subset D$. As Ω is second countable, Proposition 11.1 on page 399 implies that it contains a countable dense set E. Suppose we can find a subsequence $\{g_k\}_{k=1}^{\infty}$ of $\{f_n\}_{n=1}^{\infty}$ such that

$$g(y) = \lim_{k \to \infty} g_k(y) \tag{12.10}$$

exists for all $y \in E$. We will show that the limit in (12.10) exists for all $x \in \Omega$, that the limit function g is continuous, and that $\{g_k\}_{k=1}^{\infty}$ converges uniformly on compact subsets of Ω to g.

Let $x \in \Omega$ and $\epsilon > 0$ be given. By (b), we can choose an open set W containing x such that

$$\rho\big(g_k(x), g_k(w)\big) < \epsilon/3 \tag{12.11}$$

for each $w \in W$ and $k \in \mathcal{N}$. As E is dense in Ω, there is a $y \in E \cap W$. By (12.10), there is an $N \in \mathcal{N}$ such that $\rho\big(g_k(y), g_\ell(y)\big) < \epsilon/3$ whenever $k, \ell \geq N$. It follows that for $k, \ell \geq N$,

$$\rho\big(g_k(x), g_\ell(x)\big) \leq \rho\big(g_k(x), g_k(y)\big) + \rho\big(g_k(y), g_\ell(y)\big) + \rho\big(g_\ell(y), g_\ell(x)\big)$$
$$< \epsilon/3 + \epsilon/3 + \epsilon/3 = \epsilon. \tag{12.12}$$

Thus, we see that the sequence $\{g_k(x)\}_{k=1}^{\infty}$ is Cauchy. Because $\{g_k(x)\}_{k=1}^{\infty}$ is contained in the compact set $\overline{\{f(x) : f \in D\}}$, it follows from Theorem 11.2 that $\lim_{k \to \infty} g_k(x) = g(x)$ exists. Consequently, (12.10) continues to hold if y is replaced by any $x \in \Omega$.

We will now verify that g is continuous at each $x \in \Omega$. For each $\epsilon > 0$, we choose an open set W as in the previous paragraph. From the inequalities

$$
\begin{aligned}
\left| \rho(g(x), g(w)) - \rho(g_k(x), g_k(w)) \right| &\leq \left| \rho(g(x), g(w)) - \rho(g(x), g_k(w)) \right| \\
&\quad + \left| \rho(g(x), g_k(w)) - \rho(g_k(x), g_k(w)) \right| \\
&\leq \rho(g(w), g_k(w)) + \rho(g(x), g_k(x)),
\end{aligned}
$$

it follows that $\lim_{k \to \infty} \rho(g_k(x), g_k(w)) = \rho(g(x), g(w))$. Consequently, we can let k pass to infinity in (12.11) to obtain $\rho(g(x), g(w)) \leq \epsilon/3 < \epsilon$ for each $w \in W$.

Next we show that for each $x \in \Omega$ and $\epsilon > 0$, there is an open set O_x that contains x and an integer k_x such that

$$\rho(g_k(w), g(w)) < \epsilon, \qquad w \in O_x, \ k \geq k_x. \tag{12.13}$$

Indeed, by (b) and the continuity of g, we can choose an open set O_x containing x such that for each $w \in O_x$, $\rho(g(w), g(x)) < \epsilon/3$ and $\rho(g_k(w), g_k(x)) < \epsilon/3$ for all k. Because $g(x) = \lim_{k \to \infty} g_k(x)$, it follows that there is an integer k_x such that $\rho(g_k(x), g(x)) < \epsilon/3$ for all $k \geq k_x$. Thus, for each $w \in O_x$, we have

$$
\begin{aligned}
\rho(g_k(w), g(w)) &\leq \rho(g_k(w), g_k(x)) + \rho(g_k(x), g(x)) + \rho(g(x), g(w)) \\
&< \epsilon/3 + \epsilon/3 + \epsilon/3 = \epsilon,
\end{aligned}
$$

whenever $k \geq k_x$ and, so, (12.13) holds.

Now, let K be a compact subset of Ω and let $\epsilon > 0$ be given. Then we can cover K with finitely many open sets $\{O_{x_j}\}_{j=1}^{m}$, each of which satisfies (12.13). Let $k_0 = \max\{k_{x_j} : j = 1, 2, \ldots, m\}$. If $y \in K$, then $y \in O_{x_j}$ for some j and, hence, $\rho(g_k(y), g(y)) < \epsilon$ for $k \geq k_0$. Thus, we have shown that $\{g_k\}_{k=1}^{\infty}$ converges uniformly on compact subsets of Ω to g.

It now remains only to show that there is a subsequence $\{g_k\}_{k=1}^{\infty}$ of $\{f_n\}_{n=1}^{\infty}$ that satisfies (12.10). We shall do this by adapting the diagonalization argument used in the proof of (d) $\Rightarrow$ (c) in Theorem 11.2 on page 403.

Let $\{y_n\}_{n=1}^{\infty}$ be an enumeration of the countable dense subset E of Ω that we selected earlier. By (a) we can select a subsequence $\{f_{[1,n]}\}_{n=1}^{\infty}$ of $\{f_n\}_{n=1}^{\infty}$ such that $g(y_1) = \lim_{n \to \infty} f_{[1,n]}(y_1)$ exists. And, then again by (a), we can find a subsequence $\{f_{[2,n]}\}_{n=1}^{\infty}$ of $\{f_{[1,n]}\}_{n=1}^{\infty}$ such that $g(y_j) = \lim_{n \to \infty} f_{[2,n]}(y_j)$ exists for $j = 1, 2$. Continuing in this manner, we obtain a sequence of subsequences $\{\{f_{[k,n]}\}_{n=1}^{\infty}\}_{k=1}^{\infty}$ such that $g(y_j) = \lim_{n \to \infty} f_{[k,n]}(y_j)$ exists for $j = 1, 2, \ldots, k$. Letting $g_k = f_{[k,k]}$, for each $k \in \mathcal{N}$, we now have a subsequence of $\{f_n\}_{n=1}^{\infty}$ that satisfies (12.10). ∎

In the mathematical literature, the following variant of Theorem 12.4 is frequently cited.

□ □ □ THEOREM 12.5 Ascoli-Arzela Theorem

Let Ω be a separable topological space and D a subset of $C(\Omega, \Lambda)$ that satisfies the following conditions:
a) $\overline{\{f(x) : f \in D\}}$ is a compact subset of Λ for each $x \in \Omega$.
b) D is equicontinuous.
Then every sequence in D has a subsequence that converges uniformly on compact subsets of Ω.

PROOF In the proof of the sufficiency part of Theorem 12.4, the second countability and local compactness conditions are used only to ensure that $C(\Omega, \Lambda)$ is metrizable. If we are just trying to show that every sequence in D has a subsequence that converges uniformly on compact sets, then all that is required for the proof of Theorem 12.4 to remain valid is the existence of a countable dense subset of Ω. ■

Next we give a simple example of the use of Theorem 12.4. More elaborate applications are left to the exercises.

EXAMPLE 12.7 *Illustrates Theorem 12.4*

Let $\Omega = [a, b]$ and $\Lambda = \mathbb{C}$.
a) The closed ball $\overline{B}_r(f)$ of radius $r > 0$ in $C([a, b])$ clearly satisfies the condition (a) of Theorem 12.4. On the other hand, Exercise 12.18 shows that $\overline{B}_r(f)$ is not equicontinuous. Thus, $\overline{B}_r(f)$ is not a compact subset of $C([a, b])$.
b) Let A and B be nonnegative constants. Denote by F the collection of all functions $f \in C([a, b])$ that satisfy $|f(a)| \leq A$, f is differentiable on (a, b), and $|f'(x)| \leq B$ for all $x \in (a, b)$. We claim that $D = \overline{F}$ is compact.

Indeed, if $x \in (a, b]$, then, by the mean value theorem from elementary calculus, we have that $f(x) = f(a) + f'(t)(x - a)$ for some $t \in (a, x)$. Setting $M = A + (b - a)B$, we see that $|f(x)| \leq M$ for all $f \in F$ and $x \in [a, b]$. It follows easily that F and, hence, D satisfies Theorem 12.4(a).

If $x, y \in [a, b]$ and $x \neq y$, then another application of the mean value theorem yields $|f(x) - f(y)| \leq B|x - y|$. It follows immediately that F is equicontinuous and, hence, by Exercise 12.19, so is D. We can now conclude from Theorem 12.4 that $D = \overline{F}$ is compact. □

Exercises for Section 12.3

*Some of the exercises in this section use the concept of a compact function. Let Ω and Λ be topological spaces and $U \subset \Omega$. A function $f : U \to \Lambda$ is said to be a **compact function** if $\overline{f(U)}$ is a compact subset of Λ.*

12.18 Show that the closed ball $\overline{B}_r(f)$ in $C([a, b])$ is not equicontinuous.

12.19 Show that if $F \subset C(\Omega, \Lambda)$ is equicontinuous, then $\overline{F}$ is also equicontinuous.

12.20 Let $g \in \mathcal{L}^1([0, 1])$ and F denote the set of all functions $f \in C([0, 1])$ such that $f(0) = 0$, f is absolutely continuous, and $|f'| \leq |g|$ λ-ae. Show that $\overline{F}$ is compact.

12.21 Let $\mathcal{N}^*$ denote the one-point compactification of the space $(\mathcal{N}, \mathcal{T}_d)$, where $\mathcal{N}$ is the set of positive integers and $\mathcal{T}_d$ is the discrete topology. Prove that a closed set $D \subset C(\mathcal{N}^*)$ is compact if and only if (1) $\sup\{|f(\omega)| : f \in D\} < \infty$ and (2) there is a sequence $\{b_n\}_{n=1}^{\infty}$ such that $\lim_{n \to \infty} b_n = 0$ and $|f(n) - f(\omega)| \leq b_n$ for all $f \in D$ and $n \in \mathcal{N}$.

12.22 Let $f(x) = \sin x + \sin\sqrt{2}x$. For $t \in \mathcal{R}$, define $f_t(x) = f(x - t)$.
   a) Show that $\overline{\{f_t : t \in \mathcal{R}\}}$ is a compact subset of $C(\mathcal{R})$.
   b) Find a convergent subsequence of $\{f_{-2\pi n}\}_{n=1}^{\infty}$.

12.23 Refer to Exercise 12.22. Provide an example of a bounded function f such that $\overline{\{f_t : t \in \mathcal{R}\}}$ is not a compact subset of $C(\mathcal{R})$.

12.24 Refer to Example 12.5 on page 431. Suppose we drop the assumptions that (12.6) holds and $\alpha A < 1$, but still assume that (12.7) holds and $\alpha B \leq \beta$.
   a) Show that T still maps $\overline{B}_\beta(y_0)$ into itself.
   b) Show that T is a compact function.

Exercises 12.25–12.28 require some knowledge of the theory of functions of a complex variable. In these exercises, Ω denotes an open subset of $\mathbb{C}$ and $H(\Omega)$ the set of functions that are analytic on Ω.

12.25 Show that $H(\Omega)$ is a closed subset of $C(\Omega)$. *Hint:* Use the Cauchy integral formula.

12.26 Let $F \subset H(\Omega)$. Show that F is compact if and only if it is closed in $C(\Omega)$ and for each $z_0 \in \Omega$ there is an $r > 0$ such that $\overline{B}_r(z_0) \subset \Omega$ and

$$\sup\{\,|f(z)| : f \in F,\ z \in \overline{B}_r(z_0)\,\} < \infty.$$

12.27 Let $\Omega = B_1(0)$ and $U = \{\,f \in H(\Omega) : |f| \leq 1\,\}$. Prove that for each $z \in \Omega$, there is a $g \in U$ such that $|g'(z)| = \sup\{\,|f'(z)| : f \in U\,\}$.

12.28 Let Ω and U be as in Exercise 12.27 and $0 \leq r < 1$. Consider the function $T\colon U \to U$ defined by $T(f)(z) = f(rz)$. Prove that T is a compact function.

12.4 COMPACTNESS OF PRODUCT SPACES

The Cartesian product of two compact intervals of $\mathcal{R}$ is a compact rectangle in $\mathcal{R}^2$. Indeed, it is not hard to prove that $\Gamma \times \Lambda$ is a compact space whenever Γ and Λ are compact spaces, as you are asked to show in Exercise 12.29.

In this section, we prove a striking generalization of the forementioned fact, namely, that the Cartesian product of any collection of compact spaces is compact. As a corollary to the main result, we obtain simple sufficient conditions for compactness of spaces with weak topologies.

To begin, we briefly review two prerequisite concepts.

- If $\{A_\iota\}_{\iota \in I}$ is an indexed collection of sets, then the *Cartesian product* of the collection, denoted $\times_{\iota \in I} A_\iota$, is the set of all functions x on I such that $x(\iota) \in A_\iota$ for each $\iota \in I$. We call $x(\iota)$ the ι*th coordinate* of x and often denote it by x_ι.

- Suppose that $\{(\Omega_\iota, \mathcal{T}_\iota)\}_{\iota \in I}$ be an indexed collection of topological spaces. Let $\Omega = \times_{\iota \in I} \Omega_\iota$. The function $p_\iota\colon \Omega \to \Omega_\iota$ defined by $p_\iota(x) = x(\iota)$ is called the ι*th coordinate projection* on Ω. The weak topology on Ω determined by the family of coordinate projections $\{\,p_\iota : \iota \in I\,\}$ is called the *product topology*. Thus, the product topology is the weakest topology for which all coordinate projections are continuous.

Now we present the main result of this section, known as Tychonoff's theorem.

□ □ □ **THEOREM 12.6 Tychonoff's Theorem**

The Cartesian product of any family of compact spaces is compact. That is, if $\{\Omega_\iota\}_{\iota \in I}$ is an indexed collection of compact topological spaces, then the Cartesian product $\Omega = \times_{\iota \in I} \Omega_\iota$ is compact with respect to the product topology.

PROOF By Theorem 11.5 on page 407, it suffices to show that any collection of closed subsets of Ω that has the finite intersection property has a nonempty intersection. So, let $\mathcal{C}$ be a collection of closed subsets of Ω that has the finite intersection property. We need to prove that

$$\bigcap_{F \in \mathcal{C}} F \neq \emptyset. \tag{12.14}$$

Let $\mathfrak{A}$ denote the family of all $\mathcal{A} \subset \mathcal{P}(\Omega)$ such that $\mathcal{A}$ has the finite intersection property and $\mathcal{C} \subset \mathcal{A}$. We will use Zorn's lemma (page 15) to show that $\mathfrak{A}$ contains a maximal element with respect to the inclusion ordering $\subset$.

Suppose that $\mathfrak{C}$ is a nonempty chain in $\mathfrak{A}$. We claim that $\mathcal{U} = \bigcup_{\mathcal{A} \in \mathfrak{C}} \mathcal{A}$ is an upper bound for $\mathfrak{C}$. Clearly, $\mathcal{A} \subset \mathcal{U}$ for all $\mathcal{A} \in \mathfrak{C}$. So, we need only show that $\mathcal{U} \in \mathfrak{A}$. Because it is obvious that $\mathcal{C} \subset \mathcal{U}$, it remains to prove that $\mathcal{U}$ has the finite intersection property.

Suppose $U_1, U_2, \ldots, U_n \in \mathcal{U}$. Then, for each j, $U_j \in \mathcal{A}_j$ for some $\mathcal{A}_j \in \mathfrak{C}$. Since $\mathfrak{C}$ is a chain, there exists an $\mathcal{A} \in \mathfrak{C}$ such that $\mathcal{A}_j \subset \mathcal{A}$ for $j = 1, 2, \ldots, n$. It follows that $U_1, U_2, \ldots, U_n \in \mathcal{A}$ and, since $\mathcal{A}$ has the finite intersection property, we conclude that $\bigcap_{j=1}^n U_j \neq \emptyset$. Thus, $\mathcal{U}$ is an upper bound for $\mathfrak{C}$. Zorn's lemma now implies that $\mathfrak{A}$ has a maximal element, say, $\mathcal{A}^*$.

We claim that $\mathcal{A}^*$ has the following properties:

$$A_1, A_2, \ldots, A_n \in \mathcal{A}^* \Rightarrow \bigcap_{j=1}^n A_j \in \mathcal{A}^* \tag{12.15}$$

and

$$B \subset \Omega \text{ and } B \cap A \neq \emptyset \text{ for all } A \in \mathcal{A}^* \Rightarrow B \in \mathcal{A}^*. \tag{12.16}$$

To verify (12.15), let $A = \bigcap_{j=1}^n A_j$. We note that $\mathcal{A}^* \cup \{A\}$ has the finite intersection property and that $\mathcal{C} \subset \mathcal{A}^* \cup \{A\}$. Because $\mathcal{A}^*$ is a maximal element of $\mathfrak{A}$, we must have $\mathcal{A}^* \cup \{A\} = \mathcal{A}^*$. Hence, $A \in \mathcal{A}^*$.

To establish (12.16), we show that $\mathcal{A}^* \cup \{B\}$ has the finite intersection property. It will then follow from the maximality of $\mathcal{A}^*$ that $B \in \mathcal{A}^*$. Let $B_1, B_2, \ldots, B_m$ be distinct elements of $\mathcal{A}^* \cup \{B\}$. If $B_j \neq B$ for $j = 1, 2, \ldots, m$, then $\bigcap_{j=1}^m B_j \neq \emptyset$ because $\mathcal{A}^*$ has the finite intersection property. On the other hand, if $B_k = B$ for some k, then $\bigcap_{j \neq k} B_j \in \mathcal{A}^*$ by (12.15). Thus,

$$\bigcap_{j=1}^m B_j = B \cap \left(\bigcap_{j \neq k} B_j \right) \neq \emptyset,$$

as required.

Let $\iota \in I$. For $\mathcal{E} \subset \mathcal{A}^*$, we have $p_\iota \left(\bigcap_{A \in \mathcal{E}} A \right) \subset \bigcap_{A \in \mathcal{E}} p_\iota(A)$. Hence, the collection $\{ \overline{p_\iota(A)} : A \in \mathcal{A}^* \}$ has the finite intersection property. From the compactness of the space Ω_ι, we can conclude that $\bigcap_{A \in \mathcal{A}^*} \overline{p_\iota(A)} \neq \emptyset$. It now follows from the axiom of choice (page 14) that there is an $x \in \Omega$ such that

$$p_\iota(x) \in \bigcap_{A \in \mathcal{A}^*} \overline{p_\iota(A)}$$

for each $\iota \in I$.

We will show that $x \in \bigcap_{F \in \mathcal{C}} F$ by proving that if $F \in \mathcal{C}$ and W is an open set containing x, then

$$W \cap F \neq \emptyset. \tag{12.17}$$

It will follow that $x \in \overline{F} = F$ for each $F \in \mathcal{C}$.

We recall that the product topology is determined by the neighborhood basis that consists of finite intersections of sets of the form $p_\iota^{-1}(W_\iota)$, where W_ι is an open subset of Ω_ι. Thus, to establish (12.17), it suffices to consider the case where $W = \bigcap_{\iota \in I_0} p_\iota^{-1}(W_\iota)$ for some finite subset $I_0 \subset I$.

Let $\iota \in I_0$ and $A \in \mathcal{A}^*$. Because $p_\iota(x) \in \overline{p_\iota(A)}$ and W_ι is an open set that contains $p_\iota(x)$, it follows that $W_\iota \cap p_\iota(A) \neq \emptyset$. Consequently, $p_\iota^{-1}(W_\iota) \cap A \neq \emptyset$ for each $A \in \mathcal{A}^*$ and, hence, by (12.16), $p_\iota^{-1}(W_\iota) \in \mathcal{A}^*$. Therefore, since $F \in \mathcal{A}^*$ and $\mathcal{A}^*$ has the finite intersection property, $\left(\bigcap_{\iota \in I_0} p_\iota^{-1}(W_\iota) \right) \cap F \neq \emptyset$. ∎

Next we will apply Tychonoff's theorem to the study of compactness in the context of weak topologies. Specifically, we have the following corollary, which will be useful later when we study weak topologies on linear spaces.

□ □ □ **COROLLARY 12.2**

Suppose that Ω has the weak topology determined by a family of functions $\mathcal{F}$ and that the following conditions are satisfied:
a) *$\overline{f(\Omega)}$ is compact for each $f \in \mathcal{F}$.*
b) *If $x \neq y$, then $f(x) \neq f(y)$ for some $f \in \mathcal{F}$.*
c) *If $\{f(x_\iota)\}_{\iota \in I}$ is a convergent net for each $f \in \mathcal{F}$, then there is an $x \in \Omega$ such that $\lim f(x_\iota) = f(x)$ for each $f \in \mathcal{F}$.*
Then Ω is compact.

PROOF   Each $f \in \mathcal{F}$ maps Ω into a topological space Ω_f. From condition (a), $\Lambda_f = \overline{f(\Omega)}$ is a compact subset of Ω_f. It follows from Tychonoff's theorem that the product space $\Lambda = \times_{f \in \mathcal{F}} \Lambda_f$ is compact. Since a closed subset of a compact space is compact (Theorem 11.7 on page 408), we can establish the corollary by finding a homeomorphism from Ω onto a closed subset of Λ.

Let $h: \Omega \to \Lambda$ be defined by $h(x)(f) = f(x)$. Condition (c) is equivalent to the assertion that $h(\Omega)$ is a closed subset of Λ. That h is continuous follows from the definitions of product and weak topologies, Proposition 10.13 on page 383, and Theorem 10.1 on pages 382–383. Condition (b) says that h is one-to-one.

It remains only to show that the inverse function h^{-1} is continuous on $h(\Omega)$. If $\{h(x_\iota)\}_{\iota \in I}$ is a net in $h(\Omega)$ that converges to $h(x)$, then, by Proposition 10.13, $\{x_\iota\}_{\iota \in I}$ converges in Ω to x. It now follows from Theorem 10.1 that h^{-1} is continuous. ∎

Recall that if $A_\iota = A$ for each $\iota \in I$, where A is some set, then Cartesian products of the form $\times_{\iota \in I} A_\iota$ are usually denoted by A^I. Note that A^I is just the set of all functions from I to A. In case I is the set $\mathcal{N}$ of positive integers, A^I is the set of all sequences of elements of A. Often a typical element of $A^{\mathcal{N}}$ is written in the form $x = (x_1, x_2, \ldots)$.

EXAMPLE 12.8 Illustrates Tychonoff's Theorem

a) Consider the space $\{0, 1\}$ endowed with the discrete topology and let I be any index set. We write 2^I in place of $\{0, 1\}^I$. It follows from Tychonoff's theorem that 2^I is a compact space.

b) If $[a, b]$ is a closed bounded interval, then it follows from the Heine-Borel theorem and Tychonoff's theorem that $[a, b]^I$ is compact for any index set I. □

Exercises for Section 12.4

12.29 Give an alternative proof of Tychonoff's theorem in case the index set has only two elements. That is, without using Tychonoff's theorem, show that if Γ and Λ are compact, then so is $\Gamma \times \Lambda$ (in the product topology).

12.30 Let $(\Omega, \mathcal{T})$ be a compact Hausdorff space. Show that the topology $\mathcal{T}$ coincides with the weak topology on Ω determined by the family of functions $\mathcal{F} = C(\Omega)$.

12.31 Prove that the Cartesian product of a collection of Hausdorff spaces is a Hausdorff space.

12.32 Let Ω be a compact Hausdorff space. Show that Ω is homeomorphic to a closed subset of $[0, 1]^I$ for some set I. *Hint:* Refer to Exercise 12.31.

12.33 A topological space Ω is said to be **completely regular** if it satisfies the following two conditions:

- Ω is a T_1-space.

- Given a closed set F and a point $x \in F^c$, there is a continuous function $k\colon \Omega \to [0, 1]$ such that $k(x) = 0$ and $k(F) = \{1\}$.

Suppose that Ω is a completely regular topological space. Set $\mathcal{F} = C\big(\Omega, [0, 1]\big)$ and define $h\colon \Omega \to [0, 1]^{\mathcal{F}}$ by $h(x)(f) = f(x)$. Prove that h is a homeomorphism of Ω onto $h(\Omega)$.

12.34 This exercise continues Exercise 12.33. Let $\beta(\Omega)$ denote the closure of $h(\Omega)$ in $[0, 1]^{\mathcal{F}}$. By identifying Ω with $h(\Omega)$ via the map h we may consider Ω a dense subset of the compact Hausdorff space $\beta(\Omega)$. Thus, $\beta(\Omega)$ is a compactification of Ω, that is, a compact space that contains Ω as a dense subspace. The space $\beta(\Omega)$ is called the **Stone-Čech compactification** of Ω.

a) Prove that if $g\colon \Omega \to \mathcal{R}$ is continuous and bounded, then it has a continuous extension to $\beta(\Omega)$.

b) Does the one point compactification of Theorem 11.12 have the property in part (a)? *Note:* To appreciate the Stone-Čech compactification, it helps to consider the continuous function $\sin(1/x)$ on the interval $(0, \infty)$.

12.35 This exercise continues Exercise 12.34. Show that $\beta(\Omega)$ is the largest compactification of Ω in the following sense: If Λ is a compact Hausdorff space such that $\Omega \subset \Lambda$, $\overline{\Omega} = \Lambda$, and the topology of Ω is the same as its relative topology inherited from Λ, then there is a continuous function $f\colon \beta(\Omega) \to \Lambda$ such that $f(x) = x$ for each $x \in \Omega$.

12.36 Show that if I is uncountable, then 2^I is not metrizable.

12.37 Prove that $2^{\mathcal{N}}$ and the Cantor set are homeomorphic. *Hint:* Define the function h on $2^{\mathcal{N}}$ by $h\big((x_1, x_2, \ldots)\big) = \sum_{n=1}^{\infty} 2x_n/3^n$.

12.38 Show that if Ω is a compact metric space, then Ω is homeomorphic to a closed subset of $[0,1]^{\mathcal{N}}$.

12.39 Find a continuous function from $2^{\mathcal{N}}$ onto $[0,1]^{\mathcal{N}}$.

12.40 This exercise involves the Cantor set.
a) Let F be a nonempty closed subset of the Cantor set P. Show that there is a continuous function $r: P \to F$ such that $r(t) = t$ for each $t \in F$.
b) Use part (a) and Exercises 12.37–12.39 to show that if Ω is a compact metric space, then there is a continuous function from P onto Ω.

12.5 APPROXIMATION BY FUNCTIONS FROM A LATTICE

Recall that a real-valued function g on an interval J is **piecewise linear** if there is a partition $a_0 < a_1 < a_2 < \cdots < a_n$ of J such that on each subinterval (a_j, a_{j+1}), we have $g(t) = m_j t + b_j$ for some real numbers m_j and b_j. Consider the following two problems of approximation of continuous functions on the closed interval $[0,1]$.

Problem 1: Given $f \in C([0,1], \mathcal{R})$ and $\epsilon > 0$, find a continuous piecewise linear function g such that $|f(t) - g(t)| < \epsilon$ for each $t \in [0,1]$.

Problem 2: Given $f \in C([0,1], \mathcal{R})$ and $\epsilon > 0$, find a polynomial p such that $|f(t) - p(t)| < \epsilon$ for each $t \in [0,1]$.

Let us denote by $\mathcal{W}$ the set of continuous piecewise linear functions on $[0,1]$. And let us also denote by P_r the set of polynomials with real coefficients.[†] Then it is easy to see that Problems 1 and 2 can be solved if, respectively, $f \in \overline{\mathcal{W}}$ and $f \in \overline{P_r}$.

In this section, motivated by Problem 1, we will prove a general result from which we obtain $\overline{\mathcal{W}} = C([0,1], \mathcal{R})$ as a special case. In the next section, we will prove a theorem that has $\overline{P_r} = C([0,1], \mathcal{R})$ as a special case. It then follows that Problems 1 and 2 can be solved for each $f \in C([0,1], \mathcal{R})$.

The collection $\mathcal{W}$ of continuous piecewise linear function is a motivating example for the following definition.

DEFINITION 12.4   Lattice of Functions

A collection $\mathcal{L}$ of real-valued functions on a set Ω is called a **lattice** if it is closed under maximums and minimums. That is,
a) $f, g \in \mathcal{L}$ implies $f \vee g \in \mathcal{L}$.
b) $f, g \in \mathcal{L}$ implies $f \wedge g \in \mathcal{L}$.

[†] If we restrict a member of P_r to any subset of $\mathcal{R}$, it is continuous thereon. For convenience, we will abuse notation slightly and use P_r to denote the collection of polynomials with real coefficients considered as functions on any particular subset of $\mathcal{R}$. Context will determine the appropriate subset.

EXAMPLE 12.9 *Illustrates Definition 12.4*

a) The following collections are lattices of functions contained in $C([0,1], \mathcal{R})$:
 (i) $C([0,1], \mathcal{R})$ itself.
 (ii) The collection of nonnegative continuous functions on $[0,1]$.
 (iii) The collection $\mathcal{W}$ of continuous piecewise linear functions on $[0,1]$.
b) The collection P_r of polynomials on $[0,1]$ with real coefficients is not a lattice of functions. □

The main result of this section, the Kakutani-Krein theorem, provides a set of sufficient conditions for a lattice to be dense in $C(\Omega, \mathcal{R})$ when Ω is a compact Hausdorff space. In order to prove that result, we first establish the following theorem, which is important in its own right. You should recall the notation $\|f\|_\Omega = \sup\{|f(x)| : x \in \Omega\}$.

□ □ □ **THEOREM 12.7 Dini's Theorem**

Let Ω be a compact Hausdorff space. Suppose that $\mathcal{F} \subset C(\Omega, \mathcal{R})$ has the following two properties:
a) $f, g \in \mathcal{F}$ implies there is an $h \in \mathcal{F}$ such that $h \leq f \wedge g$.
b) The function f_0 defined by $f_0(x) = \inf\{f(x) : f \in \mathcal{F}\}$ is real-valued and continuous.
Then given $\epsilon > 0$, there exists an $f \in \mathcal{F}$ such that $\|f - f_0\|_\Omega < \epsilon$.

PROOF By the definition of f_0, we know that for each $x \in \Omega$, there is a function $f_x \in \mathcal{F}$ such that $f_x(x) < f_0(x) + \epsilon$. Because f_0 is continuous, the sets

$$U_x = \{y : f_x(y) < f_0(y) + \epsilon\}, \qquad x \in \Omega,$$

constitute an open covering of Ω. Hence, there are points $x_1, x_2, \ldots, x_n \in \Omega$ such that $\Omega = \bigcup_{j=1}^n U_{x_j}$.

Using (a), we can find an $f \in \mathcal{F}$ with $f \leq f_{x_j}$ for $j = 1, 2, \ldots, n$. Hence, for each $x \in \Omega$, we have

$$f_0(x) \leq f(x) \leq \min_{1 \leq j \leq n} f_{x_j}(x) < f_0(x) + \epsilon.$$

It follows at once that $\|f - f_0\|_\Omega < \epsilon$. ∎

DEFINITION 12.5 **Separation of Points**

A collection $\mathcal{F}$ of functions on a set Ω is said to **separate points** of Ω if whenever x and y are distinct elements of Ω, there is an $f \in \mathcal{F}$ such that $f(x) \neq f(y)$.

EXAMPLE 12.10 *Illustrates Definition 12.5*

a) If $\mathcal{F}$ contains a one-to-one function, then it separates points.
b) P_r separates the points of $[0,1]$ because it contains the identity function.

c) $\mathcal{W}$ separates the points of $[0, 1]$ because it contains the identity function.

d) Let $\mathcal{F}$ denote the polynomials on $[-1, 1]$ that contain only even powers, that is, polynomials of the form $a_0 + a_1 x^2 + \cdots + a_n x^{2n}$. Then $\mathcal{F}$ does not separate the points of $[-1, 1]$ since $f(-1) = f(1)$ for all $f \in \mathcal{F}$.

e) Suppose that Ω is a topological space with the property that there is a collection $\mathcal{F} \subset C(\Omega, \mathcal{R})$ that separates the points of Ω. Then Ω is a Hausdorff space.

f) The collection of functions $\{\sin x, \cos x\}$ separates points of $[0, 2\pi)$, but it does not separate points of $[0, 2\pi]$. □

□ □ □ **THEOREM 12.8 Kakutani-Krein Theorem**

Let Ω be a compact Hausdorff space. Suppose that $\mathcal{L} \subset C(\Omega, \mathcal{R})$ satisfies the following conditions:

a) $\mathcal{L}$ is a lattice.

b) $\mathcal{L}$ separates points of Ω.

c) $f \in \mathcal{L}$ and $c \in \mathcal{R}$ implies $cf \in \mathcal{L}$ and $f + c \in \mathcal{L}$.

Then $\overline{\mathcal{L}} = C(\Omega, \mathcal{R})$.

PROOF For $g \in C(\Omega, \mathcal{R})$, let $\mathcal{L}_g = \{ f \in \mathcal{L} : g \leq f \}$. Because $\mathcal{L}$ is a lattice, it follows that $\mathcal{L}_g$ satisfies condition (a) of Dini's theorem. Therefore, if we can show that

$$g(x) = \inf\{ f(x) : f \in \mathcal{L}_g \} \tag{12.18}$$

for each $x \in \Omega$, Dini's theorem will imply the required result. We will establish (12.18) by constructing for each $\epsilon > 0$, an $f_x \in \mathcal{L}_g$ such that $f_x(x) = g(x) + \epsilon$.

To begin, we show that for each pair of distinct points $y, z \in \Omega$ and each pair of real numbers a and b, there is an $h \in \mathcal{L}$ such that

$$h(y) = a \quad \text{and} \quad h(z) = b. \tag{12.19}$$

Using (b), we choose an $h_0 \in \mathcal{L}$ with $h_0(y) \neq h_0(z)$ and then, using (c), we conclude that the function

$$h = (a - b)\frac{h_0 - h_0(z)}{h_0(y) - h_0(z)} + b$$

belongs to $\mathcal{L}$. It is easy to see that h satisfies (12.19).

Next we consider the open set $O = \{ y : g(y) < g(x) + \epsilon \}$. For each $z \in O^c$, we can apply (12.19) to obtain an $h_z \in \mathcal{L}$ such that

$$h_z(z) = g(z) + \epsilon \quad \text{and} \quad h_z(x) = g(x) + \epsilon/2.$$

Let $V_z = \{ y : h_z(y) > g(y) \}$. Then $\{ V_z : z \in O^c \}$ is an open covering of the compact set O^c. So, there are points $z_1, z_2, \ldots, z_n \in O^c$ such that $O^c \subset \bigcup_{j=1}^n V_{z_j}$. Let

$$f_x = (g(x) + \epsilon) \vee h_{z_1} \vee h_{z_2} \vee \cdots \vee h_{z_n}.$$

It follows from (c) that $\mathcal{L}$ contains the constant function $g(x) + \epsilon$ and, hence, by (a), we have $f_x \in \mathcal{L}$. If $y \in O^c$, then $y \in V_{z_j}$ for some $j \in \{1, 2, \ldots, n\}$ and, consequently, we have that $f_x(y) \geq h_{z_j}(y) > g(y)$. On the other hand, if $y \in O$, then $f_x(y) \geq g(x) + \epsilon > g(y)$. It now follows that $f_x \in \mathcal{L}_g$. Finally, we have that

$$f_x(x) = (g(x) + \epsilon) \vee (g(x) + \epsilon/2) \vee \cdots \vee (g(x) + \epsilon/2) = g(x) + \epsilon,$$

as required. ■

EXAMPLE 12.11 *Illustrates the Kakutani-Krein Theorem*

It is easy to see that the collection $\mathcal{W}$ of continuous piecewise linear functions on $[0,1]$ satisfies the conditions of the Kakutani-Krein theorem. Consequently, $\overline{\mathcal{W}} = C([0,1], \mathcal{R})$. In other words, Problem 1 on page 442 can be solved for each $f \in C([0,1])$. □

The most important application of the Kakutani-Krein theorem comes in the next section where it is used to prove the Stone-Weierstrass theorem.

Exercises for Section 12.5

12.41 Show that Dini's theorem fails if the assumption that f_0 is continuous is dropped.

12.42 In Exercise 2.63 on page 60, we asked you to prove another version of Dini's theorem. Show that the theorem stated there is a special case of the Dini's theorem of this section (Theorem 12.7).

12.43 Verify that the collection $\mathcal{W}$ of continuous piecewise linear functions on $[0,1]$ is a lattice.

12.44 Suppose that $\mathcal{L} \subset C(\Omega, \mathcal{R})$, where Ω is a compact Hausdorff space. Show that if $\mathcal{L}$ is a lattice, then so is $\overline{\mathcal{L}}$.

12.45 Suppose that $\mathcal{L}$ is a linear subspace of $C(\Omega, \mathcal{R})$. Show that $\mathcal{L}$ is a lattice if and only if $|f| \in \mathcal{L}$ whenever $f \in \mathcal{L}$.

12.46 Give an example showing that the Kakutani-Krein theorem fails if conditions (b) and (c) are retained but (a) is dropped.

12.47 Give an example showing that the Kakutani-Krein theorem fails if conditions (a) and (c) are retained but (b) is dropped.

12.48 Give an example showing that the Kakutani-Krein theorem fails if conditions (a) and (b) are retained but (c) is dropped.

12.49 Let Ω be a compact Hausdorff space. Suppose that $\mathcal{L} \subset C(\Omega, \mathcal{R})$ satisfies the following conditions:

- $\mathcal{L}$ is closed.

- $\mathcal{L}$ is a linear subspace of $C(\Omega, \mathcal{R})$.

- $\mathcal{L}$ is a lattice.

- $f \in \mathcal{L}$, $g \in C(\Omega, \mathcal{R})$, and $0 \leq g \leq f$ imply $g \in \mathcal{L}$.

Show that either $\mathcal{L} = C(\Omega, \mathcal{R})$ or there exists a nonempty closed set $F \subset \Omega$ such that $\mathcal{L} \subset \{ f \in C(\Omega, \mathcal{R}) : f(x) = 0 \text{ for each } x \in F \}$.

12.6 APPROXIMATION BY FUNCTIONS FROM AN ALGEBRA

In our study of measure theory, we found that the concept of an algebra of functions is essential. Now we will see that this concept is also of importance in the study of approximation by functions.

DEFINITION 12.6 **Algebra of Functions**

A collection $\mathcal{A}$ of real-valued or complex-valued functions on a set Ω is called an **algebra** if it is closed under addition, scalar multiplication, and multiplication. That is, if $f, g \in \mathcal{A}$ and α is a scalar, then

a) $f + g \in \mathcal{A}$.
b) $\alpha f \in \mathcal{A}$.
c) $f \cdot g \in \mathcal{A}$.

Theorem 5.3 (page 152) and Exercise 5.32 (page 157) show, respectively, that the collection of real-valued and complex-valued functions measurable with respect to a σ-algebra of subsets of a set Ω form algebras of functions. In addition, Theorem 2.4 (page 55) shows that the collection of real-valued continuous functions on a subset of $\mathcal{R}$ constitutes an algebra of functions. It is this latter type of algebra—algebras of continuous functions—that will be important to us in this section.

EXAMPLE 12.12 *Illustrates Definition 12.6*

a) Let Ω be a topological space. Then the collection $C(\Omega, \mathcal{R})$ of real-valued continuous functions on Ω is an algebra of functions.
b) Let Ω be a topological space. Then the collection $C(\Omega)$ of complex-valued continuous functions on Ω is an algebra of functions.
c) Let P_r denote the collection of polynomials with real coefficients viewed as functions on the closed bounded interval $[a, b]$.[†] Clearly, P_r is an algebra in $C([a, b], \mathcal{R})$.
d) Let $\mathcal{U}_r$ denote the collection of trigonometric polynomials with real coefficients viewed as functions on the closed bounded interval $[a, b]$. We claim that $\mathcal{U}_r$ is also an algebra in $C([a, b], \mathcal{R})$. To see this, recall that a trigonometric polynomial u with real coefficients is a function of the form

$$u(t) = \sum_{j=0}^{n} (a_j \cos jt + b_j \sin jt),　\tag{12.20}$$

where the a_js and b_js are real numbers. That $\mathcal{U}_r$ is a linear subspace of $C([a, b], \mathcal{R})$ is clear; that it is also closed under multiplication and, hence,

---

† See the footnote on page 442.

constitutes an algebra, follows from the trigonometric identities

$$2\cos jt \cos kt = \cos(j+k)t + \cos(j-k)t,$$
$$2\sin jt \sin kt = \cos(j-k)t - \cos(j+k)t,$$
$$2\sin jt \cos kt = \sin(j+k)t + \sin(j-k)t.$$

e) By considering functions of the form (12.20) where the a_js and b_js are permitted to be complex numbers, we obtain the collection $\mathcal{U}$ of complex trigonometric polynomials. Rather than writing complex trigonometric polynomials in the form (12.20), we will usually work with the equivalent expression

$$u(t) = \sum_{j=-n}^{n} c_j e^{ijt}.$$

It is easy to check that, viewed as a subset of $C([a,b])$, $\mathcal{U}$ is an algebra of functions.

f) Suppose Ω is a compact subset of $\mathbb{C}$. Let $P(\Omega)$ denote the collection of functions in $C(\Omega)$ that are polynomials in z, that is, functions of the form $p(z) = \sum_{j=0}^{n} a_j z^j$, where the a_js are complex constants. It is clear that $P(\Omega)$ is an algebra in $C(\Omega)$. Another algebra in $C(\Omega)$, which we denote by $P^*(\Omega)$, consists of all polynomials in z and $\overline{z}$, that is, functions of the form

$$p(z,\overline{z}) = \sum_{j=0}^{n}\sum_{k=0}^{n} a_{j,k} z^j \overline{z}^k,$$

where each $a_{j,k} \in \mathbb{C}$. We will see later that the closure of $P^*(\Omega)$ is $C(\Omega)$.

g) Refer to part (f). The case $\Omega = T$, where T is the unit circle in the complex plane, $\{\, z \in \mathbb{C} : |z| = 1 \,\}$, is of particular interest. It is not hard to show that $P^*(T)$ consists of all functions of the form

$$q(z) = \sum_{j=-n}^{n} c_j z^j, \qquad z \in T,$$

where each $c_j \in \mathbb{C}$. There is a connection between $P^*(T)$ and the collection $\mathcal{U}$ of trigonometric polynomials, namely, $u \in \mathcal{U}$ if and only if $u(t) = q(e^{it})$ for some $q \in P^*(T)$. □

Motivated by Problem 2 on page 442, we now take up the question of when an algebra of functions $\mathcal{A} \subset C(\Omega, \mathcal{R})$ is dense in $C(\Omega, \mathcal{R})$. Later in this section, we will consider the same question when $\mathcal{R}$ is replaced by $\mathbb{C}$.

It is a classical result due to Karl Weierstrass that every continuous real-valued function on a closed bounded interval can be uniformly approximated arbitrarily closely by polynomials. In the notation of the preceding example, Weierstrass's theorem states that $\overline{P_r} = C([a,b], \mathcal{R})$.

Our next theorem, the Stone-Weierstrass theorem, is a far-reaching generalization of the forementioned result. The Stone-Weierstrass theorem gives a set of sufficient conditions for an algebra of functions to be dense in $C(\Omega, \mathcal{R})$ when Ω is a compact Hausdorff space. Its proof relies on the following lemma whose verification was considered in Exercise 3.2 on pages 88–89.

□ □ □   **LEMMA 12.1**

For each $\epsilon > 0$, there is a polynomial p such that $\left| |t| - p(t) \right| < \epsilon$ for all $t \in [-1, 1]$.

□ □ □   **THEOREM 12.9 Stone-Weierstrass Theorem**

Let Ω be a compact Hausdorff space. Suppose that $\mathcal{A} \subset C(\Omega, \mathcal{R})$ satisfies the following conditions:
a) $\mathcal{A}$ is an algebra.
b) $\mathcal{A}$ separates points of Ω.
c) $1 \in \mathcal{A}$.
Then $\overline{\mathcal{A}} = C(\Omega, \mathcal{R})$.

PROOF   We leave it to the reader as an exercise to show that, since $\mathcal{A}$ is an algebra of functions, so is $\overline{\mathcal{A}}$. If we can prove that $\overline{\mathcal{A}}$ is also a lattice, then the verification will be complete on account of the Kakutani-Krein theorem (page 444).

We recall that $f \vee g = (f + g + |f - g|)/2$ and $f \wedge g = (f + g - |f - g|)/2$. Thus, to prove $\overline{\mathcal{A}}$ is a lattice, it suffices to show that

$$f \in \overline{\mathcal{A}} \Rightarrow |f| \in \overline{\mathcal{A}} \tag{12.21}$$

If $f = 0$, then (12.21) is trivial. If $f \neq 0$, let $g = f/\|f\|_\Omega$ and observe that $g \in \overline{\mathcal{A}}$. Given $\epsilon > 0$, we can apply Lemma 12.1 to obtain a polynomial p such that $\left| |t| - p(t) \right| < \epsilon$ for all $t \in [-1, 1]$. Because the range of g is contained in $[-1, 1]$, it follows that

$$\left\| |g| - p \circ g \right\|_\Omega < \epsilon.$$

And because $p \circ g$ is a polynomial in powers of g and $\overline{\mathcal{A}}$ is an algebra that contains the constant functions, we conclude that $p \circ g \in \overline{\mathcal{A}}$. Thus, $|g| \in \overline{\overline{\mathcal{A}}} = \overline{\mathcal{A}}$. Finally, because f is a scalar multiple of g and $\overline{\mathcal{A}}$ is an algebra, it follows that $|f| \in \overline{\mathcal{A}}$. ■

EXAMPLE 12.13   *Illustrates the Stone-Weierstrass Theorem*

Suppose Ω is a compact subset of $\mathcal{R}^n$. Let P_r^n denote the set of polynomials in n variables with real coefficients. It is clear that, as a collection of functions on Ω, P_r^n satisfies the hypotheses of the Stone-Weierstrass theorem. It follows that any $f \in C(\Omega, \mathcal{R})$ can be approximated arbitrarily closely by polynomials in n variables with real coefficients. □

EXAMPLE 12.14   *Illustrates the Stone-Weierstrass Theorem*

The collection of real-valued trigonometric polynomials $\mathcal{U}_r$ is an algebra of functions in $C([0, \pi], \mathcal{R})$ that satisfies the hypotheses of the Stone-Weierstrass theorem. Thus, $\overline{\mathcal{U}_r} = C([0, \pi], \mathcal{R})$. As an algebra in $C([0, 2\pi], \mathcal{R})$, however, $\mathcal{U}_r$ does not satisfy condition (b) of the Stone-Weierstrass theorem because $u(0) = u(2\pi)$ for each $u \in \mathcal{U}_r$. If $g \in C([0, 2\pi], \mathcal{R})$ is such that $g(0) \neq g(2\pi)$, then g cannot be uniformly approximated arbitrarily closely by trigonometric polynomials. □

EXAMPLE 12.15   *Illustrates the Stone-Weierstrass Theorem*

Suppose $f \in C(\mathcal{R}, \mathcal{R})$ is periodic with period 2π. Then, see Example 11.11 on page 422, $f(t) = g(e^{it})$ for some function $g \in C(T, \mathcal{R})$. It is easy to verify that the algebra $P^*(T)$, defined in Example 12.12(g), is such that $P^*(T) \cap C(T, \mathcal{R})$ is an algebra in $C(T, \mathcal{R})$ that satisfies the hypotheses of the Stone-Weierstrass theorem. Consequently, for each $\epsilon > 0$, there is a $p \in P^*(T) \cap C(T, \mathcal{R})$ such that $\|g - p\|_T < \epsilon$. It follows that

$$|f(t) - p(e^{it})| < \epsilon, \qquad t \in \mathcal{R}.$$

Thus, we have proved the following important fact: *Every continuous real-valued function on $\mathcal{R}$ with period 2π can be uniformly approximated arbitrarily closely by a trigonometric polynomial.* □

Complex Version of the Stone-Weierstrass Theorem

If $C(\Omega, \mathcal{R})$ is replaced by $C(\Omega)$, the hypotheses of the Stone-Weierstrass theorem must be augmented in order to obtain an analogous result.

□ □ □ **THEOREM 12.10 Stone-Weierstrass Theorem (Complex Version)**

Let Ω be a compact Hausdorff space. Suppose that $\mathcal{A} \subset C(\Omega)$ satisfies the following conditions:

a) $\mathcal{A}$ is an algebra.
b) $\mathcal{A}$ separates points of Ω.
c) $1 \in \mathcal{A}$.
d) $f \in \mathcal{A} \Rightarrow \overline{f} \in \mathcal{A}$.
Then $\overline{\mathcal{A}} = C(\Omega, \mathcal{R})$.

PROOF   Let $\Re\mathcal{A}$ denote the set of real parts of functions in $\mathcal{A}$. Because

$$\Re\mathcal{A} = \{\, (f + \overline{f})/2 : f \in \mathcal{A} \,\},$$

it follows from the hypotheses of the theorem that $\Re\mathcal{A} \subset \mathcal{A}$ and that $\Re\mathcal{A}$ is an algebra in $C(\Omega, \mathcal{R})$.

We claim that $\Re\mathcal{A}$ separates the points of Ω. Let x and y be distinct elements of Ω. By condition (b), there is an $f \in \mathcal{A}$ such that $f(x) \neq f(y)$. We note that either $\Re f(x) \neq \Re f(y)$ or $\Re(if(x)) \neq \Re(if(y))$. Because $\mathcal{A}$ is an algebra, we have $if \in \mathcal{A}$. It follows that $\Re\mathcal{A}$ separates the points of Ω.

Because $\Re\mathcal{A}$ satisfies the hypotheses of Theorem 12.9, $\overline{\Re\mathcal{A}} = C(\Omega, \mathcal{R})$. Hence, given $f \in C(\Omega)$ and $\epsilon > 0$, we can find $g, h \in \Re\mathcal{A}$ such that $\|\Re f - g\|_\Omega < \epsilon/2$ and $\|\Im f - h\|_\Omega < \epsilon/2$. Consequently, $\|f - g - ih\|_\Omega < \epsilon$. As $\Re\mathcal{A} \subset \mathcal{A}$, it follows that $g + ih \in \mathcal{A}$. ∎

EXAMPLE 12.16   *Illustrates Theorem 12.10*

Refer to Example 12.12(f). Suppose Ω is a compact subset of $\mathbb{C}$.

a) The algebra $P^*(\Omega)$ satisfies the hypotheses of the complex version of the Stone-Weierstrass theorem. Hence, $P^*(\Omega)$ is dense in $C(\Omega)$.

b) In general, the algebra $P(\Omega)$ is not dense in $C(\Omega)$. To see this fact, consider the case $\Omega = T$ and assume that $\overline{P(T)} = C(T)$. Then there is a sequence $\{p_n\}_{n=1}^\infty \subset P(T)$ such that $\{p_n(e^{it})\}_{n=1}^\infty$ converges to e^{-it} uniformly for $t \in [0, 2\pi]$. Consequently,

$$1 = \frac{1}{2\pi} \int_0^{2\pi} e^{it} e^{-it} \, dt = \lim_{n \to \infty} \frac{1}{2\pi} \int_0^{2\pi} e^{it} p_n(e^{it}) \, dt.$$

However, as $\int_0^{2\pi} e^{ikt}\,dt = 0$ for $k \in \mathcal{N}$, it follows that the right-hand side of the previous equation equals 0, a contradiction. Thus, $P(T)$ is not dense in $C(T)$. Noting that $P(T)$ satisfies conditions (a), (b), and (c) of the complex version of the Stone-Weierstrass theorem but not condition (d), we see that this example shows that the theorem fails if condition (d) is dropped from the hypotheses. □

Exercises for Section 12.6

Note: A ★ denotes an exercise that will be subsequently referenced.

In this exercise set, we assume throughout that Ω is a compact Hausdorff space.

12.50   Let $\mathcal{A}$ be an algebra in $C(\Omega, \mathcal{R})$ or $C(\Omega)$. Prove that $\overline{\mathcal{A}}$ is also an algebra.

12.51   Verify the relations

$$f \vee g = (f + g + |f - g|)/2$$

and

$$f \wedge g = (f + g - |f - g|)/2,$$

which were used in the proof of the Stone-Weierstrass theorem.

12.52   Suppose that $\mathcal{A}$ is an algebra in $C(\Omega, \mathcal{R})$. Show that if $f \in \mathcal{A}$ and α is a positive constant, then $|f|^\alpha \in \overline{\mathcal{A}}$.

12.53   Show that Theorem 12.9 remains valid if condition (c) is replaced by the following condition: There is a $g \in \mathcal{A}$ such that $g(x) \neq 0$ for each $x \in \Omega$. *Hint:* See Exercises 12.50 and 12.52.

12.54   Show that if $\mathcal{A} \subset C(\Omega, \mathcal{R})$ satisfies conditions (a) and (b) of the Stone-Weierstrass theorem and $\overline{\mathcal{A}} \neq C(\Omega, \mathcal{R})$, then there is a point $x \in \Omega$ such that $f(x) = 0$ for each $f \in \mathcal{A}$. *Hint:* See Exercise 12.53.

12.55   A linear subspace $\mathcal{I}$ of $C(\Omega, \mathcal{R})$ is called an **ideal** if $f \in \mathcal{I}$ and $g \in C(\Omega, \mathcal{R})$ imply $g \cdot f \in \mathcal{I}$. Suppose that $\mathcal{I}$ is a proper closed ideal of $C(\Omega, \mathcal{R})$. Show that there is a nonempty closed subset $F \subset \Omega$ such that

$$\mathcal{I} = \{\, f \in C(\Omega, \mathcal{R}) : f(x) = 0 \text{ for each } x \in F \,\}.$$

Hint: By Exercise 12.54, $F = \bigcap_{f \in \mathcal{I}} f^{-1}(\{0\}) \neq \emptyset$. Show that if $g \in C(\Omega, \mathcal{R})$ vanishes on F, then there is an $f_0 \in \mathcal{I}$ such that $0 \leq f_0 \leq 1$ and $f_0(y) > 0$ for all $y \notin g^{-1}(\{0\})$. Deduce from Exercise 12.52 that $f_0^{1/n} \in \mathcal{I}$. Show that $\|g - gf_0^{1/n}\|_\Omega \to 0$ as $n \to \infty$.

12.56   Let D be a dense subalgebra of $C(\Omega)$. Show that D must separate the points of Ω.

12.57   Give an example that shows that the complex version of the Stone-Weierstrass theorem fails if condition (c) is dropped.

12.58   Show that the complex version of the Stone-Weierstrass theorem remains valid if condition (c) is replaced by the following condition: There is a $g \in \mathcal{A}$ such that $g(x) \neq 0$ for each $x \in \Omega$. *Hint:* See Exercises 12.54 and 12.55.

12.59   Suppose that $\mathcal{A} \subset C(\Omega)$ satisfies conditions (a), (c), and (d) of the complex version of the Stone-Weierstrass theorem. Also suppose that $g \in \mathcal{A}$ and that h is a complex-valued function continuous on the range of g. Show that $h \circ g \in \overline{\mathcal{A}}$.

12.60   Give an example that shows that the complex version of the Stone-Weierstrass theorem fails if instead of assuming that $\mathcal{A}$ is an algebra, we require only the weaker condition that $\mathcal{A}$ is a linear subspace of $C(\Omega)$.

12.61 A linear subspace $\mathcal{I}$ of $C(\Omega)$ is said to be **ideal** if $f \in \mathcal{I}$ and $g \in C(\Omega)$ imply $g \cdot f \in \mathcal{I}$. Suppose that $\mathcal{I}$ is a proper closed ideal of $C(\Omega)$ such that $f \in \mathcal{I}$ implies $\overline{f} \in \mathcal{I}$. Prove there is a nonempty closed subset $F \subset \Omega$ such that

$$\mathcal{I} = \{ f \in C(\Omega) : f(x) = 0 \text{ for each } x \in F \}.$$

Hint: See the hints for Exercise 12.55.

12.62 Suppose Γ and Λ are compact Hausdorff spaces. Let $\mathcal{A}$ denote the collection of functions f on $\Gamma \times \Lambda$ of the form $f(x,y) = \sum_{j=1}^{n} g_j(x)h_j(y)$, where $n \in \mathcal{N}$ and $g_j \in C(\Gamma)$ and $h_j \in C(\Lambda)$ for $j = 1, 2, \ldots, n$. Show that $\mathcal{A}$ is dense in $C(\Gamma \times \Lambda)$. Generalize this result to arbitrary Cartesian products.

12.63 Let h be a strictly increasing continuous function on $[a,b]$. Suppose that $f \in C([a,b], \mathcal{R})$ satisfies the condition $\int_a^b h^n(t)f(t)\,dt = 0$ for $n = 0, 1, \ldots$. Prove that $f = 0$.

$\star$**12.64** Let $f \in \mathcal{L}^1([0,\infty))$. Prove that if $\int_0^{\infty} e^{-tx} f(x)\,dx = 0$ for each $t \geq 0$, then $f = 0$ ae.

David Hilbert
(1862–1943)

David Hilbert was born in Königsberg, East Prussia (now Kaliningrad, Russia), on January 23, 1862. He was the first of two children of Otto and Maria Therese Hilbert.

In 1872, Hilbert entered the Friedrichskolleg Gymnasium (senior secondary school) but transferred in the fall of 1879 to the more science oriented Wilhelm Gymnasium, where he graduated in 1880. Subsequently, he attended the University of Königsberg and received his doctorate there in 1885. In 1886, Hilbert qualified as an unpaid lecturer at the University of Königsberg and acted in this capacity until 1892, when he replaced Adolf Hurwitz as assistant professor. In 1895, he was appointed to chair at the University of Göttingen, where he remained until he retired in 1930.

Influenced by Ferdinand von Lindemann (who proved the transcendence of π), Hilbert's first work was on the theory of invariants. His activity moved from algebraic forms to algebraic number theory, foundations of geometry, analysis (including the calculus of variations and integral equations), theoretical physics, and, finally, to the foundations of mathematics. The invention of the space that bears Hilbert's name grew from his work in the field of integral equations.

The treatise *Der Zahlbericht* (literally, "report on numbers") was begun in 1893 in partnership with Hermann Minkowski (who subsequently abandoned the project). In this report, Hilbert collected, reorganized, and reshaped the information of algebraic number theory into a master work of mathematical literature—for 50 years, *Der Zahlbericht* was the sacred canon of algebraic number theory. Hilbert also wrote *Grundlagen der Geometrie* (Foundations of Geometry), a text first published in 1899 and reaching its ninth edition in 1962, which put geometry in a formal axiomatic setting.

In 1925, Hilbert contracted pernicious anemia, and although he recovered from this illness, he did not resume his full scientific activity. Hilbert died in Göttingen, Germany, on February 14, 1943.

$\boxed{} \cdot 13 \cdot \boxed{}$

Hilbert Spaces and Banach Spaces

The theory of normed spaces applies ideas from linear algebra, geometry, and topology to problems of analysis. In this chapter we will study in detail the most important examples of normed spaces, namely, Hilbert spaces and the classical Banach spaces. These spaces, which are natural generalizations of Euclidean n-space $\mathcal{R}^n$ and unitary n-space $\mathbb{C}^n$, are ubiquitous in analysis. The examples we study in this chapter also serve to motivate some general theorems that appear in Chapter 14.

Section 13.1 discusses preliminaries on normed spaces; Sections 13.2 and 13.3 consider Hilbert spaces and bases and duality of Hilbert spaces; Section 13.4 examines $\mathcal{L}^p$ spaces; and Sections 13.5 and 13.6 investigate nonnegative linear functionals on $C(\Omega)$ and the dual spaces of $C(\Omega)$ and $C_0(\Omega)$.

13.1 PRELIMINARIES ON NORMED SPACES

In this section, we study some elementary properties of normed spaces. Specifically, we examine the relationship between continuity and linearity for mappings of a normed space. We also present a criterion for a normed space to be complete.

In calculus, the following properties of derivative and integral are used so often that their fundamental importance is indisputable:

$$(\alpha f + \beta g)'(t) = \alpha f'(t) + \beta g'(t)$$

$$\int_a^b (\alpha f + \beta g)(t)\,dt = \alpha \int_a^b f(t)\,dt + \beta \int_a^b g(t)\,dt.$$

These two formulas show that differentiation and integration are *linear mappings* on appropriate spaces of functions.

DEFINITION 13.1 **Linear Mappings, Operators, and Functionals**

Let Ω and Λ be linear spaces with the same scalar field. A function $L\colon \Omega \to \Lambda$ is said to be a **linear mapping** if for all $x, y \in \Omega$ and all scalars α the following two conditions are satisfied:

a) $L(x + y) = L(x) + L(y)$.

b) $L(\alpha x) = \alpha L(x)$.

Linear mappings are also referred to as **linear operators** or **linear transformations;** and in cases where Λ is the scalar field, linear mappings are usually called **linear functionals.**

It follows easily from Definition 13.1 that a linear mapping L takes the **linear combination** $\sum_{j=1}^{n} \alpha_j x_j$ to the linear combination $\sum_{j=1}^{n} \alpha_j L(x_j)$; that is, for each $n \in \mathcal{N}$,

$$L\left(\sum_{j=1}^{n} \alpha_j x_j\right) = \sum_{j=1}^{n} \alpha_j L(x_j)$$

for all $x_1, x_2, \ldots, x_n \in \Omega$ and scalars $\alpha_1, \alpha_2, \ldots, \alpha_n$.

EXAMPLE 13.1 *Illustrates Definition 13.1*

a) Let $C_1([0, 1])$ denote the collection of all complex-valued functions on $[0, 1]$ that have everywhere defined and continuous derivatives. Then the function $D\colon C_1([0, 1]) \to C([0, 1])$ defined by $D(f) = f'$ is a linear mapping.

b) The function $J\colon C([0, 1]) \to C([0, 1])$ defined by $J(f)(x) = \int_0^x f(t)\, dt$ is a linear operator.

c) The function $\ell\colon C([0, 1]) \to \mathbb{C}$ defined by $\ell(f) = \int_0^1 f(t)\, dt$ is a linear functional.

d) Let A be an $m \times n$ real matrix. Then the function $T\colon \mathcal{R}^m \to \mathcal{R}^n$ defined by $T(x) = xA$ is a linear mapping. Here xA denotes the product of x with A as matrices, where x is considered a $1 \times m$ matrix. Such mappings are the classical linear transformations studied in linear algebra. □

The next proposition, whose proof is left to the reader as Exercise 13.1, considers the relationship between continuity and linearity of mappings of normed spaces. In the statement of the proposition, as often elsewhere in the text, we use the symbol $\| \; \|$ as a generic norm, letting context determine its exact meaning.

□ □ □ **PROPOSITION 13.1**

Let Ω and Λ be normed spaces with the same scalar field and $L\colon \Omega \to \Lambda$ a linear mapping. Then the following properties are equivalent:

a) L is continuous.

b) L is continuous at some point of Ω.

c) L is continuous at 0.

d) $\sup\{\, \|L(x)\| : \|x\| \le 1 \,\} < \infty$.

e) There is a constant c such that $\|L(x)\| \le c\|x\|$ for all $x \in \Omega$.

Part (d) of Proposition 13.1 motivates the definition of a bounded linear mapping, as given in Definition 13.2.

DEFINITION 13.2   **Bounded Linear Mapping**

Suppose Ω and Λ are normed spaces with the same scalar field and that $L\colon \Omega \to \Lambda$ is a linear mapping. If

$$\|L\| = \sup\{\, \|L(x)\| : \|x\| \leq 1 \,\} < \infty,$$

then L is said to be a **bounded linear mapping.**

Proposition 13.1 shows that a linear mapping is bounded if and only if it is continuous. Note that if L is a bounded linear mapping on Ω, then we have $\|L(x)\| \leq \|L\|\|x\|$ for all $x \in \Omega$.

EXAMPLE 13.2   *Illustrates Definition 13.2*

a) Let Ω be a normed space and let $I\colon \Omega \to \Omega$ be the identity function, that is, $I(x) = x$ for all $x \in \Omega$. Then I is a bounded linear operator and we have $\|I\| = 1$; I is called the **identity operator** on Ω.

b) The linear operator J defined in Example 13.1(b) is bounded and, in fact, it is easy to show that $\|J\| = 1$.

c) The linear functional ℓ defined in Example 13.1(c) is also bounded and, again, it is easy to show that $\|\ell\| = 1$.

d) The linear mapping D defined in Example 13.1(a) is not bounded if $C_1([0,1])$ is given the norm $\|\ \|_{[0,1]}$. To establish this fact, consider the sequence of functions defined by $s_n(x) = \sin n\pi x$. Clearly, $\|s_n\|_{[0,1]} = 1$. However, because $\|D(s_n)\|_{[0,1]} = n\pi$, it follows that $\|D\| = \infty$.

e) The linear mappings discussed in Example 13.1(d) are bounded, as is implied by Exercise 13.11(b).   □

When Ω and Λ are normed spaces with the same scalar field, the collection of all bounded linear operators from Ω to Λ is denoted by $B(\Omega, \Lambda)$. If we define addition and scalar multiplication in $B(\Omega, \Lambda)$ by

$$(L_1 + L_2)(x) = L_1(x) + L_2(x) \qquad \text{and} \qquad (\alpha L_1)(x) = \alpha L_1(x),$$

then $B(\Omega, \Lambda)$ becomes a linear space. Furthermore, $\|\ \|$, as given in Definition 13.2, defines a norm on $B(\Omega, \Lambda)$. See Exercise 13.3.

From now on, unless specified otherwise, we will abbreviate the normed space $(B(\Omega, \Lambda), \|\ \|)$ by $B(\Omega, \Lambda)$. When $\Omega = \Lambda$, we usually denote $B(\Omega, \Lambda)$ by $B(\Omega)$; and when Λ is the scalar field, $B(\Omega, \Lambda)$ is denoted by Ω^* and the norm $\|\ \|$ by $\|\ \|_*$. This latter space has a special name.

DEFINITION 13.3   **Dual Space**

Let Ω be a normed space. Then the space $(\Omega^*, \|\ \|_*)$ of bounded linear functionals on Ω is called the **dual space** of Ω.

The following proposition, whose proof is left to the reader as Exercise 13.6, provides a sufficient condition for the completeness of $B(\Omega, \Lambda)$.

□ □ □ **PROPOSITION 13.2**

Let Ω and Λ be normed spaces. If Λ is complete, then so is $B(\Omega, \Lambda)$. In particular, the dual space $(\Omega^*, \| \ \|_*)$ is complete.

We will discover that in many notable cases it is possible to find a concrete description of the dual of a normed space. For example, we will prove later that $\ell \in C([0, 1])^*$ if and only if there is a unique complex Borel measure μ on $[0, 1]$ such that $\ell(f) = \int f \, d\mu$ for all $f \in C([0, 1])$.

Banach Spaces

For normed spaces, completeness is a property of such consequence that those possessing it are called *Banach spaces,* after the noted mathematician Stefan Banach. (See the biography at the beginning of Chapter 14 for more on Banach.)

DEFINITION 13.4   **Banach Space**

A complete normed space is called a **Banach space.**

EXAMPLE 13.3   *Illustrates Definition 13.4*

a) Exercises 10.59 and 10.60 on page 378 show $\mathcal{R}^n$ and $\mathbb{C}^n$ are Banach spaces.
b) By Proposition 13.2, $B(\Omega, \Lambda)$ is a Banach space whenever Λ is; in particular, Ω^* is always a Banach space.
c) If Ω is a compact topological space, then $C(\Omega)$ is a Banach space.
d) If Ω is locally compact but not compact, then Exercise 11.57(c) on page 422 shows that $C_0(\Omega)$ and $C_b(\Omega)$ are Banach spaces.
e) If $(\Omega, \mathcal{A}, \mu)$ is a measure space, then $\mathcal{L}^\infty(\mu)$ is a Banach space.   □

Our next proposition characterizes completeness in normed spaces in terms of infinite series. First let us recall some concepts from Chapter 10. If $\{x_n\}_{n=1}^\infty$ is a sequence of elements in a normed space Ω, then the expression $\sum_{n=1}^\infty x_n$ is called an **infinite series.** The sequence $\{s_n\}_{n=1}^\infty$ of elements of Ω defined by $s_n = \sum_{k=1}^n x_k$ is called the associated **sequence of partial sums.** We say the infinite series **converges** if the sequence of partial sums converges, that is, if $\lim_{n \to \infty} s_n$ exists.

Closely related to the concept of convergence of series is the concept of absolute convergence of series. If $\{x_n\}_{n=1}^\infty$ is a sequence of elements in a normed space Ω, then the infinite series $\sum_{n=1}^\infty x_n$ is said to be **absolutely convergent** or to **converge absolutely** if $\sum_{n=1}^\infty \|x_n\| < \infty$. In the normed space $\mathcal{R}$, a series of nonnegative terms converges if and only if it converges absolutely. On the other hand, the series $\sum_{n=1}^\infty (-1)^n/n$ converges but does not converge absolutely.

From calculus, we know that every absolutely convergent series of real numbers converges. The following proposition shows that this property characterizes Banach spaces.

□ □ □ PROPOSITION 13.3

A normed space Ω is a Banach space if and only if every absolutely convergent series in Ω converges.

PROOF Suppose Ω is a Banach space. Let $\sum_{n=1}^{\infty} x_n$ be an absolutely convergent series. The sequence of partial sums $s_n = \sum_{k=1}^{n} x_n$ satisfies $\|s_n - s_m\| \le \sum_{k=m+1}^{n} \|x_k\|$ for $m < n$. Therefore, it follows that $\{s_n\}_{n=1}^{\infty}$ is a Cauchy sequence. So, by completeness, $\lim_{n \to \infty} s_n$ exists.

Conversely, suppose that every absolutely convergent series in Ω converges. Let $\{y_n\}_{n=1}^{\infty}$ be a Cauchy sequence. In view of Exercise 10.79 on page 385, to prove that $\{y_n\}_{n=1}^{\infty}$ converges, we need only show that it has a convergent subsequence. By repeatedly applying the Cauchy property, we obtain a subsequence $\{y_{n_k}\}_{k=1}^{\infty}$ such that $\|y_{n_{k+1}} - y_{n_k}\| < 2^{-k}$. Now, let $x_1 = y_{n_1}$ and $x_k = y_{n_k} - y_{n_{k-1}}$ for $k \ge 2$. Then $\sum_{k=1}^{\infty} x_k$ converges absolutely. Because $y_{n_k} = \sum_{j=1}^{k} x_j$, we conclude that $\lim_{k \to \infty} y_{n_k}$ exists. ∎

Exercises for Section 13.1

13.1 Prove Proposition 13.1 on page 454.

13.2 Let $L \in B(\Omega, \Lambda)$, where Ω and Λ are normed spaces. Prove that

$$\|L\| = \sup\{ \|L(x)\| : \|x\| < 1 \} = \sup\{ \|L(x)\| : \|x\| = 1 \}.$$

13.3 Suppose that Ω and Λ are normed spaces. Prove that $\| \; \|$, as defined in Definition 13.2 on page 455, is a norm on the space $B(\Omega, \Lambda)$.

13.4 Let $g \in C([0,1])$, and consider the linear operator $L_g : C([0,1]) \to C([0,1])$ defined by $L_g(f) = gf$. Show that L_g is continuous and find $\|L_g\|$.

13.5 Show that each of the following functions is a continuous linear functional on $C([0,1])$ and find its norm:
a) $\ell(f) = f(0)$.
b) $\ell(f) = \int_0^1 f(t)\, dt$.
c) $\ell(f) = \int_0^1 f(t)h(t)\, dt$, where $h \in \mathcal{L}^1([0,1])$.

13.6 Prove Proposition 13.2.

13.7 Let $C_1([0,1])$ be defined as in Example 13.1(a) on page 454.
a) Show that $C_1([0,1])$ is not a closed subspace of $C([0,1])$.
b) Conclude that $C_1([0,1])$ equipped with the norm $\| \; \|_{[0,1]}$ is not a Banach space.

13.8 Show that the space $C_1([0,1])$ defined in Example 13.1(a) on page 454 becomes a Banach space if it is equipped with the norm $\|f\| = |f(0)| + \|f'\|_{[0,1]}$.

13.9 Refer to Example 10.6 on pages 366–367, and let Ω be a nonempty set. Show that the spaces $\ell^1(\Omega)$, $\ell^2(\Omega)$, and $\ell^\infty(\Omega)$ are all Banach spaces.

13.10 Prove that there exist discontinuous linear functionals on any infinite dimensional normed space.

13.11 This exercise shows that all linear mappings on Euclidean n-space or unitary n-space are continuous.
a) Show that all linear functionals on $\mathbb{C}^n$ or $\mathcal{R}^n$ are continuous.
b) Show that all linear mappings from $\mathbb{C}^n$ or $\mathcal{R}^n$ into a normed space are continuous.

13.12 Let S be a linear subspace of the normed space Ω. Prove that if $S^\circ \ne \emptyset$, then $S = \Omega$.

13.13 Let Γ and Λ be normed spaces. Define

$$\|(x,y)\|_1 = \|x\| + \|y\|,$$
$$\|(x,y)\|_2 = (\|x\|^2 + \|y\|^2)^{1/2},$$
$$\|(x,y)\|_\infty = \max\{\|x\|, \|y\|\}.$$

a) Prove that each of the three expressions defines a norm on the Cartesian product space $\Gamma \times \Lambda$.

b) Prove that all three norms are equivalent.

13.14 Let $\|\ \|_1$ be the norm on $C([0,1])$ defined by $\|f\|_1 = \int_0^1 |f(t)|\, dt$.

a) Show that $\|f\|_1 \le \|f\|_{[0,1]}$.

b) Are $\|\ \|_1$ and $\|\ \|_{[0,1]}$ equivalent?

13.2 HILBERT SPACES

Perhaps because they are such natural generalizations of the standard Euclidean space $(\mathcal{R}^n, \|\ \|_2)$, Hilbert spaces appear more frequently in mathematics than other Banach spaces. In addition to being intrinsically important, the theory of Hilbert spaces also merits an extensive discussion because it serves as a model for the general theory of Banach spaces. In this section, we begin our treatment of Hilbert space theory.

DEFINITION 13.5 Inner Product, Inner Product Space

Let $\mathcal{X}$ be a linear space with scalar field F either $\mathcal{R}$ or $\mathbb{C}$. An **inner product** on $\mathcal{X}$ is a function $\langle\ ,\ \rangle : \mathcal{X} \times \mathcal{X} \to F$ that satisfies the following conditions for all $x, y, z \in \mathcal{X}$ and $\alpha, \beta \in F$:

a) $\langle \alpha x + \beta y, z \rangle = \alpha \langle x, z \rangle + \beta \langle y, z \rangle$.

b) $\langle x, y \rangle = \langle y, x \rangle$ if $F = \mathcal{R}$ or $\langle x, y \rangle = \overline{\langle y, x \rangle}$ if $F = \mathbb{C}$.

c) $\langle x, x \rangle \ge 0$.

d) $\langle x, x \rangle = 0$ if and only if $x = 0$.

If $\langle\ ,\ \rangle$ is an inner product on $\mathcal{X}$, then the pair $(\mathcal{X}, \langle\ ,\ \rangle)$ is called an **inner product space**.

Note: When it is clear from context which inner product is being considered, the inner product space $(\mathcal{X}, \langle\ ,\ \rangle)$ will be indicated simply by $\mathcal{X}$. And, although we usually denote an inner product by $\langle\ ,\ \rangle$, it is sometimes convenient to have slight variations of this notation such as $\langle\ ,\ \rangle_2$ or $[\ ,\ ]$.

EXAMPLE 13.4 *Illustrates Definition 13.5*

a) $\mathbb{C}^n$ is an inner product space if we define

$$\langle z, w \rangle = \sum_{k=1}^{n} z_k \overline{w_k},$$

where $z = (z_1, \ldots, z_n)$ and $w = (w_1, \ldots, w_n)$.

b) $\mathcal{R}^n$ is an inner product space if we define

$$\langle x, y \rangle = \sum_{k=1}^{n} x_k y_k,$$

where $x = (x_1, \ldots, x_n)$ and $y = (y_1, \ldots, y_n)$. This inner product is the classical "dot product" encountered in vector-calculus courses.

Note: When we consider $\mathbb{C}^n$ or $\mathcal{R}^n$ as an inner product space, we will assume that the inner product is as in this example unless we state otherwise. □

□ □ □ **THEOREM 13.1**

Let $\mathcal{X}$ be an inner product space. Then, for all $x, y \in \mathcal{X}$,
a) $\langle x + y, x + y \rangle = \langle x, x \rangle + 2\Re\langle x, y \rangle + \langle y, y \rangle$.
b) $|\langle x, y \rangle|^2 \leq \langle x, x \rangle \langle y, y \rangle$. (Cauchy's inequality)

Moreover, if $y \neq 0$, then equality holds in (b) if and only if $x = \alpha y$ for some scalar α.

PROOF a) From Definition 13.5, we have

$$\langle x + y, x + y \rangle = \langle x, x + y \rangle + \langle y, x + y \rangle$$
$$= \langle x, x \rangle + \langle x, y \rangle + \overline{\langle x, y \rangle} + \langle y, y \rangle \qquad (13.1)$$
$$= \langle x, x \rangle + 2\Re\langle x, y \rangle + \langle y, y \rangle,$$

as required.

b) If in (13.1) we replace y by $-ty$ where t is a real scalar, then we obtain the polynomial

$$p(t) = \langle x - ty, x - ty \rangle = \gamma + \beta t + \alpha t^2,$$

where $\alpha = \langle y, y \rangle$, $\beta = -2\Re\langle x, y \rangle$, $\gamma = \langle x, x \rangle$. By Definition 13.5(c), we have $p(t) \geq 0$. It follows that $p(t)$ has at most one real root. Thus, $\beta^2 - 4\alpha\gamma \leq 0$, that is,

$$(\Re\langle x, y \rangle)^2 \leq \langle x, x \rangle \langle y, y \rangle. \qquad (13.2)$$

The proof of (b) is now complete in the case of real scalars. If the scalar field is $\mathbb{C}$, we choose $\theta \in [0, 2\pi)$ so that $e^{i\theta}\langle x, y \rangle = |\langle x, y \rangle|$ and use Definition 13.5 and (13.2) to obtain

$$|\langle x, y \rangle|^2 = (\Re\langle e^{i\theta}x, y \rangle)^2 \leq \langle e^{i\theta}x, e^{i\theta}x \rangle \langle y, y \rangle$$
$$= e^{i\theta}e^{-i\theta}\langle x, x \rangle \langle y, y \rangle = \langle x, x \rangle \langle y, y \rangle.$$

Therefore, (b) holds in any case.

Suppose now that the scalar field is $\mathcal{R}$, $y \neq 0$, and that equality holds in (b). Then the polynomial $p(t)$ has a root at $t = -\beta/(2\alpha)$. It follows from Definition 13.5(d) that $x = -(\beta/(2\alpha))y$. If the scalar field is $\mathbb{C}$, we choose θ as in the preceding paragraph. Then equality in (b) yields $e^{i\theta}x = -(\beta/(2\alpha))y$ by an argument similar to that used in the real case. ■

We have referred to the inequality in part (b) of Theorem 13.1 as *Cauchy's inequality*. But it is also known as the *Schwarz, Cauchy-Schwarz, Bunyakovski,* or *Cauchy-Bunyakovski-Schwarz (CBS)* inequality.

EXAMPLE 13.5   *Illustrates Definition 13.5 and Theorem 13.1*

Suppose $z_1, \ldots, z_n, w_1, \ldots, w_n \in \mathbb{C}$. Then it follows from Theorem 13.1 and Example 13.4 that

$$\left| \sum_{k=1}^{n} z_k \overline{w_k} \right|^2 \leq \left(\sum_{k=1}^{n} |z_k|^2 \right) \left(\sum_{k=1}^{n} |w_k|^2 \right).$$

This result is Cauchy's inequality for finite sequences of complex numbers.   □

EXAMPLE 13.6   *Illustrates Definition 13.5 and Theorem 13.1*

Refer to Example 10.6(b) on page 367. Let $(\Omega, \mathcal{A}, \mu)$ be a measure space. Recall that $\mathcal{L}^2(\mu)$ consists of all complex-valued $\mathcal{A}$-measurable functions that satisfy $\int_{\Omega} |f|^2 \, d\mu < \infty$. Also recall that we identify functions that are equal μ-ae. We will show that

$$\langle f, g \rangle = \int_{\Omega} f \overline{g} \, d\mu \tag{13.3}$$

defines an inner product on $\mathcal{L}^2(\mu)$.

Because of properties of Lebesgue integration that we established in Chapter 5, we need only prove that

$$f, g \in \mathcal{L}^2(\mu) \Rightarrow f \overline{g} \in \mathcal{L}^1(\mu). \tag{13.4}$$

But this follows immediately from the simple inequality $2|f\overline{g}| \leq |f|^2 + |g|^2$.

From now on, whenever we consider $\mathcal{L}^2(\mu)$ in the context of inner product spaces, we will always use the inner product defined by (13.3).   □

EXAMPLE 13.7   *Illustrates Definition 13.5 and Theorem 13.1*

Let $(\Omega, \mathcal{A}, P)$ be a probability space. By Example 13.6, the function $\langle \ , \ \rangle$ defined by $\langle X, Y \rangle = \mathcal{E}(XY)$ is an inner product on the space of all random variables with finite variances where, again, we identify two random variables that are equal with probability one. Note that

$$\mathrm{Cov}(X, Y) = \mathcal{E}((X - \mathcal{E}(X))(Y - \mathcal{E}(Y))) = \langle (X - \mathcal{E}(X)), (Y - \mathcal{E}(Y)) \rangle$$

and, in particular, $\mathrm{Var}(X) = \langle (X - \mathcal{E}(X)), (X - \mathcal{E}(X)) \rangle$.

The **correlation coefficient** of two random variables X and Y with finite variances is defined by

$$\rho_{X,Y} = \mathrm{Cov}(X, Y) / \sqrt{\mathrm{Var}(X)\mathrm{Var}(Y)}.$$

This quantity is used extensively in probability, statistics, and stochastic processes. From Cauchy's inequality, we see that $-1 \leq \rho_{X,Y} \leq 1$.   □

□ □ □   **COROLLARY 13.1**

Let $\mathcal{X}$ be an inner product space. Define $\| \ \|: \mathcal{X} \to \mathcal{R}$ by

$$\|x\| = \sqrt{\langle x, x \rangle}.$$

Then the following properties hold.

a) *The function $\| \ \|$ is a norm on $\mathcal{X}$.*

b) *We have*

$$\|x + y\|^2 + \|x - y\|^2 = 2\|x\|^2 + 2\|y\|^2$$

for all $x, y \in \mathcal{X}$.

c) *The inner product is continuous with respect to the product topology induced on $\mathcal{X} \times \mathcal{X}$ by the norm $\| \ \|$.*

PROOF a) Definition 10.9 on page 365 gives the three conditions for being a norm. It is easy to check that $\| \ \|$ satisfies the first two conditions. To verify the third condition, we use Theorem 13.1 to conclude that

$$\|x + y\|^2 = \|x\|^2 + 2\Re\langle x, y \rangle + \|y\|^2$$
$$\leq \|x\|^2 + 2\|x\|\|y\| + \|y\|^2$$
$$= (\|x\| + \|y\|)^2.$$

This display gives the required result.

b) Applying Theorem 13.1 again, we obtain that

$$\|x + y\|^2 = \|x\|^2 + 2\Re\langle x, y \rangle + \|y\|^2$$

and, replacing y by $-y$ in the previous equation, we get

$$\|x - y\|^2 = \|x\|^2 - 2\Re\langle x, y \rangle + \|y\|^2.$$

Adding corresponding sides of the two preceding equalities yields (b).

c) We leave the proof of part (c) to the reader as Exercise 13.15. ■

In the future, we will assume that every inner product space is also a normed space, equipped with the norm defined in Corollary 13.1. If an inner product space is complete, it is called a *Hilbert space* in honor of the mathematician David Hilbert. (See the biography at the beginning of this chapter for more about Hilbert.)

DEFINITION 13.6   **Hilbert Space**

An inner product space that is complete with respect to its norm is called a **Hilbert space.**

We already know that $\mathcal{R}^n$ and $\mathbb{C}^n$ are Hilbert spaces. Later in this chapter, we will prove that all spaces of the form $\mathcal{L}^2(\mu)$ are Hilbert spaces. But for now, we will content ourselves with knowing that $\mathcal{L}^2(\mu)$-type spaces are inner product spaces, as we showed in Example 13.6.

Nearest Points

The standard Euclidean plane $(\mathcal{R}^2, \|\ \|_2)$ serves to illustrate an essential property of Hilbert spaces that we will prove in Theorem 13.2. We know that the linear subspaces of $\mathcal{R}^2$ are $\{(0,0)\}$, $\mathcal{R}^2$, and lines passing through $(0,0)$. If L is a line through $(0,0)$ and if $x \in \mathcal{R}^2$, then the point y_0 of intersection of L and the line through x perpendicular to L is the unique point on L that is nearest to x. What is important for us is that y_0 is completely determined by the conditions

$$y_0 \in L \quad \text{and} \quad \langle x - y_0, y \rangle = 0 \quad \text{for all} \quad y \in L,$$

as seen in Fig. 13.1.

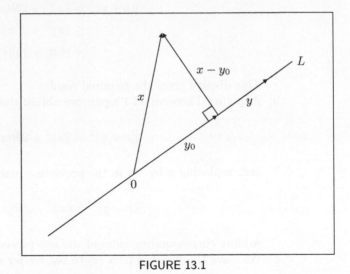

FIGURE 13.1

This property of the Euclidean plane serves to motivate the following important theorem about Hilbert spaces.

□ □ □ **THEOREM 13.2**

Let $\mathcal{H}$ be a Hilbert space and K a closed linear subspace of $\mathcal{H}$. For each $x \in \mathcal{H}$ there is a unique point $y_0 \in K$ such that

$$\|x - y_0\| = \rho(x, K),$$

where $\rho(x, K) = \inf\{\ \|x - y\| : y \in K\ \}$. Furthermore, the point y_0 is determined by the conditions

$$y_0 \in K \quad \text{and} \quad \langle x - y_0, y \rangle = 0 \quad \text{for all} \quad y \in K. \tag{13.5}$$

In other words, (13.5) determines the unique **nearest point** of K to x.

PROOF We establish the theorem when the scalar field is $\mathbb{C}$; the proof for real scalars is obtained by a slight modification. To begin, we select a sequence $\{y_n\}_{n=1}^{\infty} \subset K$ such that $\lim_{n\to\infty} \|x - y_n\| = \rho(x, K)$. We claim that $\{y_n\}_{n=1}^{\infty}$ is a Cauchy sequence. Setting $x = x - y_n$ and $y = x - y_m$ in Corollary 13.1, we obtain

$$4\|x - (y_n + y_m)/2\|^2 + \|y_n - y_m\|^2 = 2\|x - y_n\|^2 + 2\|x - y_m\|^2.$$

Since K is a linear subspace, $(y_n + y_m)/2 \in K$. It follows that

$$\|y_n - y_m\|^2 \leq 2\|x - y_n\|^2 + 2\|x - y_m\|^2 - 4\rho(x, K)^2. \tag{13.6}$$

Because the right-hand side of (13.6) tends to 0 as $n, m \to \infty$, we conclude that $\{y_n\}_{n=1}^{\infty}$ is a Cauchy sequence.

By completeness, $y_0 = \lim_{n\to\infty} y_n$ exists and, because K is closed, $y_0 \in K$. Moreover,

$$\|x - y_0\| = \lim_{n\to\infty} \|x - y_n\| = \rho(x, K).$$

To verify (13.5), it suffices to consider the case where $y \in K \setminus \{0\}$. Suppose that y_0 is a point of K nearest to x. By Theorem 13.1(a), we have

$$\|x - y_0 - \alpha y\|^2 = \|x - y_0\|^2 - 2\Re\overline{\alpha}\langle x - y_0, y \rangle + |\alpha|^2 \|y\|^2$$

for all scalars α. Choosing $\alpha = \langle x - y_0, y \rangle / \|y\|^2$, we obtain

$$\|x - y_0 - \alpha y\|^2 = \|x - y_0\|^2 - |\langle x - y_0, y \rangle|^2 / \|y\|^2.$$

Because K is a linear subspace, it follows that $y_0 + \alpha y \in K$. Hence,

$$\|x - y_0\|^2 = \rho(x, K)^2 \leq \|x - (y_0 + \alpha y)\|^2 = \|x - y_0\|^2 - |\langle x - y_0, y \rangle|^2 / \|y\|^2$$

and, consequently, $\langle x - y_0, y \rangle = 0$.

Suppose, on the other hand, that y_0 is an element of K that satisfies (13.5). Then, for every $y \in K$,

$$\begin{aligned}
\|x - y\|^2 &= \|x - y_0 + y_0 - y\|^2 \\
&= \|x - y_0\|^2 + 2\Re\langle x - y_0, y_0 - y \rangle + \|y_0 - y\|^2 \tag{13.7} \\
&= \|x - y_0\|^2 + \|y_0 - y\|^2 \geq \|x - y_0\|^2.
\end{aligned}$$

Thus, y_0 is a point of K nearest to x.

It remains to prove that y_0 is unique. Let y_1 be a point of K nearest to x. Then, by (13.7),

$$\|x - y_0\|^2 = \|x - y_1\|^2 = \|x - y_0\|^2 + \|y_0 - y_1\|^2$$

and, therefore, $\|y_0 - y_1\|^2 = 0$. It follows that $y_0 = y_1$. ∎

EXAMPLE 13.8   *Illustrates Theorem 13.2*

a) Let (x_1, y_1), (x_2, y_2), $\ldots$, (x_n, y_n) be n points in the plane. In statistics and other fields, it is important to find the straight line that best fits the n points in the sense of minimizing the sum of squared errors. That is, the problem is to find real numbers α and β that minimize

$$\sum_{j=1}^{n} \left(y_j - (\alpha + \beta x_j)\right)^2.$$

The resulting line is called the *least-squares line* or *regression line*.

We can apply Theorem 13.2 to determine the regression line by proceeding as follows. Let $x = (x_1, x_2, \ldots, x_n)$, $y = (y_1, y_2, \ldots, y_n)$, $w = (1, 1, \ldots, 1)$, and $K = \{\, aw + bx : a, b \in \mathcal{R}\,\}$. Finding the regression line is equivalent to obtaining the element y_0 of K nearest to y. Writing $y_0 = \alpha w + \beta x$, we apply (13.5) to get the equations

$$\langle \alpha w + \beta x, w \rangle = \langle y, w \rangle \qquad \text{and} \qquad \langle \alpha w + \beta x, x \rangle = \langle y, x \rangle$$

or, equivalently,

$$n\alpha + \beta \sum_{j=1}^{n} x_j = \sum_{j=1}^{n} y_j \qquad \text{and} \qquad \alpha \sum_{j=1}^{n} x_j + \beta \sum_{j=1}^{n} x_j^2 = \sum_{j=1}^{n} x_j y_j.$$

We thus have two linear equations in the two unknowns α and β. The solution, which we leave to the reader, gives the slope and y-intercept of the regression line.

b) Let μ be the measure on $[-1, 1]$ defined by $\mu(E) = \lambda(E)/2$. The quantity

$$\|f - g\|_2 = \left(\frac{1}{2} \int_{-1}^{1} |f(x) - g(x)|^2 \, dx\right)^{1/2}$$

can be thought of as the average distance between f and g. We will use Theorem 13.2 to find the function of the form $g(x) = \alpha x + \beta$ that minimizes the average distance to $f(x) = x^2$. The function g must satisfy

$$\int_{-1}^{1} (x^2 - \alpha x - \beta)(\gamma x + \delta) \, dx = 0$$

for all $\gamma, \delta \in \mathbb{C}$. A calculation shows that $2(\delta - \alpha\gamma)/3 - 2\beta\delta = 0$ for all γ and δ. It follows that $\alpha = 0$ and $\beta = 1/3$. Thus, the best approximation to x^2 of the form $\alpha x + \beta$ in the sense of the $\mathcal{L}^2(\mu)$-norm is the constant function $g(x) = 1/3$.

c) Refer to Example 13.7. Let $(\Omega, \mathcal{A}, P)$ be a probability space and X a random variable with finite variance. We will use Theorem 13.2 to determine the constant c that minimizes $\mathcal{E}((X - c)^2)$. Applying (13.5) to the subspace generated by the random variable 1, we obtain the equation $\mathcal{E}((X - c)1) = 0$. Thus, $c = \mathcal{E}(X)$ minimizes $\mathcal{E}((X - c)^2)$, and we see that the minimum value is $\text{Var}(X)$.   □

A close reading of the proof of Theorem 13.2 reveals that more than just that theorem has been established. We did not fully use the assumption that $\mathcal{H}$ is complete; rather, we only needed the completeness of the linear subspace K. The assumption that K is a linear subspace of $\mathcal{H}$ can also be relaxed.

Recall that a subset S of a linear space is said to be a **convex set** if for all $x, y \in S$ and $0 \le \alpha \le 1$, we have $\alpha x + (1 - \alpha)y \in S$; in words, whenever S contains two points, it also contains the entire line segment that connects the two points. If C is a closed convex subset, but not necessarily a linear subspace, of a Hilbert space $\mathcal{H}$, then we can still obtain a unique nearest point. However, (13.5) is in general no longer valid. (See Exercise 13.22.)

Theorem 13.2 enables us to associate with each closed linear subspace K of a Hilbert space $\mathcal{H}$ the function $P_K : \mathcal{H} \to \mathcal{H}$, where $P_K(x)$ is the point of K nearest to x. The properties of the function P_K are explored in Exercise 13.26 where, in particular, it is shown that it is a bounded linear operator on $\mathcal{H}$ with range K. The operator P_K is often referred to as the **orthogonal projection** of $\mathcal{H}$ onto K.

Orthogonality

From calculus, the ordinary dot product on $\mathcal{R}^2$ satisfies $\langle x, y \rangle = \|x\| \|y\| \cos\theta$, where θ is the angle between x and y. Thus, two vectors in $\mathcal{R}^2$ are perpendicular if and only if their dot product is 0. Similarly, the condition $\langle x, y \rangle = 0$ captures the notion of perpendicularity of two elements of a general inner product space $\mathcal{X}$. The term used for "perpendicular" in the context of inner product spaces is *orthogonal*.

DEFINITION 13.7 Orthogonality

Let $\mathcal{X}$ be an inner product space. Two elements x and y of $\mathcal{X}$ are said to be **orthogonal** if $\langle x, y \rangle = 0$. For a subset S of $\mathcal{X}$, we define the **orthogonal complement** of S, denoted $\boldsymbol{S^\perp}$, to be the set of all elements of $\mathcal{X}$ that are orthogonal to every element of S, that is,

$$S^\perp = \{\, y \in \mathcal{X} : \langle x, y \rangle = 0 \text{ for all } x \in S \,\}.$$

EXAMPLE 13.9 *Illustrates Definition 13.7*

a) The elements $(1, 0)$ and $(0, 1)$ of $\mathcal{R}^2$ are orthogonal and the orthogonal complement of $\{(1, 0)\}$ is $\{\, (0, y) : y \in \mathcal{R} \,\}$.

b) Recall that two random variables with finite variances are said to be *uncorrelated* if $\mathrm{Cov}(X, Y) = 0$. We see from Example 13.7 on page 460 that two random variables are uncorrelated if and only if $X - \mathcal{E}(X)$ and $Y - \mathcal{E}(Y)$ are orthogonal. □

It is left to the reader as Exercise 13.23 to prove that $S^\perp$ is always a closed linear subspace. Moreover, it can be shown that in Hilbert spaces, $(S^\perp)^\perp = \overline{\operatorname{span} S}$, as the reader is asked to verify in Exercise 13.25. Here we are using span S to represent the **span** of S, that is, the linear subspace of all finite linear combinations of elements of S.

Our next result is a version of Theorem 13.2 that emphasizes the role of the orthogonal complement. It also serves as the prototype for an important theorem in the general theory of normed spaces that appears in Chapter 14.

□ □ □ **THEOREM 13.3**

Let K be a proper closed linear subspace of the Hilbert space $\mathcal{H}$ and $x \in K^c$. Then there exists a unique $z_0 \in K^\perp$ such that $\|z_0\| = 1$ and

$$
\begin{aligned}
\rho(x, K) &= \inf\{\, \|x - y\| : y \in K \,\} \\
&= \sup\{\, |\langle x, z \rangle| : z \in K^\perp \text{ and } \|z\| \le 1 \,\} = \langle x, z_0 \rangle.
\end{aligned}
\tag{13.8}
$$

PROOF   Let y_0 be the nearest point of K to x. If $z \in K^\perp$ is such that $\|z\| \le 1$, then, by the definition of $K^\perp$ and Theorem 13.1, we have

$$
|\langle x, z \rangle| = |\langle x - y_0, z \rangle| \le \|x - y_0\| \|z\| \le \inf\{\, \|x - y\| : y \in K \,\}.
\tag{13.9}
$$

It follows that $\inf\{\, \|x - y\| : y \in K \,\} \ge \sup\{\, |\langle x, z \rangle| : z \in K^\perp \text{ and } \|z\| \le 1 \,\}$. Let $z_0 = (x - y_0)/\|x - y_0\|$. By (13.5), $z_0 \in K^\perp$ and, furthermore,

$$
\begin{aligned}
\inf\{\, \|x - y\| : y \in K \,\} &= \|x - y_0\| = \langle x - y_0, z_0 \rangle = \langle x, z_0 \rangle \\
&\le \sup\{\, |\langle x, z \rangle| : z \in K^\perp \text{ and } \|z\| \le 1 \,\}.
\end{aligned}
\tag{13.10}
$$

The equations in (13.8) now follow from (13.9) and (13.10). The uniqueness of z_0 is left to the reader as Exercise 13.28.   ∎

As a visual aid to understanding Theorem 13.3, we have constructed a simple illustration of the theorem in Fig. 13.2.

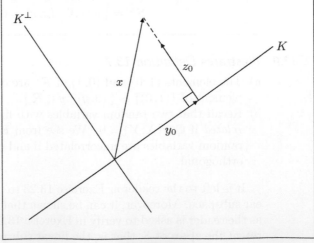

FIGURE 13.2

Exercises for Section 13.2

Note: A ★ denotes an exercise that will be subsequently referenced.

13.15 Prove part (c) of Corollary 13.1 on pages 460–461.

13.16 Let $(\mathcal{X}, \| \ \|)$ be a normed space with scalar field $\mathcal{R}$.
   a) Suppose the norm satisfies the identity in Corollary 13.1(b) on page 461. Show that there is an inner product on $\mathcal{X}$ such that $\| \ \|$ is the induced norm.
   b) Repeat part (a) in case the scalar field is $\mathbb{C}$.

13.17 A **semi inner product** on a linear space $\mathcal{X}$ is a function $\langle \ , \ \rangle \colon \mathcal{X} \times \mathcal{X} \to F$ that satisfies conditions (a), (b) and (c) of Definition 13.5 on page 458 and the following weakening of condition (d): $\langle x, x \rangle = 0$ if $x = 0$. Show that (a) and (b) of Theorem 13.1 remain valid for semi inner products.

13.18 Let $\mathcal{X}$ be a linear space with inner product $\langle \ , \ \rangle$ and $L \colon \mathcal{X} \to \mathcal{X}$ a linear operator. Show that $[x, y] = \langle L(x), L(y) \rangle$ defines a semi inner product on $\mathcal{X}$ in the sense of Exercise 13.17.

13.19 Let Ω be a nonempty set. Prove that $\ell^2(\Omega)$ is a Hilbert space with respect to the inner product given by $\langle f, g \rangle = \int_\Omega f \bar{g} \, d\mu$, where μ is counting measure on Ω.

13.20 Let $(\Omega, \mathcal{A}, \mu)$ be a measure space. Show that if $f \in \mathcal{L}^2(\mu)$, then there is a sequence of simple functions $\{r_n\}_{n=1}^\infty \subset \mathcal{L}^2(\mu)$ such that, as $n \to \infty$, $\|f - r_n\|_2 \to 0$, $\|r_n\|_2 \to \|f\|_2$, and $r_n \to f$ μ-ae.

13.21 Let $\{\mathcal{H}_n\}_{n=1}^\infty$ be a sequence of Hilbert spaces and set

$$\mathcal{H} = \left\{ x \in \bigtimes_{n=1}^\infty \mathcal{H}_n : \sum_{n=1}^\infty \|x_n\|^2 < \infty \right\}.$$

Denote by $\langle \ , \ \rangle$ the inner product for each $\mathcal{H}_n$. Show that $\mathcal{H}$ is a Hilbert space with respect to the inner product defined by $[x, y] = \sum_{n=1}^\infty \langle x_n, y_n \rangle$.

13.22 Let C be a closed convex subset of a Hilbert space $\mathcal{H}$. Show that for each $x \in \mathcal{H}$, there is a unique point $y_0 \in C$ such that $\|x - y_0\| = \rho(x, C)$.

13.23 Let S be a subset of an inner product space $\mathcal{X}$. Show that $S^\perp$ is a closed linear subspace of $\mathcal{X}$.

13.24 Verify the following properties of orthogonal complements:
   a) $A \subset B \Rightarrow B^\perp \subset A^\perp$.
   b) $A^\perp = (\text{span } A)^\perp$.
   c) $D^\perp \cap E^\perp = (D \cup E)^\perp$.

13.25 Prove that in Hilbert spaces, $(A^\perp)^\perp = \overline{\text{span } A}$.

★**13.26** Let K be a closed linear subspace of a Hilbert space $\mathcal{H}$ and P_K the associated orthogonal projection. Verify the following properties.
   a) P_K is linear.
   b) $\|P_K(x)\| \leq \|x\|$, so that P_K is continuous.
   c) $P_K \circ P_K = P_K$.
   d) $P_K^{-1}(\{0\}) = K^\perp$.
   e) The range of P_K is K.
   f) $P_{K^\perp} = I - P_K$, where I is the identity operator on $\mathcal{H}$. (See Exercise 13.25.)
   g) Deduce from part (f) that each $x \in \mathcal{H}$ can be written uniquely as $x = y + y^\perp$, where $y \in K$ and $y^\perp \in K^\perp$.

13.27 Let y_0 be a nonzero element of a Hilbert space $\mathcal{H}$ and set $K = \text{span}\{y_0\}$. Find an explicit formula for P_K.

13.28 Verify the uniqueness of z_0 in Theorem 13.3.

13.3 BASES AND DUALITY IN HILBERT SPACES

As we know, the concepts of linear independence and basis play an essential role in the theory of finite dimensional linear spaces. In the infinite dimensional case, one can use Zorn's lemma to prove the existence of a **Hamel basis** — a maximal linearly independent set B — and then show that every element of the space can be written uniquely as a finite linear combination of members of B.

Hamel bases are of little use in analysis, however, because they generally cannot be obtained by a formula or constructive process. Fortunately, in Hilbert spaces, there is an analogue of Hamel basis that is much better suited to the needs of analysis. It is this notion of basis to which we now turn our attention.

DEFINITION 13.8   Orthogonal Set; Orthonormal Set and Basis

Let $(\mathcal{X}, \langle\ ,\ \rangle)$ be an inner product space. A subset $S \subset \mathcal{X}$ is said to be an **orthogonal set** if every two distinct elements of S are orthogonal, that is, if $\langle x, y \rangle = 0$ for all $x, y \in S$ with $x \neq y$. An orthogonal set S is said to be an **orthonormal set** if $\|x\| = 1$ for each $x \in S$. If S is a nonempty orthonormal set and is contained in no strictly larger orthonormal set, then S is called an **orthonormal basis,** or simply a **basis.**

EXAMPLE 13.10   *Illustrates Definition 13.8*

a) The set of elements $\{(1, 0, 0, \ldots, 0), (0, 1, 0, \ldots, 0), \ldots, (0, 0, 0, \ldots, 1)\}$ is an orthonormal set in $\mathbb{C}^n$. Clearly, it is also a basis.

b) Let Ω be a nonempty set. For each $x \in \Omega$, let d_x denote the function that is 1 at x and 0 at all other points of Ω. Then $\{d_x : x \in \Omega\}$ is an orthonormal set in $\ell^2(\Omega)$. We will see later that it is also an orthonormal basis.

c) For each $n \in \mathcal{Z}$, define $e_n(x) = (2\pi)^{-1/2}e^{inx}$. It is easy to see that the collection of functions $\{e_n : n \in \mathcal{Z}\}$ is an orthonormal set in $\mathcal{L}^2([-\pi, \pi])$. Later we will show that it is an orthonormal basis as well.   ◻

Our next theorem provides some fundamental properties of orthonormal sets.

◻ ◻ ◻   **THEOREM 13.4**

Let $\mathcal{X}$ be an inner product space and $E = \{e_1, e_2, \ldots, e_n\}$ a finite orthonormal subset of $\mathcal{X}$. Then the following properties hold.

a) E is linearly independent.

b) $\|\sum_{j=1}^{n} \alpha_j e_j\|^2 = \sum_{j=1}^{n} |\alpha_j|^2$ for any choice of scalars $\alpha_1, \alpha_2, \ldots, \alpha_n$.

c) For each $x \in \mathcal{X}$, we have $\sum_{j=1}^{n} |\langle x, e_j \rangle|^2 \leq \|x\|^2$.

d) $x = \sum_{j=1}^{n} \langle x, e_j \rangle e_j$ for each $x \in \operatorname{span} E$.

e) $\operatorname{span} E$ is a complete subspace of $\mathcal{X}$, in particular, a closed subset of $\mathcal{X}$.

f) For each $x \in \mathcal{X}$, $y_0 = \sum_{j=1}^{n} \langle x, e_j \rangle e_j$ is the unique nearest point of $\operatorname{span} E$ to x, that is, it is the unique member y of $\operatorname{span} E$ such that $\|x - y\| = \rho(x, \operatorname{span} E)$.

PROOF The proofs of (a), (b), and (d) are left to the reader as Exercise 13.30. To prove (c), let $x \in \mathcal{X}$ and $y = \sum_{j=1}^{n} \langle x, e_j \rangle e_j$. By (b), $\|y\|^2 = \sum_{j=1}^{n} |\langle x, e_j \rangle|^2$. Also,

$$\langle x, y \rangle = \langle x, \sum_{j=1}^{n} \langle x, e_j \rangle e_j \rangle = \sum_{j=1}^{n} \overline{\langle x, e_j \rangle} \langle x, e_j \rangle = \sum_{j=1}^{n} |\langle x, e_j \rangle|^2.$$

Applying Theorem 13.1(a) on page 459, we now obtain that

$$0 \le \|x - y\|^2 = \|x\|^2 - 2\Re\langle x, y \rangle + \|y\|^2 = \|x\|^2 - \sum_{j=1}^{n} |\langle x, e_j \rangle|^2,$$

from which (c) follows immediately.

To prove (e), let $\{y_m\}_{m=1}^{\infty}$ be a Cauchy sequence in span E. From Cauchy's inequality, we have

$$|\langle y_m, e_k \rangle - \langle y_\ell, e_k \rangle| \le \|y_m - y_\ell\|.$$

Thus, $\{\langle y_m, e_k \rangle\}_{m=1}^{\infty}$ is a Cauchy sequence for $k = 1, 2, \ldots, n$. Applying part (d) and using the completeness of the scalars, we conclude that the limit

$$y = \lim_{m \to \infty} y_m = \sum_{k=1}^{n} \left(\lim_{m \to \infty} \langle y_m, e_k \rangle \right) e_k$$

exists. Clearly, $y \in$ span E. We have now shown that span E is complete. Since a complete subset of a metric space is closed, it follows that span E is closed in $\mathcal{X}$.

Next we establish (f). By Theorem 13.2 on page 462 and the defining properties of inner product, it is enough to show that $\langle x - y_0, e_k \rangle = 0$ for $k = 1, 2, \ldots, n$. Using the fact that E is an orthonormal set, we get

$$\langle x - y_0, e_k \rangle = \langle x, e_k \rangle - \sum_{j=1}^{n} \langle x, e_j \rangle \langle e_j, e_k \rangle = \langle x, e_k \rangle - \langle x, e_k \rangle = 0,$$

as required. ∎

As an immediate consequence of Theorem 13.4(c), we get the following important result, known as *Bessel's inequality*. Refer to Exercise 2.37 on page 46 for the meaning of the summation that occurs in that inequality.

□ □ □ **COROLLARY 13.2 Bessel's Inequality**

Let E be an orthonormal subset of an inner product space $\mathcal{X}$. Then

$$\sum_{e \in E} |\langle x, e \rangle|^2 \le \|x\|^2$$

for all $x \in \mathcal{X}$.

EXAMPLE 13.11   *Illustrates Theorem 13.4*

In the space $\mathcal{L}^2([-\pi, \pi])$, consider the linear subspace

$$\mathcal{U}_n = \operatorname{span}\{ e_k : -n \leq k \leq n \},$$

where $e_k(x) = (2\pi)^{-1/2}e^{ikx}$. As we noted in Example 13.10(c) on page 468, $\{ e_n : n \in \mathcal{Z} \}$ is an orthonormal set. Therefore, $\{ e_k : -n \leq k \leq n \}$ is a finite orthonormal subset of $\mathcal{L}^2([-\pi, \pi])$. It is clear that $\mathcal{U}_n$ is the space of complex trigonometric polynomials of degree at most n.

Let $f \in \mathcal{L}^2([-\pi, \pi])$. Then, from Theorem 13.4(f), the nearest member of $\mathcal{U}_n$ to f is given by

$$s_n = \sum_{|k| \leq n} \langle f, e_k \rangle e_k.$$

The number

$$\hat{f}(k) = (2\pi)^{-1/2} \langle f, e_k \rangle = \frac{1}{2\pi} \int_{-\pi}^{\pi} f(x)e^{-ikx}\, dx$$

is called the kth **Fourier coefficient** of f. Thus, the best approximation,

$$s_n(x) = \sum_{|k| \leq n} \langle f, e_k \rangle e_k = \sum_{k=-n}^{n} \hat{f}(k)e^{ikx},$$

is the nth partial sum of the **Fourier series** $\sum_{k=-\infty}^{\infty} \hat{f}(k)e^{ikx}$ associated with the function f.   □

More examples of orthonormal sets can be found by using the procedure described in the proof of the following theorem.

□ □ □   **THEOREM 13.5**

Let $\{x_m\}_{m=1}^{\infty}$ be a sequence of elements in an inner product space $\mathcal{X}$ and assume that $x_1 \neq 0$. Then there is a countable orthonormal set $\{y_1, y_2, \ldots\}$ and a nondecreasing sequence of integers $\{k(m)\}_{m=1}^{\infty}$ such that

$$\operatorname{span}\{x_1, x_2, \ldots, x_m\} = \operatorname{span}\{y_1, y_2, \ldots, y_{k(m)}\}$$

for each $m \in \mathcal{N}$.

PROOF   We outline an argument by mathematical induction, but leave the details for Exercise 13.31.

Let $y_1 = x_1/\|x_1\|$. Proceeding inductively, suppose y_1, y_2, $\ldots$, $y_{k(m)}$ have been chosen so that $\{y_1, y_2, \ldots, y_{k(m)}\}$ is an orthonormal set and

$$\operatorname{span}\{x_1, x_2, \ldots, x_m\} = \operatorname{span}\{y_1, y_2, \ldots, y_{k(m)}\}.$$

Define

$$v = x_{m+1} - \sum_{j=1}^{k(m)} \langle x_{m+1}, y_j \rangle y_j.$$

Then we find that v is orthogonal to y_j for $j = 1, 2, \ldots, k(m)$.

If $v = 0$, then $x_{m+1} \in \text{span}\{y_1, y_2, \ldots, y_{k(m)}\}$, and we let $k(m+1) = k(m)$. If $v \neq 0$, we let $k(m+1) = k(m) + 1$, and we define $y_{k(m+1)} = v/\|v\|$; then $\{y_1, y_2, \ldots, y_{k(m)}, y_{k(m+1)}\}$ is an orthonormal set. In either case, we have that

$$\text{span}\{x_1, x_2, \ldots, x_m, x_{m+1}\} = \text{span}\{y_1, y_2, \ldots, y_{k(m+1)}\},$$

as required. ∎

The following theorem provides several equivalent conditions for an orthonormal set in a Hilbert space to be a basis. It also makes clear why bases in the sense of Definition 13.8 are appropriate analogues of Hamel bases.

Before stating the theorem, we need to discuss **generalized sums** in normed spaces. Let $\{x_\iota\}_{\iota \in I}$ be an indexed collection of elements of a normed space. Then we say that the sum $\sum_{\iota \in I} x_\iota$ converges if there are only countably many nonzero terms and if for every enumeration of these terms, the resulting series converges to the same element.

□ □ □ **THEOREM 13.6**

Let $\mathcal{H}$ be a Hilbert space and E an orthonormal subset of $\mathcal{H}$. Then the following properties are equivalent:

a) E is a basis.

b) $\overline{\text{span}\, E} = \mathcal{H}$.

c) $\langle x, e \rangle = 0$ for each $e \in E$ implies $x = 0$.

d) For each $x \in \mathcal{H}$, we have $x = \sum_{e \in E} \langle x, e \rangle e$.

e) $\|x\|^2 = \sum_{e \in E} |\langle x, e \rangle|^2$ for each $x \in \mathcal{H}$.

PROOF (a) → (b): If $\overline{\text{span}\, E} \neq \mathcal{H}$, then by Theorem 13.2 on page 462, we can find a nonzero element $z \in (\overline{\text{span}\, E})^\perp$. Let $e_0 = z/\|z\|$. We note that $E \cup \{e_0\}$ is orthonormal and properly contains E. Thus, E is not a basis.

(b) ⇒ (c): Suppose that $\langle x, e \rangle = 0$ for each $e \in E$. It follows from the properties of an inner product that $\langle x, y \rangle = 0$ for each $y \in \text{span}\, E$. Using the continuity of the inner product, we conclude that x is orthogonal to every element of $\overline{\text{span}\, E}$, which by assumption equals H. Therefore, $\langle x, x \rangle = 0$ and, so, $x = 0$.

(c) ⇒ (d): It follows from Bessel's inequality that $\sum_{e \in E} |\langle x, e \rangle|^2 < \infty$. Using that fact and Exercise 2.37(c) on page 46, we see that $E_0 = \{e \in E : \langle x, e \rangle \neq 0\}$ is either countably infinite or finite. We will deal with the former case; the latter one is handled in a similar manner. Let $\{e_n\}_{n=1}^\infty$ be an enumeration of E_0 and define $x_n = \sum_{j=1}^n \langle x, e_j \rangle e_j$. If $n < m$, then Theorem 13.4(b) implies that $\|x_n - x_m\|^2 = \sum_{j=n+1}^m |\langle x, e_j \rangle|^2$. It now follows that $\{x_n\}_{n=1}^\infty$ is Cauchy and, therefore, converges to some $y \in \mathcal{H}$. We claim that $y = x$. For each $e \in E$, we have

$$\langle x - y, e \rangle = \langle x, e \rangle - \sum_{j=1}^\infty \langle x, e_j \rangle \langle e_j, e \rangle. \tag{13.11}$$

If e is not in E_0, then $\langle x, e \rangle = 0$ and $\langle e_j, e \rangle = 0$ for each j. If $e = e_k$ for some k, then the right-hand side of (13.11) reduces to $\langle x, e_k \rangle - \langle x, e_k \rangle$. Thus, $x - y$ is orthogonal to each element of E. It follows from (c) that $y = x$.

(d) $\Rightarrow$ (e): It follows from (d) and the continuity of the inner product that

$$\|x\|^2 = \langle x, x \rangle = \sum_{e \in E} \langle x, e \rangle \langle e, x \rangle = \sum_{e \in E} |\langle x, e \rangle|^2,$$

as required.

(e) $\Rightarrow$ (a): If E is not a basis, we can find an element $e_0 \in \mathcal{H}$ such that $\|e_0\| = 1$ and $\langle e_0, e \rangle = 0$ for each $e \in E$. Thus,

$$\|e_0\|^2 = 1 \neq 0 = \sum_{e \in E} |\langle e_0, e \rangle|^2.$$

The proof of the theorem is now complete. ■

EXAMPLE 13.12   *Illustrates Theorem 13.6*

Assume as known that $\mathcal{L}^2([-\pi, \pi])$ is complete, a fact that will be proved in the next section. We will show that the orthonormal set $\{ e_n : n \in \mathcal{Z} \}$, introduced in Example 13.10(c), is a basis for $\mathcal{L}^2([-\pi, \pi])$. By Theorem 13.6, it suffices to show that if $f \in \mathcal{L}^2([-\pi, \pi])$ is such that

$$\int_{-\pi}^{\pi} f(x) e^{-inx}\, dx = 0, \qquad n \in Z, \tag{13.12}$$

then $f = 0$ ae.

From (13.12), it follows immediately that $\int_{-\pi}^{\pi} f(x) p(x)\, dx = 0$ for all trigonometric polynomials p. As the reader is asked to show in Exercise 13.34, there is a sequence $\{p_n\}_{n=1}^{\infty}$ of trigonometric polynomials such that $\lim_{n \to \infty} \|\overline{f} - p_n\|_2 = 0$. Using the continuity of the inner product, we conclude that

$$\int_{-\pi}^{\pi} |f(x)|^2\, dx = \int_{-\pi}^{\pi} f(x) \overline{f(x)}\, dx = \lim_{n \to \infty} \int_{-\pi}^{\pi} f(x) p_n(x)\, dx = 0.$$

Hence, f vanishes ae.

Because $\{ e_n : n \in Z \}$ is a basis for $\mathcal{L}^2([-\pi, \pi])$, Theorem 13.6(d) implies that each function $f \in \mathcal{L}^2([-\pi, \pi])$ has the **Fourier series expansion**

$$f(x) = \sum_{n=-\infty}^{\infty} \hat{f}(n) e^{inx},$$

where the convergence is in $\mathcal{L}^2([-\pi, \pi])$. Furthermore, Theorem 13.6(e) yields

$$\frac{1}{2\pi} \int_{-\pi}^{\pi} |f(x)|^2\, dx = \sum_{n=-\infty}^{\infty} |\hat{f}(n)|^2,$$

which is called **Parseval's identity**. □

Unless we know that a Hilbert space possesses a basis, Theorem 13.6 is of little consequence. That every Hilbert space does in fact have a basis is part of our next theorem.

□ □ □ **THEOREM 13.7**

Let $\mathcal{H}$ be a Hilbert space, not $\{0\}$. Then the following properties hold.
a) $\mathcal{H}$ has a basis.
b) If E is a basis for a closed linear subspace K of $\mathcal{H}$, then there exists a basis
for $\mathcal{H}$ that contains E as a subset.
c) $\mathcal{H}$ has a countable basis if and only if $\mathcal{H}$ is separable.

PROOF We prove (a) and leave (b) and (c) to the reader as Exercises 13.35 and 13.36. Let $\mathcal{O}$ denote the collection of orthonormal subsets of H, ordered by $\subset$. Suppose that $\mathcal{C}$ is a chain of $\mathcal{O}$. Then $\bigcup_{O \in \mathcal{C}} O \in \mathcal{O}$ is an upper bound for $\mathcal{C}$. Thus, we may apply Zorn's lemma (page 15) to obtain a maximal element of $\mathcal{O}$. ∎

The Dual of a Hilbert Space

Let y be an element of the Hilbert space $\mathcal{H}$. The mapping defined by

$$\ell(x) = \langle x, y \rangle, \qquad x \in \mathcal{H}, \tag{13.13}$$

is a linear functional and satisfies $|\ell(x)| \leq \|x\|\|y\|$. Thus, ℓ belongs to the dual space $\mathcal{H}^*$. It is an important property of Hilbert spaces that all continuous linear functionals are of the form (13.13).

□ □ □ **THEOREM 13.8**

Let $\mathcal{H}$ be a Hilbert space. Then $\ell \in \mathcal{H}^*$ if and only if there is a $y \in \mathcal{H}$ such that $\ell(x) = \langle x, y \rangle$ for each $x \in \mathcal{H}$. Furthermore, $\|\ell\|_* = \|y\|$.

PROOF We have already observed that functionals of the form (13.13) belong to $\mathcal{H}^*$. Conversely, suppose that $\ell \in \mathcal{H}^*$. If ℓ is identically 0, then (13.13) holds with $y = 0$. Otherwise, $K = \ell^{-1}(\{0\})$ is a proper closed linear subspace of $\mathcal{H}$ and, consequently, $K^\perp$ contains at least one nonzero element z. For each $x \in \mathcal{H}$, we have $\ell(\ell(z)x - \ell(x)z) = 0$. Thus,

$$0 = \langle \ell(z)x - \ell(x)z, z \rangle = \ell(z)\langle x, z \rangle - \ell(x)\langle z, z \rangle.$$

It follows that $\ell(x) = \langle x, y \rangle$, where $y = (\overline{\ell(z)}/\langle z, z \rangle)z$.

To find the norm of the linear functional ℓ, we first apply Cauchy's inequality to get

$$\|\ell\|_* = \sup\{\,|\langle x, y \rangle| : \|x\| \leq 1\,\} \leq \|y\|.$$

Thus, if $y = 0$, then, trivially, $\|\ell\|_* = \|y\|$. If $y \neq 0$, we choose $w = y/\|y\|$ in order to obtain $\|y\| = \langle w, y \rangle \leq \|\ell\|_*$. ∎

Remark: If E is a basis for a Hilbert space $\mathcal{H}$, then we can write a formula for the element y given in Theorem 13.8 in terms of the basis elements. Indeed, noting that $\ell(e) = \langle e, y \rangle$, we have by Theorem 13.6 that

$$y = \sum_{e \in E} \langle y, e \rangle e = \sum_{e \in E} \overline{\ell(e)}e.$$

Theorem 13.8 is a prototype for results appearing in subsequent sections where we find explicit formulas for bounded linear functionals on various Banach spaces.

Exercises for Section 13.3

13.29 Verify the assertions of parts (b) and (c) of Example 13.10 on page 468.

13.30 Prove (a), (b), and (d) of Theorem 13.4 on page 468.

13.31 Provide the details for the proof of Theorem 13.5 on page 470.

13.32 In this exercise, E denotes an orthonormal set and $\mathcal{H}$ a Hilbert space.
a) Show that if e and e' are distinct members of E, then $\|e - e'\|^2 = 2$.
b) Show that if the closed unit ball $\overline{B}_1(0)$ of $\mathcal{H}$ is compact, then $\mathcal{H}$ is finite dimensional.

13.33 Let $[a, b]$ be a closed bounded interval.
a) Prove that the continuous functions are dense in $\mathcal{L}^2([a, b])$.
b) Formulate and prove a similar result for unbounded intervals.

13.34 Prove that the trigonometric polynomials are dense in $\mathcal{L}^2([-\pi, \pi])$. *Hint:* Refer to Exercise 13.33.

13.35 Prove part (b) of Theorem 13.7 on page 473.

13.36 Prove part (c) of Theorem 13.7 on page 473.

13.37 Let E be an orthonormal set of a Hilbert space $\mathcal{H}$. Establish the following facts.
a) $P_{\overline{\text{span } E}}(x) = \sum_{e \in E} \langle x, e \rangle e$ for all $x \in \mathcal{H}$.
b) $\rho(x, \overline{\text{span } E})^2 = \|x\|^2 - \sum_{e \in E} |\langle x, e \rangle|^2$ for all $x \in \mathcal{H}$.
c) If α is a scalar-valued function on E such that $\sum_{e \in E} |\alpha(e)|^2 < \infty$, then the sum $\sum_{e \in E} \alpha(e)e$ converges.

13.38 Refer to Theorem 13.5 on page 470.
a) Apply the technique used in the proof of that theorem to the subset of $\mathcal{L}^2([-1, 1])$ that consists of $1, x, x^2, \ldots$ to obtain an orthonormal set of polynomials $L_0, L_1, \ldots$. Show that

$$L_n(x) = (n + 1/2)^{1/2} (2^n n!)^{-1} d^n (x^2 - 1)^n / dx^n.$$

The polynomials $(2^n n!)^{-1} d^n (x^2 - 1)^n / dx^n$ are called **Legendre polynomials.**
b) Show that $\{L_0, L_1, \ldots\}$ is a basis for $\mathcal{L}^2([-1, 1])$.

13.39 The **Haar functions** are functions on $[0, 1]$ defined as follows. $H_0(t) = 1$,

$$H_1(t) = \begin{cases} 1, & t \in [0, 1/2]; \\ -1, & t \in (1/2, 1], \end{cases}$$

and

$$H_j(t) = \begin{cases} 2^{n/2} H_1(2^n t - j + 2^n), & t \in [-1 + j/2^n, -1 + (j+1)/2^n]; \\ 0, & \text{otherwise,} \end{cases}$$

for $2^n \leq j < 2^{n+1}$. Show that the Haar functions form a basis for $\mathcal{L}^2([0, 1])$.

13.40 Let $n \in \mathcal{N}$. Define a linear functional S on $\mathcal{L}^2([-\pi, \pi])$ by

$$S(f) = \sum_{k=-n}^{n} \hat{f}(k).$$

Find a function $g \in \mathcal{L}^2([-\pi, \pi])$ such that $S(f) = \int_{-\pi}^{\pi} f(x)g(x)\, dx$.

In Exercises 13.41–13.44, we will need the concepts of an isometric function and an isomorphism of normed spaces. Let Ω and Λ be normed spaces and $L: \Omega \to \Lambda$. Then L is said to be **isometric** (or to be an **isometry**) if $\|L(x)\| = \|x\|$ for each $x \in \Omega$. It is said to be an **isomorphism** if it is linear, one-to-one, onto, and continuous and L^{-1} is also continuous.

13.41   Let $\mathcal{H}$ be a separable Hilbert space. Show that there is an isometric isomorphism from $\mathcal{H}$ onto $\ell^2(\mathcal{N})$.

13.42   Let $\mathcal{H}$ be a Hilbert space. Prove there is an isometric isomorphism from $\mathcal{H}$ onto $\ell^2(S)$ for some set S.

13.43   Prove that the function $g \to \langle \cdot, \overline{g} \rangle$ defines an isometric linear mapping from $\mathcal{L}^2(\mu)$ onto $\mathcal{L}^2(\mu)^*$.

13.44   Show that there is no isometric isomorphism from $\mathcal{L}^2(\mathcal{R})$ onto $\mathcal{L}^1(\mathcal{R})$.

13.4 $\mathcal{L}^p$-SPACES

In Example 10.6 on pages 366–367, we introduced three normed spaces of measurable functions: $\mathcal{L}^1(\mu)$, $\mathcal{L}^2(\mu)$, and $\mathcal{L}^\infty(\mu)$. Now we will generalize to $\mathcal{L}^p(\mu)$, where p is any positive extended real number. These spaces are called $\mathcal{L}^p$-spaces.

We will show that for $p \geq 1$, $\mathcal{L}^p(\mu)$ is a Banach space and will describe its dual space in the spirit of Theorem 13.8 (page 473). The $\mathcal{L}^p$-spaces, along with spaces of the form $C(\Omega)$ where Ω is a compact Hausdorff space, are sometimes referred to in the literature as the *classical Banach spaces*.

DEFINITION 13.9 $\mathcal{L}^p$-Spaces

Let $(\Omega, \mathcal{A}, \mu)$ be a measure space, f a complex-valued $\mathcal{A}$-measurable function on Ω, and $0 < p \leq \infty$.

- For $0 < p < \infty$, we define

$$\sigma_p(f) = \int_\Omega |f|^p \, d\mu$$

and

$$\|f\|_p = \left(\int_\Omega |f|^p \, d\mu \right)^{1/p}.$$

- For $p = \infty$, we define

$$\|f\|_\infty = \inf\{ M : |f| \leq M \ \mu\text{-ae} \}.$$

The collection of complex-valued $\mathcal{A}$-measurable functions f such that $\|f\|_p < \infty$ is denoted $\mathcal{L}^p(\Omega, \mathcal{A}, \mu)$ or, when no confusion can arise, simply $\mathcal{L}^p(\mu)$. The spaces $\mathcal{L}^p(\Omega, \mathcal{A}, \mu)$, $0 < p \leq \infty$, are called $\mathcal{L}^p$-**spaces**.

Note: Under certain conditions, special notation is used for $\mathcal{L}^p$-spaces:

- When μ is Lebesgue measure restricted to some Lebesgue measurable subset Ω of $\mathcal{R}^n$, we write $\mathcal{L}^p(\Omega)$ for $\mathcal{L}^p(\mu)$.

- When μ is counting measure on some set Ω, we write $\ell^p(\Omega)$ for $\mathcal{L}^p(\mu)$ and, in the special case, $\Omega = \mathcal{N}$, we sometimes write simply ℓ^p.

As mentioned earlier, we identify functions that are equal μ-ae. Keeping that in mind, we will see later that $\| \ \|_p$ is a norm on the linear space $\mathcal{L}^p(\mu)$ when $1 \leq p \leq \infty$. When $0 < p < 1$, the space $\mathcal{L}^p(\mu)$ is still a linear space, but $\| \ \|_p$ is no longer a norm. Rather, in this case, $\mathcal{L}^p(\mu)$ is a metric space with metric given by $\rho_p(f, g) = \sigma_p(f - g)$. See Exercises 13.53–13.55.

EXAMPLE 13.13   Illustrates Definition 13.9

a) Let $[a, b]$ be a closed bounded interval of $\mathcal{R}$ and $0 < p < \infty$. A complex-valued Lebesgue measurable function f on $[a, b]$ is in $\mathcal{L}^p([a, b])$ if and only if $\int_a^b |f(x)|^p \, dx < \infty$.

b) Let μ be counting measure on $\{1, 2\}$. Then the space of real-valued functions in $\ell^p(\{1, 2\})$ can be identified with $\mathcal{R}^2$. We have

$$\|(x_1, x_2)\|_p = \begin{cases} (|x_1|^p + |x_2|^p)^{1/p}, & 0 < p < \infty; \\ \max\{|x_1|, |x_2|\}, & p = \infty. \end{cases}$$

Figure 13.3 shows the unit "circles" centered at $(0, 0)$ in the metric space $(\mathcal{R}^2, \rho_{0.5})$ and in the normed space $(\mathcal{R}^2, \| \ \|_p)$ for $p = 1$, 2, 3, and ∞.

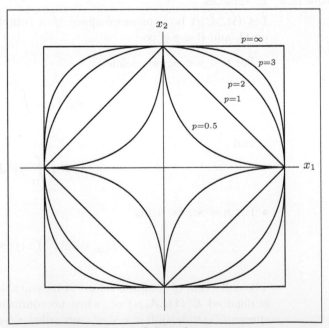

FIGURE 13.3 Selected unit circles

c) Refer to Example 7.10(c) on page 251. Let $(\Omega, \mathcal{A}, P)$ be a probability space. The random variables with finite nth moments are precisely those in $\mathcal{L}^n(P)$.

d) Let μ be counting measure on $\mathcal{N}$ and $0 < p < \infty$. A sequence $\{a_n\}_{n=1}^{\infty}$ of complex numbers is in ℓ^p if and only if $\sum_{n=1}^{\infty} |a_n|^p < \infty$. □

Our next proposition, whose proof is left to the reader as Exercise 13.45, provides some basic properties of $\mathcal{L}^p$-spaces.

□ □ □ **PROPOSITION 13.4**

Let p be a positive extended real number. Then the following properties hold.

a) $\|\alpha f\|_p = |\alpha| \|f\|_p$ *for all $f \in \mathcal{L}^p(\mu)$ and scalars α.*

b) $\mathcal{L}^p(\mu)$ *is a linear space.*

c) *For each $f \in \mathcal{L}^p(\mu)$, there exists a sequence of simple functions $\{s_n\}_{n=1}^{\infty}$ in $\mathcal{L}^p(\mu)$ such that $s_n \to f$ μ-ae, $\|f - s_n\|_p \to 0$, and $\int_{\Omega} |s_n|^p d\mu \to \int_{\Omega} |f|^p d\mu$, as $n \to \infty$.*

In Section 13.2, we used Cauchy's inequality to prove that an inner product $\langle \, , \, \rangle$ induces a norm via $\|x\| = \sqrt{\langle x, x \rangle}$. Similarly, we will use Hölder's inequality, a generalization of Cauchy's inequality, to show that $\| \ \|_p$ is a norm when $p \geq 1$.

□ □ □ **THEOREM 13.9 Hölder's Inequality**

Let $1 \leq p \leq \infty$ and q be such that $1/p + 1/q = 1$. Then for any two $\mathcal{A}$-measurable functions f and g, we have

$$\int_{\Omega} |fg| \, d\mu \leq \|f\|_p \|g\|_q. \tag{13.14}$$

Furthermore, if $1 < p < \infty$, then equality holds in (13.14) if and only if there are constants α and β not both zero such that $\alpha|f|^p = \beta|g|^q$.

PROOF Without loss of generality we can assume that $\|f\|_p$ and $\|g\|_q$ are finite and nonzero. Suppose that $1 < p < \infty$. By the concavity of the natural log function we have

$$\ln |fg| = (1/p) \ln |f|^p + (1/q) \ln |g|^q \leq \ln((1/p)|f|^p + (1/q)|g|^q).$$

Thus,

$$|fg| \leq (1/p)|f|^p + (1/q)|g|^q. \tag{13.15}$$

If $\|f\|_p = \|g\|_q = 1$, it follows from (13.15) that

$$\int_{\Omega} |fg| \, d\mu \leq (1/p) \int_{\Omega} |f|^p \, d\mu + (1/q) \int_{\Omega} |g|^q \, d\mu = 1/p + 1/q = 1 \tag{13.16}$$

and, hence, (13.14) holds in that case. In general, we can replace f and g by $f/\|f\|_p$ and $g/\|g\|_q$, respectively, and use Proposition 13.4(a) and (13.16) to obtain $(\|f\|_p \|g\|_q)^{-1} \int_{\Omega} |fg| \, d\mu \leq 1$.

We leave the cases $p = 1$ and $p = \infty$ and the "Furthermore, ..." part to the reader as Exercises 13.46–13.47. ∎

◻ ◻ ◻ **THEOREM 13.10 Minkowski's Inequality**

Let $1 \leq p \leq \infty$. Then

$$\|f + g\|_p \leq \|f\|_p + \|g\|_p$$

for all $f, g \in \mathcal{L}^p(\mu)$.

PROOF The case $p = 1$ follows immediately from the inequality $|f + g| \leq |f| + |g|$, and the case $p = \infty$ follows from the fact that if $|f| \leq M_1$ μ-ae and $|g| \leq M_2$ μ-ae, then $|f + g| \leq M_1 + M_2$ μ-ae.

Suppose that $p \in (1, \infty)$ and let q be defined via $1/p + 1/q = 1$. From

$$|f + g|^p \leq |f||f + g|^{p-1} + |g||f + g|^{p-1}$$

we get

$$\|f + g\|_p^p \leq \int_\Omega |f||f + g|^{p-1}\, d\mu + \int_\Omega |g||f + g|^{p-1}\, d\mu. \tag{13.17}$$

Noting that

$$\int_\Omega (|f + g|^{p-1})^q\, d\mu = \int_\Omega |f + g|^{qp-q}\, d\mu = \|f + g\|_p^p,$$

it follows from (13.17) and Hölder's inequality that

$$\|f + g\|_p^p \leq \|f\|_p\|f + g\|_p^{p/q} + \|g\|_p\|f + g\|_p^{p/q}.$$

Hence,

$$\|f + g\|_p^{p-p/q} \leq \|f\|_p + \|g\|_p.$$

Whereas $p - p/q = 1$, the proof is complete. ∎

It follows from Proposition 13.4 and Theorem 13.10 that $\mathcal{L}^p(\mu)$ is a normed space when $p \in [1, \infty]$. The next theorem shows that it is in fact a Banach space.

◻ ◻ ◻ **THEOREM 13.11 Riesz's Theorem**

For $1 \leq p \leq \infty$, the normed space $(\mathcal{L}^p(\mu), \|\ \|_p)$ is a Banach space, that is, a complete metric space in the metric induced by the norm $\|\ \|_p$.

PROOF We leave the case $p = \infty$ to the reader as Exercise 13.51. By Proposition 13.3 on page 457, it suffices to show that the series $\sum_{n=1}^{\infty} f_n$ converges with respect to the norm $\|\ \|_p$ whenever $\sum_{n=1}^{\infty} \|f_n\|_p < \infty$.

Consider the nondecreasing sequence of functions defined by $g_n = \sum_{k=1}^{n} |f_k|$, and set $g = \lim_{n\to\infty} g_n$. It follows immediately from Minkowski's inequality that $\int_\Omega g_n^p\, d\mu \leq (\sum_{k=1}^{n} \|f_k\|_p)^p$. Applying the monotone convergence theorem, we obtain

$$\int_\Omega g^p\, d\mu \leq \left(\sum_{n=1}^{\infty} \|f_n\|_p\right)^p < \infty.$$

Hence, g must be finite μ-ae.

It is easy to see that, whenever $g(x) < \infty$, the sequence of partial sums $s_n(x) = \sum_{k=1}^{n} f_k(x)$ is Cauchy and, hence, convergent. Let

$$s(x) = \begin{cases} \lim_{n \to \infty} s_n(x), & \text{if } g(x) < \infty; \\ 0, & \text{if } g(x) = \infty. \end{cases}$$

Then $s \in \mathcal{L}^p(\mu)$ because $\int_\Omega |s|^p \, d\mu \le \int_\Omega |g|^p \, d\mu < \infty$. Also, by using the fact that $|s - s_n|^p \le g^p$ and applying the dominated convergence theorem, we get

$$\lim_{n \to \infty} \|s - s_n\|_p^p = \lim_{n \to \infty} \int_\Omega |s - s_n|^p \, d\mu = 0.$$

We have now shown that the series $\sum_{n=1}^{\infty} f_n$ converges with respect to the norm $\| \ \|_p$. ∎

The Dual Space of $\mathcal{L}^p(\mu)$

We will now consider the problem of describing the bounded linear functionals on $\mathcal{L}^p(\mu)$. At this point, we restrict ourselves to the case where $1 < p < \infty$. To begin, we observe that for $g \in \mathcal{L}^q(\mu)$, where $1/p + 1/q = 1$, the linear functional defined by

$$\ell(f) = \int_\Omega fg \, d\mu \tag{13.18}$$

is continuous on $\mathcal{L}^p(\mu)$. Indeed, by Hölder's inequality, $|\ell(f)| \le \|f\|_p \|g\|_q$ and, therefore,

$$\|\ell\|_* \le \|g\|_q. \tag{13.19}$$

We claim that equality holds in (13.19). If $g = 0$, there is nothing to prove. So assume $\|g\|_q \ne 0$ and set

$$s(x) = \begin{cases} \overline{g(x)}/|g(x)|, & \text{if } g(x) \ne 0; \\ 0, & \text{if } g(x) = 0. \end{cases}$$

Then the function $f_0 = s|g|^{q-1}/\|g\|_q^{q-1}$ satisfies

$$\int_\Omega |f_0|^p \, d\mu = \int_\Omega |s|^p |g|^{pq-p}/\|g\|_q^{pq-p} \, d\mu = \int_\Omega |g|^q/\|g\|_q^q \, d\mu = 1.$$

Hence, $f_0 \in \mathcal{L}^p(\mu)$ and $\|f_0\|_p = 1$. Furthermore,

$$\ell(f_0) = \frac{1}{\|g\|_q^{q-1}} \int_\Omega s|g|^{q-1} g \, d\mu = \frac{1}{\|g\|_q^{q-1}} \int_\Omega |g|^q \, d\mu = \|g\|_q.$$

It follows from this last equality and (13.19) that $\|\ell\|_* = \|g\|_q$.

We have shown that functions in $\mathcal{L}^q(\mu)$ induce bounded linear functionals on $\mathcal{L}^p(\mu)$ via the formula (13.18). Now the question is whether these exhaust all bounded linear functionals on $\mathcal{L}^p(\mu)$. The following theorem shows that the answer is yes!

□ □ □ **THEOREM 13.12 Riesz Representation Theorem**

Let $1 < p < \infty$ and $1/p + 1/q = 1$. Then $\ell \in \mathcal{L}^p(\mu)^*$ if and only if there exists a unique $g \in \mathcal{L}^q(\mu)$ such that

$$\ell(f) = \int_\Omega fg\,d\mu, \qquad f \in \mathcal{L}^p(\mu).$$

Furthermore, g satisfies $\|\ell\|_* = \|g\|_q$.

PROOF In view of our discussion directly before this theorem, we need only prove necessity. So, assume that $\ell \in \mathcal{L}^p(\mu)^*$. We will work under the assumption that $(\Omega, \mathcal{A}, \mu)$ is a finite measure space and leave the general case to the reader as Exercises 13.62–13.65. We also leave the proof of the uniqueness of g for Exercise 13.59.

Define the complex measure ν on $\mathcal{A}$ by $\nu(E) = \ell(\chi_E)$. If $\mu(E) = 0$, then we have $\chi_E = 0$ μ-ae and, so, $\nu(E) = \ell(\chi_E) = 0$. Thus, ν is absolutely continuous with respect to μ. Applying the complex version of the Radon-Nikodym theorem (page 326), we conclude that there exists a function $g \in \mathcal{L}^1(\mu)$ such that

$$\ell(\chi_E) = \int_E g\,d\mu, \qquad E \in \mathcal{A}.$$

By linearity, it follows that $\ell(\phi) = \int_\Omega \phi g\,d\mu$ for all ($\mathcal{A}$-measurable) simple functions ϕ. Thus, $|\int_\Omega \phi g\,d\mu| \le \|\ell\|_* \|\phi\|_p$ for all simple functions. Let

$$s(x) = \begin{cases} \overline{g(x)}/|g(x)|, & \text{if } g(x) \ne 0; \\ 0, & \text{if } g(x) = 0. \end{cases}$$

As the reader is asked to show in Exercise 13.60, we can find a sequence of simple functions $\{\psi_n\}_{n=1}^\infty$ such that $|\psi_n| \le 1$ μ-ae and $\psi_n \to s$ μ-ae. We have

$$\left| \int_\Omega \psi_n \phi g\,d\mu \right| \le \|\ell\|_* \|\psi_n \phi\|_p \le \|\ell\|_* \|\phi\|_p$$

and, applying the dominated convergence theorem, we obtain

$$\left| \int_\Omega \phi |g|\,d\mu \right| \le \|\ell\|_* \|\phi\|_p. \tag{13.20}$$

We will use (13.20) to show that g belongs to the space $\mathcal{L}^q(\mu)$.

Let $n \in \mathcal{N}$ and $E_n = \{\, x : |g(x)| \leq n \,\}$. The function $f_0 = \chi_{E_n} |g|^{q-1}$ belongs to $\mathcal{L}^p(\mu)$. Hence, by Proposition 13.4 on page 477, there is a sequence $\{\phi_k\}_{k=1}^{\infty}$ of simple functions such that, as $k \to \infty$, $\phi_k \to f_0$ μ-ae and $\|\phi_k\|_p \to \|f_0\|_p$. Replacing ϕ_k by $\chi_{E_n} |\phi_k|$ if necessary, we may assume without loss of generality that the ϕ_ks are nonnegative and vanish outside of E_n. Using Fatou's lemma and (13.20), we obtain

$$\int_{E_n} |g|^{q-1} |g| \, d\mu \leq \liminf_{k \to \infty} \int_{\Omega} \phi_k |g| \, d\mu \leq \|\ell\|_* \liminf_{k \to \infty} \|\phi_k\|_p = \|\ell\|_* \|f_0\|_p$$

$$= \|\ell\|_* \left(\int_{E_n} |g|^{p(q-1)} \, d\mu \right)^{1/p} = \|\ell\|_* \left(\int_{E_n} |g|^q \, d\mu \right)^{1/p}$$

and, hence, that

$$\left(\int_{E_n} |g|^q \, d\mu \right)^{1/q} = \left(\int_{E_n} |g|^q \, d\mu \right)^{1-1/p} \leq \|\ell\|_*.$$

Letting $n \to \infty$ and applying the MCT, we get that $\|g\|_q \leq \|\ell\|_*$. Thus, g belongs to $\mathcal{L}^q(\mu)$.

Because $g \in \mathcal{L}^q(\mu)$, the function ℓ_g defined by $\ell_g(f) = \int_{\Omega} fg \, d\mu$ is in $\mathcal{L}^p(\mu)^*$. As ℓ and ℓ_g agree on simple functions, Proposition 13.4 implies that they are identical. ■

Remark: If $p = 1$, Theorem 13.12 remains valid under the additional assumption that $(\Omega, \mathcal{A}, \mu)$ is σ-finite, as the reader is asked to prove in Exercise 13.61. An example given in Chapter 14 shows that Theorem 13.12 fails when $p = \infty$.

In view of Theorem 13.12, we can write $\mathcal{L}^p(\mu)^* = \mathcal{L}^q(\mu)$, for $1 < p < \infty$, and, in the σ-finite case, for $p = 1$. However, for $p = \infty$, we can assert only that

$$\mathcal{L}^{\infty}(\mu)^* \supset \mathcal{L}^1(\mu). \tag{13.21}$$

See Exercise 13.58.

EXAMPLE 13.14   *Illustrates Theorem 13.12*

Refer to Example 13.11 on page 470. Let $x \in [-\pi, \pi]$ and $1 < p < \infty$. Define the linear functional ℓ_x on $\mathcal{L}^p([-\pi, \pi])$ by

$$\ell_x(f) = \sum_{k=-n}^{n} \hat{f}(k) e^{ikx}.$$

Of course, ℓ_x just gives the value at x of the nth partial sum of the Fourier series of f.

First we show that ℓ_x is bounded and then we find the function $g \in \mathcal{L}^q([-\pi, \pi])$ guaranteed by Theorem 13.12. From Hölder's inequality,

$$|\hat{f}(k)| = \left| \frac{1}{2\pi} \int_{-\pi}^{\pi} f(y) e^{-iky} \, dy \right| \le \frac{\|f\|_p}{2\pi} \left(\int_{-\pi}^{\pi} |e^{-iky}|^q \, dy \right)^{1/q} = \|f\|_p (2\pi)^{-1/p}.$$

It follows at once that $|\ell_x(f)| \le (2n + 1)(2\pi)^{-1/p} \|f\|_p$. Thus, ℓ_x is bounded. Finally, we write

$$\ell_x(f) = \sum_{k=-n}^{n} \frac{1}{2\pi} \int_{-\pi}^{\pi} f(y) e^{ik(x-y)} \, dy = \int_{-\pi}^{\pi} f(y) D_n(x - y) \, dy,$$

where

$$D_n(t) = \frac{1}{2\pi} \sum_{k=-n}^{n} e^{ikt} = \begin{cases} \dfrac{\sin((n + 1/2)t)}{2\pi \sin(t/2)}, & t \ne 0; \\[2ex] \dfrac{2n + 1}{2\pi}, & t = 0. \end{cases}$$

Thus, the function g guaranteed by Theorem 13.12 is $g(y) = D_n(x - y)$. ☐

Exercises for Section 13.4

13.45 Prove Proposition 13.4 on page 477.

13.46 Prove the "Furthermore, ..." part of Hölder's inequality (Theorem 13.9 on page 477).

13.47 Verify Hölder's inequality (Theorem 13.9 on page 477) for $p = 1$ and $p = \infty$.

13.48 Discuss the case of equality in (13.14) on page 477 when $p = 1$ or $p = \infty$.

13.49 Suppose that $p, q \in (0, \infty]$.
 a) Let r be such that $1/r = 1/p + 1/q$. Show that if $f \in \mathcal{L}^p(\mu)$ and $g \in \mathcal{L}^q(\mu)$, then $fg \in \mathcal{L}^r(\mu)$ and $\|fg\|_r \le \|f\|_p \|g\|_q$.
 b) Suppose that $(\Omega, \mathcal{A}, \mu)$ is a finite measure space. Show that if $0 < s < r \le \infty$, then $\mathcal{L}^r(\mu) \subset \mathcal{L}^s(\mu)$.

13.50 Let $(\Omega, \mathcal{A}, \mu)$ be a finite measure space. Show that for each $f \in \mathcal{L}^\infty(\mu)$, $\|f\|_p \to \|f\|_\infty$ as $p \to \infty$.

13.51 Prove that the normed space $(\mathcal{L}^\infty(\mu), \| \ \|_\infty)$ is a Banach space.

13.52 Show that $(\mathcal{L}^p([0, 1]), \| \ \|_p)$ is not an inner product space unless $p = 2$.

13.53 Show that $\| \ \|_p$ does not define a norm on $\mathcal{L}^p([0, 1])$ when $0 < p < 1$.

★13.54 Refer to Definition 13.9 on page 475.
 a) Show that if $0 < p < 1$, then $\sigma_p(f + g) \le \sigma_p(f) + \sigma_p(g)$.
 b) Deduce that $\rho_p(f, g) = \sigma_p(f - g)$ defines a metric on $\mathcal{L}^p(\mu)$ for $0 < p < 1$.

13.55 Refer to Exercise 13.54. Show that if $0 < p < 1$, then $(\mathcal{L}^p(\mu), \rho_p)$ is a complete metric space.

★13.56 Let J be a nonempty interval in $\mathcal{R}$ and $0 < p < \infty$.
 a) Show that if J is closed and bounded, then $C(J)$ is dense in $\mathcal{L}^p(J)$.
 b) Refer to Example 11.9 on page 421. Show that $C_c(J)$ is dense in $\mathcal{L}^p(J)$.
 c) Show that $C_c(J)$ is not dense in $\mathcal{L}^\infty(J)$.

13.57 Let $0 < p < \infty$. Prove that the trigonometric polynomials are dense in $\mathcal{L}^p([-\pi, \pi])$.

13.58   The result of this exercise gives meaning to the relation (13.21) on page 481. Prove that if $g \in \mathcal{L}^1(\mu)$, then $\ell(f) = \int_\Omega fg \, d\mu$ defines a bounded linear functional on $\mathcal{L}^\infty(\mu)$ and that $\|\ell\|_* = \|g\|_1$.

13.59   Prove the uniqueness of the function g in Theorem 13.12 (page 480).

★13.60   Suppose $f \in \mathcal{L}^\infty(\mu)$. Show that there exists a sequence of simple functions $\{\phi_n\}_{n=1}^\infty$ such that $|\phi_n| \le \|f\|_\infty$ μ-ae and $\lim_{n\to\infty} \phi_n = f$ μ-ae.

★13.61   Prove Theorem 13.12 (page 480) when $p = 1$ under the assumption that $(\Omega, \mathcal{A}, \mu)$ is a σ-finite measure space.

In Exercises 13.62–13.65, we complete the proof of Theorem 13.12 (page 480) by eliminating the restriction $\mu(\Omega) < \infty$.

13.62   Suppose that $(\Omega, \mathcal{A}, \mu)$ is a measure space. For $E \in \mathcal{A}$, define the measure μ_E on $\mathcal{A}$ by $\mu_E(A) = \mu(E \cap A)$.
   a) Show that $f \in \mathcal{L}^p(\mu_E)$ if and only if $\chi_E f \in \mathcal{L}^p(\mu)$.
   b) Show that if $\ell \in \mathcal{L}^p(\mu)^*$, then $\ell_E(f) = \ell(\chi_E f)$ defines a continuous linear functional on $\mathcal{L}^p(\mu_E)$ and $\|\ell_E\|_* \le \|\ell\|_*$.
   c) If $\mu(E) < \infty$, show there is a unique function $g_E \in \mathcal{L}^q(\mu)$ such that g_E vanishes outside of E, $\ell_E(f) = \int_\Omega fg_E \, d\mu$ for each $f \in \mathcal{L}^p(\mu_E)$, and $\|\ell_E\|_*^q = \int_\Omega |g_E|^q \, d\mu_E$.

13.63   Use Exercise 13.62 to prove Theorem 13.12 in case $(\Omega, \mathcal{A}, \mu)$ is σ-finite.

13.64   Let $(\Omega, \mathcal{A}, \mu)$ be an arbitrary measure space and $1 < p < \infty$. Show that if $\ell \in \mathcal{L}^p(\mu)^*$, then there exists a sequence $\{\Omega_n\}_{n=1}^\infty$ of $\mathcal{A}$-measurable sets such that $\mu(\Omega_n) < \infty$ for each $n \in \mathcal{N}$ and $\ell(\chi_A) = 0$ for each $A \in \mathcal{A}$ such that $\mu(A) < \infty$ and $A \subset (\bigcup_{n=1}^\infty \Omega_n)^c$.

13.65   Employ Exercises 13.62–13.64 to establish Theorem 13.12 for an arbitrary measure space $(\Omega, \mathcal{A}, \mu)$.

13.5 NONNEGATIVE LINEAR FUNCTIONALS ON $C(\Omega)$

We have now characterized the dual spaces of Hilbert spaces (Theorem 13.8 on page 473) and $\mathcal{L}^p$-spaces (Theorem 13.12 on page 480). Our next task, which we will begin in this section and complete in the following one, is to characterize the dual spaces of $C(\Omega)$ and $C_0(\Omega)$.

We will see that the linear functional defined on $C([0,1])$ by

$$\ell_\lambda(f) = \int_0^1 f(x) \, dx = \int_{[0,1]} f \, d\lambda$$

is typical in the sense that all bounded linear functionals on $C(\Omega)$ arise from integration with respect to some complex measure. Here we lay the foundation for the general result by characterizing those that arise from integration with respect to a (nonnegative) measure.

Borel Sets and Regular Borel Measures

In Chapter 3 we defined the collection $\mathcal{B}$ of Borel sets of $\mathcal{R}$. We showed in Theorem 3.4 that $\mathcal{B}$ is the smallest σ-algebra of subsets of $\mathcal{R}$ that contains the open sets of $\mathcal{R}$. This characterization allows us to extend the concept of Borel sets to any topological space.

DEFINITION 13.10   Borel Set, Measure, and Measurable Function

Let Ω be a topological space. The smallest σ-algebra of subsets of Ω that contains all the open sets is denoted $\mathcal{B}(\Omega)$. We use the following terminology:

- **Borel set:** a member of $\mathcal{B}(\Omega)$.

- **Borel measurable function:** a function measurable with respect to $\mathcal{B}(\Omega)$.

- **Borel measure:** a signed or complex measure on $\mathcal{B}(\Omega)$.

EXAMPLE 13.15   *Illustrates Definition 13.10*

a) $\mathcal{B}(\mathcal{R}) = \mathcal{B}$, as defined in Chapter 3.

b) $\mathcal{B}(\mathcal{R}^2) = \mathcal{B}_2 = \mathcal{B} \times \mathcal{B}$, as discussed in Exercise 6.53 on page 211. More generally, we have that $\mathcal{B}(\mathcal{R}^n) = \mathcal{B}_n = \mathcal{B} \times \mathcal{B} \times \cdots \times \mathcal{B}$, as discussed in Exercise 6.77 on page 223.

c) Let Ω be any set and $\mathcal{T} = \{\Omega, \emptyset\}$. Then $\mathcal{B}(\Omega) = \mathcal{T}$.

d) Let Ω be any set and $\mathcal{T}$ be the discrete topology on Ω. Then we have that $\mathcal{B}(\Omega) = \mathcal{T} = \mathcal{P}(\Omega)$.

e) Let $(\Omega, \mathcal{T})$ be a topological space. Then all functions in $C(\Omega)$ are Borel measurable.   □

To characterize the bounded linear functionals on $C(\Omega)$, we need the concept of a *regular Borel measure*. We recommend that the reader review the discussion of the total variation of a complex measure presented in Section 9.3 starting on page 321.

DEFINITION 13.11   Regular Borel Measure

Let Ω be a locally compact Hausdorff space. A complex Borel measure μ is said to be a **regular Borel measure** if for each $B \in \mathcal{B}(\Omega)$ and $\epsilon > 0$, there is a compact set K and an open set O such that $K \subset B \subset O$ and $|\mu|(O \setminus K) < \epsilon$.

The collection of all regular Borel measures on Ω is denoted by $M(\Omega)$; the real-valued and nonnegative regular Borel measures are denoted, respectively, by $M_r(\Omega)$ and $M_+(\Omega)$.

Remark: Definition 13.11 requires that a regular Borel measure be finite valued. Other definitions of regular Borel measure exist and some permit certain extended real-valued measures, such as Lebesgue measure, to be regular.

EXAMPLE 13.16   *Illustrates Definition 13.11*

a) Lebesgue measure on $[0, 1]$ is a regular Borel measure. In fact, Lebesgue measure on any Borel set of finite Lebesgue measure is a regular Borel measure.

b) The Lebesgue-Stieltjes measure corresponding to a distribution function on $\mathcal{R}$ is a regular Borel measure, as the reader is asked to establish in Exercise 13.68.

c) Let Ω be a locally compact Hausdorff space. For $x \in \Omega$, the Dirac measure concentrated at x, restricted to the Borel sets of Ω, is a regular Borel measure. See Exercise 13.71.   □

Suppose that Ω is a locally compact Hausdorff space. The spaces $M(\Omega)$ and $M_r(\Omega)$ are, respectively, complex and real linear spaces, where the operations of addition and scalar multiplication are defined by

$$(\mu + \nu)(B) = \mu(B) + \nu(B) \qquad \text{and} \qquad (\alpha\mu)(B) = \alpha\mu(B).$$

Referring to Exercise 9.48 on page 329, we see that the linear spaces $M(\Omega)$ and $M_r(\Omega)$ are also normed spaces, where the norm is given by the total variation, that is, $\|\mu\| = |\mu|(\Omega)$. Moreover, as the reader is asked to prove in Exercise 13.66, $M(\Omega)$ and $M_r(\Omega)$ are Banach spaces with respect to the norm $\|\ \|$.

If F is a closed subset of Ω, then any $\nu \in M(F)$ can be extended to a regular Borel measure ν' on Ω by defining

$$\nu'(B) = \nu(B \cap F), \qquad B \in \mathcal{B}(\Omega).$$

It is convenient to view ν as a measure on Ω by identifying it with ν'. In this way we can identify $M(F)$ with the linear subspace

$$\{\mu \in M(\Omega) : \mu(B) = 0 \text{ for all } B \in \mathcal{B}(\Omega) \text{ with } B \subset F^c\}.$$

Nonnegative Linear Functionals

From here on in this section, unless explicitly stated otherwise, we assume that Ω is a *compact Hausdorff space*. If $\mu \in M(\Omega)$, then μ induces a linear functional on the space $C(\Omega)$ via

$$\ell_\mu(f) = \int_\Omega f \, d\mu, \qquad f \in C(\Omega).$$

That ℓ_μ is a bounded linear functional follows from

$$|\ell_\mu(f)| \le \|f\|_\Omega |\mu|(\Omega) = \|f\|_\Omega \|\mu\|,$$

where we have applied Exercise 9.53(b) on page 330.

In this section, we will show that any linear functional on $C(\Omega)$ that satisfies a certain nonnegativity condition must be of the form ℓ_μ for some $\mu \in M_+(\Omega)$. In the next section, we will extend this result to all bounded linear functionals on $C(\Omega)$ if Ω is a compact Hausdorff space and to $C_0(\Omega)$ if Ω is a locally compact Hausdorff space.

DEFINITION 13.12   **Nonnegative Linear Functional**

A linear functional ℓ on $C(\Omega)$ is said to be **nonnegative** if $\ell(f) \ge 0$ whenever $f \ge 0$.

As the reader is asked to show in Exercise 13.75, the linear functional ℓ_μ on $C(\Omega)$ induced by a regular Borel measure μ is nonnegative if and only if μ is nonnegative.

The next theorem, whose proof is left to the reader in Exercises 13.76–13.81, presents some basic properties of nonnegative linear functionals.

□ □ □ **THEOREM 13.13**

Let Ω be a compact Hausdorff space.
a) If ℓ is a nonnegative linear functional on $C(\Omega)$, then $\ell \in C(\Omega)^*$ and, moreover, $\|\ell\|_* = \ell(1)$.
b) If $\ell \in C(\Omega)^*$ and $\ell(C(\Omega, \mathcal{R})) \subset \mathcal{R}$, then there exist nonnegative linear functionals ℓ_+ and ℓ_- such that $\|\ell\|_* = \ell_+(1) + \ell_-(1)$ and $\ell = \ell_+ - \ell_-$.

We have noted that a nonnegative regular Borel measure on Ω induces a nonnegative linear functional on $C(\Omega)$. Our next theorem shows that all nonnegative linear functionals on $C(\Omega)$ are of that type. There are two main ideas in the proof of this result. One is the use of Urysohn's lemma to obtain suitable approximations to characteristic functions of closed sets. The other is to mimic the construction of Lebesgue measure from Lebesgue outer measure.

With regard to the latter, recall that the collection $\mathcal{M}$ of Lebesgue measurable sets is defined by using the Carathéodory criterion and Lebesgue outer measure: $E \in \mathcal{M}$ if and only if

$$\lambda^*(W) = \lambda^*(W \cap E) + \lambda^*(W \cap E^c)$$

for all $W \subset \mathcal{R}$. Theorem 3.11 on page 103 shows that $\mathcal{M}$ is a σ-algebra. A careful look at the proof reveals that it uses only the properties of Lebesgue outer measure given in (a), (b), (c), and (e) of Proposition 3.1 on page 92. In other words, we have already proved the following proposition.

□ □ □ **PROPOSITION 13.5**

Let Ω be a set and ν^* an extended real-valued function on $\mathcal{P}(\Omega)$ that satisfies the following conditions:
a) $\nu^*(A) \geq 0$ for each $A \subset \Omega$.
b) $\nu^*(\emptyset) = 0$.
c) $A \subset B \Rightarrow \nu^*(A) \leq \nu^*(B)$.
d) $\{A_n\}_n \subset \mathcal{P}(\Omega) \Rightarrow \nu^* \left(\bigcup_n A_n \right) \leq \sum_n \nu^*(A_n)$.
Then the collection of subsets E of Ω that satisfy

$$\nu^*(W) = \nu^*(W \cap E) + \nu^*(W \cap E^c)$$

for all $W \subset \Omega$ is a σ-algebra whose members are called $\boldsymbol{\nu^*}$**-measurable sets.**[†]

We now state and prove the main result of this section, known as the Riesz-Markov theorem.

[†] Proposition 6.2 on page 183 shows that the outer measure ν^* induced by an appropriate set function on a semialgebra of subsets of a set Ω satisfies (a)–(d) of Proposition 13.5. Thus, the concept of ν^*-measurability given here is the same as that in Definition 6.3 on page 184.

□ □ □ **THEOREM 13.14 Riesz-Markov Theorem**

Let Ω be a compact Hausdorff space and let ℓ be a nonnegative linear functional on $C(\Omega)$. Then there exists a unique $\mu \in M_+(\Omega)$ such that

$$\ell(f) = \int_\Omega f \, d\mu, \qquad f \in C(\Omega).$$

PROOF   We start by assigning a nonnegative number $\overline{\mu}(O)$ to each open set O. If $O = \emptyset$, let $\overline{\mu}(O) = 0$; otherwise, let

$$\overline{\mu}(O) = \sup\{\ell(f) : 0 \le f \le 1 \text{ and } \operatorname{supp} f \subset O\}.$$

We note that $\overline{\mu}(O) \le \overline{\mu}(\Omega) = \ell(1)$ for all O. Next, for each $A \subset \Omega$, we define

$$\mu^*(A) = \inf\{\overline{\mu}(O) : O \text{ open and } O \supset A\}.$$

Observe that $\mu^*(O) = \overline{\mu}(O)$ whenever O is open.

We will show that μ^* satisfies the hypotheses of Proposition 13.5. Conditions (a)–(c) follow easily from the definition of μ^*. To verify condition (d), we first show that if $\{O_n\}_{n=1}^\infty$ is a sequence of open subsets of Ω, then

$$\overline{\mu}\left(\bigcup_{n=1}^\infty O_n\right) \le \sum_{n=1}^\infty \overline{\mu}(O_n). \tag{13.22}$$

Let f be a continuous function that satisfies $0 \le f \le 1$ and $\operatorname{supp} f \subset \bigcup_{n=1}^\infty O_n$. Applying Theorem 11.10 on page 412 with $K = \operatorname{supp} f$, we obtain continuous functions $f_1, f_2, \ldots, f_m$ that satisfy

- $0 \le f_j \le 1$, for each j,

- $\sum_{j=1}^m f_j(x) = 1$ for $x \in \operatorname{supp} f$,

- $\sum_{j=1}^m f_j \le 1$, and

- for each j, there is an m_j such that $\operatorname{supp} f_j \subset O_{m_j}$.

By replacing f_j by $\sum_{m_k = m_j} f_k$ if necessary, we can assume that the m_js are distinct. It is clear that $f = \sum_{j=1}^m f f_j$ and, so, $\ell(f) = \sum_{j=1}^m \ell(f f_j)$. Because $\operatorname{supp} f f_j \subset O_{m_j}$, it follows that

$$\ell(f) \le \sum_{n=1}^\infty \overline{\mu}(O_n). \tag{13.23}$$

Taking the supremum on the left-hand side of (13.23), we obtain (13.22). It is now easy to check that μ^* satisfies condition (d) of Proposition 13.5, as we ask the reader to verify in Exercise 13.82.

We complete the proof of the theorem by showing successively that

- all open sets are μ^*-measurable,

- $\mu = \mu^*_{|\mathcal{B}(\Omega)}$ is a regular Borel measure, and

- $\ell(f) = \int_\Omega f \, d\mu$ for all $f \in C(\Omega)$.

To show that an arbitrary open set O is μ^*-measurable, it suffices to prove that

$$\mu^*(A) \geq \mu^*(A \cap O) + \mu^*(A \cap O^c) \tag{13.24}$$

for all $A \subset \Omega$. Let U be an open set containing A, f a continuous function that satisfies $0 \leq f \leq 1$ and $\operatorname{supp} f \subset U \cap O$, and $V = U \cap (\operatorname{supp} f)^c$. If g is a continuous function that satisfies $0 \leq g \leq 1$ and $\operatorname{supp} g \subset V$, then

$$\operatorname{supp}(f + g) \subset \operatorname{supp} f \cup \operatorname{supp} g \subset U.$$

It follows that

$$\overline{\mu}(U) \geq \ell(f) + \ell(g). \tag{13.25}$$

From (13.25) we deduce that

$$\overline{\mu}(U) \geq \ell(f) + \overline{\mu}(V) \geq \ell(f) + \mu^*(A \cap O^c)$$

and, therefore, that

$$\overline{\mu}(U) \geq \overline{\mu}(U \cap O) + \mu^*(A \cap O^c) \geq \mu^*(A \cap O) + \mu^*(A \cap O^c).$$

As the open set U containing A was chosen arbitrarily, (13.24) holds.

Having shown that all open sets are μ^*-measurable, we can invoke Proposition 13.5 on page 486 and Proposition 6.5 on page 186 to conclude that all Borel sets are μ^*-measurable and that $\mu = \mu^*_{|\mathcal{B}(\Omega)}$ is a Borel measure. To show that μ is regular, we first observe that, by the definition of μ^*,

$$\mu(B) = \inf\{\, \mu(O) : O \text{ open and } O \supset B \,\}, \qquad B \in \mathcal{B}(\Omega). \tag{13.26}$$

Because $\mu(\Omega) = \ell(1) < \infty$, we have for each $B \in \mathcal{B}(\Omega)$ that

$$\begin{aligned}
\mu(B) &= \mu(\Omega) - \mu(B^c) = \mu(\Omega) - \inf\{\, \mu(W) : W \text{ open, } W \supset B^c \,\} \\
&= \sup\{\, \mu(W^c) : W \text{ open, } W \supset B^c \,\} \\
&= \sup\{\, \mu(F) : F \text{ closed, } F \subset B \,\}.
\end{aligned} \tag{13.27}$$

It follows at once from (13.26) and (13.27) that μ is regular.

Finally, we must show that

$$\ell(f) = \int_\Omega f \, d\mu, \qquad f \in C(\Omega). \tag{13.28}$$

Every function in $C(\Omega)$ is a linear combination of functions with values in the interval $[0, 1)$. Therefore, by the linearity of ℓ, it suffices to establish (13.28) in case $0 \leq f < 1$.

Let $n \in \mathcal{N}$. For each integer k, $0 \le k \le n$, the sets $F_k = f^{-1}([k/n, \infty))$ and $U_k = f^{-1}(((k-1)/n, \infty))$ are closed and open, respectively. Moreover, we have

$$F_{k+1} \subset U_{k+1} \subset F_k \subset U_k \qquad \text{and} \qquad \Omega = \bigcup_{k=0}^{n-1} (F_k \setminus F_{k+1}).$$

If $F_k = \emptyset$, we set $g_k = 0$. Otherwise, we first invoke the regularity of μ to choose an open set V_k such that $F_k \subset V_k \subset U_k$ and $\sum_{k=0}^{n-1} \mu(V_k \setminus F_k) < 1$ and then apply Proposition 10.14 on page 387 and Urysohn's lemma on page 388 to obtain a continuous function g_k such that $0 \le g_k \le 1$, $g_k(F_k) = \{1\}$, and supp $g_k \subset V_k$.

Let $h = (1/n) \sum_{j=0}^{n-1} g_j$. We claim $f \le h$. For each $x \in \Omega$, choose the unique k such that $x \in f^{-1}([k/n, (k+1)/n)) = F_k \setminus F_{k+1}$. If $0 \le j \le k$, then $g_j(x) = 1$ since $F_k \subset F_j$; if $j > k+1$, then $g_j(x) = 0$ since $x \in F_{k+1}^c \subset U_{k+2}^c \subset U_j^c \subset V_j^c$. It follows that

$$h(x) = (k+1)/n + g_{k+1}(x)/n \ge f(x),$$

as required. Using the fact that $f \le h$ and the nonnegativity of ℓ, we obtain

$$\ell(f) \le \ell(h) = (1/n) \sum_{j=0}^{n-1} \ell(g_j) \le (1/n) \sum_{j=0}^{n-1} \mu(V_j)$$

$$= (1/n) \sum_{j=0}^{n-1} \big(\mu(V_j \setminus F_j) + \mu(F_j)\big) \le 1/n + (1/n) \sum_{j=0}^{n-1} \mu(F_j). \tag{13.29}$$

For $j = 0, 1, \ldots, n-1$, we can write $F_j = \bigcup_{k=j}^{n-1} (F_k \setminus F_{k+1})$ and, therefore,

$$\mu(F_j) = \sum_{k=j}^{n-1} \mu(F_k \setminus F_{k+1}).$$

Applying (13.29), we get

$$\ell(f) \le 1/n + (1/n) \sum_{j=0}^{n-1} \sum_{k=j}^{n-1} \mu(F_k \setminus F_{k+1})$$

$$= 1/n + (1/n) \sum_{k=0}^{n-1} (k+1)\mu(F_k \setminus F_{k+1})$$

$$= 1/n + \mu(\Omega)/n + \sum_{k=0}^{n-1} (k/n)\mu(F_k \setminus F_{k+1})$$

$$= 1/n + \ell(1)/n + \int_\Omega \sum_{k=0}^{n-1} (k/n)\chi_{(F_k \setminus F_{k+1})} \, d\mu$$

$$\le (1 + \ell(1))/n + \int_\Omega f \, d\mu.$$

Because n was chosen arbitrarily, it follows that

$$\ell(f) \le \int_\Omega f \, d\mu. \tag{13.30}$$

We can replace f by $(1-f)/2$ in (13.30) to get

$$\ell(1) - \ell(f) \leq \mu(\Omega) - \int_\Omega f \, d\mu = \ell(1) - \int_\Omega f \, d\mu.$$

Thus (13.28) holds.

It remains only to prove the uniqueness of μ, which we leave to the reader as Exercise 13.83. ∎

Exercises for Section 13.5

13.66 Let Ω be a locally compact Hausdorff space. Show that $(M(\Omega), \|\ \|)$ and $(M_r(\Omega), \|\ \|)$ are Banach spaces, where $\|\mu\| = |\mu|(\Omega)$.

13.67 Let Ω be a locally compact Hausdorff space. Show that if $\mu \in M(\Omega)$, then $|\mu| \in M(\Omega)$.

★13.68 In this exercise, you are asked, among other things, to verify the statement of Example 13.16(b).
   a) Prove that if a locally compact metric space Ω is the countable union of compact subsets, then every complex Borel measure on Ω is regular.
   b) Show that the Lebesgue-Stieltjes measure associated with a distribution function on $\mathcal{R}$ is a regular Borel measure.

13.69 Suppose Ω is locally compact and $\mu \in M_+(\Omega)$. Prove that $C_0(\Omega)$ is dense in $\mathcal{L}^p(\mu)$ for $1 \leq p < \infty$.

13.70 Let $\mu \in M([0,1])$ satisfy

$$\int_{[0,1]} x^n \, d\mu(x) = 0$$

for $n = 0, 1, 2, \ldots$. Show that $\mu = 0$, that is, μ vanishes identically.

13.71 Suppose that Ω is a locally compact Hausdorff space. Let $x \in \Omega$ and δ_x be defined on $\mathcal{B}(\Omega)$ by

$$\delta_x(B) = \begin{cases} 1, & \text{if } x \in B; \\ 0, & \text{if } x \notin B. \end{cases}$$

   a) Show that δ_x is a regular Borel measure.
   b) Determine $\int_\Omega f \, d\delta_x$ when $f \in C(\Omega)$.

13.72 Let δ_x be as in Exercise 13.71.
   a) Show that $\|\delta_x - \delta_y\| = 1$ when $x \neq y$.
   b) Deduce that $M(\Omega)$ is not separable if Ω is uncountable.

★13.73 Show how to identify $M(\Omega)$ and $\ell^1(\Omega)$ when Ω is countable.

13.74 Let Ω be a locally compact Hausdorff space, $\mu \in M(\Omega)$, and $B \in \mathcal{B}(\Omega)$. Prove that there are sets F and G such that G is a countable intersection of open sets, F is a countable union of closed sets, $F \subset B \subset G$, and $|\mu|(G \setminus F) = 0$.

13.75 Let Ω be a compact Hausdorff space and $\mu \in M(\Omega)$. Show that $\ell_\mu(f) = \int_\Omega f \, d\mu$ defines a nonnegative linear functional on $C(\Omega)$ if and only if $\mu \in M_+(\Omega)$.

Exercises 13.76–13.81 provide the proof of Theorem 13.13 on page 486.

13.76 Show that if ℓ is a nonnegative linear functional on $C(\Omega)$, then $\ell \in C(\Omega)^*$ and, moreover, $\|\ell\|_* = \ell(1)$.

13.77 Suppose that ℓ satisfies the hypotheses of part (b) of Theorem 13.13 on page 486. For each nonnegative continuous function f on Ω, let

$$\ell_+(f) = \sup\{\,\ell(g) : 0 \le g \le f,\ g \text{ continuous}\,\}.$$

a) Show that if f_1 and f_2 are nonnegative and continuous, then

$$\ell_+(f_1 + f_2) = \ell_+(f_1) + \ell_+(f_2).$$

b) Show that $0 \le f \le g$ implies $\ell_+(f) \le \ell_+(g)$.

c) Show that $\ell_+(\alpha f) = \alpha \ell_+(f)$ whenever $f \ge 0$ and α is a nonnegative real number.

13.78 Extend the function ℓ_+ defined in Exercise 13.77 to all of $C(\Omega, \mathcal{R})$ by the formula

$$\ell_+(f) = \ell_+(\|f\| + f) - \ell_+(\|f\|),$$

where $\|f\| = \|f\|_\Omega$.

a) Prove that this new definition of $\ell_+(f)$ agrees with the old one when f is nonnegative.

b) Show that this extended ℓ_+ is linear on the space $C(\Omega, \mathcal{R})$.

13.79 Extend the function ℓ_+ defined in Exercise 13.78 to all of $C(\Omega)$ by the formula

$$\ell_+(f) = \ell_+(\Re f) + i\ell_+(\Im f).$$

a) Prove that this new definition of $\ell_+(f)$ agrees with the old one when f is real valued.

b) Show that this extended function is linear and nonnegative.

13.80 Suppose that ℓ satisfies the hypotheses of part (b) of Theorem 13.13 on page 486. Let $\ell_- = \ell_+ - \ell$, where ℓ_+ is defined as in Exercise 13.79. Show that ℓ_- is nonnegative.

13.81 Suppose that ℓ satisfies the hypotheses of part (b) of Theorem 13.13 on page 486. Let ℓ_+ and ℓ_- be defined as in Exercise 13.80. Show that $\|\ell\|_* = \ell_+(1) + \ell_-(1)$. *Hint:* If $0 \le g \le 1$, then $\|2g - 1\| \le 1$ and, so, $\|\ell\|_* \ge 2\ell(g) - 1$.

13.82 Show that the set function μ^* defined in the proof of Theorem 13.14 on page 487 satisfies condition (d) of Proposition 13.5 on page 486.

13.83 Prove the uniqueness part of Theorem 13.14 on page 487.

13.6 THE DUAL SPACES OF $C(\Omega)$ AND $C_0(\Omega)$

In this section, we extend the Riesz-Markov theorem (page 487) to arbitrary bounded linear functionals on $C(\Omega)$. We will also characterize the bounded linear functionals on $C_0(\Omega)$ when Ω is a locally compact Hausdorff space. These results show that we are justified in writing $C(\Omega)^* = M(\Omega)$ and $C_0(\Omega)^* = M(\Omega)$ in the compact and locally compact cases, respectively.

□ □ □ **LEMMA 13.1**

Suppose that Ω is a compact Hausdorff space and $\mu \in M(\Omega)$. Further suppose that ϕ is a complex-valued Borel measurable function such that $|\phi| \le 1$ $|\mu|$-ae. Then there is a sequence $\{f_n\}_{n=1}^\infty$ of continuous functions such that $\|f_n\|_\Omega \le 1$ for each n and $\int_\Omega |f_n - \phi| \, d|\mu| \to 0$ as $n \to \infty$.

PROOF By applying Exercise 13.60 on page 483, we can choose a sequence $\{\phi_n\}_{n=1}^{\infty}$ of Borel measurable simple functions such that $|\phi_n| \leq 1$ $|\mu|$-ae for all $n \in \mathcal{N}$ and $\lim_{n \to \infty} \phi_n = \phi$ $|\mu|$-ae. Applying the dominated convergence theorem, we get that

$$\lim_{n \to \infty} \int_{\Omega} |\phi_n - \phi| \, d|\mu| = 0. \tag{13.31}$$

Let $n \in \mathcal{N}$. We can write $\phi_n = \sum_{k=1}^{m} \alpha_k \chi_{E_k}$, where $|\alpha_k| \leq 1$ for each k and the E_ks are pairwise disjoint Borel sets whose union is Ω. Using the regularity of μ, we can find compact sets $F_k \subset E_k$ such that $|\mu|(E_k \setminus F_k) < 1/nm$ for $k = 1, 2, \ldots, m$.

For each k, we can write $\alpha_k = |\alpha_k| e^{i\theta_k}$, where $\theta_k \in [0, 2\pi)$. If $x \in F_k$, define $u_0(x) = |\alpha_k|$ and $v_0(x) = \theta_k$. Since the F_ks are pairwise disjoint and closed, the functions u_0 and v_0 are well-defined and continuous on $\bigcup_{k=1}^{m} F_k$ and, furthermore, $|u_0| \leq 1$.

By Tietze's extension theorem (page 389), we can extend u_0 and v_0 to continuous real-valued functions u and v on all of Ω such that $|u| \leq 1$. Let $f_n = u e^{iv}$. Then $f_n = \phi_n$ on $\bigcup_{k=1}^{m} F_k$ and $\|f_n\|_{\Omega} \leq 1$. Moreover,

$$\int_{\Omega} |\phi_n - f_n| \, d|\mu| = \sum_{k=1}^{m} \int_{E_k} |\alpha_k - f_n| \, d|\mu|$$

$$\leq \sum_{k=1}^{m} \int_{E_k \setminus F_k} (|\alpha_k| + |f_n|) \, d|\mu| \tag{13.32}$$

$$\leq \sum_{k=1}^{m} 2|\mu|(E_k \setminus F_k) \leq \sum_{k=1}^{m} 2/mn = 2/n.$$

It follows from (13.31) and (13.32) that $\lim_{n \to \infty} \int_{\Omega} |f_n - \phi| \, d|\mu| = 0$. ∎

□ □ □ THEOREM 13.15 Riesz Representation Theorem

Let Ω be a compact Hausdorff space. Then $\ell \in C(\Omega)^*$ if and only if there exists a $\mu \in M(\Omega)$ such that

$$\ell(f) = \int_{\Omega} f \, d\mu, \qquad f \in C(\Omega). \tag{13.33}$$

Furthermore, the measure μ is unique and satisfies

$$\|\ell\|_* = |\mu|(\Omega). \tag{13.34}$$

PROOF In the penultimate paragraph before Definition 13.12 (see page 485), we showed that each $\mu \in M(\Omega)$ induces a bounded linear functional on $C(\Omega)$ via the relation (13.33).

Conversely, suppose that $\ell \in C(\Omega)^*$. Define

$$\ell_{\mathrm{re}}(f) = \frac{1}{2}(\ell(f) + \overline{\ell(\overline{f})}) \qquad \text{and} \qquad \ell_{\mathrm{im}}(f) = \frac{1}{2i}(\ell(f) - \overline{\ell(\overline{f})}).$$

Then ℓ_{re} and ℓ_{im} satisfy $\ell = \ell_{\mathrm{re}} + i\ell_{\mathrm{im}}$ and the hypotheses of Theorem 13.13(b) on page 486. Therefore, by the Riesz-Markov theorem, there are measures $\mu_1, \mu_2, \mu_3, \mu_4 \in M_+(\Omega)$ such that

$$\ell_{\mathrm{re}}(f) = \int_\Omega f \, d\mu_1 - \int_\Omega f \, d\mu_2 \qquad \text{and} \qquad \ell_{\mathrm{im}}(f) = \int_\Omega f \, d\mu_3 - \int_\Omega f \, d\mu_4$$

for all $f \in C(\Omega)$. Thus, the measure $\mu = \mu_1 - \mu_2 + i(\mu_3 - \mu_4)$ belongs to $M(\Omega)$ and satisfies (13.33).

To verify (13.34), we note first that

$$\|\ell\|_* = \sup\left\{\left|\int_\Omega f \, d\mu\right| : \|f\|_\Omega \le 1\right\} \le \sup\{\|f\|_\Omega |\mu|(\Omega) : \|f\|_\Omega \le 1\} = |\mu|(\Omega).$$

To prove the reverse inequality, we first apply Exercise 9.53 on page 330 to obtain a Borel measurable complex-valued function ϕ such that $|\phi| = 1$ $|\mu|$-ae and $\int_\Omega v \, d\mu = \int_\Omega v\phi \, d|\mu|$ for all $v \in \mathcal{L}^1(|\mu|)$. Now applying Lemma 13.1 to $\overline{\phi}$, we choose a sequence $\{f_n\}_{n=1}^\infty$ of continuous functions such that $\|f_n\|_\Omega \le 1$ and $\int_\Omega |f_n - \overline{\phi}| \, d|\mu| \to 0$. We have

$$\left|\int_\Omega f_n \, d\mu - |\mu|(\Omega)\right| = \left|\int_\Omega \phi(f_n - \overline{\phi}) \, d|\mu|\right| \le \int_\Omega |f_n - \overline{\phi}| \, d|\mu|.$$

It follows that $|\mu|(\Omega) \le \|\ell\|_*$ and, hence, (13.34) holds. The proof of uniqueness is left to the reader as Exercise 13.84. ∎

The Case Ω Locally Compact

Next we extend Theorem 13.15 to locally compact, noncompact Hausdorff spaces. In this case, we work with $C_0(\Omega)$ rather than $C(\Omega)$ because $\| \ \|_\Omega$ is no longer a norm on $C(\Omega)$.

□ □ □ **THEOREM 13.16 Riesz Representation Theorem**

Let Ω be a locally compact, noncompact Hausdorff space. Then $\ell \in C_0(\Omega)^*$ if and only if there exists a $\mu \in M(\Omega)$ such that

$$\ell(f) = \int_\Omega f \, d\mu, \qquad f \in C_0(\Omega). \tag{13.35}$$

Furthermore, the measure μ is unique and satisfies $\|\ell\|_* = |\mu|(\Omega)$.

PROOF   Let $\ell \in C_0(\Omega)^*$. We prove the existence of the measure μ that satisfies (13.35), but leave the proofs of the remainder of the assertions to the reader as Exercise 13.86.

Let $\Omega^* = \Omega \cup \{\omega\}$ be the one-point compactification of Ω, as described in Theorem 11.12 on page 414. Define the function L on $C(\Omega^*)$ by $L(g) = \ell(g_{|\Omega} - g(\omega))$. Clearly L is linear. That it is also bounded, follows from

$$|L(g)| = |\ell(g_{|\Omega} - g(\omega))| \le \|\ell\|_* \|g_{|\Omega} - g(\omega)\|_\Omega \le 2\|\ell\|_* \|g\|_{\Omega^*}.$$

Hence, by Theorem 13.15, there is a measure $\mu^* \in M(\Omega^*)$ such that

$$L(g) = \int_{\Omega^*} g \, d\mu^*, \qquad g \in C(\Omega^*).$$

Letting $\mu = \mu^*_{|\mathcal{B}(\Omega)}$, we obtain

$$L(g) = \int_\Omega g \, d\mu + g(\omega)\mu^*(\{\omega\}), \qquad g \in C(\Omega^*). \tag{13.36}$$

Now let $f \in C_0(\Omega)$. By defining $f^*(x) = f(x)$ for $x \in \Omega$ and $f^*(\omega) = 0$, we can extend f to a function $f^* \in C(\Omega^*)$ with the same norm; indeed, $C_0(\Omega)$ is the collection of restrictions to Ω of functions in $C(\Omega^*)$ that vanish at ω. We have by (13.36) that $\ell(f) = L(f^*) = \int_\Omega f \, d\mu$. The regularity of μ follows from Exercise 13.85.   ∎

Two simple but instructive illustrations of Theorem 13.16 are provided in Example 13.17. In the next chapter, we will see more elaborate applications of the results of this section.

EXAMPLE 13.17   *Illustrates Theorem 13.16*

a) When it is given the discrete topology, the set of positive integers $\mathcal{N}$ becomes a locally compact space. $C_0(\mathcal{N})$ is simply the collection of all sequences $\{a_n\}_{n=1}^\infty$ of complex numbers such that $\lim_{n \to \infty} a_n = 0$. Applying Exercise 13.73 on page 490, we can identify $M(\mathcal{N})$ with $\ell^1(\mathcal{N})$ and, consequently, we can write $C_0(\mathcal{N})^* = \ell^1(\mathcal{N})$. It follows from Theorem 13.16 that each bounded linear functional ℓ on $C_0(\mathcal{N})$ is of the form $\ell(a) = \sum_{n=1}^\infty a_n b_n$ for some $b \in \ell^1(\mathcal{N})$ and, furthermore, that $\|\ell\|_* = \sum_{n=1}^\infty |b_n|$.

b) Let Ω be a locally compact Hausdorff space and let $x_0 \in \Omega$. Define the function ℓ on $C_0(\Omega)$ by $\ell(f) = f(x_0)$. Clearly, $\ell \in C_0(\Omega)^*$ and $\|\ell\|_* \le 1$. Since $f(x_0) = \int_\Omega f \, d\delta_{x_0}$, it follows from the uniqueness part of Theorem 13.16 that $\mu = \delta_{x_0}$. Moreover, $\|\ell\|_* = |\delta_{x_0}|(\Omega) = \delta_{x_0}(\Omega) = 1$.   □

Exercises for Section 13.6

13.84   Verify the uniqueness assertion in Theorem 13.15 on page 492.

13.85   Let Ω be a locally compact, noncompact Hausdorff space and $\Omega^* = \Omega \cup \{\infty\}$ its one point compactification.
a) Show that $\mathcal{B}(\Omega) \subset \mathcal{B}(\Omega^*)$.

b) Show that $\mu \in M(\Omega)$ if and only if there exists $\mu^* \in M(\Omega^*)$ such that $\mu^*(B) = \mu(B)$ for all $B \in \mathcal{B}(\Omega)$.

13.86 Verify the assertions in Theorem 13.16 (page 493) that we did not prove.

13.87 Refer to Exercises 11.33 and 11.34 on page 410. Let Ω be a compact Hausdorff space, g a lower-semicontinuous function on Ω, and $\mu \in M_+(\Omega)$. Prove that

$$\int_\Omega g \, d\mu = \sup \left\{ \int_\Omega f \, d\mu : f \in C(\Omega) \text{ and } f \le g \right\}.$$

13.88 Let Ω and Λ be compact Hausdorff spaces, $\mu \in M(\Omega)$, and $G: \Omega \to \Lambda$ be continuous.
a) Show that there is a $\nu \in M(\Lambda)$ such that $\int_\Lambda f \, d\nu = \int_\Omega f \circ G \, d\mu$ for all $f \in C(\Lambda)$.
b) Verify that $\nu = \mu \circ G^{-1}$, the measure induced by μ and G.

13.89 Define the linear functional ℓ on $C([0,1] \times [0,1])$ by $\ell(f) = \int_0^1 f(x,x) \, dx$. Describe explicitly the measure μ that satisfies $\ell(f) = \int_{[0,1] \times [0,1]} f \, d\mu$. *Hint:* Refer to Exercise 13.88.

13.90 In Exercise 6.64 on page 221, we defined the convolution product of two nonnegative σ-finite Borel measures on $\mathcal{R}$. An alternative definition that holds for any two (complex) Borel measures on $\mathcal{R}$ is given as follows. For $\mu, \nu \in M(\mathcal{R})$, define the convolution product of μ and ν to be the unique measure $\mu * \nu \in M(\mathcal{R})$ that satisfies

$$\int_\mathcal{R} f \, d\mu * \nu = \int_\mathcal{R} \int_\mathcal{R} f(x+y) \, d\mu(x) \, d\nu(y), \qquad f \in C_0(\mathcal{R}).$$

Show that for $\mu, \nu \in M_+(\mathcal{R})$, this definition agrees with the one presented in Exercise 6.64(d) on page 221.

13.91 Refer to Exercise 13.90. For $\mu \in M(\mathcal{R})$, find $\mu * \delta_0$.

13.92 Let Ω be a locally compact Hausdorff space and $\nu \in M_+(\Omega)$. Denote by $\mathcal{AC}(\nu)$ the collection of measures in $M(\Omega)$ that are absolutely continuous with respect to ν. Prove that $\mathcal{AC}(\nu)$ is a closed subspace of $M(\Omega)$.

13.93 Refer to Exercise 13.92. Show that $\mathcal{L}^1(\nu)$ is isometrically isomorphic to $\mathcal{AC}(\nu)$ via the correspondence $f \to \nu_f$, where $\nu_f(B) = \int_B f \, d\nu$.

Stefan Banach
(1892–1945)

Stefan Banach was born in Kraków, Austria-Hungary (now Poland), on March 30, 1892, to Stefan Greczek and Katarzyna Banach, both of whom were natives of the Podhale region. Unusually, his surname was that of his mother instead of his father (although he received his father's given name).

Banach attended primary school in Kraków. He began his secondary education in 1902 at the Henryk Sienkiewicz Gymnasium No. 4, also in Kraków, where he became known as a prodigy. Although Banach attended the University of Lwów, he was awarded his doctorate in mathematics in 1919 under the unusual circumstance of not completing a university education. In his 1920 dissertation, published in *Fundamenta mathematicae* in 1922, he defined what is now called a Banach space.

A professor at the University of Lwów from 1927, Banach was also a member of the Polish and Ukrainian Academies of Science. With his friend Hugo Steinhaus, one of the renowned mathematicians of the age, he founded the journal *Studia Mathematica*.

Through his writings and through his students, many of whom, like Stanisław Mazur, Władysław Orlicz, Juliusz Schauder, and Stanisław Ulam, became famous researchers, Banach exerted tremendous influence on mathematics. In his classic monograph, "Theorie des opérations linéaires" (Theory of linear operators), he laid the foundations of modern functional analysis.

Banach also made important contributions to the theory of measure and integration, to orthogonal series, and to general topology. Besides *Banach space*, mathematical concepts named after him include the Banach-Tarski paradox, the Hahn-Banach theorem, the Banach-Steinhaus theorem, and the Banach-Mazur game.

During the three-year Nazi occupation of Lwów from 1941–1944, Banach was forced to work in a German infectious disease institute where his health deteriorated. In January 1945, he was diagnosed with lung cancer and died in Lwów on August 31, 1945.

14

Normed Spaces and Locally Convex Spaces

In this chapter, we will develop the basic theory of both normed spaces and locally convex spaces. Included is an examination of the following topics: the Hahn-Banach theorem, linear operators on Banach spaces, compact self-adjoint operators on Hilbert spaces, topological linear spaces, weak and weak* topologies, and compact convex sets.

14.1 THE HAHN-BANACH THEOREM

We begin by introducing notation for a normed space that is suggested by the duality theory of Hilbert spaces — each bounded linear functional ℓ on a Hilbert space $\mathcal{H}$ is of the form $\ell(x) = \langle x, y \rangle$ for some $y \in \mathcal{H}$. (See Theorem 13.8 on page 473.)

Let $(\Omega, \| \ \|)$ be a normed space. For $x \in \Omega$ and $x^* \in \Omega^*$, we define

$$\langle x, x^* \rangle = x^*(x).$$

And when $A \subset \Omega$, we let

$$A^\perp = \{\, x^* \in \Omega^* : \langle x, x^* \rangle = 0 \text{ for all } x \in A \,\}.$$

As the reader is asked to verify in Exercise 14.1, $A^\perp$ is a closed linear subspace of Ω^*.

The notation we have just introduced and Theorem 13.3 on page 466 suggest the following conjecture:

Let K be a closed linear subspace of the normed space Ω and let $x \in \Omega$. Then there is an $x_0^* \in K^\perp$ such that $\|x_0^*\|_* \leq 1$ and

$$\rho(x, K) = \inf\{ \|x - y\| : y \in K \}$$
$$= \sup\{ |\langle x, x^* \rangle| : x^* \in K^\perp \text{ and } \|x^*\|_* \leq 1 \}$$
$$= \langle x, x_0^* \rangle.$$

Verifying this conjecture depends on being able to extend a linear functional on $\operatorname{span}(\{x\} \cup K)$ to all of Ω without increasing its norm. And that extension requires a fundamental result that is the main topic of this section—the Hahn-Banach theorem.

□ □ □ **THEOREM 14.1 Hahn-Banach Theorem**

Let V be a linear space with real scalars and p a real-valued function on V such that

$$p(u + v) \leq p(u) + p(v), \qquad u, v \in V$$

and

$$p(\alpha v) = \alpha p(v), \qquad v \in V, \ \alpha \geq 0.$$

Suppose V_0 is a linear subspace of V and ℓ_0 is a linear functional on V_0 such that

$$\ell_0(v) \leq p(v), \qquad v \in V_0.$$

Then there exists a linear functional ℓ on V such that $\ell(v) = \ell_0(v)$ for each $v \in V_0$ and $\ell(u) \leq p(u)$ for each $u \in V$.

PROOF If $V_0 = V$, there is nothing to prove. So, assume V_0 is a proper subspace of V. We begin by enlarging only slightly the domain of the functional ℓ_0. Let $v_1 \in V_0^c$ and consider the linear subspace

$$V_1 = \{ \alpha v_1 + v : \alpha \in \mathcal{R}, \ v \in V_0 \}.$$

To define a linear functional ℓ_1 on V_1 that agrees with ℓ_0 on V_0 and satisfies

$$\ell_1(\alpha v_1 + v) \leq p(\alpha v_1 + v), \ \alpha \in \mathcal{R}, \ v \in V_0, \tag{14.1}$$

it suffices to assign a value β to $\ell_1(v_1)$ such that

$$\alpha\beta \leq p(\alpha v_1 + v) - \ell_0(v), \ \alpha \in \mathcal{R}, \ v \in V_0. \tag{14.2}$$

Indeed, if we can determine such a β, then the mapping $\ell_1 : V_1 \to \mathcal{R}$, defined by $\ell_1(\alpha v_1 + v) = \alpha\beta + \ell_0(v)$, will give the required extension, as the reader can easily verify.

If $\alpha = 0$, then, by hypothesis, (14.2) holds for any choice of β. If $\alpha > 0$, then (14.2) holds if and only if

$$\beta \leq \alpha^{-1} p(\alpha v_1 + v) - \alpha^{-1} \ell_0(v) = p(v_1 + \alpha^{-1} v) - \ell_0(\alpha^{-1} v)$$

for each $v \in V_0$.

As v varies over all of V_0, so does $\alpha^{-1}v$. Hence (14.2) holds for $\alpha > 0$ if and only if

$$\beta \leq \inf\{\, p(v_1 + u) - \ell_0(u) : u \in V_0 \,\}.$$

Similarly, if $\alpha < 0$, then (14.2) holds if and only if

$$-p(-v_1 - \alpha^{-1}v) + \ell_0(-\alpha^{-1}v) = \alpha^{-1}p(\alpha v_1 + v) - \alpha^{-1}\ell_0(v) \leq \beta$$

for each $v \in V_0$. Hence (14.2) holds for $\alpha < 0$ if and only if

$$\sup\{\, -p(-v_1 + w) + \ell_0(w) : w \in V_0 \,\} \leq \beta.$$

It follows that we can choose a suitable value for $\ell_1(v_1)$ if

$$\sup\{\, -p(-v_1 + w) + l_0(w) : w \in V_0 \,\} \leq \inf\{\, p(v_1 + u) - \ell_0(u) : u \in V_0 \,\}. \quad (14.3)$$

We will now verify (14.3). For $u, w \in V_0$ we have

$$\ell_0(u) + \ell_0(w) = \ell_0(u + w) \leq p(u + v_1 - v_1 + w) \leq p(u + v_1) + p(-v_1 + w).$$

Thus,

$$-p(-v_1 + w) + \ell_0(w) \leq p(v_1 + u) - \ell_0(u).$$

Since u and w are arbitrary members of V_0, it follows that (14.3) is valid.[†]

Consider the collection $\mathcal{E}$ of pairs of the form (L, W), where W is a linear subspace of V that contains V_0, and L is a linear functional on W that agrees with ℓ_0 on V_0 and satisfies $L(w) \leq p(w)$ for each $w \in W$. We define an order relation $\prec$ on $\mathcal{E}$ by $(L_1, W_1) \prec (L_2, W_2)$ if $W_1 \subset W_2$ and L_2 agrees with L_1 on W_1.

As the reader is asked to verify in Exercise 14.2, $\prec$ is a partial ordering and each chain in $\mathcal{E}$ has an upper bound. It follows from Zorn's lemma (page 15) that $\mathcal{E}$ has a maximal element (ℓ_ω, V_ω) with respect to the ordering $\prec$.

To complete the proof of the theorem, it suffices to show that $V_\omega = V$. Suppose to the contrary that $V_\omega^c \neq \emptyset$ and let $v_\omega \in V_\omega^c$. Then we can apply the argument used in the beginning of the proof with V_ω replacing V_0, ℓ_ω replacing ℓ_0, and v_ω replacing v_1 to obtain a linear functional $\hat{\ell}_\omega$ on $\mathrm{span}(V_\omega \cup \{v_\omega\})$ such that $(\hat{\ell}_\omega, \mathrm{span}(V_\omega \cup \{v_\omega\})) \in \mathcal{E}$ and

$$(\ell_\omega, V_\omega) \prec (\hat{\ell}_\omega, \mathrm{span}(V_\omega \cup \{v_\omega\})).$$

Thus, we have reached a contradiction to the maximality of (ℓ_ω, V_ω). It follows that $V_\omega = V$. ∎

[†] Having succeeded in finding a suitable extension of the linear functional ℓ_0 to the subspace $V_1 = \mathrm{span}(V_0 \cup \{v_1\})$, we could now proceed inductively to find a sequence of subspaces of the form $V_n = \mathrm{span}(V_0 \cup \{v_1, v_2, \ldots, v_n\})$ and corresponding linear functionals ℓ_n that extend ℓ_0 and satisfy $\ell_n(u) \leq p(u)$, in the hope that the V_ns would exhaust V. That this approach cannot work in general can be seen by considering a space V that is not the span of a countable set. Nevertheless, as the following argument shows, this idea becomes effective if we replace the inductive procedure by "transfinite induction" based on Zorn's lemma.

EXAMPLE 14.1   *Illustrates the Hahn-Banach Theorem*

Let $\ell_r^\infty(\mathcal{N})$ or, more briefly, ℓ_r^∞, denote the real linear space consisting of all bounded sequences of real numbers and set

$$E = \{\, x \in \ell_r^\infty : \lim_{n \to \infty} x_n \text{ exists} \,\}.$$

Consider the linear functional L_0 on E defined by $L_0(x) = \lim_{n \to \infty} x_n$. We will use the Hahn-Banach theorem to extend L_0 to all of ℓ_r^∞.

Define $p: \ell_r^\infty \to \mathcal{R}$ by $p(x) = \limsup_{n \to \infty} n^{-1} \sum_{k=1}^n x_k$. Then p satisfies the hypotheses of the Hahn-Banach theorem and, according to part (b) of Exercise 2.35 on page 46, $p(x) = L_0(x)$ for all $x \in E$. It follows that there is a linear functional L on ℓ_r^∞ that agrees with L_0 on E and satisfies $L(x) \le p(x)$ for all $x \in \ell_r^\infty$. The functional L shares with L_0 the following properties:

$$\liminf_{n \to \infty} x_n \le L(x) \le \limsup_{n \to \infty} x_n \quad \text{and} \quad L(x) = L(x^{(1)}), \tag{14.4}$$

where $x_n^{(1)} = x_{n+1}$ for each n. (See Exercise 14.3.) For this reason L can be thought of as assigning a "generalized limit" to any bounded sequence of real numbers. Linear functionals on ℓ_r^∞ that satisfy (14.4) are called **Banach limits**.   □

Next we present a version of the Hahn-Banach theorem that is valid in the case of complex scalars.

□ □ □   **THEOREM 14.2 Hahn-Banach Theorem (Complex Version)**

Let V be a linear space with complex scalars and p a real-valued function on V such that

$$p(u + v) \le p(u) + p(v), \qquad u, v \in V$$

and

$$p(\alpha v) = |\alpha| p(v), \qquad v \in V, \ \alpha \in \mathbf{C}.$$

Suppose V_0 is a linear subspace of V and ℓ_0 is a linear functional on V_0 such that

$$|\ell_0(v)| \le p(v), \qquad v \in V_0.$$

Then there exists a linear functional ℓ on V such that $\ell(v) = \ell_0(v)$ for each $v \in V_0$ and $|\ell(u)| \le p(u)$ for each $u \in V$.

PROOF   Because $\mathcal{R} \subset \mathbf{C}$, it follows that V and V_0 are also linear spaces with respect to $\mathcal{R}$. Furthermore, $\Re\ell_0$ is linear with respect to real scalars and satisfies $\Re\ell_0(v) \le |\ell_0(v)| \le p(v)$ for each $v \in V_0$. Hence, we can apply Theorem 14.1 to obtain a function ℓ_r on V that is linear with respect to real scalars and satisfies $\ell_r(v) = \Re\ell_0(v)$ for each $v \in V_0$ and $\ell_r(u) \le p(u)$ for each $u \in V$.

We note that $\Re\ell_0(iv) = -\Im\ell_0(v)$ and, so, $\ell_0(v) = \Re\ell_0(v) - i\Re\ell_0(iv)$. Thus, the function ℓ defined on V by $\ell(u) = \ell_r(u) - i\ell_r(iu)$ agrees with ℓ_0 on V_0 and it is easy to see that ℓ is linear with respect to complex scalars.

We will complete the proof by showing that $|\ell(u)| \le p(u)$ for each $u \in V$. Choosing a complex number α such that $|\alpha| = 1$ and $\alpha\ell(u) = |\ell(u)|$, we have

$$|\ell(u)| = \ell(\alpha u) = \ell_r(\alpha u) \le p(\alpha u) = |\alpha| p(u) = p(u),$$

as required.   ■

□ □ □   **COROLLARY 14.1**

Let S be a linear subspace of the normed space Ω and let $\ell \in S^*$. Then there exists an $L \in \Omega^*$ such that $L_{|S} = \ell$ and $\|L\|_* = \|\ell\|_*$.

PROOF   Apply Theorem 14.2 with $p(x) = \|\ell\|_* \|x\|$.   ■

Armed with Theorem 14.2, we can now prove the conjecture made on page 498 in the paragraph prior to the statement of the Hahn-Banach theorem.

□ □ □   **THEOREM 14.3**

Let K be a closed linear subspace of the normed space Ω and let $x \in \Omega$. Then there is an $x_0^* \in K^\perp$ such that $\|x_0^*\|_* \leq 1$ and

$$\rho(x, K) = \inf\{ \|x - y\| : y \in K \}$$
$$= \sup\{ |\langle x, x^* \rangle| : x^* \in K^\perp \text{ and } \|x^*\|_* \leq 1 \}$$
$$= \langle x, x_0^* \rangle.$$

PROOF   We will prove the theorem under the additional assumption that $x \in K^c$, but will leave to the reader the case $x \in K$ as Exercise 14.4.

Let $V_0 = \{ \alpha x + y : \alpha \in \mathbb{C}, \ y \in K \}$ and ℓ_0 the linear functional defined on V_0 by $\ell_0(\alpha x + y) = \alpha \rho(x, K)$. Then $|\ell_0(\alpha x + y)| \leq \|\alpha x + y\|$. Thus, ℓ_0 satisfies the hypotheses of Theorem 14.2 with $p = \| \ \|$. It follows that there is a linear functional x_0^* on Ω that satisfies

$$x_0^*(\alpha x + y) = \alpha \rho(x, K), \qquad \alpha \in \mathbb{C}, \ y \in K \tag{14.5}$$

and

$$|x_0^*(z)| \leq \|z\|, \qquad z \in \Omega. \tag{14.6}$$

The relations (14.5) and (14.6) show $x_0^* \in K^\perp$, $\|x_0^*\|_* \leq 1$, and $x_0^*(x) = \rho(x, K)$. Finally, let $x^* \in K^\perp$ with $\|x^*\|_* \leq 1$. If $y \in K$, then we have

$$|\langle x, x^* \rangle| = |x^*(x) - x^*(y)| \leq \|x^*\|_* \|x - y\| \leq \|x - y\|.$$

Taking the infimum over $y \in K$, we get that $|\langle x, x^* \rangle| \leq \rho(x, K)$. Therefore, we have

$$\sup\{ |\langle x, x^* \rangle| : x^* \in K^\perp \text{ and } \|x^*\|_* \leq 1 \} \leq \rho(x, K) = \langle x, x_0^* \rangle.$$

The reverse inequality is trivial.   ■

As our first application of Theorem 14.3, we use it to prove an attractive and useful symmetry between the norms $\| \ \|$ and $\| \ \|_*$.

□ □ □   **COROLLARY 14.2**

Let $(\Omega, \| \ \|)$ be a normed space, $x_0 \in \Omega$, and $x_0^* \in \Omega^*$. Then

$$\|x_0\| = \sup\{ |\langle x_0, x^* \rangle| : \|x^*\|_* \leq 1 \} \tag{14.7}$$

and

$$\|x_0^*\|_* = \sup\{ |\langle x, x_0^* \rangle| : \|x\| \leq 1 \}, \tag{14.8}$$

and, moreover, the supremum in (14.7) is attained. In particular, if $x_0 \neq 0$, then there is an $x^* \in \Omega^*$ with norm 1 such that $\|x_0\| = \langle x_0, x^* \rangle$.

PROOF   Equation (14.8) is just the definition of the norm of the bounded linear functional x_0^*. To obtain (14.7), we apply Theorem 14.3 with $x = x_0$ and $K = \{0\}$ to get an $x_1^* \in \Omega^*$ with the properties $\|x_1^*\|_* \leq 1$ and $x_1^*(x_0) = \rho(x_0, \{0\}) = \|x_0\|$. It follows that

$$\|x_0\| \leq \sup\{\,|\langle x_0, x^*\rangle| : \|x^*\|_* \leq 1\,\}. \tag{14.9}$$

On the other hand, we have $|\langle x_0, x^*\rangle| \leq \|x_0\|$ whenever $\|x^*\|_* \leq 1$. Consequently, the reverse of (14.9) is also valid. Finally, since $x_1^*(x_0/\|x_0\|) = 1$, we see that $\|x_1^*\|_* = 1$.   ∎

EXAMPLE 14.2   *Illustrates Theorem 14.3*

Let Ω be a compact Hausdorff space and F a nonempty closed subset of Ω. We consider the linear subspace of $C(\Omega)$ defined by

$$\mathcal{I}_F = \{\, f \in C(\Omega) : f(x) = 0 \text{ for each } x \in F \,\}.$$

Theorem 13.15 (page 492) shows that we can identify $C(\Omega)^*$ with the space $M(\Omega)$ of regular Borel measures on Ω.

We claim that

$$\mathcal{I}_F^\perp = \{\, \mu \in M(\Omega) : |\mu|(F^c) = 0 \,\}. \tag{14.10}$$

That $\mathcal{I}_F^\perp$ contains the right-hand side of (14.10) follows easily from the equation $\int_\Omega f\, d\mu = \int_F f\, d\mu + \int_{F^c} f\, d\mu$.

To show that $\mathcal{I}_F^\perp$ is contained in the right-hand side of (14.10), it suffices to prove that $\mu(E) = 0$ whenever $\mu \in \mathcal{I}_F^\perp$ and E is a closed subset of F^c. Let $\epsilon > 0$ be given. By the regularity of the measure μ, there is an open set O such that $E \subset O$ and $|\mu|(O \setminus E) < \epsilon$. Furthermore, we can choose O so that $O \subset F^c$.

Next, we apply Urysohn's lemma to obtain a continuous function g such that $0 \leq g \leq 1$, $g(E) = \{1\}$, and $g(O^c) = \{0\}$. Then $g \in \mathcal{I}_F$ and, hence,

$$0 = \int_\Omega g\, d\mu = \mu(E) + \int_{O \setminus E} g\, d\mu.$$

It follows that

$$|\mu(E)| = \left|\int_{O \setminus E} g\, d\mu\right| \leq |\mu|(O \setminus E) < \epsilon.$$

Thus, $\mu(E) = 0$.

Having established (14.10), we can now use Theorem 14.3 to assert that for each $f \in C(\Omega)$, there is a regular Borel measure μ_0 such that $|\mu_0|(\Omega) \leq 1$, $|\mu_0|(F^c) = 0$, and

$$\rho(f, \mathcal{I}_F) = \sup\left\{\left|\int_\Omega f\, d\mu\right| : |\mu|(\Omega) \leq 1 \text{ and } |\mu|(F^c) = 0\right\} = \int_\Omega f\, d\mu_0$$

for some measure μ_0. (See Exercise 14.12.)

It is also clear from (14.10) that $\mathcal{I}_F^\perp$ can be identified with $M(F)$ since it consists of measures in $M(\Omega)$ that vanish on Borel subsets of F^c.   □

EXAMPLE 14.3   *Illustrates Corollary 14.2*

a) Consider the Banach space $\mathcal{L}^p(\mu)$, where $1 < p < \infty$, and let q be such that $1/p + 1/q = 1$. Applying the Riesz representation theorem for $\mathcal{L}^p$-spaces (Theorem 13.12 on page 480) and Corollary 14.2, we obtain the following fact: If $f \in \mathcal{L}^p(\mu)$, then there is a function $g \in \mathcal{L}^q(\mu)$ such that $\|g\|_q \leq 1$ and $\|f\|_p = \int_\Omega fg \, d\mu$.

b) We can use Corollary 14.2 to show that Theorem 13.12 cannot be extended to the case $p = \infty$. To do that, consider the space ℓ^∞ and define $\{x_n\}_{n=1}^\infty \in \ell^\infty$ by $x_n = n/(n+1)$. Note that $\|x\|_\infty = 1$. If we could extend Theorem 13.12 to hold for $p = \infty$, then we could apply Corollary 14.2 to find a $y \in \ell^1$ such that

$$\sum_{n=1}^\infty |y_n| = \|y\|_1 = 1 = \|x\|_\infty = \sum_{n=1}^\infty x_n y_n = \sum_{n=1}^\infty \frac{n}{n+1} y_n.$$

However, this result is impossible because the quantity on the right-hand side is in modulus strictly less than the quantity on the left. □

More on Duality

As we mentioned at the beginning of this section, the notation $\langle x, x^* \rangle$ is motivated by Hilbert space theory. Specifically, if $\mathcal{H}$ is a Hilbert space, then each bounded linear functional on $\mathcal{H}$ is of the form $\ell_y(x) = \langle x, y \rangle$ for some $y \in \mathcal{H}$ and, moreover, $\|\ell_y\|_* = \|y\|$.

The mapping $j: \mathcal{H} \to \mathcal{H}^*$ defined by $j(y) = \ell_y$ is, therefore, onto and isometric. It is also almost, but not quite, linear. Indeed, we have $j(y + z) = j(y) + j(z)$ and $j(\alpha y) = \overline{\alpha} j(y)$. Due to these properties of j, we can use it to identify $\mathcal{H}^*$ and $\mathcal{H}$. When this identification is made, it becomes clear that the notation

$$A^\perp = \{ x^* \in \Omega^* : \langle x, x^* \rangle = 0 \text{ for all } x \in A \}$$

has the same meaning for Hilbert spaces as the notation for the orthogonal complement of a set introduced in Definition 13.7 on page 465.

When Ω is a normed space, but not a Hilbert space, and $A \subset \Omega$, then $A^\perp$ no longer resides in Ω but, rather, in Ω^*. Consequently, whereas the notation $(A^\perp)^\perp$ makes sense in Hilbert spaces, it does not in general normed spaces.

One way to generalize the notation for a double orthogonal complement from inner product spaces to general normed spaces is to define, for $B \subset \Omega^*$,

$$^\perp B = \{ x \in \Omega : \langle x, x^* \rangle = 0 \text{ for all } x^* \in B \}.$$

Then an analogue of the notation $(A^\perp)^\perp$ that makes sense in arbitrary normed spaces is $^\perp(A^\perp)$.

□ □ □   **THEOREM 14.4**

Let Ω be a normed space, $A \subset \Omega$, and $B \subset \Omega^*$. Then the following properties hold:

a) $^\perp B$ *is a closed linear subspace of* Ω.

b) $^\perp(A^\perp) = \overline{\operatorname{span} A}$.

c) *If* $A^\perp = \{0\}$, *then* $\operatorname{span} A$ *is dense in* Ω.

PROOF   The proof of (a) is left to the reader as Exercise 14.11 and (c) follows immediately from (b). Thus, we move on to the verification of (b). From (a), we know that $^\perp(A^\perp)$ is a closed linear subspace of Ω. Therefore, because $A \subset {}^\perp(A^\perp)$, we have $\overline{\operatorname{span} A} \subset {}^\perp(A^\perp)$.

To establish the reverse inclusion, let $x \in {}^\perp(A^\perp)$. By applying Theorem 14.3 with $K = \overline{\operatorname{span} A}$, there is an $x^* \in \overline{\operatorname{span} A}^\perp$ such that $\rho(x, \overline{\operatorname{span} A}) = \langle x, x^* \rangle$. Because $\overline{\operatorname{span} A}^\perp \subset A^\perp$, it follows that $\langle x, x^* \rangle = 0$. Therefore, $x \in \overline{\operatorname{span} A}$.   ∎

EXAMPLE 14.4   *Illustrates Theorem 14.4*

In this example we make use of the following result from the theory of analytic functions: Let g be a function that is analytic on a connected open subset O of $\mathbb{C}$. If there is a sequence $\{b_n\}_{n=1}^\infty$ of distinct elements of O such that $b = \lim_{n\to\infty} b_n$ exists and belongs to O and $g(b_n) = 0$ for each $n \in \mathcal{N}$, then g vanishes identically on all of O.

Consider the space $C([a,b])$, where $a < b$ and $0 \notin [a,b]$. For $\alpha \in \mathbb{C}$, define $f_\alpha(x) = 1/(x - \alpha)$ and let $\{a_n\}_{n=1}^\infty$ be a sequence of distinct elements of $\mathbb{C} \setminus [a,b]$ with $\lim_{n\to\infty} a_n = 0$. We will prove that $\operatorname{span}\{\, f_{a_n} : n \in \mathcal{N} \,\}$ is dense in $C([a,b])$ by showing that $\{\, f_{a_n} : n \in \mathcal{N} \,\}^\perp = \{0\}$.

Suppose $\mu \in \{\, f_{a_n} : n \in \mathcal{N} \,\}^\perp$ and let

$$g(z) = \int_{[a,b]} \frac{1}{x - z}\, d\mu(x).$$

Because g is analytic on the open connected set $\mathbb{C} \setminus [a,b]$ and $g(a_n) = 0$ for $n \in \mathcal{N}$, we conclude that g vanishes identically on $\mathbb{C} \setminus [a,b]$. This fact, in turn, implies that the nth derivative

$$g^{(n)}(z) = \int_{[a,b]} \frac{n!}{(x - z)^{n+1}}\, d\mu(x)$$

vanishes identically on $\mathbb{C} \setminus [a,b]$. In particular, then, $\int_{[a,b]} x^{-n}\, d\mu(x) = 0$ for each $n \in \mathcal{N}$.

From the complex version of the Stone-Weierstrass theorem (Theorem 12.10 on page 449), it follows that the span of the functions $1, x^{-1}, x^{-2}, \ldots$ is dense in $C([a,b])$. From this fact, we can conclude that $\int_{[a,b]} x^{-1} f(x)\, d\mu(x) = 0$ for each $f \in C([a,b])$ and, consequently, $\mu = 0$. We leave the details for the reader as Exercise 14.13.   ❑

Exercises for Section 14.1

Note: A ★ denotes an exercise that will be subsequently referenced.

14.1   Let A be a subset of the normed space Ω. Show that $A^\perp$ is a closed linear subspace of Ω^*.

14.2   In the proof of Theorem 14.1, verify that $\prec$ is a partial ordering and that each chain of $\mathcal{E}$ has a $\prec$-upper bound.

14.3   Verify (14.4) in Example 14.1. *Hint:* See Exercise 2.35 on page 46.

14.4   Prove Theorem 14.3 in the case where $x \in K$.

★14.5 For a normed space Ω, let Ω^{**} denote the dual space of the dual space, that is, $(\Omega^*)^*$. Let $J\colon\Omega \to \Omega^{**}$ be defined by $\langle x^*, J(x)\rangle = \langle x, x^*\rangle$. Show that J is a linear isometry.

14.6 Show that the mapping J defined in Exercise 14.5 is onto if Ω is a Hilbert space.

14.7 Show that the mapping J defined in Exercise 14.5 is onto if $\Omega = \mathcal{L}^p(\mu)$, $1 < p < \infty$.

14.8 Show that the mapping J defined in Exercise 14.5 is not onto if $\Omega = \ell^1$.

14.9 Prove that Ω is a separable space if Ω^* is separable.

14.10 Show that the converse of the assertion of Exercise 14.9 is false.

14.11 Prove part (a) of Theorem 14.4 on page 503.

14.12 Show that in Example 14.2 (page 502), the measure μ_0 can be chosen to be $\alpha\delta_x$ for some $x \in F$ and some constant α with $|\alpha| = 1$.

14.13 Refer to Example 14.4.
a) Show that the function g is analytic on $\mathbb{C} \setminus [a, b]$.
b) Verify the formula $g^{(n)}(z) = \int_{[a,b]} n!/(x-z)^{n+1}\, d\mu(x)$.
c) Prove that $\operatorname{span}\{\, x^{-(n-1)} : n \in \mathcal{N}\,\}$ is dense in $C([a,b])$.
d) Use part (c) and the fact that $\int_{[a,b]} x^{-n}\, d\mu(x) = 0$ for each $n \in \mathcal{N}$ to show that $\int_{[a,b]} x^{-1} f(x)\, d\mu(x) = 0$ for all $f \in C([a,b])$.
e) Use part (d) to conclude that $\int_B x^{-1}\, d\mu(x) = 0$ for each $B \in \mathcal{B}([a,b])$.
f) Deduce from part (e) that $\mu = 0$.

14.2 LINEAR OPERATORS ON BANACH SPACES

Linear operators (mappings) appear in most branches of mathematics, but especially in analysis, where differentiation and integration are basic processes. In this section, we present some important general results about continuous (bounded) linear operators on Banach spaces.

The Open Mapping Theorem

Let T be a continuous function from a topological space Ω to a topological space Λ. Then, by definition, the inverse image of an open set under T is open, that is, $T^{-1}(U)$ is open in Ω whenever U is open in Λ.

On the other hand, as the reader is asked to verify in Exercise 14.14, it is not generally true that the image of an open set under T is open, that is, that $T(O)$ is open in Λ whenever O is open in Ω. However, if Ω and Λ are Banach spaces and T is linear, continuous, and onto, then our next theorem, called the **open mapping theorem,** shows that T does carry open sets to open sets.

We will employ the following notation. Suppose A and B are subsets of a linear space and α is a scalar. Then

$$A + B = \{\, x + y : x \in A,\ y \in B\,\}$$

and

$$\alpha A = \{\, \alpha x : x \in A\,\}.$$

Furthermore, we define $A - B = A + (-1)B$ and $x + B = \{x\} + B$.

Note that, in a normed space Ω, we have

$$B_r(x) = x + rB_1(0)$$

for all $x \in \Omega$ and $r > 0$.

□ □ □ **THEOREM 14.5 Open Mapping Theorem**

Let Ω and Λ be Banach spaces and $T: \Omega \to \Lambda$ be continuous, linear, and onto. Then $T(O)$ is open in Λ whenever O is open in Ω.

PROOF We claim that it suffices to prove that there is an $\epsilon > 0$ such that

$$B_\epsilon(0) \subset T(B_1(0)). \tag{14.11}$$

Indeed, suppose that (14.11) holds. If $O \subset \Omega$ is open, then for each $x \in O$, there is a $\delta > 0$ such that $x + \delta B_1(0) = B_\delta(x) \subset O$. Hence, by (14.11),

$$B_{\delta\epsilon}(T(x)) = T(x) + \delta\epsilon B_1(0) = T(x) + \delta B_\epsilon(0)$$
$$\subset T(x) + \delta T(B_1(0)) = T(x + \delta B_1(0)) \subset T(O).$$

As $T(x)$ is an arbitrary point of $T(O)$, it follows that $T(O)$ is open.

Therefore, to establish the theorem, we need only show that (14.11) is valid. As a first step, we will prove that there is an $\epsilon > 0$ such that

$$B_\epsilon(0) \subset \overline{T(B_{1/2}(0))}. \tag{14.12}$$

Since $\Omega = \bigcup_{n=1}^\infty nB_{1/2}(0)$, we have

$$\Lambda = T(\Omega) = \bigcup_{n=1}^\infty nT(B_{1/2}(0)).$$

It follows from the Baire category theorem (Theorem 12.1 on page 426) that there exist $m \in \mathcal{N}$, $y_0 \in \Lambda$, and $\alpha > 0$ such that $B_\alpha(y_0) \subset m\overline{T(B_{1/2}(0))}$ and, hence, that $y_0/m + B_{\alpha/m}(0) \subset \overline{T(B_{1/2}(0))}$. Moreover, because $y_0/m \in \overline{T(B_{1/2}(0))}$, we also have that $-y_0/m \in (-1)\overline{T(B_{1/2}(0))} = \overline{T(B_{1/2}(0))}$. Thus,

$$B_{\alpha/2m}(0) = (1/2)B_{\alpha/m}(0) = -y_0/2m + (1/2)(y_0/m + B_{\alpha/m}(0))$$
$$\subset (1/2)\overline{T(B_{1/2}(0))} + (1/2)\overline{T(B_{1/2}(0))}$$
$$\subset \overline{T((1/2)B_{1/2}(0)) + (1/2)B_{1/2}(0)} \subset \overline{T(B_{1/2}(0))}.$$

(See Exercise 14.15.) We have now verified (14.12) with $\epsilon = \alpha/2m$.

Finally, we will derive (14.11) from (14.12). Let $y \in B_\epsilon(0)$. By (14.12) we can find an $x_1 \in B_{1/2}(0)$ such that $\|y - T(x_1)\| < \epsilon/2$. As

$$y - T(x_1) \in (1/2)B_\epsilon(0) \subset (1/2)\overline{T(B_{1/2}(0))} = \overline{T(B_{1/4}(0))},$$

it follows that there exists $x_2 \in B_{1/4}(0)$ such that $\|y - T(x_1) - T(x_2)\| < \epsilon/4$. Proceeding in this fashion and using mathematical induction, we obtain a sequence $\{x_n\}_{n=1}^\infty$ of elements of Ω such that

$$\|x_n\| < 2^{-n} \quad \text{and} \quad \left\|y - \sum_{j=1}^n T(x_j)\right\| < \epsilon/2^n.$$

Because Ω is a Banach space, Proposition 13.3 on page 457 implies that the series $\sum_{n=1}^\infty x_n$ converges, say, to x. Noting that

$$\|x\| \le \sum_{n=1}^\infty \|x_n\| < \sum_{n=1}^\infty 2^{-n} = 1,$$

we conclude that $x \in B_1(0)$. By the continuity of T, we have $T(x) = y$ and, so, $y \in T(B_1(0))$. We have shown that $B_\epsilon(0) \subset T(B_1(0))$. ∎

□ □ □ **COROLLARY 14.3**

Suppose that the operator T satisfies the hypotheses of the open mapping theorem and is also one-to-one. Then the following properties hold:
a) T^{-1} is linear and continuous.
b) $\|T\|^{-1} \leq \|T^{-1}\|$.
c) We have

$$\|T^{-1}\|^{-1}\|x\| \leq \|T(x)\| \leq \|T\|\|x\|$$

for all $x \in \Omega$.

PROOF It is easy to see that the inverse function $T^{-1}: \Lambda \to \Omega$ is linear. Moreover, as $(T^{-1})^{-1}(O) = T(O)$ is open in Λ whenever O is open in Ω, T^{-1} is continuous. Thus, (a) holds.

By (a), we know that $\|T^{-1}\| < \infty$. For each $y \in \Lambda$, we have

$$\|y\| = \|T(T^{-1}(y))\| \leq \|T\|\|T^{-1}(y)\| \leq \|T\|\|T^{-1}\|\|y\|,$$

from which (b) follows immediately.

To obtain (c), we need only prove the first inequality. We observe that, for each $x \in \Omega$,

$$\|x\| = \|T^{-1}(T(x))\| \leq \|T^{-1}\|\|T(x)\|,$$

as required. ∎

EXAMPLE 14.5 *Illustrates Corollary 14.3*

It follows from Exercises 4.26 and 12.64 on pages 128 and 451, respectively, that the Laplace transform L defined by

$$L(f)(s) = \int_0^\infty e^{-sx} f(x)\, dx, \qquad s \geq 0,$$

is a one-to-one linear operator from the Banach space $\left(\mathcal{L}^1([0,\infty)), \|\ \|_1\right)$ into the Banach space $\left(C_0([0,\infty)), \|\ \|_{[0,\infty)}\right)$. Because

$$|L(f)(s)| \leq \int_0^\infty e^{-sx}|f(x)|\, dx \leq \int_0^\infty |f(x)|\, dx,$$

we have $\|L(f)\|_{[0,\infty)} \leq \|f\|_1$. It follows that L is continuous and $\|L\| \leq 1$.

We will use Corollary 14.3 to show that L is not onto, that is, there are functions in $C_0([0,\infty))$ that are not Laplace transforms of functions in $\mathcal{L}^1([0,\infty))$. Suppose to the contrary that L is onto. By Corollary 14.3(c), there is a positive constant c such that

$$c\|f\|_1 \leq \|L(f)\|_{[0,\infty)} \tag{14.13}$$

for all $f \in \mathcal{L}^1([0,\infty))$. Let $n \in \mathcal{N}$ and define

$$f_n = \chi_{[n,n+1)} - \chi_{[n+1,n+2)}.$$

Then $\|f_n\|_1 = 2$. Moreover,

$$L(f_n)(s) = e^{-ns}(e^{-2s} - 2e^{-s} + 1)/s = se^{-ns}((e^{-s} - 1)/s)^2.$$

It is easy to check that the maximum of $((e^{-s} - 1)/s)^2$ is 1 and that the maximum of se^{-ns} is $1/ne$. Thus, using (14.13), we obtain $2c \leq 1/ne$. As $n \in \mathcal{N}$ was chosen arbitrarily, we conclude that $c = 0$, which is a contradiction. □

One can get a better appreciation for the power of the open mapping theorem by trying to explicitly construct a function in $C_0([0, \infty))$ that is not a Laplace transform of a function in $\mathcal{L}^1([0, \infty))$. We leave that to the reader and, instead, present another interesting corollary of the open mapping theorem.

□ □ □ **COROLLARY 14.4**

Let Ω be a linear space. Suppose that $\|\ \|$ and $\|\ \|_0$ are norms on Ω such that $(\Omega, \|\ \|)$ and $(\Omega, \|\ \|_0)$ are Banach spaces. If there is a positive constant α such that

$$\|x\| \leq \alpha\|x\|_0, \qquad x \in \Omega, \tag{14.14}$$

then there is a positive constant β such that

$$\|x\|_0 \leq \beta\|x\|, \qquad x \in \Omega. \tag{14.15}$$

PROOF   From (14.14), we see that the identity map

$$I : (\Omega, \|\ \|_0) \to (\Omega, \|\ \|)$$

is continuous. The relation (14.15) now follows from Corollary 14.3. ■

We note that Corollary 14.4 shows that the topology of a Banach space $(\Omega, \|\ \|)$ cannot be strictly weaker than the topology induced by a norm $\|\ \|_0$ for which $(\Omega, \|\ \|_0)$ is complete. And, as the reader is asked to verify in the exercises, Corollary 14.4 can also be used to prove that any finite dimensional normed space is isomorphic to either $\mathbb{C}^n$ or $\mathcal{R}^n$ for some $n \in \mathcal{N}$.

The Closed Graph Theorem

Another application of Corollary 14.4 provides a condition equivalent to continuity for linear operators on Banach spaces. Let Ω and Λ be normed spaces and $T : \Omega \to \Lambda$ a linear operator. Then, as we know from Exercise 10.53 on page 378, T is continuous if and only if it satisfies

$$\lim_{n \to \infty} x_n = x \quad \Rightarrow \quad \lim_{n \to \infty} T(x_n) = T(x).$$

A weaker condition on T is that

$$\lim_{n \to \infty} x_n = x \quad \text{and} \quad \lim_{n \to \infty} T(x_n) = y \quad \Rightarrow \quad y = T(x). \tag{14.16}$$

A linear operator that satisfies (14.16) is said to be **closed.** We use that terminology because (14.16) is equivalent to the condition that the graph of T — the set $\{(x, T(x)) : x \in \Omega\}$ — is a closed subset of the product space $\Omega \times \Lambda$.

In our next theorem, called the **closed graph theorem,** we will prove that, when Ω and Λ are Banach spaces, not only is being closed a necessary condition for a linear operator to be continuous, but it is also sufficient. First, however, we present an example showing that, in general, a closed linear operator need not be continuous.

EXAMPLE 14.6 *A Discontinuous Closed Linear Operator*

Let $\Omega = \{ f : f' \in C([0,1]) \}$ and $\Lambda = C([0,1])$, both equipped with the sup-norm, and let $D: \Omega \to \Lambda$ be the differentiation operator, $D(f) = f'$. Suppose that $\{f_n\}_{n=1}^{\infty}$ is a sequence of functions in Ω such that $f_n \to f$ and $D(f_n) \to g$. It follows from these assumptions and the second fundamental theorem of calculus that $f(x) = f(0) + \int_0^x g(t)\, dt$. This, in turn, implies that $D(f) = g$ and hence that D is closed. But D is not continuous, as can be seen by considering the sequence $f_n(x) = \sin(nx)/n$. □

□ □ □ **THEOREM 14.6 Closed Graph Theorem**

Let Ω and Λ be Banach spaces and $T: \Omega \to \Lambda$ a linear operator. If T is closed, then it is continuous.

PROOF We define a second norm on Ω by

$$\|x\|_0 = \|x\| + \|T(x)\|$$

and show that $(\Omega, \| \ \|_0)$ is a Banach space. Let $\{x_n\}_{n=1}^{\infty}$ be a Cauchy sequence with respect to the norm $\| \ \|_0$. Because $\|x_n - x_m\| \le \|x_n - x_m\|_0$ and $\|T(x_n) - T(x_m)\| \le \|x_n - x_m\|_0$, it follows that $\{x_n\}_{n=1}^{\infty}$ and $\{T(x_n)\}_{n=1}^{\infty}$ are Cauchy sequences in $(\Omega, \| \ \|)$ and Λ, respectively. Consequently, $x = \lim x_n$ and $y = \lim T(x_n)$ both exist. Because, by assumption, T is closed, we have $y = T(x)$. It follows that

$$\lim_{n \to \infty} \|x_n - x\|_0 = \lim_{n \to \infty} \|x_n - x\| + \lim_{n \to \infty} \|T(x_n) - T(x)\| = 0.$$

Thus, $(\Omega, \| \ \|_0)$ is a Banach space.

Since $\|x\| \le \|x\|_0$ for all $x \in \Omega$, it follows from Corollary 14.4 that there is a positive constant β such that $\|T(x)\| \le \|x\|_0 \le \beta\|x\|$ for all $x \subset \Omega$. Hence, T is continuous. ■

The Uniform Boundedness Principle for Linear Operators

A look at the proof of the open mapping theorem reveals that it is a consequence of the Baire category theorem. We conclude this section with another application of the Baire category theorem to the theory of linear operators on Banach spaces.

□ □ □ **THEOREM 14.7 Uniform Boundedness Principle for Linear Operators**

Suppose that $\mathcal{F}$ is a collection of continuous linear operators from a Banach space Ω into a normed space Λ. If $\sup\{ \|T(x)\| : T \in \mathcal{F} \} < \infty$ for each $x \in \Omega$, then $\sup\{ \|T\| : T \in \mathcal{F} \} < \infty$.

PROOF By Theorem 12.2 on page 428, there exists an $x_0 \in \Omega$ and a $\delta > 0$ such that

$$M = \sup\{ \|T(x)\| : T \in \mathcal{F}, \ \|x - x_0\| \le \delta \} < \infty.$$

If $\|u\| \le 1$, then

$$\|T(u)\| \le \delta^{-1}(\|T(x_0 + \delta u)\| + \|T(x_0)\|) \le 2M\delta^{-1}.$$

It follows that $\|T\| \le 2M\delta^{-1}$ for each $T \in \mathcal{F}$. ■

Exercises for Section 14.2

14.14   Let Ω and Λ be Banach spaces.

  a) Provide an example of a linear and continuous mapping from Ω into Λ that does not take open sets to open sets.

  b) Provide an example of a continuous mapping from Ω onto Λ that does not take open sets to open sets.

14.15   Show that if A and B are subsets of a normed linear space and α is a nonzero scalar, then $\overline{\alpha A} = \alpha \overline{A}$ and $\overline{x + A} = x + \overline{A}$, $\overline{A} + \overline{B} \subset \overline{A + B}$, and $\overline{A} - \overline{B} \subset \overline{A - B}$.

14.16   Provide an example of a continuous linear operator $T : \Omega \to \Lambda$, where Ω and Λ are normed spaces, such that T is one-to-one and onto but T^{-1} is not continuous.

Exercises 14.17 and 14.18 show that every finite dimensional normed space is isomorphic to either $\mathbb{C}^n$ or $\mathcal{R}^n$ for some $n \in \mathcal{N}$. Recall that the dimension of a linear space is the number of elements in a Hamel basis.

14.17   Let Ω be a finite dimensional normed space.

  a) Show that the dimension of Ω^* is at most the dimension of Ω. *Hint:* What is the dimension of the space of all linear functionals on Ω?

  b) Use Exercise 14.5 on page 505 and Proposition 13.2 on page 456 to deduce that Ω is complete.

14.18   Let Ω be a finite dimensional normed space and $\{x_1, x_2, \ldots, x_n\}$ a Hamel basis for Ω. Recall that each $x \in \Omega$ can be written uniquely in the form $x = \sum_{j=1}^{n} a_j(x) x_j$, where the $a_j(x)$s are scalars. For $x \in \Omega$, define

$$p(x) = \left(\sum_{j=1}^{n} |a_j(x)|^2 \right)^{1/2}.$$

  a) Show that p is a norm on Ω.

  b) Show that there exist positive constants α and β such that for each $x \in \Omega$, we have $\alpha p(x) \leq \|x\| \leq \beta p(x)$.

  c) Deduce that Ω is isomorphic to $\mathbb{C}^n$ or to $\mathcal{R}^n$ in the case of complex or real scalars, respectively.

14.19   Let Ω be a normed linear space such that $\overline{B}_1(0)$ is compact. Prove that Ω is finite dimensional. *Hint:* Find $x_1^*, x_2^*, \ldots, x_n^* \in \Omega^*$ such that

$$\{ x : \|x\| = 1 \} \subset \bigcup_{j=1}^{n} \{ x : |\langle x, x_j^* \rangle| > 1/2 \}$$

and consider the mapping $L(x) = (\langle x, x_1^* \rangle, \langle x, x_2^* \rangle, \ldots, \langle x, x_n^* \rangle)$.

*In Exercises 14.20–14.22, we consider projections of a normed space. Let Ω be a normed space. A linear operator $P : \Omega \to \Omega$ is called a **projection** if both the range of P and $P^{-1}(\{0\})$ are closed and $P \circ P = P$. Exercise 13.26 on page 467 shows that orthogonal projections, as defined in Section 13.2 on page 465, are projections in the sense defined here.*

14.20   Show that if Ω is a Banach space, then all projections of Ω are continuous.

14.21   Show that if P is a projection on Ω, then $\|x - P(x)\| \geq \rho(x, \text{range } P)$ for each $x \in \Omega$.

14.22   Let K be a finite dimensional subspace of Ω. Show that there is a projection of Ω with range equal to K.

Exercises 14.23–14.27 elaborate on Example 13.11 on page 470. Let S_n denote the linear operator defined on $C([-\pi, \pi])$ by $S_n(f) = s_n$.

14.23 Refer to the definition of a *projection* given in the paragraph prior to Exercise 14.20. Show that S_n is a projection of $C([-\pi, \pi])$ with range $\mathcal{U}_n = \text{span}\{ e_k : -n \le k \le n \}$.

14.24 Show that

$$\|S_n(f) - f\|_{[-\pi,\pi]} \ge \rho(f, \mathcal{U}_n) \ge |f(\pi) - f(-\pi)|/2.$$

Deduce that $f(\pi) = f(-\pi)$ is a necessary condition for the uniform convergence of the Fourier series $\sum_{k=-n}^{n} \hat{f}(k)e^{ikx}$ to the function $f(x)$.

14.25 Show that for each $f \in C([-\pi, \pi])$,

$$S_n(f)(x) = \frac{1}{2\pi} \int_{-\pi}^{\pi} f(t) \frac{\sin((n + 1/2)(x - t))}{\sin((x - t)/2)}\, dt.$$

14.26 Let $C_p([-\pi, \pi]) = \{ f \in C([-\pi, \pi]) : f(\pi) = f(-\pi) \}$. Show that

$$\sup \left\{ |S_n(f)(x)| : f \in C_p([-\pi, \pi]),\ \|f\|_{[-\pi,\pi]} \le 1 \right\}$$

$$= \frac{1}{2\pi} \int_{-\pi}^{\pi} \left| \frac{\sin((n + 1/2)(x - t))}{\sin((x - t)/2)} \right|\, dt.$$

★14.27 Show that there is a function in $C_p([-\pi, \pi])$ whose Fourier series diverges at some point.

14.3 COMPACT SELF-ADJOINT OPERATORS

We first note that, *throughout this section, $T: \mathcal{H} \to \mathcal{H}$ denotes a continuous linear operator on a Hilbert space $\mathcal{H}$, not $\{0\}$, with inner product $\langle\ ,\ \rangle$.*

From Exercise 14.28, for fixed $y \in \mathcal{H}$, the mapping $x \to \langle T(x), y \rangle$ defines a continuous linear functional on $\mathcal{H}$. Hence, by Theorem 13.8 on page 473, there is a (unique) element of $\mathcal{H}$, which we denote $T^*(y)$, such that $\langle T(x), y \rangle = \langle x, T^*(y) \rangle$ for all $x \in \mathcal{H}$.

We leave it to the reader as Exercise 14.29 to show that T^* is a continuous linear operator on $\mathcal{H}$; it is called the **adjoint** of T. If $T^* = T$, then T is said to be a **self-adjoint operator.**

EXAMPLE 14.7 *Self-Adjoint Operators*

a) Clearly, the identity operator, defined by $I(x) = x$ for $x \in \mathcal{H}$, is self-adjoint.

b) If T is self-adjoint and $\lambda \in \mathcal{R}$, then λT is self-adjoint. See Exercise 14.30.

c) Let $T: \mathbb{C}^n \to \mathbb{C}^n$ be a (continuous) linear operator and let $[t_{ij}]_{i,j}$ be the $n \times n$ matrix that corresponds to T with respect to some basis. Then T is self-adjoint if and only if $[t_{ij}]_{i,j}$ is a Hermitian matrix, that is, if and only if $\overline{t_{ji}} = t_{ij}$ for all $1 \le i, j \le n$. See Exercise 14.31.

d) Define $T: \ell^2(\mathcal{N}) \to \ell^2(\mathcal{N})$ by $T(f)(n) = \lambda_n f(n)$, where $\{\lambda_n\}_{n=1}^{\infty}$ is a bounded sequence of real numbers. Then T is self-adjoint. See Exercise 14.32. □

Let $\mathcal{H}$ be a Hilbert space of finite dimension n (i.e., $\mathcal{H}$ has a finite basis that consists of n elements) and let $T: \mathcal{H} \to \mathcal{H}$ be self-adjoint. An important result of linear algebra is that there are real numbers $\lambda_1, \lambda_2, \ldots, \lambda_n$ and an orthonormal basis $\{u_1, u_2, \ldots, u_n\}$ for $\mathcal{H}$ such that

$$T(x) = \sum_{j=1}^{n} \lambda_j \langle x, u_j \rangle u_j, \qquad x \in \mathcal{H}. \tag{14.17}$$

In this section, we will derive a version of (14.17) that holds as well for infinite dimensional Hilbert spaces. We begin with the following definition.

DEFINITION 14.1   Compact Linear Operator

A linear operator $T: \mathcal{H} \to \mathcal{H}$ is said to be a **compact linear operator** if the closure of $\{T(x) : \|x\| \leq 1\}$ is compact.

The following proposition provides an equivalent condition for compactness of a linear operator. Its proof is left to the reader as Exercise 14.35.

□ □ □   **PROPOSITION 14.1**

A linear operator is compact if and only if for every bounded sequence $\{x_n\}_{n=1}^{\infty}$ of $\mathcal{H}$, the sequence $\{T(x_n)\}_{n=1}^{\infty}$ has a convergent subsequence.

It is easy to see that a compact linear operator is continuous (bounded). Here are some examples of compact linear operators.

EXAMPLE 14.8   *Illustrates Definition 14.1*

a) If $\mathcal{H}$ is finite dimensional, then every linear operator $T: \mathcal{H} \to \mathcal{H}$ is compact. This result follows from Exercise 14.18(c), the Heine-Borel theorem, and the fact that T is continuous. Alternatively, it follows from Exercise 14.36.

b) Let $u_1, u_2, \ldots, u_n, w_1, w_2, \ldots, w_m \in \mathcal{H}$. Then the linear operator $T: \mathcal{H} \to \mathcal{H}$ defined by

$$T(x) = \sum_{i=1}^{n} \sum_{j=1}^{m} \langle x, u_i \rangle w_j$$

is compact. This result follows from Exercise 14.36 and the fact that the range of T is a finite-dimensional space.

c) Suppose that $\{T_n\}_{n=1}^{\infty}$ is a sequence of compact linear operators on $\mathcal{H}$ such that $\lim_{n\to\infty} \|T - T_n\| = 0$. Then T is a compact linear operator. See Exercise 14.37.   □

EXAMPLE 14.9   *Illustrates Definition 14.1*

Let $(\Omega, \mathcal{A}, \mu)$ be a measure space and let $K \in \mathcal{L}^2(\mu \times \mu)$. For $f \in \mathcal{L}^2(\mu)$, define $T_K(f)$ on Ω by

$$T_K(f)(x) = \int_{\Omega} K(x, y) f(y) \, d\mu(y), \qquad x \in \Omega.$$

Then, as you are asked to verify in Exercise 14.38(a), T_K is a continuous linear operator that maps $\mathcal{L}^2(\mu)$ into itself. We claim that T_K is compact.

For $n = 1, 2, \ldots$, let $K_n \in \mathcal{L}^2(\mu \times \mu)$ be a simple function of the form

$$K_n(x, y) = \sum_{i=1}^{N_n} \sum_{j=1}^{M_n} k_{ij}^{(n)} \chi_{A_{in}}(x) \chi_{B_{jn}}(y),$$

chosen so that $\|K - K_n\|_2 < 1/n$. (See Exercise 14.38(b)). We have that

$$T_{K_n}(f)(x) = \int_\Omega K_n(x, y) f(y) d\mu(y) = \sum_{i=1}^{N_n} \sum_{j=1}^{M_n} k_{ij}^{(n)} \langle f, \chi_{B_{jn}} \rangle \chi_{A_{in}}(x).$$

Referring to Example 14.8(b), we see that each T_{K_n} is a compact linear operator on $\mathcal{L}^2(\mu)$.

Applying Cauchy's inequality, we find that, for each $x \in \Omega$,

$$|T_K(f)(x) - T_{K_n}(f)(x)|^2 \leq \|f\|_2^2 \int_\Omega |K(x, y) - K_n(x, y)|^2 \, d\mu(y) \quad \mu\text{-ae.}$$

Using the preceding inequality and Fubini's theorem, we get that

$$\|T_K(f) - T_{K_n}(f)\|_2^2 = \int_\Omega |T_K(f)(x) - T_{K_n}(f)(x)|^2 \, d\mu(x)$$

$$\leq \|f\|_2^2 \int_\Omega \left(\int_\Omega |K(x, y) - K_n(x, y)|^2 \, d\mu(y) \right) d\mu(x)$$

$$= \|f\|_2^2 \int_{\Omega \times \Omega} |K(x, y) - K_n(x, y)|^2 \, d(\mu \times \mu)(x, y)$$

$$= \|f\|_2^2 \|K - K_n\|_2^2 \leq \frac{1}{n^2} \|f\|_2^2.$$

Therefore, we see that $\|T_K - T_{K_n}\| \leq 1/n$. Referring now to Example 14.8(c), we conclude that T_K is a compact linear operator. □

Note: An important special case of the preceding example is the **Volterra operator** on $\mathcal{L}^2([0, 1])$ given by $V(f)(x) = \int_0^x f(y) \, dy$.

Eigenvalues, Eigenvectors, and Eigenspaces

In linear algebra, where finite-dimensional linear spaces are studied, the concepts of *eigenvalues* (aka *characteristic values, proper values, latent values*), *eigenvectors* (aka *characteristic vectors, proper vectors, latent vectors*), and *eigenspaces* play a significant role. These concepts also apply to infinite-dimensional linear spaces, where they are important as well.

DEFINITION 14.2 Eigenvalue, Eigenvector, and Eigenspace

A number λ is called an **eigenvalue** of the linear operator T if there is a nonzero element x such that $T(x) = \lambda x$. Any such x is called an **eigenvector** of T with eigenvalue λ. The subspace $(T - \lambda I)^{-1}(\{0\})$, consisting of 0 and all eigenvectors of T with eigenvalue λ, is called the **eigenspace** of λ.

Note: The set of all eigenvalues of a compact linear operator T is called its **spectrum**.

EXAMPLE 14.10   *Illustrates Definition 14.2*

a) Define $T: \ell^2(\mathcal{N}) \to \ell^2(\mathcal{N})$ by $T(f)(n) = \lambda_n f(n)$, where $\{\lambda_n\}_{n=1}^{\infty}$ is a bounded sequence of complex numbers. Let $e_n \in \ell^2(\mathcal{N})$ be such that $e_n(n) = 1$ and $e_n(j) = 0$ for $j \neq n$. Then each λ_n is an eigenvalue of T, and each e_n is an eigenvector of T with eigenvalue λ_n.

b) We claim that the aforementioned Volterra operator V has no eigenvalues. First we show that 0 is not an eigenvalue of V. Indeed, if $V(f) = 0$, then, by Lemma 8.2 on page 289, we have that $f = 0$ ae. Next suppose that $\lambda \neq 0$ and that $V(f) = \lambda f$ for some $f \in \mathcal{L}^2([0,1])$. Then

$$\int_0^x f(y)\, dy = \lambda f(x) \text{ ae.}$$

From Proposition 8.2 on page 286, the left side of the preceding display is continuous on $[0,1]$ and, hence, we can take f to be likewise. Applying the first fundamental theorem of calculus for Riemann integration, we find that f is differentiable and $f(x) = \lambda f'(x)$ on $[0,1]$. Consequently, $f(x) = ce^{x/\lambda}$. It follows that $0 = \lambda f(0) = \lambda c$; so, $c = 0$. Hence, $f = 0$. Therefore, λ is not an eigenvalue of V. □

The following proposition provides some essential properties of eigenvalues, eigenvectors, and eigenspaces for self-adjoint and compact operators.

□ □ □   **PROPOSITION 14.2**

Let T be a self-adjoint operator on $\mathcal{H}$. Then the following properties hold:

a) *$\langle T(x), x \rangle$ is real for all $x \in \mathcal{H}$.*

b) *All eigenvalues of T are real.*

c) *If λ_1 and λ_2 are distinct eigenvalues of T and x_1 and x_2 are corresponding eigenvectors, then $\langle x_1, x_2 \rangle = 0$.*

d) *If λ is an eigenvector of T, then $|\lambda| \leq \|T\|$.*

e) *If T is also compact, then the eigenspace corresponding to a nonzero eigenvalue of T is finite dimensional.*

PROOF   a) Let $x \in \mathcal{H}$. Because T is self-adjoint, $\langle T(x), x \rangle = \langle x, T(x) \rangle = \overline{\langle T(x), x \rangle}$. Consequently, $\langle T(x), x \rangle$ is real.

b) Suppose that λ is an eigenvalue of T. Let x be an eigenvector of T with eigenvalue λ. Then

$$\lambda \|x\|^2 = \lambda \langle x, x \rangle = \langle \lambda x, x \rangle = \langle T(x), x \rangle,$$

which is real by part (a). Hence, λ is real.

c) Using the facts that T is self-adjoint and λ_2 is real, we get that

$$\lambda_1 \langle x_1, x_2 \rangle = \langle \lambda_1 x_1, x_2 \rangle = \langle T(x_1), x_2 \rangle = \langle x_1, T(x_2) \rangle$$
$$= \langle x_1, \lambda_2 x_2 \rangle = \overline{\lambda_2} \langle x_1, x_2 \rangle = \lambda_2 \langle x_1, x_2 \rangle.$$

As λ_1 and λ_2 are distinct, it follows that $\langle x_1, x_2 \rangle = 0$.

d) Let x be an eigenvector of T with eigenvalue λ. Then

$$|\lambda|\|x\| = \|\lambda x\| = \|T(x)\| \le \|T\|\|x\|.$$

Therefore, $|\lambda| \le \|T\|$.

e) Let $K = (T - \lambda I)^{-1}(\{0\})$. Because $T - \lambda I$ is continuous and $\{0\}$ is closed, we have that K is closed. We claim that K is finite dimensional. Suppose to the contrary. Then K contains an infinite orthonormal sequence, say, $\{u_1, u_2, \ldots\}$. From Theorem 13.4(b) on page 468, we get that, for $i \ne j$,

$$\|T(u_i) - T(u_j)\|^2 = \|\lambda u_i - \lambda u_j\|^2 = \lambda^2\|u_i - u_j\|^2 = 2\lambda^2 > 0.$$

It follows that $\{T(u_n)\}_{n=1}^{\infty}$ has no convergent subsequence, which, in view of Proposition 14.1 on page 512, contradicts the fact that T is a compact linear operator. Hence, K is finite dimensional. ■

We will also need the following two lemmas. The proof of the second one is left to the reader as Exercise 14.42.

□ □ □ **LEMMA 14.1**

Let T be a compact self-adjoint operator. Then either $\|T\|$ or $-\|T\|$ is an eigenvalue of T.

PROOF If T is identically zero, then $\|T\| = 0$ is an eigenvalue of T. So, we can assume that T is not identically zero.

By Exercise 13.2 on page 457, we can find a sequence $\{x_n\}_{n=1}^{\infty}$ of members of $\mathcal{H}$ such that $\|x_n\| = 1$ for all $n \in \mathcal{N}$ and $\lim_{n \to \infty} \|T(x_n)\| = \|T\|$. Letting T^2 denote the composition of T with itself and using properties of inner product, we get that

$$\|T^2(x_n) - \|T(x_n)\|^2 x_n\|^2 = \|T^2(x_n)\|^2 - 2\|T(x_n)\|^2\Re\langle T^2(x_n), x_n\rangle + \|T(x_n)\|^4.$$

However, because T is self-adjoint,

$$\langle T^2(x_n), x_n\rangle = \langle T(x_n), T(x_n)\rangle = \|T(x_n)\|^2.$$

Thus,

$$\|T^2(x_n) - \|T(x_n)\|^2 x_n\|^2 = \|T^2(x_n)\|^2 - \|T(x_n)\|^4.$$

But,

$$\|T^2(x_n)\| = \|T(T(x_n))\| \le \|T\|\|T(x_n)\|,$$

and, so,

$$\|T^2(x_n) - \|T(x_n)\|^2 x_n\|^2 \le \|T(x_n)\|^2\left(\|T\|^2 - \|T(x_n)\|^2\right).$$

Hence,

$$\lim_{n \to \infty} \|T^2(x_n) - \|T(x_n)\|^2 x_n\| = 0. \tag{14.18}$$

It is easy to verify that T^2 is compact. Hence, by Proposition 14.1 on page 512, there is a subsequence $\{x_{n_k}\}_{k=1}^{\infty}$ of $\{x_n\}_{n=1}^{\infty}$ such that $\lim_{k \to \infty} T^2(x_{n_k}) = v$ for some $v \in \mathcal{H}$. Because

$$\|v - \|T(x_{n_k})\|^2 x_{n_k}\| \leq \|v - T^2(x_{n_k})\| + \|T^2(x_{n_k}) - \|T(x_{n_k})\|^2 x_{n_k}\|,$$

it follows from (14.18) that

$$v = \lim_{k \to \infty} \|T(x_{n_k})\|^2 x_{n_k} = \|T\|^2 \lim_{k \to \infty} x_{n_k}, \tag{14.19}$$

and, consequently,

$$\|v\| = \|T\|^2 \lim_{k \to \infty} \|x_{n_k}\| = \|T\|^2 \cdot 1 = \|T\|^2 > 0.$$

Thus, $v \neq 0$.

Referring again to (14.19), we get that

$$T^2(v) = \|T\|^2 \lim_{k \to \infty} T^2(x_{n_k}) = \|T\|^2 v.$$

Consequently,

$$(T + \|T\|I)((T - \|T\|I)(v)) = T^2(v) - \|T\|^2 v = 0.$$

Let $w = (T - \|T\|I)(v)$. If $w = 0$, then $\|T\|$ is an eigenvalue of T because $v \neq 0$. If $w \neq 0$, then $-\|T\|$ is an eigenvalue of T. ■

□ □ □ **LEMMA 14.2**

Let T be self-adjoint, and let K be a subspace of $\mathcal{H}$ such that $T(K) \subset K$. Then $T(K^{\perp}) \subset K^{\perp}$ and $T_{|K^{\perp}}$ is a self-adjoint continuous linear operator on $K^{\perp}$. Moreover, we have that $\|T_{|K}\| \leq \|T\|$ and $\|T_{|K^{\perp}}\| \leq \|T\|$.

The Spectral Theorem

Let $K_1,\ K_2,\ \ldots,\ K_N$ be subspaces of $\mathcal{H}$ such that $\mathrm{span}\left(\bigcup_{i=1}^{N} K_i\right) = \mathcal{H}$ and that $\langle x, y \rangle = 0$ whenever $x \in K_i$ and $y \in K_j$ with $i \neq j$. In this case, we write

$$\mathcal{H} = K_1 \oplus K_2 \oplus \cdots \oplus K_N.$$

We note that each $x \in \mathcal{H}$ can be expressed uniquely as

$$x = x_1 + x_2 + \cdots + x_N,$$

where $x_i \in K_i$ for $i = 1,\ 2,\ \ldots,\ N$.

We can now establish a generalization of (14.17) on page 512 to any Hilbert space. This result is known as the **spectral theorem.**

□ □ □ **THEOREM 14.8 Spectral Theorem for Compact Self-Adjoint Operators**

Let $\mathcal{H}$ be a Hilbert space and let $T: \mathcal{H} \to \mathcal{H}$ be a compact self-adjoint operator. Then there is a countable orthonormal set $\{u_n\}_n$ such that

$$T(x) = \sum_n \lambda_n \langle x, u_n \rangle u_n, \qquad x \in \mathcal{H}, \qquad (14.20)$$

where the λ_ns are the nonzero (not necessarily distinct) eigenvalues of T.

Note: The spectral theorem is often phrased in the following equivalent way: *The orthogonal complement of the kernel (null space) of T has a countable orthonormal basis that consists of eigenvectors of T; in other words, a compact self-adjoint operator can be unitarily diagonalized.*

PROOF If T is identically zero, then the theorem is trivially true. So, we assume that T is not identically zero. From Lemma 14.1, T has an eigenvalue ξ_1 such that $|\xi_1| = \|T\| > 0$. Let K_1 be the eigenspace of ξ_1. From Proposition 14.2(e) on page 514, K_1 has finite dimension, say, n_1. Let $H_1 = K_1^\perp$. We have that $\mathcal{H} = K_1 \oplus H_1$, $T(K_1) \subset K_1$, and, in view of Lemma 14.2, $T(H_1) \subset H_1$. Let $T_1 = T_{|H_1}$.

If $T_1 = 0$ (i.e., T_1 is identically zero), then we say that the process has terminated. In this case, let $\{u_1, u_2, \ldots, u_{n_1}\}$ be an orthonormal basis for K_1. Each $x \in \mathcal{H}$ can be written in the form $\sum_{j=1}^{n_1} \langle x, u_j \rangle u_j + z$, where $z \in H_1$. Thus,

$$T(x) = \sum_{j=1}^{n_1} \lambda_j \langle x, u_j \rangle u_j,$$

where $\lambda_j = \xi_1$ for $1 \le j \le n_1$, as required.

If $T_1 \ne 0$, then it has an eigenvalue ξ_2 such that $|\xi_2| = \|T_1\| > 0$. Moreover, by Lemma 14.2,

$$|\xi_2| = \|T_1\| \le \|T\| = |\xi_1|.$$

Let K_2 be the eigenspace of ξ_2. From Proposition 14.2(e), K_2 has finite dimension, say, n_2. Let $H_2 = K_2^\perp$. Then

$$\mathcal{H} = K_1 \oplus H_1 = K_1 \oplus K_2 \oplus H_2,$$

$T(K_2) \subset K_2$, and $T(H_2) \subset H_2$. Let $T_2 = T_{|H_2}$.

If $T_2 = 0$ (i.e., T_2 is identically zero), then we say that the process has terminated. In this case, let $\{u_{n_1+1}, u_{n_1+2}, \ldots, u_{n_1+n_2}\}$ be an orthonormal basis for K_2. Each $x \in \mathcal{H}$ can be written in the form $\sum_{j=1}^{n_1+n_2} \langle x, u_j \rangle u_j + z$, where $z \in H_2$. Therefore,

$$T(x) = \sum_{j=1}^{n_1+n_2} \lambda_j \langle x, u_j \rangle u_j,$$

where $\lambda_j = \xi_1$ for $1 \le j \le n_1$ and $\lambda_j = \xi_2$ for $n_1 + 1 \le j \le n_1 + n_2$, as required.

If $T_2 \neq 0$, then it has an eigenvalue ξ_3 such that $|\xi_3| = \|T_2\| > 0$. Moreover, by Lemma 14.2,

$$|\xi_3| = \|T_2\| \leq \|T_1\| = |\xi_2|.$$

Let K_3 be the eigenspace of ξ_3. From Proposition 14.2(e), K_3 has finite dimension, say, n_3. Let $H_3 = K_3^\perp$. Then

$$\mathcal{H} = K_1 \oplus K_2 \oplus H_2 = K_1 \oplus K_2 \oplus K_3 \oplus H_3,$$

$T(K_3) \subset K_3$, and $T(H_3) \subset H_3$. Let $T_3 = T_{|H_3}$.

Continuing in this manner, either the process eventually terminates, in which case the theorem is proved, or it does not eventually terminate. In the latter case, we obtain an infinite sequence of eigenspaces K_1, K_2, ..., and corresponding eigenvalues ξ_1, ξ_2, ..., such that $|\xi_{n+1}| = \|T_n\| > 0$ for all $n \in \mathcal{N}$ and

$$|\xi_1| \geq |\xi_2| \geq \cdots, \tag{14.21}$$

where $T_n = T_{|H_n}$ and $H_n = K_n^\perp$.

We claim that $\lim_{n\to\infty} \xi_n = 0$. Suppose to the contrary. Then, in view of (14.21), there is an $\epsilon > 0$ such that $|\xi_n| \geq \epsilon$ for all $n \in \mathcal{N}$. For each $n \in \mathcal{N}$, we choose a $v_n \in K_n$ with unit norm. Then $\{v_1, v_2, \ldots\}$ is an orthonormal set and

$$\|T(v_m) - T(v_n)\|^2 = \|\xi_m v_m - \xi_n v_n\|^2 = |\xi_m|^2 + |\xi_n|^2 \geq 2\epsilon^2$$

for all $m, n \in \mathcal{N}$ with $m \neq n$. It follows that $\{T(v_n)\}_{n=1}^\infty$ has no convergent subsequence, which contradicts the compactness of T (see Proposition 14.1 on page 512). Hence, $\lim_{n\to\infty} \xi_n = 0$.

Note that $H_1 \supset H_2 \supset \cdots$. Let $H_\infty = \bigcap_{n=1}^\infty H_n$ and let $x \in H_\infty$. Then, for all $n \in \mathcal{N}$,

$$\|T(x)\| = \|T_{|H_n}(x)\| = \|T_n(x)\| \leq \|T_n\|\|x\| = |\xi_{n+1}|\|x\|.$$

Therefore, because $\lim_{n\to\infty} \xi_n = 0$, we have that $T(x) = 0$. Consequently, we see that T is identically zero on H_∞.

If $H_\infty = \{0\}$, let $B_\infty = \emptyset$; otherwise, let B_∞ be an orthonormal basis for H_∞. Moreover, for each $n \in \mathcal{N}$, let B_n be an orthonormal basis for K_n. We claim that $B = \left(\bigcup_{n=1}^\infty B_n\right) \cup B_\infty$ is an orthonormal basis for $\mathcal{H}$. Indeed, suppose that $\langle x, u \rangle = 0$ for all $u \in B$. Then $x \in K_n^\perp = H_n$ for all $n \in \mathcal{N}$ and, so, $x \in H_\infty$. If $H_\infty = \{0\}$, then $x = 0$; otherwise, because B_∞ is an orthonormal basis for H_∞ and $\langle x, u \rangle = 0$ for all $u \in B_\infty$, we again have that $x = 0$. Consequently, we have shown that $\langle x, u \rangle = 0$ for each $u \in B$ implies $x = 0$. Hence, B is an orthonormal basis for $\mathcal{H}$.

Now, as we have seen, each B_n is finite. Consequently, $\bigcup_{n=1}^\infty B_n$ is countable (actually, countably infinite), say, $\{u_1, u_2, \ldots\}$. Then each $x \in \mathcal{H}$ can be written in the form $x = \sum_{n=1}^\infty \langle x, u_n \rangle u_n + z$, where $z \in H_\infty$. Hence,

$$T(x) = \sum_{n=1}^\infty \langle x, u_n \rangle T(u_n) + T(z) = \sum_{n=1}^\infty \langle x, u_n \rangle \lambda_n u_n + 0 = \sum_{n=1}^\infty \lambda_n \langle x, u_n \rangle u_n,$$

where, for each $n \in \mathcal{N}$, λ_n is the eigenvalue for u_n. ∎

From Theorem 13.7(c) on page 473, we know that $\mathcal{H}$ has a countable basis if and only if it is separable. In that case, it follows that the set B_∞ in the proof of Theorem 14.8 is countable. Recalling from that proof that T is identically zero on H_∞, we see that either $H_\infty = \{0\}$ or it is the eigenspace of the eigenvalue 0. In either case, we get the following corollary of Theorem 14.8.

□ □ □ **COROLLARY 14.5**

Suppose that $\mathcal{H}$ is a separable Hilbert space, not $\{0\}$, and let $T:\mathcal{H} \to \mathcal{H}$ be a compact self-adjoint operator. Then $\mathcal{H}$ has a countable orthonormal basis consisting of eigenvectors of T.

Exercises for Section 14.3

14.28 Let $T:\mathcal{H} \to \mathcal{H}$ be a continuous linear operator on a Hilbert space $\mathcal{H}$. Show that, for fixed $y \in \mathcal{H}$, the mapping $x \to \langle T(x), y \rangle$ defines a continuous linear functional on $\mathcal{H}$.

14.29 Prove that T^*, the adjoint of a continuous linear operator T, is a continuous linear operator.

14.30 Show that, if T is self-adjoint and $\lambda \in \mathcal{R}$, then λT is self-adjoint.

14.31 Let $T:\mathbb{C}^n \to \mathbb{C}^n$ be a linear operator and let $[t_{ij}]_{i,j}$ be the $n \times n$ matrix that corresponds to T with respect to some basis. Show that T is self-adjoint if and only if $\overline{t_{ji}} = t_{ij}$ for all $1 \le i, j \le n$, that is, if and only if $[t_{ij}]_{i,j}$ is a Hermitian matrix.

14.32 Suppose that $\{\lambda_n\}_{n=1}^{\infty}$ is a bounded sequence of real numbers. Define $T:\ell^2(\mathcal{N}) \to \ell^2(\mathcal{N})$ by $T(f)(n) = \lambda_n f(n)$. Show that T is self-adjoint.

14.33 Let $\mathcal{H}$ be an n-dimensional Hilbert space, $\{v_1, v_2, \ldots, v_n\}$ an orthonormal basis for $\mathcal{H}$, and $T:\mathcal{H} \to \mathcal{H}$ a linear operator.
   a) Explain why we can write $T(v_i) = \sum_{j=1}^{n} a_{ij} v_j$ for $1 \le i \le n$. Identify each a_{ij} in terms of T and $v_1, v_2, \ldots, v_n$.
   b) Deduce from part (a) that

$$T(x) = \sum_{i=1}^{n} \sum_{j=1}^{n} a_{ij} \langle x, v_i \rangle v_j, \qquad x \in \mathcal{H}.$$

   c) Suppose that $\mathcal{H}$ has an orthonormal basis $\{u_1, u_2, \ldots, u_n\}$ such that $T(u_i) = \lambda_i u_i$ for $1 \le i \le n$, where $\lambda_1, \lambda_2, \ldots, \lambda_n$ are real numbers. Show that (14.17) on page 512 holds.
   d) In part (c), describe the matrix for T relative to the basis $\{u_1, u_2, \ldots, u_n\}$.

14.34 Show that a compact linear operator is continuous.

14.35 Prove Proposition 14.1 on page 512.

14.36 Show that if the linear operator $T:\mathcal{H} \to \mathcal{H}$ has a finite-dimensional range, then it is compact.

14.37 Suppose that $\{T_n\}_{n=1}^{\infty}$ is a sequence of compact linear operators on a Hilbert space $\mathcal{H}$ such that $\lim_{n \to \infty} \|T - T_n\| = 0$. Prove that T is a compact linear operator.

14.38 Let $(\Omega, \mathcal{A}, \mu)$ be a measure space and let $K \in \mathcal{L}^2(\mu \times \mu)$. For $f \in \mathcal{L}^2(\mu)$, define $T_K(f)$ to be the function on Ω given by

$$T_K(f)(x) = \int_{\Omega} K(x,y) f(y) \, d\mu(y), \qquad x \in \Omega.$$

   a) Show that T_K is a continuous linear operator that maps $\mathcal{L}^2(\mu)$ into itself.
   b) Show that for each $n \in \mathcal{N}$, there is a simple function $K_n \in \mathcal{L}^2(\mu \times \mu)$ of the form

$$K_n(x,y) = \sum_{i=1}^{N_n} \sum_{j=1}^{M_n} k_{ij}^{(n)} \chi_{A_{in}}(x) \chi_{B_{jn}}(y)$$

   such that $\|K - K_n\|_2 < 1/n$.
   c) Show that T_K is self-adjoint if and only if $K(y,x) = \overline{K(x,y)}$ for $\mu \times \mu$-almost all $(x,y) \in \Omega \times \Omega$.

14.39   The notion of a compact linear operator makes sense in any normed space. Show that the Volterra operator is compact as a mapping of $C([0,1])$ into itself.

14.40   Show that the eigenspace of an eigenvalue λ is a (linear) subspace of $\mathcal{H}$.

14.41   Show that the linear operator $T\colon \ell^2(\mathcal{N}) \to \ell^2(\mathcal{N})$ in Example 14.10(a) on page 514 is compact if and only if $\lim_{n\to\infty} \lambda_n = 0$.

14.42   Prove Lemma 14.2 on page 516.

14.43   Determine the adjoint of the Volterra operator V.

14.44   Let V be the Volterra operator. Discuss the eigenvalues of $V + V^*$.

14.4   TOPOLOGICAL LINEAR SPACES

Let Ω be a locally compact Hausdorff space. We recall from Section 11.5 (see page 416) that, for $S \subset \Omega$,

$$\rho_S(f,g) = \sup\{\,|f(x) - g(x)| : x \in S\,\}, \qquad f,g \in C(\Omega).$$

And we also recall (see page 417) that the weak topology on $C(\Omega)$ determined by the family of functions $\{\,\rho_K(\cdot,g) : K \text{ compact}, \ g \in C(\Omega)\,\}$ is called the topology of uniform convergence on compact sets and is denoted by $\mathcal{T}(\Omega)$.

As we know, if Ω is compact, then

$$\|f\|_\Omega = \rho_\Omega(f,0) = \sup\{\,|f(x)| : x \in \Omega\,\}, \qquad f \in C(\Omega),$$

is a norm on the linear space $C(\Omega)$ that induces the topology $\mathcal{T}(\Omega)$.

On the other hand, if Ω is not compact, then the topology $\mathcal{T}(\Omega)$ on the linear space $C(\Omega)$ is not induced by a norm. Indeed, suppose to the contrary that $\mathcal{T}(\Omega)$ is induced by a norm $\|\ \|$. Because $\{\,f : \|f\| < 1\,\}$ is an open set that contains 0, it follows from Proposition 10.7 on page 371 that there exist $n \in \mathcal{N}$, compact sets $K_1, K_2, \ldots, K_n$, and positive numbers $\epsilon_1, \epsilon_2, \ldots, \epsilon_n$ such that

$$\bigcap_{j=1}^{n} \{\,f : \rho_{K_j}(f,0) < \epsilon_j\,\} \subset \{\,f : \|f\| < 1\,\}. \tag{14.22}$$

Since Ω is not compact, there is an $x_0 \notin \bigcup_{j=1}^{n} K_j$. Applying Theorem 11.9 on page 411, we choose $g \in C(\Omega)$ such that $g(x_0) = 1$ and $g(x) = 0$ for $x \in \bigcup_{j=1}^{n} K_j$. It now follows that from (14.22) that $|\alpha|\|g\| = \|\alpha g\| < 1$ for each $\alpha \in \mathbb{C}$. On the other hand, because $g \neq 0$, we have $\|g\| \neq 0$. Having reached a contradiction, we conclude that there is no norm inducing the topology $\mathcal{T}(\Omega)$.

Nevertheless, it is clear that the topology and the linear structure on $C(\Omega)$ are related. In this section, we develop a theory of *topological linear spaces* that encompasses not only normed spaces, but interesting spaces like $(C(\Omega), \mathcal{T}(\Omega))$ as well.

DEFINITION 14.3   Topological Linear Space

Let Ω be a linear space with scalar field F and let $\mathcal{T}$ be a topology on Ω. Then we say that $(\Omega, \mathcal{T})$ is a **topological linear space** if the operations of addition and scalar multiplication are continuous, that is, if the functions $A: \Omega \times \Omega \to \Omega$ and $M: F \times \Omega \to \Omega$ defined by

$$A(x, y) = x + y \qquad \text{and} \qquad M(\alpha, x) = \alpha x$$

are continuous.

EXAMPLE 14.11   *Illustrates Definition 14.3*

a) Any normed space is a topological linear space.

b) If Ω is a locally compact Hausdorff space, then $(C(\Omega), \mathcal{T}(\Omega))$ is a topological linear space, as the reader is asked to verify in Exercise 14.45.

c) Let $(\Omega, \mathcal{A}, \mu)$ be a measure space. For $0 < p < 1$, the space $\mathcal{L}^p(\mu)$ is a topological linear space with respect to the topology induced by the metric ρ_p, where

$$\rho_p(f, g) = \int_\Omega |f - g|^p \, d\mu, \qquad f, g \in \mathcal{L}^p(\mu).$$

See Exercise 14.57. □

Unless there is a danger of ambiguity, we will write Ω for the topological linear space $(\Omega, \mathcal{T})$. In the remainder of this section, we assume that the scalar field of Ω is $\mathbb{C}$ unless we state otherwise. Our results are easily adapted to the case of real scalars.

The next two propositions provide some basic properties of topological linear spaces. Note that the second one shows that the topology of a topological linear space is determined by a neighborhood basis at 0. We leave the proofs of both propositions to the reader as Exercises 14.46 and 14.47.

□ □ □ **PROPOSITION 14.3**

Let Ω be a topological linear space.

a) *Suppose $y \in \Omega$ and α is a nonzero scalar. Then the mappings $T(x) = x + y$ and $S(x) = \alpha x$ are homeomorphisms of Ω onto itself.*

b) *Suppose U is an open subset of Ω, $A \subset \Omega$, and α is a nonzero scalar. Then $A + U$ and αU are open subsets of Ω.*

□ □ □ **PROPOSITION 14.4**

Let Ω be a topological linear space. Then there is a collection $\mathcal{W}$ of open sets each containing 0 with the following properties:

a) $W_1, W_2 \in \mathcal{W} \Rightarrow W_3 \subset W_1 \cap W_2$ *for some $W_3 \subset \mathcal{W}$.*

b) $W \in \mathcal{W}$ *and* $x \in W \Rightarrow$ *there is a $W_1 \in \mathcal{W}$ such that $x + W_1 \subset W$.*

c) $W \in \mathcal{W} \Rightarrow$ *there is a $W_1 \in \mathcal{W}$ such that $W_1 + W_1 \subset W$.*

d) $W \in \mathcal{W} \Rightarrow$ *there is a $W_1 \in \mathcal{W}$ and an $\epsilon > 0$ such that $\alpha W_1 \subset W$ whenever $|\alpha| < \epsilon$.*

e) $\{x + W : x \in \Omega, \ W \in \mathcal{W}\}$ *is a basis for the open sets of Ω.*

Conversely, if $\mathcal{W}$ is a collection of subsets of a linear space Ω that satisfies (a)–(d) and $0 \in W$ for each $W \in \mathcal{W}$, then $\{x + W : x \in \Omega, \ W \in \mathcal{W}\}$ is a basis for a topology $\mathcal{T}$ on Ω and $(\Omega, \mathcal{T})$ is a topological linear space.

Locally Convex Topological Linear Spaces and Seminorms

When Ω is a normed space, we can take the collection $\mathcal{W}$ in Proposition 14.4 to be $\{B_r(0) : r > 0\}$. Thus, in the important case of a normed space, the sets in $\mathcal{W}$ can be assumed convex. This convexity property turns out to be the key to a significant generalization of normed spaces, called *locally convex topological linear spaces*.

DEFINITION 14.4   Locally Convex Topological Linear Space

A topological linear space Ω is said to be **locally convex** if there is a collection $\mathcal{W}$ of convex open sets each containing 0 such that

a) $W_1, W_2 \in \mathcal{W} \Rightarrow W_3 \subset W_1 \cap W_2$ for some $W_3 \in \mathcal{W}$.

b) $W \in \mathcal{W}$ and $x \in W \Rightarrow$ there is a $W_1 \in \mathcal{W}$ such that $x + W_1 \subset W$.

c) $W \in \mathcal{W} \Rightarrow$ there is a $W_1 \in \mathcal{W}$ such that $W_1 + W_1 \subset W$.

d) $W \in \mathcal{W} \Rightarrow$ there is a $W_1 \in \mathcal{W}$ and an $\epsilon > 0$ such that $\alpha W_1 \subset W$ whenever $|\alpha| < \epsilon$.

e) $\{x + W : x \in \Omega,\ W \in \mathcal{W}\}$ is a basis for the open sets of Ω.

EXAMPLE 14.12   Illustrates Definition 14.4

a) Any normed space is a locally convex topological linear space.

b) As we will see, if Ω is a locally compact Hausdorff space, then $(C(\Omega), \mathcal{T}(\Omega))$ is a locally convex topological linear space.

c) Exercise 14.58 shows that the space in Example 14.11(c) is not a locally convex topological linear space.   □

In practice, locally convex topological linear spaces are often defined in terms of collections of objects called *seminorms*. A seminorm has the defining properties of a norm except that the seminorm of a nonzero element may be 0.

DEFINITION 14.5   Seminorm

Let Ω be a linear space that has as its scalar field F either $\mathcal{R}$ or $\mathbb{C}$. A function $\sigma \colon \Omega \to \mathcal{R}$ is said to be a **seminorm** on Ω if it satisfies the following conditions for all $x, y \in \Omega$ and $\alpha \in F$:

a) $\sigma(x) \geq 0$.

b) $\sigma(0) = 0$.

c) $\sigma(\alpha x) = |\alpha|\sigma(x)$.

d) $\sigma(x + y) \leq \sigma(x) + \sigma(y)$.

Remark: Although condition (b) follows from condition (c), we have included the former to retain the resemblance of a seminorm to a norm.

Let σ be a seminorm on a linear space Ω. Then for each $x \in \Omega$ and $r > 0$, we define

$$B_r^\sigma(x) = \{\, y : \sigma(x - y) < r \,\}.$$

It is important to note that, by the defining properties of a seminorm, sets of the form $B_r^\sigma(x)$ are convex.

EXAMPLE 14.13 *Illustrates Definition 14.5*

a) Any norm is a seminorm.

b) Let Ω be locally compact, noncompact, Hausdorff space and K a compact subset of Ω. The function $\|\ \|_K$ defined by

$$\|f\|_K = \rho_K(f, 0) = \sup\{\, |f(x)| : x \in K \,\}, \qquad f \in C(\Omega),$$

is an example of a seminorm that is not a norm.

c) If ℓ is a linear functional on a linear space V, then $|\ell|$ is a seminorm on V. □

Let $\mathcal{S}$ be a collection of seminorms defined on a linear space Ω. For $\sigma \in \mathcal{S}$ and $x \in \Omega$, define σ_x by

$$\sigma_x(y) = \sigma(x + y), \qquad y \in \Omega.$$

Then we define the **topology induced by $\mathcal{S}$** to be the weak topology on Ω determined by the family of functions $\{\, \sigma_x : x \in \Omega,\ \sigma \in \mathcal{S} \,\}$.

□ □ □ **PROPOSITION 14.5**

Let Ω be a linear space that has the topology $\mathcal{T}_\mathcal{S}$ induced by a family of seminorms $\mathcal{S}$ and let $\mathcal{W}$ denote the collection of subsets of Ω that consist of intersections of finitely many sets of the form $B_r^\sigma(0)$. Then
a) $\{\, x + W : x \in \Omega,\ W \in \mathcal{W} \,\}$ *is a basis for $\mathcal{T}_\mathcal{S}$.*
b) $(\Omega, \mathcal{T}_\mathcal{S})$ *is a locally convex topological linear space.*

PROOF It can be shown that $\mathcal{W}$ satisfies the conditions (a)–(d) of Proposition 14.4. The details are left to the reader as Exercise 14.48. ∎

EXAMPLE 14.14 *Topologies Induced by a Collection of Seminorms*

a) Let $(\Omega, \|\ \|)$ be a normed space. Then the topology induced by the single-element collection $\mathcal{S} = \{\|\ \|\}$ is the same as the topology induced by the norm $\|\ \|$.

b) Let Ω be a locally compact Hausdorff space. Then it follows from Example 14.13(b) and Proposition 14.5 that $(C(\Omega), \mathcal{T}(\Omega))$ is a locally convex topological linear space. □

□ □ □ **PROPOSITION 14.6**

Let Ω be a linear space with a topology induced by a collection of seminorms $\mathcal{S}$. Then the following results hold:
a) Ω *is Hausdorff if and only if $x \neq 0 \Rightarrow \sigma(x) \neq 0$ for some $\sigma \in \mathcal{S}$.*
b) Suppose $\{x_\iota\}_{\iota \in I}$ is a net in Ω. Then $\lim x_\iota = x$ if and only if $\lim \sigma(x_\iota - x) = 0$ for each $\sigma \in \mathcal{S}$.

PROOF a) Suppose Ω is Hausdorff and let $x \neq 0$. Then, by Proposition 14.5, there is an $\epsilon > 0$ and a $\sigma \in \mathcal{S}$ such that $x \notin B_\epsilon^\sigma(0)$. Hence, $\sigma(x) \geq \epsilon$. To prove the converse, let x and y be distinct elements of Ω. Then, by assumption, there is an $\epsilon > 0$ and $\sigma \in \mathcal{S}$ such that $\sigma(x - y) = \epsilon$. It follows that $x + B_{\epsilon/2}^\sigma(0)$ and $y + B_{\epsilon/2}^\sigma(0)$ are disjoint open sets that contain x and y, respectively. Hence, $(\Omega, \mathcal{T}_\mathcal{S})$ is Hausdorff.

b) By Proposition 10.13 on page 383, we know that $\lim x_\iota = x$ if and only if $\lim \sigma_y(x_\iota) = \sigma_y(x)$ for each $y \in \Omega$ and $\sigma \in \mathcal{S}$. Suppose that $\lim x_\iota = x$ and let $\sigma \in \mathcal{S}$. Then, setting $y = -x$, we get

$$\lim \sigma(x_\iota - x) = \lim \sigma_{-x}(x_\iota) = \sigma_{-x}(x) = \sigma(0) = 0.$$

Conversely, suppose that $\lim \sigma(x_\iota - x) = 0$ for each $\sigma \in \mathcal{S}$. Then, using condition (d) of Definition 14.5, we get

$$\lim |\sigma_y(x_\iota) - \sigma_y(x)| \leq \lim \sigma(x_\iota - x) = 0$$

for all $y \in \Omega$ and $\sigma \in \mathcal{S}$. Thus, $\lim x_\iota = x$. ■

For topologies induced by collections of seminorms, there is a nice analogue of Proposition 13.1 on page 454. We present this as Proposition 14.7 and leave the proof to the reader as Exercise 14.49.

□ □ □ PROPOSITION 14.7

Let Ω and Λ be linear spaces with the same scalar fields and with topologies induced, respectively, by the collections of seminorms $\mathcal{S}_1$ and $\mathcal{S}_2$. For a linear mapping $T: \Omega \to \Lambda$, the following properties are equivalent:

a) T is continuous.

b) T is continuous at 0.

c) For each $\sigma \in \mathcal{S}_2$ there exist $\sigma_1, \sigma_2, \ldots, \sigma_n \in \mathcal{S}_1$ and a constant α such that $\sigma(T(x)) \leq \alpha \max\{\sigma_j(x) : j = 1, 2, \ldots n\}$ for all $x \in \Omega$.

We observe that if $\Lambda = \mathbb{C}$ (with the usually topology), then the seminorm σ in Proposition 14.7 is just the modulus of a complex number.

Linear Functionals and Separation by Hyperplanes

Because continuous linear functionals play an important role in the theory of normed spaces, one might expect that they would also be significant in the theory of topological linear spaces. Surprisingly, though, there are naturally arising examples of topological linear spaces that have no continuous linear functionals other than the one that is identically 0. (See Exercise 14.58.)

The situation becomes much more agreeable, however, if local convexity is assumed. We will devote the remainder of this section to showing that, in the locally convex case, there are an abundance of continuous linear functionals; indeed there are enough to separate elements from closed convex subsets.

DEFINITION 14.6   Internal Point, Support Function

Let V be a linear space and A a convex subset of V.

a) An element $u \in A$ is said to be an **internal point** of A if for each $v \in V$, there is an $\epsilon > 0$ such that $u + \alpha v \in A$ for all scalars α such that $|\alpha| < \epsilon$.

b) If 0 is an internal point of A, then the function s_A defined on V by

$$s_A(v) = \inf\{\, r : r^{-1}v \in A, \ r > 0 \,\}$$

is called the **support function** of A.

EXAMPLE 14.15   *Illustrates Definition 14.6*

Let $(\Omega, \|\ \|)$ be a normed space. Then 0 is an internal point of $B_1(0)$ and, as the reader is asked to verify in Exercise 14.51, $s_{B_1(0)} = \|\ \|$.   □

Our next proposition, Proposition 14.8, shows that support functions behave much like norms.

□ □ □   PROPOSITION 14.8

Let V be a linear space and 0 an internal point of the convex set A. Then

a) $s_A(\alpha v) = \alpha s_A(v)$ for all $v \in V$ and $\alpha \geq 0$.

b) $s_A(v_1 + v_2) \leq s_A(v_1) + s_A(v_2)$ for all $v_1, v_2 \in V$.

c) $\{\, v : s_A(v) < 1 \,\} \subset A \subset \{\, v : s_A(v) \leq 1 \,\}$.

PROOF   We leave the proofs of (a) and (c) to the reader in Exercise 14.52. To prove (b), let r_1 and r_2 be positive numbers such that $r_1^{-1}v_1, \ r_2^{-1}v_2 \in A$. As A is convex, we have that

$$(r_1 + r_2)^{-1}(v_1 + v_2) = (r_1/(r_1 + r_2))r_1^{-1}v_1 + (r_2/(r_1 + r_2))r_2^{-1}v_2 \in A.$$

Hence, $s_A(v_1 + v_2) \leq r_1 + r_2$. Taking infimums with respect to r_1 and r_2, we obtain (b).   ■

We will use support functions to study the phenomenon of **separation by a hyperplane.** To introduce that topic, let E and F be disjoint closed convex subsets of $\mathcal{R}^2$. Then, as shown in Fig. 14.1 at the top of the next page, E and F can be *separated* by a line L in the following sense: Associated with L, there is a nontrivial linear functional ℓ on $\mathcal{R}^2$ and a real number α such that $L = \ell^{-1}(\{\alpha\})$, $E \subset \ell^{-1}((-\infty, \alpha])$, and $F \subset \ell^{-1}([\alpha, \infty))$. Note that

$$\sup\{\, \ell(v) : v \in E \,\} \leq \inf\{\, \ell(u) : u \in F \,\}$$

is a necessary condition for such a separation.

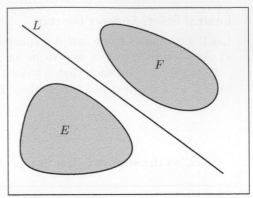

FIGURE 14.1 Separation by a hyperplane

Similarly, disjoint closed convex subsets of $\mathcal{R}^3$ can be separated by a plane. We now proceed to generalize these two simple examples to locally convex topological linear spaces.

◻ ◻ ◻ **THEOREM 14.9**

Let V be a linear space with real scalars. Suppose that A_1 and A_2 are nonempty disjoint convex subsets of V and that A_1 has an internal point. Then there is a nontrivial linear functional ℓ on V such that

$$\sup\{\,\ell(v) : v \in A_1\,\} \leq \inf\{\,\ell(u) : u \in A_2\,\}. \tag{14.23}$$

PROOF Let v_1 be an internal point of A_1, v_2 be any point of A_2, and $v_0 = v_2 - v_1$. Then it is easy to check that the set

$$A = v_0 + A_1 - A_2$$

is convex and contains 0 as an internal point.

We define a linear functional ℓ_0 on the subspace $V_0 = \{\,\alpha v_0 : \alpha \in \mathcal{R}\,\}$ by $\ell_0(\alpha v_0) = \alpha$ and will show that ℓ_0 is dominated on V_0 by s_A. Since $A_1 \cap A_2 = \emptyset$, we have $v_0 \notin A$. Hence, by Proposition 14.8, $s_A(v_0) \geq 1$. Using Proposition 14.8 again, we conclude that for $\alpha \geq 0$,

$$\ell_0(\alpha v_0) = \alpha \leq \alpha s_A(v_0) = s_A(\alpha v_0).$$

On the other hand, if $\alpha < 0$, then $\ell_0(\alpha v_0) \leq s_A(\alpha v_0)$ is trivially true.

We can now invoke the Hahn-Banach theorem (Theorem 14.1 on page 498) to obtain a linear functional ℓ on V such that $\ell_{|V_0} = \ell_0$ and

$$\ell(v) \leq s_A(v), \qquad v \in V.$$

Because $\ell(v_0) = 1$, ℓ is not identically 0. Also, if $v \in A_1$ and $u \in A_2$, then $v_0 + v - u \in A$ and, so, by Proposition 14.8,

$$1 + \ell(v) - \ell(u) = \ell(v_0 + v - u) \leq s_A(v_0 + v - u) \leq 1.$$

The inequality (14.23) now follows immediately. ∎

Next, we will prove a version of Theorem 14.3 (page 501) for topological linear spaces. To do so, we will need the following lemma.

□ □ □ **LEMMA 14.3**

Let ℓ be a linear functional on the topological linear space Ω. If there is an open set W containing 0 such that $\sup\{\Re\ell(x) : x \in W\} < \infty$, then ℓ is continuous.

PROOF It suffices to show that if $x_0 \in \Omega$ and $\epsilon > 0$, then there is an open set U containing 0 such that

$$\ell(x_0 + U) \subset \{z : |\ell(x_0) - z| < \epsilon\}.$$

Let $b_0 = \sup\{\Re\ell(x) : x \in W\}$. Since $0 \in W$, it follows that $b_0 \geq 0$. Applying Proposition 14.4(d) on page 521, we choose a $\delta > 0$ and an open set O containing 0 such that $\alpha O \subset W$ whenever $|\alpha| < \delta$. Let $b > b_0$ and set

$$U = \epsilon b^{-1} \bigcup_{|\alpha| < \delta} \alpha O.$$

Then U is open and $e^{it}U \subset U$ for all $t \in \mathcal{R}$.

Let $y \in U$. By selecting t so that $|\ell(y)| = e^{it}\ell(y)$, we get

$$|\ell(y)| = e^{it}\ell(y) = \Re\ell(e^{it}y) = \epsilon b^{-1}\Re\ell(\epsilon^{-1}be^{it}y) \leq \epsilon b^{-1}b_0 < \epsilon.$$

Hence, $|\ell(x_0) - \ell(x_0 + y)| < \epsilon$. ∎

□ □ □ **THEOREM 14.10**

Let F be a nonempty closed convex subset of a locally convex topological linear space Ω. If $x_0 \in F^c$, then there is a continuous linear functional ℓ on Ω such that $\Re\ell(x_0) < \inf\{\Re\ell(x) : x \in F\}$.

PROOF The set $-x_0 + F$ is convex and closed and does not contain 0. Since Ω is locally convex, there is a convex open set U that contains 0 and is disjoint from $-x_0 + F$. And because scalar multiplication is continuous, the point 0 is an internal point of U. Thus, by Theorem 14.9, there is a nontrivial linear functional ℓ on Ω such that

$$\sup\{\Re\ell(u) : u \in U\} \leq \inf\{\Re\ell(v) : v \in -x_0 + F\}.$$

Applying Lemma 14.3, we conclude that ℓ is continuous.

Because $0 \in U$, we have $\sup\{\Re\ell(u) : u \in U\} \geq 0$. We claim that this inequality is strict. Suppose to the contrary and let $z \in \Omega$. Then by choosing $\epsilon > 0$ small enough, we get $e^{it}\epsilon z \in U$ for all $t \in \mathcal{R}$ and, hence, $\Re\ell(e^{it}\epsilon z) \leq 0$ for all $t \in \mathcal{R}$. Upon selecting t so that $e^{it}\ell(z) = |\ell(z)|$, it follows that $\ell(z) = 0$. We have shown that ℓ is identically 0, a contradiction.

From $\sup\{\Re\ell(u) : u \in U\} > 0$ and the preceding displayed inequality, we conclud that $\inf\{\Re\ell(v) : v \in -x_0 + F\} > 0$. So, $\Re\ell(x_0) < \inf\{\Re\ell(x) : x \in F\}$. ∎

Remaining consistent with the notation used for normed spaces, we will write Ω^* for the space of all continuous linear functionals on the topological linear space Ω and $\langle x, x^*\rangle$ for $x^*(x)$ when $x \in \Omega$ and $x^* \in \Omega^*$. Theorem 14.10 shows that when Ω is locally convex and Hausdorff, Ω^* has enough members to separate the points of Ω. We refer the reader to Exercises 14.57–14.58 for an example of a topological linear space where the only continuous linear functional is identically zero.

When the convex set in Theorem 14.10 is a linear subspace of Ω, we have the following refinement, which is also a generalization of Theorem 14.3 on page 501.

□ □ □ **COROLLARY 14.6**

Let K be a closed linear subspace of the locally convex topological linear space Ω and let $x \in K^c$. Then there is an $x_0^* \in K^\perp$ such that $\langle x, x_0^* \rangle > 0$.

PROOF   From Theorem 14.10, there exists a continuous linear functional x_1^* such that $\Re\langle x, x_1^* \rangle < \inf\{\Re\langle y, x_1^* \rangle : y \in K\}$. Let $y \in K$ and $b > 0$, and choose t so that $e^{it}\langle y, x_1^* \rangle = |\langle y, x_1^* \rangle|$. Because $-be^{it}y \in K$, we have

$$\Re\langle x, x_1^* \rangle < \Re\langle -be^{it}y, x_1^* \rangle = \Re(-be^{it}\langle y, x_1^* \rangle) = -b|\langle y, x_1^* \rangle|.$$

As b is an arbitrary positive number, it follows that $\langle y, x_1^* \rangle = 0$. If s is chosen so that $e^{is}\langle x, x_1^* \rangle = |\langle x, x_1^* \rangle|$, then the functional $x_0^* = e^{is}x_1^*$ satisfies the assertions of the corollary.   ■

The next corollary shows that the point x_0 in Theorem 14.10 can be replaced by a compact convex set disjoint from F.

□ □ □ **COROLLARY 14.7**

Let Ω be a locally convex topological linear space. Suppose F and K are, respectively, nonempty closed and compact convex subsets of Ω such that $F \cap K = \emptyset$. Then there is an $x^* \in \Omega^*$ such that

$$\sup\{\Re\langle y, x^* \rangle : y \in K\} < \inf\{\Re\langle x, x^* \rangle : x \in F\}.$$

PROOF   Clearly $F - K$ is convex; we claim that it is also closed. Let $\{x_\iota - y_\iota\}_{\iota \in I}$ be a net in $F - K$ such that $\lim(x_\iota - y_\iota) = z$. Because K is compact, there is a subnet $\{y_{\iota_\upsilon}\}_{\upsilon \in \Upsilon}$ that converges to an element $y \in K$. It follows that $\{x_{\iota_\upsilon}\}_{\upsilon \in \Upsilon}$ is a net in F that converges to $z + y$. Because F is closed, $z + y \in F$; hence, $z \in F - K$. We have now shown that $F - K$ is closed.

Because F and K are disjoint, $0 \notin F - K$. Hence, we can apply Theorem 14.10 to obtain a continuous linear functional x^* such that

$$0 = \langle 0, x^* \rangle < \inf\{\Re\langle x - y, x^* \rangle : x \in F, \; y \in K\}$$
$$= \inf\{\Re\langle x, x^* \rangle : x \in F\} - \sup\{\Re\langle y, x^* \rangle : y \in K\},$$

as required.   ■

In the remaining two sections of this chapter, we will see many applications of Theorem 14.10 and its corollaries. Here we content ourselves with the following simple examples.

EXAMPLE 14.16   *Illustrates Theorem 14.10*

a) Suppose that F is a nonempty closed convex subset of the normed space Ω. If $0 \notin F$, then Theorem 14.10 implies that there exists an $x^* \in \Omega^*$ such that $\inf\{\Re\langle x, x^* \rangle : x \in F\} > 0$.

b) Let $F = \{f \in C([0, 1]) : \Re f(t) \geq t\}$. As $0 \notin F$, part (a) guarantees the existence of an $x^* \in C([0, 1])^*$ such that $\inf\{\Re\langle f, x^* \rangle : f \in F\} > 0$. The continuous linear functional on $C([0, 1])$ defined by $\langle g, x^* \rangle = \int_0^1 g(t)\, dt$ satisfies that condition. In fact, the infimum is $1/2$.

c) If we replace $C([0, 1])$ by $C(\mathcal{R})$ in part (b), but use the same F and x^*, then we obtain an illustration of Theorem 14.10 for the case of a locally convex topological linear space that is not a normed space.   □

Exercises for Section 14.4

14.45 Let Ω be a locally compact Hausdorff space. Prove that $(C(\Omega), \mathcal{T}(\Omega))$ is a topological linear space.

14.46 Prove Proposition 14.3 on page 521.

14.47 Prove Proposition 14.4 on page 521.

14.48 Prove Proposition 14.5 on page 523.

14.49 Prove Proposition 14.7 on page 524.

14.50 Let U be an open convex subset of a topological linear space. Show that all points of U are internal points.

14.51 Show that in a normed space, the support function of the open unit ball around 0 is equal to the norm.

14.52 Prove (a) and (c) of Proposition 14.8 on page 525.

14.53 Let ℓ be a linear functional on a topological linear space Ω. Prove that the following conditions on ℓ are equivalent:
a) ℓ is continuous.
b) ℓ is continuous at some point of Ω.
c) $\sup\{\,\Re\ell(u) : u \in U\,\} < \infty$ for some nonempty open set U.
d) $\inf\{\,\Re\ell(u) : u \in U\,\} > -\infty$ for some nonempty open set U.
e) $\sup\{\,|\ell(u)| : u \in U\,\} < \infty$ for some nonempty open set U.

14.54 Show that Corollary 14.7 fails if the compactness assumption on K is replaced by the assumption that K is closed.

14.55 Let A and B be subsets of a topological linear space Ω. Show that if A is closed and B is compact, then $A + B$ is closed.

14.56 Show that if a topological linear space is locally convex and T_1 (i.e., single-element subsets are closed), then it is Hausdorff.

14.57 Consider the space $\mathcal{L}^p([0,1])$, where $0 < p < 1$. By Exercise 13.54 on page 482, the function ρ_p defined by

$$\rho_p(f,g) = \sigma_p(f-g) = \int_{[0,1]} |f-g|^p \, d\lambda, \qquad f, g \in \mathcal{L}^p([0,1])$$

is a metric on $\mathcal{L}^p([0,1])$. Show that $\mathcal{L}^p([0,1])$ is a topological linear space with respect to the topology induced by ρ_p.

14.58 Refer to Exercise 14.57.
a) Show that $(\mathcal{L}^p([0,1]))^*$ contains only the functional that is identically 0.
b) Deduce that $\mathcal{L}^p([0,1])$ is not locally convex when $0 < p < 1$.

In Exercises 14.59–14.61, $C^\infty(\mathcal{R})$ denotes the space of complex-valued functions that have derivatives of all orders at each point of $\mathcal{R}$. For each nonnegative integer n, define

$$\sigma_n(f) = \max_{0 \le m \le n} \sup\left\{\,\left(1+t^2\right)^n \left|f^{(m)}(t)\right| : t \in \mathcal{R}\,\right\}, \qquad f \in C^\infty(\mathcal{R}).$$

We will consider the space

$$\mathcal{S}(\mathcal{R}) = \{\,f \in C^\infty(\mathcal{R}) : \sigma_n(f) < \infty, \ n = 0, 1, 2, \ldots\,\}.$$

14.59 Let the notation be as in the preceding paragraph.
a) Show that $\mathcal{S}(\mathcal{R})$ is a linear space.

    b) Show that functions of the form $p(x)e^{-x^2}$, where $p(x)$ is a polynomial, belong to $\mathcal{S}(\mathcal{R})$.

14.60   Let the notation be as in the paragraph preceding Exercise 14.59.

    a) Show that $\{\,\sigma_n : n = 0, 1, \ldots\,\}$ is a family of seminorms that induces a Hausdorff topology on $\mathcal{S}(\mathcal{R})$.

    b) Show that the linear operators $D(f) = f'$ and $M(f) = pf$, where p is a polynomial, are continuous with respect to this topology.

14.61   Call a subset $\mathcal{F} \subset S(\mathcal{R})$ *bounded* if $\sup\{\,\sigma_n(f) : f \in \mathcal{F}\,\} < \infty$ for each nonnegative integer n. Let $\mathcal{S}(\mathcal{R})$ have the topology defined in Exercise 14.60. Prove the following version of the Heine-Borel theorem: $\mathcal{F} \subset \mathcal{S}(\mathcal{R})$ is compact if and only if it is closed and bounded.

14.5 WEAK AND WEAK* TOPOLOGIES

In this section, we will introduce and discuss the main properties of the weak topology on a normed space and the weak* topology on its dual space. We will include an investigation of weak convergence and weak boundedness of sequences in a normed space and an important result about weak* compactness.

    Let ℓ be a linear functional on a linear space V. Then, as we have seen, $|\ell|$ is a seminorm on V. Therefore, according to Proposition 14.5 on page 523, if $\mathcal{F}$ is a family of linear functionals on V, the collection of seminorms $\{\,|\ell| : \ell \in \mathcal{F}\,\}$ induces a locally convex topology on V. The cases where V is a normed space or its dual are important enough to warrant a special definition.

DEFINITION 14.7   Weak and Weak* Topologies

Let Ω be normed space. For each $x \in \Omega$, define the linear functional ℓ_x on Ω^* by $\ell_x(x^*) = x^*(x) = \langle x, x^* \rangle$. Then we use the following terminology:

- **Weak topology:** the topology on Ω induced by the collection of seminorms $\{\,|x^*| : x^* \in \Omega^*\,\}$.

- **Weak* topology:** the topology on Ω^* induced by the collection of seminorms $\{\,|\ell_x| : x \in \Omega\,\}$.

Because we work with the norm topologies on Ω and Ω^* as well as the weak and weak* topologies, it is useful to have a convenient way to distinguish these topologies. When no modifier is used, we assume that the topology is the norm topology. So, for instance, when we say that a function is continuous, we mean that it is continuous with respect to the norm topology, and when we say that a set is closed, we mean that it is closed with respect to the norm topology.

    On the other hand, we will employ the words *weak* and *weakly* to indicate "with respect to the weak topology" and, similarly, use the term *weak** to indicate "with respect to the weak* topology." Thus, for example, a function that is continuous with respect to the weak* topology is called weak* continuous and a set that is closed in the weak topology is called weak closed or weakly closed.

Let $\{x_\iota\}_{\iota \in I}$ and $\{x_\iota^*\}_{\iota \in I}$ be nets in Ω and Ω^*, respectively. Then we use the notation

$$\text{wlim} \, x_\iota = x \qquad \text{and} \qquad \text{w}^*\text{lim} \, x_\iota^* = x^* \qquad (14.24)$$

to denote, respectively, that $\{x_\iota\}_{\iota \in I}$ weak converges (or converges weakly) to x and $\{x_\iota^*\}_{\iota \in I}$ weak* converges to x^*. We observe that (14.24) holds if and only if

$$\lim \langle x_\iota, x^* \rangle = \langle x, x^* \rangle, \qquad x^* \in \Omega^*,$$

and

$$\lim \langle x, x_\iota^* \rangle = \langle x, x^* \rangle, \qquad x \in \Omega,$$

respectively.

EXAMPLE 14.17 *Illustrates Definition 14.7*

a) Let $\mathcal{H}$ be a Hilbert space. By Theorem 13.8 on page 473, each element of $\mathcal{H}^*$ is of the form $\langle \cdot, y \rangle$ for some $y \in \mathcal{H}$. Hence, by associating y with $\langle \cdot, y \rangle$, we can identify $\mathcal{H}$ with its dual space. A consequence of this identification is that the weak and weak* topologies of a Hilbert space coincide.

Now suppose that $\mathcal{H}$ contains an infinite orthonormal sequence $\{e_n\}_{n=1}^\infty$. Then, according to Bessel's inequality (Corollary 13.2 on page 469), we have $\sum_{n=1}^\infty |\langle x, e_n \rangle|^2 \leq \|x\|^2 < \infty$ for each $x \in \mathcal{H}$ and, so, $\lim_{n \to \infty} \langle x, e_n \rangle = 0$ for each $x \in \mathcal{H}$. It follows that $\text{wlim} \, e_n = 0$; but, on the other hand, because $\|e_n\| = 1$ for all n, the sequence $\{e_n\}_{n=1}^\infty$ cannot converge to 0 with respect to the norm topology on $\mathcal{H}$.

This simple example shows that the norm topology on a normed space can be strictly stronger than the weak topology. Indeed, as the reader is asked to verify in Exercise 14.65, the weak (weak*) topology coincides with the norm topology on Ω (Ω^*) if and only if Ω is finite dimensional.

b) Suppose that Ω is a compact Hausdorff space. In view of the Riesz representation theorem for compact spaces (Theorem 13.15 on page 492), we know that the dual space of $C(\Omega)$ can be identified with the space $M(\Omega)$ of regular Borel measures on Ω. Thus, for a sequence $\{f_n\}_{n=1}^\infty$ of $C(\Omega)$, we have $\text{wlim} \, f_n = f$ if and only if

$$\lim_{n \to \infty} \int_\Omega f_n \, d\mu = \int_\Omega f \, d\mu, \qquad \mu \in M(\Omega).$$

And, for a sequence $\{\mu_n\}_{n=1}^\infty$ of $M(\Omega)$, we have $\text{w}^*\text{lim} \, \mu_n = \mu$ if and only if

$$\lim_{n \to \infty} \int_\Omega f \, d\mu_n = \int_\Omega f \, d\mu, \qquad f \in C(\Omega).$$

c) Suppose that Ω is a locally compact Hausdorff space. In view of the Riesz representation theorem for locally compact spaces (Theorem 13.16 on page 493), we know that the dual space of $C_0(\Omega)$ can be identified with the space $M(\Omega)$ of regular Borel measures on Ω. Thus, the same results hold as in part (b), provided we replace $C(\Omega)$ by $C_0(\Omega)$.

d) A sequence $\{X_n\}_{n=1}^\infty$ of random variables (not necessarily defined on the same probability space) is said to **converge in distribution** to the random variable X if the sequence $\{\mu_{X_n}\}_{n=1}^\infty$ converges to μ_X in the weak* topology.

We write $X_n \overset{d}{\to} X$ to indicate convergence in distribution. Therefore, we have $X_n \overset{d}{\to} X$ if and only if

$$\lim_{n \to \infty} \int_{\mathcal{R}} f \, d\mu_{X_n} = \int_{\mathcal{R}} f \, d\mu_X, \qquad f \in C_0(\mathcal{R}).$$

Actually, it can be shown that the previous limit holds for all $f \in C_b(\mathcal{R})$. (See Exercise 14.76). An equivalent condition for $X_n \overset{d}{\to} X$ is that for each $x \in \mathcal{R}$ at which F_X is continuous, $F_{X_n}(x) \to F_X(x)$ as $n \to \infty$. The reader is asked to verify this in Exercise 14.78. In that exercise, we also ask the reader to show that convergence in probability implies convergence in distribution.

e) We will show that a familiar example of convergence is really weak* convergence in disguise. Consider the sequence of measures in $M([0,1])$ defined by $\mu_n = (1/n) \sum_{j=1}^n \delta_{j/n}$. Then, for each $f \in C([0,1])$, we have

$$\lim_{n \to \infty} \int_{[0,1]} f \, d\mu_n = \lim_{n \to \infty} (1/n) \sum_{j=1}^n f(j/n) = \int_0^1 f(x) \, dx.$$

It follows that the sequence $\{\mu_n\}_{n=1}^\infty$ converges in the weak* topology to Lebesgue measure on $[0,1]$. □

Our next theorem provides some fundamental properties of the weak topology on a normed space.

□ □ □ **THEOREM 14.11**

Let Ω be a normed space.

a) *With respect to the weak topology, Ω is a locally convex Hausdorff topological linear space.*

b) *Weakly closed subsets of Ω are closed.*

c) *Convex closed subsets of Ω are weakly closed.*

PROOF   a) By definition, the weak topology is induced by the collection of seminorms $\{ |x^*| : x^* \in \Omega^* \}$. Therefore, it follows from Proposition 14.5 on page 523 that Ω is locally convex with respect to the weak topology. That the weak topology is Hausdorff follows from Corollary 14.2 on page 501 and Proposition 14.6 on page 523.

b) This result follows immediately because the weak topology on Ω is weaker than the norm topology on Ω.

c) Let F be a nonempty closed convex subset of Ω. We will prove that F^c is weakly open by showing that for each $x \in F^c$, there is a weakly open set W such that

$$x \in W \subset F^c. \tag{14.25}$$

By Theorem 14.10 on page 527, there exists an $x^* \in \Omega^*$ such that

$$\Re\langle x, x^* \rangle < d = \inf\{ \Re\langle y, x^* \rangle : y \in F \}.$$

It follows, therefore, that the weakly open set $W = \{ w \in \Omega : \Re\langle w, x^* \rangle < d \}$ satisfies (14.25). ∎

Bounded and Weakly Bounded Sets

A subset E of a normed space Ω is said to be **bounded** if

$$\sup\{\,\|x\| : x \in E\,\} < \infty.$$

Our next definition provides a less restrictive notion of boundedness.

DEFINITION 14.8 Weakly Bounded Set

A subset E of a normed space Ω is said to be **weakly bounded** if

$$\sup\{\,|\langle x, x^* \rangle| : x \in E\,\} < \infty$$

for each $x^* \in \Omega^*$.

The inequality $|\langle x, x^* \rangle| \leq \|x^*\|_*\|x\|$ implies that bounded sets are weakly bounded. Although not obvious, it is nevertheless true that weakly bounded sets are bounded.

□ □ □ **THEOREM 14.12**

A subset of a normed space is weakly bounded if and only if it is bounded.

PROOF We have already proved sufficiency. To prove necessity, we first note that each $x \in E$ determines a continuous linear functional ℓ_x on Ω^* via $\ell_x(x^*) = \langle x, x^* \rangle$. Because E is weakly bounded,

$$\sup\{\,|\ell_x(x^*)| : x \in E\,\} = \sup\{\,|\langle x, x^* \rangle| : x \in E\,\} < \infty$$

for each $x^* \in \Omega^*$. Recalling that Ω^* is a Banach space, we apply the uniform boundedness principle for linear operators (Theorem 14.7 on page 509) to conclude that $\sup\{\,\|\ell_x\| : x \in E\,\} < \infty$. However, from Corollary 14.2 on page 501, we know that $\|\ell_x\| = \|x\|$. ∎

EXAMPLE 14.18 *Illustrates Theorem 14.12*

In this example we will use Theorem 14.12 to characterize the weakly convergent sequences in the space $C(\Omega)$, where Ω is a compact Hausdorff space. Specifically, we will show that a sequence of functions in $C(\Omega)$ converges weakly if and only if it converges pointwise and is uniformly bounded.

Suppose that $\{f_n\}_{n=1}^{\infty}$ is a sequence in $C(\Omega)$ converging weakly to f. Then

$$\lim_{n \to \infty} \int_\Omega f_n \, d\mu = \int_\Omega f \, d\mu \tag{14.26}$$

for each $\mu \in M(\Omega)$. Setting $\mu = \delta_x$, we obtain

$$\lim_{n \to \infty} f_n(x) = f(x), \qquad x \in \Omega. \tag{14.27}$$

Also, since weakly convergent sequences are weakly bounded, it follows from Theorem 14.12 that

$$\sup \{ \, \|f_n\|_\Omega : n \in \mathcal{N} \, \} < \infty. \tag{14.28}$$

Thus, (14.27) and (14.28) are necessary conditions for the weak convergence of $\{f_n\}_{n=1}^\infty$ to f.

Next, we show that (14.27) and (14.28) together are sufficient conditions for the weak convergence of $\{f_n\}_{n=1}^\infty$ to f. Let $\mu \in M(\Omega)$. Then, because of (14.27), (14.28), and $|\mu|(\Omega) < \infty$, we can apply the Lebesgue dominated convergence theorem to obtain (14.26). □

Compactness in the Weak* Topology

One of the most important properties of the weak* topology is that the closed unit ball is always weak* compact. This famous result is known as Alaoglu's theorem.

□ □ □ **THEOREM 14.13 Alaoglu's Theorem**

In the dual space Ω^* of a normed space Ω, the closed unit ball,

$$\overline{B}_1^*(0) = \{ \, x^* \in \Omega^* : \|x^*\|_* \leq 1 \, \},$$

is weak* compact.

PROOF In Exercise 14.5 on page 505, we introduced the linear operator $J : \Omega \to \Omega^{**}$ defined by $J(x)(x^*) = \langle x, x^* \rangle$. The relative weak* topology on $\overline{B}_1^*(0)$ is just the weak topology induced by the family $\mathcal{F}$ of restrictions of functions in $J(\Omega)$. We will establish the theorem by showing that the family $\mathcal{F}$ satisfies the hypotheses of Corollary 12.2 on page 440.

If $x \in \Omega$ and $x^* \in \overline{B}_1^*(0)$, then $|J(x)(x^*)| = |\langle x, x^* \rangle| \leq \|x\|$. Consequently, $\overline{J(x)(\overline{B}_1^*(0))}$ is a compact subset of $\mathbb{C}$ for each $x \in \Omega$. Thus, $\mathcal{F}$ satisfies condition (a) of Corollary 12.2. If x_1^* and x_2^* are distinct elements of $\overline{B}_1^*(0)$, then for some $y \in \Omega$,

$$J(y)(x_1^*) = \langle y, x_1^* \rangle \neq \langle y, x_2^* \rangle = J(y)(x_2^*).$$

Hence, condition (b) of Corollary 12.2 holds.

To verify condition (c) of Corollary 12.2, let $\{x_\iota^*\}_{\iota \in I}$ be a net in $\overline{B}_1^*(0)$ such that $\lim \langle x, x_\iota^* \rangle = \ell(x)$ exists for each $x \in \Omega$. Whereas

$$\ell(\alpha x + \beta y) = \lim \langle \alpha x + \beta y, x_\iota^* \rangle = \lim (\alpha \langle x, x_\iota^* \rangle + \beta \langle y, x_\iota^* \rangle) = \alpha \ell(x) + \beta \ell(y)$$

and $|\ell(x)| = \lim |\langle x, x_\iota^* \rangle| \leq \|x\|$, it follows that ℓ is a continuous linear functional on Ω with norm at most 1. Furthermore, ℓ is the weak* limit of the net $\{x_\iota^*\}_{\iota \in I}$. Hence, condition (c) of Corollary 12.2 is satisfied. ∎

□ □ □ **COROLLARY 14.8**

Every bounded net in the dual space Ω^* of a normed space Ω has a weak* convergent subnet.

PROOF A bounded net is contained in a closed ball $\overline{B}_r^*(0)$ for sufficiently large r. Since $\overline{B}_r^*(0) = r\overline{B}_1^*(0)$, it follows that $\overline{B}_r^*(0)$ is also weak* compact. An application of Theorem 11.5 on page 407 completes the proof. ∎

In practice, Corollary 14.8 is often applied in an effort to obtain a weak* convergent subsequence of a bounded sequence. But unless we know that $\overline{B}_1^*(0)$ is metrizable with respect to the weak* topology, all that we can assert is that a bounded sequence has a weak* convergent sub*net*. Our next theorem shows that $\overline{B}_1^*(0)$ is weak* metrizable if Ω is separable.

□ □ □   THEOREM 14.14

Let Ω be a separable normed space. Then the following properties hold:
a) $\overline{B}_1^*(0)$ is weak* metrizable.
b) Every bounded sequence in Ω^* has a weak* convergent subsequence.

PROOF We note, in view of Alaoglu's theorem and Theorem 11.2 on page 403, that (b) follows from (a). We now outline the proof of (a), leaving the details to the reader in Exercise 14.71.

Let $\{x_n\}_{n=1}^\infty$ be a sequence that is dense in the closed ball $\overline{B}_1(0)$. We define a metric ρ on $\overline{B}_1^*(0)$ by

$$\rho(x^*, y^*) = \sum_{n=1}^\infty 2^{-n} |\langle x_n, x^* \rangle - \langle x_n, y^* \rangle|.$$

For each x^*, the function $\rho(x^*, \cdot)$ is weak* continuous. It follows that $B_r^\rho(x^*)$ is weak* open for each $r > 0$ and $x^* \in \overline{B}_1^*(0)$. Thus, the topology induced by ρ is weaker than the relative weak* topology. Since $\overline{B}_1^*(0)$ is weak* compact, we conclude from Corollary 11.5 on page 409 that the topology induced by ρ coincides with the weak* topology. ∎

EXAMPLE 14.19 *Illustrates Theorem 14.14*

Let $g \in \mathcal{L}^2([0,1] \times [0,1])$ and define $L: \mathcal{L}^2([0,1]) \to \mathcal{L}^2([0,1])$ by

$$L(f)(x) = \int_0^1 g(x,y) f(y) \, dy.$$

Then L is a linear operator and satisfies

$$\|L\| \leq \left(\int_0^1 \int_0^1 |g(x,y)|^2 \, dx \, dy \right)^{1/2}.$$

We will prove there is an $f_0 \in \mathcal{L}^2([0,1])$ such that $\|L(f_0)\|_2 = \|L\|$. Let $\{f_n\}_{n=1}^\infty$ be a sequence with $\|f_n\|_2 \leq 1$ and $\lim_{n \to \infty} \|L(f_n)\|_2 = \|L\|$. By Theorem 14.14, there is a subsequence $\{f_{n_j}\}_{j=1}^\infty$ that converges weakly to some f_0. Hence,

$$\lim_{j \to \infty} L(f_{n_j})(x) = \lim_{j \to \infty} \langle f_{n_j}, \overline{g(x, \cdot)} \rangle = \langle f_0, \overline{g(x, \cdot)} \rangle = L(f_0)(x)$$

for almost all x. By Cauchy's inequality,

$$|L(f_{n_j})(x)|^2 \leq \int_0^1 |g(x,y)|^2 \, dy.$$

Thus, the Lebesgue dominated convergence theorem implies that

$$\|L\|^2 = \lim_{j \to \infty} \|L(f_{n_j})\|_2^2 = \|L(f_0)\|_2^2,$$

as required. □

Exercises for Section 14.5

14.62 Let V be a linear space and $\mathcal{F}$ a family of linear functionals on V. Then the collection of seminorms $\mathcal{S} = \{\,|\ell| : \ell \in \mathcal{F}\,\}$ induces a locally convex topology $\mathcal{T}_{\mathcal{S}}$ on V. Show that $\mathcal{T}_{\mathcal{S}}$ is the same as the weak topology $\mathcal{T}_{\mathcal{F}}$ determined by the family $\mathcal{F}$ in the sense of Definition 10.10 on page 370.

14.63 Construct a sequence in $C([0,1])$ such that $\|f_n\|_{[0,1]} \equiv 1$ for all $n \in \mathcal{N}$ and wlim $f_n = 0$.

14.64 Let Ω be an infinite dimensional normed space.

a) Show that if W is a weak open set that contains 0, then W contains an infinite dimensional linear subspace of Ω. *Hint:* Consider the linear mapping $L: \Omega \to \mathbb{C}^n$ defined by $L(x) = (\langle x, x_1^* \rangle, \langle x, x_2^* \rangle, \ldots, \langle x, x_n^* \rangle)$ for appropriate linear functionals $x_1^*, x_2^*, \ldots, x_n^* \in \Omega^*$.

b) Show that if U is a weak* open set that contains 0, then U contains an infinite dimensional linear subspace of Ω^*. *Hint:* Consider an appropriate analogue of the hint for part (a).

14.65 Use Exercise 14.64 to prove the following facts.

a) The norm and weak topologies are equal only for finite dimensional spaces.

b) The norm and weak* topologies are equal only for finite dimensional spaces.

14.66 Consider the normed space ℓ^1.

a) Prove that if a sequence converges weakly in ℓ^1, then it converges in the norm.

b) Deduce from part (a) and Exercise 14.65 that ℓ^1 with the weak topology is not metrizable.

14.67 In the space ℓ^2, let e_n denote the sequence which is 1 at the nth position and 0 elsewhere, and set $E = \{\,e_n + ne_m : m > n \geq 1\,\}$.

a) Show that E is closed.

b) Show that 0 is in the weak closure of E. Deduce that there is a net in E that converges weakly to 0.

c) Show that there is no sequence in E that is weakly convergent to 0.

d) Deduce from parts (b) and (c) that ℓ^2 with the weak topology is not metrizable.

★**14.68** Show that if Ω is a compact or locally compact Hausdorff space, then $M_+(\Omega)$ is weak* closed.

14.69 Suppose that Ω is a compact or locally compact Hausdorff space and that $M(\Omega)$ is endowed with the weak* topology. Let $\Delta: \Omega \to M(\Omega)$ be defined by $\Delta(x) = \delta_x$. Prove that Δ is continuous.

★**14.70** Let Ω be a Hausdorff space and set $P(\Omega) = \{\,\mu \in M_+(\Omega) : \mu(\Omega) = 1\,\}$, the collection of probability measures in $M(\Omega)$.

a) Show that if Ω is compact, then $P(\Omega)$ is weak* compact.

b) Show that if Ω is locally compact but not compact, then $P(\Omega)$ is not weak* compact.

14.71 Provide the details of the proof of Theorem 14.14 on page 535.

14.72 Prove that in a separable Hilbert space, every bounded sequence has a weakly convergent subsequence.

14.73 Consider the space $\mathcal{L}^p([a,b])$, where $1 < p < \infty$. Prove that every bounded sequence has a weakly convergent subsequence.

14.74 Refer to Example 14.19 on pages 535–536.

a) Show that L maps $\mathcal{L}^2([0,1])$ into $\mathcal{L}^2([0,1])$.

b) Verify that

$$\|L\| \le \left(\int_0^1 \int_0^1 |g(x,y)|^2 \, dx \, dy \right)^{1/2}.$$

14.75 Recall that $C_0(\mathcal{R})^* = M(\mathcal{R})$. Let $\{\mu_n\}_{n=1}^\infty$ be a sequence in $M_+(\mathcal{R})$ and $\mu \in M_+(\mathcal{R})$.

a) Show that if $\sup\{\mu_n(\mathcal{R}) : n \in \mathcal{N}\} < \infty$ and $\lim_{n\to\infty} F_{\mu_n}(x) = F_\mu(x)$ at every x where F_μ is continuous, then $\text{w}^*\lim \mu_n = \mu$.

b) Show by example that the converse of the statement in part (a) is false.

c) Show that if $\text{w}^*\lim \mu_n = \mu$ and $\lim_{x\to-\infty} \sup\{F_{\mu_n}(x) : n \in \mathcal{N}\} = 0$, then we have $\lim_{n\to\infty} F_{\mu_n}(x) = F_\mu(x)$ at every x where F_μ is continuous.

★*14.76* Suppose that $\{\mu_n\}_{n=1}^\infty$ is a sequence in $M_+(\mathcal{R})$. Recall that $\text{w}^*\lim \mu_n = \mu$ means

$$\int_\mathcal{R} f \, d\mu_n \to \int_\mathcal{R} f \, d\mu, \qquad f \in C_0(\mathcal{R}).$$

Show that if $\text{w}^*\lim \mu_n = \mu$ and $\mu_n(\mathcal{R}) \to \mu(\mathcal{R})$, then

$$\int_\mathcal{R} f \, d\mu_n \to \int_\mathcal{R} f \, d\mu, \qquad f \in C_b(\mathcal{R}).$$

Hint: First consider the case where $0 \le f \le 1$.

14.77 Suppose that $\{F_n\}_{n=1}^\infty$ is a sequence of distribution functions on $\mathcal{R}$. Further suppose that $\sup\{F_n(\infty) : n \in \mathcal{N}\} < \infty$ and $\lim_{x\to-\infty} \sup\{F_n(x) : n \in \mathcal{N}\} = 0$. Show that there is a subsequence $\{F_{n_k}\}_{k=1}^\infty$ of $\{F_n\}_{n=1}^\infty$ and a distribution function F such that $\lim_{k\to\infty} F_{n_k}(x) = F(x)$ at every x where F is continuous. This result is a version of what is known as **Helly's selection principle.** *Hint:* Observe that $C_0(\mathcal{R})$ is separable. Use Exercise 14.75(c).

14.78 Refer to Example 14.17(d) on pages 531–532. Let X be a random variable and $\{X_n\}_{n=1}^\infty$ a sequence of random variables.

a) Show that $X_n \xrightarrow{d} X$ if and only if $F_{X_n}(x) \to F_X(x)$ at every x where F_X is continuous. *Hint:* Use Exercises 14.75 and 14.76.

b) Suppose $\{X_n\}_{n=1}^\infty$ and X are all defined on the same probability space. Prove that if $\{X_n\}_{n=1}^\infty$ converges to X in probability, then $X_n \xrightarrow{d} X$. *Hint:* Use the fact that functions in $C_0(\mathcal{R})$ are uniformly continuous and Theorem 9.10 on page 344.

14.6 COMPACT CONVEX SETS

In this section, we will study subsets of locally convex topological linear spaces that are both convex and compact. We will prove the Krein-Milman theorem, a result that describes how compact convex sets are generated by their irreducible elements. Additionally, we will give an application of the Krein-Milman theorem to the trigonometric moment problem.

To begin, we introduce some simple geometric ideas. Let $v_1, v_2, \ldots, v_n$ be elements of a linear space and $\alpha_1, \alpha_2, \ldots, \alpha_n$ nonnegative scalars that sum to 1. Then the sum

$$v = \alpha_1 v_1 + \alpha_2 v_2 + \cdots + \alpha_n v_n \tag{14.29}$$

is called a **convex combination** of the v_is. When $n = 2$, we see that v must lie on the line segment connecting v_1 and v_2. Thus, a set is convex if it contains all convex combinations of any two of its elements. It is not hard to show that a convex set contains all convex combinations of any finite subset of its elements. (See Exercise 14.79.)

Some convex combinations are trivial, such as $v = \alpha_1 v + \alpha_2 v + \cdots + \alpha_n v$ or $v = 1v + 0v_2 + \cdots + 0v_n$. We say that a convex combination of the form (14.29) is **proper** if there are at least two distinct indices i and j such that α_i and α_j are positive, and either $v \neq v_i$ or $v \neq v_j$.

An element of a convex set C is either a proper convex combination of elements of C or it is not. The latter case, where the element is "irreducible" with respect to convex combinations, is important enough to warrant the following definition.

DEFINITION 14.9   **Extreme Point**

An element of a convex set C is called an **extreme point** of C if it is not a proper convex combination of elements of C. We use **ex** C to denote the set of all extreme points of C.

There are various useful equivalent conditions for an element of a convex set to be an extreme point. In describing these conditions, we let $[u, v]$ denote the *closed line segment* $\{ (1 - \alpha)u + \alpha v : \alpha \in [0, 1] \}$ and (u, v) denote the *open line segment* $\{ (1 - \alpha)u + \alpha v : \alpha \in (0, 1) \}$. We leave it to the reader to show that each of the following conditions is equivalent to x being an extreme point of the convex set C.

- If $x \in (u, v)$, where $u, v \in C$, then $x = u = v$.

- If $x \in [u, v]$, where $u, v \in C$, then $x = u$ or $x = v$.

Closely related to the concept of an extreme point is that of a *face*, as described in our next definition.

DEFINITION 14.10   **Face of a Convex Set**

Let C be a convex set. A set $F \subset C$ is said to be a **face** of C if it satisfies the following two conditions:
a) F is convex.
b) $\alpha u + (1 - \alpha)v \in F$, where $0 < \alpha < 1$ and $u, v \in C$, implies $u, v \in F$.

EXAMPLE 14.20   *Illustrates Definitions 14.9 and 14.10*

Figure 14.2 shows a closed triangular region T in the plane $\mathcal{R}^2$. The extreme points of the region are the vertices, p_1, p_2, and p_3; in symbols $\mathrm{ex}\, T = \{p_1, p_2, p_3\}$. The single element sets $\{p_1\}$, $\{p_2\}$, $\{p_3\}$, the edges $[p_1, p_2]$, $[p_2, p_3]$, $[p_3, p_1]$, the set T itself, and (vacuously) the empty set are faces of T.   ◻

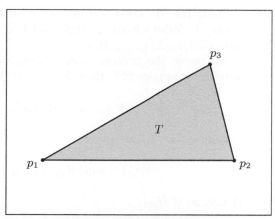

FIGURE 14.2 A triangular region

We observe that in Example 14.20, every element of the triangular region T is a convex combination of the extreme points. As we will see shortly, this is close to being typical of compact convex subsets of Hausdorff locally convex topological linear spaces in that every element of such a set is "approximately" a convex combination of the extreme points.

Let A be a subset of a linear space. The set of all possible convex combinations of elements of A is called the **convex hull** of A and is denoted **cov**(A). Referring to Example 14.20, we see that the line segment $[p_1, p_2]$ is the convex hull of $\{p_1, p_2\}$ and that the triangular region T is the convex hull of $\{p_1, p_2, p_3\}$.

In topological linear spaces, we write $\overline{\mathrm{cov}}(A)$ for $\overline{\mathrm{cov}(A)}$ and call this set the **closed convex hull** of A. Proposition 14.9 provides some basic properties of the convex hull and closed convex hull.

□ □ □ **PROPOSITION 14.9**

Let Ω be a topological linear space and $A, B \subset \Omega$. Then the following properties hold:

a) cov A is convex.
b) If B is convex and $A \subset B$, then $\mathrm{cov}(A) \subset B$.
c) If B is convex, then $\overline{B}$ is convex.
d) If B is closed and convex and $A \subset B$, then $\overline{\mathrm{cov}}(A) \subset B$.
e) If Ω is locally convex, then

$$\overline{\mathrm{cov}}(A) = \left\{\, x \in \Omega : \Re\langle x, x^* \rangle \leq \sup\{\, \Re\langle y, x^* \rangle : y \in A \,\} \text{ for all } x^* \in \Omega^* \,\right\}.$$

PROOF We leave the proofs of parts (a), (b), (c), and (d) to the reader as Exercise 14.81. To prove part (e), we let, for each $x^* \in \Omega^*$,

$$b(x^*) = \sup\{\, \Re\langle y, x^* \rangle : y \in A \,\},$$

and

$$H_{x^*} = \{\, x \in \Omega : \Re\langle x, x^* \rangle \leq b(x^*) \,\}.$$

We must show that $\overline{\mathrm{cov}}(A) = \bigcap_{x^* \in \Omega^*} H_{x^*}$.

Each H_{x^*} is closed, because x^* is continuous, and is convex, because x^* is linear. It follows from (d) that $\overline{\text{cov}}(A) \subset H_{x^*}$ for each $x^* \in \Omega^*$. Therefore, we have $\overline{\text{cov}}(A) \subset \bigcap_{x^* \in \Omega^*} H_{x^*}$.

To prove the reverse containment, suppose that $x_0 \notin \overline{\text{cov}}(A)$. By Theorem 14.10 on page 527, there is an $x_1^* \in \Omega^*$ such that

$$\Re\langle x_0, x_1^* \rangle < \inf\{\, \Re\langle y, x_1^* \rangle : y \in \overline{\text{cov}}(A) \,\}.$$

Letting $x_0^* = -x_1^*$, we obtain

$$b(x_0^*) \leq \sup\{\, \Re\langle y, x_0^* \rangle : y \in \overline{\text{cov}}(A) \,\} < \Re\langle x_0, x_0^* \rangle.$$

Hence, $x_0 \notin H_{x_0^*}$. ∎

We note the following consequences of Proposition 14.9.

- $\text{cov}(A)$ is the smallest convex set that contains A.

- $\overline{\text{cov}}(A)$ is the intersection of all closed half-spaces that contain A.

The Krein-Milman Theorem

We are now ready to prove the main result of this section — the Krein-Milman theorem.

□ □ □ **THEOREM 14.15 Krein-Milman Theorem**

Let Ω be a Hausdorff locally convex topological linear space. If K is a nonempty compact convex subset of Ω, then $K = \overline{\text{cov}}(\text{ex } K)$. In particular, $\text{ex } K \neq \emptyset$.

PROOF It suffices to consider the case of real scalars. Let $\mathcal{F}(K)$ denote the collection of nonempty closed faces of K. Then the following assertions hold, as the reader is asked to verify in Exercise 14.82:

$$\mathcal{F} \subset \mathcal{F}(K) \quad \text{and} \quad \bigcap_{F \in \mathcal{F}} F \neq \emptyset \quad \Rightarrow \quad \bigcap_{F \in \mathcal{F}} F \in \mathcal{F}(K) \tag{14.30}$$

$$F \in \mathcal{F}(K) \quad \Rightarrow \quad \mathcal{F}(F) \subset \mathcal{F}(K) \tag{14.31}$$

$$F \in \mathcal{F}(K) \quad \Rightarrow \quad \text{ex } F \subset \text{ex } K \tag{14.32}$$

$$x^* \in \Omega^* \quad \Rightarrow \quad \{\, x \in K : \langle x, x^* \rangle = \inf x^*(K) \,\} \in \mathcal{F}(K). \tag{14.33}$$

The collection $\mathcal{F}(K)$ is partially ordered by reverse inclusion $\supset$. If $\mathcal{F}$ is a chain in $\mathcal{F}(K)$, then, because $\mathcal{F}$ has the finite intersection property, the intersection $F_1 = \bigcap_{F \in \mathcal{F}} F$ is nonempty; hence, by (14.30), $F_1 \in \mathcal{F}(K)$. Since $F \supset F_1$ for each $F \in \mathcal{F}$, we see that $\mathcal{F}$ has a $\supset$-upper bound. Thus, by Zorn's lemma, there is a $\supset$-maximal nonempty closed face F_0.

We claim that F_0 has only one element. Suppose to the contrary. Then Theorem 14.10 (page 527) implies the existence of an $x^* \in \Omega^*$ that is nonconstant on F_0. It follows from (14.31) and (14.33) that $\{\, x \in F_0 : \langle x, x^* \rangle = \inf x^*(F_0) \,\}$ is a nonempty closed face of K that is properly contained in F_0. This contradiction shows that $F_0 = \{x\}$ for some $x \in K$. Because F_0 is a face, x must be an extreme point of K.

Each $F \in \mathcal{F}(K)$ is also a compact convex set and, consequently, by the preceding argument, ex $F \neq \emptyset$. It follows from (14.32) that each $F \in \mathcal{F}(K)$ contains an extreme point of K.

We are now in a position to show that $K = \overline{\mathrm{cov}}(\mathrm{ex}\, K)$. Since K is closed and convex, we have $\overline{\mathrm{cov}}(\mathrm{ex}\, K) \subset K$. To prove the reverse inclusion, suppose that $K \setminus \overline{\mathrm{cov}}(\mathrm{ex}\, K) \neq \emptyset$. Then Theorem 14.10 implies that there is a $y^* \in \Omega^*$ such that $\inf y^*(K) < \inf y^*(\overline{\mathrm{cov}}(\mathrm{ex}\, K))$. Hence, $\{\, x \in K : \langle x, y^* \rangle = \inf y^*(K)\,\}$ is a nonempty closed face of K that is disjoint from ex K. Since every nonempty closed face of K contains an extreme point of K, we have reached a contradiction. Thus, $K = \overline{\mathrm{cov}}(\mathrm{ex}\, K)$. ■

Some of the most important applications of the Krein-Milman theorem involve optimization of linear functionals. Corollary 14.9 provides some particulars.

□ □ □ **COROLLARY 14.9**

Suppose Ω is a Hausdorff locally convex topological linear space and let $x^* \in \Omega^*$. If K is a nonempty compact convex subset of Ω, then there exist $x_1, x_2, x_3 \in \mathrm{ex}\, K$ such that

a) $\Re\langle x_1, x^* \rangle = \inf\{\, \Re\langle y, x^* \rangle : y \in K\,\}$,
b) $\Re\langle x_2, x^* \rangle = \sup\{\, \Re\langle y, x^* \rangle : y \in K\,\}$,
c) $|\langle x_3, x^* \rangle| = \sup\{\, |\langle y, x^* \rangle| : y \in K\,\}$.

PROOF We prove part (a) and leave parts (b) and (c) to the reader as Exercise 14.84. Let $\alpha = \inf\{\, \Re\langle y, x^* \rangle : y \in K\,\}$. From (14.33), $F = \{\, x \in K : \Re\langle x, x^* \rangle = \alpha\,\}$ is a nonempty closed face of K. So, the Krein-Milman theorem implies that ex $F \neq \emptyset$. Applying (14.32), we conclude that F contains at least one element of ex K. ■

EXAMPLE 14.21 *Illustrates the Krein-Milman Theorem*

a) Suppose that K is a compact convex subset of R^n. Given a real-valued continuous function g on K, it is often a difficult problem to find the maximum value of g. In case g is linear, however, this problem is simplified by the relation $\sup g(K) = \sup g(\mathrm{ex}\, K)$. For example, if K is the triangle in Fig. 14.2, then the maximum on K of a real-valued linear functional is attained at one of the points of $\{p_1, p_2, p_3\}$. This property of linear functionals on finite-dimensional compact convex sets is fundamental to the subject of linear programming.

b) Let $D = \{\, z \in \mathbb{C} : |z| < 1\,\}$ and $H(D)$ the collection of analytic functions on D, equipped with the relative topology inherited from $C(D)$. Each function $f \in H(D)$ has a power series expansion $f(z) = \sum_{n=0}^{\infty} a_n(f)z^n$. As the reader is asked to show in Exercise 14.85, $a_n \colon H(D) \to \mathbb{C}$ is a continuous linear functional for each nonnegative integer n. Consequently, if K is a compact convex subset of $H(D)$ and $n \in \mathcal{N}$, then there exists a $g \in \mathrm{ex}\, K$ such that $|a_n(g)| = \sup\{\, |a_n(f)| : f \in K\,\}$. This observation is useful in the theory of complex variables. □

Next we give a measure-theoretic version of the Krein-Milman theorem. To begin, we define the concept of a *representing measure*.

DEFINITION 14.11 **Representing Measure**

Let K be a compact convex subset of a topological linear space Ω. A measure $\mu \in M(K)$ is said to be a **representing measure** for the element $x \in \Omega$ if μ is nonnegative, $\mu(K) = 1$, and

$$\langle x, x^* \rangle = \int_K \langle y, x^* \rangle \, d\mu(y)$$

for each $x^* \in \Omega^*$.

It is easy to see that in a Hausdorff topological linear space, x can be written as a convex combination

$$x = \alpha_1 x_1 + \alpha_2 x_2 + \cdots + \alpha_n x_n$$

if and only if the measure

$$\mu = \alpha_1 \delta_{x_1} + \alpha_2 \delta_{x_2} + \cdots + \alpha_n \delta_{x_n}$$

is a representing measure for x. This result suggests that a representing measure is a kind of generalized convex combination. As we will now see, the Krein-Milman theorem shows that every element of a compact convex subset of a Hausdorff locally convex topological linear space is, in that sense, a generalized convex combination of extreme points.

□ □ □ **THEOREM 14.16 Krein-Milman Theorem (Measure-Theoretic Version)**

Let Ω be a Hausdorff locally convex topological linear space. If K is a nonempty compact convex subset of Ω, then each $x \in K$ has a representing measure μ such that

$$\mu(K \setminus \overline{\operatorname{ex} K}) = 0. \tag{14.34}$$

PROOF Let $x \in K$. It follows from our first version of the Krein-Milman theorem (Theorem 14.15) that x is the limit of a net $\{x_\iota\}_{\iota \in I}$ contained in $\operatorname{cov}(\operatorname{ex} K)$. Now each x_ι is a convex combination

$$x_\iota = \alpha_{\iota,1} x_{\iota,1} + \alpha_{\iota,2} x_{\iota,2} + \cdots + \alpha_{\iota,n_\iota} x_{\iota,n_\iota},$$

where $x_{\iota,1}, x_{\iota,2}, \ldots, x_{\iota,n_\iota} \in \operatorname{ex} K$. Hence, x_ι has the representing measure

$$\mu_\iota = \alpha_{\iota,1} \delta_{x_{1,\iota}} + \alpha_{\iota,2} \delta_{x_{2,\iota}} + \cdots + \alpha_{\iota,n_\iota} \delta_{x_{\iota,n_\iota}}.$$

From Alaoglu's theorem (page 534), we know that the net of measures $\{\mu_\iota\}_{\iota \in I}$ has a subnet $\{\mu_{\iota_\eta}\}_{\eta \in J}$ that is weak* convergent to a measure μ in $M(K)$. And, by Exercise 14.70 on page 536, μ is a probability measure. That μ is a representing measure for x follows from the fact that

$$\langle x, x^* \rangle = \lim \langle x_{\iota_\eta}, x^* \rangle = \lim \int_K \langle y, x^* \rangle \, d\mu_{\iota_\eta}(y) = \int_K \langle y, x^* \rangle \, d\mu(y)$$

for each $x^* \in \Omega^*$.

It remains to verify (14.34). By the regularity of μ, it is enough to show that $\mu(F) = 0$ whenever F is a compact subset of $K \setminus \overline{\text{ex}\,K}$. We apply Urysohn's lemma to obtain a continuous function $g\colon K \to [0, 1]$ such that g vanishes on $\overline{\text{ex}\,K}$ and is constantly 1 on F. Since each of the measures μ_{ι_η} satisfies (14.34), it follows that $\mu(F) \leq \int_K g\,d\mu = \lim \int_K g\,d\mu_{\iota_\eta} = 0$. ■

Remark: We emphasize that the essential point of Theorem 14.16 is not simply the existence of a representing measure for each point of K, but the existence of a representing measure concentrated on $\overline{\text{ex}\,K}$, in the sense of (14.34).

The Trigonometric Moment Problem

The remainder of this section is devoted to a rather elaborate application of the measure-theoretic version of the Krein-Milman theorem. We will use that theorem to solve the classical trigonometric moment problem:[†]

Given a doubly infinite sequence of complex numbers $\{c_n\}_{n=-\infty}^{\infty}$, find necessary and sufficient conditions for the existence of a measure μ such that

$$c_n = \frac{1}{2\pi} \int_{[-\pi, \pi)} e^{-inx}\,d\mu(x), \qquad n \in \mathcal{Z}. \tag{14.35}$$

Note the following:

- If (14.35) holds, then the c_ns are called the **Fourier coefficients** of the measure μ.

- If (14.35) holds and $\mu \ll \lambda$, then $c_n = \widehat{(d\mu/d\lambda)}(n)$. That is, the Fourier coefficients of the measure μ are the Fourier coefficients of the Radon-Nikodym derivative of μ with respect to λ.

We assume that the space $\ell^\infty(\mathcal{Z})$ of bounded sequences is equipped with the weak* topology, where we recall that $\ell^\infty(\mathcal{Z})$ is the dual space of $\ell^1(\mathcal{Z})$. (Refer to Exercise 13.61 on page 483.)

We first derive a necessary condition for the existence of a measure μ that satisfies (14.35). So, suppose that (14.35) holds. Then, for each $n \in \mathcal{N}$ and complex numbers $\{\lambda_j\}_{j=-n}^n$, we have

$$\sum_{j,k=-n}^{n} \lambda_j \overline{\lambda_k} c_{j-k} = \frac{1}{2\pi} \int_{[-\pi, \pi)} \sum_{j,k=-n}^{n} \lambda_j \overline{\lambda_k} e^{-i(j-k)x}\,d\mu(x)$$

$$= \frac{1}{2\pi} \int_{[-\pi, \pi)} \left| \sum_{k=-n}^{n} \lambda_k e^{-ikx} \right|^2 d\mu(x).$$

[†] See "Moments in Mathematics," H.J. Landau (ed.), *Proceedings of Symposia in Applied Mathematics*, **37** (American Mathematical Society, 1987), for an introduction to the voluminous literature on problems of this type.

Hence, we see that a necessary condition for the existence of a measure μ that satisfies (14.35) is that

$$\sum_{j,k=-n}^{n} \lambda_j \overline{\lambda_k} c_{j-k} \geq 0, \qquad \{\lambda_j\}_{j=-n}^{n} \subset \mathbb{C}, \ n \in \mathcal{N}. \tag{14.36}$$

Sequences satisfying (14.36) are called **nonnegative definite.** Consequently, we have proved the necessity part of the following theorem.

□ □ □ **THEOREM 14.17**

Let $\{c_n\}_{n=-\infty}^{\infty}$ be a doubly infinite sequence of complex numbers. A necessary and sufficient condition for the existence of a measure μ such that

$$c_n = \frac{1}{2\pi} \int_{[-\pi,\pi)} e^{-inx} \, d\mu(x), \qquad n \in \mathcal{Z},$$

is that $\{c_n\}_{n=-\infty}^{\infty}$ is nonnegative definite.

To establish sufficiency in Theorem 14.17, we first introduce some notation. Let D and D_1 denote, respectively, the collection of nonnegative definite sequences and the collection of nonnegative definite sequences that are 1 at the 0th position. Verification of the following basic properties of D and D_1 are left to the reader as Exercise 14.89.

(P1) If $f \in D$, then $|f(n)| \leq f(0)$ for all $n \in \mathcal{Z}$.

(P2) If $f \in D$, then $f(-n) = \overline{f(n)}$ for all $n \in \mathcal{Z}$.

(P3) $f \in D$ and $a \geq 0 \Rightarrow af \in D$.

(P4) If $w \in \mathbb{C}$ with $|w| = 1$, then the sequence defined by $e_w(n) = w^n$ is in D_1.

(P5) If $w \in \mathbb{C}$ with $|w| = 1$, then $e_w \in \text{ex} \, D_1$.

(P6) D is a weak* closed subset of $\ell^\infty(\mathcal{Z})$.

(P7) D_1 is a convex, weak* compact subset of $\ell^\infty(\mathcal{Z})$.

(P8) If $T = \{\, z \in \mathbb{C} : |z| = 1 \,\}$ and D_1 has the weak* topology, then the function $E: T \to D_1$ defined by $E(w) = e_w$ is continuous.

We note that (P1) and (P3) imply that every element of D is a nonnegative scalar multiple of an element of D_1. Therefore, we need only establish sufficiency in Theorem 14.17 for elements of D_1.

The crucial assertion for what follows is that

$$\text{ex} \, D_1 = E(T). \tag{14.37}$$

Suppose, for the moment, that we have established (14.37), and let $f \in D_1$. By (P8) and (14.37), $\text{ex} \, D_1$ is weak* closed. Applying Theorem 14.16, we obtain a representing measure ν_f for f such that $\nu_f(D_1 \setminus \text{ex} \, D_1) = 0$. Thus, we have

$$\ell(f) = \int_{\text{ex} \, D_1} \ell(g) \, d\nu_f(g)$$

for each weak* continuous linear functional ℓ.

Define $\hat{E}: [-\pi, \pi) \to \mathrm{ex}\, D_1$ by $\hat{E}(x) = E(e^{-ix})$. Then, as you are asked to show in Exercise 14.90, the set function μ_f defined by

$$\mu_f(A) = 2\pi \nu_f(\hat{E}(A)), \qquad A \in \mathcal{B}([-\pi, \pi)),$$

is a measure. As you are also asked to show in Exercise 14.90, it now follows from Theorem 9.10 on page 344 that

$$\ell(f) = \frac{1}{2\pi} \int_{[-\pi, \pi)} \ell(\hat{E}(x))\, d\mu_f(x).$$

If, for each $n \in \mathcal{Z}$, we apply the previous equation to the linear functional that evaluates sequences at the integer n, we obtain

$$f(n) = \frac{1}{2\pi} \int_{[-\pi, \pi)} \hat{E}(x)(n)\, d\mu_f(x) = \frac{1}{2\pi} \int_{[-\pi, \pi)} e^{-inx}\, d\mu_f(x), \qquad n \in \mathcal{Z},$$

as required.

It remains to show that (14.37) is valid. In view of (P5), we can conclude that $E(T) \subset \mathrm{ex}\, D_1$, but the reverse inclusion is more difficult to prove. Let $f \in \mathrm{ex}\, D_1$. For a fixed but arbitrary integer m, we define the sequence

$$h(k) = i(f(k+m)f(-m) - f(k-m)f(m)), \qquad k \in \mathcal{Z}.$$

We claim that

$$f \pm h/2 \in D_1. \tag{14.38}$$

By (P3) and the fact that $h(0) = 0$, we see that proving (14.38) is equivalent to showing that $2f \pm h \in D$.

Let $n \in \mathcal{N}$ and $\{\lambda_j\}_{j=-n}^{n} \subset \mathbb{C}$. Setting $\lambda_j = 0$ for $|j| > n$, the condition for nonnegative definiteness can be expressed as

$$\sum_{j,k} \lambda_j \overline{\lambda_k} (2f(j-k) \pm h(j-k)) \geq 0, \tag{14.39}$$

where $\sum_{j,k}$ indicates summation over all integers j and k. Let S denote the left-hand side of (14.39) and note that S is real since $\overline{h(k)} = h(-k)$ for all $k \in \mathcal{Z}$. Then, by using (14.36), (P1), and (P2), we obtain

$$\begin{aligned} S \geq &\sum_{j,k} \lambda_j \overline{\lambda_k} f(j-k) \\ &\pm \sum_{j,k} i\lambda_j \overline{\lambda_k f(m)}\, f(j-k+m) \\ &\mp \sum_{j,k} i\lambda_j f(m)\overline{\lambda_k} f(j-k-m) \\ &+ \sum_{j,k} f(m)\lambda_j \overline{f(m)\lambda_k}\, f(j-k). \end{aligned} \tag{14.40}$$

On the right-hand side of (14.40), we replace k by $k + m$ in the second sum, j by $j + m$ in the third sum, and both j by $j + m$ and k by $k + m$ in the fourth sum to obtain

$$S \geq \sum_{j,k} \lambda_j \overline{\lambda_k} f(j-k) \pm \sum_{j,k} i\lambda_j \overline{\lambda_{k+m} f(m)}\, f(j-k)$$

$$\mp \sum_{j,k} i\lambda_{j+m} f(m) \overline{\lambda_k} f(j-k) + \sum_{j,k} f(m) \lambda_{j+m} \overline{f(m)\lambda_{k+m}}\, f(j-k)$$

$$= \sum_{j,k} (\lambda_j \mp i\lambda_{j+m} f(m)) \overline{(\lambda_k \mp i\lambda_{k+m} f(m))} f(j-k).$$

Because f is nonnegative definite, it follows that $S \geq 0$. Therefore, $2f \pm h$ is nonnegative definite and, so, (14.38) holds.

Now, we have

$$f = (1/2)(f + h/2) + (1/2)(f - h/2).$$

Because $f \in \operatorname{ex} D_1$, it follows that $h = 0$ and, therefore,

$$f(k+m)f(-m) = f(k-m)f(m), \qquad k, m \in \mathcal{Z}. \tag{14.41}$$

We will use (14.41) to show that $f \in E(T)$.

First we observe that there is a smallest positive integer n such that $f(n) \neq 0$. For otherwise, it follows from (P2) that

$$f(k) = \begin{cases} 1, & \text{if } k = 0; \\ 0, & \text{if } k \neq 0. \end{cases}$$

But, as the reader is asked to verify in Exercise 14.91, this sequence is not an extreme point of D_1, a contradiction.

Let $b = |f(n)|$ and $f(n) = be^{it}$. Any integer m has the form $m = sn + r$, where s and r are integers and $0 \leq r < n$. Thus, from (P2) and (14.41), we have $f(sn + r)be^{-it} = f((s-2)n + r)be^{it}$ or, equivalently,

$$f(sn + r) = f((s-2)n + r)e^{2it}.$$

It follows that, when $s \geq 0$,

$$f(sn + r) = \begin{cases} e^{ist} f(r), & \text{if } s \text{ is even}; \\ e^{i(s+1)t}\, \overline{f(n-r)}, & \text{if } s \text{ is odd}. \end{cases}$$

Recalling that $f(q) = 0$ when $0 < q < n$ and using (P2) again, we obtain

$$f(sn + r) = \begin{cases} e^{ist}, & \text{if } s \text{ is even and } r = 0; \\ 0, & \text{if } 0 < r < n; \\ be^{ist}, & \text{if } s \text{ is odd and } r = 0 \end{cases} \tag{14.42}$$

for an arbitrary integer s.

Next we make use of a well-known property of roots of unity, namely,

$$\frac{1}{n}\sum_{k=0}^{n-1}\omega^{km} = \begin{cases} 1, & \text{if } n \text{ divides } m; \\ 0, & \text{if } n \text{ does not divide } m \end{cases} \tag{14.43}$$

where $\omega = e^{2\pi i/n}$. It follows from (14.42) and (14.43) that

$$f(m) = \frac{1+b}{2n}\sum_{k=0}^{n-1}\omega^{km}e^{imt/n} + \frac{1-b}{2n}\sum_{k=0}^{n-1}e^{im\pi/n}\omega^{km}e^{imt/n}.$$

Consequently, we have written f as a convex combination of sequences of the forms $(\omega^k e^{it/n})^m$ and $(e^{i\pi/n}\omega^k e^{it/n})^m$. Since $f \in \operatorname{ex} D_1$, it follows that $b = n = 1$ and $f(m) = e^{imt}$. We have shown that $f \in E(T)$. Hence (14.37) is valid.

Exercises for Section 14.6

14.79 Show that a convex set contains any convex combination of its elements.

14.80 Show that x is an extreme point of the convex set C if and only if it satisfies the following condition: $x = \alpha u + (1-\alpha)v$, where $0 < \alpha < 1$ and $u, v \in C$, implies $u = v = x$.

14.81 Prove parts (a)–(d) of Proposition 14.9 on page 539.

14.82 Prove assertions (14.30)–(14.33) on page 540.

14.83 Why does it suffice to prove the Krein-Milman theorem (Theorem 14.15 on page 540) in the case of real scalars?

14.84 Prove parts (b) and (c) of Corollary 14.9 on page 541.

14.85 Verify that the coefficient functionals in Example 14.21(b) on page 541 are continuous.

14.86 Suppose that Ω is a compact Hausdorff space. Let $P(\Omega)$ denote the collection of probability measures on Ω, that is, the set $\{\mu \in M_1(\Omega) : \mu(\Omega) = 1\}$. By Exercise 14.70 on page 536, $P(\Omega)$ is weak* compact, and it is easy to see that $P(\Omega)$ is convex. Show that $\operatorname{ex} P(\Omega) = \{\delta_x : x \in \Omega\}$.

14.87 Let K be a compact convex subset of a locally convex topological linear space and set $P(K) = \{\mu \in M_+(K) : \mu(K) = 1\}$. Show that each $\mu \in P(K)$ is the representing measure for some point of K. *Hint:* See Exercise 14.86.

14.88 Consider the space $\mathcal{L}^1(\mathcal{R})$.
a) Show that $\overline{B}_1(0)$ has no extreme points.
b) Deduce that Theorem 13.12 on page 480 fails in the case $p = \infty$.

14.89 Prove assertions (P1)–(P8) on page 544.

14.90 In the proof of Theorem 14.17, we defined $\hat{E}:[-\pi,\pi) \to \operatorname{ex} D_1$ by $\hat{E}(x) = E(e^{-ix})$. We also defined the set function $\mu_f:\mathcal{B}([-\pi,\pi)) \to \mathcal{R}$ by $\mu_f(A) = 2\pi\nu_f(\hat{E}(A))$.
a) Show that $\hat{E}$ is continuous, one-to-one, and onto.
b) Show that μ_f is well-defined, that is, if $A \in \mathcal{B}([-\pi,\pi))$, then $\hat{E}(A) \in \mathcal{B}(\operatorname{ex} D_1)$. *Hint:* First show that this result holds true if A is a closed interval contained in $[-\pi,\pi)$.
c) Show that μ_f is a measure on $\mathcal{B}([-\pi,\pi))$.
d) Show that $\ell(f) = (2\pi)^{-1}\int_{[-\pi,\pi)}\ell(\hat{E}(x))\,d\mu_f(x)$ for each weak* continuous linear functional ℓ.

14.91 Define

$$f(n) = \begin{cases} 1, & \text{if } n = 0; \\ 0, & \text{if } n \neq 0. \end{cases}$$

Show that $f \notin \operatorname{ex} D_1$.

14.92 Let Ω be a compact Hausdorff space and $f: \Omega \to \Omega$ be continuous. A measure $\mu \in M(\Omega)$ is said to be **invariant** with respect to f if

$$\mu(f^{-1}(A)) = \mu(A), \qquad A \in \mathcal{B}(\Omega).$$

Let $\mathcal{I}_f$ denote the collection of all probability measures in $M(\Omega)$ that are invariant with respect to f.
a) Show that $\mathcal{I}_f$ is convex and weak* compact.
b) Suppose $\mu \in \mathrm{ex}\,\mathcal{I}_f$. Prove that if $A \in \mathcal{B}(\Omega)$ and $f^{-1}(A) \subset A$, then $\mu(A) \in \{0, 1\}$. Measures that satisfy this condition are called **ergodic**. *Hint:* Refer to Theorem 9.10 on page 344.

PART FOUR

□

Harmonic Analysis, Dynamical Systems, and Hausdorff Measure

Ingrid Daubechies
(1954–)

Ingrid Daubechies was born in Houthalen, Belgium, on August 17, 1954, and is now a naturalized U.S. citizen. Her father, Marcel Daubechies, is a retired civil mining engineer who encouraged her to pursue her interest in science, and her mother, Simone Daubechies (nee Duran), is a retired criminologist who taught her the value of being your own person.

In 1980, Daubechies received her Ph.D. in physics from Vrije Universiteit Brussel (Free University in Brussels, Belgium) and remained there in a research position until 1987. From 1987 to 1994, she was a member of the technical staff at AT&T Bell Laboratories, taking leaves to spend time at the University of Michigan and Rutgers University.

The research interests of Daubechies focus on the mathematical aspects of time-frequency analysis (in particular, wavelets) and its applications. The simplest example of what is now known as a wavelet family was discovered in 1909 by Alfréd Haar. The usefulness of Haar's wavelets is limited, however, because they are discontinuous. Daubechies made a major contribution to wavelet theory when she found generalizations of Haar wavelets that, in addition to being highly regular, are considerably more effective for representing functions.

Daubechies has an extensive list of awards and honors. For instance, in 1994, she received the American Mathematical Society (AMS) Steele Prize for Exposition for her book *Ten Lectures on Wavelets*; in 1997, she was awarded the AMS Ruth Lyttle Satter prize; in 1998, she was elected to the United States National Academy of Sciences. Daubechies became, in 2000, the first woman to receive the National Academy of Sciences Award in Mathematics for excellence in published mathematical research. The award honored her "for fundamental discoveries on wavelets and wavelet expansions and for her role in making wavelets methods a practical basic tool of applied mathematics."

Since 1994, Daubechies has been a professor at Princeton University in the Mathematics Department and the Program in Applied and Computational Mathematics and, since 2004, William R. Kenan, Jr. Professor of Mathematics.

15

Elements of Harmonic Analysis

Much of the subject matter of real analysis emerged from attempts by various mathematicians to deal with problems associated with the idea, developed by Jean Baptiste Joseph Fourier, of expanding a function in a series of the form

$$f(x) = a_0 + \sum_{n=1}^{\infty}(a_n \cos nx + b_n \sin nx). \tag{15.1}$$

This expansion can be thought of as a decomposition of f into an infinite sum of harmonic (oscillatory) terms. Thus one speaks of (15.1) as a *harmonic analysis* of the function f.

In this chapter, we will investigate the meaning of (15.1) and some of its many variations by using ideas that we have explored and results that we have obtained in previous chapters. Sections 15.1 and 15.2 deal with properties of Fourier series, Sections 15.3–15.5 investigate the Fourier transform, and Sections 15.6 and 15.7 are devoted to wavelet expansions, analogues of Fourier expansions that have received much attention in recent years.

15.1 INTRODUCTION TO FOURIER SERIES

Recall that a complex-valued function f on $\mathcal{R}$ is said to be **periodic** with **period** p if

$$f(x + p) = f(x), \qquad x \in \mathcal{R}.$$

In most cases of interest, there is a smallest positive period, called the **basic period** of f. (See Exercise 15.1.) The reciprocal of the basic period is called the **frequency** of the periodic function. For convenience, and because it involves no real loss of generality, we restrict our treatment to functions having period 2π. (See Exercise 15.2.)

In harmonic analysis, often called Fourier analysis after its main founder, researchers attempt to understand complicated periodic functions in terms of simple ones. Specifically, it was Fourier's idea to try to represent a function with period 2π as a series of the form (15.1). The formula (15.1) yields the function f as a sum of simple oscillating terms, that is, sine and cosine terms whose frequencies form the arithmetic progression $1/2\pi, 2/2\pi, 3/2\pi, \ldots$. The main purpose of this section and the next section is to explore the meaning and ramifications of (15.1).

Using the formulas

$$2\cos x = e^{ix} + e^{-ix} \qquad \text{and} \qquad 2i\sin x = e^{ix} - e^{-ix},$$

we can recast (15.1) in the more compact form

$$f(x) = \sum_{n=-\infty}^{\infty} c_n e^{inx}. \tag{15.2}$$

Assuming that the series converges rapidly enough so that the integral and sum can be interchanged, we obtain

$$\int_{-\pi}^{\pi} f(x)e^{-inx}\,dx = \sum_{k=-\infty}^{\infty} c_k \int_{-\pi}^{\pi} e^{i(k-n)x}\,dx$$

and, hence, that

$$c_n = \frac{1}{2\pi}\int_{-\pi}^{\pi} f(x)e^{-inx}\,dx.$$

This result shows how the coefficients c_n, $n = 0, \pm1, \pm2, \ldots$, can be calculated explicitly from the function f and serves as motivation for the following definition.

DEFINITION 15.1   Fourier Coefficients, Transform, and Series
For $f \in \mathcal{L}^1([-\pi,\pi])$, the function $\hat{f}: \mathcal{Z} \to \mathbb{C}$ defined by

$$\hat{f}(n) = \frac{1}{2\pi}\int_{-\pi}^{\pi} f(x)e^{-inx}\,dx$$

is called the **Fourier transform** of f. We refer to the number $\hat{f}(n)$ as the **nth Fourier coefficient** of f and to the expression

$$\sum_{n=-\infty}^{\infty} \hat{f}(n)e^{inx}$$

as the **Fourier series** of f.

The Fourier series of f is said to **converge** at x if the sequence of partial sums

$$\sum_{k=-n}^{n} \hat{f}(k)e^{ikx}, \qquad n \in \mathcal{N},$$

has a finite limit $s(x)$. Convergence to $s(x)$ will often be indicated by the equation $s(x) = \sum_{n=-\infty}^{\infty} \hat{f}(n)e^{inx}$. The Fourier series of f is said to **converge in the norm** $\| \ \|$ to the function s if the forementioned partial sums converge to s with respect to $\| \ \|$.

EXAMPLE 15.1 *Illustrates Definition 15.1*

a) Let $f(x) = \sin mx$, where $m \in \mathcal{N}$. Then $\hat{f}(\pm m) = \pm 1/2i$ and $\hat{f}(n) = 0$ otherwise. The Fourier series of f converges to $f(x)$ for all x. In fact, the partial sums at x equal $f(x)$ as soon as $n \geq m$.

b) Consider the function f defined by

$$f(x) = \begin{cases} -\frac{1}{2}, & \text{if } x \in [-\pi, 0); \\ \frac{1}{2}, & \text{if } x \in [0, \pi]. \end{cases}$$

An easy calculation shows $\hat{f}(0) = 0$ and $\hat{f}(n) = i((-1)^n - 1)/2\pi n$ if $n \neq 0$. We will see later that the Fourier series of f converges to $f(x)$ at points of $(-\pi, 0) \cup (0, \pi)$ and to 0 at $-\pi$, 0, and π. □

The following theorem provides some basic properties of Fourier transforms of functions in $\mathcal{L}^1([-\pi, \pi])$.

□ □ □ **THEOREM 15.1**

Let $f \in \mathcal{L}^1([-\pi, \pi])$. Then

a) $\hat{f} \in C_0(\mathcal{Z})$.

b) $\|\hat{f}\|_{\mathcal{Z}} \leq \|f\|_1/2\pi$.

c) $\hat{f} = 0 \Rightarrow f = 0$ ae.

PROOF We leave the proof of part (a) to the reader as Exercise 15.3. Part (b) follows from the inequality

$$|\hat{f}(n)| \leq \frac{1}{2\pi} \int_{-\pi}^{\pi} |f(x)e^{-inx}| \, dx = \frac{1}{2\pi}\|f\|_1.$$

Example 13.12 on page 472 shows that part (c) holds whenever f is a function in $\mathcal{L}^2([-\pi, \pi])$. The same argument remains valid, however, if $\mathcal{L}^2([-\pi, \pi])$ is replaced by $\mathcal{L}^1([-\pi, \pi])$ and $\| \ \|_2$ is replaced by $\| \ \|_1$. ■

Any function with period 2π is completely determined by its values on the interval $(-\pi, \pi]$. Conversely, a function f defined on $(-\pi, \pi]$ extends uniquely to a periodic function on all of $\mathcal{R}$ via

$$f(x) = f(x - 2\pi k), \qquad x \in ((2k-1)\pi, \ (2k+1)\pi], \quad k \in \mathcal{Z}. \tag{15.3}$$

A continuous function f on $[-\pi, \pi]$ extends to a continuous periodic function via (15.3) if and only if $f(-\pi) = f(\pi)$.

We will use $C_{2\pi}$ to denote the space of continuous complex-valued functions with period 2π. And, unless stated otherwise, we assume that $C_{2\pi}$ is equipped with the supremum norm,

$$\|f\|_{\mathcal{R}} = \|f\|_{[-\pi,\pi]} = \|f\|_\infty = \sup\{\,|f(x)| : x \in [-\pi,\pi]\,\}.$$

The normed space $C_{2\pi}$ is identified via (15.3) with a closed subspace of the Banach space $C([-\pi,\pi])$ and, hence, it is a Banach space.

Similarly, for $1 \le p \le \infty$, $\mathcal{L}_{2\pi}^p$ denotes the space of complex-valued Lebesgue measurable functions on $\mathcal{R}$ with period 2π whose restrictions to $[-\pi,\pi]$ are in $\mathcal{L}^p([-\pi,\pi])$. We assume that $\mathcal{L}_{2\pi}^p$ has the norm

$$\|f\|_p = \begin{cases} (\int_{-\pi}^{\pi} |f(x)|^p\,dx)^{1/p}, & \text{if } 1 \le p < \infty; \\ \inf\{\,M : |f| \le M \text{ ae}\,\}, & \text{if } p = \infty. \end{cases}$$

Since $\mathcal{L}_{2\pi}^p$ is identified via (15.3) with the space $\mathcal{L}^p([-\pi,\pi])$, it is, according to Riesz's theorem (Theorem 13.11 on page 478), a Banach space.

For a function f on $\mathcal{R}$, let f_y denote the **translated function** defined by $f_y(x) = f(x-y)$. If $f \in \mathcal{L}_{2\pi}^p$, then so is f_y and, likewise, if $f \in C_{2\pi}$, then so is f_y. Thus we say that the spaces $C_{2\pi}$ and $\mathcal{L}_{2\pi}^p$ are **translation invariant.** Furthermore, it follows from Exercise 15.2 that $\|f_y\|_p = \|f\|_p$.

Some important properties of the spaces $\mathcal{L}_{2\pi}^p$ and $C_{2\pi}$ are given in the following proposition. Its proof is left to the reader as Exercise 15.4.

□ □ □ **PROPOSITION 15.1**

a) For $p \ge 1$, we have $C_{2\pi} \subset \mathcal{L}_{2\pi}^\infty \subset \mathcal{L}_{2\pi}^p \subset \mathcal{L}_{2\pi}^1$.

b) For $1 \le p < \infty$, $C_{2\pi}$ is dense in $\mathcal{L}_{2\pi}^p$ with respect to the norm $\|\ \|_p$.

c) For $1 \le p < \infty$, we have $(\mathcal{L}_{2\pi}^p)^* = \mathcal{L}_{2\pi}^q$, where $1/p + 1/q = 1$. Specifically, ℓ is a continuous linear functional on $\mathcal{L}_{2\pi}^p$ if and only if there exists a function $g \in \mathcal{L}_{2\pi}^q$ such that

$$\ell(f) = \int_{-\pi}^{\pi} f(x)g(x)\,dx, \qquad f \in \mathcal{L}_{2\pi}^p.$$

Furthermore, $\|\ell\|_* = \|g\|_q$.

d) For $f \in \mathcal{L}_{2\pi}^1$ and $y \in \mathcal{R}$, we have

$$\int_y^{y+2\pi} f(x)\,dx = \int_{-\pi}^{\pi} f(x)\,dx.$$

e) Each function in $C_{2\pi}$ is uniformly continuous on $\mathcal{R}$.

f) For $f \in \mathcal{L}_{2\pi}^1$ and $y \in \mathcal{R}$, we have

$$\widehat{f_y}(n) = e^{-iny}\hat{f}(n), \qquad n \in Z.$$

For $f \in \mathcal{L}_{2\pi}^1$, let $S_n(f)$ denote the nth partial sum of the Fourier series of f:

$$S_n(f)(x) = \sum_{k=-n}^{n} \hat{f}(k)e^{ikx}. \tag{15.4}$$

It is easy to see that (15.4) defines a linear operator S_n on $\mathcal{L}_{2\pi}^1$ with range in $C_{2\pi}$. Because $\mathcal{L}_{2\pi}^p \subset \mathcal{L}_{2\pi}^1$, it follows that S_n also maps $\mathcal{L}_{2\pi}^p$ into $C_{2\pi}$. The following proposition describes some essential properties of S_n.

□ □ □ PROPOSITION 15.2

Define

$$D_n(x) = \sum_{k=-n}^{n} e^{ikx}, \qquad n = 0, 1, 2, \dots.$$

Then the following properties hold.

a) We have

$$D_n(x) = \begin{cases} \dfrac{\sin((n + 1/2)x)}{\sin(x/2)}, & \text{if } x/2\pi \notin \mathcal{Z}; \\[2ex] 2n + 1, & \text{if } x/2\pi \in \mathcal{Z}. \end{cases}$$

b) $(2\pi)^{-1} \int_{-\pi}^{\pi} D_n(t)\, dt = 1.$

c) $D_n(-x) = D_n(x)$ *for each* $x \in \mathcal{R}.$

d) If $f \in \mathcal{L}_{2\pi}^1$, *then* $S_n(f)(x) = (2\pi)^{-1} \int_{-\pi}^{\pi} f(t) D_n(x - t)\, dt$ *for each* $x \in \mathcal{R}.$

e) If f *is a trigonometric polynomial, then* $S_n(f) = f$ *whenever* $n \geq \deg f.$

PROOF   We prove part (a) and leave the remaining parts to the reader as Exercise 15.5. If $x/2\pi \in \mathcal{Z}$, then $e^{ikx} = 1$ for each integer k and, consequently, $D_n(x) = 2n + 1$. If $x/2\pi \notin \mathcal{Z}$, then, by using $\overline{e^{ikx}} = e^{-ikx}$ and the formula for a geometric sum, we have

$$D_n(x) = \sum_{k=0}^{n} e^{ikx} + \sum_{k=1}^{n} e^{-ikx} = 2\Re\left(\sum_{k=0}^{n} e^{ikx}\right) - 1 = \Re 2\, \frac{e^{i(n+1)x} - 1}{e^{ix} - 1} - 1$$

$$= \Re 2\, \frac{e^{i(n+1/2)x} - e^{-ix/2}}{e^{ix/2} - e^{-ix/2}} - 1 = \frac{\sin((n + 1/2)x)}{\sin(x/2)},$$

as required.   ■

The function D_n, often called the **Dirichlet kernel,** changes sign more and more rapidly as n increases. This fact is a main reason why so little can be said about the behavior of the sequence $\{S_n(f)\}_{n=1}^{\infty}$ of partial sums of the Fourier series of a function f unless special conditions are imposed. However, the corresponding sequence of averages

$$A_n(f) = \frac{1}{n} \sum_{k=0}^{n-1} S_k(f), \qquad n \in \mathcal{N}, \tag{15.5}$$

satisfies a formula similar to Proposition 15.2(d), where D_n is replaced by a nonnegative function. This fact tends to make $\{A_n(f)\}_{n=1}^{\infty}$ a more tractable sequence than $\{S_n(f)\}_{n=1}^{\infty}$.

Clearly, (15.5) defines a linear operator on $\mathcal{L}_{2\pi}^1$ with range in $C_{2\pi}$. Proposition 15.3 presents some basic properties of A_n.

□ □ □ PROPOSITION 15.3

Define

$$F_n(x) = \frac{1}{n} \sum_{k=0}^{n-1} D_k(x), \qquad n \in \mathcal{N}.$$

Then the following results hold.

a) We have

$$F_n(x) = \begin{cases} \dfrac{1}{n} \left(\dfrac{\sin(nx/2)}{\sin(x/2)} \right)^2, & \text{if } x/2\pi \notin \mathcal{Z}; \\ n, & \text{if } x/2\pi \in \mathcal{Z}. \end{cases}$$

b) $(2\pi)^{-1} \int_{-\pi}^{\pi} F_n(t)dt = 1$.

c) $F_n(-x) = F_n(x)$ *for each* $x \in \mathcal{R}$.

d) For each $\delta \in (0, \pi)$, $\lim_{n \to \infty} \sup\{ F_n(x) : \delta \le |x| \le \pi \} = 0$.

e) If $f \in \mathcal{L}_{2\pi}^1$, *then* $A_n(f)(x) = (2\pi)^{-1} \int_{-\pi}^{\pi} f(t) F_n(x - t)\, dt$ *for each* $x \in \mathcal{R}$.

f) If $1 \le p \le \infty$ *and* $f \in \mathcal{L}_{2\pi}^p$, *then* $\| A_n(f) \|_p \le \| f \|_p$.

PROOF The proofs of parts (a)–(e) are left to the reader as Exercise 15.6. To prove part (f), we argue as follows. If $p = \infty$, then, by part (e),

$$|A_n(f)(x)| \le \frac{1}{2\pi} \int_{-\pi}^{\pi} |f(t)| F_n(x - t)\, dt \le \frac{\| f \|_\infty}{2\pi} \int_{-\pi}^{\pi} F_n(x - t)\, dt, \qquad x \in \mathcal{R}.$$

Applying Proposition 15.2, parts (c) and (b), we obtain $|A_n(f)(x)| \le \| f \|_\infty$ for each $x \in \mathcal{R}$ and, so, $\| A_n(f) \|_\infty \le \| f \|_\infty$.

Now suppose that $1 \le p < \infty$ and let $1/p + 1/q = 1$. For $x \in \mathcal{R}$, define the Borel measure μ on $[-\pi, \pi]$ by $\mu(B) = (2\pi)^{-1} \int_B F_n(x - t)\, d\lambda(t)$. It follows from parts (a), (b), (c), and Exercise 15.2 that μ is nonnegative and $\mu([-\pi, \pi]) = 1$. Moreover, by part (e) and Exercise 5.61 on page 164, $A_n(f)(x) = \int_{-\pi}^{\pi} f(t)\, d\mu(t)$. Hence, by Hölder's inequality (Theorem 13.9 on page 477),

$$|A_n(f)(x)| \le \left(\int_{[-\pi, \pi]} |f(t)|^p\, d\mu(t) \right)^{1/p}.$$

Applying Exercise 5.61 again and using Fubini's theorem, we obtain

$$\| A_n(f) \|_p^p \le \frac{1}{2\pi} \int_{-\pi}^{\pi} |f(t)|^p \int_{-\pi}^{\pi} F_n(x - t)\, dx\, dt = \| f \|_p^p.$$

Thus, part (f) is established. ∎

The function F_n introduced in Proposition 15.3, often referred to as **Fejér's kernel,** plays an essential role in Fourier analysis.

We conclude this section by introducing another important function in harmonic analysis, namely, the **sinc function,** defined by

$$\operatorname{sinc} x = \begin{cases} \dfrac{\sin x}{x}, & \text{for } x \ne 0; \\ 1, & \text{for } x = 0. \end{cases} \tag{15.6}$$

As Exercise 4.40 on page 138 shows, sinc is not Lebesgue integrable over $\mathcal{R}$, yet its improper Riemann integral exists. In particular, we can assert that

$$\lim_{b \to \infty} \int_b^{\infty} \operatorname{sinc} x\, dx = 0. \tag{15.7}$$

We will find (15.7) useful in the next section.

Exercises for Section 15.1

Note: A ★ denotes an exercise that will be subsequently referenced.

15.1 Let f be a complex-valued measurable function on $\mathcal{R}$ such that there are arbitrarily small positive numbers p that satisfy $f(x + p) = f(x)$ for almost all $x \in \mathcal{R}$. Show that f is constant ae.

★15.2 Let f be a complex-valued function on $\mathcal{R}$.
a) Show that f has period p if and only if the function defined by $f(px/2\pi)$ has period 2π.
b) Show that if f has period $p > 0$ and is integrable over every bounded interval of $\mathcal{R}$, then $\int_x^{x+p} f(t)\, dt$ is independent of x.

15.3 Prove Theorem 15.1(a) on page 553, often referred to as the **Riemann-Lebesgue lemma**. *Hint:* Start with the case where f is the characteristic function of a subinterval of $[-\pi, \pi]$.

15.4 Prove Proposition 15.1 on page 554. *Hint:* For part (b), see Exercise 13.56 on page 482 and, for part (d), refer to Exercise 15.2.

15.5 Complete the proof of Proposition 15.2 on page 555.

15.6 Prove parts (a)–(e) of Proposition 15.3 on pages 555–556.

15.7 Let f be a complex-valued function defined on $\mathcal{R}$ that satisfies the following conditions: f is not identically 0, is continuous at 0, has period 2π, and $f(x + y) = f(x)f(y)$ for all $x, y \in \mathcal{R}$. Show that $f(x) = e^{inx}$ for some integer n.

★15.8 For $f, g \in \mathcal{L}_{2\pi}^1$, let the function $f * g$ be defined by

$$(f * g)(x) = \frac{1}{2\pi} \int_{-\pi}^{\pi} f(x - y)g(y)\, dy.$$

Show that $f * g$ is well-defined ae and belongs to $\mathcal{L}_{2\pi}^1$. $f * g$ is called the **convolution** of f and g. Observe that $S_n(f) = f * D_n$ and $A_n(f) = f * F_n$.

In Exercises 15.9–15.12, $f * g$ denotes the convolution product introduced in Exercise 15.8.

★15.9 Verify that the convolution product is commutative and associative, that is, prove that
a) $f * g = g * f$.
b) $(f * g) * h = f * (g * h)$.

★15.10 Show that if $g \in \mathcal{L}_{2\pi}^\infty$, then $f * g \in C_{2\pi}$.

★15.11 Show that if $f \in \mathcal{L}_{2\pi}^1$ has a derivative at all points of $\mathcal{R}$, $f' \in \mathcal{L}_{2\pi}^\infty$, and $g \in \mathcal{L}_{2\pi}^1$, then $(f * g)'$ exists and equals $(f') * g$.

★15.12 Prove that $\widehat{f * g} = \hat{f}\hat{g}$.

*In Exercises 15.13–15.16, you are asked to consider Fourier coefficients of measures. A measure $\mu \in M([-\pi, \pi])$ is said to be **periodic** if $\mu(\{-\pi\}) = \mu(\{\pi\})$. The set $M_{2\pi}$ of all periodic measures is a closed subspace of the Banach space $M([-\pi, \pi])$. The **Fourier coefficients** of a measure $\mu \in M_{2\pi}$ are defined by the formula*

$$\hat{\mu}(n) = \frac{1}{2\pi} \int_{[-\pi,\pi]} e^{-inx}\, d\mu(x), \qquad n \in \mathcal{Z}.$$

In Exercises 15.13–15.16, we assume that μ and ν are members of $M_{2\pi}$.

★15.13 Show that if $\mu \ll \lambda$ with Radon-Nikodym derivative g, then $\hat{\mu} = \hat{g}$.

15.14   Prove that $|\hat{\mu}(n)| \le |\mu|([-\pi, \pi])/2\pi$. Does $\hat{\mu}$ always lie in $C_0(\mathcal{Z})$? (See part (a) of Theorem 15.1.)

15.15   Set $S_n(\mu)(x) = \sum_{k=-n}^{n} \hat{\mu}(k)e^{ikx}$. Verify that $S_n(\delta_0) = D_n/2\pi$.

15.16   Show that if $\hat{\mu} = \hat{\nu}$, then $\mu = \nu$. *Hint:* See Example 12.15 on pages 448–449.

15.2 CONVERGENCE OF FOURIER SERIES

For a particular function $f \in \mathcal{L}_{2\pi}^1$ and a number $x \in \mathcal{R}$, we pose the following two questions:

Question 1: Does the Fourier series of f converge at x?

Question 2: If the answer to Question 1 is yes, does the Fourier series of f at x converge to $f(x)$?

Because these two questions are so broadly posed, one has to give the answer "not always" to both. Nevertheless, they serve as motivators for a host of interesting and useful results. In this section, we present samples of two approaches to answering Questions 1 and 2, namely:

- Narrow the class of functions under consideration.

- Modify the convergence requirement.

The following theorem shows, among other things, that if the Fourier coefficients of a function converge to 0 rapidly enough, then the Fourier series converges to the function for almost all x.

□ □ □   **THEOREM 15.2**

Let $f \in \mathcal{L}_{2\pi}^1$. If $\hat{f} \in \ell^1(\mathcal{Z})$, then the Fourier series of f converges uniformly to a continuous function g such that $f = g$ ae. In particular, the Fourier series of f converges to f almost everywhere.

PROOF   Because $\sum_{n=-\infty}^{\infty} |\hat{f}(n)| < \infty$, we deduce from Exercise 10.89 on page 386 that the series

$$g(x) = \sum_{n=-\infty}^{\infty} \hat{f}(n)e^{inx}$$

converges uniformly on $\mathcal{R}$ and that g is continuous and has period 2π. It follows that $\hat{g} = \hat{f}$ and, consequently, from Theorem 15.1 on page 553 that $g = f$ ae. ∎

EXAMPLE 15.2   *Illustrates Theorem 15.2*

Let f be the function defined on $\mathcal{R}$ with period 2π that satisfies $f(x) = 1 - |x|/\pi$ for $x \in [-\pi, \pi]$. An easy calculation shows that $\hat{f}(0) = 1/2$ and

$$\hat{f}(n) = \frac{1 - (-1)^n}{n^2 \pi^2}, \qquad n \ne 0.$$

Because $\hat{f} \in \ell^1(\mathcal{Z})$, Theorem 15.2 implies that we have the expansion

$$1 - \frac{|x|}{\pi} = \sum_{n=-\infty}^{\infty} \hat{f}(n)e^{inx} = \frac{1}{2} + \sum_{n=1}^{\infty} \frac{1 - (-1)^n}{n^2\pi^2}(e^{inx} + e^{-inx})$$

$$= \frac{1}{2} + \frac{4}{\pi^2}\sum_{n=0}^{\infty} \frac{\cos((2n+1)x)}{(2n+1)^2},$$

where the series converges uniformly for $x \in [-\pi, \pi]$. Setting $x = 0$, we obtain the formula

$$\frac{\pi^2}{8} = \sum_{n=0}^{\infty} \frac{1}{(2n+1)^2}$$

as a special case. ☐

Our next theorem, due to Dirichlet, shows that the Fourier series of a function of bounded variation converges everywhere and that it converges almost everywhere to the function. The proof of Dirichlet's theorem requires the following lemma.

☐ ☐ ☐ **LEMMA 15.1**

Suppose that F is a right-continuous function of bounded variation defined on $[a, b]$. Then there exists a $\mu \in M([a, b])$ such that $F(t) - F(a) = \mu((a, t])$ for each $t \in [a, b]$.

PROOF Suppose first that F is nondecreasing and nonnegative. Then we can extend F to a distribution function on $\mathcal{R}$ by defining $F(t) = 0$ for $t < a$ and $F(t) = F(b)$ for $t > b$. Applying Theorem 6.3 on page 196 to the extended version of F, we see that there is a finite Borel measure μ on $\mathcal{R}$ such that $F(t) - F(a) = \mu((a, t])$ for each $t \in (a, b]$. By Exercise 13.68 on page 490, μ is regular and, hence, the restriction of μ to Borel measurable subsets of $[a, b]$ is a regular Borel measure that satisfies the assertion of the lemma.

To continue, we next assume that F is real valued. Then, according to Theorem 8.3 on page 283, we can write $F = g_1 - g_2$ where g_1 and g_2 are nondecreasing functions on $[a, b]$. Letting $\beta = g_1(a) \wedge g_2(a)$, we define $F_1(b) = g_1(b) - \beta$, $F_2(b) = g_2(b) - \beta$, and for $t \in [a, b)$, $F_1(t) = g_1(t+) - \beta$ and $F_2(t) = g_2(t+) - \beta$. Then F_1 and F_2 are nonnegative, nondecreasing, and right continuous on $[a, b]$, and we have $F = F_1 - F_2$. Therefore, there exist $\mu_1, \mu_2 \in M([a, b])$ such that $F(t) - F(a) = \mu_1((a, t]) - \mu_2((a, t])$ for each $t \in [a, b]$. It follows that the signed measure $\mu = \mu_1 - \mu_2$ satisfies the assertion of the lemma.

It remains to establish the lemma in case F is complex valued. This is done by first noting that the real and imaginary parts of F are real-valued, right-continuous functions of bounded variation and then applying what we just proved to the real and imaginary parts of F. ■

□ □ □ **THEOREM 15.3 Dirichlet's Theorem**

Suppose that $f \in \mathcal{L}_{2\pi}^1$ and is of bounded variation on the interval $[-\pi, \pi]$. Then

$$\frac{1}{2}(f(x+) + f(x-)) = \sum_{n=-\infty}^{\infty} \hat{f}(n)e^{inx} \tag{15.8}$$

for each $x \in \mathcal{R}$. In particular, the Fourier series of f converges to f almost everywhere.

PROOF Because $f(x+) = f(x)$ for all but countably many x, if we replace $f(x)$ by $f(x+)$ at each x, the Fourier coefficients of the function are unaltered. Therefore, without loss of generality, we can assume that f is right continuous on $\mathcal{R}$.

We first show that (15.8) holds when $x = 0$. Using Proposition 15.2 on page 555, we obtain

$$\begin{aligned}
S_n(f)(0) &= \frac{1}{2\pi} \int_{-\pi}^{\pi} f(t) D_n(t)\, dt \\
&= \frac{1}{2\pi} \int_0^{\pi} (f(t) - f(0)) D_n(t)\, dt \\
&\quad + \frac{1}{2\pi} \int_{-\pi}^0 (f(t) - f(0-)) D_n(t)\, dt \\
&\quad + \frac{f(0)}{2\pi} \int_0^{\pi} D_n(t)\, dt + \frac{f(0-)}{2\pi} \int_{-\pi}^0 D_n(t)\, dt \tag{15.9} \\
&= \frac{1}{2\pi} \int_0^{\pi} (f(t) - f(0)) D_n(t)\, dt \\
&\quad + \frac{1}{2\pi} \int_0^{\pi} (f(-t) - f(0-)) D_n(t)\, dt \\
&\quad + \frac{1}{2}(f(0) + f(0-)).
\end{aligned}$$

We will show that

$$\lim_{n \to \infty} \int_0^{\pi} (f(t) - f(0)) D_n(t)\, dt = 0. \tag{15.10}$$

Set $g(t) = (t/2)f(t)/\sin(t/2)$ for $t \in (0, \pi]$ and $g(0) = f(0)$. Clearly, g is right continuous and, referring to Exercise 8.28 on page 285, we see that it has bounded variation over $[0, \pi]$. Hence, by Lemma 15.1, there is a regular Borel measure μ on $[0, \pi]$ such that

$$g(t) - g(0) = \mu((0, t]) = \int_{(0,t]} d\mu(x)$$

for $t \in [0, \pi]$.

Applying Fubini's theorem, we obtain

$$\int_0^\pi (f(t) - f(0))D_n(t)\, dt$$

$$= \int_0^\pi (g(t) - g(0))\frac{\sin((n+1/2)t)}{t/2}\, dt = \int_0^\pi \int_{(0,t]} \frac{\sin((n+1/2)t)}{t/2}\, d\mu(x)\, dt$$

$$= \int_{(0,\pi]} \int_x^\pi \frac{\sin((n+1/2)t)}{t/2}\, dt\, d\mu(x) = 2\int_{(0,\pi]} \int_{(n+1/2)x}^{(n+1/2)\pi} \operatorname{sinc} v\, dv\, d\mu(x).$$

It follows from (15.7) on page 556 that the sequence

$$\int_{(n+1/2)x}^{(n+1/2)\pi} \operatorname{sinc} v\, dv$$

is uniformly bounded and tends to 0 as $n \to \infty$ for $x \in (0, \pi]$. Therefore, by the dominated convergence theorem, (15.10) is satisfied. A similar argument applied to the function $g(-t)$ shows that

$$\lim_{n\to\infty} \int_0^\pi (f(-t) - f(0-))D_n(t)\, dt = 0.$$

In view of (15.9), we conclude that

$$\lim_{n\to\infty} S_n(f)(0) = \frac{1}{2}(f(0) + f(0-)). \tag{15.11}$$

To complete the proof of (15.8), we apply (15.11) to the translated function f_{-x}. Using Proposition 15.2 on page 555 and Exercise 15.2 on page 557, we obtain

$$S_n(f_{-x})(0) = \frac{1}{2\pi}\int_{-\pi}^\pi f(t+x)D_n(t)\, dt = \frac{1}{2\pi}\int_{-\pi+x}^{\pi+x} f(t)D_n(t-x)\, dt = S_n(f)(x).$$

Thus, from (15.11) and Exercise 15.17, we have

$$\lim_{n\to\infty} S_n(f)(x) = \frac{1}{2}(f_{-x}(0) + f_{-x}(0-)) = \frac{1}{2}(f(x) + f(x-)),$$

as required. The last sentence in the statement of this theorem follows from the fact that a function of bounded variation has only countably many points of discontinuity. ∎

EXAMPLE 15.3   *Illustrates Dirichlet's Theorem*

Refer to Example 15.1(b) on page 553. Clearly, f is a right-continuous function of bounded variation and $f(0) = 1/2$ and $f(0-) = -1/2$. Considering the Fourier series of f at $x = 0$, we have

$$\sum_{n=-\infty}^\infty \hat{f}(n) = \lim_{n\to\infty}\left(\sum_{k=1}^n \frac{i((-1)^k - 1))}{2\pi k} + \sum_{k=-n}^{-1} \frac{i((-1)^k - 1))}{2\pi k} \right)$$

$$= 0 = \frac{1}{2}(f(0) + f(0-)),$$

as predicted by Dirichlet's theorem.   □

Theorems 15.2 and 15.3 are about pointwise convergence of Fourier series. Other interesting results can be obtained if we allow alternate modes of convergence. For example, if $f \in \mathcal{L}_{2\pi}^2$, then it follows from Example 13.12 on page 472 that

$$\lim_{n \to \infty} \|f - S_n(f)\|_2 = 0, \tag{15.12}$$

that is, the Fourier series of f converges to f with respect to the norm $\| \ \|_2$. The weaker formula

$$\lim_{n \to \infty} \|f - A_n(f)\|_2 = 0, \tag{15.13}$$

which follows immediately from (15.12), suggests still another way of looking at the convergence of Fourier series.

Intuitively, because averaging tends to diminish fluctuations, it should tend to make partial sums easier to handle. That this intuition is correct is borne out by the following theorem of which (15.13) is a special case.

□ □ □ **THEOREM 15.4**

a) If $f \in C_{2\pi}$, then $\lim_{n \to \infty} \|f - A_n(f)\|_\infty = 0$.

b) For $1 \le p < \infty$, if $f \in \mathcal{L}_{2\pi}^p$, then $\lim_{n \to \infty} \|f - A_n(f)\|_p = 0$.

c) If $f \in \mathcal{L}_{2\pi}^\infty$, then the sequence $\{A_n(f)\}_{n=1}^\infty$ converges to f in the weak* topology of $\mathcal{L}_{2\pi}^\infty$.

PROOF   We show first that for $f \in C_{2\pi}$,

$$\lim_{n \to \infty} A_n(f)(0) = f(0). \tag{15.14}$$

Later in the proof, we will generalize the argument to establish part (a). By Proposition 15.3 on pages 555–556, we have

$$|A_n(f)(0) - f(0)| \le \frac{1}{2\pi} \int_{-\pi}^{\pi} |f(t) - f(0)| F_n(t) \, dt.$$

Given $\epsilon > 0$, there exists a $\delta \in (0, \pi)$ such that $|f(t) - f(0)| < \epsilon/2$ for $t \in (-\delta, \delta)$. Hence,

$$\begin{aligned} |A_n(f)(0) - f(0)| &\le \frac{\epsilon}{2} \cdot \frac{1}{2\pi} \int_{-\delta}^{\delta} F_n(t) \, dt \\ &+ \frac{1}{2\pi} \int_{[-\pi,\pi]\setminus(-\delta,\delta)} |f(t) - f(0)| F_n(t) \, dt. \end{aligned} \tag{15.15}$$

Since, by Proposition 15.3, $(2\pi)^{-1} \int_{-\delta}^{\delta} F_n(t) dt \le 1$, we have that

$$|A_n(f)(0) - f(0)| \le \frac{\epsilon}{2} + 2\|f\|_\infty \sup\{F_n(t) : \delta \le |t| \le \pi\}.$$

Equation (15.14) now follows from Proposition 15.3(d).

To complete the proof of part (a), we observe that, as f is uniformly continuous on $\mathcal{R}$, the δ in the foregoing argument can be chosen so that whenever $t \in (-\delta, \delta)$, $|f_{-x}(t) - f_{-x}(0)| < \epsilon/2$ for all $x \in \mathcal{R}$. It follows that the inequality (15.15) is satisfied when f is replaced by f_{-x}. As $A_n(f_{-x})(0) = A_n(f)(x)$ and $\|f_{-x}\|_\infty = \|f\|_\infty$, we deduce that

$$|A_n(f)(x) - f(x)| \leq \frac{\epsilon}{2} + 2\|f\|_\infty \sup\{F_n(t) : \delta \leq |t| \leq \pi\}.$$

Hence, the sequence $\{A_n(f)\}_{n=1}^\infty$ converges uniformly to f. This result verifies part (a) of the theorem.

To establish part (b), we make use of the density of $C_{2\pi}$ in $\mathcal{L}_{2\pi}^p$ and the inequality $\|g\|_p \leq (2\pi)^{1/p}\|g\|_\infty$ for functions in $C_{2\pi}$. For $\epsilon > 0$, choose $g \in C_{2\pi}$ such that $\|f - g\|_p < \epsilon/3$. Then we have

$$\|A_n(f) - f\|_p \leq \|A_n(f - g)\|_p + \|A_n(g) - g\|_p + \|f - g\|_p$$

$$< \frac{2}{3}\epsilon + (2\pi)^{1/p}\|A_n(g) - g\|_\infty.$$

It now follows from part (a) that $\|A_n(f) - f\|_p < \epsilon$ for sufficiently large n.

For part (c), we must show that

$$\lim_{n \to \infty} \int_{-\pi}^\pi A_n(f)(x)\, g(x)\, dx = \int_{-\pi}^\pi f(x)g(x)\, dx \tag{15.16}$$

for all $g \in \mathcal{L}_{2\pi}^1$. By Proposition 15.3 and Fubini's theorem,

$$\int_{-\pi}^\pi A_n(f)(x)g(x)\, dx = \frac{1}{2\pi} \int_{-\pi}^\pi \int_{-\pi}^\pi f(t)g(x)F_n(x - t)\, dt\, dx$$

$$= \int_{-\pi}^\pi f(t)A_n(g)(t)\, dt.$$

Consequently,

$$\left| \int_{-\pi}^\pi A_n(f)(x)\, g(x)\, dx - \int_{-\pi}^\pi f(x)g(x)\, dx \right|$$

$$= \left| \int_{-\pi}^\pi f(t)(A_n(g)(t) - g(t))\, dt \right| \leq \|f\|_\infty \|A_n(g) - g\|_1$$

and, hence, in view of part (b), we see that (15.16) holds. ∎

EXAMPLE 15.4 *Illustrates Theorem 15.4*

It follows from Exercise 14.27 on page 511 that there exists a function $f \in C_{2\pi}$ such that the sequence $\{S_n(f)(x_0)\}_{n=1}^\infty$ diverges for some x_0. Theorem 15.4 shows that by averaging the $S_n(f)$s we can remove the divergence at x_0 and get uniform convergence for all x. □

Exercises for Section 15.2

15.17 Show that $f_{-x}(0+) = f(x+)$ and $f_{-x}(0-) = f(x-)$.

15.18 Explain why part (b) of Theorem 15.4 (page 562) is false when $p = \infty$.

15.19 **Localization of Fourier series.**
a) Suppose $f \in \mathcal{L}_{2\pi}^1$ and that f vanishes on some open interval $J \subset [-\pi, \pi]$. Show that the Fourier series of f converges uniformly to 0 on compact subsets of J. *Hint:* Start with the case where f is the characteristic function of an interval disjoint from J.
b) Deduce from part (a) that if $f, g \in \mathcal{L}_{2\pi}^1$ and $f = g$ on some open interval $J \subset [-\pi, \pi]$, then $\sum_{n=-\infty}^{\infty} \hat{f}(n)e^{inx}$ converges at a point $x \in J$ if and only if $\sum_{n=-\infty}^{\infty} \hat{g}(n)e^{inx}$ converges.

15.20 Suppose $f^{(j)} \in C_{2\pi}$ for $j = 0, 1, \ldots, m-1$ and that $f^{(m-1)}$ is absolutely continuous.
a) Show that $|\hat{f}(n)| \leq \|f^{(m)}\|_1 / (2\pi n^m)$.
b) Deduce that if $m \geq 2$, then the Fourier series of f converges uniformly to f.

★15.21 In this exercise, we evaluate $\int_0^\infty \operatorname{sinc} x \, dx$.
a) Consider the function f with period 2π that satisfies

$$f(x) = \begin{cases} -(\pi + x)/2, & \text{if } -\pi \leq x < 0; \\ (\pi - x)/2, & \text{if } 0 \leq x \leq \pi. \end{cases}$$

Show that

$$\hat{f}(n) = \begin{cases} 0, & \text{if } n = 0; \\ (2ni)^{-1}, & \text{if } n \neq 0. \end{cases}$$

Deduce that

$$\sum_{n=1}^{\infty} \frac{\sin nx}{n} = \begin{cases} -(\pi + x)/2, & \text{if } -\pi \leq x < 0; \\ 0, & \text{if } x = 0; \\ (\pi - x)/2, & \text{if } 0 < x \leq \pi. \end{cases}$$

b) Show that for $x \in [-\pi, \pi]$,

$$S_n(f)(x) = -\frac{1}{2}x + \frac{1}{2}\int_0^x D_n(t) \, dt.$$

c) Show that for $x \in [-\pi, \pi]$,

$$S_n(f)(x) = -\frac{1}{2}x + \int_0^x \frac{\sin((n+1/2)t)}{t} \, dt + q_n(x),$$

where $q_n(x)$ tends to 0 uniformly for $x \in [-\pi, \pi]$ as $n \to \infty$.
d) Deduce from part (c) that $\int_0^\infty \operatorname{sinc} v \, dv = \pi/2$.

15.22 In this exercise, we study the behavior near 0 of the sequence of partial sums of the Fourier series of a function f. It is assumed that f is right continuous and of bounded variation on $[-\pi, \pi]$. For $0 < \delta < \pi$, let

$$w_{n,\delta}(f) = \sup\{ S_n(f)(x) - S_n(f)(-x) : 0 < x < \delta \}$$

and $w_\delta(f) = \limsup_{n \to \infty} w_{n,\delta}(f)$.
a) For each nonnegative integer n, verify that the function

$$G_n(x) = \int_0^x \frac{\sin(n+1/2)t}{t} \, dt$$

has a local max and local min at $\pi/(n+1/2)$ and $-\pi/(n+1/2)$, respectively, and, moreover, that

$$G_n(\pi/(n+1/2)) = \int_0^\pi \operatorname{sinc} t \, dt$$

and

$$G_n(-\pi/(n+1/2)) = -\int_0^\pi \operatorname{sinc} t \, dt.$$

b) Deduce that $\lim_{\delta \to 0+} w_\delta(f_0) = 2 \int_0^\pi \text{sinc}\, t \, dt$, where f_0 denotes the function defined in Exercise 15.21(a).

c) Show that if f is continuous at 0, then $\lim_{\delta \to 0+} w_\delta(f) = 0$. *Hint:* Study the proof of Dirichlet's theorem carefully. Use Exercise 15.21 to show that there is a constant c such that $\left| \int_x^y D_n(t) \, dt \right| \le c$ for all n and for all $x, y \in [-\pi, \pi]$.

d) Show that if f is discontinuous at 0, then

$$\lim_{\delta \to 0+} w_\delta(f) = \frac{2}{\pi}(f(0) - f(0-)) \int_0^\pi \text{sinc}\, t \, dt.$$

Hint: Consider $f - af_0$ for a suitable constant a and use part (c). This property of the sequence of partial sums near a jump discontinuity is known as **Gibbs' phenomenon**.

15.23 Let $M_{2\pi}$ denote the space of periodic measures on $[-\pi, \pi]$, as defined in the paragraph preceding Exercise 15.13 on page 557. Verify that $M_{2\pi}$ can be identified with the dual space of $C_{2\pi}$ in the sense that every continuous linear functional on $C_{2\pi}$ is of the form $f \to \int f \, d\mu$ for some $\mu \in M_{2\pi}$.

15.24 Refer to Exercise 15.23. Suppose that $\mu \in M_{2\pi}$ and set

$$A_n(\mu)(x) = \frac{1}{n} \sum_{k=0}^{n-1} S_k(\mu)(x), \qquad n \in \mathcal{N},$$

where $S_n(\mu)(x) = \sum_{k=-n}^n \hat{\mu}(k) e^{ikx}$ and $\hat{\mu}(k) = (2\pi)^{-1} \int_{[-\pi, \pi]} e^{-ikx} \, d\mu(x)$. Define ν_n on $\mathcal{B}([-\pi, \pi])$ by $\nu_n(B) = \int_B A_n(\mu)(x) \, dx$. Prove that the sequence $\{\nu_n\}_{n=1}^\infty$ converges in the weak* topology of $M_{2\pi}$ to Lebesgue measure on $[-\pi, \pi]$.

Let I be a bounded interval with a nonempty interior. A sequence $\{a_n\}_{n=1}^\infty$ of elements of I is said to be **uniformly distributed** *in I if*

$$\lim_{n \to \infty} \frac{1}{n} N(\{k \in \mathcal{N} : 1 \le k \le n, \ a_k \in J\}) = \frac{\ell(J)}{\ell(I)}$$

for each subinterval $J \subset I$. Here $N(E)$ denotes the number of elements of a set E and ℓ denotes length. The idea of uniform distribution is that the relative frequency of points of the sequence that lie in an interval is proportional to the length of the interval.

15.25 The object of this exercise is to establish **Weyl's criteria** for uniform distribution, which we do for $I = [-\pi, \pi]$. Prove that the following statements are equivalent:

a) $\{a_n\}_{n=1}^\infty$ is uniformly distributed in $[-\pi, \pi]$.

b) $\lim_{n \to \infty} n^{-1} \sum_{k=1}^n f(a_k) = (2\pi)^{-1} \int_{-\pi}^\pi f(x) \, dx$ for each $f \in C([-\pi, \pi])$.

c) $\lim_{n \to \infty} n^{-1} \sum_{k=1}^n e^{ia_k m} = 0$ for every nonzero integer m. *Hint:* If part (b) holds, consider continuous functions f and g with the property that $0 \le g \le \chi_J \le f$.

★**15.26** Show that the sequence $\{nb - [nb]\}_{n=1}^\infty$ is uniformly distributed in $[0, 1]$ if and only if b is irrational. Here $[\ ]$ denotes the greatest integer function.

15.3 THE FOURIER TRANSFORM

Fourier series expansions express 2π-periodic functions in terms of the oscillatory functions e^{inx}, $n \in \mathcal{Z}$, whose basic periods constitute the discrete set $\{2\pi, 2\pi/2, 2\pi/3, \ldots\}$. In this section, we discuss an analogous expansion for certain nonperiodic functions in terms of oscillatory functions of the form e^{itx}, where the parameter t is continuous rather than discrete and summation is replaced by integration. Specifically, we have the following definition.

DEFINITION 15.2 **Fourier Transform**

For $f \in \mathcal{L}^1(\mathcal{R})$, the function $\hat{f} \colon \mathcal{R} \to \mathbb{C}$ defined by

$$\hat{f}(t) = \frac{1}{\sqrt{2\pi}} \int_{-\infty}^{\infty} f(x) e^{-itx}\, dx$$

is called the **Fourier transform** of f.

We should point out the following facts:

- Definition 15.2 deviates from the one given in Exercise 5.81 on page 173 by the factor $(2\pi)^{-1/2}$. In fact, slightly different definitions appear in various mathematical subfields, mostly for aesthetic reasons.

- The term *Fourier transform* is used in both Definition 15.1 (for periodic functions) and Definition 15.2 (for $\mathcal{L}^1$-functions). There is little room for confusion, however, because the Fourier transform of a function in $\mathcal{L}^1_{2\pi}$ is a function on $\mathcal{Z}$, whereas that of a function in $\mathcal{L}^1(\mathcal{R})$ is a function on $\mathcal{R}$. Moreover, the only function common to both $\mathcal{L}^1_{2\pi}$ and $\mathcal{L}^1(\mathcal{R})$ is the zero function. The advantage of using common terminology is that it suggests many important analogies between properties of the two transforms.

The following theorem, whose proof is left to the reader as Exercise 15.27, provides some basic properties of the Fourier transform. One of the properties employs the notation $f_{a,b}$ to represent the **translation-dilation** of the function f. That is, we write

$$f_{a,b}(x) = \frac{1}{\sqrt{|a|}} f((x-b)/a)$$

for $a, b \in \mathcal{R}$ and $a \neq 0$.

□ □ □ **THEOREM 15.5**

Let $f \in \mathcal{L}^1(\mathcal{R})$. Then the following results hold:

a) $\hat{f} \in C_0(\mathcal{R})$.

b) $\|\hat{f}\|_{\mathcal{R}} \leq \|f\|_1 / \sqrt{2\pi}$.

c) The function $F \colon \mathcal{L}^1(\mathcal{R}) \to C_0(\mathcal{R})$ defined by $F(f) = \hat{f}$ is a continuous linear mapping.

d) For $a, b \in \mathcal{R}$ and $a \neq 0$, we have

$$\widehat{f_{a,b}}(t) = \sqrt{|a|}\, e^{-ibt} \hat{f}(at), \qquad t \in \mathcal{R}.$$

e) If $\int_{-\infty}^{\infty} |x f(x)|\, dx < \infty$, then $\hat{f}\,'$ exists for all $t \in \mathcal{R}$ and

$$\hat{f}\,'(t) = \frac{1}{\sqrt{2\pi}} \int_{-\infty}^{\infty} (-ix) f(x) e^{-itx}\, dx.$$

f) If f' exists ae and $f' \in \mathcal{L}^1(\mathcal{R})$, then $\widehat{f'}(t) = it\hat{f}(t)$.

We observe that parts (a) and (b) of Theorem 15.5 are, respectively, the analogues of parts (a) and (b) of Theorem 15.1 on page 553. Moreover, for $a = 1$, part (d) of Theorem 15.5 is the analogue of part (f) of Proposition 15.1 on page 554. The properties of the Fourier transform described in parts (e) and (f) of Theorem 15.5 have numerous applications to many fields, including differential equations and probability theory.

EXAMPLE 15.5 *Illustrates Definition 15.2 and Theorem 15.5*

For $c > 0$, we have

$$\widehat{\chi_{[-c,c]}}(t) = \frac{1}{\sqrt{2\pi}} \int_{-c}^{c} e^{-itx}\, dx = \sqrt{2/\pi}\, c \operatorname{sinc}(ct).$$

Using this fact and Theorem 15.5(d), it is easy to obtain the Fourier transform of any integrable step function. □

EXAMPLE 15.6 *Illustrates Definition 15.2 and Theorem 15.5*

The function $g(x) = e^{-x^2/2}$, often called the **Gaussian function,** arises in many areas of mathematics, including harmonic analysis, probability theory, and statistics. We will prove that the Gaussian function is its own Fourier transform, that is, $\hat{g} = g$. To accomplish this proof, we first note that g satisfies the condition of Theorem 15.5(e) and, consequently,

$$\hat{g}'(t) = \frac{1}{\sqrt{2\pi}} \int_{-\infty}^{\infty} (-ix)e^{-x^2/2} e^{-itx}\, dx.$$

Applying integration by parts, we find that $\hat{g}'(t) = -t\hat{g}(t)$. This differential equation has the solution $\hat{g}(t) = \hat{g}(0)e^{-t^2/2}$. As the reader is asked to verify in Exercise 15.28,

$$\sqrt{2\pi}\, \hat{g}(0) = \int_{-\infty}^{\infty} e^{-x^2/2}\, dx = \sqrt{2\pi}.$$

Hence, $\hat{g}(0) = 1$, as required. □

Convolution Products

In the theory of Fourier series, we find frequent appearances of convolution products, that is, integrals of the form

$$(f * g)(x) = \frac{1}{2\pi} \int_{-\pi}^{\pi} f(x - y)\, g(y)\, dy,$$

where $f, g \in \mathcal{L}^1_{2\pi}$. Some basic properties of convolution are examined in Exercises 15.8–15.12 on page 557. For instance, Exercise 15.12 shows that convolution multiplication of periodic functions corresponds to ordinary multiplication of Fourier coefficients.

In the theory of the Fourier transform, a similar notion of convolution product for $\mathcal{L}^1(\mathcal{R})$-functions plays an essential role. We begin with the following definition.

DEFINITION 15.3   Convolution of Functions

For $f, g \in \mathcal{L}^1(\mathcal{R})$, the function $f * g$ defined by

$$(f * g)(x) = \frac{1}{\sqrt{2\pi}} \int_{-\infty}^{\infty} f(x - y)\, g(y)\, dy, \qquad x \in \mathcal{R},$$

is called the **convolution** of f and g.

As with the definition of the Fourier transform, minor modifications of the definition of convolution given in Definition 15.3 appear in various mathematical subfields. In particular, the factor $(2\pi)^{-1/2}$ is often omitted, as was done in Exercise 6.63(d) on page 221. From this point on, we will use Definition 15.3.

The following theorem summarizes some basic properties of the convolution product. Its proof is left to the reader as Exercise 15.31.

□ □ □   THEOREM 15.6

Let $f, g, h \in \mathcal{L}^1(\mathcal{R})$. Then the following results hold:
a) $f * g \in \mathcal{L}^1(\mathcal{R})$ and, in fact, $\|f * g\|_1 \leq \|f\|_1 \|g\|_1 / \sqrt{2\pi}$.
b) $f * g = g * f$.
c) $(f * g) * h = f * (g * h)$.
d) $f * (g + h) = f * g + f * h$.
e) $\widehat{f * g} = \hat{f}\hat{g}$.

EXAMPLE 15.7   *Illustrates Definition 15.3*

The integrals

$$I_T(f)(x) = \frac{1}{\sqrt{2\pi}} \int_{-T}^{T} \hat{f}(t) e^{itx}\, dt, \qquad T > 0,$$

are analogous to the partial sums of a Fourier series. We will show how to express $I_T(f)$ as a convolution product. Using Fubini's theorem, we have

$$I_T(f)(x) = \frac{1}{2\pi} \int_{-T}^{T} e^{itx} \int_{-\infty}^{\infty} f(y) e^{-ity}\, dy\, dt$$

$$= \frac{1}{2\pi} \int_{-\infty}^{\infty} f(y) \int_{-T}^{T} e^{-it(y-x)}\, dt\, dy = (f * D_T)(x),$$

where $D_T(x) = \sqrt{2/\pi}\, T \operatorname{sinc}(Tx)$. The function D_T is the continuous analogue of the Dirichlet kernel.   □

Uniqueness and Inversion

Based on an analogy with Fourier series, we might expect the formula

$$\lim_{T \to \infty} I_T(f)(x) = f(x) \tag{15.17}$$

to hold, at least under some reasonable conditions on f. Indeed, the following heuristic argument suggests that (15.17) is valid when f is continuous at x. By

Exercise 15.21 on page 564, we have

$$f(x) = \frac{1}{\pi} \int_{-\infty}^{\infty} f(x) \operatorname{sinc} y \, dy$$

and, by Example 15.7,

$$I_T(f)(x) = \frac{1}{\pi} \int_{-\infty}^{\infty} f(y) \operatorname{sinc}(T(x-y)) \, d(Ty) = \frac{1}{\pi} \int_{-\infty}^{\infty} f(x-y/T) \operatorname{sinc} y \, dy.$$

It follows that

$$I_T(f)(x) - f(x) = \frac{1}{\pi} \int_{-\infty}^{\infty} \left(f(x-y/T) - f(x) \right) \operatorname{sinc} y \, dy.$$

Hence,

$$\lim_{T \to \infty} I_T(f)(x) - f(x) = \frac{1}{\pi} \int_{-\infty}^{\infty} \lim_{T \to \infty} \left(f(x-y/T) - f(x) \right) \operatorname{sinc} y \, dy = 0.$$

The obstacle to making this argument rigorous is that the function sinc does not belong to $\mathcal{L}^1(\mathcal{R})$ and so the dominated convergence theorem cannot be applied. A way around this obstruction is to pass from the integals $I_T(f)$ to their averages.

For $f \in \mathcal{L}^1(\mathcal{R})$, let

$$J_T(f) = \frac{1}{T} \int_0^T I_t(f) \, dt.$$

Integrals of the form $J_T(f)$, which are analogous to averages of partial sums of Fourier series, make tractable substitutes for $I_T(f)$. Like $I_T(f)$, the integral $J_T(f)$ is also a convolution product, as can be seen as follows. By Fubini's theorem,

$$J_T(f)(x) = \frac{1}{T} \int_0^T I_t(f)(x) \, dt = \frac{1}{\pi} \frac{1}{T} \int_0^T \int_{-\infty}^{\infty} f(y) t \operatorname{sinc}(t(x-y)) \, dy \, dt$$

$$= \frac{1}{\pi} \int_{-\infty}^{\infty} f(y) \frac{1}{T} \int_0^T t \operatorname{sinc}(t(x-y)) \, dt \, dy$$

$$= \frac{1}{\pi} \int_{-\infty}^{\infty} f(y) \frac{1 - \cos(T(x-y))}{T(x-y)^2} \, dy = (f * G_T)(x),$$

where

$$G_T(x) = \sqrt{\frac{2}{\pi}} \frac{1 - \cos(Tx)}{Tx^2} = \sqrt{\frac{2}{\pi}} \frac{\sin^2(Tx/2)}{Tx^2/2}.$$

The function G_T is the continuous analogue of the Fejér kernel. Three of its essential properties are presented in Lemma 15.2. Parts (a) and (c) of the lemma are obvious; part (b) is left to the reader as Exercise 15.32.

□ □ □ **LEMMA 15.2**

The function G_T defined by the previous equation has the following properties.

a) $G_T \geq 0$.

b) $(2\pi)^{-1/2} \int_{-\infty}^{\infty} G_T(x)\,dx = 1$.

c) For each $\delta > 0$, $\lim_{T\to\infty} \int_{\mathcal{R}\setminus(-\delta,\delta)} G_T(x)\,dx = 0$.

Our next theorem presents results analogous to those given for Fourier series in Theorem 15.4 on page 562.

□ □ □ **THEOREM 15.7**

a) If $f \in C_0(\mathcal{R}) \cap \mathcal{L}^1(\mathcal{R})$, then $\lim_{T\to\infty} \|f - J_T(f)\|_{\infty} = 0$.

b) If $f \in \mathcal{L}^1(\mathcal{R})$, then $\lim_{T\to\infty} \|f - J_T(f)\|_1 = 0$.

PROOF To prove part (a), let $\epsilon > 0$. As f is uniformly continuous, we can choose $\delta > 0$ so that $|f(x-y) - f(x)| < \epsilon$ for all $y \in (-\delta, \delta)$ and $x \in \mathcal{R}$. Then, by Lemma 15.2, we have

$$J_T(f)(x) - f(x) = \frac{1}{\sqrt{2\pi}} \int_{-\infty}^{\infty} f(y) G_T(x-y)\,dy - f(x)$$

$$= \frac{1}{\sqrt{2\pi}} \int_{-\infty}^{\infty} (f(x-y) - f(x)) G_T(y)\,dy.$$

Thus,

$$|J_T(f)(x) - f(x)| \leq \frac{1}{\sqrt{2\pi}} \int_{-\infty}^{\infty} |f(x-y) - f(x)| G_T(y)\,dy$$

$$\leq \frac{\epsilon}{\sqrt{2\pi}} \int_{(-\delta,\delta)} G_T(y)\,dy + \sqrt{\frac{2}{\pi}}\, \|f\|_{\infty} \int_{\mathcal{R}\setminus(-\delta,\delta)} G_T(y)\,dy$$

$$\leq \epsilon + \sqrt{\frac{2}{\pi}}\, \|f\|_{\infty} \int_{\mathcal{R}\setminus(-\delta,\delta)} G_T(y)\,dy.$$

Applying Lemma 15.2(c), we conclude that

$$\lim_{T\to\infty} \sup_{x\in\mathcal{R}} |J_T(f)(x) - f(x)| \leq \epsilon.$$

As $\epsilon > 0$ was arbitrarily chosen, we see that part (a) holds.

To establish part (b), we begin by using Fubini's theorem to conclude that

$$\int_{-\infty}^{\infty} |J_T(f)(x) - f(x)|\,dx \leq \frac{1}{\sqrt{2\pi}} \int_{-\infty}^{\infty} \int_{-\infty}^{\infty} |f(x-y) - f(x)| G_T(y)\,dy\,dx$$

$$= \frac{1}{\sqrt{2\pi}} \int_{-\infty}^{\infty} \int_{-\infty}^{\infty} |f(x-y) - f(x)| G_T(y)\,dx\,dy$$

$$= \frac{1}{\sqrt{2\pi}} \int_{-\infty}^{\infty} \|f_y - f\|_1 G_T(y)\,dy.$$

The function $h(y) = \|f_y - f\|_1$ is bounded by $2\|f\|_1$ and, as we ask the reader to verify in Exercise 15.34, is continuous at 0. It follows by the argument used in part (a) that

$$\lim_{T \to \infty} \int_{-\infty}^{\infty} \|f_y - f\|_1 G_T(y)\, dy = 0.$$

Thus, part (b) is proved. ∎

□ □ □ **COROLLARY 15.1 Uniqueness Property of Fourier Transforms**

If $f, g \in \mathcal{L}^1(\mathcal{R})$ and $\hat{f} = \hat{g}$, then $f = g$ ae.

PROOF   If $\hat{f} = 0$, then $I_T(f)$ and, hence, $J_T(f)$ vanishes for every $T > 0$. Applying Theorem 15.7(b), we conclude that $f = 0$ ae. The corollary now follows from the linearity of the Fourier transform. ∎

Corollary 15.1 implies that an $\mathcal{L}^1(\mathcal{R})$-function is determined by its Fourier transform in the sense that two functions with the same transform must be identical almost everywhere. Theorem 15.8 gives a recipe for recovering a function from its Fourier transform. Such recipes are referred to as **inversion theorems.**

□ □ □ **THEOREM 15.8**

Suppose that both f and $\hat{f}$ belong to $\mathcal{L}^1(\mathcal{R})$. Then

$$f(x) = \frac{1}{\sqrt{2\pi}} \int_{-\infty}^{\infty} \hat{f}(t) e^{itx}\, dt$$

for almost all $x \in \mathcal{R}$.

PROOF   Let

$$g(x) = \frac{1}{\sqrt{2\pi}} \int_{-\infty}^{\infty} \hat{f}(t) e^{itx}\, dt.$$

Because, by assumption, $\hat{f} \in \mathcal{L}^1(\mathcal{R})$, it follows that g is well defined for all $x \in \mathcal{R}$. Using the dominated convergence theorem, we can write

$$g(x) = \lim_{T \to \infty} \frac{1}{\sqrt{2\pi}} \int_{-T}^{T} \hat{f}(t) e^{itx}\, dt = \lim_{T \to \infty} I_T(f)(x).$$

From this result, we can conclude that $g(x) = \lim_{T \to \infty} J_T(f)(x)$, as the reader is asked to verify in Exercise 15.35. In particular, we have shown that the sequence $\{J_n(f)\}_{n=1}^{\infty}$ converges to g pointwise on $\mathcal{R}$.

By Theorem 15.7(b), the sequence $\{J_n(f)\}_{n=1}^{\infty}$ converges to f in the $\mathcal{L}^1(\mathcal{R})$-norm. Applying Exercise 5.84 on page 176 and Proposition 5.12 on page 175, we deduce that there is a subsequence of $\{J_n(f)\}_{n=1}^{\infty}$ that converges to f almost everywhere. Consequently, $f = g$ ae. ∎

Applying Theorem 15.8 and Theorem 15.5 on page 566, we obtain the following corollary.

◻ ◻ ◻ **COROLLARY 15.2**

If both f and $\hat{f}$ belong to $\mathcal{L}^1(\mathcal{R})$, then $\hat{\hat{f}}(x) = f(-x)$ for almost all $x \in \mathcal{R}$. Furthermore, f is equal to a continuous function almost everywhere.

Although Theorem 15.8 is adequate for handling functions that satisfy certain mild restrictions, such as the ones given in Exercise 15.36, it is by no means the last word on inversion of the Fourier transform. Indeed, Example 15.5 on page 567 shows that the Fourier transform of the characteristic function of an interval fails to be Lebesgue integrable.

The Schwartz Class

Having studied the action of the Fourier transform on the space $\mathcal{L}^1(\mathcal{R})$, we now consider its behavior on a class of smooth functions that decay rapidly at $\pm\infty$.

We denote by $C^\infty(\mathcal{R})$ the space of complex-valued functions that have derivatives of all orders at each point of $\mathcal{R}$. For $f \in C^\infty(\mathcal{R})$, let

$$\sigma_n(f) = \max_{0 \le k \le n} \sup_{x \in \mathcal{R}} \left(1 + x^2\right)^n \left|f^{(k)}(x)\right|, \qquad n = 0, 1, 2, \ldots.$$

If $\sigma_n(f) < \infty$ for all $n \ge 0$, then we say that f is **rapidly decreasing.**

The collection of all rapidly decreasing functions is called the **Schwartz class** and is denoted $\mathcal{S}(\mathcal{R})$. We note that $\mathcal{S}(\mathcal{R})$ is a linear space and that $\{\sigma_n\}_{n=0}^\infty$ is a family of seminorms defining a locally convex topology on $\mathcal{S}(\mathcal{R})$. The following proposition provides some basic properties of $\mathcal{S}(\mathcal{R})$.

◻ ◻ ◻ **PROPOSITION 15.4**

Let $f \in C^\infty(\mathcal{R})$. Then
a) $f \in \mathcal{S}(\mathcal{R})$ if and only if $pf^{(m)}$ is bounded on $\mathcal{R}$ for every polynomial p and every integer $m \ge 0$.
b) $f \in \mathcal{S}(\mathcal{R})$ implies that $pf^{(m)} \in \mathcal{S}(\mathcal{R})$ for every polynomial p and every integer $m \ge 0$.
c) $\mathcal{S}(\mathcal{R}) \subset \mathcal{L}^1(\mathcal{R})$.
d) $f \in \mathcal{S}(\mathcal{R})$ implies that $\hat{f} \in C^\infty(\mathcal{R})$.

PROOF Parts (a), (b), and (c) are left to the reader as Exercise 15.38. Part (d) follows from parts (a), (b), and (c) and repeated application of Theorem 15.5(e) on page 566. ∎

EXAMPLE 15.8 *Illustrates the Schwartz Class*

The verification of each part below is left to the reader as Exercise 15.39.
a) The function $f(x) = x^n e^{-bx^2}$, where n is a nonnegative integer and b is a positive real number, is rapidly decreasing—that is, $f \in \mathcal{S}(\mathcal{R})$.
b) A function $f \colon \mathcal{R} \to \mathcal{R}$ is called a **bump function** if $f \in C^\infty(\mathcal{R})$ and has compact support. For instance, let $a < b$ and define $f(x) = e^{-1/(x-a)(b-x)}$ if $a < x < b$, and 0 elsewhere. Then f is a bump function. Any bump function is rapidly decreasing and, hence, is in $\mathcal{S}(\mathcal{R})$.

c) Let f be a rational function (quotient of two polynomials) whose denominator has no real zeros. Then $f \in C^\infty(\mathcal{R})$ but is not rapidly decreasing; in other words, $f \in C^\infty(\mathcal{R}) \setminus \mathcal{S}(\mathcal{R})$. □

In the next theorem, we give some properties of functions in the Schwartz class with regard to the Fourier transform.

□ □ □ **THEOREM 15.9**

Let $f \in \mathcal{S}(\mathcal{R})$. Then

a) $\hat{f} \in \mathcal{S}(\mathcal{R})$.

b) $\hat{\hat{f}}(x) = f(-x)$ for all $x \in \mathcal{R}$.

c) the correspondence $f \to \hat{f}$ maps $\mathcal{S}(\mathcal{R})$ one-to-one onto itself.

PROOF   a) By repeated application of Theorem 15.5(e), we get that

$$\hat{f}^{(m)}(t) = \frac{1}{\sqrt{2\pi}} \int_{-\infty}^{\infty} (-ix)^m f(x) e^{-itx} \, dx$$

for all nonnegative integers m. Multiplying both sides of the preceding display by $(it)^\ell$, where ℓ is a nonnegative integer, and integrating the right side by parts ℓ times yields

$$(it)^\ell \hat{f}^{(m)}(t) = \frac{1}{\sqrt{2\pi}} \int_{-\infty}^{\infty} \frac{d^\ell((-ix)^m f(x))}{dx^\ell} e^{-itx} \, dx. \qquad (15.18)$$

Parts (b) and (c) of Proposition 15.4 give $d^\ell((-ix)^m f(x))/dx^\ell \in \mathcal{L}^1(\mathcal{R})$. This fact and (15.18) imply that $|t|^\ell |\hat{f}^{(m)}(t)|$ is bounded for all nonnegative integers ℓ and m. It now follows from parts (d) and (a) of Proposition 15.4 that $\hat{f} \in \mathcal{S}(\mathcal{R})$.

b) Referring to part (a) of this theorem, Proposition 15.4(c), and Theorem 15.8, we conclude that

$$\hat{\hat{f}}(x) = \frac{1}{\sqrt{2\pi}} \int_{-\infty}^{\infty} \hat{f}(t) e^{-ixt} \, dt = \frac{1}{\sqrt{2\pi}} \int_{-\infty}^{\infty} \hat{f}(t) e^{it(-x)} \, dt = f(-x),$$

as required.

c) For $f \in \mathcal{S}(\mathcal{R})$, define $F(f) = \hat{f}$. From part (a), $F: \mathcal{S}(\mathcal{R}) \to \mathcal{S}(\mathcal{R})$. We want to show that F is one-to-one and onto. Suppose that $F(f) = F(g)$. Then, by part (b), we have for each $x \in \mathcal{R}$,

$$f(x) = \hat{\hat{f}}(-x) = \widehat{F(f)}(-x) = \widehat{F(g)}(-x) = \hat{\hat{g}}(-x) = g(x).$$

Consequently, F is one-to-one. Now suppose that $f \in \mathcal{S}(\mathcal{R})$. Define h on $\mathcal{R}$ by $h(x) = \hat{f}(-x)$. By part (a), $h \in \mathcal{S}(\mathcal{R})$. Moreover, by part (b),

$$F(h)(x) = \hat{h}(x) = \hat{\hat{f}}(-x) = f(x)$$

for all $x \in \mathcal{R}$. Thus, F is onto. ∎

In our next theorem, we show that the Schwartz class is well-behaved with respect to convolution.

☐☐☐ **THEOREM 15.10**

Let $f, g \in \mathcal{S}(\mathcal{R})$. Then
a) fg and $f * g$ are in $\mathcal{S}(\mathcal{R})$.
b) $\widehat{fg} = \hat{f} * \hat{g}$.

PROOF   We leave the verification of part (a) to the reader as Exercise 15.40. For part (b), we refer to Theorem 15.6(e) on page 568 and Theorem 15.9(b) to get

$$\widehat{\widehat{fg}}(x) = fg(-x) = f(-x)g(-x) = \hat{\hat{f}}(x)\hat{\hat{g}}(x) = \widehat{\hat{f}\hat{g}}(x) = \widehat{\hat{f} * \hat{g}}(x).$$

Hence, we see that $\widehat{fg}$ and $\hat{f} * \hat{g}$ have the same Fourier transform and, consequently, they are equal by Theorem 15.9(c). ∎

Exercises for Section 15.3

15.27 Prove Theorem 15.5 on page 566.

15.28 Show that $\int_{-\infty}^{\infty} e^{-x^2/2} \, dx = \sqrt{2\pi}$. *Hint:* Use polar coordinates to evaluate the double integral $\int_{-\infty}^{\infty} \int_{-\infty}^{\infty} e^{-(x^2+y^2)/2} \, dx \, dy$.

15.29 Calculate the Fourier transform of the function $f(x) = e^{-(x-b)^2/a}$, where a and b are real constants with $a > 0$.

15.30 Convolution products also appear naturally in probability theory. Suppose that X and Y are independent random variables with probability density functions f_X and f_Y, respectively.
a) Establish that the random variable $X + Y$ has probability density function given by $f_{X+Y} = \sqrt{2\pi} f_X * f_Y$.
b) Explain the discrepancy between the result in part (a) and the one obtained in Exercise 7.56(c) on page 247.

15.31 Prove Theorem 15.6 on page 568.

15.32 Prove that there is no identity for the convolution product, that is, there does not exist a function $h \in \mathcal{L}^1(\mathcal{R})$ such that $f = f * h$ for all $f \in \mathcal{L}^1(\mathcal{R})$. Note, however, that Theorem 15.7 shows that $\lim_{T \to \infty} f * G_T = f$ for all $f \in \mathcal{L}^1(\mathcal{R})$.

15.33 Show that

$$\int_{-\infty}^{\infty} \frac{1 - \cos(Tx)}{Tx^2} \, dx = \pi.$$

Hint: Use

$$\frac{1 - \cos(Tx)}{Tx^2} = \frac{1}{T} \int_0^T \frac{\sin(tx)}{x} \, dt$$

and Exercise 15.21 on page 564.

15.34 Let $1 \le p < \infty$ and $f \in \mathcal{L}^p(\mathcal{R})$. Show that the function h defined by $h(y) = \|f_y - f\|_p$ is continuous on $\mathcal{R}$.

15.35 Let $f \in \mathcal{L}^1(\mathcal{R})$ and $x \in \mathcal{R}$. Suppose that $\lim_{T \to \infty} I_T(f)(x)$ exists and equals, say, L. Prove that $\lim_{T \to \infty} J_T(f)(x)$ also exists and equals L.

15.36 Suppose that f'' exists and is finite everywhere and that $f, f', f'' \in \mathcal{L}^1(\mathcal{R})$. Prove that $\hat{f} \in \mathcal{L}^1(\mathcal{R})$.

15.37 Suppose that $f, f', f'' \in \mathcal{L}^1(\mathcal{R}) \cap C(\mathcal{R})$ and that there is a constant M such that

$$|f(x)| \vee |f'(x)| \vee |f''(x)| \le \frac{M}{1+x^2}, \qquad x \in \mathcal{R}.$$

a) Prove the **Poisson summation formula:**

$$\sum_{k=-\infty}^{\infty} \hat{f}(k) = \sqrt{2\pi} \sum_{n=-\infty}^{\infty} f(2\pi n).$$

Hint: Consider the function $g(x) = \sum_{n=-\infty}^{\infty} f(x + 2\pi n)$.

b) Use the Poisson summation formula to verify the **Jacobi theta function identity:**

$$\sum_{n=-\infty}^{\infty} e^{-n^2/2t} = \sqrt{2\pi t} \sum_{n=-\infty}^{\infty} e^{-2\pi^2 n^2 t}, \qquad t > 0.$$

Hint: Refer to Exercise 15.29.

15.38 Prove parts (a), (b), and (c) of Proposition 15.4 on page 572.

15.39 Verify parts (a), (b), and (c) of Example 15.8 on pages 572–573.

15.40 Establish part (a) of Theorem 15.10.

15.41 Prove that $f \in \mathcal{S}(\mathcal{R})$ and $g \in \mathcal{L}^1(R)$ imply $f * g \in \mathcal{S}(\mathcal{R})$.

15.42 Prove that the linear operator $F \colon S(\mathcal{R}) \to S(\mathcal{R})$ defined by $F(f) = \hat{f}$ is continuous with respect to the topology induced by the seminorms $\{\sigma_n\}_{n=0}^{\infty}$.

15.4 FOURIER TRANSFORMS OF MEASURES

In this section, we will extend the concept of Fourier transform from functions in $\mathcal{L}^1(\mathcal{R})$ to measures in $M(\mathcal{R})$. As an application of Fourier transforms of measures, we will obtain several interesting and important results in probability theory, including the celebrated central limit theorem.

DEFINITION 15.4   Fourier Transform of a Measure

For $\mu \in M(\mathcal{R})$, the function $\hat{\mu} \colon \mathcal{R} \to \mathbb{C}$ defined by

$$\hat{\mu}(t) = \frac{1}{\sqrt{2\pi}} \int_{\mathcal{R}} e^{-itx} \, d\mu(x)$$

is called the **Fourier transform** of μ.

EXAMPLE 15.9   *Illustrates Definition 15.4*

a) The Radon-Nikodym theorem for complex measures (page 326) and Exercise 9.51 on page 330 imply that if $\mu \ll \lambda$, then $\hat{\mu} = \widehat{d\mu/d\lambda}$.

b) If $a \in \mathcal{R}$ and $\mu = \delta_a$, then $\hat{\mu}(t) = (2\pi)^{-1/2} e^{-iat}$.   □

Our next proposition, whose proof is left for Exercise 15.43, provides some basic properties of Fourier transforms of measures.

☐ ☐ ☐ **PROPOSITION 15.5**

Let $\mu \in M(\mathcal{R})$. Then the following results hold:

a) $\hat{\mu} \in C(\mathcal{R})$.

b) $|\hat{\mu}(t)| \leq |\mu|(\mathcal{R})/\sqrt{2\pi}$.

c) If $\mu(B) = \int_B f(x) \, d\lambda(x)$ for some $f \in \mathcal{L}^1(\mathcal{R})$, then $\hat{\mu} = \hat{f}$.

The integrals $I_T(f)$ and $J_T(f)$, defined in Section 15.3, play an important role in the theory of Fourier transforms of $\mathcal{L}^1(\mathcal{R})$-functions. They have natural analogues when f is replaced by a measure: If $\mu \in M(\mathcal{R})$, we let

$$I_T(\mu)(x) = \frac{1}{\sqrt{2\pi}} \int_{-T}^{T} \hat{\mu}(t) e^{itx} \, dt, \qquad T > 0,$$

and

$$J_T(\mu)(x) = \frac{1}{T} \int_0^T I_t(\mu)(x) \, dt, \qquad T > 0.$$

Using these integrals, we can show that a measure is determined by its Fourier transform. We begin with the following theorem whose proof is left to the reader as Exercise 15.44.

☐ ☐ ☐ **THEOREM 15.11**

For $\mu \in M(\mathcal{R})$, define

$$\mu_T(B) = \int_B J_T(\mu)(x) \, d\lambda(x), \qquad B \in \mathcal{B}.$$

Then the following results hold:

a) $\mu_T \in M(\mathcal{R})$.

b) $|\mu_T|(\mathcal{R}) \leq |\mu|(\mathcal{R})$.

c) The net $\{\mu_T\}_{T \in (0,\infty)}$ converges in the weak* topology to μ.

☐ ☐ ☐ **COROLLARY 15.3 Uniqueness Property of Fourier Transforms**

If $\mu, \nu \in M(\mathcal{R})$ and $\hat{\mu} = \hat{\nu}$, then $\mu = \nu$.

PROOF   If $\hat{\mu} = 0$, then $J_T(\mu)$ and, hence, μ_T vanishes for every $T > 0$. Applying part (c) of Theorem 15.11, we conclude that $\mu = 0$. The corollary now follows from the linearity of the Fourier transform of a measure. ∎

Corollary 15.3 implies that a measure is determined by its Fourier transform in the sense that two measures with the same transform must be identical. When μ is a probability measure, we can get still more information about its relationship with $\hat{\mu}$.

☐ ☐ ☐ **LEMMA 15.3**

Suppose $\mu \in M_+(\mathcal{R})$ and $\mu(\mathcal{R}) = 1$. Then, for each $c > 0$, we have

$$\mu([-2c, 2c]) \geq \sqrt{2\pi} \, c \left| \int_{-1/c}^{1/c} \hat{\mu}(t) \, dt \right| - 1.$$

PROOF   Let $b > 0$. Then

$$\frac{\sqrt{2\pi}}{2b} \int_{-b}^{b} \hat{\mu}(t)\, dt = \frac{1}{2b} \int_{-b}^{b} \int_{\mathcal{R}} e^{-itx}\, d\mu(x)\, dt$$

$$= \int_{\mathcal{R}} \frac{1}{2b} \int_{-b}^{b} e^{-itx}\, dt\, d\mu(x) = \int_{\mathcal{R}} \mathrm{sinc}(bx)\, d\mu(x).$$

It is easy to see that $|\mathrm{sinc}(bx)| \leq 1$ for all x and that $|\mathrm{sinc}(bx)| \leq (2bc)^{-1}$ when $|x| > 2c$. It follows that

$$\frac{\sqrt{2\pi}}{2b} \left| \int_{-b}^{b} \hat{\mu}(t)\, dt \right| \leq \mu([-2c, 2c]) + \frac{1}{2bc} \mu(\mathcal{R} \setminus [-2c, 2c]).$$

Taking $b = 1/c$ and using $\mu(\mathcal{R}) = 1$, we get

$$\frac{c\sqrt{2\pi}}{2} \left| \int_{-1/c}^{1/c} \hat{\mu}(t)\, dt \right| \leq \frac{1}{2} + \frac{1}{2} \mu([-2c, 2c]),$$

from which the assertion of the lemma follows immediately.   ∎

Just as convolution of functions plays an important role in the theory of Fourier transforms of functions, convolution of measures figures prominently in the theory of Fourier transforms of measures.

DEFINITION 15.5   **Convolution of Measures**

For $\mu, \nu \in M(\mathcal{R})$, the Borel measure $\mu * \nu$ defined by

$$(\mu * \nu)(B) = \frac{1}{\sqrt{2\pi}} \int_{\mathcal{R}} \mu(B - x)\, d\nu(x), \qquad B \in \mathcal{B},$$

is called the **convolution** of μ and ν.

Note: As with the definition of convolution of functions, minor modifications of the definition of convolution of measures given in Definition 15.5 appear in various mathematical subfields. In particular, the factor $(2\pi)^{-1/2}$ is often omitted, as was done in Exercise 6.64(d) on page 221. From this point on, we will use Definition 15.5.

Our next proposition, whose proof is left for the reader as Exercise 15.47, shows that convolution of measures (Definition 15.5) is consistent with convolution of functions (Definition 15.3 on page 568).

◻ ◻ ◻ **PROPOSITION 15.6**

Let $f, g \in \mathcal{L}^1(\mathcal{R})$. Define measures μ and ν by

$$\mu(B) = \int_B f \, d\lambda \quad \text{and} \quad \nu(B) = \int_B g \, d\lambda, \qquad B \in \mathcal{B}.$$

Then

$$(\mu * \nu)(B) = \int_B (f * g) \, d\lambda, \qquad B \in \mathcal{B}.$$

Equivalently, if μ and ν are absolutely continuous with respect to Lebesgue measure, then so is $\mu * \nu$ and, moreover, $d(\mu * \nu)/d\lambda = (d\mu/d\lambda) * (d\nu/d\lambda)$.

Proposition 15.6 shows that convolution of functions corresponds to a special case of convolution of measures. More examples of convolution of measures are contained in Example 15.10.

EXAMPLE 15.10 *Illustrates Definition 15.5*

a) For each $\mu \in M(\mathcal{R})$, we have

$$(\delta_0 * \mu)(B) = \frac{1}{\sqrt{2\pi}} \int_{\mathcal{R}} \delta_0(B - x) \, d\mu(x) = \frac{1}{\sqrt{2\pi}} \mu(B), \qquad B \in \mathcal{B}.$$

In other words, $\delta_0 * \mu = (2\pi)^{-1/2} \mu$ for all $\mu \in M(\mathcal{R})$.

b) Let X and Y be independent random variables. Then Exercise 7.56 on page 247 implies that $\mu_{X+Y} = \sqrt{2\pi} \, \mu_X * \mu_Y$. ◻

The following analog of Theorem 15.6 on page 568 gives some basic properties of convolution of measures. Its proof is left for Exercise 15.48.

◻ ◻ ◻ **THEOREM 15.12**

Let $\mu, \nu, \gamma \in M(\mathcal{R})$. Then the following results hold:
a) $\mu * \nu \in M(\mathcal{R})$.
b) $\mu * \nu = \nu * \mu$.
c) $(\mu * \nu) * \gamma = \mu * (\nu * \gamma)$.
d) $\mu * (\nu + \gamma) = \mu * \nu + \mu * \gamma$.
e) $\widehat{\mu * \nu} = \hat{\mu}\hat{\nu}$.

Fourier Transforms in Probability Theory

Let X be a random variable. Then the function $\psi_X(t) = \mathcal{E}(e^{itX})$ is called **characteristic function** of the random variable X, not to be confused with the characteristic function of a set. It is easy to see that

$$\psi_X(t) = \int_{\mathcal{R}} e^{itx} d\mu_X(x) = \sqrt{2\pi} \, \widehat{\mu_X}(-t).$$

Instead of using the characteristic function ψ_X, as is usually done in probability theory, we will work with the essentially equivalent Fourier transform $\widehat{\mu_X}$.

Recall that a sequence $\{X_n\}_{n=1}^\infty$ of random variables is said to converge in distribution to the random variable X, written $X_n \overset{d}{\to} X$, if the sequence $\{\mu_{X_n}\}_{n=1}^\infty$ of measures converges to μ_X in the weak* topology. Exercise 14.76 on page 537 shows that $X_n \overset{d}{\to} X$ if and only if

$$\lim_{n\to\infty} \int_{\mathcal{R}} f \, d\mu_{X_n} = \int_{\mathcal{R}} f \, d\mu_X, \qquad f \in C_b(\mathcal{R}).$$

Suppose that $X_n \overset{d}{\to} X$. Then it follows immediately from the previous equation that

$$\lim_{n\to\infty} \widehat{\mu_{X_n}}(t) = \widehat{\mu_X}(t), \qquad t \in \mathcal{R}. \tag{15.19}$$

In other words, convergence in distribution of a sequence of random variables implies pointwise convergence of the Fourier transforms of the corresponding distributions. The following important theorem, due to Paul Lévy, provides a partial converse to this result.

□ □ □   **THEOREM 15.13 Lévy's Theorem**

Let $\{X_n\}_{n=1}^\infty$ be a sequence of random variables such that the sequence $\{\widehat{\mu_{X_n}}\}_{n=1}^\infty$ of Fourier transforms converges pointwise to a function h that is continuous at $t = 0$. Then there is a random variable X such that $X_n \overset{d}{\to} X$ and $h = \widehat{\mu_X}$.

PROOF   Let $\{X_{n_k}\}_{k=1}^\infty$ be any subsequence of $\{X_n\}_{n=1}^\infty$ and $\mu_k = \mu_{X_{n_k}}$. We will first show that $\{\mu_k\}_{k=1}^\infty$ has a subsequence $\{\mu_{k_j}\}_{j=1}^\infty$ that converges in the weak* topology of $M(\mathcal{R})$ to a probability measure μ. Because the μ_ks are probability measures, it follows from Theorem 14.14 on page 535 that there is a subsequence $\{\mu_{k_j}\}_{j=1}^\infty$ that converges in the weak* topology of $M(\mathcal{R})$ to a regular Borel measure μ and, by Exercise 14.68 on page 536, $\mu \in M_+(\mathcal{R})$.

We will show that μ is a probability measure. Let $\epsilon > 0$. Because h is continuous at 0, there is a $\delta > 0$ such that $|h(t) - h(0)| < \epsilon$ for $|t| < \delta$. Let c be a positive real number such that $c^{-1} < \delta$. Select a continuous function g that satisfies $0 \le g \le 1$, $g(x) = 1$ for $|x| \le 2c$, and $g(x) = 0$ for $|x| \ge 2c + 1$. Applying Lemma 15.3, we have

$$\int_{\mathcal{R}} g(x) \, d\mu_{k_j}(x) \ge \mu_{k_j}([-2c, 2c]) \ge \sqrt{2\pi}\, c \left| \int_{-1/c}^{1/c} \widehat{\mu_{k_j}}(t) \, dt \right| - 1. \tag{15.20}$$

Letting $j \to \infty$ and using weak* convergence on the left-hand side of (15.20) and dominated convergence on the right-hand side of (15.20), we obtain

$$\int_{\mathcal{R}} g(x) \, d\mu(x) \ge \sqrt{2\pi}\, c \left| \int_{-1/c}^{1/c} h(t) \, dt \right| - 1$$

$$\ge \sqrt{2\pi}\, c \left| \int_{-1/c}^{1/c} h(0) \, dt \right| - \sqrt{2\pi}\, c \left| \int_{-1/c}^{1/c} (h(t) - h(0)) \, dt \right| - 1.$$

Using $[-1/c, 1/c] \subset (-\delta, \delta)$ and $h(0) = (2\pi)^{-1/2}$, we conclude that

$$\mu(\mathcal{R}) \ge \int_{\mathcal{R}} g(x) \, d\mu(x) \ge 1 - 2\sqrt{2\pi}\, \epsilon.$$

Because ϵ is an arbitrary positive number, it follows that $\mu(\mathcal{R}) \geq 1$. On the other hand, if $f \in C_0(\mathcal{R})$ with $|f| \leq 1$, then

$$\left| \int_{\mathcal{R}} f \, d\mu \right| = \lim_{j \to \infty} \left| \int_{\mathcal{R}} f \, d\mu_{k_j} \right| \leq 1.$$

Applying the Riesz representation theorem (Theorem 13.16 on page 493), we deduce that $\mu(\mathcal{R}) \leq 1$. Thus, we have shown that $\mu(\mathcal{R}) = 1$.

Next we apply Exercise 14.76 on page 537 to assert that for each $f \in C_b(\mathcal{R})$, we have $\int_{\mathcal{R}} f \, d\mu_{k_j} \to \int_{\mathcal{R}} f \, d\mu$ as $j \to \infty$. Letting $f(x) = (2\pi)^{-1/2} e^{-itx}$, we obtain that $h(t) = \hat{\mu}(t)$.

Now suppose that $\{X_{m_k}\}_{k=1}^\infty$ is another subsequence of $\{X_n\}_{n=1}^\infty$. By the preceding argument, there is a subsequence of $\{\mu_{X_{m_k}}\}_{k=1}^\infty$ that converges in the weak* topology to a probability measure $\nu \in M(\mathcal{R})$ with $\hat{\nu}(t) = h(t)$. Invoking the uniqueness property of Fourier transforms of measures (Corollary 15.3), we conclude that $\nu = \mu$. Thus, we have shown that every subsequence of $\{\mu_{X_n}\}_{n=1}^\infty$ has a subsequence that converges weak* to the probability measure μ.

In a metric space, a sequence converges to a limit L if every subsequence has a subsequence that converges to L. Because the set of probability measures in $M(\mathcal{R})$ is weak* metrizable, it follows that $\{\mu_{X_n}\}_{n=1}^\infty$ converges in the weak* topology to the probability measure μ.

To complete the proof, let X be the identity function on $\mathcal{R}$. Then, as a random variable on the probability space $(\mathcal{R}, \mathcal{B}, \mu)$, we have that $\mu = \mu_X$ and, because w*$\lim \mu_{X_n} = \mu$, we conclude that $X_n \xrightarrow{d} X$. ∎

In view of Proposition 15.5(a) on page 576, the uniqueness property of Fourier transforms, and Lévy's theorem, we obtain the following corollary.

□ □ □ **COROLLARY 15.4**

Let $X, X_1, X_2, \ldots$ be random variables. Then $X_n \xrightarrow{d} X$ if and only if $\widehat{\mu_{X_n}} \to \widehat{\mu_X}$ pointwise as $n \to \infty$.

The Central Limit Theorem

The strong law of large numbers for sequences of independent and identically distributed (iid) random variables, Theorem 7.9 on page 263, is one of the two most important theorems in probability theory. The other is the *central limit theorem*. This latter remarkable and useful result states that the partial sums of any sequence of iid random variables is asymptotically normally distributed, provided only that the random variables have finite variance.

We will use Lévy's theorem to prove the central limit theorem. But first we require a lemma, the verification of which is left to the reader as Exercise 15.52.

□ □ □ **LEMMA 15.4**

Let $\mu \in M_+(\mathcal{R})$ satisfy $\mu(\mathcal{R}) = 1$ and $\int_{\mathcal{R}} x^2 \, d\mu(x) < \infty$. Set $m_1 = \int_{\mathcal{R}} x \, d\mu(x)$ and $m_2 = \int_{\mathcal{R}} x^2 \, d\mu(x)$. Then

$$\sqrt{2\pi} \, \hat{\mu}(t) = 1 - im_1 t - m_2 t^2/2 + \alpha(t),$$

where $\lim_{t \to 0} \alpha(t)/t^2 = 0$.

□ □ □ **THEOREM 15.14 Central Limit Theorem**

Suppose $X_1, X_2, \ldots$, are mutually independent and identically distributed random variables with mean m and finite variance σ^2. Let $S_n = \sum_{k=1}^{n} X_k$. Then

$$\lim_{n \to \infty} P\left(a < \frac{S_n - nm}{\sqrt{n}\sigma} \leq b\right) = \frac{1}{\sqrt{2\pi}} \int_a^b e^{-x^2/2} \, dx$$

uniformly for all $-\infty \leq a < b \leq \infty$.

PROOF The reader is asked in Exercise 15.53 to show that we can, without loss of generality, assume that $m = 0$ and $\sigma^2 = 1$. It follows from Example 15.10(b) on page 578 that

$$\mu_{S_n} = (2\pi)^{(n-1)/2} \underbrace{\mu * \mu * \cdots * \mu}_{n \text{ times}},$$

where μ denotes the common distribution of the X_ns. Let $Z_n = S_n/\sqrt{n}$. Then, by Theorem 15.12(e) and Exercise 15.45,

$$\widehat{\mu_{Z_n}}(t) = \widehat{\mu_{S_n}}(t/\sqrt{n}) = \frac{1}{\sqrt{2\pi}} \left[\sqrt{2\pi}\, \hat{\mu}(t/\sqrt{n})\right]^n.$$

Using Lemma 15.4, we get

$$\widehat{\mu_{Z_n}}(t) = \frac{1}{\sqrt{2\pi}} \left[1 - \frac{t^2}{2n} + \alpha\left(\frac{t}{\sqrt{n}}\right)\right]^n.$$

Consequently,

$$\lim_{n \to \infty} \widehat{\mu_{Z_n}}(t) = \frac{1}{\sqrt{2\pi}} e^{-t^2/2} = \frac{1}{\sqrt{2\pi}} \hat{g}(t),$$

where g is the Gaussian function discussed in Example 15.6 on page 567.

By Lévy's theorem, the sequence $\{Z_n\}_{n=1}^{\infty}$ converges in distribution to a random variable with the distribution $\nu(B) = (2\pi)^{-1/2} \int_B e^{-x^2/2} \, dx$. Applying Exercise 14.76 on page 537, we conclude that

$$\lim_{n \to \infty} \int_{\mathcal{R}} f \, d\mu_{Z_n} = \int_{\mathcal{R}} f \, d\nu, \qquad f \in C_b(\mathcal{R}).$$

Let $0 < \epsilon < (b - a)/2$. Choose a continuous function f_1 such that $0 \leq f_1 \leq 1$, $f_1(x) = 0$ for $x \notin (a, b)$, and $f_1(x) = 1$ for $x \in [a + \epsilon, b - \epsilon]$. And choose a continuous function f_2 such that $0 \leq f_2 \leq 1$, $f_2(x) = 1$ for $x \in [a, b]$, and $f_2(x) = 0$ for $x \notin (a - \epsilon, b + \epsilon)$. Then

$$\int_{\mathcal{R}} f_1(x) \, d\mu_{Z_n}(x) \leq \int_{(a,b]} d\mu_{Z_n}(x) < \int_{\mathcal{R}} f_2(x) \, d\mu_{Z_n}(x).$$

Thus,

$$\frac{1}{\sqrt{2\pi}} \int_{a+\epsilon}^{b-\epsilon} e^{-x^2/2} \, dx \leq \int_{\mathcal{R}} f_1(x) \, d\nu(x) \leq \liminf_{n \to \infty} \mu_{Z_n}((a,b])$$

and

$$\limsup_{n \to \infty} \mu_{Z_n}((a,b]) \leq \int_{\mathcal{R}} f_2(x) \, d\nu(x) \leq \frac{1}{\sqrt{2\pi}} \int_{a-\epsilon}^{b+\epsilon} e^{-x^2/2} \, dx.$$

Because ϵ can be made arbitrarily small, it follows that

$$\frac{1}{\sqrt{2\pi}} \int_a^b e^{-x^2/2}\, dx = \lim_{n\to\infty} \mu_{Z_n}((a,b]) = \lim_{n\to\infty} P\left(a < \frac{S_n}{\sqrt{n}} \le b\right),$$

as required. The uniformity in a and b follows from Exercise 15.54. ∎

EXAMPLE 15.11 *Illustrates the Central Limit Theorem*

As a consequence of the strong law of large numbers for iid random variables, we proved, in Corollary 7.5, Borel's strong law of large numbers: Suppose that E is an event associated with some random experiment and let p be its probability. Denote by $n(E)$ the number of times that event E occurs in n independent repetitions of the experiment. Then, with probability one, $\lim_{n\to\infty} n(E)/n = p$.

Similarly, we can obtain as a special case of the central limit theorem, the following result known as the **DeMoivre-Laplace theorem:**

$$\lim_{n\to\infty} P\left(a < \frac{n(E) - np}{\sqrt{np(1-p)}} \le b\right) = \frac{1}{\sqrt{2\pi}} \int_a^b e^{-x^2/2}\, dx$$

uniformly for all $-\infty \le a < b \le \infty$. To prove this result, define for each $n \in \mathcal{N}$, $X_n = 1$ or 0 according to whether event E occurs or does not occur on the nth repetition of the experiment. Then X_1, X_2, ... are iid and have common mean p and variance $p(1-p)$. Noting that $n(E) = X_1 + \cdots + X_n$, we obtain the DeMoivre-Laplace theorem from the central limit theorem. □

Exercises for Section 15.4

15.43 Prove Proposition 15.5 on page 576.

15.44 Prove Theorem 15.11 on page 576.

15.45 Establish the following facts.
a) Let $\mu \in M(\mathcal{R})$ and $\nu(B) = \mu(a^{-1}(B - b))$, where a and b are constants with $a \ne 0$. Show that $\hat{\nu}(t) = e^{-ibt} \hat{\mu}(at)$.
b) Let X be a random variable and set $Y = aX + b$, where a and b are constants with $a \ne 0$. Show that $\widehat{\mu_Y}(t) = e^{-ibt} \hat{\mu}(at)$.

15.46 Let $\mu \in M(\mathcal{R})$. Show that if $\int_{\mathcal{R}} |x|^k d|\mu|(x) < \infty$, where k is a positive integer, then the kth derivative $\hat{\mu}^{(k)}$ exists and

$$\hat{\mu}^{(k)}(t) = \frac{1}{\sqrt{2\pi}} \int_{\mathcal{R}} (-ix)^k e^{-itx}\, d\mu(x), \qquad t \in \mathcal{R}.$$

15.47 Prove Proposition 15.6 on page 578.

15.48 Prove Theorem 15.12 on page 578.

15.49 Let $\mu \in M(\mathcal{R})$. Show that $\hat{\mu}$ has period 2π if and only if $|\mu|(\mathcal{Z}^c) = 0$.

*In Exercises 15.50 and 15.51, we borrow some terminology from communications engineering. A measure $\mu \in M(\mathcal{R})$ is said to be **time limited** if it vanishes on Borel subsets of $[-a, a]^c$ for some $a > 0$, and it is said to be **band limited** if $\hat{\mu}$ vanishes outside of $[-b, b]$ for some $b > 0$.*

15.50   Show that if μ is band limited, then there is an $f \in \mathcal{L}^1(\mathcal{R})$ such that $\mu(B) = \int_B f d\lambda$ for all $B \in \mathcal{B}$. In other words, prove that every band-limited measure is absolutely continuous with respect to Lebesgue measure.

15.51   Show that a measure that is both band limited and time limited must vanish identically. *Hint:* Use the fact that if a function analytic on $\mathbb{C}$ vanishes on a nonempty open interval, then it vanishes identically.

15.52   Prove Lemma 15.4 on page 580.

15.53   Show that it suffices to prove the central limit theorem (Theorem 15.14 on page 581) in the case of zero mean and unit variance.

15.54   Let $\{X_n\}_{n=1}^{\infty}$ be a sequence of random variables.
   a) Show that if $X_n \xrightarrow{d} X$, where X is a continuous random variable, then $F_{X_n} \to F_X$ uniformly on $\mathcal{R}$. *Hint:* Show that for each $\epsilon > 0$, there is a $T > 0$ such that $\mu_{X_n}\left([-T, T]^c\right) < \epsilon$ for all n. Use the uniform continuity of F_X.
   b) Use part (a) to deduce the uniformity in the central limit theorem.

15.5 $\mathcal{L}^2$-THEORY OF THE FOURIER TRANSFORM

Because $\mathcal{L}^2_{2\pi} \subset \mathcal{L}^1_{2\pi}$, Fourier coefficients are defined for functions in $\mathcal{L}^2_{2\pi}$. Indeed, the theory of Fourier series for $\mathcal{L}^2_{2\pi}$-functions is particularly well understood. In the sense of convergence in the norm of $\mathcal{L}^2_{2\pi}$, we have, for $f \in \mathcal{L}^2_{2\pi}$, that

$$f(x) - \sum_{n=-\infty}^{\infty} \hat{f}(n) e^{inx} \tag{15.21}$$

and, furthermore,

$$\|f\|_2^2 = 2\pi \sum_{n=-\infty}^{\infty} |\hat{f}(n)|^2. \tag{15.22}$$

Given the strong analogy between Fourier coefficients and the Fourier transform, we would expect similar results to hold for functions in $\mathcal{L}^2(\mathcal{R})$ provided, of course, that the sums in (15.21) and (15.22) are replaced by suitable integrals.

However, there is an immediate problem: Since $\mathcal{L}^2(\mathcal{R}) \not\subset \mathcal{L}^1(\mathcal{R})$, the Fourier transform is not defined for all functions in $\mathcal{L}^2(\mathcal{R})$. To proceed, we must therefore first provide an appropriate definition of the Fourier transform of such functions. In this section, we will see that the "correct" definition leads naturally to extensions of (15.21) and (15.22).

We begin by studying the integral

$$\int_{-\infty}^{\infty} |\hat{f}(t)|^2 \, dt, \qquad f \in \mathcal{L}^1(\mathcal{R}). \tag{15.23}$$

In particular, we would like to know when this integral is finite. By the MCT, the finiteness of (15.23) is equivalent to that of $\lim_{T \to \infty} \int_{-T}^{T} |\hat{f}(t)|^2 \, dt$.

Referring to Example 15.5 on page 567 and applying Fubini's theorem yields

$$\int_{-T}^{T} |\hat{f}(t)|^2 \, dt = \frac{1}{2\pi} \int_{-\infty}^{\infty} \int_{-\infty}^{\infty} f(x)\overline{f(y)} \int_{-T}^{T} e^{-i(x-y)t} \, dt \, dx \, dy$$

$$= \frac{1}{\pi} \int_{-\infty}^{\infty} \int_{-\infty}^{\infty} f(x)\overline{f(y)} T \, \mathrm{sinc}(T(x-y)) \, dx \, dy.$$

The presence of the term $\mathrm{sinc}(T(x-y))$ makes these integrals hard to handle. Consequently, we will employ the averaging technique used in previous sections.

We first note that

$$\frac{1}{T} \int_0^T \int_{-t}^{t} |\hat{f}(s)|^2 \, ds \, dt = \frac{1}{\sqrt{2\pi}} \int_{-\infty}^{\infty} \int_{-\infty}^{\infty} f(x)\overline{f(y)} G_T(x-y) \, dx \, dy, \quad (15.24)$$

where

$$G_T(x) = \sqrt{\frac{2}{\pi}} \frac{1 - \cos(Tx)}{Tx^2} = \sqrt{\frac{2}{\pi}} \frac{\sin^2(Tx/2)}{Tx^2/2}.$$

Because, by Exercise 15.55,

$$\lim_{T \to \infty} \int_{-T}^{T} |\hat{f}(t)|^2 \, dt = \lim_{T \to \infty} \frac{1}{T} \int_0^T \int_{-t}^{t} |\hat{f}(s)|^2 \, ds \, dt,$$

it follows that finiteness in (15.23) is equivalent to that of the right-hand side of the previous equation. Our strategy therefore is to examine finiteness in (15.23) by working with the right-hand side of (15.24).

□ □ □ **LEMMA 15.5**

If $f \in \mathcal{L}^2(\mathcal{R}) \cap \mathcal{L}^1(\mathcal{R})$, then $\hat{f} \in \mathcal{L}^2(\mathcal{R})$ and $\|\hat{f}\|_2 = \|f\|_2$.

PROOF   By Lemma 15.2 on page 570,

$$\|f\|_2^2 = \frac{1}{\sqrt{2\pi}} \int_{-\infty}^{\infty} \|f\|_2^2 G_T(x) \, dx = \frac{1}{\sqrt{2\pi}} \int_{-\infty}^{\infty} \int_{-\infty}^{\infty} f(y)\overline{f(y)} G_T(x) \, dx \, dy.$$

On the other hand, from (15.24) we have

$$\frac{1}{T} \int_0^T \int_{-t}^{t} |\hat{f}(s)|^2 \, ds \, dt = \frac{1}{\sqrt{2\pi}} \int_{-\infty}^{\infty} \int_{-\infty}^{\infty} f(x+y)\overline{f(y)} G_T(x) \, dx \, dy.$$

Using Fubini's theorem, we get

$$\frac{1}{T} \int_0^T \int_{-t}^{t} |\hat{f}(s)|^2 \, ds \, dt - \|f\|_2^2 = \frac{1}{\sqrt{2\pi}} \int_{-\infty}^{\infty} \int_{-\infty}^{\infty} (f(x+y) - f(y))\overline{f(y)} G_T(x) \, dy \, dx.$$

and then applying Cauchy's inequality, we get

$$\left| \frac{1}{T} \int_0^T \int_{-t}^{t} |\hat{f}(s)|^2 \, ds \, dt - \|f\|_2^2 \right| \leq \frac{1}{\sqrt{2\pi}} \|f\|_2 \int_{-\infty}^{\infty} \|f_{-x} - f\|_2 G_T(x) \, dx.$$

We now proceed as in the proof of Theorem 15.7(b) on pages 570–571 to show that the right side of the preceding inequality tends to 0 as $T \to \infty$. Thus,

$$\int_{-\infty}^{\infty} |\hat{f}(y)|^2 \, dy = \lim_{T \to \infty} \frac{1}{T} \int_0^T \int_{-t}^{t} |\hat{f}(s)|^2 \, ds \, dt = \|f\|_2^2,$$

as required.   ■

□ □ □ **THEOREM 15.15 Plancherel's Theorem**

There exists a unique linear operator $\mathcal{F}: \mathcal{L}^2(\mathcal{R}) \to \mathcal{L}^2(\mathcal{R})$ *with the following properties:*

a) *For each* $f \in \mathcal{L}^2(\mathcal{R}) \cap \mathcal{L}^1(\mathcal{R})$, $\mathcal{F}(f) = \hat{f}$ *ae.*

b) *For each* $f \in \mathcal{L}^2(\mathcal{R})$, $\lim_{M \to \infty} \|\mathcal{F}(f) - \widehat{\tau_M f}\|_2 = 0$, *where* $\tau_M = \chi_{[-M,M]}$.

c) $\|\mathcal{F}(f)\|_2 = \|f\|_2$ *for each* $f \in \mathcal{L}^2(\mathcal{R})$.

d) $\langle \mathcal{F}(f), \mathcal{F}(g) \rangle = \langle f, g \rangle$ *for* $f, g \in \mathcal{L}^2(\mathcal{R})$.

e) $\mathcal{F}(\mathcal{F}(f))(x) = f(-x)$ *ae for each* $f \in \mathcal{L}^2(\mathcal{R})$.

PROOF Let $f \in \mathcal{L}^2(\mathcal{R})$ and let $\{M_n\}_{n=1}^\infty$ be a sequence of positive numbers tending to ∞. Then $\{\tau_{M_n} f\}_{n=1}^\infty \subset \mathcal{L}^2(\mathcal{R}) \cap \mathcal{L}^1(\mathcal{R})$ and $\lim_{n \to \infty} \|\tau_{M_n} f - f\|_2 = 0$. Now let $f_n = \tau_{M_n} f$. From Lemma 15.5, it follows that $\|\widehat{f_n} - \widehat{f_m}\|_2 = \|f_n - f_m\|_2$. Consequently, the sequence $\{\widehat{f_n}\}_{n=1}^\infty$ is Cauchy. Using the completeness of $\mathcal{L}^2(\mathcal{R})$, we now define

$$\mathcal{F}(f) = \lim_{n \to \infty} \widehat{f_n} = \lim_{n \to \infty} \widehat{\tau_{M_n} f}. \tag{15.25}$$

As the reader is asked to show in Exercise 15.56, the limit in (15.25) is independent of the particular sequence of M_ns.

If $f \in \mathcal{L}^2(\mathcal{R}) \cap \mathcal{L}^1(\mathcal{R})$, then the dominated convergence theorem implies that the sequence $\{\widehat{f_n}\}_{n=1}^\infty$ converges pointwise to $\hat{f}$. Thus, part (a) holds.

That $\mathcal{F}$ is a linear operator can be seen as follows: Let $f, g \in \mathcal{L}^2(\mathcal{R})$ and $\alpha, \beta \in \mathbb{C}$, and, for each $n \in \mathcal{N}$, set $h_n = \tau_{M_n} \alpha f + \tau_{M_n} \beta g$. Using the linearity of the Fourier transform on $\mathcal{L}^1(\mathcal{R})$, we have $\widehat{h_n} = \alpha \widehat{\tau_{M_n} f} + \beta \widehat{\tau_{M_n} g}$. Passing to the limit on n, we get $\mathcal{F}(\alpha f + \beta g) = \alpha \mathcal{F}(f) + \beta \mathcal{F}(g)$.

Part (b) follows from the definition of $\mathcal{F}(f)$ and Exercise 15.56. We obtain part (c) from Lemma 15.5 via

$$\|\mathcal{F}(f)\|_2 = \lim_{n \to \infty} \|\widehat{\tau_{M_n} f}\|_2 = \lim_{n \to \infty} \|\tau_{M_n} f\|_2 = \|f\|_2.$$

The uniqueness of a linear operator that satisfies parts (a) and (c) is a consequence of the fact that $\mathcal{L}^1(\mathcal{R}) \cap \mathcal{L}^2(\mathcal{R})$ is a dense subset of $\mathcal{L}^2(\mathcal{R})$. Part (d) is left to the reader as Exercise 15.57.

It remains to prove part (e). As $\mathcal{L}^2(\mathcal{R}) \cap \mathcal{L}^1(\mathcal{R})$ is dense in $\mathcal{L}^2(\mathcal{R})$, it suffices to prove that $\mathcal{F}(\mathcal{F}(f)) = R(f)$ for all $f \in \mathcal{L}^2(\mathcal{R}) \cap \mathcal{L}^1(\mathcal{R})$, where $R(f)(x) = f(-x)$. So, assume that $f \in \mathcal{L}^2(\mathcal{R}) \cap \mathcal{L}^1(\mathcal{R})$. Then, as the reader is asked to show in Exercise 15.57(b),

$$f * G_T \in \mathcal{L}^2(\mathcal{R}) \cap \mathcal{L}^1(\mathcal{R}) \qquad \text{and} \qquad \widehat{f * G_T} \in \mathcal{L}^2(\mathcal{R}) \cap \mathcal{L}^1(\mathcal{R}) \tag{15.26}$$

and

$$\lim_{T \to \infty} \|f * G_T - f\|_2 = 0. \tag{15.27}$$

Applying parts (a) and (c), (15.26), (15.27), and Theorem 15.8 on page 571, we obtain

$$0 = \lim_{T \to \infty} \|f * G_T - f\|_2 = \lim_{T \to \infty} \|\widehat{f * G_T} - \mathcal{F}(\mathcal{F}(f))\|_2$$

$$= \lim_{T \to \infty} \|R(f * G_T) - \mathcal{F}(\mathcal{F}(f))\|_2 = \|R(f) - \mathcal{F}(\mathcal{F}(f))\|_2.$$

The proof of Plancherel's theorem is now complete. ∎

The operator $\mathcal{F}$ given in Plancherel's theorem extends the definition of the Fourier transform to the space $\mathcal{L}^2(\mathcal{R})$. From now on, we will call $\mathcal{F}(f)$ the **Fourier transform** of an $\mathcal{L}^2(\mathcal{R})$-function f and write

$$\mathcal{F}(f) = \hat{f}, \qquad f \in \mathcal{L}^2(\mathcal{R}).$$

Although, strictly speaking, this expression is an abuse of notation, we observe from part (a) of Plancherel's theorem that this notation is consistent with previous usage.

EXAMPLE 15.12 *Illustrates Plancherel's Theorem*

The function sinc is not in $\mathcal{L}^1(\mathcal{R})$, but it is in $\mathcal{L}^2(\mathcal{R})$. By Example 15.5 on page 567, we have $\operatorname{sinc} t = \sqrt{\pi/2}\,\widehat{\chi_{[-1,1]}}(t)$. Applying part (e) of Plancherel's theorem, we deduce that

$$\widehat{\operatorname{sinc}}\, x = \sqrt{\pi/2}\,\chi_{[-1,1]}(-x) = \sqrt{\pi/2}\,\chi_{[-1,1]}(x)$$

for almost all x. In particular, $\widehat{\operatorname{sinc}}$ is not continuous. □

The Fourier transform on $\mathcal{L}^2(\mathcal{R})$ retains some of the properties that it has on $\mathcal{L}^1(\mathcal{R})$ — but others are lost. Specifically, our next theorem shows that part (d) of Theorem 15.5 (page 566) remains valid for $\mathcal{L}^2(\mathcal{R})$-functions, as do parts (e) and (f), provided we modify the notion of derivative. On the other hand, the Fourier transform of an $\mathcal{L}^2(\mathcal{R})$-function need not be continuous, as Example 15.12 shows.

Let $f \in \mathcal{L}^2(\mathcal{R})$. If $\|(f_{-h} - f)/h - \phi\|_2 \to 0$ as $h \to 0$, then we say that ϕ is the **derivative of f in the $\mathcal{L}^2$-sense**, and write $\phi = f'$. That this definition of derivative is close to the usual one can be seen from Exercise 15.59(a).

□ □ □ **THEOREM 15.16**

Let $f \in \mathcal{L}^2(\mathcal{R})$. Then the following results hold:

a) For $a, b \in \mathcal{R}$ and $a \neq 0$, we have $\widehat{f_{a,b}}(t) = \sqrt{|a|}\,e^{-ibt}\hat{f}(at)$ ae.

b) If $\int_{-\infty}^{\infty} x^2|f(x)|^2\,dx < \infty$, then $\hat{f}' = \hat{g}$ in the $\mathcal{L}^2$-sense, where $g(x) = -ixf(x)$.

c) If f' exists in the $\mathcal{L}^2$-sense, then $\widehat{f'}(t) = it\hat{f}(t)$ ae.

PROOF See Exercise 15.59. ■

Exercises for Section 15.5

15.55 Show that

$$\lim_{T \to \infty} \int_{-T}^{T} |\hat{f}(t)|^2\,dt = \lim_{T \to \infty} \frac{1}{T}\int_0^T \int_{-t}^t |\hat{f}(s)|^2\,ds\,dt.$$

15.56 Prove that the limit in (15.25) on page 585 is independent of the particular sequence of M_ns tending to ∞.

15.57 Refer to Plancherel's theorem (page 585).
a) Prove part (d) of the theorem.
b) Verify (15.26) and (15.27).

15.58 Show that the Fourier transform is onto $\mathcal{L}^2(\mathcal{R})$.

15.59 Establish the following facts.
a) If $f \in \mathcal{L}^2(\mathcal{R})$ is an absolutely continuous function such that $f' \in \mathcal{L}^2(\mathcal{R})$, then f' is also the derivative of f in the $\mathcal{L}^2$-sense.
b) Theorem 15.16.

15.60 Let f be a continuous function in $\mathcal{L}^2(\mathcal{R})$ such that $\sum_{n=-\infty}^{\infty} |f(n)| < \infty$ and $\hat{f}(t) = 0$ for $|t| \geq \pi$.
a) Show that, considered a function on $[-\pi, \pi]$, $\hat{f}$ has the Fourier series expansion $\hat{f}(t) = \sum_{n=-\infty}^{\infty} c_n e^{int}$, where $c_n = (2\pi)^{-1/2} f(-n)$.
b) Show that $f(x) = \sum_{n=-\infty}^{\infty} f(n) \operatorname{sinc}(\pi(x - n))$ for each $x \in \mathcal{R}$.
c) Use part (b) to prove the following result, which is known as the **Shannon sampling theorem.** Let L be a positive constant. Suppose that g is a continuous function in $\mathcal{L}^2(\mathcal{R})$ such that $\sum_{n=-\infty}^{\infty} |g(n\pi/L)| < \infty$ and $\hat{g}(t) = 0$ for $|t| \geq L$. Then

$$g(x) = \sum_{n=-\infty}^{\infty} g(n\pi/L) \operatorname{sinc}(Lx - \pi n).$$

The Shannon sampling theorem is used extensively in communications engineering.

In Exercises 15.61–15.68, we consider an important class of special functions that are closely related to the Fourier transform.

15.61 Let μ be the measure defined by $\mu(B) = \int_B e^{-x^2} d\lambda(x)$ and let $\langle \,, \rangle_\mu$ be the inner product induced by μ.
a) Verify that the space $\mathcal{L}^2(\mu)$ contains all polynomials.
b) Apply Theorem 13.5 on page 470 to the sequence $1, x, x^2, \ldots$ to obtain a sequence of polynomials $H_0, H_1, \ldots$, where H_n is of degree n, that are orthonormal with respect to $\langle \,, \rangle_\mu$. The H_ns are often referred to as **Hermite polynomials.**
c) Deduce that any polynomial p of degree n can be expressed as $p = \sum_{k=0}^{n} \langle p, H_k \rangle_\mu H_k$.

15.62 Refer to Exercise 15.61.
a) Deduce that the functions $h_n(x) = H_n(x)e^{-x^2/2}$ constitute an orthonormal sequence in $\mathcal{L}^2(\mathcal{R})$.
b) Show that $k_n = \hat{h}_n$ takes the form $k_n(t) = K_n(t)e^{-t^2/2}$, where K_n is a polynomial of degree at most n.

15.63 Prove that, for each $n \in \mathcal{N}$, we have $K_n = a_n H_n$, where $a_n \in \{1, -1, i, -i\}$.

15.64 Let α_n denote the leading coefficient of H_n. Show that

$$h'_n(x) - xh_n(x) = -2\frac{\alpha_n}{\alpha_{n+1}} h_{n+1}(x).$$

15.65 Prove that

$$e^{-x^2/2} h_n(x) = c_n \frac{d^n}{dx^n} e^{-x^2},$$

where $c_n = (-1)^n \alpha_n 2^{-n}$.

15.66 Referring to Exercise 15.63, verify that $a_n = (-i)^n$.

15.67 Show that $\{h_0, h_1, h_2, \ldots\}$ form an orthonormal basis for $\mathcal{L}^2(\mathcal{R})$.

15.68 Establish that the Fourier transform of a function $f \in \mathcal{L}^2(\mathcal{R})$ can be written in the form $\hat{f} = \sum_{n=0}^{\infty} (-i)^n \langle f, h_n \rangle h_n$.

15.6 INTRODUCTION TO WAVELETS

The theory of Fourier series seeks expansions of the form

$$f(x) = \sum_{n=-\infty}^{\infty} \hat{f}(n) e^{inx}$$

that express the function f as an infinite linear combination of dilations of the basic oscillating function $E(x) = e^{ix}$.

Similarly, **wavelet theory** is concerned with expansions of the form

$$f(x) = \sum_{n,m=-\infty}^{\infty} c_{nm} \psi(a_n x + b_m) \tag{15.28}$$

that express f as an infinite linear combination of translations-dilations of a single function ψ called a *wavelet*. Wavelet theory, however, unlike the theory of Fourier series, emphasizes the case where ψ is localized, that is, ψ vanishes or decays rapidly outside of some bounded interval.

This and the following section provide a brief introduction to the burgeoning theory of wavelets. We begin with a discussion of the family of Haar wavelets. Motivated by the example of Haar wavelets, we will then introduce the concept of a multiresolution analysis of $\mathcal{L}^2(\mathcal{R})$.

In our discussion of wavelets, we will restrict ourselves to functions in the Hilbert space $\mathcal{L}^2(\mathcal{R})$. And when we consider convergence of expansions of the form (15.28), we will always do so in the context of the usual $\mathcal{L}^2(\mathcal{R})$-norm. It will therefore be unambiguous to drop the subscript on that norm and to write $\langle \, , \, \rangle$ for the usual inner product on $\mathcal{L}^2(\mathcal{R})$.

As a further restriction, we will only investigate expansions of the form (15.28) in case $a_n = 2^{-n}$ and $b_m = -m$, where n and m vary over the set $\mathcal{Z}$ of all integers. Double sums of the form $\sum_{n,m=-\infty}^{\infty}$ will be denoted by $\sum_{n,m}$.

Wavelets and Haar Wavelets

We will employ the notation

$$f_{(n,m)}(x) = f_{2^n, 2^n m}(x) = 2^{-n/2} f(2^{-n} x - m).$$

It is important to note that if $f \in \mathcal{L}^2(\mathcal{R})$ and $\|f\| = 1$, then $\|f_{(n,m)}\| = \|f\| = 1$ for all $n, m \in \mathcal{Z}$.

DEFINITION 15.6   Orthonormal Wavelet Basis, Wavelet

Let $\psi \in \mathcal{L}^2(\mathcal{R})$. If the collection of functions $\{\, \psi_{(n,m)} : n, m \in \mathcal{Z} \,\}$ is an orthonormal basis for $\mathcal{L}^2(\mathcal{R})$, then it is called an **orthonormal wavelet basis** and the function ψ is called a **basic wavelet** or, more simply, a **wavelet.**

The following example introduces an important orthonormal wavelet basis and illustrates some basic ideas of wavelet theory.

EXAMPLE 15.13   The Haar Wavelet

For each $n \in \mathcal{Z}$, let $\mathcal{V}_n$ denote the set of all functions in $\mathcal{L}^2(\mathcal{R})$ that are constant on every interval of the form $[\ell\, 2^n, (\ell+1)2^n)$, where $\ell \in \mathcal{Z}$. Clearly, we have $\mathcal{V}_n \subset \mathcal{V}_{n-1}$ for each $n \in \mathcal{Z}$. Moreover, as the reader is asked to verify in Exercise 15.69, we have

$$\mathcal{V}_n \text{ is a closed linear subspace of } \mathcal{L}^2(\mathcal{R}), \tag{15.29}$$

$$\overline{\bigcup_{n \in \mathcal{Z}} \mathcal{V}_n} = \mathcal{L}^2(\mathcal{R}), \tag{15.30}$$

$$\bigcap_{n \in \mathcal{Z}} \mathcal{V}_n = \{0\}. \tag{15.31}$$

Let $\varphi = \chi_{[0,1)}$. From Exercise 15.70,

$$\mathcal{V}_n = \overline{\operatorname{span}\{\,\varphi_{(n,m)} : m \in \mathcal{Z}\,\}}. \tag{15.32}$$

Applying (15.30), it follows that

$$\overline{\operatorname{span}\{\,\varphi_{(n,m)} : m, n \in \mathcal{Z}\,\}} = \mathcal{L}^2(\mathcal{R}). \tag{15.33}$$

Also, for each $n \in \mathcal{Z}$, the family $\{\,\varphi_{(n,m)} : m \in \mathcal{Z}\,\}$ is orthonormal.

But although $\{\,\varphi_{(n,m)} : m, n \in \mathcal{Z}\,\}$ resembles an orthonormal wavelet basis, it is not because it lacks orthogonality. Indeed, we have, for example, that $\langle \varphi_{(n,0)}, \varphi_{(0,0)} \rangle \neq 0$. The problem is that the $\varphi_{(n,m)}$s are nonnegative-valued. To produce an orthonormal wavelet basis, we will modify φ.

To that end, we define

$$h(x) = \begin{cases} 1, & \text{if } 0 \le x < 1/2; \\ -1, & \text{if } 1/2 \le x < 1; \\ 0, & \text{otherwise.} \end{cases}$$

Members of the family $\mathcal{B}_h = \{\, h_{(n,m)} : n, m \in \mathcal{Z}\,\}$ are called **Haar functions.**

It is not difficult to show that $\mathcal{B}_h$ is orthonormal; but verifying that it is a basis for $\mathcal{L}^2(\mathcal{R})$ is somewhat more challenging. Suppose we can prove that

$$\varphi \in \overline{\operatorname{span} \mathcal{B}_h}. \tag{15.34}$$

Then, because

$$f \in \overline{\operatorname{span} \mathcal{B}_h} \quad \Rightarrow \quad f_{(n,m)} \in \overline{\operatorname{span} \mathcal{B}_h}, \ n, m \in \mathcal{Z}, \tag{15.35}$$

(see Exercise 15.72), it follows that

$$\operatorname{span}\{\,\varphi_{(n,m)} : m, n \in \mathcal{Z}\,\} \subset \overline{\operatorname{span} \mathcal{B}_h}.$$

Thus, by (15.33), $\overline{\operatorname{span} \mathcal{B}_h} = \mathcal{L}^2(\mathcal{R})$ and, hence, $\mathcal{B}_h$ is a basis.

It remains to verify (15.34), which we will do by proving that

$$\varphi = \sum_{m,n} \langle \varphi, h_{(n,m)} \rangle h_{(n,m)}. \qquad (15.36)$$

As the reader is asked to show in Exercise 15.73,

$$\langle \varphi, h_{(n,m)} \rangle = \begin{cases} 2^{-n/2}, & \text{if } n > 0 \text{ and } m = 0; \\ 0, & \text{otherwise.} \end{cases}$$

To establish (15.36), we first prove that

$$\varphi(x) = \sum_{n=1}^{\infty} 2^{-n} h(2^{-n}x), \qquad x \in \mathcal{R}. \qquad (15.37)$$

Clearly, both sides of (15.37) are 0 if $x < 0$. If $x \in [0,1)$, $2^{-n}x \in [0,1/2)$ for all $n \geq 1$ and, consequently, both the left and right sides of (15.37) equal 1.

If $x \in [1,\infty)$, select $k \in \mathcal{N}$ such that $2^{k-1} \leq x < 2^k$. Then $2^{-n}x \in [0,1/2)$ for $n > k$, $2^{-k}x \in [1/2,1)$, and $2^{-n}x \in [1,\infty)$ for $n < k$. Therefore, the right side of (15.37) is $-2^{-k} + \sum_{n=k+1}^{\infty} 2^{-n} = 0$ which, of course, equals the left side of (15.37). It now follows from Proposition 13.3 on page 457 and the DCT that (15.36) holds.

We have now shown that the family of Haar functions constitutes an orthonormal basis for $\mathcal{L}^2(\mathcal{R})$. Hence, it forms an orthonormal wavelet basis and h is a wavelet, called the **Haar wavelet**. □

Multiresolution Analysis

Guided by the essential features of Example 15.13, we can establish a general framework for constructing orthonormal wavelet bases. Specifically, we will work with a sequence $\{\mathcal{V}_n\}_{n=-\infty}^{\infty}$ of closed subspaces of $\mathcal{L}^2(\mathcal{R})$ that satisfies the following conditions.

(M1) $\cdots \subset \mathcal{V}_2 \subset \mathcal{V}_1 \subset \mathcal{V}_0 \subset \mathcal{V}_{-1} \subset \mathcal{V}_{-2} \subset \cdots$.

(M2) $\overline{\bigcup_{n \in \mathcal{Z}} \mathcal{V}_n} = \mathcal{L}^2(\mathcal{R})$.

(M3) $\bigcap_{n \in \mathcal{Z}} \mathcal{V}_n = \{0\}$.

(M4) $f \in \mathcal{V}_n$ if and only if $f_{(-n,0)} \in \mathcal{V}_0$.

(M5) $f \in \mathcal{V}_0$ if and only if $f_{(0,m)} \in \mathcal{V}_0$ for all $m \in \mathcal{Z}$.

(M6) There is a function $\varphi \in \mathcal{V}_0$ such that $\{\varphi_{(0,m)} : m \in \mathcal{Z}\}$ is an orthonormal basis for $\mathcal{V}_0$.

A sequence $\{\mathcal{V}_n\}_{n=-\infty}^{\infty}$ of closed subspaces of $\mathcal{L}^2(\mathcal{R})$ that satisfies (M1)–(M6) is said to be a **multiresolution analysis** of $\mathcal{L}^2(\mathcal{R})$.

As noted in Example 15.13, if $\mathcal{V}_n$ denotes the collection of $\mathcal{L}^2(\mathcal{R})$-functions that are constant on every interval of the form $[\ell 2^n, (\ell+1)2^n)$, where $\ell \in \mathcal{Z}$, then $\{\mathcal{V}_n\}_{n=-\infty}^{\infty}$ is a multiresolution analysis with $\varphi = \chi_{[0,1)}$.

In the general framework described by (M1)–(M6), the family of functions $\{\varphi_{(n,m)} : m \in \mathcal{Z}\}$ is an orthonormal basis for $\mathcal{V}_n$ for each n, but $\varphi_{(n,m)}$ and $\varphi_{(j,k)}$ may not be orthogonal if $n \neq j$. Rather, an orthonormal wavelet basis can be constructed by using φ, as we will show in the next section.

For the remainder of this chapter, we will assume that $\{\mathcal{V}_n\}_{n=-\infty}^{\infty}$ is a multiresolution analysis. The orthogonal projection of $\mathcal{L}^2(\mathcal{R})$ onto $\mathcal{V}_n$ will be denoted P_n.

Our next lemma, whose proof is left to the reader as Exercise 15.74, will be needed in our development of the theory of wavelets.

□ □ □   **LEMMA 15.6**

Let $\mathcal{W}_0 = \mathcal{V}_0^{\perp} \cap \mathcal{V}_{-1}$, $\mathcal{W}_n = \{f_{(n,0)} : f \in \mathcal{W}_0\}$, and $Q_n : \mathcal{L}^2(\mathcal{R}) \to \mathcal{W}_n$ be the orthogonal projection of $\mathcal{L}^2(\mathcal{R})$ onto $\mathcal{W}_n$. Then the following results hold.

a) If $\ell \neq n$, then $\langle f, g \rangle = 0$ for all $f \in \mathcal{W}_\ell$ and $g \in \mathcal{W}_n$.

b) $P_{n-1} = P_n + Q_n$.

c) For each $f \in \mathcal{L}^2(\mathcal{R})$, we have $f = \sum_{n=-\infty}^{\infty} Q_n(f)$, where the series converges absolutely with respect to the $\mathcal{L}^2(\mathcal{R})$-norm.

If $\{\psi_{(0,m)} : m \in \mathcal{Z}\}$ is an orthonormal basis for $\mathcal{W}_0$, then $\{\psi_{(n,m)} : m \in \mathcal{Z}\}$ is an orthonormal basis for $\mathcal{W}_n$. Consequently, by Theorem 13.6 on page 471, we have that $Q_n(f) = \sum_{m=-\infty}^{\infty} \langle Q_n(f), \psi_{(n,m)} \rangle \psi_{(n,m)}$ for each $f \in \mathcal{L}^2(\mathcal{R})$. It follows from Lemma 15.6(c) that $f = \sum_{n,m} \langle Q_n(f), \psi_{(n,m)} \rangle \psi_{(n,m)}$. On the other hand, by Lemma 15.6(a), the family $\{\psi_{(n,m)} : n, m \in \mathcal{Z}\}$ is orthonormal. Thus, $\{\psi_{(n,m)} : n, m \in \mathcal{Z}\}$ is an orthonormal wavelet basis for $\mathcal{L}^2(\mathcal{R})$.

We have shown that if $\{\psi_{(0,m)} : m \in \mathcal{Z}\}$ is an orthonormal basis for $\mathcal{W}_0$, then $\{\psi_{(n,m)} : n, m \in \mathcal{Z}\}$ is an orthonormal wavelet basis for $\mathcal{L}^2(\mathcal{R})$. We conclude this section by giving sufficient conditions for $\{\psi_{(0,m)} : m \in \mathcal{Z}\}$ to be an orthonormal basis for $\mathcal{W}_0$.

□ □ □   **PROPOSITION 15.7**

Suppose that $\psi \in \mathcal{W}_0$ is such that $\|\psi\| = 1$ and also satisfies the following two conditions:

a) *For each $n \in \mathcal{Z} \setminus \{0\}$, we have $\int_{-\infty}^{\infty} e^{int} |\hat{\psi}(t)|^2 \, dt = 0$.*

b) *For each $f \in \mathcal{W}_0$, there exists $F \in \mathcal{L}_{2\pi}^2$ such that $\hat{f} = F\hat{\psi}$.*

Then $\{\psi_{(0,m)} : m \in \mathcal{Z}\}$ is an orthonormal basis for $\mathcal{W}_0$ and, consequently, we have that $\{\psi_{(n,m)} : n, m \in \mathcal{Z}\}$ is an orthonormal wavelet basis for $\mathcal{L}^2(\mathcal{R})$.

PROOF   Applying condition (a), Plancherel's theorem (page 585), and Theorem 15.16 (page 586), we get that

$$\langle \psi_{(0,m)}, \psi_{(0,p)} \rangle = \int_{-\infty}^{\infty} e^{i(p-m)t} |\hat{\psi}(t)|^2 \, dt = \begin{cases} 0, & \text{if } p \neq m; \\ 1, & \text{if } p = m. \end{cases}$$

Thus, $\{\psi_{(0,m)} : m \in \mathcal{Z}\}$ is orthonormal. We will show that it forms a basis for $\mathcal{W}_0$ by applying Theorem 13.6(c) on page 471.

Suppose that $f \in \mathcal{W}_0$ and $\langle f, \psi_{(0,m)} \rangle = 0$ for each $m \in \mathcal{Z}$. Then, by condition (b) and, again, Plancherel's theorem and Theorem 15.16,

$$\int_{-\infty}^{\infty} e^{imt} F(t) |\hat{\psi}(t)|^2 \, dt = \langle F\hat{\psi}, \psi_{(0,m)}^{\frown} \rangle = 0 \tag{15.38}$$

for each $m \in \mathcal{Z}$. Now, we have

$$\int_{-\infty}^{\infty} e^{imt} F(t) |\hat{\psi}(t)|^2 \, dt = \sum_{\ell=-\infty}^{\infty} \int_{[2\ell\pi, 2(\ell+1)\pi)} e^{imt} F(t) |\hat{\psi}(t)|^2 \, dt$$

$$= \int_{-\pi}^{\pi} e^{imt} F(t) \sum_{\ell=-\infty}^{\infty} |\hat{\psi}(t + (2\ell+1)\pi)|^2 \, dt.$$

The function $g(t) = F(t) \sum_{\ell=-\infty}^{\infty} |\hat{\psi}(t + (2\ell+1)\pi)|^2$ belongs to $\mathcal{L}_{2\pi}^1$ since, by Cauchy's inequality,

$$\int_{-\pi}^{\pi} |g(t)| \, dt = \int_{-\pi}^{\pi} |F(t)| \sum_{\ell=-\infty}^{\infty} |\hat{\psi}(t + (2\ell+1)\pi)|^2 \, dt$$

$$= \int_{-\infty}^{\infty} |F(t)| |\hat{\psi}(t)|^2 \, dt \leq \|\hat{f}\| \|\hat{\psi}\| = \|f\| < \infty.$$

In view of (15.38), all Fourier coefficients of g vanish and, consequently, Theorem 15.1 on page 553 implies that $g = 0$ ae. It follows that $\hat{f} = F\hat{\psi}$ vanishes ae on $\mathcal{R}$ and hence that $f = 0$ ae. ∎

Exercises for Section 15.6

15.69 Verify (15.29)–(15.31) on page 589.

15.70 Refer to Example 15.13 on pages 589–590.
a) Prove that $\mathcal{V}_n = \text{span}\{ \varphi_{(n,m)} : m \in \mathcal{Z} \}$.
b) Show that (a) holds for any multiresolution analysis of $\mathcal{L}^2(\mathcal{R})$.

15.71 Prove that the Haar functions form an orthonormal family.

15.72 Verify (15.35) on page 589.

15.73 Show that

$$\langle \chi_{[0,1)}, h_{(n,m)} \rangle = \begin{cases} 2^{-n/2}, & \text{if } n > 0 \text{ and } m = 0; \\ 0, & \text{otherwise.} \end{cases}$$

15.74 Prove Lemma 15.6 on page 591.

15.75 Calculate the Fourier transforms of the Haar functions.

15.76 For a multiresolution analysis $\{\mathcal{V}_n\}_{n=-\infty}^{\infty}$, let P_n denote the orthogonal projection of $\mathcal{L}^2(\mathcal{R})$ onto $\mathcal{V}_n$. Is $P_n \circ P_{n-1} = P_n$? *Hint:* See Exercise 13.26 on page 467.

15.77 Show that the Haar functions do not span a dense subspace of $\mathcal{L}^1(\mathcal{R})$.

15.78 Let $\{ h_{(n,m)} : n, m \in \mathcal{Z} \}$ be the family of Haar functions. Define

$$f(x) = \begin{cases} 2x, & \text{for } 0 \leq x < 1/2; \\ 2 - 2x, & \text{for } 1/2 \leq x < 1; \\ 0, & \text{for } x \notin [0, 1). \end{cases}$$

Sketch the graph of the partial sum $\sum_{m=-1}^{1} \sum_{n=-1}^{1} \langle f, h_{n,m} \rangle h_{n,m}$.

15.7 ORTHONORMAL WAVELET BASES; THE WAVELET TRANSFORM

Working with a multiresolution analysis $\{\mathcal{V}_n\}_{n=-\infty}^{\infty}$, we will construct a function ψ that satisfies the conditions of Proposition 15.7 on page 591 and thereby obtain an orthonormal wavelet basis for $\mathcal{L}^2(\mathcal{R})$. We will also introduce a continuous version of the wavelet expansion.

Scaling Functions

Recall that a sequence $\{\mathcal{V}_n\}_{n=-\infty}^{\infty}$ of closed subspaces of $\mathcal{L}^2(\mathcal{R})$ is called a multiresolution analysis of $\mathcal{L}^2(\mathcal{R})$ if it satisfies (M1)–(M6) on page 590. By (M6) there is a function φ such that $\{\varphi_{(0,m)} : m \in \mathcal{Z}\}$ is an orthonormal basis for $\mathcal{V}_0$. We will call φ a **scaling function** of the multiresolution analysis $\{\mathcal{V}_n\}_{n=-\infty}^{\infty}$. Properties of φ are developed in the following lemmas.

□ □ □ **LEMMA 15.7**

Let φ be a scaling function of the multiresolution analysis $\{\mathcal{V}_n\}_{n=-\infty}^{\infty}$. Then

$$\sum_{\ell=-\infty}^{\infty} |\hat{\varphi}(t + 2\pi\ell)|^2 = \frac{1}{2\pi}$$

for almost all $t \in \mathcal{R}$.

PROOF We outline the proof here and leave the details to the reader as Exercise 15.79. Let $g(t) = \sum_{\ell=-\infty}^{\infty} |\hat{\varphi}(t + 2\pi\ell)|^2$. Then g is an extended real-valued function with period 2π. That g is finite almost everywhere is a consequence of the fact that $\int_{-\pi}^{\pi} g(t)\,dt = \|\hat{\varphi}\|^2 = 1$. We now see that $g \in \mathcal{L}_{2\pi}^1$.

Using Plancherel's theorem (page 585) and Theorem 15.16 on page 586, it can be shown that the Fourier coefficients $\hat{g}(k)$ vanish for $k \neq 0$. Consequently, g has the same Fourier coefficients as the function that is constantly equal to $1/2\pi$. Applying Theorem 15.1(c) on page 553 now yields the required result. ∎

□ □ □ **LEMMA 15.8**

Let φ be a scaling function of the multiresolution analysis $\{\mathcal{V}_n\}_{n=-\infty}^{\infty}$. Then the following results hold.

a) For almost all $x \in \mathcal{R}$,

$$\varphi(x) = \sqrt{2} \sum_{n=-\infty}^{\infty} \varphi_n \varphi(2x - n),$$

where $\varphi_n = \langle \varphi, \varphi_{(-1,n)} \rangle$. Moreover, $\sum_{n=-\infty}^{\infty} |\varphi_n|^2 = 1$.

b) For almost all $t \in \mathcal{R}$,

$$\hat{\varphi}(t) = m_0(t/2)\,\hat{\varphi}(t/2),$$

where $m_0(t) = (1/\sqrt{2}) \sum_{n=-\infty}^{\infty} \varphi_n e^{-int}$.

PROOF Again, we only outline the proof, leaving the details for Exercise 15.80. To prove (a), it suffices to show that $\{\varphi_{(-1,n)} : n \in \mathcal{Z}\}$ is an orthonormal basis for the space $\mathcal{V}_{-1}$. Applying the Fourier transform to both sides of the equation for φ given in part (a), leads to the verification of (b). ∎

□ □ □ **LEMMA 15.9**

Let φ be a scaling function of the multiresolution analysis $\{\mathcal{V}_n\}_{n=-\infty}^{\infty}$ and let m_0 be as in Lemma 15.8(b). Then $|m_0(t)|^2 + |m_0(t + \pi)|^2 = 1$ ae.

PROOF By Lemmas 15.7 and 15.8, we have

$$\frac{1}{2\pi} = \sum_{\ell=-\infty}^{\infty} |\hat{\varphi}(t + 2\pi\ell)|^2 = \sum_{\ell=-\infty}^{\infty} |\hat{\varphi}(t + 4\pi\ell)|^2 + \sum_{\ell=-\infty}^{\infty} |\hat{\varphi}(t + 2\pi(2\ell + 1))|^2$$

$$= \sum_{\ell=-\infty}^{\infty} |\hat{\varphi}(t/2 + 2\pi\ell)|^2 |m_0(t/2 + 2\pi\ell)|^2$$

$$+ \sum_{\ell=-\infty}^{\infty} |\hat{\varphi}(t/2 + \pi + 2\pi\ell)|^2 |m_0(t/2 + \pi + 2\pi\ell)|^2$$

for almost all t. Using Lemma 15.7 and the fact that m_0 has period 2π, we obtain that

$$\frac{1}{2\pi} = |m_0(t/2)|^2 \sum_{\ell=-\infty}^{\infty} |\hat{\varphi}(t/2 + 2\pi\ell)|^2$$

$$+ |m_0(t/2 + \pi)|^2 \sum_{\ell=-\infty}^{\infty} |\hat{\varphi}(t/2 + \pi + 2\pi\ell)|^2$$

$$= \frac{1}{2\pi} |m_0(t/2)|^2 + \frac{1}{2\pi} |m_0(t/2 + \pi)|^2.$$

The assertion of the lemma now follows immediately. ∎

Our next lemma characterizes the action of the Fourier transform on the space $\mathcal{V}_{-1}$.

□ □ □ **LEMMA 15.10**

Let φ be a scaling function of the multiresolution analysis $\{\mathcal{V}_n\}_{n=-\infty}^{\infty}$. For $f \in \mathcal{V}_{-1}$, set $f_n = \langle f, \varphi_{(-1,n)} \rangle$. Then $\hat{f}(t) = m_f(t/2)\,\hat{\varphi}(t/2)$, where

$$m_f(t) = \frac{1}{\sqrt{2}} \sum_{n=-\infty}^{\infty} f_n e^{-int}. \tag{15.39}$$

PROOF Recall that $\mathcal{W}_0 = \mathcal{V}_0^{\perp} \cap \mathcal{V}_{-1}$. Because $\{\varphi_{(-1,n)} : n \in \mathcal{Z}\}$ is an orthonormal basis for $\mathcal{V}_{-1}$, every $f \in \mathcal{W}_0$ has the expansion

$$f = \sum_{n=-\infty}^{\infty} \langle f, \varphi_{(-1,n)} \rangle \varphi_{(-1,n)} = \sum_{n=-\infty}^{\infty} f_n \varphi_{(-1,n)}.$$

The required result now follows from a straightforward application of the Fourier transform by using Theorem 15.16 on page 586. ∎

When the function f belongs to the space $\mathcal{W}_0$, more can be said about the function m_f given in (15.39).

□ □ □ **LEMMA 15.11**

Let φ be a scaling function of the multiresolution analysis $\{V_n\}_{n=-\infty}^{\infty}$ and let m_0 and m_f be as in Lemmas 15.8 and 15.10, respectively. If $f \in W_0$, then

$$\overline{m_0(t)}\, m_f(t) + \overline{m_0(t+\pi)}\, m_f(t+\pi) = 0$$

for almost all t.

PROOF   The proof is similar to that of Lemma 15.7 on page 593. We sketch it here and leave the details to the reader as Exercise 15.81. Let

$$G(t) = \sum_{\ell=-\infty}^{\infty} \hat{f}(t + 2\pi\ell)\, \overline{\hat{\varphi}(t + 2\pi\ell)}.$$

Then $G \in \mathcal{L}_{2\pi}^1$ and all of its Fourier coefficients vanish; so, $G = 0$ ae. Applying Lemmas 15.8 and 15.10, we have, for almost all t, that

$$0 = \sum_{\ell=-\infty}^{\infty} m_f(t/2 + \pi\ell)\, \overline{m_0(t/2 + \pi\ell)}\, |\hat{\varphi}(t/2 + \pi\ell)|^2.$$

The proof is completed by an argument similar to that used in the proof of Lemma 15.9. ∎

Next we have a formula that characterizes the Fourier transform of a function in the space W_0.

□ □ □ **LEMMA 15.12**

Let φ be a scaling function of the multiresolution analysis $\{V_n\}_{n=-\infty}^{\infty}$ and let m_0 be as in Lemma 15.8 on page 593. Then $f \in W_0$ if and only if there is a function $F \in \mathcal{L}_{2\pi}^2$ such that

$$\hat{f}(t) = e^{it/2}\, \overline{m_0(t/2 + \pi)}\, \hat{\varphi}(t/2)\, F(t). \tag{15.40}$$

PROOF   Suppose that $f \in W_0$. Let

$$L(t) = \begin{cases} m_f(t)/\overline{m_0(t+\pi)}, & \text{if } m_0(t+\pi) \neq 0; \\ 0, & \text{if } m_0(t+\pi) = 0. \end{cases}$$

It follows from Lemmas 15.9 and 15.11 that

$$L(t) = -L(t + \pi) \tag{15.41}$$

and

$$m_f(t) = \overline{m_0(t+\pi)}\, L(t). \tag{15.42}$$

Now let $F(t) = e^{-it/2}L(t/2)$. Applying Lemma 15.10 and (15.42), we deduce that

$$\hat{f}(t) = e^{it/2}\, \overline{m_0(t/2 + \pi)}\, \hat{\varphi}(t/2)\, F(t).$$

From (15.41), we see that F has period 2π. Also, it follows from the definition of F that $|F(t)|^2 |m_0(t/2 + \pi)|^2 = |m_f(t/2)|^2$. Hence, by Lemma 15.9,

$$|F(t)|^2 = |F(t)|^2 |m_0(t/2)|^2 + |F(t)|^2 |m_0(t/2 + \pi)|^2 = |m_f(t/2 + \pi)|^2 + |m_f(t/2)|^2.$$

Consequently, by Theorem 13.6 on page 471, we have

$$\int_{-\pi}^{\pi} |F(t)|^2\, dt = \int_{-\pi}^{\pi} \left(|m_f(t/2 + \pi)|^2 + |m_f(t/2)|^2 \right) dt$$

$$= 2 \int_{-\pi}^{\pi} |m_f(t)|^2\, dt = 4\pi \sum_{n=-\infty}^{\infty} |f_n|^2 = 4\pi \|f\|^2 < \infty.$$

This result shows that $F \in \mathcal{L}_{2\pi}^2$.

Conversely, suppose that f satisfies (15.40) for some $F \in \mathcal{L}_{2\pi}^2$. We then have a Fourier series expansion $F(t) = \sum_{n=-\infty}^{\infty} \widehat{F}(n) e^{int}$, where $\sum_{n=-\infty}^{\infty} |\widehat{F}(n)|^2 < \infty$. Thus,

$$\hat{f}(t) = \sum_{n=-\infty}^{\infty} \widehat{F}(n) e^{int} e^{it/2} \overline{m_0(t/2 + \pi)}\, \hat{\varphi}(t/2). \qquad (15.43)$$

Applying Theorem 15.16 on page 586, we have

$$e^{it/2} \overline{m_0(t/2 + \pi)}\, \hat{\varphi}(t/2) = \frac{1}{\sqrt{2}} \sum_{k=-\infty}^{\infty} \overline{\varphi_k} e^{ik(t/2 + \pi)} e^{it/2} \hat{\varphi}(t/2)$$

$$= \sum_{k=-\infty}^{\infty} (-1)^{k-1} \overline{\varphi_{k-1}} \widehat{\varphi_{(-1,-k)}}(t),$$

where we recall from Lemma 15.8 that $\varphi_k = \langle \varphi, \varphi_{(-1,k)} \rangle$.

The series

$$\psi = \sum_{k=-\infty}^{\infty} (-1)^{k-1} \overline{\varphi_{k-1}} \varphi_{(-1,-k)}$$

defines a function in the space $\mathcal{V}_{-1}$. By the continuity of the Fourier transform, we have

$$\hat{\psi}(t) = e^{it/2} \overline{m_0(t/2 + \pi)}\, \hat{\varphi}(t/2). \qquad (15.44)$$

Hence, we can use Plancherel's theorem and Theorem 15.16 to rewrite (15.43) as

$$\hat{f}(t) = \sum_{n=-\infty}^{\infty} \widehat{F}(n) e^{int} \hat{\psi}(t).$$

Therefore, $\hat{f}$ is also the Fourier transform of the function $\sum_{n=-\infty}^{\infty} \widehat{F}(n) \psi_{(0,-n)}$. Applying the uniqueness property of Fourier transforms, we conclude that

$$f = \sum_{n=-\infty}^{\infty} \widehat{F}(n) \psi_{(0,-n)}.$$

Because $\psi_{(0,-n)} \in \mathcal{V}_{-1}$ for each $n \in \mathcal{Z}$, it follows that $f \in \mathcal{V}_{-1}$. To complete the proof, we must verify that $f \in \mathcal{V}_0^\perp$. This verification will be accomplished if we can prove that

$$\langle \psi_{(0,-n)}, \varphi_{(0,-n+m)} \rangle = \langle \psi, \varphi_{(0,m)} \rangle = 0 \tag{15.45}$$

for all $m \in \mathcal{Z}$. However, we have by Plancherel's theorem, Lemmas 15.7 and 15.8, and (15.44) that

$$\langle \psi, \varphi_{(0,m)} \rangle = \langle \hat{\psi}, \hat{\varphi}_{(0,m)} \rangle$$

$$= \int_{-\infty}^{\infty} e^{i(m+1/2)t} \, \overline{m_0(t/2+\pi)} \, \overline{m_0(t/2)} \, |\hat{\varphi}(t/2)|^2 \, dt$$

$$= \sum_{\ell=-\infty}^{\infty} 2 \int_{0}^{2\pi} e^{i(2m+1)t} \, \overline{m_0(t+\pi)} \, \overline{m_0(t)} \, |\hat{\varphi}(t+2\pi\ell)|^2 \, dt$$

$$= \frac{1}{\pi} \int_{0}^{2\pi} e^{i(2m+1)t} \, \overline{m_0(t+\pi)} \, \overline{m_0(t)} \, dt$$

$$= \frac{1}{\pi} \int_{0}^{\pi} e^{i(2m+1)t} \, \overline{m_0(t+\pi)} \, \overline{m_0(t)} \, dt$$

$$\qquad + \frac{1}{\pi} \int_{0}^{\pi} e^{i(2m+1)(t+\pi)} \, \overline{m_0(t+2\pi)} \, \overline{m_0(t+\pi)} \, dt$$

$$= 0.$$

This result verifies (15.45) and completes the proof of the lemma. ■

Construction of Orthonormal Wavelet Bases

In the course of the proof of Lemma 15.12, we constructed the function

$$\psi = \sum_{k=-\infty}^{\infty} (-1)^{k-1} \overline{\varphi_{k-1}} \varphi_{(-1,-k)}.$$

As the next theorem shows, ψ is a wavelet.

□ □ □ **THEOREM 15.17**

Let $\{\mathcal{V}_n\}_{n=-\infty}^{\infty}$ be a multiresolution analysis of $\mathcal{L}^2(\mathcal{R})$ with scaling function φ. Define

$$\psi = \sum_{k=-\infty}^{\infty} (-1)^{k-1} \overline{\varphi_{k-1}} \varphi_{(-1,-k)}, \tag{15.46}$$

where $\varphi_k = \langle \varphi, \varphi_{(-1,k)} \rangle$. Then $\{ \psi_{(n,m)} : n, m \in \mathcal{Z} \}$ is an orthonormal wavelet basis for $\mathcal{L}^2(\mathcal{R})$.

PROOF We will prove the theorem by verifying that ψ satisfies the hypotheses of Proposition 15.7 on page 591. To begin, we note that $\|\psi\| = 1$ since $\sum_{n=-\infty}^{\infty} |\varphi_n|^2 = 1$. Also, in the course of proving Lemma 15.12, we actually established that $\psi \in \mathcal{W}_0$. It now follows from (15.44) and Lemma 15.12 that condition (b) of Proposition 15.7 is satisfied.

It remains to show that condition (a) of Proposition 15.7 holds. To that end, we apply (15.44), Lemma 15.7, and Lemma 15.9 to obtain that

$$\int_{-\infty}^{\infty} e^{int} |\hat{\psi}(t)|^2 \, dt = \int_{-\infty}^{\infty} e^{int} |m_0(t/2 + \pi)|^2 \, |\hat{\varphi}(t/2)|^2 \, dt$$

$$= \sum_{\ell=-\infty}^{\infty} 2 \int_0^{2\pi} e^{i2nt} |m_0(t + \pi)|^2 \, |\hat{\varphi}(t + 2\pi\ell)|^2 \, dt$$

$$= \frac{1}{\pi} \int_0^{2\pi} e^{i2nt} |m_0(t + \pi)|^2 \, dt$$

$$= \frac{1}{\pi} \int_0^{\pi} e^{i2nt} |m_0(t + \pi)|^2 \, dt + \frac{1}{\pi} \int_0^{\pi} e^{i2nt} |m_0(t + 2\pi)|^2 \, dt$$

$$= \frac{1}{\pi} \int_0^{\pi} e^{i2nt} \, dt = 0,$$

for $n \neq 0$. ∎

EXAMPLE 15.14 *Illustrates Theorem 15.17*

Refer to Example 15.13 on pages 589–590. If we apply Formula (15.46) to the scaling function $\varphi = \chi_{[0,1)}$, we obtain the wavelet

$$\psi(x) = \varphi(2x + 1) - \varphi(2x + 2).$$

This wavelet is quite similar to the basic Haar wavelet h. In fact, we have $\psi_{(0,1)} = -h$. It follows that, in this case, the orthonormal wavelet basis determined by ψ consists of the Haar functions multiplied by the factor -1. □

The Wavelet Transform

Next we introduce a continuous version of the discrete wavelet expansion. To begin, we recall that for $a, b \in \mathcal{R}$ and $a \neq 0$, the function $\psi_{a,b}$ is the *translation-dilation* of the function ψ:

$$\psi_{a,b}(x) = \frac{1}{\sqrt{|a|}} \, \psi((x - b)/a).$$

Here now is the definition of the *wavelet transform*.

DEFINITION 15.7 Wavelet Transform

Let ψ be a fixed function in $\mathcal{L}^2(\mathcal{R}) \setminus \{0\}$. Then, for each $f \in \mathcal{L}^2(\mathcal{R})$, the function $Wf: (0, \infty) \times \mathcal{R} \to \mathbb{C}$ defined by

$$Wf(a, b) = \langle f, \psi_{a,b} \rangle$$

is called the **wavelet transform** of f.

Note: Although the wavelet transform depends on the fixed function ψ, we have retained the terminology found in the literature by writing W instead of W_ψ and by using the terminology "the wavelet transform" instead of, say, "the wavelet transform with respect to ψ."

EXAMPLE 15.15 *Illustrates Definition 15.7*

Let $T > 0$ and set $\psi = \chi_{[-T,T]}/2T$. From Plancherel's theorem, we have, for each $f \in \mathcal{L}^2(\mathcal{R})$, that $Wf(a,b) = \langle \hat{f}, \widehat{\psi_{a,b}} \rangle$. Referring now to Example 15.5 on page 567, we conclude that

$$Wf(a,b) = \sqrt{\frac{a}{2\pi}} \int_{-\infty}^{\infty} \hat{f}(t) \operatorname{sinc}(atT) e^{ibt}\, dt.$$

Replacing f by $\hat{f}$ and again applying Plancherel's theorem, we obtain that

$$W\hat{f}(a,b) = \sqrt{\frac{a}{2\pi}} \int_{-\infty}^{\infty} \hat{\hat{f}}(t) \operatorname{sinc}(atT) e^{ibt}\, dt = \sqrt{\frac{a}{2\pi}} \int_{-\infty}^{\infty} f(x) \operatorname{sinc}(axT) e^{-ibx}\, dx.$$

If f is also in $\mathcal{L}^1(\mathcal{R})$, then we can use the dominated convergence theorem to conclude that

$$\lim_{T \to 0^+} W\hat{f}(1,t) = \frac{1}{\sqrt{2\pi}} \int_{-\infty}^{\infty} f(x) e^{-itx}\, dx = \hat{f}(t).$$

Consequently, we obtain the Fourier transform of f as a limiting case of a wavelet transform. □

If ψ is a wavelet, that is, if $\{\, \psi_{(n,m)} : n, m \in \mathcal{Z} \,\}$ happens to be an orthonormal wavelet basis for $\mathcal{L}^2(\mathcal{R})$, then f can be recovered from its wavelet transform via

$$f = \sum_{n,m} \langle f, \psi_{(n,m)} \rangle \psi_{(n,m)} = \sum_{n,m} Wf(2^n, 2^n m) \psi_{2^n, 2^n m}.$$

This result suggests, in general, the heuristic formula

$$f = \int_{(0,\infty) \times \mathcal{R}} Wf(a,b) \psi_{a,b}\, d\mu(a,b) \tag{15.47}$$

for recovering a function from its wavelet transform. We will show how sense can be made of (15.47) by choosing the measure μ appropriately and imposing mild restrictions on ψ.

We begin by deriving the measure μ. By Plancherel's theorem and Theorem 15.16 on page 586, we have

$$Wf(a,b) = \langle \hat{f}, \psi_{a,b} \rangle = \sqrt{a} \int_{-\infty}^{\infty} \hat{f}(t) \overline{\hat{\psi}(at)}\, e^{ibt}\, dt = \widehat{F}(-b),$$

where $F(t) = \sqrt{2\pi a}\, \hat{f}(t) \overline{\hat{\psi}(at)}$. Again applying Plancherel's theorem, we obtain, for each $a > 0$, that

$$\int_{-\infty}^{\infty} |Wf(a,b)|^2\, db = \int_{-\infty}^{\infty} |\widehat{F}(-b)|^2\, db$$

$$= \int_{-\infty}^{\infty} |F(t)|^2\, dt = 2\pi a \int_{-\infty}^{\infty} |\hat{f}(t)|^2 |\hat{\psi}(at)|^2\, dt.$$

Multiplying by a^{-2} and integrating over $(0, \infty)$ yields

$$\int_0^\infty \int_{-\infty}^\infty |Wf(a,b)|^2 \frac{1}{a^2} \, db \, da$$

$$= \int_{-\infty}^\infty 2\pi |\hat{f}(t)|^2 \int_0^\infty \frac{1}{a} |\hat{\psi}(at)|^2 \, da \, dt \qquad (15.48)$$

$$= \int_0^\infty 2\pi |\hat{f}(t)|^2 \int_0^\infty \frac{1}{s} |\hat{\psi}(s)|^2 \, ds \, dt - \int_{-\infty}^0 2\pi |\hat{f}(t)|^2 \int_{-\infty}^0 \frac{1}{s} |\hat{\psi}(s)|^2 \, ds \, dt.$$

We are now ready to impose a restriction on ψ, namely, that

$$-\int_{-\infty}^0 \frac{1}{s} |\hat{\psi}(s)|^2 \, ds = \int_0^\infty \frac{1}{s} |\hat{\psi}(s)|^2 \, ds = C_\psi < \infty.$$

With this restriction on ψ, we can prove a theorem for the wavelet transform that is analogous to Plancherel's theorem.

☐ ☐ ☐ **THEOREM 15.18**

Suppose that $\psi \in \mathcal{L}^2(\mathcal{R}) \setminus \{0\}$ and that

$$-\int_{-\infty}^0 \frac{1}{s} |\hat{\psi}(s)|^2 \, ds = \int_0^\infty \frac{1}{s} |\hat{\psi}(s)|^2 \, ds = C_\psi < \infty. \qquad (15.49)$$

Define the Borel measure μ_0 on $(0, \infty)$ via

$$\mu_0(B) = \frac{1}{2\pi C_\psi} \int_B a^{-2} \, da, \qquad B \in \mathcal{B},$$

and let $\mu = \mu_0 \times \lambda$. Then the wavelet transform is a linear operator from $\mathcal{L}^2(\mathcal{R})$ to $\mathcal{L}^2(\mu)$ that satisfies

$$\|Wf\|_{2,\mu} = \|f\|, \qquad f \in \mathcal{L}^2(\mathcal{R}), \qquad (15.50)$$

where $\| \ \|_{2,\mu}$ denotes the $\mathcal{L}^2$-norm on $\mathcal{L}^2(\mu)$.

PROOF It follows from (15.48), (15.49), and Plancherel's theorem that

$$\int_{(0,\infty)\times\mathcal{R}} |Wf(a,b)|^2 \, d\mu(a,b) = \frac{1}{2\pi C_\psi} \int_0^\infty \int_{-\infty}^\infty |Wf(a,b)|^2 \frac{1}{a^2} \, db \, da$$

$$= \int_{-\infty}^\infty |\hat{f}(t)|^2 \, dt = \int_{-\infty}^\infty |f(x)|^2 \, dx.$$

Thus, (15.50) is valid. ■

Theorem 15.18 provides a likely candidate for the measure μ appearing in the heuristic formula (15.47). Still, the problem of correctly interpreting (15.47) remains. One possible approach, explored in Exercise 15.85, is to show that under appropriate conditions

$$f(x) = \int_{(0,\infty)\times\mathcal{R}} Wf(a,b) \psi_{a,b}(x) \, d\mu(a,b)$$

for almost all x. A more subtle, but easier to prove, interpretation is based on the following theorem whose verification is left to the reader as Exercise 15.88.

□ □ □ THEOREM 15.19

Suppose that $\psi \in \mathcal{L}^2(\mathcal{R}) \setminus \{0\}$ and

$$-\int_{-\infty}^0 \frac{1}{s}|\hat{\psi}(s)|^2\, ds = \int_0^\infty \frac{1}{s}|\hat{\psi}(s)|^2\, ds = C_\psi < \infty.$$

Then (15.47) is valid in the sense that

$$\langle f, g \rangle = \int_{(0,\infty)\times\mathcal{R}} Wf(a,b)\langle \psi_{a,b}, g \rangle\, d\mu(a,b), \qquad f, g \in \mathcal{L}^2(\mathcal{R}),$$

where μ is defined as in Theorem 15.18.

The theory of wavelets is an important and active research area. As a starting point for the interested reader, we recommend the paper "Wavelet transforms and orthonormal wavelet bases" by I. Daubechies (*Proceedings of Symposia in Applied Mathematics*, Vol. 47, American Math. Soc., Providence, RI, 1993).

Exercises for Section 15.7

15.79 Provide the details of the proof of Lemma 15.7 on page 593.

15.80 Provide the details of the proof of Lemma 15.8 on page 593.

15.81 Provide the details of the proof of Lemma 15.11 on page 595.

15.82 Show that (15.49) is satisfied if ψ is real-valued and $\hat{\psi}$ vanishes in some open interval that contains 0.

15.83 Show that (15.49) is satisfied if ψ is real-valued, $\hat{\psi}$ is continuous in some open interval that contains 0, $\hat{\psi}(0) = 0$, and $\hat{\psi}'(0)$ exists.

15.84 Show that (15.49) is satisfied by the Haar wavelet h discussed in Example 15.13 on pages 589–590. Find C_ψ in this case.

15.85 Suppose ψ satisfies (15.49), μ is defined as in Theorem 15.18, and $f, \hat{f} \in \mathcal{L}^2(\mathcal{R}) \cap \mathcal{L}^1(\mathcal{R})$. Prove that

$$f(x) = \int_{(0,\infty)\times\mathcal{R}} Wf(a,b)\psi_{a,b}(x)\, d\mu(a,b)$$

for almost all $x \in \mathcal{R}$.

15.86 Consider the Hermite function h_1 discussed in Exercises 15.61–15.68 on page 587.
a) Find C_{h_1}.
b) Determine Wh_0.

15.87 Find a formula for $Wf_{c,d}$ in terms of Wf.

15.88 Prove Theorem 15.19.

Claude Elwood Shannon (1916–2001)

Claude Elwood Shannon was born in Gaylord, Michigan, on April 30, 1916. He was the only child of Claude Elwood Shannon, Sr., a self-made businessman and, for a time, a probate judge, and Mabel Wolf Shannon, a language teacher and, for several years, principal of Gaylord High School .

In 1932, Shannon graduated from Gaylord High School; in 1936, he earned a bachelor's degree at the University of Michigan; and, in 1940, he was awarded both a master's degree and a doctorate in mathematics at the Massachusetts Institute of Technology.

After working as a National Research Fellow at Princeton University for a year, Shannon joined the staff at Bell Labs in 1941. His charge there was to determine the most efficient method of transmitting information. Because of his success in presenting the transmission of information as a precise mathematical theory, he is considered one of the founders of information theory. Shannon related the relaying of information to a binary system of yes/no choices, represented by a 1/0 binary code, a representation that is still integral to computer design today.

In 1956, Shannon accepted the position of Visiting Professor of Electronic Communication at MIT; in 1957, Professor of Communications Science and Mathematics; and in 1958, Donner Professor of Science.

Shannon received many awards for his work. For instance, in 1940, he was honored with the Alfred Nobel American Institute of American Engineers Award; in 1966, the National Medal of Science; in 1985, the Audio Engineering Society Gold Medal; and, also in 1985, the Kyoto Prize. He was also honored with the Marconi Lifetime Achievement Award by the Guglielmo Marconi International Fellowship Foundation in 2000.

In addition to communications engineering, Shannon's methods have profoundly influenced many other fields, including mathematics, computer science, statistics, biology, physics, cryptography, linguistics, and phonetics. Shannon passed away on February 24, 2001, in Medford, Massachusetts.

16

Measurable Dynamical Systems

In this chapter, we discuss the theory of measurable dynamical systems. Section 16.1 introduces the theory by providing a motivating heuristic illustration, stating the definition of a measurable dynamical system, and presenting several examples.

Section 16.2 discusses ergodicity and presents a proof of the famous pointwise ergodic theorem. Section 16.3 examines isomorphisms of measurable dynamical systems and introduces the concept of entropy. And Section 16.4 investigates the entropy of a Bernoulli shift.

16.1 INTRODUCTION AND EXAMPLES

To introduce this chapter, we construct a simple heuristic model that illustrates the idea of a measurable dynamical system. Imagine a particle p confined in some compact region $\Omega \subset \mathcal{R}^3$.

Suppose that p moves around inside Ω according to the following rule: If p is at x at time n, then it moves to $\psi(x)$ at time $n+1$, where $\varphi \colon \Omega \to \Omega$ is a function that is independent of n. Although, according to this rule, the particle is always moving in Ω, the law that governs its movement remains constant for all time.

For $A \subset \Omega$, let

$$\mu_n(A) = \begin{cases} 1, & \text{if } p \in A \text{ at time } n; \\ 0, & \text{otherwise.} \end{cases}$$

Then the expression

$$\mu(A) = \lim_{n \to \infty} \frac{1}{n} \sum_{k=0}^{n-1} \mu_k(A)$$

represents an average over time of the number of visits of the particle to the set A, that is, the number of visits to A per unit time by the particle.

Let $\mathcal{A}$ denote the collection of subsets of Ω such that the previous limit exists. Clearly, $\emptyset$ and Ω belong to $\mathcal{A}$, and it is easy to see that μ satisfies the following conditions:

- $\mu(A) \geq 0$ for all $A \in \mathcal{A}$.

- $\mu(\emptyset) = 0$.

- $\mu(\Omega) = 1$.

- If $A, B \in \mathcal{A}$ are disjoint, then $\mu(A \cup B) = \mu(A) + \mu(B)$.

Consequently, we see that the triple $(\Omega, \mathcal{A}, \mu)$ resembles a probability space. Suppose that, indeed,

$$(\Omega, \mathcal{A}, \mu) \quad \text{is a probability space.} \tag{16.1}$$

Because the particle is in A at time n if and only if it is in $\varphi^{-1}(A)$ at time $n - 1$, we have $\mu_{n-1}(\varphi^{-1}(A)) = \mu_n(A)$. It follows that

$$A \in \mathcal{A} \quad \Rightarrow \quad \varphi^{-1}(A) \in \mathcal{A} \tag{16.2}$$

and

$$A \in \mathcal{A} \quad \Rightarrow \quad \mu(A) = \mu(\varphi^{-1}(A)). \tag{16.3}$$

Thus, a quadruple $(\Omega, \mathcal{A}, \mu, \varphi)$ that satisfies (16.1)–(16.3) models the average behavior of the aforementioned simple particle motion. Formally, we have the following definition.

DEFINITION 16.1 **Invariant Measure, Measurable Dynamical System**

Let $(\Omega, \mathcal{A}, \mu)$ be a measure space. Suppose that $\varphi \colon \Omega \to \Omega$ and that $\varphi^{-1}(A) \in \mathcal{A}$ for all $A \in \mathcal{A}$. Then μ is said to be **invariant** with respect to φ if

$$\mu(A) = \mu(\varphi^{-1}(A)), \qquad A \in \mathcal{A},$$

that is, if $\mu \circ \varphi^{-1} = \mu$ as measures on $\mathcal{A}$. If μ is invariant with respect to φ and is also a probability measure, then the quadruple $(\Omega, \mathcal{A}, \mu, \varphi)$ is referred to as a **measurable dynamical system.**

In the remainder of this section, we will present a variety of examples of measurable dynamical systems, showing their relevance and importance.

EXAMPLE 16.1 *Addition Modulo One*

Let the operation $\overset{\circ}{+}$ be defined on $[0,1)$ by

$$x \overset{\circ}{+} y = (x+y) \bmod 1 = \begin{cases} x+y, & \text{if } x+y < 1; \\ x+y-1, & \text{if } x+y \geq 1. \end{cases}$$

For fixed $b \in [0,1)$, let $\varphi_b(x) = x \overset{\circ}{+} b$. Then $([0,1), \mathcal{M}_{[0,1)}, \lambda_{[0,1)}, \varphi_b)$ is a measurable dynamical system. □

EXAMPLE 16.2 *Rotation Through an Angle*

Let E be the map from $[0,1)$ onto the unit circle T in the complex plane defined by $E(x) = e^{2\pi i x}$ and let $\mathcal{A} = \{ A \subset T : E^{-1}(A) \in \mathcal{M} \}$. Define the measure μ on $\mathcal{A}$ by $\mu(A) = \lambda(E^{-1}(A))$, so that μ is normalized arc-length measure on T. Also, for fixed $b \in [0,1)$, define $\psi_b : T \to T$ by $\psi_b(z) = e^{2\pi i b} z$. Then $(T, \mathcal{A}, \mu, \psi_b)$ is a measurable dynamical system. In a sense that will be made precise later, this example is the same as the previous one. □

EXAMPLE 16.3 *Multiplication by 2 Mod One*

Let the mapping φ_b in Example 16.1 be replaced by

$$\varphi(x) = 2x \bmod 1 = x \overset{\circ}{+} x.$$

As the reader is asked to verify in Exercise 16.1, φ is measurable with respect to $\mathcal{M}_{[0,1)}$ and $\lambda_{[0,1)}$ is invariant with respect to φ. Hence, $([0,1), \mathcal{M}_{[0,1)}, \lambda_{[0,1)}, \varphi)$ is a measurable dynamical system. It is interesting to note that if $x \in [0,1)$ has the binary expansion $x = 0.x_1 x_2 x_3 \ldots$, then we have $\varphi(x) = 0.x_2 x_3 \ldots$ (2). □

EXAMPLE 16.4 *Bernoulli Schemes*

Let $S = \{1, 2, \ldots, N\}$, where $N \geq 2$, and let $p = (p_1, p_2, \ldots, p_N)$, where $p_j \geq 0$ for each $j \in S$ and $\sum_{j=1}^{N} p_j = 1$. The vector p defines a probability measure μ_0 on S via $\mu_0(\{j\}) = p_j$.

Recall that the Cartesian product $\Omega = S^{\mathcal{Z}}$ consists of all functions on the integers with values in S, or, alternatively, all doubly infinite sequences of elements of S. From μ_0, we will construct a probability measure μ on Ω by extending the development of product measure given in Theorem 6.10 on page 219.

To begin, let F be a finite set of integers and a a function from F into S. Then we define

$$C_{F,a} = \{ f \in \Omega : f(j) = a(j) \text{ for } j \in F \}.$$

Let $\mathcal{C}$ be the collection of subsets of Ω that consists of $\emptyset$, Ω, and all sets of the form $C_{F,a}$.

Next, we define a set function ι on $\mathcal{C}$ by letting $\iota(\emptyset) = 0$, $\iota(\Omega) = 1$, and

$$\iota(C_{F,a}) = \prod_{j \in F} \mu_0(\{a(j)\}).$$

In Exercise 16.2, we ask the reader to show that $\mathcal{C}$ is a semialgebra of subsets of Ω and that ι satisfies conditions (E1)–(E3) on page 180 and condition (E4) of Theorem 6.2 on page 188. Consequently, by that theorem, ι extends uniquely to a probability measure μ on the σ-algebra $\mathcal{A}$ generated by $\mathcal{C}$. We now have a probability space $(\Omega, \mathcal{A}, \mu)$.

Next, we define the function $\varphi \colon \Omega \to \Omega$ by $\varphi(f)(j) = f(j+1)$. If we consider the elements of Ω doubly infinite sequences, then the effect of φ is to move each term of a sequence f one place to the left. For this reason, the mapping φ is often called a **Bernoulli shift.**

It is easy to see that

$$\varphi^{-1}(C_{F,a}) = C_{F^*, a^*}, \tag{16.4}$$

where $F^* = \{\, j + 1 : j \in F \,\}$ and $a^*(j+1) = a(j)$. It follows that the σ-algebra $\{\, A \subset \Omega : \varphi^{-1}(A) \in \mathcal{A} \,\}$ contains $\mathcal{C}$ and, hence, $\varphi^{-1}(A) \in \mathcal{A}$ for each $A \in \mathcal{A}$.

We claim that the measure μ is invariant with respect to φ. Let the measure ν be defined on $\mathcal{A}$ by $\nu(A) = \mu(\varphi^{-1}(A))$. By (16.4) we have

$$\nu(C_{F,a}) = \mu(C_{F^*, a^*}) = \prod_{j \in F^*} \mu_0(a^*(j)) = \prod_{j \in F} \mu_0(a(j)) = \mu(C_{F,a}).$$

Thus ν and μ agree on $\mathcal{C}$ and so, by Theorem 6.2, $\nu = \mu$. This means that μ is invariant with respect to φ.

We have shown that $(\Omega, \mathcal{A}, \mu, \varphi)$ is a measurable dynamical system. This system is known in the literature as a **Bernoulli scheme** and is often denoted by $B(p_1, p_2, \ldots, p_N)$. ◻

EXAMPLE 16.5 Continued Fraction Expansions

Decimal and binary expansions are familiar ways of representing real numbers. Less familiar but, nevertheless, useful and interesting is the expansion of a real number as a continued fraction. This expansion is based on iteration of the function defined on $[0, 1)$ by

$$\varphi(x) = \begin{cases} 1/x - \lfloor 1/x \rfloor, & \text{for } x \neq 0; \\ 0, & \text{for } x = 0, \end{cases}$$

where $\lfloor\ \rfloor$ denotes the greatest integer function.

We can express x in terms of $\varphi(x)$ by

$$x = \frac{1}{\alpha(x) + \varphi(x)}, \tag{16.5}$$

where

$$\alpha(x) = \begin{cases} \lfloor 1/x \rfloor, & \text{if } x \neq 0; \\ \infty, & \text{if } x = 0. \end{cases}$$

Replacing x by $\varphi(x)$ and substituting into the right-hand side of (16.5), we obtain

$$x = \cfrac{1}{\alpha(x) + \cfrac{1}{\alpha(\varphi(x)) + \varphi(\varphi(x))}}.$$

Repeating this procedure, we obtain

$$x = \cfrac{1}{\alpha(x) + \cfrac{1}{\alpha(\varphi(x)) + \cfrac{1}{\alpha(\varphi(\varphi(x))) + \cfrac{1}{\ddots + \cfrac{1}{\alpha(\varphi^{(n)}(x)) + \varphi^{(n+1)}(x)}}}}},$$

where $\varphi^{(n)}$ indicates the nth iterate of φ. As the reader is asked to show in Exercise 16.3, the sequence of quotients

$$x_n = \cfrac{1}{\alpha(x) + \cfrac{1}{\alpha(\varphi(x)) + \cfrac{1}{\alpha(\varphi(\varphi(x))) + \cfrac{1}{\ddots + \cfrac{1}{\alpha(\varphi^{(n)}(x))}}}}}$$

converges to x. Thus, we have the continued fraction expansion

$$x = \cfrac{1}{\alpha(x) + \cfrac{1}{\alpha(\varphi(x)) + \cfrac{1}{\alpha(\varphi(\varphi(x))) + \cfrac{1}{\ddots}}}}. \tag{16.6}$$

Clearly, the mapping φ is the key element for obtaining (16.6). We will relate a measurable dynamical system to the continued fraction by finding a probability measure μ on Borel subsets of $[0, 1)$ that is invariant with respect to φ. To obtain μ, we begin by deriving a necessary condition for a Borel measure on $[0, 1)$ to be both invariant with respect to φ and absolutely continuous with respect to $\lambda_{[0,1)}$.

Suppose then that μ is a Borel measure on $[0, 1)$ that is invariant with respect to φ and absolutely continuous with respect to $\lambda_{[0,1)}$. Set $g = d\mu/d\lambda$. Then, for each $t \in (0, 1)$, we have

$$\int_{(0,t)} g(x) \, d\lambda(x) = \int_{\varphi^{-1}((0,t))} g(x) \, d\lambda(x).$$

Using

$$\varphi^{-1}((0,t)) = \bigcup_{k=1}^{\infty} \{x : \lfloor 1/x \rfloor = k \text{ and } \varphi(x) < t\} = \bigcup_{k=1}^{\infty} \left(\frac{1}{k+t}, \frac{1}{k}\right],$$

we obtain

$$\int_{(0,t)} g(x) \, d\lambda(x) = \sum_{k=1}^{\infty} \int_{\frac{1}{k+t}}^{\frac{1}{k}} g(x) \, dx. \tag{16.7}$$

Ignoring questions of convergence, we differentiate both sides of (16.7) to get the equation

$$g(t) = \sum_{k=1}^{\infty} g\left(\frac{1}{t+k}\right) \frac{1}{(t+k)^2}. \tag{16.8}$$

Equation (16.8) looks formidable. To find a solution, it is helpful to recast it as a functional equation:

$$g(t) = g\left(\frac{1}{t+1}\right) \frac{1}{(t+1)^2} + \sum_{k=1}^{\infty} g\left(\frac{1}{t+k+1}\right) \frac{1}{(t+k+1)^2}$$

$$= g\left(\frac{1}{t+1}\right) \frac{1}{(t+1)^2} + g(t+1). \tag{16.9}$$

The form of (16.9) suggests that we try solutions of the type $g(t) = (t+1)^a$. Substituting for $g(t)$ in (16.9) gives

$$(t+1)^a = \left(\frac{1}{t+1}+1\right)^a \frac{1}{(t+1)^2} + (t+2)^a. \tag{16.10}$$

It is not hard to see that (16.10) is satisfied for all $t \in [0,1)$ if $a = -1$.

The preceding informal argument suggests that measures on $[0,1)$ of the form

$$\mu(B) = \int_B \frac{c}{x+1}\, d\lambda(x)$$

are invariant with respect to the transformation φ. It is left for Exercise 16.4 to verify this suggestion formally. The choice $c = (\log 2)^{-1}$ yields an invariant probability measure on $[0,1)$. ☐

EXAMPLE 16.6 *Hamiltonian Systems*

Consider the system of differential equations

$$\frac{dq_j}{dt} = \frac{\partial H}{\partial p_j}, \quad \frac{dp_j}{dt} = -\frac{\partial H}{\partial q_j}, \quad j = 1, 2, \ldots, 3n, \tag{16.11}$$

where H is a function on $\mathcal{R}^{6n}$ of the form

$$H(p,q) = H(p_1, p_2, \ldots, p_{3n}, q_1, q_2, \ldots, q_{3n}) = \frac{1}{2} \sum_{j=1}^{3n} \frac{p_j^2}{m_j} + V(q_1, q_2, \ldots, q_{3n}).$$

Such systems of differential equations are important in mechanics, where the vectors $(q_{3j-2}, q_{3j-1}, q_{3j})$ and $(p_{3j-2}, p_{3j-1}, p_{3j})$, $1 \le j \le n$, represent the position and momentum, respectively, of the jth of n particles that are moving in $\mathcal{R}^3$. The term $\frac{1}{2}\sum_{j=1}^{3n} p_j^2/m_j$ gives the total kinetic energy of the n particles, and $V(q_1, q_2, \ldots, q_{3n})$ is the energy associated with interactions of the n particles.

Assuming that V is reasonably well behaved, it follows from the general theory of differential equations that, for each

$$(x,y) = (x_1, x_2, \ldots, x_{3n}, y_1, y_2, \ldots, y_{3n}) \in \mathcal{R}^{6n},$$

there is a unique solution

$$\alpha(t, x, y) = (p_1(t, x, y), \ldots, p_{3n}(t, x, y), q_1(t, x, y), \ldots, q_{3n}(t, x, y))$$

to the system that is defined for all t and satisfies $\alpha(0, x, y) = (x, y)$.

Also, under appropriate hypotheses on the function V, it can be shown that, for $j = 1, 2, \ldots, 3n$, all second-order partial derivatives of the functions $p_j(t, x, y)$ and $q_j(t, x, y)$ with respect to each of the variables $t, x_1, \ldots, y_{3n}$ exist and are continuous.

For fixed t, define $\varphi_t : \mathcal{R}^{6n} \to \mathcal{R}^{6n}$ by $\varphi_t(x, y) = \alpha(t, x, y)$. As you are asked to prove in Exercise 16.6,

$$\lambda_{6n}(E) = \lambda_{6n}(\varphi_t^{-1}(E)), \qquad E \in \mathcal{M}_{6n}. \tag{16.12}$$

This result is known as **Liouville's theorem.**

Next, we will combine (16.12) with an invariance property of H to produce a measurable dynamical system. The property of H that we need is

$$H \circ \varphi_t = H. \tag{16.13}$$

To obtain (16.13), we use the chain rule and (16.11). We have

$$
\begin{aligned}
\frac{d}{dt} H(\varphi_t(x, y)) &= \sum_{j=1}^{3n} \frac{\partial H}{\partial p_j}(\varphi_t(x, y)) \frac{dp_j(t, x, y)}{dt} + \frac{\partial H}{\partial q_j}(\varphi_t(x, y)) \frac{dq_j(t, x, y)}{dt} \\
&= \sum_{j=1}^{3n} \left[\frac{\partial H}{\partial p_j}(\varphi_t(x, y)) \left(-\frac{\partial H}{\partial q_j}(\varphi_t(x, y)) \right) \right. \\
&\qquad \left. + \frac{\partial H}{\partial q_j}(\varphi_t(x, y)) \frac{\partial H}{\partial p_j}(\varphi_t(x, y)) \right] \\
&= 0.
\end{aligned}
$$

It follows that

$$H(\varphi_t(x, y)) = H(\varphi_0(x, y)) = H(x, y)$$

and, hence, (16.13) holds.

For each $c \in \mathcal{R}$, let $\Omega_c = H^{-1}((-\infty, c))$. Then, in view of (16.13), we have that $\varphi_t(\Omega_c) \subset \Omega_c$. Assuming, as in many applications, that Ω_c is a bounded set with positive Lebesgue measure, we can define a probability measure on Lebesgue measurable subsets of Ω_c by

$$\mu(E) = \frac{\lambda_{6n}(E)}{\lambda_{6n}(\Omega_c)}. \tag{16.14}$$

It follows from (16.12) that μ is invariant with respect to φ_t and, consequently, $(\Omega_c, \mathcal{M}_{\Omega_c}, \mu, \varphi_t)$ is a measurable dynamical system. □

Exercises for Section 16.1

Note: A ★ denotes an exercise that will be subsequently referenced.

16.1 Prove that the mapping φ in Example 16.3 on page 605 is measurable with respect to $\mathcal{M}_{[0,1)}$ and that $\lambda_{[0,1)}$ is invariant with respect to φ.

16.2 Refer to Example 16.4 on pages 605–606.
a) Show that the collection $\mathcal{C}$ is a semialgebra.
b) Show that the set function ι defined on the collection $\mathcal{C}$ satisfies conditions (E1)–(E3) on page 180 and condition (E4) on page 188.

16.3 Prove that the sequence $\{x_n\}_{n=1}^{\infty}$, defined in Example 16.5 on page 607, converges to x. *Hint:* If $x \in Q$, show that $x_n = x$ for sufficiently large n. If $x \notin Q$, show that $x_n = p_n/q_n$ where $\{p_n\}_{n=1}^{\infty}$ and $\{q_n\}_{n=1}^{\infty}$ are sequences of integers defined recursively by: $p_{-1} = 0$, $p_0 = 1$, and $p_n = a_n p_{n-1} + p_{n-2}$ for $n \geq 1$; $q_{-1} = 1$, $q_0 = \alpha(x)$, and $q_n = a_n q_{n-1} + q_{n-2}$ for $n \geq 1$. Here $a_n = \alpha(\varphi^{(n)}(x))$.

16.4 Prove that the measure

$$\mu(B) = \frac{1}{\log 2} \int_B \frac{1}{x+1} \, d\lambda(x), \qquad B \in \mathcal{B}_{[0,1)},$$

is invariant with respect to the mapping φ defined in Example 16.5 on page 606.

16.5 Suppose that T is a one-to-one continuously differentiable transformation from $\mathcal{R}^n$ onto $\mathcal{R}^n$ with Jacobian identically 1. Prove that T is a measurable transformation from $(\mathcal{R}^n, \mathcal{M}_n)$ to $(\mathcal{R}^n, \mathcal{M}_n)$ and that $\lambda_n = \lambda_n \circ T^{-1}$ as measures on $\mathcal{M}_n$. In solving this exercise, you may want to refer to Exercise 6.19 on page 191. Proceed as follows.
a) Show that, as measures on $\mathcal{B}_n$, $\lambda_n = \lambda_n \circ T^{-1}$. *Hint:* Use Theorem 9.11 on page 351.
b) Apply part (a) to show that T is a measurable transformation from $(\mathcal{R}^n, \mathcal{M}_n)$ to $(\mathcal{R}^n, \mathcal{M}_n)$.
c) Use parts (a) and (b) to show that, as measures on $\mathcal{M}_n$, $\lambda_n = \lambda_n \circ T^{-1}$.

16.6 Refer to Example 16.6, which begins on page 608.
a) Prove that $\det J_{\varphi_t} = 1$. *Hint:* Show that $\partial \det J_{\varphi_t}(x, y)/\partial t = 0$.
b) Verify (16.12). *Hint:* Use part (a) and Exercise 16.5. You may assume as known that $\varphi_t \colon \mathcal{R}^{6n} \to \mathcal{R}^{6n}$ is one-to-one and onto; this result follows, for instance, from the problem given at the bottom of page 68 in V.I. Arnold's *Mathematical Methods of Classical Mechanics*, 2nd edition (New York: Springer-Verlag, 1989).

★16.7 Suppose that Ω is a compact Hausdorff space and let $\varphi \colon \Omega \to \Omega$ be continuous. Show that there is a regular Borel probability measure μ on Ω such that $\mu(\varphi^{-1}(B)) = \mu(B)$ for all Borel subsets of Ω. *Hint:* Fix $\omega \in \Omega$ and apply the Hahn-Banach theorem (page 498) with the subadditive function $\sigma \colon C(\Omega, \mathcal{R}) \to \mathcal{R}$ defined by

$$\sigma(f) = \limsup_{n \to \infty} \frac{1}{n} \sum_{k=0}^{n-1} f(\varphi^{(k)}(\omega)),$$

where $\varphi^{(k)}$ is the kth iterate of φ.

★16.8 Let Ω be a compact Hausdorff space and $\varphi \colon \Omega \to \Omega$ be continuous. Show that the collection $I(\varphi)$ of regular Borel probability measures on Ω that are invariant with respect to φ is weak* compact and convex.

16.9 Let $\varphi(x) = x^2$. Show that the only regular Borel probability measures on $[0, 1]$ that are invariant with respect to the function φ are those of the form $c\delta_0 + (1 - c)\delta_1$, where $0 \leq c \leq 1$.

16.10 Suppose that $\varphi: [0,1] \to [0,1]$ is absolutely continuous, strictly increasing, and onto. Set $\psi = \varphi^{-1}$. Show that if μ is absolutely continuous with respect to $\lambda_{[0,1]}$ and invariant with respect to φ, then

$$\frac{d\mu}{d\lambda}(x) = \frac{d\mu}{d\lambda}(\psi(x))\psi'(x)$$

for almost all $x \in [0,1]$.

16.11 Suppose that $\varphi: [0,1] \to [0,1]$ is continuously differentiable and that for each $x \in [0,1]$, $\varphi^{-1}(\{x\})$ is a finite set. Let μ be absolutely continuous with respect to $\lambda_{[0,1]}$ and set $g = d\mu/d\lambda$. Show that μ is invariant with respect to φ if and only if

$$g(x) = \sum_{y \in \varphi^{-1}(\{x\})} \frac{g(y)}{|\varphi'(y)|}$$

for almost all $x \in [0,1]$.

16.2 ERGODIC THEORY

Let $(\Omega, \mathcal{A}, \mu, \varphi)$ be a measurable dynamical system. Recall that for $n \in \mathcal{N}$, $\varphi^{(n)}$ denotes the nth iterate of φ. We also define $\varphi^{(0)}$ to be the identity function on Ω.

For $x \in \Omega$, the sequence

$$x, \; \varphi(x), \; \varphi(\varphi(x)), \; \ldots, \; \varphi^{(n)}(x), \; \ldots,$$

called the **orbit** of x, describes the path of the point x as it moves in Ω under iterations of the mapping φ. In ergodic theory, we try to find out as much as possible about this sequence.

Oftentimes, in applications, orbits cannot be observed directly, but rather data are obtained in the form of numerical sequences

$$f(x), \; f(\varphi(x)), \; f(\varphi(\varphi(x))), \; \ldots, \; f(\varphi^{(n)}(x)), \; \ldots,$$

where f is some function defined on Ω. In this section, we prove some general results about the average behavior of the sequence $\left\{f(\varphi^{(n-1)}(x))\right\}_{n=1}^{\infty}$. Specifically, we will first establish that for each $f \in \mathcal{L}^1(\mu)$, the limit

$$f^* = \lim_{n \to \infty} \frac{1}{n} \sum_{k=0}^{n-1} f \circ \varphi^{(k)}$$

exists μ-ae. Then we will investigate the important case where f^* is constant μ-ae for all $f \in \mathcal{L}^1(\mu)$.

□ □ □ THEOREM 16.1 Pointwise Ergodic Theorem

For each $f \in \mathcal{L}^1(\mu)$, the limit

$$f^* = \lim_{n \to \infty} \frac{1}{n} \sum_{k=0}^{n-1} f \circ \varphi^{(k)} \tag{16.15}$$

exists μ-ae. Furthermore, $f^* \in \mathcal{L}^1(\mu)$ and satisfies

$$\int_{\Omega} f^* \, d\mu = \int_{\Omega} f \, d\mu. \tag{16.16}$$

PROOF   We will prove the theorem in the special case $f = \chi_B$, leaving the proof of the general case for Exercises 16.13 and 16.14.[†]

We begin by considering the number of visits to the set B among the first n terms of the orbit of x, that is, $S_n(x) = \sum_{k=0}^{n-1} \chi_B \circ \varphi^{(k)}(x)$, and the average number of visits $A_n(x) = S_n(x)/n$.

Suppose that we can show that the functions $\overline{A}(x) = \limsup_{n \to \infty} A_n(x)$ and $\underline{A}(x) = \liminf_{n \to \infty} A_n(x)$ satisfy

$$\int_\Omega \overline{A} \, d\mu \leq \mu(B) \quad \text{and} \quad \int_\Omega \underline{A} \, d\mu \geq \mu(B). \tag{16.17}$$

Then we would have $\int_\Omega (\overline{A} - \underline{A}) \, d\mu \leq 0$ and, because $\overline{A} - \underline{A} \geq 0$, it would follow that the limit (16.15) exists μ-ae and that (16.16) holds.

We proceed to verify (16.17). Our arguments will make use of the following properties of the functions $\underline{A}$ and $\overline{A}$:

$$0 \leq \underline{A} \leq \overline{A} \leq 1 \tag{16.18}$$

and

$$\underline{A} \circ \varphi = \underline{A} \quad \text{and} \quad \overline{A} \circ \varphi = \overline{A} \tag{16.19}$$

(See Exercise 16.12.)

To understand the proof of (16.17), it helps to think of the parameter n as time. Then $A_n(x)$ represents the average number of visits of the orbit of x to the set B by time $n - 1$.

Let $\epsilon > 0$ and let $\tau_\epsilon(x)$ denote the first time that the average number of visits exceeds $\overline{A}(x) - \epsilon$. Symbolically, we have

$$\tau_\epsilon(x) = \min\{ n \in \mathcal{N} : A_n(x) > \overline{A}(x) - \epsilon \}.$$

We observe that by (16.18), $\tau_\epsilon(x)$ is always a positive integer. From

$$\{ x : \tau_\epsilon(x) \geq c \} = \bigcap_{n < c} \{ x : A_n(x) \leq \overline{A}(x) - \epsilon \},$$

it follows that τ_ϵ is $\mathcal{A}$-measurable.

Either τ_ϵ is essentially bounded or it is not, that is, either

$$\tau_\epsilon \in \mathcal{L}^\infty(\mu) \tag{16.20}$$

or

$$\tau_\epsilon \notin \mathcal{L}^\infty(\mu). \tag{16.21}$$

Suppose first that (16.20) holds. Then we can choose an integer M such that

$$\mu\big(\tau_\epsilon^{-1}((M, \infty))\big) = 0. \tag{16.22}$$

[†] This proof is adapted from one given by M. Keene, "Ergodic Theory and Subshifts of Finite Type," in *Ergodic Theory, Symbolic Dynamics and Hyperbolic Spaces*, edited by T. Bedford, M. Keene, and C. Series (Oxford, UK: Oxford University Press, 1991). Keene's argument is based on ideas in Y. Kamae, "A Simple Proof of the Ergodic Theorem Using Non-Standard Analysis" (*Israel J. of Math*, **42**, pp 284–290, 1982).

For each $x \in \Omega$, we consider the sequence of integers

$$\tau_1(x) = \tau_\epsilon(x), \ \tau_2(x) = \tau_\epsilon(\varphi^{(\tau_1(x))}(x)), \ \tau_3(x) = \tau_\epsilon(\varphi^{(\tau_1(x)+\tau_2(x))}(x)), \ \dots .$$

It follows from (16.22) and the invariance of μ with respect to φ that, for μ-almost all x, we have

$$\tau_j(x) \le M, \qquad j \in \mathcal{N}. \tag{16.23}$$

Suppose that x satisfies (16.23). In the remainder of the proof, we will suppress the dependence of τ_j on x. Let n be a positive integer greater than M and let q be such that

$$\sigma_q \le n < \sigma_{q+1},$$

where we are using the notation $\sigma_q = \tau_1 + \tau_2 + \cdots + \tau_q$. Then

$$S_n(x) \ge S_{\sigma_q}(x)$$

$$= \sum_{k=0}^{\sigma_1-1} \chi_B \circ \varphi^{(k)}(x) + \sum_{k=\sigma_1}^{\sigma_2-1} \chi_B \circ \varphi^{(k)}(x) + \cdots + \sum_{k=\sigma_{q-1}}^{\sigma_q-1} \chi_B \circ \varphi^{(k)}(x)$$

$$= S_{\tau_1}(x) + S_{\tau_2}(\varphi^{(\sigma_1)}(x)) + \cdots + S_{\tau_q}(\varphi^{(\sigma_{q-1})}(x)).$$

It follows from (16.19) and the definition of τ_ϵ that

$$S_{\tau_1}(x) \ge \tau_1(\overline{A}(x) - \epsilon),$$

$$S_{\tau_2}(\varphi^{(\sigma_1)}(x)) \ge \tau_2(\overline{A}(\varphi^{(\sigma_1)}(x)) - \epsilon) = \tau_2(\overline{A}(x) - \epsilon)$$

$$\vdots$$

$$S_{\tau_q}(\varphi^{(\sigma_{q-1})}(x)) \ge \tau_q(\overline{A}(\varphi^{(\sigma_{q-1})}(x)) - \epsilon) = \tau_q(\overline{A}(x) - \epsilon).$$

Hence,

$$S_n(x) \ge \sigma_q(\overline{A}(x) - \epsilon) \ge (n - \tau_{q+1})(\overline{A}(x) - \epsilon).$$

Applying (16.23), we conclude that, for μ-almost all xs, we have the inequality

$$S_n(x) \ge (n - M)(\overline{A}(x) - \epsilon). \tag{16.24}$$

We have shown that (16.24) holds for μ-almost all $x \in \Omega$. Integrating both sides of (16.24) and using the invariance of μ with respect to φ, we obtain

$$n\mu(B) = \sum_{k=0}^{n-1} \mu((\varphi^{(k)})^{-1}(B)) = \int_\Omega S_n(x) \, d\mu(x) \ge \int_\Omega (n - M)(\overline{A}(x) - \epsilon) \, d\mu(x).$$

Dividing by n and letting $n \to \infty$, we get

$$\mu(B) \ge \int_\Omega \overline{A}(x) \, d\mu(x) - \epsilon.$$

A similar argument shows that

$$\mu(B) \le \int_{\Omega} \underline{A}(x)\,d\mu(x) + \epsilon.$$

As $\epsilon > 0$ was chosen arbitrarily, we obtain (16.17). Thus, the proof of the theorem is complete in case (16.20) holds.

It remains to establish (16.17) in case (16.21) holds, that is, when τ_ϵ is not essentially bounded. The idea is to reduce the proof to the case where (16.20) holds by slightly enlarging the set B.

Because τ_ϵ is finite-valued, we can choose a positive integer M such that $\mu(\tau_\epsilon^{-1}((M,\infty))) < \epsilon$. Now we set $B^\epsilon = B \cup \tau_\epsilon^{-1}((M,\infty))$,

$$S_n^\epsilon(x) = \sum_{k=0}^{n-1} \chi_{B^\epsilon} \circ \varphi^{(k)}(x),$$

$A_n^\epsilon(x) = S_n^\epsilon(x)/n$, and $\tau^\epsilon(x) = \min\{\, n \in \mathcal{N} : A_n^\epsilon(x) > \overline{A}(x) - \epsilon \,\}$.

It follows immediately that $\tau^\epsilon \le \tau_\epsilon$. We claim that

$$\tau^\epsilon(x) \le M, \qquad x \in \Omega. \tag{16.25}$$

For, if $\tau^\epsilon(x) > M$, then $\tau_\epsilon(x) > M$. Hence, $A_1^\epsilon(x) = 1 > \overline{A}(x) - \epsilon$, but this implies that $\tau^\epsilon(x) = 1 \le M$, a contradiction.

We can now apply the arguments that we used in the case (16.20) to obtain $\mu(B^\epsilon) \ge \int_\Omega \overline{A}(x)\,d\mu(x) - \epsilon$. Therefore,

$$\mu(B) + \epsilon > \mu(B) + \mu\left(\tau_\epsilon^{-1}((M,\infty))\right) \ge \mu(B^\epsilon) \ge \int_\Omega \overline{A}(x)\,d\mu(x) - \epsilon. \tag{16.26}$$

By similar arguments, we obtain that

$$\mu(B) - \epsilon \le \int_\Omega \underline{A}(x)\,d\mu(x) + \epsilon. \tag{16.27}$$

From (16.26) and (16.27), we deduce that (16.17) holds. ∎

EXAMPLE 16.7 Illustrates the Pointwise Ergodic Theorem

Consider the Bernoulli scheme of Example 16.4 on pages 605–606. Let $k \in \mathcal{N}$. Define $F: \Omega \to \mathcal{R}$ by $F(f) = f(k)$. Because

$$F^{-1}(\{m\}) = \{\, f \in \Omega : f(k) = m \,\} = C_{\{k\},m},$$

F is $\mathcal{A}$-measurable. Applying the pointwise ergodic theorem, we conclude that the average

$$F^*(f) = \lim_{n \to \infty} \frac{1}{n} \sum_{j=0}^{n-1} f(k+j)$$

exists for almost all $f \in \Omega$. We also have

$$\int_\Omega F^*\,d\mu = \int_\Omega F\,d\mu = \sum_{m=1}^{N} m\mu(C_{\{k\},m}) = \sum_{m=1}^{N} m p_m,$$

as is easily verified. □

Ergodicity

Many interesting measurable dynamical systems have the property that for each $f \in \mathcal{L}^1(\mu)$, the average f^* in the pointwise ergodic theorem is constant almost everywhere. In Theorem 16.2, we will see that this property is characterized by the following condition:

$$E \in \mathcal{A} \quad \& \quad E = \varphi^{-1}(E) \quad \Rightarrow \quad \mu(E) = 0 \quad \text{or} \quad \mu(E) = 1. \tag{16.28}$$

To understand the meaning of (16.28), it helps to consider its negation. Suppose that $\Omega_1 \in \mathcal{A}$, $\Omega_1 = \varphi^{-1}(\Omega_1)$, and $0 < \mu(\Omega_1) < 1$. Let $\Omega_2 = \Omega \setminus \Omega_1$. Then we also have $\Omega_2 \in \mathcal{A}$, $\Omega_2 = \varphi^{-1}(\Omega_2)$, and $0 < \mu(\Omega_2) < 1$.

For $j = 1$, 2, we define the σ-algebra $\mathcal{A}_j = \{ A \cap \Omega_j : A \in \mathcal{A} \}$ and a corresponding probability measure $\mu_j(A \cap \Omega_j) = \mu(A \cap \Omega_j)/\mu(\Omega_j)$. Denoting by φ_j the restriction of the mapping φ to Ω_j, we obtain the two measurable dynamical systems $(\Omega_j, \mathcal{A}_j, \mu_j, \varphi_j)$, $j = 1$, 2.

For $x \in \Omega$, the orbit $\{\varphi^{(n-1)}(x)\}_{n=1}^{\infty}$ is contained in either Ω_1 or Ω_2. Indeed, that orbit equals either $\{\varphi_1^{(n-1)}(x)\}_{n=1}^{\infty}$ or $\{\varphi_2^{(n-1)}(x)\}_{n=1}^{\infty}$. Thus, we have complete information about the orbits of $(\Omega, \mathcal{A}, \mu, \varphi)$ if we have it for each of the two smaller systems $(\Omega_j, \mathcal{A}_j, \mu_j, \varphi_j)$, $j = 1$, 2.

DEFINITION 16.2 **Ergodicity**

A measurable dynamical system $(\Omega, \mathcal{A}, \mu, \varphi)$ is called **ergodic** if

$$E \in \mathcal{A} \quad \& \quad E = \varphi^{-1}(E) \quad \Rightarrow \quad \mu(E) = 0 \quad \text{or} \quad \mu(E) = 1.$$

Exercise 16.21 shows that the measurable dynamical system in Example 16.3 on page 605 is ergodic. Example 16.8, which we will present shortly, shows that the measurable dynamical system in Example 16.2 on page 605 is ergodic if and only if b is irrational.

In the proof of our next theorem, we will need to know that ergodicity is equivalent to

$$E \in \mathcal{A} \quad \& \quad E \subset \varphi^{-1}(E) \quad \Rightarrow \quad \mu(E) = 0 \quad \text{or} \quad \mu(E) = 1. \tag{16.29}$$

We leave the verification of this fact to the reader as Exercise 16.15.

□ □ □ **THEOREM 16.2**

Let $(\Omega, \mathcal{A}, \mu, \varphi)$ be a measurable dynamical system. Then the following properties are equivalent:

a) $(\Omega, \mathcal{A}, \mu, \varphi)$ is ergodic.

b) For each $f \in \mathcal{L}^1(\mu)$, the average

$$f^* = \lim_{n \to \infty} \frac{1}{n} \sum_{k=0}^{n-1} f \circ \varphi^{(k)}$$

is constant μ-ae.

c) If $f \in \mathcal{L}^1(\mu)$ and $f \circ \varphi = f$ μ-ae, then f is constant μ-ae.

PROOF   The equivalence of (b) and (c) is left for Exercise 16.16. Suppose (a) holds and that $f \in \mathcal{L}^1(\mu)$ is such that $f \circ \varphi = f$ μ-ae. To show that f is constant μ-ae, it suffices to assume that f is real-valued. Let $D = \{\, x \in \Omega : f(x) \neq f \circ \varphi(x) \,\}$. Then $\mu(D) = 0$. Letting $\varphi^{-k} = (\varphi^{(k)})^{-1}$, we have from the invariance of μ that $\mu(\varphi^{-k}(D)) = 0$ for all k. Hence,

$$\mu\left(\bigcup_{k=0}^{\infty} \varphi^{-k}(D) \right) \leq \sum_{k=0}^{\infty} \mu(\varphi^{-k}(D)) = 0.$$

Now, let $b \in \mathcal{R}$ and set $E = f^{-1}((-\infty, b)) \setminus \bigcup_{k=0}^{\infty} \varphi^{-k}(D)$. Then we have that $\mu(E) = \mu(f^{-1}((-\infty, b)))$ and $E \subset \varphi^{-1}(E)$. By (a) and (16.29), we know that $\mu(f^{-1}((-\infty, b)))$ equals either 0 or 1. It is now immediate that f is constant μ-ae. Consequently, we see that $(a) \Rightarrow (c)$.

Conversely, suppose that (c) holds. Let $E \in \mathcal{A}$ be such that $E = \varphi^{-1}(E)$. Then $\chi_E \circ \varphi = \chi_{\varphi^{-1}(E)} = \chi_E$. Hence, by (c), χ_E is constant μ-ae. It follows that $\mu(E)$ is either 0 or 1. Thus, we have shown that $(c) \Rightarrow (a)$ ■

EXAMPLE 16.8   *Illustrates Theorem 16.2*

Using Theorem 16.2, we will now show that the measurable dynamical system of Example 16.2 on page 605 is ergodic if and only if b is an irrational number. Suppose $f \in \mathcal{L}^1(\mu)$ is such that $f \circ \varphi = f$. Then the Fourier coefficients of the function $g(x) = f(e^{ix})$ must satisfy $\hat{g}(n) = e^{2\pi i n b} \hat{g}(n)$.

If b is irrational, it follows that $\hat{g}(n) = 0$ for all nonzero integers n. Thus, f is constant by Theorem 15.1 on page 553. Consequently, we see that $(T, \mathcal{A}, \mu, \psi_b)$ is ergodic.

On the other hand, if b is rational, say, $b = p/q$, where p and q are integers, then the function $f(z) = z^q$ is nonconstant and satisfies $f \circ \psi_b = f$. Hence, $(T, \mathcal{A}, \mu, \psi_b)$ is not ergodic.   ☐

From the pointwise ergodic theorem and Theorem 16.2, we obtain the following important corollary.

☐ ☐ ☐   COROLLARY 16.1

If $(\Omega, \mathcal{A}, \mu, \varphi)$ is ergodic, then for each $f \in \mathcal{L}^1(\mu)$,

$$\lim_{n \to \infty} \frac{1}{n} \sum_{k=0}^{n-1} f \circ \varphi^{(k)}(x) = \int_{\Omega} f \, d\mu$$

for almost all $x \in \Omega$.

Exercises for Section 16.2

16.12   Verify (16.19) on page 612.

16.13   Let $(\Omega, \mathcal{A}, \mu, \varphi)$ be a measurable dynamical system. Suppose that $f \in \mathcal{L}^1(\mu)$ and that $f \geq 0$ μ-ae. Let $N \in \mathcal{N}$ and let $\epsilon > 0$ be given. Set $S_n(f) = \sum_{k=0}^{n-1} f \circ \varphi^{(k)}$, $A_n(f) = S_n(f)/n$, $\overline{A}(f) = \limsup A_n(f)$, and

$$\tau_\epsilon = \min\{\, n \in \mathcal{N} : A_n > \min\{N, \overline{A}(f) - \epsilon\} \,\}.$$

Show that

$$\int_\Omega f\,d\mu \geq \int_\Omega \min\{N, \overline{A}(f) - \epsilon\}\,d\mu,$$

if $\tau_\epsilon \in \mathcal{L}^\infty(\mu)$, and

$$\int_\Omega f\,d\mu \geq \int_\Omega \min\{N, \overline{A}(f) - \epsilon\}\,d\mu - \epsilon,$$

if $\tau_\epsilon \notin \mathcal{L}^\infty(\mu)$.

16.14   Use Exercise 16.13 to complete the proof of the pointwise ergodic theorem.

16.15   Prove the equivalence of ergodicity and condition (16.29) on page 615.

16.16   Prove the equivalence of (b) and (c) in Theorem 16.2 on page 615.

Exercises 16.17–16.20 are devoted to proving an $\mathcal{L}^2$-version of the pointwise ergodic theorem.

16.17   Let V denote the collection of all $f \in \mathcal{L}^2(\mu)$ such that

$$f^* = \lim_{n \to \infty} \frac{1}{n} \sum_{k=0}^{n-1} f \circ \varphi^{(k)}$$

exists in the sense of convergence in the $\mathcal{L}^2(\mu)$-norm. Prove that V is a closed linear subspace of $\mathcal{L}^2(\mu)$.

16.18   Let V be as in Exercise 16.17 and let

$$Y = \{\, f \in \mathcal{L}^2(\mu) : f \circ \varphi - f \,\}.$$

Show that $Y \subset V$ and $P(f) = f^*$ for all $f \in V$, where $P \colon \mathcal{L}^2(\mu) \to Y$ is the orthogonal projection.

16.19   Refer to Exercises 16.17 and 16.18. Let $Z = \{\, f \circ \varphi - f : f \in \mathcal{L}^2(\mu)\,\}$. Show that $Z \subset V$ and $P(Z) = \{0\}$.

16.20   Refer to Exercises 16.17–16.19. Show that $(Y + Z)^\perp = \{0\}$ and deduce that $V = \mathcal{L}^2(\mu)$. This result proves the $\mathcal{L}^2$-**ergodic theorem:** For each $f \in \mathcal{L}^2(\mu)$, the limit

$$f^* = \lim_{n \to \infty} \frac{1}{n} \sum_{k=0}^{n-1} f \circ \varphi^{(k)}$$

exists in the sense of convergence in the $\mathcal{L}^2(\mu)$-norm. *Hint:* Show that $h \in (Y + Z)^\perp$ implies $\langle h \circ \varphi, f \rangle = \langle h, f \circ \varphi \rangle$.

16.21   Show that the measurable dynamical system of Example 16.3 on page 605 is ergodic by employing the following argument.
a) Show that if $\varphi^{-1}(A) = A$, then $\lambda(A \cap I) = \lambda(A)\lambda(I)$, whenever I is a subinterval of $[0, 1)$ of the form $I = [p/2^n, q/2^n)$ for integers p and q.
b) Extend the result in part (a) to arbitrary subintervals of $[0, 1)$.
c) Show that $\lambda(A) = \lambda(A)^2$.

16.22   In Exercise 16.21, you were asked to verify that the measurable dynamical system of Example 16.3 on page 605 is ergodic. Provide an alternative verification by using the Fourier coefficients $c_n = \int_0^1 e^{-2\pi i n x} f(x)\,dx$, $n \in \mathcal{Z}$.

16.23 Show that if $(\Omega, \mathcal{A}, \mu_1, \varphi)$ and $(\Omega, \mathcal{A}, \mu_2, \varphi)$ are both ergodic, then either $\mu_1 = \mu_2$ or $\mu_1 \perp \mu_2$.

16.24 Suppose that Ω is a compact Hausdorff space. Let $\mathcal{A}$ be the collection of Borel subsets of Ω and let $\varphi \colon \Omega \to \Omega$ be a continuous function. Consider

$$I(\Omega) = \{\, \mu \in P(\Omega) : \mu(\varphi^{-1}(A)) = \mu(A) \text{ for all } A \in \mathcal{A} \,\}.$$

Show that μ is an extreme point of $I(\Omega)$ if and only if $(\Omega, \mathcal{A}, \mu, \varphi)$ is ergodic. Refer to Exercises 16.7 and 16.8 on page 610.

16.25 Let Ω, $\mathcal{A}$, and φ be as in Exercise 16.24. Show that for each $\nu \in I(\Omega)$, there is a regular Borel measure Σ_ν on the weak* closure of ex $I(\Omega)$ such that

$$\int_{\overline{\mathrm{ex}\, I(\Omega)}} \left(\int_\Omega f \, d\mu \right) d\Sigma_\nu(\mu) = \int_\Omega f \, d\nu, \qquad f \in C(\Omega).$$

Hint: See Theorem 14.16 on page 542.

16.3 ISOMORPHISM OF MEASURABLE DYNAMICAL SYSTEMS; ENTROPY

This section is an introduction to some ideas motivated by the question: "When are two measurable dynamical systems essentially the same?" First we will give a definition of what it means for measurable dynamical systems to be *isomorphic*. Then we will present a powerful tool for deciding when two measurable dynamical systems are isomorphic, namely, *entropy*.

DEFINITION 16.3 **Isomorphism of Measurable Dynamical Systems**

Two measurable dynamical systems $(\Omega, \mathcal{A}, \mu, \varphi)$ and $(\Lambda, \mathcal{S}, \nu, \psi)$ are said to be **isomorphic** if there are mappings $J \colon \Omega \to \Lambda$ and $K \colon \Lambda \to \Omega$ such that
a) $J^{-1}(B) \in \mathcal{A}$ for each $B \in \mathcal{S}$,
b) $K^{-1}(A) \in \mathcal{S}$ for each $A \in \mathcal{A}$,
c) $\mu(J^{-1}(B)) = \nu(B)$ for each $B \in \mathcal{S}$,
d) $\nu(K^{-1}(A)) = \mu(A)$ for each $A \in \mathcal{A}$,
e) $J \circ \varphi = \psi \circ J$ μ-ae,
f) $K \circ \psi = \varphi \circ K$ ν-ae,
g) $K \circ J(x) = x$ μ-ae,
h) $J \circ K(y) = y$ ν-ae.
Each of the mappings J and K is called an **isomorphism**.

As the reader is asked to verify in Exercise 16.27, the measurable dynamical systems given in Examples 16.1 (page 605) and 16.2 (page 605) are isomorphic via the mapping $E(x) = e^{2\pi i x}$ defined in the latter example.

A more complicated example of a pair of isomorphic measurable dynamical systems is obtained by considering a so-called one-sided variation of the Bernoulli scheme $B(1/2, 1/2)$.

EXAMPLE 16.9 *Illustrates Definition 16.3*

Refer to Example 16.4 on pages 605–606. The construction of the Bernoulli scheme is unaffected if the space $\Omega = S^{\mathcal{Z}}$ is replaced by $\Omega_+ = S^{\mathcal{N}}$. If $S = \{0, 1\}$ and $(p_1, p_2) = (1/2, 1/2)$, the measure μ is replaced by the measure μ_+ that satisfies $\mu_+(C_{F,a}) = 2^{-N(F)}$ and the function φ is replaced by the function φ_+ given by $\varphi_+((x_1, x_2, x_3, \ldots)) = (x_2, x_3, \ldots)$. It can be shown that the mapping $J: \Omega_+ \to [0, 1)$ defined by

$$J((x_1, x_2, x_3, \ldots)) = \begin{cases} \sum_{j=1}^{\infty} x_j 2^{-j}, & \text{if } x_j = 0 \text{ for some } j; \\ 0, & \text{if } x_j = 1 \text{ for all } j. \end{cases}$$

is an isomorphism of $(\Omega_+, \mathcal{A}_+, \mu_+, \varphi_+)$ onto the measurable dynamical system $([0, 1), \mathcal{B}_{[0,1)}, \lambda_{[0,1)}, \varphi)$ of Example 16.3 on page 605. See Exercise 16.28. □

The idea of isomorphism immediately suggests the following problem: Given two measurable dynamical systems, determine whether they are isomorphic. A natural approach to this problem is to seek *invariants* of measurable dynamical systems. An **invariant** of a measurable dynamical system $(\Omega, \mathcal{A}, \mu, \varphi)$ is a number or property $\mathcal{I}(\Omega, \mathcal{A}, \mu, \varphi)$ such that if $(\Omega, \mathcal{A}, \mu, \varphi)$ and $(\Lambda, \mathcal{S}, \nu, \psi)$ are isomorphic, then $\mathcal{I}(\Omega, \mathcal{A}, \mu, \varphi)$ and $\mathcal{I}(\Lambda, \mathcal{S}, \nu, \psi)$ are identical.

Here is a simple illustration of the use of invariants. As the reader is asked to verify in Exercise 16.29, the property of being ergodic is an invariant of a measurable dynamical system. From Example 16.8 on page 616, we know that if $b \in Q$ and $c \notin Q$, then $(T, \mathcal{A}, \mu, \psi_b)$ is not ergodic and $(T, \mathcal{A}, \mu, \psi_c)$ is ergodic. Therefore, those two measurable dynamical systems are not isomorphic.

Entropy

The remainder of this section is devoted to a discussion of numerical measures of information. To motivate the pertinent ideas, we consider the following "thought experiment."

Let $(\Omega, \mathcal{A}, \mu)$ be probability space. Suppose that the distribution of the location of a particle p in Ω is given by the probability measure μ; that is, for each $A \in \mathcal{A}$, the probability that p is in A equals $\mu(A)$. The object of our experiment is to locate the position of p as closely as possible.

Let $\mathfrak{P}$ be a measurable partition of $(\Omega, \mathcal{A})$. Suppose that we can extract information about the location of p by answering, for each $A \in \mathfrak{P}$, the question: "Is p in A?" In other words, we can ascertain which element of $\mathfrak{P}$ contains p.

Some partitions tell us more than others about the location of p. For example, for the probability space $([0, 1), \mathcal{M}_{[0,1)}, \lambda_{[0,1)})$, we expect more information from $\mathfrak{P} = \{[0, 1/2), [1/2, 1)\}$ than $\mathfrak{Q} = \{[0, 1/100), [1/100, 1)\}$. The reason is because we are guaranteed that $\mathfrak{P}$ will reduce by 50% the measure of the set where we have to look for p, whereas, unless we are lucky, $\mathfrak{Q}$ will reduce it by only 1%.

To proceed rigorously, we need to assign a number to the amount of information gained from a measurable partition. That number is called the *entropy* of the measurable partition.

DEFINITION 16.4 **Entropy of a Measurable Partition**

Let $(\Omega, \mathcal{A}, \mu)$ be a probability space and $\mathfrak{P}$ a measurable partition of $(\Omega, \mathcal{A})$. Then the **entropy** of $\mathfrak{P}$, denoted by $H(\mathfrak{P})$, is defined by

$$H(\mathfrak{P}) = -\sum_{A \in \mathfrak{P}} \mu(A) \log \mu(A),$$

where we use the convention that $0 \log 0 = 0$.

At the end of this section, we will derive the formula for $H(\mathfrak{P})$ from some plausible properties of a measure of information. For the present, we content ourselves with the intuitively satisfying observation that in the case of the probability space $([0,1), \mathcal{M}_{[0,1)}, \lambda_{[0,1)})$, the entropy of a two element partition

$$H(\{A, A^c\}) = -\lambda(A) \log \lambda(A) - (1 - \lambda(A)) \log(1 - \lambda(A))$$

is maximized when $\lambda(A) = \lambda(A^c) = 1/2$.

To determine the basic properties of entropy, we introduce the concept of the refinement of a measurable partition. We say that a measurable partition $\mathfrak{Q}$ is a **refinement** of the measurable partition $\mathfrak{P}$ and write $\mathfrak{P} \ll \mathfrak{Q}$ if every element of $\mathfrak{P}$ is a union of elements of $\mathfrak{Q}$. For two measurable partitions $\mathfrak{P}$ and $\mathfrak{R}$, there is a smallest common refinement given by $\mathfrak{P} \vee \mathfrak{R} = \{ A \cap B : A \in \mathfrak{P}, B \in \mathfrak{R} \}$.

□ □ □ **PROPOSITION 16.1**

Let $(\Omega, \mathcal{A}, \mu)$ be a probability space and $\mathfrak{P}, \mathfrak{Q}$, and $\mathfrak{R}$ be measurable partitions of $(\Omega, \mathcal{A})$. Then the following properties hold:

a) $\mathfrak{P} \ll \mathfrak{Q} \Rightarrow H(\mathfrak{P}) \leq H(\mathfrak{Q})$.

b) $H(\mathfrak{P} \vee \mathfrak{R}) \leq H(\mathfrak{P}) + H(\mathfrak{R})$.

PROOF To prove (a), we start by observing that each $A \in \mathfrak{P}$ is a disjoint union of members of $\mathfrak{Q}$. Thus, for $\mu(A) > 0$, we have

$$-\mu(A) \log \mu(A) = -\sum_{\substack{B \subset A \\ B \in \mathfrak{Q}}} \mu(B) \log \mu(A)$$

$$= -\sum_{\substack{B \subset A \\ B \in \mathfrak{Q}}} \mu(B) \log \mu(B) + \sum_{\substack{B \subset A \\ B \in \mathfrak{Q}}} \mu(B) \log \frac{\mu(B)}{\mu(A)}$$

$$\leq -\sum_{\substack{B \subset A \\ B \in \mathfrak{Q}}} \mu(B) \log \mu(B).$$

Summing over $A \in \mathfrak{P}$, we obtain that

$$H(\mathfrak{P}) = -\sum_{A \in \mathfrak{P}} \mu(A) \log \mu(A) \leq -\sum_{A \in \mathfrak{P}} \sum_{\substack{B \subset A \\ B \in \mathfrak{Q}}} \mu(B) \log \mu(B) = H(\mathfrak{Q}).$$

The proof of (b) is based on the fact that the function $g(t) = -t \log t$ is concave on $[0, 1]$; that is, g satisfies

$$\sum_{j=1}^{n} c_j g(t_j) \leq g\left(\sum_{j=1}^{n} c_j t_j\right) \qquad (16.30)$$

for all convex combinations of elements of $[0, 1]$. Without loss of generality, we can assume that $\mu(C) > 0$ for all $C \in \mathfrak{R}$. Thus, we can write

$$\mu(A) = \sum_{C \in \mathfrak{R}} \left(\frac{\mu(A \cap C)}{\mu(C)}\right) \mu(C) \qquad (16.31)$$

for each $A \in \mathfrak{P}$. It follows from (16.30) and (16.31) that

$$-\mu(A) \log \mu(A) \geq -\sum_{C \in \mathfrak{R}} \mu(C) \left(\frac{\mu(A \cap C)}{\mu(C)}\right) \log \left(\frac{\mu(A \cap C)}{\mu(C)}\right)$$

$$= -\sum_{C \in \mathfrak{R}} \mu(A \cap C) \log \mu(A \cap C) + \sum_{C \in \mathfrak{R}} \mu(A \cap C) \log \mu(C).$$

Summing over $A \in \mathfrak{P}$, we get

$$H(\mathfrak{P}) \geq -\sum_{A \in \mathfrak{P}} \sum_{C \in \mathfrak{R}} \mu(A \cap C) \log \mu(A \cap C) + \sum_{C \in \mathfrak{R}} \sum_{A \in \mathfrak{P}} \mu(A \cap C) \log \mu(C)$$

$$= -\sum_{A \in \mathfrak{P}} \sum_{C \in \mathfrak{R}} \mu(A \cap C) \log \mu(A \cap C) + \sum_{C \in \mathfrak{R}} \mu(C) \log \mu(C)$$

$$= H(\mathfrak{P} \vee \mathfrak{R}) - H(\mathfrak{R}).$$

Thus, (b) is proved. ∎

Entropy and Measurable Dynamical Systems

Up to this point, we have defined entropy for probability spaces $(\Omega, \mathcal{A}, \mu)$. Now we introduce a dynamical aspect by considering the measurable dynamical system $(\Omega, \mathcal{A}, \mu, \varphi)$. Suppose that we modify the "thought experiment" introduced on page 619 by allowing the particle p to move according to the following rule: If p is at x at time 0, then its position at time 1 is $\varphi(x)$, its position at time 2 is $\varphi(\varphi(x))$, etc.

If we use a measurable partition $\mathfrak{P}$ to obtain information about the location of p at time 0, then the measurable partition

$$\varphi^{-n}\mathfrak{P} = \{\, (\varphi^{(n)})^{-1}(A) : A \in \mathfrak{P} \,\}$$

yields corresponding information about the particle's location at time n, and the measurable partition

$$\mathfrak{P}^{(n)} = \mathfrak{P} \vee \varphi^{-1}\mathfrak{P} \vee \cdots \vee \varphi^{-(n-1)}\mathfrak{P}$$

yields corresponding information about the path of successive positions of p at times 0 through $n - 1$ as it moves in Ω under the action of φ.

□ □ □ PROPOSITION 16.2

Let $(\Omega, \mathcal{A}, \mu, \varphi)$ be a measurable dynamical system and $\mathfrak{P}$ a measurable partition of $(\Omega, \mathcal{A})$. Then the following properties hold:

a) $H(\varphi^{-k}\mathfrak{P}) = H(\mathfrak{P})$.

b) $H((\varphi^{-k}\mathfrak{P})^{(n)}) = H(\mathfrak{P}^{(n)})$.

c) $H(\mathfrak{P}^{(n+m)}) \leq H(\mathfrak{P}^{(n)}) + H(\mathfrak{P}^{(m)})$.

PROOF Parts (a) and (b) follow immediately from the definition of the entropy of a measurable partition and the invariance of μ with respect to φ. To obtain (c), we begin with the observation $\mathfrak{P}^{(n+m)} = \mathfrak{P}^{(n)} \vee (\varphi^{-n}\mathfrak{P})^{(m)}$. It follows from Proposition 16.1 that

$$H(\mathfrak{P}^{(n+m)}) \leq H(\mathfrak{P}^{(n)}) + H((\varphi^{-n}\mathfrak{P})^{(m)}).$$

The assertion (c) is now an immediate consequence of (b). ∎

Using Proposition 16.2(c), it can be shown that the limit

$$H(\mathfrak{P}, \varphi) = \lim_{n \to \infty} \frac{H(\mathfrak{P}^{(n)})}{n}$$

exists. (See Exercise 16.30.) We can think of $H(\mathfrak{P}, \varphi)$ as the time average for the entropies associated with the measurable partitions $\mathfrak{P}^{(n)}$. The quantity

$$h(\varphi) = \sup\{ H(\mathfrak{P}, \varphi) : \mathfrak{P} \text{ a partition of } (\Omega, \mathcal{A}) \},$$

which can be viewed as the maximum amount of information that can be extracted from the dynamical system per unit time, is called the **entropy of the measurable dynamical system** $(\Omega, \mathcal{A}, \mu, \varphi)$. As the reader is asked to show in Exercise 16.34, h is an invariant of $(\Omega, \mathcal{A}, \mu, \varphi)$.

Calculation of the entropy of a measurable dynamical system is often not an easy task. In the next section, though, we will find a formula for the entropy of the Bernoulli scheme $B(p_1, p_2, \ldots, p_N)$.

Motivating the Formula for the Entropy of a Partition

We will now motivate the formula for the entropy of a measurable partition given in Definition 16.4 on page 620. Let us return to the "thought experiment" discussed previously in this section. Recall that a particle is located in Ω according to the probability measure μ. That is, for each $A \in \mathcal{A}$, the probability that

$$p \text{ is in } A \tag{16.32}$$

equals $\mu(A)$.

We would like to assign a numerical value $I(A)$ to the information contained in the event (16.32). It seems reasonable to require that $I(A)$ be a decreasing function of $\mu(A)$. In other words, the smaller the probability of A, the greater the information that is obtained from the knowledge that p is in A. Thus, we should have a decreasing function f defined on $[0, 1]$ such that

$$I(A) = f(\mu(A)). \tag{16.33}$$

Another plausible condition on $I(A)$ is that it should assign the value 0 to the sure event:

$$I(\Omega) = f(1) = 0. \tag{16.34}$$

Equation (16.34) reflects the fact that knowing p is in Ω provides no information.

Our final condition on I concerns the total information in two independent events, say, A and B. Knowing that one of the events occurs provides no probabilistic information regarding the occurrence of the other event. Therefore, there is no redundancy in the information imparted by knowing that p is in A and the information imparted by knowing that p is in B. Therefore, the total information imparted by knowing that p is in $A \cap B$ is the aggregate of the individual information:

$$I(A \cap B) = I(A) + I(B). \tag{16.35}$$

Combining (16.33–16.35), we obtain a decreasing function f defined on probabilities such that

$$f(1) = 0 \qquad \text{and} \qquad f(st) = f(s) + f(t), \quad s, t \in [0, 1]. \tag{16.36}$$

As the reader is asked to verify in Exercise 16.35, the only decreasing functions on $[0, 1]$ that satisfy (16.36) are those of the form

$$f(t) = -a \log t,$$

where a is a positive constant.

For convenience, we choose $a = 1$ to arrive at the following definition of the information content of a single event:

$$I(A) = -\log \mu(A).$$

Now consider a measurable partition $\mathfrak{P} = \{A_1, A_2, \ldots, A_n\}$. The discrete random variable

$$X = \sum_{j=1}^{n} I(A_j) \chi_{A_j}$$

gives the information gained by knowing which element of $\mathfrak{P}$ contains p. The expected value of X is

$$\mathcal{E}(X) = \int_{\Omega} X \, d\mu = \sum_{j=1}^{n} I(A_j) \mu(A_j) = -\sum_{j=1}^{n} \mu(A_j) \log \mu(A_j) = H(\mathfrak{P}).$$

Thus, we see that the entropy of a measurable partition is the expected amount of information gained by knowing which element of the measurable partition contains the particle p.

Exercises for Section 16.3

16.26 Prove that isomorphism of measurable dynamical systems is an equivalence relation.

16.27 Show that the measurable dynamical systems in Examples 16.1 and 16.2 on page 605 are isomorphic.

16.28 Refer to Example 16.9 on page 619. Show that the measurable dynamical system $(\Omega_+, \mathcal{A}_+, \mu_+, \varphi_+)$ is isomorphic to $([0,1), \mathcal{B}_{[0,1)}, \lambda_{[0,1)}, \varphi)$, where φ is the mapping defined in Example 16.3 on page 605.

16.29 Prove that ergodicity is an invariant of a measurable dynamical system.

16.30 Suppose that $\{a_n\}_{n=1}^{\infty}$ is a sequence of real numbers that satisfies the subadditivity condition $a_{n+m} \leq a_n + a_m$. Show that $\lim_{n \to \infty} a_n/n$ exists as a real number or, possibly, $-\infty$. *Hint:* Let $m \in \mathcal{N}$ be fixed, but arbitrary. Each $n \in \mathcal{N}$ can be written as $n = \ell m + r$, where $\ell \geq 0$ and $0 \leq r < m$. Thus, $a_n \leq \ell a_m + a_r$.

16.31 Consider the probability space $([0,1), \mathcal{M}_{[0,1)}, \lambda_{[0,1)})$. Show that, among all measurable partitions of $([0,1), \mathcal{M}_{[0,1)})$ with n members, entropy is maximized by the measurable partition $\{[(j-1)/n, j/n)\}_{j=1}^{n}$.

16.32 Let $(\Omega, \mathcal{A}, \mu)$ be a probability space. Show that if $\mathfrak{P}$ is a measurable partition with n elements, then $H(\mathfrak{P}) \leq \log n$.

16.33 Let φ be the identity function. Calculate the entropy of $(\Omega, \mathcal{A}, \mu, \varphi)$.

16.34 Prove that if $(\Omega, \mathcal{A}, \mu, \varphi)$ is isomorphic to $(\Lambda, \mathcal{S}, \nu, \psi)$, then $h(\varphi) = h(\psi)$.

16.35 Show that if $f: [0,1] \to [0, \infty]$ is nonincreasing and satisfies (16.36), then it must be of the form $f(t) = -a \log t$.

In the remaining exercises of this section, we consider an alternative approach to the concept of the information in an event. As previously in this chapter, $(\Omega, \mathcal{A}, \mu)$ is a probability space.

16.36 Let J be an information function on $\mathcal{A}$ of the form $J(A) = g(\mu(A))$, where g is a function defined on $[0,1]$. Suppose that there is also a conditional information function defined by

$$J(A \mid B) = \begin{cases} \mu(B)g(\mu(A \mid B)), & \text{if } \mu(B) > 0; \\ 0, & \text{if } \mu(B) = 0. \end{cases}$$

Define the joint information function of A and B to be the sum of the information in B and the information in A given that B does not occur, that is,

$$J(A, B) = J(B) + J(A \mid B^c).$$

Suppose that the following conditions are satisfied:

- $\{\mu(B) : B \subset A\} = [0, \mu(A)]$.
- $J(\Omega) = 0$.
- $J(A, B) = J(B, A)$.

Show that g satisfies the functional equation

$$g(x) + (1-x)g\left(\frac{y}{1-x}\right) = g(y) + (1-y)g\left(\frac{x}{1-y}\right)$$

for $x, y \in [0,1]$ and $x + y \leq 1$.

16.37 Let g be as in Exercise 16.36.

    a) Show that $g(x) = g(1-x)$.

    b) Deduce that $J(A) = J(A^c)$, that is, the information in A is the same as that in A^c. Observe that the information function I discussed at the end of this section fails to have this property.

16.38 Let g be as in Exercise 16.36.

    a) Assuming that g is twice continuously differentiable, show that it must have the form

$$g(x) = c(x \log x + (1-x) \log(1-x))$$

    for some constant c. *Hint:* Differentiate both sides of the equation in Exercise 16.36, first with respect to x and then with respect to y, and then use the substitutions $u = y/(1-x)$ and $v = x/(1-y)$.

    b) Deduce that, in the case of two element partitions, using $J(A)$ as a measure of the information content of an event A leads to the same definition of entropy as given in Definition 16.4.

16.4 THE KOLMOGOROV-SINAI THEOREM; CALCULATION OF ENTROPY

Our goal in this section is to prove a theorem, due to Kolmogorov and Sinai, that will enable us to calculate the entropy of the Bernoulli scheme $B(p_1, p_2, \ldots, p_N)$.

We will need the following natural extension of the notion of the entropy of a measurable partition. Suppose that $\mathfrak{P}$ and $\mathfrak{Q}$ are measurable partitions of $(\Omega, \mathcal{A}, \mu)$. The **conditional entropy** of $\mathfrak{P}$ relative to $\mathfrak{Q}$ is defined by

$$H(\mathfrak{P} \mid \mathfrak{Q}) = -\sum_{B \in \mathfrak{Q}} \sum_{A \in \mathfrak{P}} \mu(B)\mu(A \mid B) \log \mu(A \mid B),$$

where we assign the value 0 to a summand in which $\mu(B) = 0$.

□ □ □ **PROPOSITION 16.3**

Let $(\Omega, \mathcal{A}, \mu, \varphi)$ be a measurable dynamical system and let $\mathfrak{P}$, $\mathfrak{Q}$, and $\mathfrak{R}$ be measurable partitions of $(\Omega, \mathcal{A})$. Then the following properties hold:

    a) $H(\mathfrak{P} \mid \mathfrak{Q}) \leq H(\mathfrak{P})$.

    b) $H(\mathfrak{P} \vee \mathfrak{Q}) = H(\mathfrak{Q}) + H(\mathfrak{P} \mid \mathfrak{Q})$.

    c) $H(\mathfrak{P} \vee \mathfrak{Q} \mid \mathfrak{R}) \leq H(\mathfrak{P} \mid \mathfrak{R}) + H(\mathfrak{Q} \mid \mathfrak{R})$.

    d) $\mathfrak{P} \ll \mathfrak{Q} \Rightarrow H(\mathfrak{P} \mid \mathfrak{R}) \leq H(\mathfrak{Q} \mid \mathfrak{R})$.

    e) $\mathfrak{Q} \ll \mathfrak{R} \Rightarrow H(\mathfrak{P} \mid \mathfrak{R}) \leq H(\mathfrak{P} \mid \mathfrak{Q})$.

    f) $H(\varphi^{-1}\mathfrak{P} \mid \varphi^{-1}\mathfrak{Q}) = H(\mathfrak{P} \mid \mathfrak{Q})$.

    g) $H(\mathfrak{Q}, \varphi) \leq H(\mathfrak{Q} \mid \mathfrak{P}) + H(\mathfrak{P}, \varphi)$.

PROOF The proofs of (a)–(f) are left for Exercises 16.39–16.40. To obtain (g), we argue as follows. By Proposition 16.1 on page 620 and (b) and (c), we have

$$H(\mathfrak{Q}^{(n)}) \leq H(\mathfrak{Q}^{(n)} \vee \mathfrak{P}^{(n)}) = H(\mathfrak{P}^{(n)}) + H(\mathfrak{Q}^{(n)} \mid \mathfrak{P}^{(n)})$$

$$\leq H(\mathfrak{P}^{(n)}) + \sum_{j=0}^{n-1} H(\varphi^{-j}\mathfrak{Q} \mid \mathfrak{P}^{(n)}).$$

Using (e) and (f), we conclude that

$$H(\mathfrak{Q}^{(n)}) \leq H(\mathfrak{P}^{(n)}) + \sum_{j=0}^{n-1} H(\varphi^{-j}\mathfrak{Q} \mid \varphi^{-j}\mathfrak{P}) \leq H(\mathfrak{P}^{(n)}) + nH(\mathfrak{Q} \mid \mathfrak{P}).$$

Recalling from page 622 that

$$H(\mathfrak{Q}, \varphi) = \lim_{n \to \infty} \frac{H(\mathfrak{Q}^{(n)})}{n}, \tag{16.37}$$

we see that (g) holds. ∎

Next, we need a lemma about approximating σ-algebras by algebras of sets. In stating the lemma, we recall the notation for the symmetric difference of two sets: $E \triangle F = (E \setminus F) \cup (F \setminus E)$.

◻ ◻ ◻ **LEMMA 16.1**

Let $(\Omega, \mathcal{A}, \mu)$ be a probability space, $\mathcal{F} \subset \mathcal{A}$ an algebra of sets, and $\mathcal{E}$ the smallest σ-algebra that contains $\mathcal{F}$. Then, given $E \in \mathcal{E}$ and $\epsilon > 0$, there exists an $F \in \mathcal{F}$ such that $\mu(E \triangle F) < \epsilon$.

PROOF Let $\mathcal{G}$ denote the collection of all $G \in \mathcal{A}$ with the property that there is a sequence $\{F_n\}_{n=1}^{\infty} \subset \mathcal{F}$ such that $\lim_{n \to \infty} \mu(G \triangle F_n) = 0$. As the reader is asked to prove in Exercise 16.41, $\mathcal{G}$ is an algebra of sets.

The lemma will be established if we can show that $\mathcal{G}$ is actually a σ-algebra. Let $\{G_n\}_{n=1}^{\infty}$ be a sequence of sets in $\mathcal{G}$. We must prove that $\bigcup_{n=1}^{\infty} G_n \in \mathcal{G}$. First we disjointize the G_ns. Let $E_1 = G_1$ and, for $n \geq 2$, let $E_n = G_n \setminus \bigcup_{k=1}^{n-1} G_k$. Because $\mathcal{G}$ is an algebra, $\{E_n\}_{n=1}^{\infty} \subset \mathcal{G}$; moreover, we have $\bigcup_{n=1}^{\infty} E_n = \bigcup_{n=1}^{\infty} G_n$. Let $E = \bigcup_{n=1}^{\infty} E_n$.

Because $\mathcal{G}$ is an algebra, $\bigcup_{j=1}^{n} E_j \in \mathcal{G}$. It follows that for each $n \in \mathcal{N}$, there is an $F_n \in \mathcal{F}$ such that $\mu\left(\left(\bigcup_{j=1}^{n} E_j\right) \triangle F_n\right) < 1/n$. Now, we have

$$E \triangle F_n \subset \left(\bigcup_{j=n+1}^{\infty} E_j\right) \cup \left(\left(\bigcup_{j=1}^{n} E_j\right) \triangle F_n\right).$$

Hence,

$$\mu(E \triangle F_n) \leq \mu\left(\bigcup_{j=n+1}^{\infty} E_j\right) + \mu\left(\left(\bigcup_{j=1}^{n} E_j\right) \triangle F_n\right) \leq \sum_{j=n+1}^{\infty} \mu(E_j) + \frac{1}{n}.$$

Because $\sum_{n=1}^{\infty} \mu(E_n) \leq 1$, we conclude that $\lim_{n \to \infty} \mu(E \triangle F_n) = 0$. Consequently, $E \in \mathcal{G}$. ∎

For $A, B \in \mathcal{A}$, $\mu(A \mid B) \log \mu(A \mid B)$ will be close to 0 if $\mu(A \mid B)$ is either close to 0 or close to 1. In other words, $\mu(A \mid B) \log \mu(A \mid B)$ will be close to 0 if A and B are either nearly disjoint or nearly equal. This observation makes it reasonable to consider the conditional entropy $H(\mathfrak{P} \mid \mathfrak{Q})$ to be a measure of closeness of the measurable partitions $\mathfrak{P}$ and $\mathfrak{Q}$. From this viewpoint, our next lemma concerns approximating one measurable partition by another.

□ □ □ **LEMMA 16.2**

Let $\mathcal{F} \subset \mathcal{A}$ be an algebra of sets, $\mathcal{E}$ the smallest σ-algebra that contains $\mathcal{F}$, and $\mathfrak{P} \subset \mathcal{E}$ a measurable partition. Then, for each $\epsilon > 0$, there is a measurable partition $\mathfrak{Q} \subset \mathcal{F}$ such that $H(\mathfrak{P} \mid \mathfrak{Q}) < \epsilon$.

PROOF We sketch the proof, using imprecise terms such as "small" and "close," leaving the details for Exercise 16.42. Let $\mathfrak{P} = \{A_1, A_2, \ldots, A_n\}$. The main idea of the proof is to use Lemma 16.1 to approximate each A_j by a $C_j \in \mathcal{F}$.

Let δ be a small positive number. By Lemma 16.1, we can find, for each j, a set $C_j \in \mathcal{F}$ such that $\mu(A_j \triangle C_j) < \delta$. We will use the C_js to construct a measurable partition of Ω. First, we disjointize the C_js by defining $B_j = C_j \setminus \bigcup_{k \neq j} C_k$. Then we obtain a measurable partition $\mathfrak{Q} = \{B_1, B_2, \ldots, B_n, B_{n+1}\}$ by letting $B_{n+1} = \Omega \setminus \bigcup_{j=1}^{n} B_j$. Because $\mathcal{F}$ is an algebra, it follows that $B_j \in \mathcal{F}$ for all j.

Now we consider the conditional entropy

$$H(\mathfrak{P} \mid \mathfrak{Q}) = -\sum_{j=1}^{n} \sum_{k=1}^{n+1} \mu(B_k) \mu(A_j \mid B_k) \log \mu(A_j \mid B_k).$$

On the right-hand side of the previous equation, the sum of the terms for which $k = n + 1$ is dominated by $n\mu(B_{n+1}) \log 2/2$. This latter expression can be made small by choosing δ appropriately, because $|\mu(A_j) - \mu(B_j)|$ is small for $1 \leq j \leq n$ and $\sum_{j=1}^{n} \mu(A_j) = 1$.

We use

$$-\mu(B_k)\mu(A_j \mid B_k) \log \mu(A_j \mid B_k) \leq -\mu(A_j \mid B_k) \log \mu(A_j \mid B_k)$$

and the observation that $\mu(A_j \mid B_k)$ is close to 0 when $j \neq k$ and close to 1 when $j = k$, to assert that the sum of the remaining terms of $H(\mathfrak{P} \mid \mathfrak{Q})$ is small when δ is sufficiently small. ∎

In the remainder of this section, we assume that $(\Omega, \mathcal{A}, \mu, \varphi)$ is a measurable dynamical system. We also continue to use the convention that $\varphi^{(0)}$ denotes the identity on Ω.

□ □ □ **LEMMA 16.3**

Let $\mathfrak{P}$ be a measurable partition. Then $H(\mathfrak{P}^{(k)}, \varphi) = H(\mathfrak{P}, \varphi)$ for all $k \geq 1$.

PROOF It is easy to check that $(\mathfrak{P}^{(k)})^{(n)} = \mathfrak{P}^{(k+n-1)}$. Hence,

$$H(\mathfrak{P}^{(k)}, \varphi) = \lim_{n \to \infty} \frac{H((\mathfrak{P}^{(k)})^{(n)})}{n} = \lim_{n \to \infty} \frac{H(\mathfrak{P}^{(k+n-1)})}{n}$$

$$= \lim_{m \to \infty} \frac{H(\mathfrak{P}^{(m)})}{m} = H(\mathfrak{P}, \varphi),$$

as required. ∎

If φ is a 1-1 correspondence and $(\Omega, \mathcal{A}, \mu, \varphi^{-1})$ is a measurable dynamical system, then we say that φ is **invertible**. In such cases, the notation

$$\mathfrak{P}^{(m,n)} = \varphi^{-m}\mathfrak{P} \vee \varphi^{-m-1}\mathfrak{P} \vee \cdots \vee \varphi^{-n}\mathfrak{P}$$

is meaningful for each pair of integers n, m with $m \leq n$.

◻ ◻ ◻ **LEMMA 16.4**

If φ is invertible and $\mathfrak{P}$ is a measurable partition, then

$$H(\mathfrak{P}^{(m,n)}, \varphi) = H(\mathfrak{P}, \varphi)$$

for each pair of integers n, m with $m \leq n$.

PROOF It is easy to see that $\mathfrak{P}^{(m,n)} = (\varphi^{-m}\mathfrak{P})^{(n-m+1)}$. Hence, by Lemma 16.3, we have $H(\mathfrak{P}^{(m,n)}, \varphi) = H(\varphi^{-m}\mathfrak{P}, \varphi)$. Since μ is invariant with respect to both φ and φ^{-1}, it follows that $H(\varphi^{-m}\mathfrak{P}, \varphi) = H(\mathfrak{P}, \varphi)$. ∎

Next we discuss the relationship between measurable partitions and algebras of sets. Specifically, if φ is invertible and $\mathfrak{P}$ is a measurable partition, then for each $n \in \mathcal{N}$, the collection

$$\mathcal{A}_n(\mathfrak{P}) = \left\{ B \in \mathcal{A} : B \text{ is a union of members of } \mathfrak{P}^{(-n,n)} \right\}$$

is an algebra of subsets of Ω. Because we have $\mathcal{A}_n(\mathfrak{P}) \subset \mathcal{A}_{n+1}(\mathfrak{P})$, the collection $\mathcal{A}_\infty(\mathfrak{P}) = \bigcup_{n=1}^{\infty} \mathcal{A}_n(\mathfrak{P})$ is also an algebra of subsets of Ω. (See Exercise 16.43.)

We are now ready to state and prove the main result of this section, which is known as the **Kolmogorov-Sinai theorem.** In doing so, we recall that the entropy of the measurable dynamical system $(\Omega, \mathcal{A}, \mu, \varphi)$ is defined by

$$h(\varphi) = \sup\{ H(\mathfrak{P}, \varphi) : \mathfrak{P} \text{ a partition of } (\Omega, \mathcal{A}) \}.$$

◻ ◻ ◻ **THEOREM 16.3 Kolmogorov-Sinai Theorem**

Let $(\Omega, \mathcal{A}, \mu, \varphi)$ be a measurable dynamical system and assume that φ is invertible. Suppose that $\mathfrak{P}$ is a measurable partition of $(\Omega, \mathcal{A})$ such that $\mathcal{A}$ is the smallest σ-algebra that contains $\mathcal{A}_\infty(\mathfrak{P})$. Then $h(\varphi) = H(\mathfrak{P}, \varphi)$.

PROOF By the definition of $h(\varphi)$, it suffices to prove that

$$H(\mathfrak{Q}, \varphi) \leq H(\mathfrak{P}, \varphi) \tag{16.38}$$

for each measurable partition $\mathfrak{Q}$. It follows from Proposition 16.3(g) on page 625 that

$$H(\mathfrak{Q}, \varphi) \leq H(\mathfrak{Q} \mid \mathfrak{P}^{(-n,n)}) + H(\mathfrak{P}^{(-n,n)}, \varphi)$$

for all $n \in \mathcal{N}$. Hence, by Lemma 16.4, we have

$$H(\mathfrak{Q}, \varphi) \leq H(\mathfrak{Q} \mid \mathfrak{P}^{(-n,n)}) + H(\mathfrak{P}, \varphi). \tag{16.39}$$

Given $\epsilon > 0$, we can apply Lemma 16.2 to find a measurable partition $\mathfrak{R}$ such that $\mathfrak{R} \subset \mathcal{A}_\infty(\mathfrak{P})$ and $H(\mathfrak{Q} \mid \mathfrak{R}) < \epsilon$. Because $\mathfrak{R}$ is a finite collection, it follows that $\mathfrak{R} \subset \mathcal{A}_n(\mathfrak{P})$ for some n. In particular, we have $\mathfrak{R} \ll \mathfrak{P}^{(-n,n)}$. Applying Proposition 16.3(e), we get $H(\mathfrak{Q} \mid \mathfrak{P}^{(-n,n)}) \leq H(\mathfrak{Q} \mid \mathfrak{R}) < \epsilon$. Hence, by (16.39), we have $H(\mathfrak{Q}, \varphi) \leq \epsilon + H(\mathfrak{P}, \varphi)$. Since ϵ is an arbitrary positive number, the assertion (16.38) follows and the proof is complete. ∎

There is a version of the Kolmogorov-Sinai theorem that is valid when φ is not necessarily invertible. For a measurable partition $\mathfrak{P}$, let

$$\tilde{\mathcal{A}}_n(\mathfrak{P}) = \left\{ B \in \mathcal{A} : B \text{ is a union of members of } \mathfrak{P}^{(n)} \right\}$$

and let $\tilde{\mathcal{A}}_\infty(\mathfrak{P}) = \bigcup_{n=1}^\infty \tilde{\mathcal{A}}_n(\mathfrak{P})$. If, in the proof of the Kolmogorov-Sinai theorem, we replace $\mathcal{A}_n(\mathfrak{P})$ and $\mathcal{A}_\infty(\mathfrak{P})$ by $\tilde{\mathcal{A}}_n(\mathfrak{P})$ and $\tilde{\mathcal{A}}_\infty(\mathfrak{P})$, respectively, we obtain a proof of the following theorem.

□ □ □ **THEOREM 16.4**

Let $(\Omega, \mathcal{A}, \mu, \varphi)$ be a measurable dynamical system. Suppose $\mathfrak{P}$ is a measurable partition of $(\Omega, \mathcal{A})$ such that $\mathcal{A}$ is the smallest σ-algebra that contains $\tilde{\mathcal{A}}_\infty(\mathfrak{P})$. Then $h(\varphi) = H(\mathfrak{P}, \varphi)$.

EXAMPLE 16.10 *Entropy of a Bernoulli Scheme*

We apply the Kolmogorov-Sinai theorem to obtain the entropy of the Bernoulli scheme $B(p_1, p_2, \ldots, p_N)$, introduced in Example 16.4 on pages 605–606. Consider the measurable partition of $(\Omega, \mathcal{A})$ given by $\mathfrak{P} = \{ C_{\{0\},k} : k = 1, 2, \ldots, N \}$. The entropy of $\mathfrak{P}$ is

$$H(\mathfrak{P}) = -\sum_{k=1}^N \mu(C_{\{0\},k}) \log \mu(C_{\{0\},k}) = -\sum_{k=1}^N p_k \log p_k.$$

We now show that $\mathfrak{P}$ satisfies the hypothesis of the Kolmogorov-Sinai theorem. It is easy to see that φ is invertible. We have $\varphi^{-1}(C_{\{0\},k}) = C_{\{1\},k}$ and, more generally, $\varphi^{-\ell}(C_{\{0\},k}) = C_{\{\ell\},k}$ for every integer ℓ. Therefore, a typical element of $\mathfrak{P}^{(-m,m)}$ is of the form $\bigcap_{\ell=-m}^m C_{\{\ell\},k_\ell} = C_{\{-m,-m+1,\ldots,m\},b}$ where $b(\ell) = k_\ell$ for $-m \le \ell \le m$.

We recall that, in this example, $\mathcal{A}$ is the σ-algebra generated by sets of the form $C_{F,a}$, where F is a finite set of integers and a is a function from F to $\{1, 2, \ldots, N\}$. By choosing m large enough, we can assume $F \subset \{-m, \ldots, m\}$. Hence, we can write $C_{F,a} = \bigcup_b C_{\{-m,\ldots,m\},b}$, where the union is over all functions $b: \{-m, \ldots, m\} \to \{1, \ldots, N\}$ such that $b(\ell) = a(\ell)$ for all $\ell \in F$.

It follows that $C_{F,a}$ belongs to $\mathcal{A}_m(\mathfrak{P})$ and this in turn implies that the algebra $\mathcal{A}_\infty(\mathfrak{P})$ contains all sets of the form $C_{F,a}$. Thus, $\mathcal{A}$ is the smallest σ-algebra that contains $\mathcal{A}_\infty(\mathfrak{P})$.

Next, we calculate $H(\mathfrak{P}, \varphi)$. The entropy of $\mathfrak{P}^{(m)}$ is

$$H(\mathfrak{P}^{(m)}) = -\sum_{k_0=1}^N \sum_{k_1=1}^N \cdots \sum_{k_{m-1}=1}^N \prod_{\ell=0}^{m-1} \mu(C_{\{\ell\},k_\ell}) \log \prod_{\ell=0}^{m-1} \mu(C_{\{\ell\},k_\ell})$$

$$= -\sum_{k_0=1}^N \sum_{k_1=1}^N \cdots \sum_{k_{m-1}=1}^N \prod_{\ell=0}^{m-1} p_{k_\ell} \log \prod_{\ell=0}^{m-1} p_{k_\ell}.$$

As the reader is asked to verify in Exercise 16.44, by using $\sum_{k=1}^{N} p_k = 1$, it can be shown that

$$\sum_{k_0=1}^{N} \sum_{k_1=1}^{N} \cdots \sum_{k_{m-1}=1}^{N} \prod_{\ell=0}^{m-1} p_{k_\ell} \log \prod_{\ell=0}^{m-1} p_{k_\ell} = m \sum_{k=1}^{N} p_k \log p_k.$$

Applying the Kolmogorov-Sinai theorem, we conclude that

$$h(\varphi) = H(\mathfrak{P}, \varphi) = \lim_{m \to \infty} \frac{H(\mathfrak{P}^{(m)})}{m} = - \sum_{k=1}^{N} p_k \log p_k.$$

Consequently, we see that the entropy of the Bernoulli scheme $B(p_1, p_2, \ldots, p_N)$ equals $-\sum_{k=1}^{N} p_k \log p_k$. ☐

Using Example 16.10 and the fact that entropy is an invariant of a measurable dynamical system, we obtain the following result: If two Bernoulli schemes $B(p_1, p_2, \ldots, p_N)$ and $B(q_1, q_2, \ldots, q_M)$ are isomorphic, then

$$\sum_{k=1}^{N} p_k \log p_k = \sum_{\ell=1}^{M} q_\ell \log q_\ell. \tag{16.40}$$

Thus, for example, we see that $B(1/2, 1/2)$ and $B(1/3, 1/3, 1/3)$ are not isomorphic because $\log 2 \neq \log 3$.

Actually, a stronger result exists regarding Bernoulli schemes: A necessary and sufficient condition for $B(p_1, p_2, \ldots, p_N)$ and $B(q_1, q_2, \ldots, q_M)$ to be isomorphic is that (16.40) holds.[†]

Exercises for Section 16.4

16.39 Prove (a), (b), and (c) of Proposition 16.3 on page 625.

16.40 Prove (d), (e), and (f) of Proposition 16.3 on page 625.

16.41 Suppose $(\Omega, \mathcal{A}, \mu)$ is a probability space. Let $\mathcal{G}$ denote the collection of all $E \in \mathcal{A}$ with the property that there is a sequence $\{F_n\}_{n=1}^{\infty} \subset \mathcal{F}$ such that $\lim_{n \to \infty} \mu(E \triangle F_n) = 0$. Prove that $\mathcal{G}$ is an algebra of subsets of Ω.

16.42 Provide the details for the proof of Lemma 16.2 on page 627.

16.43 Prove the following fact: If $\{\mathcal{A}_n\}_{n=1}^{\infty}$ is a sequence of algebras of subsets of some set Ω such that $\mathcal{A}_n \subset \mathcal{A}_{n+1}$, then $\bigcup_{n=1}^{\infty} \mathcal{A}_n$ is also an algebra of subsets of Ω.

16.44 Using $\sum_{k=1}^{N} p_k = 1$, show that

$$\sum_{k_0=1}^{N} \sum_{k_1=1}^{N} \cdots \sum_{k_{m-1}=1}^{N} \prod_{\ell=0}^{m-1} p_{k_\ell} \log \prod_{\ell=0}^{m-1} p_{k_\ell} = m \sum_{k=1}^{N} p_k \log p_k.$$

---

† For a proof of this result, see M. Keane and M. Smorodinsky, "Bernoulli Schemes of the Same Entropy are Finitarily Isomorphic" (*Annals of Math.*, **109**, pp 397–406, 1979).

16.45 Show that if $(\Omega, \mathcal{A}, \mu, \varphi)$ has entropy $h(\varphi)$, then $h(\varphi^{(k)}) = kh(\varphi)$ when $k \in \mathcal{N}$ and, if φ is invertible, then $h(\varphi^{(k)}) = |k|h(\varphi)$ for all $k \in \mathcal{Z}$.

16.46 Refer to Example 16.1 on page 605. Show that $h(\varphi_b) = 0$ if b is rational. *Hint:* See Exercise 16.45.

16.47 Let $(\Omega, \mathcal{A}, \mu, \varphi)$ be a measurable dynamical system and $\mathfrak{P}$ a measurable partition of $(\Omega, \mathcal{A})$. Show that $H(\mathfrak{P}, \varphi) = \lim_{n \to \infty} H(\mathfrak{P} \,|\, (\varphi^{-1}\mathfrak{P})^{(n)})$. *Hint:* Use Proposition 16.3 on page 625 to show that

$$H(\mathfrak{P}^{(k+1)}) = H(\mathfrak{P} \,|\, (\varphi^{-1}\mathfrak{P})^{(k)}) + H(\mathfrak{P}^{(k)}).$$

16.48 Consider the measurable dynamical system in Example 16.1 on page 605 and assume b is irrational. Let $\mathfrak{P} = \{[0, b), [b, 1)\}$,

$$\widetilde{\mathcal{A}}_n = \left\{ B \in \mathcal{A} : B \text{ is a union of members of } (\varphi^{-1}\mathfrak{P})^{(n)} \right\}, \qquad n \in \mathcal{N},$$

and $\widetilde{\mathcal{A}}_\infty = \bigcup_{n=1}^{\infty} \widetilde{\mathcal{A}}_n$. Establish that the smallest σ-algebra that contains $\mathcal{A}_\infty(\mathfrak{P})$ is the σ-algebra of Borel subsets of $[0, 1)$. *Hint:* See Exercise 15.26 on page 565.

16.49 Refer to Example 16.1 on page 605. Show that $h(\varphi_b) = 0$ if b is irrational. *Hint:* Use Exercises 16.47 and 16.48.

Benoit B. Mandelbrot
(1924–2010)

Benoit Mandelbrot was born in Warsaw, Poland, on November 20, 1924, into a family with a strong academic tradition. His mother was a physician and, although his father made a living by buying and selling clothes, Mandelbrot described him as "a very scholarly person."

In 1936, anticipating the threat posed by Nazi Germany, Mandelbrot's family, being Jewish, fled from Poland to France. Until the beginning of World War II, Mandelbrot attended the Lycée Rolin in Paris. His family then moved to Tulle, France. In 1944, Mandelbrot returned to Paris, where he studied at the Lycée du Parc in Lyon. From 1945–1947, he attended the École Polytechnique, where he worked under Gaston Julia and Paul Lévy. And from 1947–1949, he studied at the California Institute of Technology, where he received a master's degree in aeronautics. Subsequently, he returned to France, where he obtained his Ph.D. in Mathematical Sciences at the University of Paris in 1952.

From 1949–1957, Mandelbrot was a staff member at the Centre National de la Recherche Scientifique. In 1958, he and his wife moved to the United States, where he became a member of the research staff at the IBM Thomas J. Watson Research Center in Yorktown Heights, New York. He stayed at IBM until 1987, when IBM ended pure research in his division. Mandelbrot then joined the Department of Mathematics at Yale, where he remained until his retirement in 2005, when he was Sterling Professor of Mathematical Sciences.

Although Mandelbrot is most noted as the founder of fractal geometry (he coined the term *fractal*), he did research in many areas other than mathematics, such as information theory, economics, and fluid dynamics. For his outstanding contributions, Mandelbrot received many honors and prizes: the Franklin Medal (1986), the Wolf prize for physics (1993), and the Japan Prize for Science and Technology (2003), to name just a few.

Mandelbrot died in Cambridge, Massachusetts, on October 14, 2010. He was 85 years old.

17

Hausdorff Measure and Fractals

This chapter provides an introduction to a circle of ideas about measure and dimension. It is natural to assign dimension n to any (Lebesgue-measurable) set $E \subset \mathcal{R}^n$ with $\lambda_n(E) > 0$. We will consider an extension of n-dimensional Lebesgue measure called Hausdorff measure that assigns non-integral values to the dimensions of certain sets.

Section 17.1 generalizes the definition of outer measure given in Definition 6.2 on page 182 to settings without a semialgebra. Section 17.2 defines Hausdorff measure and develops some of its properties. Section 17.3 introduces Hausdorff dimension and topological dimension. And Section 17.4 discusses fractal sets.

17.1 OUTER MEASURE AND MEASURABILITY

Previously, we discussed Lebesgue outer measure and, more generally, the outer measure induced by a set function on a semialgebra. However, we can define outer measure in much broader contexts and do so in the next definition.

DEFINITION 17.1 **Outer Measure**

Let Ω be a set. An **outer measure** ν^* on $\mathcal{P}(\Omega)$ is an extended real-valued function that satisfies the following conditions for all subsets A, B, A_1, A_2, ... of Ω:

a) $\nu^*(A) \geq 0$.   **(nonnegativity)**
b) $\nu^*(\emptyset) = 0$.
c) $A \subset B \Rightarrow \nu^*(A) \leq \nu^*(B)$.   **(monotonicity)**
d) $\nu^*(\bigcup_n A_n) \leq \sum_n \nu^*(A_n)$.   **(countable subadditivity)**

EXAMPLE 17.1 *Illustrates Definition 17.1*

    a) Proposition 3.1 on page 92 shows that Lebesgue outer measure λ^* is an outer measure in the sense of Definition 17.1.

    b) Let Ω be a set, $\mathcal{C}$ a semialgebra of subsets of Ω, and ι a nonnegative extended real-valued set function on $\mathcal{C}$ that satisfies Conditions (E1)–(E3) on page 180. Then, in view of Proposition 6.2 on page 183, the outer measure μ^* induced by ι and $\mathcal{C}$, as defined in Definition 6.2 on page 182, is an outer measure in the sense of Definition 17.1. In particular, then, n-dimensional Lebesgue outer measure λ_n^* is an outer measure in that sense.   □

An outer measure ν^* admits application of the Carathéodory criterion, as defined for Lebesgue outer measure on page 102 and used again in Definition 6.3 on page 184.

DEFINITION 17.2   Measurable Sets

    Let Ω be a set and ν^* an outer measure on $\mathcal{P}(\Omega)$. A set $E \subset \Omega$ is said to be $\boldsymbol{\nu^*}$**-measurable** if

$$\nu^*(W) = \nu^*(W \cap E) + \nu^*(W \cap E^c) \tag{17.1}$$

for all subsets W of Ω. The collection of all ν^*-measurable sets is denoted by $\mathcal{A}$.

□ □ □   THEOREM 17.1

    The collection $\mathcal{A}$ of all ν^*-measurable sets is a σ-algebra. Moreover, ν^* is a measure on $\mathcal{A}$; that is, $\nu^*|_{\mathcal{A}}$ is a measure.

PROOF   We can prove that $\mathcal{A}$ is a σ-algebra by using the same techniques as those in the proof of Theorem 3.11 on page 103 and replacing $\mathcal{R}$ by Ω and $\mathcal{M}$ by $\mathcal{A}$.

    Now, let $\nu = \nu^*|_{\mathcal{A}}$. We want to prove that ν is a measure. Because ν^* is an outer measure, it suffices to establish countable additivity. We establish countable additivity by mimicking the proof of Theorem 3.12 on page 105 and again replacing $\mathcal{R}$ by Ω and $\mathcal{M}$ by $\mathcal{A}$.   ■

Metric Outer Measure

In this chapter, Ω is usually a metric space, which we tacitly assume is separable (see Definition 11.1 on page 397). Let Ω be a metric space with metric ρ. For nonempty sets $A, B \subset \Omega$, we define the **distance** between A and B, denoted $\boldsymbol{\rho(A, B)}$, by

$$\rho(A, B) = \inf\{\rho(x, y) : x \in A, y \in B\}.$$

Also, we define the **diameter** of a nonempty set $A \subset \Omega$, denoted **diam A**, by

$$\operatorname{diam} A = \sup\{\rho(x, y) : x, y \in A\}.$$

By convention, we take $\operatorname{diam} \emptyset = 0$.

From Theorem 3.11 on page 103, we know that the collection $\mathcal{M}$ of Lebesgue-measurable sets (i.e., λ^*-measurable sets) contains the Borel sets of $\mathcal{R}$. However, as we will see later in this section, it is not necessarily true that the collection $\mathcal{A}$ of ν^*-measurable sets contains the Borel sets of a metric space Ω. To ensure that $\mathcal{A} \supset \mathcal{B}(\Omega)$, further conditions must be imposed, as stated in the following definition.

DEFINITION 17.3 **Metric Outer Measure**

Let (Ω, ρ) be a metric space. An outer measure ν^* on $\mathcal{P}(\Omega)$ is said to be a **metric outer measure** if
$$\nu^*(A \cup B) = \nu^*(A) + \nu^*(B)$$
for all $A, B \subset \Omega$ such that $\rho(A, B) > 0$.

Note: Theorem 3.8 on page 96 shows that Lebesgue outer measure λ^* is a metric outer measure.

□ □ □ **PROPOSITION 17.1**

Suppose that ν^* is a metric outer measure on $\mathcal{P}(\Omega)$. Then every closed set of Ω is ν^*-measurable.

PROOF Let F be a closed subset of Ω and let W be any subset of Ω. It is easy to see that the proof of Theorem 3.9 on page 99 remains valid when $(\mathcal{R}, d)$ is replaced by (Ω, ρ) and λ^* is replaced by ν^*. Applying that result with $O = F^c$, $A = W \cap F^c$, and $B = W \cap F$, we get that

$$\nu^*(W) = \nu^*(A \cup B) = \nu^*(A) + \nu^*(B) = \nu^*(W \cap F^c) + \nu^*(W \cap F).$$

Hence, F is ν^*-measurable. ∎

□ □ □ **THEOREM 17.2**

Suppose that ν^* is a metric outer measure on $\mathcal{P}(\Omega)$. Then every Borel set is ν^*-measurable; that is, $\mathcal{B}(\Omega) \subset \mathcal{A}$.

PROOF From Theorem 17.1 and Proposition 17.1, the collection $\mathcal{A}$ of ν^*-measurable sets is a σ-algebra that contains all closed sets of Ω. Hence, $\mathcal{A}$ must contain all open sets of Ω. As $\mathcal{B}(\Omega)$ is the smallest σ-algebra that contains all open sets of Ω, it follows that $\mathcal{A} \supset \mathcal{B}(\Omega)$. ∎

Outer Measures Induced by Set Functions on Coverings

Our next theorem shows that the definition for μ^* (given in Definition 6.2 on page 182) can be significantly generalized to produce a large class of outer measures. Before stating and proving that theorem, we should recall the following concepts from Chapter 11 (page 399): Let $E \subset \Omega$. A collection $\mathcal{S}$ of subsets of Ω such that $E \subset \bigcup_{S \in \mathcal{S}} S$ is called a *covering* of E. A subcollection of $\mathcal{S}$ that is also a covering of E is called a *subcovering*.

□ □ □ **THEOREM 17.3**

Let Ω be a set, $\mathcal{S}$ a covering of Ω, and $h\colon \mathcal{S} \to [0, \infty]$. Then the set function m^* defined on $\mathcal{P}(\Omega)$ by

$$m^*(A) = \inf \left\{ \sum_n h(S_n) : \{S_n\}_n \subset \mathcal{S}, \bigcup_n S_n \supset A \right\} \tag{17.2}$$

is an outer measure. Moreover, m^* is the unique outer measure on Ω that satisfies the following two conditions:

a) *$m^*(S) \leq h(S)$ for all $S \in \mathcal{S}$.*

b) *If ν^* is an outer measure on $\mathcal{P}(\Omega)$ such that $\nu^*(S) \leq h(S)$ for all $S \in \mathcal{S}$, then $\nu^*(A) \leq m^*(A)$ for all $A \subset \Omega$.*

PROOF   Since h is nonnegative, it follows from (17.2) that m^* satisfies the nonnegativity condition of an outer measure. And, since the empty set $\emptyset$ is covered by the empty covering and the empty sum has value 0, we have that $m^*(\emptyset) = 0$.

Suppose now that A and B are subsets of Ω with $A \subset B$. Then any countable subcovering of B by sets from $\mathcal{S}$ is also a countable subcovering of A by sets from $\mathcal{S}$. Hence, $m^*(A) \leq m^*(B)$, so that m^* satisfies the monotonicity condition of an outer measure.

Next, suppose that $\{A_n\}_{n=1}^\infty$ is a sequence of subsets of Ω. If $m^*(A_n) = \infty$ for some $n \in \mathcal{N}$, then, clearly, (d) of Definition 17.1 holds. So, assume $m^*(A_n) < \infty$ for all $n \in \mathcal{N}$. Let $\epsilon > 0$ be given. For each $n \in \mathcal{N}$, we can, by (17.2), choose a sequence $\{S_{nk}\}_k$ of members of $\mathcal{S}$ such that $\bigcup_k S_{nk} \supset A_n$ and

$$\sum_k h(S_{nk}) < m^*(A_n) + \frac{\epsilon}{2^n}.$$

The collection $\{S_{nk}\}_{n,k}$ is then a countable subcovering of $\bigcup_n A_n$ by members of $\mathcal{S}$, and we have

$$m^*\left(\bigcup_{n=1}^\infty A_n \right) \leq \sum_{n,k} h(S_{nk}) \leq \sum_{n=1}^\infty \left(m^*(A_n) + \frac{\epsilon}{2^n} \right) = \sum_{n=1}^\infty m^*(A_n) + \epsilon.$$

As $\epsilon > 0$ was arbitrary, we see that m^* satisfies the countable subaddivity condition of an outer measure.

We have now shown that m^* defined by (17.2) is an outer measure on $\mathcal{P}(\Omega)$. The remaining parts of the proof are left to the reader as Exercise 17.2.   ■

EXAMPLE 17.2   *Illustrates Theorem 17.3*

a) Let Ω be a set, $\mathcal{C}$ a semialgebra of subsets of Ω, and ι a nonnegative extended real-valued set function on $\mathcal{C}$ that satisfies Conditions (E1)–(E3) on page 180. Let $\mathcal{S} = \mathcal{C}$ and $h = \iota$. We note that $\mathcal{S}$ is a covering of Ω (see the second bulleted item on page 181). In view of (17.2) and Definition 6.2 on page 182, we see that $m^* = \mu^*$, the outer measure induced by ι and $\mathcal{C}$. And, by Proposition 6.2(a) on page 183, the inequality in Theorem 17.3(a) is, in this case, equality; in other words, $m^*(S) = h(S)$ for all $S \in \mathcal{S}$.

b) Let $\Omega = \mathcal{R}$, $\mathcal{S} = \{\,[a,b) : -\infty < a < b < \infty\,\}$, and $h([a,b)) = (b-a)^3$. Note that $\mathcal{S}$ is a covering of $\mathcal{R}$. From Theorem 17.3(a), $m^*(S) \leq h(S)$ for all $S \in \mathcal{S}$. We claim that strict inequality must hold for some $S \in \mathcal{S}$. Suppose, to the contrary, that $m^*(S) = h(S)$ for all $S \in \mathcal{S}$. Applying the countable-subadditivity property of an outer measure, we get

$$8 = m^*([0,2)) = m^*([0,1) \cup [1,2)) \leq m^*([0,1)) + m^*([1,2)) = 1 + 1 = 2,$$

a contradiction. Therefore, strict inequality in $m^*(S) \leq h(S)$ must hold for some $S \in \mathcal{S}$.   □

As we stated earlier, when Ω is a metric space, it is not necessarily true that the collection $\mathcal{A}$ of measurable sets contains the collection $\mathcal{B}(\Omega)$ of Borel sets. We now provide an example of a nonmeasurable Borel set.

EXAMPLE 17.3   A Nonmeasurable Borel Set

Let $\Omega = \mathcal{R}$, $\mathcal{S} = \{\,[a,b) : -\infty < a < b < \infty\,\}$, and $h([a,b)) = \sqrt{b-a}$. We claim that $m^*([0,1)) = 1$. Indeed, if $[0,1) \subset \bigcup_n [a_n, b_n)$, then $\sum_n (b_n - a_n) \geq 1$. Thus,

$$\left(\sum_n \sqrt{b_n - a_n} \right)^2 = \sum_n \left(\sqrt{b_n - a_n} \right)^2 + 2 \sum_{i<j} \sum \left(\sqrt{b_i - a_i} \sqrt{b_j - a_j} \right)$$

$$\geq \sum_n (b_n - a_n) \geq 1.$$

Hence, $\sum_n \sqrt{b_n - a_n} \geq 1$ and, so, from (17.2), it follows that $m^*([0,1)) \geq 1$. But, by Theorem 17.3(a), $m^*([0,1)) \leq h([0,1)) = 1$. Therefore, $m^*([0,1)) = 1$. A similar argument shows that $m^*([-1,0)) = 1$.

Referring again to Theorem 17.3(a), we see that $m^*([-1,1)) \leq \sqrt{2}$. Thus,

$$m^*([-1,1)) \leq \sqrt{2}$$
$$< 2 = 1 + 1 = m^*([0,1)) + m^*([-1,0))$$
$$= m^*([-1,1) \cap [0,1]) + m^*([-1,1) \cap [0,1]^c).$$

So, (17.1) on page 634 fails for $E = [0,1]$; that is, $[0,1]$ is not m^*-measurable. Thus, we have found a Borel set of $\mathcal{R}$ that is not m^*-measurable.   □

Suppose now that Ω is a metric space, $\mathcal{S}$ is a covering of Ω, and $h\colon \mathcal{S} \to [0, \infty]$. Further suppose that, for each $x \in \Omega$ and $\epsilon > 0$, there is an $S \in \mathcal{S}$ such that $x \in S$ and $\operatorname{diam} S < \epsilon$. Then, for fixed $\epsilon > 0$, the collection

$$S_\epsilon = \{\, S \in \mathcal{S} : \operatorname{diam} S < \epsilon \,\} \tag{17.3}$$

is a subcovering of Ω by sets from $\mathcal{S}$.

Let m^*_ϵ be the outer measure defined by (17.2), but with $\mathcal{S}_\epsilon$ in place of $\mathcal{S}$. As the reader is asked to show in Exercise 17.6, if $0 < \epsilon_1 < \epsilon_2$, then $m^*_{\epsilon_1} \geq m^*_{\epsilon_2}$, that is, $m^*_{\epsilon_1}(A) \geq m^*_{\epsilon_2}(A)$ for all $A \subset \Omega$. As a consequence, we can define the function m^*_0 on $\mathcal{P}(\Omega)$ by

$$m^*_0(A) = \lim_{\epsilon \to 0} m^*_\epsilon(A) = \sup_{\epsilon > 0} m^*_\epsilon(A), \qquad A \subset \Omega. \tag{17.4}$$

□ □ □ **THEOREM 17.4**

Let (Ω, ρ) be a metric space, $\mathcal{S}$ a covering of Ω, and $h: \mathcal{S} \to [0, \infty]$. Suppose that, for each $x \in \Omega$ and $\epsilon > 0$, there is an $S \in \mathcal{S}$ such that $x \in S$ and $\operatorname{diam} S < \epsilon$. Then the function m_0^* on $\mathcal{P}(\Omega)$, defined in (17.4), is a metric outer measure.

PROOF That m_0^* is an outer measure is left for Exercise 17.7. Hence, it remains to show that, if $A, B \subset \Omega$ with $\rho(A, B) > 0$, then $m_0^*(A \cup B) = m_0^*(A) + m_0^*(B)$. Because m_0^* is an outer measure, it suffices to prove that

$$m_0^*(A \cup B) \geq m_0^*(A) + m_0^*(B). \tag{17.5}$$

To that end, choose $\epsilon > 0$ such that $\epsilon < \rho(A, B)$, and let $\mathcal{T}$ be any countable covering of $A \cup B$ by sets from $\mathcal{S}_\epsilon$. We note that no set in $\mathcal{T}$ can intersect both A and B. Let

$$\mathcal{T}_A = \{ T \in \mathcal{T} : T \cap A \neq \emptyset \} \quad \text{and} \quad \mathcal{T}_B = \{ T \in \mathcal{T} : T \cap B \neq \emptyset \}.$$

Then $\mathcal{T}_A$ and $\mathcal{T}_B$ are disjoint countable coverings of A and B, respectively. Therefore,

$$\sum_{T \in \mathcal{T}} h(T) \geq \sum_{T \in \mathcal{T}_A} h(T) + \sum_{T \in \mathcal{T}_B} h(T) \geq m_\epsilon^*(A) + m_\epsilon^*(B).$$

Taking the infimum over $\mathcal{T}$, we deduce that $m_\epsilon^*(A \cup B) \geq m_\epsilon^*(A) + m_\epsilon^*(B)$. Passing to the limit as $\epsilon \to 0$, we get (17.5). ∎

Exercises for Section 17.1

17.1 Suppose that Ω is compact and that ν^* is a metric outer measure on $\mathcal{P}(\Omega)$ such that $\nu^*(\Omega) < \infty$. Prove that $\nu^*{}_{|\mathcal{B}(\Omega)}$ is a regular Borel measure. (Refer to Definition 13.11 on page 484.)

17.2 Complete the proof of Theorem 17.3 on page 636 by showing that parts (a) and (b) and uniqueness are satisfied.

17.3 Refer to Theorem 17.3 on page 636. Suppose that $\Omega = \mathcal{R}$, $\mathcal{S} = \{ [n, n + 1) : n \in \mathcal{Z} \}$, and $h([n, n + 1)) = 1$ for all $n \in \mathcal{Z}$. Observe that $\mathcal{S}$ is a covering of $\mathcal{R}$.
a) Show that $m^*(S) = h(S)$ for all $S \in \mathcal{S}$.
b) Show that m^* is not a measure.

17.4 Refer to Theorem 17.3 on page 636. Let $\Omega = \mathcal{R}$, $\mathcal{S} = \{ [a, b) : -\infty < a < b < \infty \}$, and $h([a, b)) = (b - a)^2$. Find $m^*([0, 1])$.

17.5 Refer to Theorem 17.3 on page 636. Suppose that $\mathcal{T}$ is a subcovering of Ω by sets from $\mathcal{S}$, and let $g = h_{|\mathcal{T}}$. Denote the outer measure induced by $\mathcal{T}$ and g by n^*. Show that $m^* \leq n^*$; that is, that $m^*(A) \leq n^*(A)$ for all $A \subset \Omega$.

17.6 Refer to page 637. Show that, if $0 < \epsilon_1 < \epsilon_2$, then $m_{\epsilon_1}^* \geq m_{\epsilon_2}^*$, that is, $m_{\epsilon_1}^*(A) \geq m_{\epsilon_2}^*(A)$ for all $A \subset \Omega$.

17.7 Show that the function m_0^* on $\mathcal{P}(\Omega)$, defined in (17.4), is an outer measure.

17.2 HAUSDORFF MEASURE

In this section, we introduce Hausdorff measure and examine some of its properties. We assume throughout that Ω is a separable metric space.

Let r and δ be positive real numbers. As the reader is asked to show in Exercise 17.8, the collection $\{\, S \subset \Omega : \operatorname{diam} S < \delta \,\}$ is a covering of Ω. It then follows immediately from Theorem 17.3 on page 636 that the set function $H^*_{r,\delta}$ defined on $\mathcal{P}(\Omega)$ by

$$H^*_{r,\delta}(A) = \inf\left\{ \sum_n (\operatorname{diam} S_n)^r : \bigcup_n S_n \supset A, \ \operatorname{diam} S_n < \delta \right\}$$

is an outer measure.

In the notation of (17.3) on page 637, $\{\, S \subset \Omega : \operatorname{diam} S < \delta \,\} = \mathcal{P}(\Omega)_\delta$. Thus, applying Theorem 17.4 with $\mathcal{S} = \mathcal{P}(\Omega)$ and $h(S) = (\operatorname{diam} S)^r$, we conclude that the set function H^*_r defined on $\mathcal{P}(\Omega)$ by

$$H^*_r(A) = \lim_{\delta \to 0} H^*_{r,\delta}(A) \tag{17.6}$$

is a metric outer measure.

Now, we know from Theorems 17.1 and 17.2 on pages 634 and 635, respectively, that the H^*_r-measurable sets form a σ-algebra that contains all the Borel sets of Ω. Moreover, the restriction of H^*_r to the σ-algebra of H^*_r-measurable sets is a measure, which we denote H_r. That measure is called **r-dimensional Hausdorff measure.**

Hausdorff Measure on $\mathcal{R}^n$

Our next objective is to establish that, by using an appropriate metric on $\mathcal{R}^n$, n-dimensional Hausdorff measure and n-dimensional Lebesgue measure agree on all Borel sets. The metric on $\mathcal{R}^n$ that we use is defined by

$$\rho_n(x,y) = \max\{\, |x_k - y_k| : k = 1, \ldots, n \,\}. \tag{17.7}$$

Exercise 10.25 on page 369 implies that the collection of Borel sets of $(\mathcal{R}^n, \rho_n)$ is $\mathcal{B}_n$; that is, equipped with the metric ρ_n, we have that $\mathcal{B}(\mathcal{R}^n) = \mathcal{B}_n$. Also, with the metric ρ_n, balls are n-dimensional cubes (i.e., sets of the form $I_1 \times \cdots \times I_n$, where the I_ks are finite intervals of the same length).

□ □ □ **LEMMA 17.1**

We have $\lambda^*_n(S) \leq (\operatorname{diam} S)^n$ for all $S \subset \mathcal{R}^n$.

PROOF Let $S \subset \mathcal{R}^n$. We can, without loss of generality, assume that $\operatorname{diam} S < \infty$. For $k = 1, \ldots, n$, let

$$a_k = \inf\{\, x_k : x \in S \,\} \qquad \text{and} \qquad b_k = \sup\{\, x_k : x \in S \,\}.$$

Now, suppose $x, y \in S$. For $1 \leq k \leq n$, we have $|x_k - y_k| \leq \rho_n(x,y) \leq \operatorname{diam} S$ and, hence, $b_k - a_k \leq \operatorname{diam} S$. As $S \subset [a_1, b_1] \times \cdots \times [a_n, b_n]$, it follows that

$$\lambda^*_n(S) \leq (b_1 - a_1) \cdots (b_n - a_n) \leq (\operatorname{diam} S)^n,$$

as required. ■

We will also need the following lemma whose proof we leave to the reader as Exercise 17.10.

□ □ □ **LEMMA 17.2**

Let B be a ball in $(\mathcal{R}^n, \rho_n)$. Then $\lambda_n(B) = (\operatorname{diam} B)^n$.

In the proof of our next lemma, we use the following observation: The Vitali covering theorem (Theorem 8.1 on page 275) remains valid for $\mathcal{R}^n$ if "intervals" are replaced by "n-dimensional intervals" and d (the usual metric on $\mathcal{R}$) is replaced by ρ_n.

□ □ □ **LEMMA 17.3**

We have $H_n(O) \leq \lambda_n(O)$ for all open sets $O \subset \mathcal{R}^n$.

PROOF   We can assume without loss of generality that $\lambda_n(O) < \infty$. Let $\epsilon > 0$ be given.

Now, as the reader is asked to show in Exercise 17.11, for each $\delta > 0$, there is a collection $\mathcal{V}$ of closed balls, each with positive radius, such that $\mathcal{V}$ is countable, $\bigcup_{B \in \mathcal{V}} B = O$, $\mathcal{V}$ is a Vitali cover of O, and $\operatorname{diam} B < \delta$ for all $B \in \mathcal{V}$.

Applying the aforementioned $\mathcal{R}^n$-version of the Vitali covering theorem, there is a finite disjoint collection $\{B_1, \ldots, B_{k_1}\} \subset \mathcal{V}$ such that

$$\lambda_n\left(O \setminus \bigcup_{j=1}^{k_1} B_j\right) < 1.$$

Let $O_1 = O \setminus \bigcup_{j=1}^{k_1} B_j$. We observe that O_1 is open in $\mathcal{R}^n$ and that the collection $\mathcal{V}_1 = \{B \in \mathcal{V} : B \subset O_1\}$ is a Vitali cover of O_1. Hence, there is a finite disjoint collection $\{B_{k_1+1}, \ldots, B_{k_2}\} \subset \mathcal{V}_1$ such that

$$\lambda_n\left(O_1 \setminus \bigcup_{j=k_1+1}^{k_2} B_j\right) < \frac{1}{2}.$$

By continuing in this manner, we obtain a nested sequence of open sets $O = O_0 \supset O_1 \supset O_2 \supset \cdots$, a pairwise disjoint sequence $B_1, B_2, \ldots$ of sets from $\mathcal{V}$, and an increasing sequence of integers $0 = k_0 < k_1 < k_2 < \cdots$ such that

$$O_{m+1} = O_m \setminus \bigcup_{j=k_m+1}^{k_{m+1}} B_j \quad \text{and} \quad \lambda_n(O_{m+1}) < \frac{1}{m+1}$$

for $m = 0, 1, 2, \ldots$.

For $m = 0, 1, 2, \ldots$, set $W_m = \bigcup_{j=k_m+1}^{k_{m+1}} B_j$. Then $O = \left(\bigcup_{j=0}^m W_j\right) \cup O_{m+1}$ for each m. Also, set

$$W = \bigcup_{m=0}^{\infty} W_m = \bigcup_{j=1}^{\infty} B_j.$$

Since $O \setminus W \subset O_{m+1}$, $\lambda_n(O \setminus W) \leq 1/(m+1)$ for each nonnegative integer m. So, we have $\lambda_n(O \setminus W) = 0$ and, because $W \subset O$, we also have $\lambda_n(O) = \lambda_n(W)$.

From Exercise 17.12, we can choose balls A_1, A_2, $\ldots$, each with diameter less than δ, that cover $O \setminus W$ and satisfy $\sum_{k=1}^{\infty} \lambda_n(A_k) < \epsilon$. Then the A_ks and B_js together cover O. Consequently, by Lemma 17.2 and the definition of $H_{n,\delta}^*$, we have that

$$\lambda_n(O) + \epsilon = \lambda_n(W) + \epsilon = \sum_{j=1}^{\infty} \lambda_n(B_j) + \epsilon > \sum_{j=1}^{\infty} \lambda_n(B_j) + \sum_{k=1}^{\infty} \lambda_n(A_k)$$

$$= \sum_{j=1}^{\infty} (\operatorname{diam} B_j)^n + \sum_{k=1}^{\infty} (\operatorname{diam} A_k)^n \geq H_{n,\delta}^*(O).$$

Letting δ tend to 0 and using (17.6) on page 639 and the fact that H_n is a metric outer measure, we conclude that

$$\lambda_n(O) + \epsilon \geq \lim_{\delta \to 0} H_{n,\delta}^*(O) = H_n^*(O) = H_n(O).$$

As $\epsilon > 0$ was arbitrary, the lemma is proved. ■

□ □ □ **THEOREM 17.5**

Let $n \in \mathcal{N}$ and let H_n be n-dimensional Hausdorff measure on $(\mathcal{R}^n, \rho_n)$, where ρ_n is the metric given by (17.7) on page 639. Then $H_n = \lambda_n$ as measures on $\mathcal{B}_n$.

PROOF Suppose that $B \in \mathcal{B}_n$. For $\delta > 0$, let $\{S_k\}_k$ be any sequence of subsets of $\mathcal{R}^n$ such that $\bigcup_k S_k \supset B$ and $\operatorname{diam} S_k < \delta$ for all k. Applying the monotonicity and countable-subadditivity properties of an outer measure and then applying Lemma 17.1 (page 639), we get that

$$\lambda_n(B) = \lambda_n^*(B) \leq \lambda_n^*\left(\bigcup_k S_k\right) \leq \sum_k \lambda_n^*(S_k) \leq \sum_k (\operatorname{diam} S_k)^n.$$

It follows that $\lambda_n(B) \leq H_{n,\delta}^*(B)$. Referring now to (17.6) on page 639 and using the fact that all Borel sets of $\mathcal{R}^n$ are H_n^*-measurable, we conclude that

$$\lambda_n(B) \leq \lim_{\delta \to 0} H_{n,\delta}^*(B) = H_n^*(B) = H_n(B).$$

Hence, $\lambda_n(B) \leq H_n(B)$.

To obtain the reverse inequality, we first observe that

$$\lambda_n(B) = \inf\{\lambda_n(O) \,; O \supset B,\ O \text{ open}\},$$

a result that follows from Exercise 17.13. Consequently, in view of Lemma 17.3, we have that

$$\lambda_n(B) = \inf\{\lambda_n(O) : O \supset B,\ O \text{ open}\}$$
$$\geq \inf\{H_n(O) : O \supset B,\ O \text{ open}\} \geq H_n(B).$$

Hence, $\lambda_n(B) \geq H_n(B)$. ■

Metric Relations for Hausdorff Measures

Next, we examine some relations among Hausdorff measures induced by different metrics on the same set. In doing so, we will use the following notation and conventions. Let Ω be a set, c a positive real number, and ρ and σ metrics on Ω.

- By $c\rho$, we mean the metric on Ω defined by $(c\rho)(x, y) = c\rho(x, y)$ for all $x, y \in \Omega$.
- By $\rho \leq \sigma$, we mean that $\rho(x, y) \leq \sigma(x, y)$ for all $x, y \in \Omega$.
- By $\mathcal{B}_\rho(\Omega)$, we mean the collection of Borel sets induced by ρ, that is, the smallest σ-algebra of subsets of Ω that contains the topology induced by ρ.

We will need the following lemma whose proof is left to the reader as Exercise 17.14.

□ □ □ **LEMMA 17.4**

Suppose that ρ and σ are metrics on a set Ω.
a) If $\rho \leq \sigma$, then $\mathcal{B}_\rho(\Omega) \subset \mathcal{B}_\sigma(\Omega)$.
b) If $\sigma = c\rho$ for some positive real number c, then $\mathcal{B}_\sigma(\Omega) = \mathcal{B}_\rho(\Omega)$.
c) If $a\rho \leq \sigma \leq b\rho$ for some positive real numbers a and b, then $\mathcal{B}_\sigma(\Omega) = \mathcal{B}_\rho(\Omega)$.

Note: If we are considering two or more metrics on a set Ω and the collections of Borel sets induced by those metrics are identical, then we use the phrase "Borel sets of Ω" to refer to those identical collections of Borel sets.

The following proposition provides some relations among Hausdorff measures induced by metrics on the same set. We leave its proof to the reader as Exercise 17.15.

□ □ □ **PROPOSITION 17.2**

Suppose that ρ and σ are metrics on Ω. Let H_r^ρ and H_r^σ be the corresponding r-dimensional Hausdorff measures, where r is a positive real number.
a) If $\sigma = c\rho$ for some positive real number c, then $H_r^\sigma(B) = c^r H_r^\rho(B)$ for all Borel sets B of Ω.
b) If $a\rho \leq \sigma \leq b\rho$ for some positive real numbers a and b, then

$$a^r H_r^\rho(B) \leq H_r^\sigma(B) \leq b^r H_r^\rho(B)$$

for all Borel sets B of Ω.

EXAMPLE 17.4 *Illustrates Proposition 17.2*

Let $\Omega = \mathcal{R}^n$. Consider the metrics ρ_n and σ_n defined by

$$\rho_n(x, y) = \max\{\, |x_k - y_k| : 1 \leq k \leq n \,\} \quad \text{and} \quad \sigma_n(x, y) = \left(\sum_{k=1}^n (x_k - y_k)^2 \right)^{1/2}.$$

Then

$$\rho_n(x, y) \leq \sigma_n(x, y) \leq \sqrt{n}\, \rho_n(x, y)$$

for all $x, y \in \mathcal{R}^n$. Hence, from Proposition 17.2(b), for each $r > 0$,

$$H_r^{\rho_n}(B) \leq H_r^{\sigma_n}(B) \leq (\sqrt{n})^r H_r^{\rho_n}(B), \qquad B \in \mathcal{B}_n. \tag{17.8}$$

We note that (17.8) shows that the measures $H_r^{\rho_n}$ and $H_r^{\sigma_n}$ are mutually absolutely continuous. We also note that, in the special case when $r = n$ (a positive integer), Theorem 17.5 on page 641 and (17.8) yield

$$\lambda_n(B) \le H_n^{\sigma_n}(B) \le (\sqrt{n})^n \lambda_n(B) \tag{17.9}$$

for all $B \in \mathcal{B}_n$.

From (17.9), we see that $H_1^d(B) = \lambda(B)$ for all $B \in \mathcal{B}$, where $d(x, y) = |x - y|$. More generally, it can be shown that, for each $n \in \mathcal{N}$,

$$H_n^{\sigma_n}(B) = \frac{2^n \Gamma(\frac{n}{2} + 1)}{\pi^{n/2}} \lambda_n(B), \qquad B \in \mathcal{B}_n. \tag{17.10}$$

Thus, equipped with the usual Euclidean metric, n-dimensional Hausdorff measure is a constant multiple of n-dimensional Lebesgue measure. It is interesting to note that the constant equals the reciprocal of the volume of a ball of diameter 1 in $\mathcal{R}^n$.[†]   □

Exercises for Section 17.2

17.8 Suppose that (Ω, ρ) is a metric space. Let δ be a positive real number. Prove that the collection $\{ S \subset \Omega : \text{diam } S < \delta \}$ is a covering of Ω.

17.9 Suppose that (Ω, ρ) is a metric space. Let $r > 0$. Use the definition of r-dimensional Hausdorff measure to show that $H_r(\{x\}) = 0$ for all $x \in \Omega$. Deduce that $H_r(C) = 0$ for all countable subsets of Ω.

17.10 Prove Lemma 17.2 on page 640.

17.11 Consider the metric space $(\mathcal{R}^n, \rho_n)$, where ρ_n is given by (17.7) on page 639. Suppose that $O \subset \mathcal{R}^n$ is an open set with $\lambda_n(O) < \infty$. Show that, for each $\delta > 0$, there is a collection $\mathcal{V}$ of closed balls, each with positive radius, such that $\mathcal{V}$ is countable, $\bigcup_{B \in \mathcal{V}} B = O$, $\mathcal{V}$ is a Vitali cover of O, and diam $B < \delta$ for all $B \in \mathcal{V}$.

17.12 Consider the metric space $(\mathcal{R}^n, \rho_n)$, where ρ_n is given by (17.7) on page 639. Suppose that $A \subset \mathcal{R}^n$ and $\lambda_n(A) = 0$. Show that, for each $\epsilon > 0$ and $\delta > 0$, there is a sequence of balls $\{B_k\}_{k=1}^\infty$ such that diam $B_k < \delta$ for all k, $\bigcup_{k=1}^\infty B_k \supset A$, and $\sum_{k=1}^\infty \lambda_n(B_k) < \epsilon$. *Hint:* Use the definition of n-dimensional Lebesgue outer measure and the fact that, with the metric ρ_n, balls are cubes.

17.13 Show that $\lambda_n(E) = \inf\{ \lambda_n(O) : O \supset E,\ O \text{ open} \}$ for all $E \in \mathcal{M}_n$.

17.14 Prove Lemma 17.4.

17.15 Prove Proposition 17.2.

17.16 Verify all details of Example 17.4 other than (17.10).

17.3 HAUSDORFF DIMENSION AND TOPOLOGICAL DIMENSION

In this section, we define both the Hausdorff dimension and the topological dimension of a subset of a metric space. We begin by studying the dependence of Hausdorff measure H_r on the parameter r.

[†] For a proof of (17.10), see, for instance, pages 157–161 of the text *Measure Theory and Integration* (Graduate Studies in Mathematics, vol. 76) by Michael E. Taylor (Providence, RI: American Mathematical Society, 2006).

□ □ □ **THEOREM 17.6**

Let Ω be a metric space and let $B \in \mathcal{B}(\Omega)$. Then the following properties hold.
a) $H_r(B)$ is a nonincreasing function of r.
b) If $H_r(B) < \infty$, then $H_q(B) = 0$ for all $q > r$.

PROOF We leave the proof of part (a) to the reader as Exercise 17.17. For part (b), let $r > 0$ and $B \in \mathcal{B}(\Omega)$. Suppose that $H_r(B) < \infty$ and that $q > r$. Recall that, for each $p > 0$ and $\delta > 0$,

$$H_{p,\delta}^*(B) = \inf\left\{\sum_k (\operatorname{diam} S_k)^p : \bigcup_k S_k \supset B, \ \operatorname{diam} S_k < \delta\right\}.$$

For $\delta > 0$, let $\{S_k\}_k$ be any sequence of subsets of Ω such that $\bigcup_k S_k \supset B$ and $\operatorname{diam} S_k < \delta$ for all k. Then

$$H_{q,\delta}^*(B) \le \sum_k (\operatorname{diam} S_k)^q = \sum_k (\operatorname{diam} S_k)^{q-r}(\operatorname{diam} S_k)^r \le \delta^{q-r}\sum_k (\operatorname{diam} S_k)^r.$$

It follows that $H_{q,\delta}^*(B) \le \delta^{q-r}H_{r,\delta}^*(B)$. Letting $\delta \to 0$ and noting that $q - r > 0$, we conclude by referring to (17.6) on page 639 that $H_q(B) = 0$. ∎

The proof of the following corollary of Theorem 17.6 is left to the reader as Exercise 17.18.

□ □ □ **COROLLARY 17.1**

Let Ω be a metric space and let $B \in \mathcal{B}(\Omega)$. Then exactly one of the following statements holds true.
a) $H_r(B) = 0$ for all $r > 0$.
b) $H_r(B) = \infty$ for all $r > 0$.
c) There is a unique positive real number r_0 such that $H_r(B) = \infty$ for $r < r_0$ and $H_r(B) = 0$ for $r > r_0$.

With Corollary 17.1 in mind, we now make the following definition.

DEFINITION 17.4 **Hausdorff Dimension**

Let Ω be a metric space. For $B \in \mathcal{B}(\Omega)$, we define the **Hausdorff dimension of B,** denoted $\dim_H(B)$, to be 0, ∞, and r_0 in the cases (a), (b), and (c), respectively, of Corollary 17.1. Equivalently,

$$\dim_H(B) = \sup\{r > 0 : H_r(B) = \infty\} = \inf\{r > 0 : H_r(B) = 0\},$$

where we let $\sup \emptyset = 0$ and $\inf \emptyset = \infty$.

In general, the Hausdorff dimension of a Borel set depends on the metric with which Ω is equipped. Nonetheless, we have the following proposition, which follows immediately from Proposition 17.2(b) on page 642 and the definition of Hausdorff dimension.

□ □ □ **PROPOSITION 17.3**

Suppose ρ and σ are metrics on a set Ω. Let $\dim_H^\rho$ and $\dim_H^\sigma$ denote Hausdorff dimensions relative to ρ and σ, respectively. If $a\rho \leq \sigma \leq b\rho$ for some positive real numbers a and b, then $\dim_H^\sigma(B) = \dim_H^\rho(B)$ for all Borel sets B of Ω.

In Section 17.4, we will calculate the Hausdorff dimensions of some interesting sets. Here, we simply point out that the Hausdorff dimension of a countable set is zero, as you are asked to verify in Exercise 17.19.

Topological Dimension

Hausdorff dimension is a measure-theoretic construct. We now discuss a notion of dimension, called *topological dimension*, that depends only on the topology of a space.

To define topological dimension, we need the concept of the *boundary* of a set. Let E be a subset of a topological space. The **boundary of E,** denoted ∂E, is defined by $\partial E = \overline{E} \setminus E^\circ$. Basic properties of the boundary of a set are given in Exercise 10.64 on page 378. Additionally, we note that a set has empty boundary if and only if it is both open and closed.

We note that the (geometrical) dimension of a line segment is 1; its boundary consists of its two endpoints, which have dimension 0. The dimension of a (solid) square is 2; its boundary consists of four line segments, which have dimension 1. The dimension of a (solid) cube is 3; its boundary consists of six squares, which have dimension 2. The pattern then is that the dimension of the object equals 1 plus the dimension of its boundary. Because the concept of boundary makes sense in any topological space, as noted in the preceding paragraph, we can apply this pattern to topological spaces to define a topological dimension.

Although topological dimension can be defined for any topological space, we will, as previously, concentrate on (separable) metric spaces. Here now is the definition of topological dimension that we will use.[†]

DEFINITION 17.5 **Topological Dimension**

Let Ω be a metric space. The **topological dimension of Ω,** denoted $\dim_T(\Omega)$, is defined inductively as follows:

a) $\dim_T(\emptyset) = -1$.

b) $\dim_T(\Omega) = 0$ if Ω has a neighborhood basis that consists of sets with empty boundaries.

c) $\dim_T(\Omega) \leq k$ if Ω has a neighborhood basis that consists of sets S such that $\dim_T(\partial S) \leq k - 1$.

d) $\dim_T(\Omega) = k$ if $\dim_T(\Omega) \leq k$ but $\dim_T(\Omega) \not\leq k - 1$.

e) $\dim_T(\Omega) = \infty$ if $\dim_T(\Omega) \not\leq k$ for all k.

[†] There are three commonly used definitions for topological dimension: *small inductive dimension (ind), large inductive dimension (Ind),* and *Lebesgue covering dimension (dim).* The definition that we give in Definition 17.5 is for small inductive dimension. For separable metric spaces, the three notions of dimensions are the same; that is, if Ω is a separable metric space, then $\text{ind}(\Omega) = \text{Ind}(\Omega) = \dim(\Omega)$.

Note: Let (Ω, ρ) be a metric space and let $E \subset \Omega$. Then, by $\dim_T(E)$, we mean the topological dimension of the metric space $(E, \rho_{|E \times E})$.

EXAMPLE 17.5 Topological Dimension of a Finite Set

Let F be a finite nonempty subset of a metric space, say, $F = \{x_1, \ldots, x_n\}$. We leave it to the reader as Exercise 17.25 to show that F has the discrete topology. In particular, each single-element set is both open and closed and, hence, has empty boundary. As $\{\{x_j\} : 1 \le j \le n\}$ constitutes a neighborhood basis for F, it follows from Definition 17.5(b) that $\dim_T(F) = 0$. □

EXAMPLE 17.6 Topological Dimension of the Cantor Set

Refer to the construction of the Cantor set on page 62. We have $P = \bigcap_{n=1}^{\infty} P_n$, where each P_n is the union of 2^n disjoint closed intervals $\{J_{nk}\}_{k=1}^{2^n}$ each of length 3^{-n}. We note that the distance between two distinct J_{nk}s is at least 3^{-n}.

For each $n \in \mathcal{N}$ and $1 \le k \le 2^n$, let $L_{nk} = J_{nk} \cap P$. Then each L_{nk} is closed in P. In addition, let I_{nk} be the open interval with the same center as J_{nk} but with length 3^{-n+1}. Then $L_{nk} = I_{nk} \cap P$ so that each L_{nk} is also open in P. Hence, each L_{nk} has empty boundary in P.

Suppose now that $x \in P$. For each $\epsilon > 0$, choose $n \in \mathcal{N}$ so that $3^{-n+1} < \epsilon$ and choose $k \in \{1, 2, \ldots, 2^n\}$ such that $x \in L_{nk}$. Then $L_{nk} \subset (x - \epsilon, x + \epsilon)$. It follows that $\{L_{nk} : 1 \le k \le 2^n, n \in \mathcal{N}\}$ constitutes a neighborhood basis for P. Thus, by Definition 17.5(b), $\dim_T(P) = 0$. □

EXAMPLE 17.7 Topological Dimension of the Real Line

As we know, the collection of all intervals of the form $(x - \epsilon, x + \epsilon)$, where $x \in \mathcal{R}$ and $\epsilon > 0$, is a neighborhood basis for $\mathcal{R}$. The boundary of $(x - \epsilon, x + \epsilon)$ is the finite set $\{x - \epsilon, x + \epsilon\}$, which, by Example 17.5, has topological dimension 0. Hence, by Definition 17.5(c), $\dim_T(\mathcal{R}) \le 1$. However, because the only subsets of $\mathcal{R}$ that are both open and closed are $\emptyset$ and $\mathcal{R}$, it follows that $\dim_T(\mathcal{R}) \not\le 0$. Hence, by Definition 17.5(d), we have that $\dim_T(\mathcal{R}) = 1$. □

Next, we show that topological dimension is indeed a topological property; that is, that topological dimension is preserved by homeomorphisms.

□ □ □ **THEOREM 17.7**

If Γ and Λ are homeomorphic metric spaces, then $\dim_T(\Gamma) = \dim_T(\Lambda)$.

PROOF Let $h : \Gamma \to \Lambda$ be a homeomorphism from Γ onto Λ.

We proceed by induction. If Γ is empty, then so is Λ and, hence, both spaces have topological dimension -1.

Assume that the theorem is valid for all spaces Γ_0 with $\dim_T(\Gamma_0) \le n$. Let Γ have topological dimension $n + 1$. Then, by definition, there is a neighborhood basis $\mathcal{U}$ of Γ that consists of open sets U with $\dim_T(\partial U) \le n$.

Now, because h is a homeomorphism, $\{h(U) : U \in \mathcal{U}\}$ is a neighborhood basis for Λ. Also, if $U \in \mathcal{U}$, then $h(\partial U) = \partial h(U)$. Since the restriction of h to ∂U is a homeomorphism onto $\partial h(U)$, it follows from the induction assumption that $\dim_T(\partial h(U)) \le n$. Therefore, Λ has a neighborhood basis of sets S such that $\dim_T(\partial S) \leq n$, which means that $\dim_T(\Lambda) \leq n + 1$. On the other hand, if $\dim_T(\Lambda) \leq n$, then, by the induction assumption, we have that $\dim_T(\Gamma) \leq n$, which is a contradiction of the fact that $\dim_T(\Gamma) = n + 1$. Consequently, we have that $\dim_T(\Lambda) = n + 1$. Thus, the theorem has been proved in all cases where Γ has finite topological dimension.

Assume now that $\dim_T(\Gamma) = \infty$. We claim that $\dim_T(\Lambda) = \infty$. Suppose to the contrary that $\dim_T(\Lambda) = n$ for some integer n. But then we would have that $\dim_T(\Gamma) = n$, a contradiction. ∎

We conclude our treatment of topological dimension by stating the following theorem.[†]

□ □ □   **THEOREM 17.8**

The topological dimension of Euclidean n-space is n; that is, $\dim_T(\mathcal{R}^n) = n$.

Exercises for Section 17.3

Note: A ★ denotes an exercise that will be subsequently referenced.

17.17   Prove part (a) of Theorem 17.6 on page 644.

17.18   Prove Corollary 17.1 on page 644.

17.19   Show that countable subsets of a metric space have Hausdorff dimension zero.

17.20   Show that, if $0 < H_r(B) < \infty$, then $\dim_H(B) = r$.

17.21   Prove that $\dim_H(E) \leq \dim_H(F)$ for all subsets E and F of a metric space with $E \subset F$.

17.22   Show that $\dim_H(\mathcal{R}^n) = n$.

★17.23   Prove that any Borel set of $\mathcal{R}^n$ with positive n-dimensional Lebesgue measure has Hausdorff dimension n; that is, if $B \in \mathcal{B}_n$ and $\lambda_n(B) > 0$, then $\dim_H(B) = n$.

17.24   Let E be a subset of a topological space. Show that $\partial E = \emptyset$ if and only if E is both open and closed.

17.25   Show that a finite nonempty subset of a metric space has the discrete topology and that its single-element sets constitute a neighborhood basis for the set.

17.26   Prove that $\dim_T(E) \leq \dim_T(F)$ for all subsets E and F of a metric space with $E \subset F$.

17.27   Use Theorem 17.7 and Exercise 17.26 to prove that $\dim_T(I) = 1$ for all nondegenerate intervals $I \subset \mathcal{R}$.

★17.28   For $0 < \alpha < 1$, let P_α be a modified Cantor set obtained in a manner similar to that of the ordinary Cantor set except that, at the nth step, open intervals of length $\alpha/3^n$, instead of length $1/3^n$, are removed. Determine $\dim_T(P_\alpha)$.

17.29   Let C be a circle in $\mathcal{R}^2$. Show that $\dim_T(C) = 1$.

17.30   Prove that, if Ω is an infinite, connected metric space, then $\dim_T(\Omega) > 0$.

[†] For a proof, see, for instance, page 66 of the text *Modern Dimension Theory* (Berlin: Heldermann Verlag, 1983) by Jun-iti Nagata.

17.4 FRACTALS

The term **fractal** was coined by Benoit Mandelbrot to describe a set whose Hausdorff dimension exceeds its topological dimension. We will use that definition of fractal, but note that other definitions are also employed.

A result of Edward Marczewski, also known as Edward Szpilrajn, is that (for separable metric spaces) *the Hausdorff dimension of a set is always greater than or equal to its topological dimension.*

Consequently, we can define a fractal to be a set whose Hausdorff and topological dimensions differ. We observe that, because the topological dimension of a set is either a nonnegative integer or infinity, a sufficient condition for a set to be a fractal is that its Hausdorff dimension is a non-integer real number.

Fractal sets are highly irregular, yet frequently possess underlying symmetries. Due to Mandelbrot's efforts, it is now known that fractals are ubiquitous in nature and society.[†]

In this section, we present some illustrative examples of fractals. For purposes of contrast, we begin with a simple example of a non-fractal.

EXAMPLE 17.8 *Unit Cube in Euclidean n-Space*

Let $Q = \{ x \in \mathcal{R}^n : 0 < x_k < 1,\ 1 \leq k \leq n \}$ be an open unit cube in $\mathcal{R}^n$. Because Q has positive Lebesgue measure, we see from Exercise 17.23 on page 647 that $\dim_H(Q) = n$.

Moreover, as the reader is asked to show in Exercise 17.31, Q is homeomorphic to $\mathcal{R}^n$. Therefore, from Theorem 17.7 on page 646 and Theorem 17.8 on page 647, we have that $\dim_T(Q) = \dim_T(\mathcal{R}^n) = n$.

We have thus shown that the Hausdorff and topological dimensions of Q are equal, namely, to n. Therefore, Q is not a fractal. □

EXAMPLE 17.9 *The Cantor Set*

In Example 17.6 on page 646, we showed that the topological dimension of the Cantor set P is 0. As you are asked to show in Exercise 17.36, the Hausdorff dimension of the Cantor set is $\log 2 / \log 3$ (≈ 0.6309). Hence, we have that

$$\dim_T(P) = 0 < \frac{\log 2}{\log 3} = \dim_H(P).$$

Because the Hausdorff dimension of the Cantor set exceeds its topological dimension, that set is a fractal. □

In our next example, we discuss the **Sierpinski gasket** (also called the **Sierpinski triangle** or **Sierpinski sieve**), named after the Polish mathematician Wacław Sierpiński (1882–1969) who described it in 1915.

[†] For an excellent overview of many applications of fractals, see the book *Fractals, Chaos, Power Laws: Minutes from an Infinite Paradise* (Mineola, NY: Dover Publications, Inc., 2009) by Manfred Schroeder.

EXAMPLE 17.10   *The Sierpinski Gasket*

To construct the Sierpinski gasket, we begin with a closed equilateral triangle (including boundary and interior). At the first step, we remove an open equilateral triangle T, as depicted in Fig. 17.1. The result is three congruent closed equilateral triangles, S_{11}, S_{12}, and S_{13}, each with one-half the side-length of the original triangle. Let $S_1 = \bigcup_{k=1}^{3} S_{1k}$.

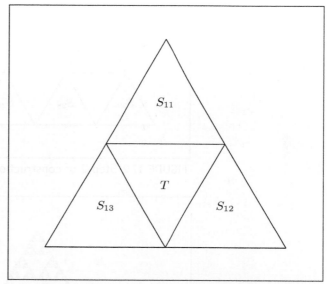

FIGURE 17.1 Step 1 of construction of Sierpinski gasket

At the second step, we remove an open equilateral triangle from each of the three closed equilateral triangles obtained at the first step, as depicted in Fig. 17.2 at the top of the following page. The result is nine congruent closed equilateral triangles S_{2k}, $1 \leq k \leq 9$, each with one-fourth the side-length of the original triangle. Let $S_2 = \bigcup_{k=1}^{9} S_{2k}$.

Continuing in this manner, we obtain a nested sequence of unions of closed equilateral triangles $S_1 \supset S_2 \supset \cdots$, where

$$S_n = \bigcup_{k=1}^{3^n} S_{nk}, \qquad n \in \mathcal{N},$$

and where, for each $n \in \mathcal{N}$, the S_{nk}, $1 \leq k \leq 3^n$, are congruent closed equilateral triangles, each with side-length $1/2^n$ of the original triangle.

The set

$$S = \bigcap_{n=1}^{\infty} S_n$$

is the Sierpinski gasket. We picture that set in Fig. 17.3 on the next page.

As you are asked to show in Exercise 17.37, the Sierpinski gasket has topological dimension 1 and Hausdorff dimension $\log 3/\log 2$ (≈ 1.5850). Thus, S is a fractal with $\dim_T(S) = 1$ and $\dim_H(S) = \log 3/\log 2$. □

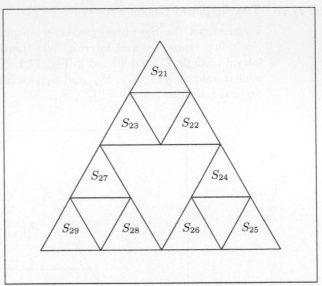

FIGURE 17.2 Step 2 of construction of Sierpinski gasket

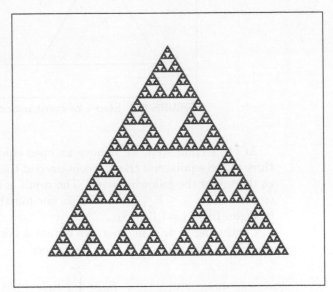

FIGURE 17.3 The Sierpinski gasket

EXAMPLE 17.11 The Mandelbrot Set

For each $c \in \mathbb{C}$, define $P_c\colon \mathbb{C} \to \mathbb{C}$ by $P_c(z) = z^2 + c$. Also, denote by P_c^n the nth iterate of P_c, that is, the composition of P_c with itself n times. Then the **Mandelbrot set** is defined as follows:

$$M = \left\{ c \in \mathbb{C} : \sup_{n \in \mathcal{N}} |P_c^n(0)| < \infty \right\}.$$

The Mandelbrot set is depicted in Fig. 17.4.

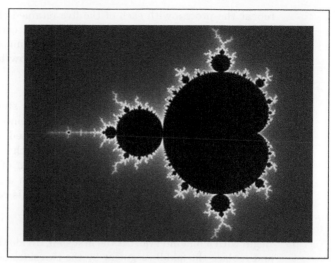

FIGURE 17.4 The Mandelbrot set

It is known that the topological dimension of the Mandelbrot set is 2. From this fact, it follows that the Hausdorff dimension of the Mandelbrot set is also 2 because

$$2 = \dim_T(M) \le \dim_H(M) \le \dim_H(\mathcal{R}^2) = 2.$$

Hence, by the definition of fractal that we are using, the Mandelbrot set is not a fractal.

On the other hand, a celebrated result of Mitsuhiro Shishikura is that the boundary of the Mandelbrot set also has Hausdorff dimension 2 and, thus, exceeds its topological dimension of 1. Therefore, the boundary of the Mandelbrot set is a fractal. □

In our brief introduction to fractals here, we have barely scratched the surface of that extensive and important topic. For more details, you can refer to the book *Measure, Topology, and Fractal Geometry, 2/e* (New York: Springer, 2008) by Gerald A. Edgar. You might also want to examine the Wikipedia page "List of fractals by Hausdorff dimension."

Exercises for Section 17.4

17.31 Let $Q = \{ x \in \mathcal{R}^n : 0 < x_k < 1, \ 1 \le k \le n \}$ be an open unit cube in $\mathcal{R}^n$. Show that Q is homeomorphic to $\mathcal{R}^n$.

17.32 Refer to Exercise 17.28 on page 647. For $0 < \alpha < 1$, let P_α be a modified Cantor set obtained in a manner similar to that of the ordinary Cantor set except that, at the nth step, open intervals of length $\alpha/3^n$, instead of length $1/3^n$, are removed.
a) Determine $\dim_H(P_\alpha)$.
b) Is P_α a fractal? Justify your answer.

17.33 Begin with a closed unit square. At the first step, decompose the square into nine congruent subsquares with a 3×3 grid; retain the middle closed square and the four closed squares at the corners, removing the other squares; call the resulting set V_1, which is a union of five closed squares, each with edge length $1/3$. At the second step,

repeat the process with each of the five retained closed squares and call the resulting set V_2, which is a union of 25 closed squares, each with edge length 1/9. Continue in this manner to obtain a nested sequence of unions of closed squares $V_1 \supset V_2 \supset \cdots$. The set $V = \bigcap_{n=1}^{\infty} V_n$ is called the **Vicsek snowflake.**
a) Construct graphs of V_1, V_2, and V_3.
b) Try to imagine the Vicsek snowflake, V.

17.34 Begin with a closed unit square. At the first step, decompose the square into nine congruent subsquares with a 3×3 grid and remove the middle open square; call the resulting set C_1, which is a union of eight closed squares, each with edge length 1/3. At the second step, repeat the process with each of the eight remaining closed squares and call the resulting set C_2, which is a union of 64 closed squares, each with edge length 1/9. Continue in this manner to obtain a nested sequence of unions of closed squares $C_1 \supset C_2 \supset \cdots$. The set $C = \bigcap_{n=1}^{\infty} C_n$ is called the **Sierpinski carpet.**
a) Construct graphs of C_1, C_2, and C_3.
b) Try to imagine the Sierpinski carpet, C.

17.35 Begin with a unit interval. At the first step, divide the interval into three equal segments and replace the middle segment by the two sides of an equilateral triangle of the same length as the segment being removed to make a "tent" (so that, if the middle segment remained, it and the two added segments would form an upward-pointed equilateral triangle); call the resulting set K_1, which consists of four line segments, each with length 1/3. At the second step, repeat the process with each of the four line segments and call the resulting set K_2, which consists of 16 line segments, each with length 1/9. Continue in this manner indefinitely to obtain a set K, called the **Koch curve.** *Note:* We will make the existence of K rigorous in Exercise 17.40.
a) Construct graphs of K_1, K_2, and K_3.
b) Try to imagine the Koch curve, K.
c) Show that K has infinite length.

Log-ratio formula for Hausdorff dimension: Under certain conditions, the Hausdorff dimension of a set in $\mathcal{R}^n$ can be determined by using a simple formula. We consider compact sets C with the property that there exist integers $k \geq 2$ and m such that C can be divided into k congruent subsets each of which when magnified by a factor of m is congruent to C. Such sets are called (exactly) **self-similar.**

To motivate the log-ratio formula for the Hausdorff dimension of a compact set, we consider the closed unit cube C in $\mathcal{R}^n$. We note that C can be divided into 2^n congruent closed cubes, each of which when magnified by 2 is congruent to C. Hence, C is self-similar, and we can take $k = 2^n$ and $m = 2$. Because $\dim_H(C) = n$, we have $m^{\dim_H(C)} = k$ or, equivalently,

$$\dim_H(C) = \frac{\log k}{\log m}. \tag{17.11}$$

This formula for Hausdorff dimension holds for all self-similar compact sets. You are to apply (17.11) in Exercises 17.36–17.40.

17.36 Refer to Example 17.9 on page 648.
a) Use (17.11) to verify that the Hausdorff dimension of the Cantor set is $\log 2/\log 3$.
b) Use part (a) to deduce that the topological dimension of the Cantor set is 0.
c) Use part (a) to show that $\lambda(P) = 0$.

17.37 Refer to Example 17.10 on page 649.
a) Apply (17.11) to verify that the Hausdorff dimension of the Sierpinski gasket is indeed $\log 3/\log 2$.

b) Show that the topological dimension of the Sierpinski gasket is 1.

c) Show that $\lambda_2(S) = 0$.

17.38 Refer to Exercise 17.33.

a) Use (17.11) to determine the Hausdorff dimension of the Vicsek snowflake.

b) Without computing the topological dimension of the Vicsek snowflake, explain why it is a fractal.

c) Determine the topological dimension of the Vicsek snowflake.

17.39 Refer to Exercise 17.34.

a) Use (17.11) to determine the Hausdorff dimension of the Sierpinski carpet.

b) Without computing the topological dimension of the Sierpinski carpet, explain why it is a fractal.

c) Find the topological dimension of the Sierpinski carpet.

17.40 Refer to Exercise 17.35.

a) For each $n \in \mathcal{N}$, divide the interval $[0, 1]$ into 4^n equal-length subintervals. Then define $f_n: [0, 1] \to \mathcal{R}^2$ to be the piecewise linear continuous function that, going from left to right, maps the jth subinterval onto the jth line segment of K_n. Observe that $f_n([0, 1]) = K_n$. Prove that $\{f_n\}_{n=1}^{\infty}$ converges uniformly. *Hint:* First prove that $\{f_n\}_{n=1}^{\infty}$ is uniformly Cauchy.

b) The *Koch curve* K is defined to be the range of the limit of the sequence $\{f_n\}_{n=1}^{\infty}$ constructed in part (a). Show that K is connected.

c) Use (17.11) to determine the Hausdorff dimension of the Koch curve.

d) Without computing the topological dimension of the Koch curve, explain why it is a fractal.

e) Find the topological dimension of the Koch curve.

Index

Symbol Index

Symbol Index (cont.)

Symbol Index (cont.)

Symbol Index (cont.)

Symbol Index (cont.)

Printed and bound by CPI Group (UK) Ltd, Croydon, CR0 4YY

03/10/2024

01040320-0001